CAT

# Stochastic Modeling and Geostatistics: Principles, Methods, and Case Studies, Volume II

*Edited by*

T. C. Coburn

J. M. Yarus

R. L. Chambers



## AAPG Computer Applications in Geology 5

Published by

**The American Association of Petroleum Geologists**

Tulsa, Oklahoma

ISBN: 978-0-89181-704-8
0-89181-704-2



ON THE COVER: Three-dimensional representations of: simulated facies for the combined Wustro, Ebstorf, and Dethlingen intervals; simulated porosity of the Wustro interval; and simulated permeability for the combined Wustro, Ebstorf, and Dethlingen intervals in the Rotliegend Gas Field, northwestern Germany. For more details see the chapter by R. K. Shrestha, S. Faruqi, M. Boeckmann, and M.-K. Wang beginning on page 237 of this volume.

AAPG Editor: Ernest A. Mancini
Geoscience Director: James B. Blankenship

This publication is available from:

The AAPG Bookstore
P.O. Box 979
Tulsa, OK U.S.A. 74101-0979
Phone: 1-918-584-2555
or 1-800-364-AAPG (U.S.A. only)
Fax: 1-918-260-2652
or 1-800-898-2274 (U.S.A. only)
E-mail: bookstore@aapg.org
www.aapg.org

Geological Society Publishing House
Unit 7, Brassmill Enterprise Centre
Brassmill Lane, Bath BA13JN
U.K.
Phone: +44-1225-445046
Fax: +44-1225-442836
E-mail: sales@geolsoc.org.uk
www.geolsoc.org.uk

Canadian Society of Petroleum Geologists
600, 640-8th Avenue S.W.
Calgary, Alberta T2P 1G7
Canada
Phone: 1-403-264-5610
Fax: 1-403-264-5898
E-mail: Jaime.croft@cspg.org
www.cspg.org

Affiliated East-West Press Private Ltd.
G-1/16 Ansari Road, Darya Ganj
New Delhi 110 002
India
Phone: +91-11-23279113
Fax: +91-11-23260538
E-mail: affiliat@vsnl.com

# *Preface*

The basic ideas of geostatistics have been known since the earliest writings of Krige, Matheron, and their contemporaries; but it was not until the mid- to late-1980s that geostatistical methodology would undergo a major expansion that would ultimately change the face of data analysis in all disciplines that use spatial data. While many of the new developments initially took form in academic centers like Stanford University, the University of Texas, and Ecole des Mines in Fontainbleau, they were given life in the research centers of companies like BP/Amoco, Marathon Oil, and ExxonMobil through practical application to applied geoscience and petroleum-related problems.

By the early 1990s, geostatistics had reached a certain level of maturity and it was beginning to be integrated into a number of reservoir studies. The idea for a book on geostatistics began to germinate at Amoco and Marathon Oil; and in 1994, the American Association of Petroleum Geologists (AAPG) published Stochastic Modeling and Geostatistics: Principles, Methods, and Case Studies (SMG) as the third contribution to its series on computer applications in geology. The book presented both a reflective review of developments in geostatistics up to that time and a forward look at the application of a number of new developments which were just beginning to be come to the forefront.

Unbeknownst to the editors and to AAPG, SMG appeared at a time that coincided with the chronological cusp of an even greater explosion in spatial thinking, stochastic modeling, and geostatistical data analysis. Whether intended or not, from that point forward the practice of geoscience experienced a rather remarkable evolution, as did many other scientific disciplines. Since its publication, SMG has become one of the most highly regarded, oft-quoted, and visible texts on geostatistical methodology, and it has served to stimulate countless new ideas, applications, and techniques.

It has now been more than ten years since SMG was published, and the discipline of geostatistics has undergone tremendous expansion and development since that time. In particular, the fundamental ideas of geostatistics have become even more intertwined with the broader notions of stochastic modeling and spatial statistics that, in their own right, have undergone unparalleled expansion and growth. In this same timeframe, computational power has mushroomed and new information technology (e.g., geographic information systems) has literally transformed the practice of geoscience.

Because much has happened in the realm of geostatistics since 1994, the task of summarizing and characterizing the signal accomplishments would be an almost insurmountable task. Yet, we believe it is paramount that the geostatistics community—in particular, the earth science part of that community—once again step back, reflect on its past achievements, and look forward to further developments and enhancements. Like sequels to good movies, sequels to good books are difficult to create; and so it is with considerable trepidation that we have once again undertaken the assignment of compiling a body of information about geostatistics and stochastic modeling that is both representative of achievements throughout the past decade and forward-looking enough to catapult the discipline to a new level. The challenge of preparing yet another high-profile book on geostatistics is quite daunting when so many good ones have been published in recent years.

The idea for this second volume of SMG took root late in 2000 at an AAPG-sponsored Hedberg conference held at The Woodlands Conference Center outside Houston, Texas. Many of the presentations from that conference were subsequently revised and formalized, and are included as part of this volume. The Hedberg conference papers have been supplemented by other selected reports and essays that span the continuum of methodology and applications from recent years, as well as some relatively new ideas and technologies that will likely experience further development and implementation in the near future.

The volume was initially compiled in the summers of 2001–2003 while the senior editor served as visiting scientist at the U.S. Geological Survey in Reston, Virginia, and at Sandia National Laboratories in Albuquerque, New Mexico. This was a challenging, transitional timeframe marked by changes in thought, practice, and profession from which we, like many others, were not exempt, and so this effort has required much more time to complete than originally envisioned.

The principal title of the book is once again Stochastic Modeling and Geostatistics. Whereas "geostatistics" is a fairly well defined notion, "stochastic modeling" has become a somewhat more nebulous term due to the corresponding development of probabilistic approaches that are not specifically spatial in nature. Consequently, in this second volume of SGM, as in the first, stochastic modeling is assumed to involve geostatistical or related spatial techniques. We acknowledge both the interrelatedness of the two disciplines and the diverging lines of thought and practice.

In addition to the application of some relatively new methodologies, readers of SMG II will no doubt recognize the use of several familiar ideas—possibly even some early vintages of such ideas—in both well-known and novel situations. One recurring theme is the integration of seismic information, reflecting the significant growth that has occurred in this area in recent years. In addition, the book includes material that is not necessarily petroleum-related and information that is not solely intended for use in the earth sciences. We have purposely broadened the scope of the text to underscore the expanding influence of geostatistics in disciplines other than the geosciences.

In recent years the field of geostatistics has been blessed with the addition of numerous new researchers and practitioners—both young and old alike—and readers will note the expanded collection of authors represented here. These men and women constitute a remarkable cross-section of the human capital devoted to geostatistical endeavors in academia, industry, and government, and our hats go off to them for their continued commitment to excellence and creativity, and to their patience with us as editors.

The book is divided into five sections. Each section is accompanied by its own short introduction. As a general introduction, Section I contains reflective essays pertaining to the state of geostatistics and stochastic modeling, and addressing their interrelatedness and their connection to one another. Section II is devoted to theoretical developments and the advancement of fundamental principles. Section III focuses on novel applications of methodology in a variety of settings and situations. Fully developed case studies are grouped together in Section IV. While there is a fine line between methodological applications and case studies, the chapters in this section represent more comprehensive investigations encompassing the technical, practical, and operational aspects of conducting a geostatistical investigation. Section V contains papers on systems and resources that are pertinent to the continuing evolution of geostatistics and stochastic modeling. The resulting collection of chapters represents what we believe to be a good mix of theory, methods, applications, and philosophy.

While embracing diversity of ideas, applications, methodologies, and situations, we felt it necessary to impose some editorial consistency on the final version. Good communication is essential to the successful completion of any geostatistical project, and the use of common terminology promotes better understanding among all parties involved. Hence, we have made an extensive effort to standardize language and word usage throughout the book, and to impose a consistent presentation format. For example, we have chosen to use the word "semivariogram" as opposed to "variogram," we have chosen "data" to be plural rather than singular, and we have required some rigor in the use of the terms "model" and "modeling." In the editing process we have made a conscious effort to enforce these rules, but in an effort of this size, complete adherence is essentially impossible. No doubt disparities remain. Finally, we have made every effort to eliminate spelling and grammatical mistakes, but some will have gone undetected.

Throughout the book there are references to vendors and products that are of a proprietary nature. We have made every attempt to give credit where credit is due, and we apologize in advance to those who may not have received proper acknowledgement. Reference to specific commercial or governmental entities by various authors, or mention of products or services available from such organizations, does not necessarily represent our views or opinions, or those of the publisher, nor does it in any way constitute an endorsement.

From the beginning of this project we have had three main objectives: (1) to advance the general knowledge of geostatistics; (2) to benchmark the state-of-the-art in geostatistical thought and practice, principally within the geoscience community at the beginning of the 21$^{st}$ century; and (3) to foster an understanding in the commercial world that employing geostatistical technologies can positively impact the bottom line. As the old adage goes, "nothing succeeds like success;" and so we hope the readers see the successful implementation of geostatistical tools in every chapter of this book. Whether or not we have achieved our objectives remains an open question—one that only our readers can address. We also hope SMG II will be as informative and enduring as the first volume has been, and that it will be a springboard for greater expansion of the geostatistics discipline.

Timothy C. Coburn, Abilene, Texas
Jeffrey M. Yarus, Houston, Texas
Richard L. Chambers, Tulsa, Oklahoma

# *Acknowledgments*

There are so many books on the shelves of stores and libraries that it sometimes seems they simply appear overnight. The truth of the matter is that an enormous amount of planning and effort goes into every one of them. Books don't just happen. It sometimes takes years to complete one, even with the involvement of numerous individuals.

The completion of a book like this one is solely dependent on the countless hours of volunteers. Every editor, author, and reviewer is a volunteer, and they freely contribute their time and effort because they believe in the project and they get a sense of satisfaction in seeing it through to completion.

We are particularly grateful to the staff at AAPG for their patience, guidance, and helpfulness throughout the publication process. We are also indebted to our colleagues in our respective companies who, throughout the duration of this project, have tolerated our quirkiness, our bouts of anxiety, and sometimes even our short tempers, and who have admirably covered for us when we just couldn't quite get it right.

Special thanks go to Abilene Christian University, the U.S. Geological Survey in Reston, the University of Nebraska, George Mason University, Quantitative Geosciences, and Sandia National Laboratories for permitting us the use of their facilities, libraries, and computing resources to prepare various aspects of the final draft.

In addition, we gratefully acknowledge the tireless service of the following individuals who assisted us in providing expert peer reviews so that the final product would be of the highest quality:

Hans Wackernagel, Centre de Geostatistique, Ecole des Mines de Paris
Alain Galli, Centre d'Économie Industrielle, Ecole des Mines de Paris
Ricardo Olea, formerly of the Kansas Geological Survey
Steven Carle, Lawrence Livermore National Laboratory
Eileen Poeter, Colorado School of Mines
Mary Harris, Savannah River Site
Greg Glach, Savannah River Site
Carlos Torres-Verdin, University of Texas-Austin
Michael Hohn, West Virginia Geological Survey
Petter Abrahamsen, Roxar
Raj Shrestha, Veritas DGC, Inc.
Carol Gotway, Centers for Disease Control and Prevention
Seiichi Nagihara, Texas Tech University
Marc Genton, North Carolina State University
Dale Zimmerman, University of Iowa
Karl Kramer, Roxar
Joost Herweijer, Reservoir Team Ltd.
Christopher Murray, Pacific Northwest National Laboratory
Ting Ting Yao, ExxonMobil
Didier Renard, Centre de Geostatistique, Ecole des Mines de Paris
Lars Holden, Norwegian Computing Center
Dean Oliver, University of Oklahoma
Jim Jennings, Texas Bureau of Economic Geology
Jeffrey May, EOG Resources
Alister MacDonald, Roxar
Jerry Jensen, Texas A&M University
Christopher White, Louisiana State University
Olivier Dubrule, Total-Elf-Fina
Donald Myers, University of Arizona

Mohan Kelkar, University of Tulsa
Sangar Ronghe, University of Brunei
Henning Omre, Norwegian University of Science and Technology
Jim Carr, University of Nevada-Reno
Roland Froidevaux, FSS Consultants
Pierre Delfiner, Total-Elf-Fina
Alexandre Desbarats, Canadian Geological Survey
Dennis Prezybindowski, ChevronTexaco
Yves Touffait, Geovariances
Steve Clawson, iReservoir, Inc.
Henning Trappe, TEEC
Tapan Mukerji, Stanford University

Last, but by no means least, we have racked up a huge debt to our wives, children, and other family members because of the countless hours this project has taken us away from them, not only in terms of our physical absence but also our reduced attention to their own needs. We salute them for their continuing patience, their tireless love, and their unending encouragement of our professional pursuits through endeavors such as this.

From Tim: Thank you Nancy, Claire, and Kevin

From Jeffrey: Thank you Mary, Jordan, David, and Rachel

From Rich: Thank you Verda, Kris, Kelli, and Keri

# TABLE OF CONTENTS

# About the Editors

**Timothy C. Coburn**

Tim Coburn is Professor of Statistics at Abilene Christian University in Abilene, Texas, where he teaches courses in applied statistics, information and decision support systems, and knowledge management. He holds a Ph.D. in statistics from Oklahoma State University. Along with his teaching duties, Coburn is actively engaged in statistical consulting and applied research across several segments of the energy complex, including petroleum resource estimation, subsurface characterization, energy economics and market analysis, development of alternative transportation fuels, and environmental restoration. He has also served as Visiting Professor/Researcher in the Mathematical Geology Group at the Kansas Geological Survey in Lawrence, Kansas; the Eastern Energy Resources Division of the U.S. Geological Survey in Reston, Virginia; and the Energy and Environment Division at Sandia National Laboratory in Albuquerque, New Mexico.

Prior to beginning his tenure at ACU, Coburn held a number of technical and administrative positions at Phillips Petroleum Company, Marathon Oil Company, and the National Renewable Energy Laboratory. At Phillips, he served as Research Statistician at the Bartlesville Research Center, and as a consuting statistician in the Western Division Drilling and Production Office in Denver, where he was assigned to the Prudhoe Bay equity redetermination on behalf of the Mobil-Phillips-Chevron consortium. At Marathon's Petroleum Technology Center in Littleton, Colorado, he was first employed as Senior Scientist in the Geophysical Technology Department and then, for many years, as Supervisor of Quantitative Geosciences. He also served as Supervisor of Mapping and Database in the Exploration Services organization at Marathon's Headquarters Office in Houston.

Coburn is active in many professional and industry organizations, including the International Association for Mathematical Geology and the American Association of Petroleum Geologists. He was a long-time member of AAPG's Publications and Geologic Computing committees, and he currently serves on the Editorial Board of *Natural Resources Research* and as the Book Review Editor for *Mathematical Geology*.

**Jeffrey M. Yarus**

Jeffrey Yarus is co-owner of Quantitative Geosciences, LLP, and lives in Houston, Texas. While attending the College of Wooster, he received a scholarship to study geology for one year at the University of Durham, England, before earning his B.A. at Wooster. He earned his M.S. and Ph.D. degrees in geology from the University of South Carolina, where he refined his interests in computer mapping and numerical and statistical analysis. He joined Amoco Production Company in New Orleans, Louisiana, as a production geologist for the Gulf Coast Region. He left Amoco, to work as an independent for eight years on a variety of domestic basins in the Rockies, mid-continent, and Appalachian regions. He joined Marathon Oil Company's Petroleum Technology Center in 1988, where he was instrumental in instituting desktop computer mapping and geostatistical technology. Before co-founding Quantitative Geosciences with Richard Chambers, he held various managerial positions with GeoGraphix, Geomath, Roxar (formerly Smedvig Technologies), and Knowledge Reservoir. His professional activities include publishing, consulting, lecturing, and teaching applied training courses in reservoir characterization and geostatistical applications. Yarus currently serves as Adjunct Professor in the Petroleum Engineering Department at the University of Houston. He has also served as Adjunct Professor at the University of Colorado, Denver, and as Visiting Professor at the Colorado School of Mines. He is the current Chairman of the AAPG Reservoir Development Committee.

**Richard L. Chambers**

Rich Chambers is co-owner of Quantitative Geosciences, LLP, and lives in Bixby, Oklahoma. He earned B.A. and M.S. degrees in geology from the University of Montana, and then attended Michigan State University, earning a Ph.D. in geology with an emphasis on quantitative sedimentology and statistical applications. He worked for the Great Lakes Environmental Research Laboratory, National Oceanic and Atmospheric Administration (NOAA), U.S. Department of Commerce for eight years before joining Phillips Petroleum Company, where for eight years his activities included research in high-resolution P- and S-wave seismic imaging and developing exploration plays and lease acquisition strategies as a member of a regional basin studies team. In 1988, he joined the Amoco Production Company Research Center where he worked for nine years in various technical and management capacities related to the application of geostatistical methods for exploration and production problems. In 1997, he became a consultant providing training and geostatistical modeling expertise to the petroleum industry. His professional activities include publishing, consulting, lecturing, and teaching applied training courses in reservoir characterization and geostatistical applications. In late 2001, he co-founded Quantitative Geosciences, LLP, with Jeffrey Yarus, which has offices in Broken Arrow and Houston.

# *Dedication*

This book is dedicated to the men and women of the Mathematical Geology Group at the Kansas Geological Survey. The application of quantitative and computational techniques to geoscience problems arguably got its beginning at KGS in the mid-1960s; and in the succeeding decades, the Mathematical Geology Group served as the philosophical and intellectual home base for this evolving discipline. As a result of internal management conflict within the Survey and the University of Kansas, the Mathematical Geology Group was surreptitiously disbanded in 2002 and several of its most distinguished researchers were fired, depriving the international community of a focal point for interaction and collaboration on many important issues. Although this turn of events was unfortunate, the work and reputation of the group continues to live through other venues and situations, and its influence will no doubt persist for years to come. Our hats go off to all of those individuals who gave so much of their lives to make this small troop of geoscientists a quality, respected organization within the scientific world. We thank you for your dedication to the discipline and for your many technical contributions that have had such a large emphasis on our own lives and careers, as well as those of many others.

We also dedicate this book as a memorial to Gladys Srivastava, whose love of language and learning captivated many within the mathematical geology community. She was an inspiration to her family, her friends, and everyone she met. Her hospitality was the source of many happy times and spirited conversations, and she was part of the emotional and intellectual glue that still binds us together.

# *Acknowledgments*

AAPG wishes to thank the following
for their generous contributions to

## Stochastic Modeling and Geostatistics: Principles, Methods, and Case Studies, Volume II

**ExxonMobil Upstream Research**

Contributions are applied toward the production cost of publication,
thus directly reducing the book's purchase price
and making the volume available to a larger readership.

*Section I*

# Introduction—Looking Back and Moving Forward

We begin this volume with a deep appreciation of the role that geostatistics and stochastic modeling have played in the geosciences over the past several decades, as well as a full awareness of their evolving influence in recent years. In this first section we revisit some of the historical developments in these disciplines that now loom large in the legacy of geoscientific investigation, and we look forward to future advances that may have an equally significant impact.

In the first chapter, the editors briefly lay the groundwork for the book. They specifically focus on developments that have occurred since the first volume of *Stochastic Modeling and Geostatistics* appeared, and they engage in a bit of crystal-ball gazing about what might transpire in the future. In the second chapter, Don Myers, who has been a giant in the area of geostatistics and stochastic modeling for many years, provides some additional historical perspective, reflects on progress that has been made throughout the time of his own career (as well as some nagging difficulties), and speculates about the future. This is one of Don's finest essays spanning a wide variety of ideas and perspectives. The third chapter consists of a virtual conversation between Tim Coburn, Patrick Corbett (Heriot-Watt University), Jerry Jensen (Texas A&M University), Mohan Kelkar (University of Tulsa), Dean Oliver (University of Oklahoma), Akhil Datta-Gupta (Texas A&M University), and Chris White (Louisiana State University). The topic of this conversation is the impact of geostatistics in petroleum, production, and reservoir engineering. Some of the most important developments in geostatistics have had their beginnings on the engineering side of the petroleum business, and readers will find this to be an extremely informed and enlightening discussion about the past, present, and future of the discipline by some of the world's leading experts.

Coburn, T. C., J. M. Yarus, and R. L. Chambers, 2006, Geostatistics and stochastic modeling: Bridging into the 21st century, *in* T. C. Coburn, J. M. Yarus, and R. L. Chambers, eds., Stochastic modeling and geostatistics: Principles, methods, and case studies, volume II: AAPG Computer Applications in Geology 5, p. 3–9.

Chapter 1

# Geostatistics and Stochastic Modeling: Bridging into the 21st Century

**T. C. Coburn**
*Abilene Christian University*
*Abilene, Texas, U.S.A.*

**J. M. Yarus**
*Quantitative Geosciences, LLP*
*Houston, Texas, U.S.A.*

**R. L. Chambers**
*Quantitative Geosciences, LLP*
*Broken Arrow, Oklahoma, U.S.A.*

## ABSTRACT

There has been a remarkable expansion in the fields of geostatistics and spatiostochastic modeling during the decades spanning the advent of the new millennium. The nearly ubiquitous nature of spatial relationships across all disciplines and all walks of life is one of the most important scientific, social, and philosophical realizations of the era; and the evolution of thought and practice in geostatistics and spatiostochastic model reflects this newfound understanding. This chapter traces some of the major developments in these fields during this time frame, particularly in the geoscience context. It also presents a glimpse into the future regarding what geostatistics and spatiostochastic modeling might look like in the coming years.

In 1994, when the first volume of *Stochastic Modeling and Geostatistics* (Yarus and Chambers, 1994) appeared, geostatistics was still a developing discipline, and its relationship to the broader field of stochastic modeling was not well defined. In fact, in one of the early chapters of that volume, Coburn (1994) offered the following observation:

> Geostatistics, as an art and science, is growing up within the petroleum industry, particularly within exploration and production. After approximately 10 years of determined effort by a relatively small number of individuals at a handful of industry and university research centers, geostatistics finally, although cautiously, seems to be catching on.

and caught on it has. Since 1994, so much has happened, and so much development has occurred that


DOI:10.1306/10631037CA51306

it is hard to know where to begin. The decade of the 1990s brought so much change in terms of improved theory and novel applications that geostatistics hardly even seems like the same discipline it was back then; and the pace of development into the 21st century has certainly not slowed. This expansion of interest in all things geostatistical is evidenced by a swelling body of literature (see the last chapter in this volume), higher attendance at an increasing number of technical conferences, and extensive commercialization of computational tools.

What is fueling all of this activity and interest? There is, of course, a natural maturing of the geostatistics discipline that is occurring; but there are other important factors at play as well. First and foremost is the power of computing. The effect of the revolution in computational power and the availability of computational resources at relatively cheap prices probably cannot be measured or overemphasized. Before computing grew up, the promise of geostatistics, like so many other good quantitative ideas, seemed likely to fade under the burden of mathematical tedium and complexity; but the availability of essentially unlimited computational power has almost literally brought geostatistics to life from a practitioner's point of view.

Second, geostatistics and stochastic modeling became inexorably intertwined. The first volume of *Stochastic Modeling and Geostatistics* represents one of the earliest times that the two ideas were formally linked together, particularly in the earth sciences. In fact, Srivastava's (1994) opening chapter on stochastic methods for reservoir characterization may have consummated the marriage. Since that time, however, the broader field of stochastic modeling has come to embrace the ideas of geostatistics in ways that were previously not possible, nor perhaps even desirable. This linkage has grown out of improved theoretical and methodological work in both camps that is grounded in the ideas of applied probability. Clearly, not all stochastic modeling is linked to geostatistics, and so, over time, this specific subdiscipline has come to be better characterized as spatiostochastic modeling.

Third, an intense interest in spatial ideas developed during the course of the 1990s throughout the scientific community as societal relationships changed and economies became more global. Spatial thinking, which had been developing on several fronts began to coalesce into a more unified body of knowledge. The work of Noel Cressie (first at Iowa State University and later at Ohio State University) and others during this time frame, and specifically, the publication of *Statistics for Spatial Data* (Cressie, 1993), helped spur this fusion of thought and approach, which had begun earlier with contributions from Ripley (1981, 1988), Diggle (1983), and others. In particular, many ideas from the field of quantitative geography, which had previously not been directly tied to geostatistics and stochastic modeling, began to exert a strong influence (note the simultaneous rapid expansion of geographic information systems [GIS] during this same time frame). The Department of Geography at the University of California–Santa Barbara has been one of several particularly influential academic centers.

Fourth, the geostatistics community, and the broader spatial statistics domain as a whole, has been blessed and enriched by an influx of young researchers who are extremely bright and eager to embrace new challenges. Organizations like the International Association for Mathematical Geology have helped stimulate interest among these individuals by offering to provide funds to support research activities; and additional training programs have evolved beyond academic centers that have historically been linked with geostatistics and stochastic modeling, particularly outside the United States. Several examples can be given, but the move of Clayton Deutsch from Stanford University to the University of Alberta is particularly noteworthy in the geoscience arena. If anything, the development of geostatistics and stochastic modeling, particularly within the geosciences, has taken on an even more international flavor than before, with a great deal of important work being done in Norway, France, the United Kingdom, China, and elsewhere.

Here at the midpoint of the first decade in the 21st century, it is interesting to reflect on some of the ways in which geostatistics, spatial statistics, stochastic modeling, and related disciplines have penetrated and impacted the scientific community. Seemingly, almost every discipline has been influenced, including areas of business, education, and the social sciences. Precision agriculture; estimation of natural resources; precipitation mapping; ecological modeling; reservoir characterization; groundwater-flow patterns; environmental remediation; disease tracking; market analysis; remote sensing, forensic analysis, and satellite imaging; national security; and human and animal population dynamics are just some of the many ideas and tasks that can be found linked to geostatistics and stochastic modeling in a simple Google™ search of the Internet. Evidence of the growing influence of spatial and stochastic thinking in general problem solving is also readily available in the literature and on the technical conference and training circuit. Since 1995, at least 25 new or revised texts pertaining to geostatistics, spatial statistics, or stochastic modeling have appeared, two of the most recent of which are authored by Moore (2001) and Schabenberger and Gotway (2004). Throughout this time frame, the American Statistical Association (ASA) and affiliated societies have regularly hosted or sponsored meetings and conferences focused on applications of spatial statistics, such as Spatial Statistics: Integrating Statistics, GIS, and Statistical Graphics presented by the ASA Section on Statistics in the Environment in 2002.

The expansion of geostatistics and stochastic modeling into the environmental arena is particularly noteworthy, with numerous applications related to cleanup and remediation, waste disposal, subsurface pollutant transport, air quality, climate change, and the like (e.g., see Jean et al., 2004). This heightened concentration of activity can be partly attributed to increased awareness of environmental issues, worldwide, which were brought into focus by the Kyoto Protocol, an amendment to the United Nations Framework Convention on Climate Change opened for signature in December 1997. Several important conferences specifically focused on geostatistics, spatial statistics, and stochastic modeling in environmental contexts have occurred in the decade since Kyoto, such as the 2005 International Workshop on Modeling and Data Analysis in Environmetrics, Geostatistics, and Related Areas and the continuing biennial GeoENV series begun in 1996. There have been abundant contributions to the literature as well. Along with the many scholarly articles published in journals such as *Stochastic Environmental Research and Risk Assessment* and *Environmental Geosciences*, the texts of Walden and Guttorp (1992), Mowrer et al. (1996), Webster and Oliver (2001), and Renard et al. (2005) are notable examples of the emphasis on environmental concerns during this period of time.

Perhaps the clearest manifestation of geostatistics as a discipline has been the quadrennial series of meetings of the International Geostatistics Congress, the most recent of which was held in Banff in 2004. The North American Council on Geostatistics, a similar but much smaller group, has also met on a periodic basis since the mid-1990s. Several other important meetings and conferences specifically devoted to geostatistics have been conducted during this same general time frame, including the series of CODATA (Committee on Data for Science and Technology) conferences on geomathematics and geostatistics. Many of these events have had a conspicuous focus on mineral and petroleum exploration and development, as well as other segments of the geosciences because the origin of geostatistics can be traced, in part, to these disciplines. Many new and important ideas in geostatistics continue to be conceived and developed at centers of excellence such as the Stanford Center for Reservoir Forecasting, Centre de Géostatistique de l'Ecole des Mines de Paris, and the Norwegian Computing Center, and some of the newest textbooks on geostatistics are decidedly aimed at the mineral and petroleum industries (e.g., Deutsch, 2002; Kelkar et al., 2002; Journel and Kyriakidis, 2004; Caers, 2005).

The broader concept of stochastic modeling (beyond its ties to geostatistics) has also experienced an increasing level of adaptation and application in numerous disciplines, including fields such as finance that were previously more focused on deterministic methods. Some of the most promising theoretical and methodological developments have involved various flavors of stochastic simulation, the Markov chain Monte Carlo (MCMC), and the empirical Bayes modeling techniques (see Gilks et al., 1996; Gamerman, 1997; Kendall et al., 2005). Although quite a bit of work had already been done with stochastic modeling in the late 1980s and early 1990s (in the earth sciences, most notably in the area of hydrogeology), the texts of Ripley (1987) and Guttorp (1995) were two of several that helped propel ideas about stochastic modeling into the broader social, scientific, and economic realms.

As the foregoing discussion suggests, the decades bridging the new millennium have been a time of extensive development and maturation for geostatistics and stochastic modeling, both as separate yet highly interrelated disciplines. Although crystal-ball gazing is commonly dangerous, it seems likely that this expansion of knowledge, application, and influence will continue unabated for some time to come. Although it is doubtful that anyone will be doing geostatistics or stochastic modeling on their iPods® or cell phones anytime soon, even greater access to data and information via the Internet and through wireless capabilities; more powerful computational algorithms and visualization tools; and cheaper, faster, and ever smaller computers (along with parallel processing) will no doubt provide a further boost to what is currently believed to be the state of the art in these and related disciplines. In addition, as interconnectedness increases, questions about spatial relationships and spatial uncertainty will naturally follow, which will, in turn, promote even greater interest in geostatistical and spatiostochastic problem-solving approaches. As with all analytical and modeling tools, there are, and will continue to be, limitations, particularly because not all methods and models are uniformly applicable to all situations. Hack (2005) provides one such assessment of limitations relative to the characterization of sand and gravel resources.

If the current trend persists, there should be even further convergence of ideas in spatial thinking from different disciplines, which, in turn, will affect developments in geostatistics and stochastic modeling. This move to a more holistic approach to spatial data analysis that incorporates best practices from several viewpoints is long overdue. In particular, for much of their history, geostatisticians have largely worked beyond the realm of geographers and more classically trained statisticians who view spatial relationships from somewhat different perspectives. In recent years, however, there has been more determined outreach from all camps and a greater willingness to incorporate alternative ideas and approaches. To a large degree, such cooperation has been driven by the maturation of computational technologies and the advent of tools such as GIS, which facilitate more rapid and extensive data processing and better visualization and physical conceptualization of spatial relationships and modeling results. Because GIS is one of the fastest growing areas of information technology, its dominance in the management, delivery, presentation, analysis, and interpretation of spatial data is likely to drive even greater scientific collaboration and additional convergence of analytical and modeling approaches (see Coburn and Yarus, 2000; Dubois, 2000). Commercial software products such as those provided by the Environmental Systems Research Institute already blend typical GIS functionality with geostatistical modeling. The growing convergence of spatial thinking has also been the hallmark of a continuing series of international conferences on spatial accuracy assessment (Mowrer et al., 1996, and others), and additional evidence of the cross-fertilization of ideas is appearing in the literature of several disciplines (e.g., Kyriakidis, 2001).

In the specific areas of petroleum exploration and development, there is likely to be even more emphasis on methods to effectively incorporate data from multiple sources, particularly seismic and well-log data. This comes as no surprise because exploration and development activities rely heavily on these kinds of data, and the quality of such information is constantly being enhanced. However, with the recent upturn in oil and gas prices, and with continuing strain on supply and demand, there will be increased pressure to incorporate additional indicators whose quality and reliability may not be as well established. Improving techniques for upscaling petrophysical properties, particularly permeability, has been an important focus of reservoir engineering research throughout this time frame, and that emphasis is likely to continue and perhaps even become more intense. Further, Dubrule and Damsleth (2001) have suggested that the establishment of decision support workflows will be given greater prominence in the geoscience sector (also see Herweijer et al., 2006).

Another focal point will be the development of faster and more efficient computational algorithms that facilitate the processing and analysis of massive data sets, such as those produced through finer cell discretization of subsurface features or through higher resolution data collection. Such algorithms are likely to increasingly rely on soft computing techniques (e.g., fuzzy logic and neural networks) and artificial intelligence.

Along similar lines, a significant focus on three-dimensional visualization and earth modeling that first emerged in the late 1970s and early 1980s and came to fruition in the 1990s is likely to attract even more attention and support, particularly in the petroleum industry. Gocad, a comprehensive modeling and visualization package originally developed by the computer science group within the national school of geology (École Nationale des Sciences Géographiques; ENSG) in Nancy, France, is one of the best known products of this type. Schlumberger, Roxar AS, and other commercial vendors continue to develop related technology.

As suggested previously, in recent years, there has been considerable scientific interest in the use of empirical Bayesian approaches to explore high-dimensional probability distributions and the use of MCMC methods (including Metropolis-Hastings sampling, Gibbs sampling, and related tools) to estimate unknown parameters and construct bounds on the uncertainty of these estimates. Such approaches represent alternatives to geostatistically based stochastic simulation, which will no doubt receive considerable attention. There is also likely to be continued interest in other data-driven spatial estimation procedures that are closely tied to the notions of bootstrapping, jackknifing, and cross-validation.

Much more work is needed on the selection, ranking, and comparison of modeling approaches; evaluation of competing model realizations, arising from different scenarios; and determination of the optimum number of realizations. Research has already been conducted on several fronts and from the perspectives of a variety of disciplines (e.g., Zimmerman et al., 1998, 1999; Kupfersberger and Deutsch, 1999; Brandsaeter et al., 2001; Gilman et al., 2002; Dubois and Galmarini, 2005; Patriarche et al., 2005; Hoeting et al., 2006), but a unified approach is still not available. Software implementations of some model ranking procedures are already beginning to appear in some commercial products. Clearly, each specific problem or objective should and will drive the selection of an optimum approach; yet establishment of a set of clear-cut best practices would be a major contribution. Any attempt to construct broad-based comparisons must be based on well-designed computational experiments.

Finally, given the extent to which computer technology already permeates all of science and business, it seems readily apparent that there will be more extensive and more thorough software implementation of new analytical and modeling approaches and algorithms, and in turn, more widespread commercialization of them. Commercially available software packages have steadily evolved since the 1980s and continue to do so even today, with the rollout time for new functionality being progressively shortened. The metamorphosis of the venerable Bluepack geostatistical software, which originated at Centre de Géostatistique de Fontainbleau, into Isatis, currently marketed by Geovariances, is but one example.

The number of low-cost and publicly available software packages for geostatistics and stochastic modeling has also increased during this same time frame, and their functionality has greatly improved (see Corps of Engineers, 1997, for a nice comparison of the STATPAC, GeoEas, GSLIB, and GMS2.0 packages; other comparisons may be found at www.ai-geostats.org and www.ibmpcug.co.uk/~bedrock/gsd). Yet, although cost is an advantage of these products, most parties will find themselves drawn to the larger commercially available packages for which maintenance and support, as well as resources to add functionality and upgrade the user interface, are more readily available. Except, perhaps, for GSLIB, which continues to be developed by the Stanford Center for Reservoir Characterization and associated organizations, and which has a loyal user base, most of the low-cost and publicly available software packages are not likely to be improved much beyond their current versions simply because of economics.

In summary, the decades bridging the advent of the new millennium have brought significant expansion in the theory and practice of geostatistics and stochastic modeling. Such expansion has resulted, in part, from a more comprehensive understanding that spatial relationships exist across nearly all disciplines and walks of life. Although perhaps originally regarded as individual domains of thought, geostatistics and stochastic modeling actually have many themes in common, and distinct lines of convergence are emerging. Such convergence should result in a more holistic approach to spatial data analysis in the future, along with even greater integration of methods originating from several different disciplines, better and faster algorithms capable of processing much larger and more diverse data sets, increased emphasis on best practices, and faster operationalization of new techniques driven by the commercial interests of software vendors and service organizations.

## REFERENCES CITED

Brandsaeter, I., H. T. Wist, A. Naess, O. Lia, O. J. Arntzen, P. S. Ringrose, A. W. Martinius, and T. R. Lerdahl, 2001, Ranking of stochastic realizations of complex tidal reservoirs using streamline simulation criteria: Petroleum Geoscience, v. 7, supplement, p. 53–63.

Caers, J., 2005, Petroleum geostatistics: Richardson, Texas, Society of Petroleum Engineers, 96 p.

Coburn, T. C., 1994, Reflections on the proliferation of

geostatistics in petroleum exploration and production, *in* J. M. Yarus and R. L. Chambers, eds., Stochastic modeling and geostatistics: Principles, methods, and case studies: AAPG Computer Applications in Geology 3, p. 21–23.

Coburn, T. C., and J. M. Yarus, 2000, GIS and geostatistics— A marriage of technologies, *in* T. C. Coburn and J. M. Yarus, eds., Geographic information systems for petroleum exploration and development: AAPG Computer Contributions 4, p. 295–298.

Corps of Engineers, 1997, Practical aspects of applying geostatistics at hazardous, toxic, and radioactive waste sites: Washington, U.S. Department of the Army, Technical Letter 1110-1-175, 95 p.: www.usace.army.mil/usace-docs /eng-tech-ltrs/etl110-1-175/toc.html (accessed June 21, 2006).

Cressie, N. A. C., 1993, Statistics for spatial data, revised ed.: New York, John Wiley & Sons, 900 p.

Deutsch, C. V., 2002, Geostatistical reservoir modeling: New York, Oxford University Press, 376 p.

Diggle, P., 1983, Statistical analysis of spatial point processes: New York, Academic Press, 148 p.

Dubois, G., 2000, Interfacing GIS and geostatistics, *in* J. Pilz, ed., Special Volume on Conference Proceedings of STATGIS'99: Dresden, CPress-Verlag, 11 p.

Dubois, G., and S. Galmarini, 2005, Introduction to the spatial interpolation comparison (SIC) 2004 exercise and presentation of the data sets: Applied Geographic Information Systems, v. 1, p. 09-01–09-11.

Dubrule, O., and E. Damsleth, 2001, Achievements and challenges in petroleum geostatistics: Petroleum Geoscience, v. 7, supplement, p. 1–7.

Gamerman, D., 1997, Markov chain Monte Carlo: Stochastic simulation for Bayesian inference: London, Chapman & Hall, 245 p.

Gilks, W. R., S. Richardson, and D. J. Spiegelhalter, 1996, Markov chain Monte Carlo in practice: London, Chapman & Hall, 486 p.

Gilman, J. R., H.-Z. Meng, M. J. Uland, P. J. Dzurman, and S. Cosic, 2002, Statistical ranking of stochastic geomodels using streamline simulation: A field application approach: Richardson, Texas, Society of Petroleum Engineers Annual Technical Conference and Exhibition, SPE Paper 77374, 11 p.

Guttorp, P., 1995, Stochastic modeling of scientific data: New York, Chapman & Hall, 372 p.

Hack, D. R., 2005, Issues and challenges in the application of geostatistics and spatial-data analysis to the characterization of sand-and-gravel resources, *in* J. D. Bliss, P. R. Moyle, and K. R. Long, Contributions to industrial-minerals research: U.S. Geological Survey Bulletin, v. 2209-J, 14 p.

Herweijer, J. C., J. M. Yarus, and M. J. V. Suana, 2006, Process-based knowledge management: Overview and application of iterative 3-D modeling workflows, *in* T. C. Coburn, J. M. Yarus, and R. L. Chambers, eds., Stochastic modeling and geostatistics: Principles, methods, and case studies, vol. II: AAPG Computer Applications in Geology 5, p. 313–324.

Hoeting, J. A., R. A. Davis, A. A. Merton, and S. E. Thompson, 2006, Model selection for geostatistical models: Contemporary Statistics and Ecology, v. 16, p. 87–98.

Jean, J. A., J. M. Yarus, G. P. Flach, M. R. Millings, M. K. Harris, R. L. Chambers, and F. H. Syms, 2004, Three-dimensional geologic model of southeastern Tertiary coastal-plain sediments, Savannah River Site, South Carolina: An applied geostatistical approach for environmental applications: AAPG/DEG Environmental Geosciences, v. 11, p. 205–220.

Journel, A. G., and P. C. Kyriakidis, 2004, Evaluation of mineral reserves: A simulation approach: New York, Oxford University Press, 216 p.

Kelkar, M., G. Perez, and A. Chopra, 2002, Applied geostatistics for reservoir characterization: Richardson, Texas, Society of Petroleum Engineers, 264 p.

Kendall, W. S., F. Liang, and J.-S. Wang, 2005, Markov chain Monte Carlo: Innovations and applications: Hackensack, New Jersey, World Scientific, 220 p.

Kupfersberger, H., and C. V. Deutsch, 1999, Ranking stochastic realizations for improved aquifer response uncertainty assessment: Journal of Hydrology, v. 223, p. 54–65.

Kyriakidis, P. C., 2001, Geostatistical models of uncertainty for spatial data, *in* C. T. Hunsaker, M. F. Goodchild, M. A. Friedl, and T. J. Case, eds., Spatial uncertainty in ecology: Implications for remote sensing and GIS applications: New York, Springer-Verlag, p. 175–213.

Moore, M., 2001, Spatial statistics: Methodological aspects and applications: New York, Springer, 282 p.

Mowrer, H. T., R. L. Czaplewski, and R. H. Hamre, 1996, Spatial accuracy assessment in natural resources and environmental sciences: Second International Symposium: Ft. Collins, Colorado, U.S. Department of Agriculture Forest Service, General Technical Report RM-GTR-277, 728 p.

Patriarche, D., M. Castro, and P. Goovaerts, 2005, Estimating regional hydraulic conductivity fields— A comparative study of geostatistical methods: Mathematical Geology, v. 37, p. 587–613.

Renard, P., H. Demougeot-Renard, and R. Froidevoix, 2005, Geostatistics for environmental applications: Proceedings of the 5th European Conference on Geostatistics for Environmental Applications: Berlin, Springer, 480 p.

Ripley, B. D., 1981, Spatial statistics: New York, John Wiley & Sons, 252 p.

Ripley, B. D., 1987, Stochastic simulation: New York, John Wiley & Sons, 237 p.

Ripley, B. D., 1988, Statistical inference for spatial processes: Cambridge, Cambridge University Press, 148 p.

Schabenberger, O., and C. A. Gotway, 2004, Statistical methods for spatial data analysis: Boca Raton, Florida, Chapman Hall-CRC, 488 p.

Srivastava, R. M., 1994, An overview of stochastic methods for reservoir characterization, *in* J. M. Yarus and R. L. Chambers, eds., Stochastic modeling and geostatistics: Principles, methods, and case studies: AAPG Computer Applications in Geology 3, p. 3–16.

Walden, A. T., and P. Guttorp, 1992, Statistics in the environmental and earth sciences: London, E. Arnold, 306 p.

Webster, R., and M. A. Oliver, 2001, Geostatistics for environmental scientists: New York, John Wiley & Sons, 271 p.

Yarus, J. M., and R. L. Chambers, eds., 1994, Stochastic modeling and geostatistics: Principles, methods, and case studies: AAPG Computer Applications in Geology 3, 379 p.

Zimmerman, D., et al., 1998, A comparison of seven geostatistically based inverse approaches to estimate transmissivities for modeling advective transport by groundwater flow: Water Resources Research, v. 34, p. 1373–1413.

Zimmerman, D., C. Pavlik, A. Ruggles, and M. P. Armstrong, 1999, An experimental comparison of ordinary and universal kriging and inverse distance weighting: Mathematical Geology, v. 31, p. 375–390.

Myers, D. E., 2006, Reflections on geostatistics and stochastic modeling, *in* T. C. Coburn, J. M. Yarus, and R. L. Chambers, eds., Stochastic modeling and geostatistics: Principles, methods, and case studies, volume II: AAPG Computer Applications in Geology 5, p. 11–22.

*Chapter 2*

# Reflections on Geostatistics and Stochastic Modeling

**D. E. Myers**
*University of Arizona*
*Tucson, Arizona, U.S.A.*

## ABSTRACT

In recent years, geostatistics and stochastic modeling have made a tremendous impact on scientific investigation. This chapter describes the relationship between these two ideas, provides a historical perspective on their development, and discusses the ways in which they have evolved, both separately and in concert with each other. Important issues impacting future development are also addressed.

## INTRODUCTION

In recent years, geostatistics and stochastic modeling have found their way into several scientific endeavors where they have been used in significant ways to address a diverse array of problems that are important to the well-being of mankind. Interest in these particular problem-solving approaches has expanded from rather modest beginnings to the point that individuals across almost all disciplines recognize their value and actively incorporate them into research and applications. This has become increasingly more evident as the importance of space and geography have become recognized in science and industry, and as scientists have come to embrace the ideas of spatial analysis and spatial statistics.

The increased interest in geostatistics and stochastic modeling has also roughly tracked a series of conferences on these topics that have been conducted throughout the last 30 yr. In 1975, for example, there was a North Atlantic Treaty Organization Advanced Study Institute entitled Advanced Geostatistics in the Mining Industry (although it included applications in petroleum and hydrology as well). A 1993 conference entitled Geostatistics for the Next Century provided a look at how geostatistics might impact the succeeding decades. Finally, the quadrennial International Geostatistics Congress was established nearly 30 yr ago, with the most recent installment (2004) being conducted in Banff, Alberta, Canada.

Throughout the years, many have reflected on the future of geostatistics and stochastic modeling, with at least one entire conference being dedicated to the subject (Dimitrakopoulos, 1994). At the 1996 International Geostatistics Congress held in Wollongong, New South Wales, Australia, Srivastava (1997) posed a timely question, asking, "Where are we going?" Although the question might not have been completely answered, it certainly inspired considerable debate; and since that time, there seems to have been a greater effort to give the discipline more overall focus. Clearly, not all the problems have been solved, nor all the issues addressed. The objective of the present discussion is to underscore some of the situations, both technical and philosophical, that, from the author's perspective, represent ongoing distractions


DOI:10.1306/1063803CA53224

in the field. Obviously, this will not be the last time someone reflects on the state of geostatistics and stochastic modeling; but perhaps this exchange will help to stimulate still more lively conversation, more open communication, and possibly even resolution of some of the ongoing difficulties.

It is necessary to make a few general comments at the outset of this dialog to set the stage for what is to follow. First, the distinction between stochastic methods (which for now will encompass both geostatistics and other stochastic modeling approaches) and deterministic methods is, today, somewhat blurred. In fact, the distinction is somewhat more of a perception or convenience instead of a reality. In their book entitled An Introduction to Stochastic Modeling, Karlin and Taylor (1998, p. 492) stated the following:

> A quantitative description of a natural phenomenon is called a mathematical model of that phenomenon ... A deterministic model predicts a single outcome for a given set of circumstances. A stochastic model predicts a set of possible outcomes weighted by their likelihoods or probabilities... However, phenomena are not in and of themselves inherently stochastic. Rather, to model the phenomenon as stochastic or deterministic is the choice of the observer.

Although the title of the book is An Introduction to Stochastic Modeling, it is, in fact, really about stochastic processes, which is a different topic with a different focus than is implied in the present context. Stochastic modeling, then, might be understood as somewhat more general than geostatistics, although both emphasize stochastic phenomena. However, in stochastic modeling, more emphasis is placed on modeling, whereas in geostatistics, more emphasis is placed on data analysis.

Second, most developments in geostatistics and stochastic modeling have been, and likely always will be, strongly driven by applications. Note that Holden et al. (2003) interpreted stochastic modeling as simply pertaining to the modeling of a petroleum reservoir. In fact, it is probably fair to say that not many really new ideas in this field have been promulgated through fundamental theoretical research, in the same sense that new ideas in mathematics and traditional statistics are developed from first principles. The applications-driven nature of geostatistics and stochastic modeling is broadly evident by the number and diversity of examples appearing in an ever-widening literature. As a result, some of the issues that have arisen throughout the years really stem from the proliferation of methodologies beyond the conventional boundaries of earth science in which geostatistics and stochastic modeling have evolved. Such issues are both theoretical and practical in nature and sometimes simply reflect the kind of conflict that emerges at the interface of history, language, terminology, and culture. Therefore, to understand some of the difficulties currently plaguing the discipline of geostatistics and stochastic modeling, the best place to start may be in the past.

## A BIT OF HISTORY

Geostatistics is a relatively new discipline, and much of its development has occurred throughout the last 30–40 yr. Through its flagship journal, *Mathematical Geology*, the International Association for Mathematical Geology (IAMG) has largely been responsible for disseminating many of the theoretical advances in geostatistics, with other organizations, corporations, and academic institutions making many important contributions in both theory and applications. The IAMG dates from 1968, and almost from its inception, the association recognized the significance of this emerging discipline. However, even well before that time, there were many examples of probability and statistics being applied to earth science investigations. For example, the crucial work of Gandin (1963), Matheron (1965), and Matérn (1986) all predate the establishment of the IAMG. Perhaps because of language differences (Swedish in the case of Matérn and Russian in the case of Gandin) as well as Matheron's affiliation with Ecole des Mines in Paris, his work and those of his students became better known.

In retrospect, Michel David's move from Fontainebleau to the University of Montreal in 1968 (see Dimitrakopoulos and Dagbert, 2001) and Andre Journel's move from Fontainebleau to Stanford University in 1978 were watershed events that greatly increased the interest in geostatistics, particularly in the United States and Canada. Together, David (1977) and Journel and Huijbregts (1978) profoundly influenced the theoretical and methodological development of geostatistics for several decades and fundamentally altered the way scientists view the physical world. The impact of these two scholars on spatial thinking and the practice and mechanics of stochastic modeling is well known, and their names are commonly associated with the expansion of geostatistics from the earth sciences into other disciplines, including medicine, public health, business, and the environment.

Despite the extensive influence of these and other individuals, there was not a large number of people using geostatistical methods in the 1970s and into the 1980s. The knowledge base and number of practitioners have grown tremendously since then, but the community of geostatisticians and stochastic modelers remains comparatively small even today. In fact, formal academic training programs in these disciplines are still not widely available, with only a handful of universities in the United States offering such programs.

One particular problem has to do with an ongoing conflict between the practitioners of geostatistics and traditional statistical methods. Geostatistics is viewed in some circles as a reinvention and repackaging of statistical principles that were already well known; however, the most devoted of geostatisticians contend that traditional statistical methods are totally ineffectual at incorporating spatial variability. One thing seems true: the emergence of geostatistics has forced practitioners of traditional statistical methods to embrace the importance of spatial variation. In fact, interest in spatial statistics has exploded in the statistical community in the last decade. In turn, geostatisticians have come to embrace more of the traditional ideas of statistics. The two camps have certainly not yet become one; but there does seem to be a greater level of cooperation and mutual appreciation than in years gone by. Such a convergence of ideas can only be good for quantitative problem solving in general because it diminishes the distrust and misunderstanding of the techniques harbored by those who are peripheral to the conflict and who are in need of real solutions to their problems.

## GEOSTATISTICS VS. STOCHASTIC MODELING

As noted above, stochastic modeling is perhaps more general than geostatistics, but other differences exist. Stochastic differential equations are models, Markov chains are models, there are models for time series, and fractals are commonly used for models. In contrast, kriging in its various forms is not really modeling. Although kriging is closely linked to modeling of the variogram or covariance function, the kriging process itself is not quite the same as modeling in the traditional sense.

Although the term geostatistics has became synonymous with the stochastic approach to spatial estimation, there are those who contend that this view is far too narrow. Given the breadth of work in spatial statistics and spatial estimation in recent years (see Anselin, 1988; Davis, 2002; Haining, 2003), this complaint could certainly be afforded some credence.

Stochastic modeling, to the extent that it is distinct from geostatistics, has, perhaps, had stronger mathematical ties. The link to mathematics is readily apparent in its many applications (e.g., turbulence problems; see Batchelor, 1953; Lumley, 1970; and the work of Kolmogorov as summarized by Hunt et al., 1991; Frisch and Kolmogorov, 1995). The principal upshot of stochastic modeling research has been to replace deterministic differential equations with stochastic differential equations, which are especially important when considering transport problems in the subsurface.

Although the breadth of applications for geostatistics has been steadily increasing, stochastic modeling is probably better known in disciplines such as hydrology and petroleum engineering. The principles of stochastic modeling are also known beyond the realm of earth and environmental science, with applications in fields such as mathematical finance and actuarial science (Actuarial Foundation, 2003). Geostatistics, however, finds most of its applications in the exploration and characterization of natural resources, with a particular historical link to the mining industry.

A natural question that might be asked, then, is whether sufficient cross-fertilization is occurring among disciplines; that is, whether the ideas and results generated in one discipline or area of application are being sufficiently used in other areas. Although it would be hard to give a definitive answer, simply asking the question raises the level of interest and consideration. Unquestionably, more interdisciplinary interaction is needed throughout all the sciences, along with more integration of science and business; and so, acknowledging the need for increased cooperation and communication could only serve to enhance the understanding of geostatistical and stochastic modeling approaches and their applications in scientific investigations.

## COMPUTING AND SOFTWARE CONSIDERATIONS

To a considerable extent, it is possible to link and trace the development of geostatistics and stochastic modeling, as well as growth in their applications, to the advent of inexpensive and accessible computing (e.g., interactive multiuser systems such Digital Equipment's virtual address extension (VAX) machines and the personal computer). Ready access to computational resources has given rise to several individualized and customized computer programs, because individuals and groups have tended to develop their own software throughout the years. This is in strong contrast to the way in which software for performing more traditional statistical computation. Many of today's geostatistical algorithms were developed during the same time frame that computing capabilities were expanding, and so the algorithms had to be refined in concert with computational enhancements. Traditional statistical methods and algorithms, however, are somewhat older, and some of them predate the modern computer.

Because many of the procedures and routines used to perform traditional statistical computations were already on the shelf, it was fairly straightforward when the computer came along to operationalize and compile them into integrated packages that could be further developed and commercially distributed. Consequently, as computing organizations began to flourish in businesses and universities, statistical software packages such as the Statistical Analysis System and the Statistical Package for the Social Sciences, although in their infancy, were already available for distribution; and so, as demand increased, it was natural for these packages to be routinely acquired. In most cases, the purchase and acquisition decision, along with the provision of subsequent maintenance and support, was assigned to central computing organizations, and the code was commonly accessible only on mainframe machines. At the same time, because many organizations were obtaining and using the same software, there was a strong move toward standardization (of the algorithms) and widespread testing of the code, thereby increasing their appeal. As a direct consequence, a strong market for statistical software evolved that persists even today, with continued enhancement of the procedures and codes being almost totally commercially driven.

As suggested above, the manner in which software evolved to perform traditional statistical computations is quite different from the evolution of geostatistical programs and software for stochastic modeling. Although there is, indeed, some commercially available geostatistical software, such as the venerable Bluepack and its successor Isatis, as well as geostatistical add-ons for comprehensive statistical packages like S-Plus and geographic information systems such as ArcView developed by the Environmental Systems Research Institute, commercialization has not been the primary driver, and the market for these products is considerably smaller. For petroleum and mining companies, cost has been less of an impediment; but for many practitioners, there is a strong reliance on free software such as the aging GeoEas, GSLIB, and more recently, Gstat and GeoR, both of which have been ported to the freeware platform R. Although there are obvious advantages to this approach, there are also distinct disadvantages. For example, there is little in the systems, or by way of geostatistical practice, to ensure that the same data processed by two different software packages will produce the same results. Further, the options and features of the different implementations are not likely to be the same. In general, the algorithms are moderately well understood, but there may be extensive differences in their implementation. The advent of Fortran-based GSLIB perhaps set some standards; but GSLIB has not been systematically updated as a package. Now, many algorithms are not included, and the use of the Fortran code in batch operations is becoming outdated.

This situation is complicated, of course, by the wide variety of people, groups, and disciplines that use geostatistics and stochastic modeling. Few are exposed to the broad range of journals that now publish papers whose results are based on geostatistical analysis or stochastic modeling. In addition, companies and businesses are reluctant to divulge proprietary codes and systems because they want to maintain their competitive advantages.

Such circumstances suggest the need for increased standardization. One possible solution would be to establish a formal mechanism by which similar codes or packages could be operationally and numerically compared, with the results of such comparisons being widely disseminated. This is a common practice with regard to commercially available statistical (traditional) analysis packages (see the regular reviews that appear in publications such as *The American Statistician*). Without diminishing the uniqueness of individual codes, the objective might be to establish some common performance benchmarks that could be recognized and accepted throughout the geostatistical community.

## SIMULATION

Simulation is a term that evokes different meanings for individuals working in different disciplines. It may be deterministic in character (e.g., numerical solution of a differential equation), or it may be probabilistic (e.g., Monte Carlo methods). Simulation is used for a variety of problems and applications, particularly when it is difficult or costly to obtain live data. For example, the U.S. Geological Survey has developed several routines to simulate fluid flow in the subsurface that have been widely adopted. In geostatistics and stochastic modeling, simulation commonly refers to the process of generating multiple realizations of a random function to obtain an acceptable numerical solution to a problem. One or more of these realizations may serve as the input for other computer programs (e.g., MODFLOW, the modular, three-dimensional [3-D], finite-difference ground-water–flow model developed by the U.S. Geological Survey uses simulated cell values for hydraulic conductivity to generate alternative flow patterns).

By its very nature, simulation cannot produce an answer that is absolutely correct. Although the process may yield a very good approximation that is altogether admissible, uncertainty is always associated with the result. This uncertainty arises in several ways. Obviously, assumptions that are improperly imposed, or imperfections in the estimates of one or more process parameters, can lead to questionable results; but there is also uncertainty in knowing which of the many results (or realizations) to choose from among all those that can be produced through simulation. In addition, there can be many different ways to simulate the same process or phenomenon, and so the choice of an approach, or algorithm, can also contribute to uncertainty about the result.

Several different algorithms are associated with the geostatistical or stochastic approach to simulation, including the turning-bands algorithm (which is really a procedure to generate 3-D realizations from multiple one-dimensional [1-D] realizations), covariance matrix decomposition (variously called Cholesky decomposition, LU decomposition, etc.), sequential Gaussian simulation (and multiple variants thereof), and simulated annealing (which is really based on an optimization approach of the same name). Some of these implicitly rely on assumptions pertaining to the multivariate Gaussian distribution, and all of them essentially require use of a known covariance function (i.e., only second-order properties of the random function are reproduced). Each has been developed because of perceived weaknesses or difficulties with competing or alternative approaches; but unfortunately, in many cases, it is not immediately clear how to choose among them. Little effort has been devoted to theoretical comparisons, and empirical examinations are generally complicated by the several restrictions previously suggested.

First, it is commonly the case in practice that only a small number of realizations can be generated (because of cost, time, or other constraints), with each realization encompassing only a finite (although perhaps large) number of locations. In this situation, it would be essentially impossible to conduct any meaningful numerical comparisons among the results of several competing algorithms. Second, little work has been done on the problem of selecting the best realization from among all those that can be produced through simulation. Because process parameters are commonly assumed to be characterized by statistical distributions (e.g., Gaussian), any number of different realizations can be randomly produced, and identification of one or more that appear to be optimum is largely a subjective process. Finally, on an even more fundamental level, the initial choice of the algorithm(s) itself is important if the simulation results (i.e., the realizations) are to be used for further analysis or for decision making because those results are likely to change if a different algorithm is selected. Note that, with the exception of simulated annealing, the various algorithms only reproduce the distributional characteristics of the quantity of interest in an average sense (i.e., averaged across realizations), and so, direct comparisons of individual realizations produced by different algorithms using essentially the same inputs would not be completely valid anyway.

An additional aspect of simulation that has received inadequate attention is random-number generation. All geostatistical-simulation algorithms rely on random-number generators when producing various results, and yet there is a tendency in the literature to ignore the possible effects arising from inappropriate or inadequate procedures (see Van Niel and Laffan, 2001). This situation begs for more theoretical and empirical investigation, but virtually no work has been done on it within the geostatistical community. A solid research effort to examine and substantiate the quality of random-number generation algorithms could yield great dividends and broaden the collective understanding of simulation in the geostatistical context.

## THE NOTION OF SUPPORT

One of the crucial distinctions between spatial statistics (which could be interpreted as including both geostatistics and stochastic modeling) and what might be called classical or traditional statistics is the explicit recognition of the importance of the support of the data. Support has to do with the idea that the value of some quantity of interest is related to the physical size (and possibly the dimensions) of the unit on which it is recorded. For example, in the context of ore reservation estimation, it is commonly the case that assay values are associated with the volume and shape of a core, and average grades are associated with the size and shape of mining blocks. A fundamental understanding of this notion has developed over the years in various disciplines. For example, in geography, it is widely known as the "modifiable areal unit" problem; and in a classic paper, Smith (1938) recognizes its implications for agriculture. For a discussion of Smith's results in the context of geostatistics, see Zhang et al. (1990). Although much of the investigation of support predates the more recent expansion of geostatistical methodology, the ideas of block kriging and regularized variograms are, in fact, tools for incorporating the idea of support into geostatistical analyses.

The theory pertaining to these tools is fairly well known, yet practical problems of application still need to be solved. For example, computing point-to-block and block-to-block covariances is commonly accomplished through numerical integration, which is imbedded in software. However, the software options often do not accommodate the irregularly shaped regions that are commonly found in practice, and so, the notion of support is incompletely or inaccurately addressed. Furthermore, whereas a regularized variogram or covariance can be theoretically related to a point support model, it is difficult in practice to obtain such a model from nonpoint support data. Such difficulties arise, for example, in the contexts of upscaling and downscaling geological, petrophysical, or engineering properties. Still other complications may occur in practice when determining the actual support of the data (e.g., when hydraulic conductivity is measured using a pumping test), resulting in restricted application of the currently known theory.

## BAYES, ENTROPY, AND MULTIPOINT CORRELATION

As suggested above, an even more diverse array of ideas from traditional statistics have made their way into geostatistical thinking and research in recent years. Three such ideas, in particular, have captured the interest and imagination of practitioners.

The first has to do with multipoint correlation. Both the variogram and the covariance function are two-point functions (i.e., each quantifies the similarity or dissimilarity of the values at a pair of locations in space). Each is a second-moment function. Geostatistics and, to some extent, stochastic modeling are strongly based on the assumption that knowledge of second-order moments is sufficient. The kriging equations depend only on the variogram or covariance (with appropriate assumptions about the mean) and not on other properties or characteristics of the random function. However, it is also known that second-order moments are far from adequate in characterizing even a second-order stationary random function. The variogram is somewhat analogous to a derivative in the sense that both filter out constants because both are based on first-order differences. In contrast, higher order generalized functions are based on higher order differences. First-order differences are essentially dimension-free, whereas higher order differences are not. As seen in Delfiner (1976), to generate sample-generalized covariances, it is necessary to construct higher order differences. To obtain acceptable multiple first-order differences, one may rigidly translate any pair of points. The coefficients in the difference (+1,−1) remain unchanged; but the same is not true for a second-order difference. In 1-D space, one can take the trio of points ($s - h, s, s + h$) with coefficients (1,−2,1), respectively. Moreover, if the pattern of points is rigidly translated, then the coefficients will likely change. Hence, what might seem to be an obvious extension to two-dimensional space does not work.

Although a multiple-point correlation function is not quite the same as a generalized covariance, some analogies are present. Under a second-order stationarity assumption, $\text{Cov}(Z(s + h),Z(s))$ is a function of $h$ alone. Even without this assumption, the geometrical pattern determined by the pair of points is the same (a line segment; although its orientation might change) and is not dependent on the magnitude of $h$. In contrast, the assumption that $\text{Cov}(Z(s), Z(s + h_1), Z(s + h_1),\ldots, Z(s + h_K))$ is only a function of $(h_1,\ldots, h_K)$ is much stronger. Moreover, the geometrical pattern of the points $(s, s + h_1,\ldots, s + h_K)$ can change greatly as the relative magnitudes and the orientations of the $(h_1,\ldots, h_K)$ change. Clearly, $\text{Cov}(Z(s), Z(s + h_1), Z(s + h_1),\ldots, Z(s + h_K))$ captures more information than $\text{Cov}(Z(s + h), Z(s))$.

Various authors (e.g., Guardano and Srivastava, 1993; Krishman and Journel, 2003) have proposed some form of multipoint correlation function that would characterize the random function to a greater degree. This idea does not seem to have progressed very far, and practical difficulties still exist. In particular, estimating and modeling such functions would likely require large data sets. There is also the question of how to actually apply such functions.

The second issue concerns the evolution of geostatistical methods that have been developed from the Bayesian point of view (see Diggle et al., 2003). In the traditional statistics literature, there are commonly heated disputes about whether the Bayesian or frequentist approach is better. Both the successes of Bayesian statistics and the advent of greater computing power have led to an interest in Bayesian geostatistical methods. The geoR package for R incorporates basic Bayesian geostatistical tools (also, see Diggle and Tawn, 1998). Whether such approaches continue to be developed likely depends on the availability of appropriate software, and currently, such software is not widely available.

The third issue has to do with an evolving understanding about the results obtained with kriging, which, of course, is a widely used spatial estimation procedure. The usual kriging equations are obtained by minimizing the estimation variance (with the unbiasedness constraint); and, yet, it is now well known that the resulting kriging variance is not exactly a variance in the usual sense of the term. The kriging variance does not directly depend on the data and, hence, provides only a relative measure of reliability. This has led to an interest in entropy, best exemplified by the work of G. Christakos (e.g., Journel and Deutsch 1993; Christakos and Li, 1998; Hristopulos and Christakos, 2001). Several definitions of entropy exist, and one must be careful to distinguish between the discrete and the continuous case (see Cover and Thomas, 1991, especially chapters 9 and 11). First, consider the discrete case. Suppose there are outcomes $E_1,\ldots, E_n$ with associated probabilities $p(E_1),\ldots, p(E_n)$. Then, the information-theoretic entropy is given by

$$-\sum p(E_i)\ln p(E_i)$$

This is also known as Shannon's (1948) entropy but is also found in Pauli (1933). It can be interpreted as the average loss of ignorance or gain in knowledge.

For many applications, however, one must consider continuous distributions. If $f(x)$ is the density function of a univariate continuous random variable, then the entropy is defined as

$$H(f) = -\int f(x)\ln f(x)\mathrm{d}x$$

This expression is different from the entropy for a discrete random variable. The values of the density function are not probabilities, and in particular, they are not bounded by the interval [0,1]. Thus, the entropy may not exist (i.e., the [improper] integral may not converge to a finite value). Moreover, it need not be positive.

To make this definition consistent with the previous one, it is necessary to introduce a reference density $g(x)$ and consider the following integral:

$$-\int f(x)\ln[f(x)/g(x)]\mathrm{d}x$$

The justification for maximizing the entropy can be made from various perspectives, but even then, the solution is not unique. It is commonly noted or claimed that the normal distribution has maximum entropy. This is not quite complete; a question about additional constraints must be addressed. If the density function is only nonzero on an interval $[a,b]$, then the uniform distribution has maximum entropy. If the density is only nonzero for $[0, \infty)$ and the expected value is fixed, then the exponential distribution has maximum entropy. If the density function is nonzero on $(-\infty,\infty)$ with fixed expected value and fixed variance, then the normal distribution has maximum entropy. The actual maximal value will depend on the variance. In Bayesian maximum entropy, it is the posterior distribution for which the entropy is maximized.

Unfortunately, this higher level of mathematical complexity and a lack of readily available software have precluded a formal test of whether the entropy results are really better than those obtained with the usual forms of kriging. Additional investigation is clearly needed, but such work may simply have to wait until practice and operational implementation catches up more with theory.

With regard to each of these issues (as well as others), an impediment to further development and expanded application seems to be the absence of software, especially at the commercial or semicommercial level. That such software is not yet readily available likely reflects a lack of easily translatable algorithms and a general immaturity of the scientific principles. Although this situation is expected to change, the extent to which such ideas will permeate geostatistical thinking in practice is yet to be determined.

## DESCRIPTIVE OR INFERENTIAL?

Traditional statistics operates at two levels: descriptive and inferential. If the objective is to summarize the measurements represented by a given data set, then the task is a descriptive one (hence, the term "descriptive statistics"). However, if the data are assumed to be a random sample from a larger population and the objective is to use such sample data to draw conclusions or make inferences about the entire population, the task is one of inference (hence, the term "inferential statistics"). Both geostatistics and stochastic modeling are somewhat closer to descriptive statistics than to inferential statistics. That is, drawing conclusions about the specific data set and/or the specific source of the data is more common than drawing conclusions about the entire population in question. In particular, the geostatistical literature is almost void of references to tests of hypotheses, which is a fundamental approach to traditional statistical inference, particularly from the frequentist viewpoint. However, hypothesis testing could be a valuable factor. For example, it might be desirable to test the underlying assumptions of the modeling approach (such as second-order or intrinsic stationarity) or to evaluate the goodness of fit of the variogram and covariance function. The book by Stein (1999) is perhaps one of the few texts that devotes any space to such ideas. Whitten (2003) raises a more general question about the function of hypothesis testing and questions why this has received less attention. Pardo-Igúzquiza and Dowd (2004) provide one example of applying hypothesis testing in the context of geostatistics.

## UNCERTAINTY AND RELIABILITY

Statistics, by its very nature, is intended to deal with problems and data in a manner that acknowledges conclusions will be couched in terms of uncertainty (e.g., probabilities of types I and II errors associated with hypothesis testing; confidence level and margin of error associated with confidence interval estimates of population parameters). When kriging was first introduced and promoted as a superior estimation technique (i.e., superior to the nearest neighbor technique commonly in use at the time), one of the claimed advantages was that the estimates have minimum variance (i.e., the kriging variance is minimum). As suggested above, it was subsequently recognized that the kriging variance is more a function of the data location pattern and the variogram model than it is of the data themselves. At best, it is a relative measure of reliability because it can be artificially increased or decreased without changing the estimated values. Moreover, as what has been pointed out by several authors, the kriging variance does not truly incorporate the uncertainty associated with estimating and modeling the variogram. This point is addressed, at least in part, by Stein (1999) but under rather strong assumptions. Consequently, the question might be asked as to whether there are more adequate ways to quantify the uncertainties associated with spatial estimation; and if so, how can they be used in a practical problem?

Although interest in quantifying the uncertainty associated with variogram modeling goes back at least to Davis and Borgman (1979, 1982), there have been a series of more recent papers (e.g., Pardo-Igúzquiza and Dowd, 2001; Ortiz and Deutsch, 2002; Marchant and Lark, 2004). One important point is commonly ignored: the sample variogram estimates the values of the variogram but does not directly estimate the variogram itself (i.e., the function). In practice, then, one must choose a family of variograms (e.g., Matérn, spherical, and power with one or more parameters). Then, the sample variogram or the variogram cloud is used to estimate the parameters. The problem is even more complicated in the case of a nested model variogram. Neither maximum likelihood nor weighted least squares do a good of job of detecting the different components in a nested model (or even the need for one). Although Matheron (1973) gives an integral representation for variograms, it is more difficult to translate this into practical use. Several extant results also make use of a multivariate normality assumption, and hence, they are most applicable in the case of variograms linked to covariance functions. Again, this is an area in which more work is needed.

## COLLABORATIVE EFFORTS IN RESEARCH AND SOFTWARE DEVELOPMENT

A great deal has already been made of the importance of software to the proliferation of geostatistical methodology and stochastic modeling approaches. As already suggested, it appears that future developments will be strongly intertwined with the creation of software packages that implement the various ideas, algorithms, and approaches. Such an effort requires substantial financial and intellectual resources. The geostatistical community has greatly benefited in this respect from collaboration between industry and academia. One successful approach, which became popular in the 1980s, has been for the consortia of companies to provide financial backing of academic research programs in the form of a participation fee to obtain proprietary access to research results and computational code. At least four such collaborative efforts are worth noting, each involving one or more academic groups and one or more segments of the petroleum industry.

At the top of this list is the Stanford Center for Reservoir Forecasting (SCRF), which is well known among, and well supported by, oil companies. Over the years of its existence, SCRF has given birth to many new ideas in geostatistics, producing numerous research publications. Most of the actual details of the algorithms, as well as specialized codes, are reserved, of course, for the financial supporters and participants.

The gOcad project at the University of Nancy, which is focused on 3-D Earth modeling, is another such collaboration between industry and academia. The consortium has resulted in the development of the well-known gOcad software package, which provides an alternative to traditional computer-aided drawing of complex geological surfaces based on discrete smooth interpolation. Although the actual software is reserved for supporters and group members, the theory is well documented in the book entitled Geomodeling (Mallet, 2002), and both petroleum and environmental applications have been reported.

A third example is Petbool, which is both a research project and a software package originating out of the collaboration between the Pontifical Catholic University in Rio de Janeiro and Petrobras (see Tavares et al., 2001). The acronym Petbool stands for the combination of Petrobras and Boolean, and the software provides 3-D visualization capabilities, along with object-based geological modeling of oil reservoirs.

Finally, the Statistical Analysis of Natural Resources group at the Norwegian Computing Center has developed multiple software packages largely targeting the petroleum industry, which are summarized and discussed in a recent article by Holden et al. (2003).

The collaborative approach has been both good and bad for the geostatistical community. Although it has resulted in many new discoveries and developments that have greatly expanded the scientific and computational boundaries of the discipline, the proprietary nature of programs has somewhat restricted their application. Individuals and groups without the financial resources to participate are left struggling to devise alternative computational approaches on their own, which has led to unnecessary tension between those who have access to the best algorithms and code and those who do not, as well as incomplete understanding of the solutions to problems that can be obtained. Consequently, both the industry and the scientific community might now be better served by more open communication of geostatistical knowledge and greater accessibility to software than what has been available in the past. Such a suggestion, of course, requires a different kind and level of cooperation and a great deal of leadership and effort to make it work. More immediately, a survey article providing a more detailed summary and comparison of the different analytical and modeling approaches, as well as the software capabilities, would be very useful. The practices common in the broader field of statistics may be relevant here. STATLIB (http://lib.stat

.cmu.edu/) is an archive of algorithms and program codes. As noted previously, *The American Statistician* has a regular section devoted to the review of statistical packages. In addition, a section of the American Statistical Association is devoted to statistical software and graphics.

## LANGUAGE DIFFICULTIES

As in the case of all scientific disciplines, there have been many controversies within the geostatistical community over the years. Interestingly, argument about the meaning and intent of some of the fundamental terminology still exists. For example, the term "variogram" was originally used to denote the quantity

$$\mathrm{Var}[Z(s+h) - Z(s)] = 2\gamma(h)$$

under the assumption that it was finite for all values of $s$ and $h$ and did not depend on $s$ (see Matheron, 1971). However, under the second-order stationarity assumption, it is easy to show that

$$0.5\ \mathrm{Var}[Z(s+h) - Z(s)] = \mathrm{Var}[Z(s)] - \mathrm{Cov}[Z(s+h), Z(s)]$$

or

$$\gamma(h) = C(0) - C(h)$$

where $\gamma(h) = 0.5\mathrm{Var}[Z(s+h) - Z(s)]$ and $C(h) = \mathrm{Cov}[Z(s+h), Z(s)]$. Thus, it was natural to focus more on half of the variogram; hence, the term "semivariogram." However, it soon became apparent that there were few, if any, instances in geostatistics where it was really necessary or even useful to consider the (original) variogram instead of the semivariogram (as an example, the kriging equations are easily derived and written in terms of the semivariogram). In the 1980s, many authors began using the term variogram to denote the semivariogram, omitting any reference to the original quantity. There were two principal advantages of this shift: (1) it simplified the language usage in all written and oral communications (e.g., experimental variogram seemed easier to say and communicate than experimental semivariogram); and (2) it avoided the confusion created when the two terms were used interchangeably or incorrectly, even when a distinction might have been technically correct.

Clearly, the shift in terminology was not and is not universal, and there are those who insist on using the original usage, which is certainly their prerogative. It is difficult to say which term is more common today; but, as an example, an examination of the proceedings of the 1988 International Geostatistics Congress in Avignon, France, suggests that most authors (or perhaps the editor) used variogram exclusively, a few used semivariogram exclusively, and a few others used the two terms interchangeably. A similar pattern can be observed in many later texts and compilations. The documentation for some software packages (e.g., GeoEas, http://www.epa.gov/nerlesd1/databases/geo-eas/access.htm) only uses variogram, and several authors (e.g., Chiles and Delfiner, 1999) only acknowledge semivariogram as an older, unused term. The issue is not, and likely never will be, resolved.

Unfortunately, this is not the only inconsistency in geostatistical terminology. At least two other words (or terms) that appear with some frequency in the literature do not always have precise meanings. "Robust" (or robustness) is one example. Kendall and Stuart (1979, p. 492) wrote that "A statistical procedure which is insensitive to departures from the assumptions which underlie it is called robust." This definition is originally attributed to Box (1953). The problem is that the assumptions underlying a particular procedure may not be clearly stated, or their relevance may not be clearly understood; and so, too commonly, the term robust is used as a general catch-all adjective. In the context of geostatistics, it is generally thought that the ordinary kriging estimator is robust with respect to the values of the variogram parameters. However, it may not be so robust with respect to the underlying distribution of the random function or the intrinsic stationarity assumption; and so, as a broadly descriptive term, robustness may not be an appropriate characterization.

Another word that presents some interpretive difficulties is "representativeness." Sometimes, it will be claimed that a sample is representative, and occasionally, other quantities or statistics are called representative. Unfortunately, it is commonly unclear in what sense the characteristic of representativeness applies. Intuitively, representative is a word that sounds desirable, and consequently, it is tempting to claim that some quantity or procedure is representative. It is easy, in fact, to think of ways in which samples might be considered representative. For example, if the empirical distribution of the sample is the same or nearly the same as the distribution of the population, then the sample might be called representative. In this case, the sample mean and variance might be expected to be close to the population mean and variance, respectively. Unfortunately, none of these attributes can be known in advance, and in fact, the population may not have a finite mean or finite variance. Obviously, random samples need not be representative at all. Hence, to remove the ambiguity and to avoid confusion, it would be extremely helpful in all communications of this nature for geostatisticians and their colleagues to explicitly state the sense in which representativeness applies.

Although language differences might not be the primary cause of divisions among practitioners, they still represent a source of irritation and confusion for those outside the immediate geostatistical community, and they do not serve to place the field in the best scientific light. To ensure the future of the discipline, geostatisticians would do well to engage in a conversation aimed at standardizing language and terminology and in making geostatistical communications more effective, more understandable, and more accessible to a wider range of potential users.

## FINAL THOUGHTS

It seems obvious that geostatistics and stochastic modeling are alive and well, and that they will continue to be adapted and exploited for the foreseeable future. As suggested earlier, neither discipline has evolved by altogether conventional means through largely theoretical academic endeavors, but instead, through extensive experience, practice, and applied problem solving in the context of a rich and diverse array of applications. It is this focus on applications that makes the tools of these disciplines so attractive in many areas of investigation.

In 50 or even 20 yr, geostatistics and stochastic modeling will no doubt look different. They may continue to converge, or they may diverge along entirely new or different paths. However, if history is a strong indicator of the future, it seems certain they will further evolve within an applications and problem-solving framework. There will be theoretical enhancements, to be sure, and perhaps some truly astonishing breakthroughs; but the need to address both simple and thorny questions from an applied point of view is likely to remain the primary driver.

In a very real sense, the world is becoming smaller with the relentless advances of technology. As a result, science, business, industry, medicine, politics, and the like are becoming increasingly focused on spatial relationships. It is the spatial domain in which stochastic modeling and geostatistics found their beginnings, and it is within this same spatial domain that they will surely continue to flourish.

Obviously, the future is unknown; but stochastic modeling and geostatistics seem destined to exert even greater influence on the way people think about the world around them. Although their past contributions will persist, there will likely be many new applications and developments that will have profound influence on global thinking and well-being. This book catalogs some of the many examples illustrating progress and enhancements throughout the last 10 yr, specifically in the rich areas of earth and petroleum science that have been traditional strongholds for geostatistics and stochastic modeling since their early beginnings. The next decade and beyond promises to be an equally productive and exciting time in which geostatistics and stochastic modeling impact not only the geosciences but important areas of investigation far beyond this traditional base.

## REFERENCES CITED

Actuarial Foundation, 2003, Stochastic modeling Symposium, September 4–5, Toronto: http://www.actuaries.ca/meetings/archive_stochasticsymposium_2003_e.html (accessed April 15, 2005).

Anselin, L., 1988, Spatial econometrics: Methods and models: Dordrecht, Kluwer Academic Publishers, 284 p.

Batchelor, G. K., 1953, The theory of homogeneous turbulence: New York, Cambridge University Press, 197 p.

Box, G. E. P., 1953, Non-normality and tests on variances: Biometrika, v. 40, p. 318.

Chiles, J.-P., and P. Delfiner, 1999, Geostatistics: Modeling spatial uncertainty: New York, John Wiley & Sons, 695 p.

Christakos, G., and X. Li, 1998, Bayesian maximum entropy analysis and mapping: A farewell to kriging estimators?: Mathematical Geology, v. 30, p. 435–462.

Cover, T. M., and J. A. Thomas, 1991, Elements of information theory: New York, John Wiley & Sons, 542 p.

David, M., 1977, Geostatistical ore reserve estimation: Amsterdam, Elsevier, 364 p.

Davis, B. M., and L. E. Borgman, 1979, Some exact sampling

distributions for variogram estimators: Mathematical Geology, v. 11, p. 643–653.
Davis, B. M., and L. E. Borgman, 1982, A note on the asymptotic distribution of the sample variogram: Mathematical Geology, v. 14, p. 189–193.
Davis, J., 2002, Statistics and data analysis in geology, 3d ed.: New York, John Wiley & Sons, 638 p.
Delfiner, P., 1976, Linear estimation of non-stationary spatial phenomena, *in* M. Guarascio, M. David, and C. J. Huijbregts, eds., Advanced geostatistics in the mining industry: Dordrecht, D. Reidel Publishing Company, p. 49–68.
Diggle, P., P. J. Ribeiro Jr., and O. F. Christensen, 2003, An introduction to model-based geostatistics, *in* J. Møller, ed., Spatial statistics and computational methods: New York, Springer, p. 43–86.
Diggle, P. J., and J. A. Tawn, 1998, Model based geostatistics: Applied Statistics, v. 47, p. 299–350.
Dimitrakopoulos, R., 1994, Geostatistics for the next century: Dordrecht, Kluwer Academic Publishing, 497 p.
Dimitrakopoulos, R., and M. Dagbert, 2001, Farewell to Michel David (1945–2000): Mathematical Geology, v. 33, p. 241–244.
Frisch, U., and A. N. Kolmogorov, 1995, Turbulence: The legacy of A. N. Kolmogorov: New York, Cambridge University Press, 310 p.
Gandin, L. S., 1963, Objective analysis of meteorological fields: Leningrad, Gidrometeorologicheskoe Izsatel'stvo (GIMIZ), 242 p. (reprinted by Israel Program for Scientific Translations, Jerusalem, 1965).
Guardano, F., and M. Srivastava, 1993, Beyond bivariate moments, *in* A. Soares, ed., Geostatistics Troia '92: Dordrecht, Kluwer Academic Publishing, p. 133–144.
Haining, R., 2003, Spatial data analysis: Theory and practice: New York, Cambridge University Press, 432 p.
Holden, L., P. Mostad, B. Nielsen, J. Gjerde, C. Townsend, and S. Ottesen, 2003, Stochastic structural modeling: Mathematical Geology, v. 35, p. 899–914.
Hristopulos, D. T., and C. Christakos, 2001, Practical calculation of non-Gaussian multivariate moments in spatiotemporal Bayesian maximum entropy analysis: Mathematical Geology, v. 33, p. 543–568.
Hunt, J. C. R., O. M. Phillips, and D. Williams, 1991, Turbulence and stochastic processes: Kolmogorov's ideas 50 years on: London, Royal Society, 240 p.
Journel, A. G., and C. V. Deutsch, 1993, Entropy and spatial disorder: Mathematical Geology, v. 25, p. 329–356.
Journel, A. G., and C. J. Huijbregts, 1978, Mining geostatistics: New York, Academic Press, 600 p.
Karlin, S., and H. Taylor, 1998, An introduction to stochastic modeling, 3d ed.: San Diego, Academic Press, 631 p.
Kendall, M., and A. Stuart, 1979, The advanced theory of statistics, 4th ed.: New York, Macmillan Publishing, v. 2, 748 p.
Krishman, S., and A. G. Journel, 2003, Spatial connectivity: From variograms to multiple-point measures: Mathematical Geology, v. 35, p. 915–926.
Lumley, J. L., 1970, Stochastic tools in turbulence: New York, Academic Press, 194 p.
Mallet, J.-L., 2002, Geomodeling: New York, Oxford University Press, 599 p.
Marchant, B. P., and R. M. Lark, 2004, Estimating variogram uncertainty: Mathematical Geology, v. 36, p. 867–898.
Matérn, B., 1986, Spatial variation, 2d ed.: Berlin, Springer-Verlag, 151 p.
Matheron, G., 1965, Les variables regionalisees et leur estimation: Paris, Masson et Cie, 305 p.
Matheron, G., 1971, The theory of regionalized variables and its applications: Fontainebleau, Ecole Nationale Supérieure des Mines de Paris, Les Cahiers du Centre de Morphologie Mathématique, Fascicule 5, 211 p.
Matheron, G., 1973, The intrinsic random functions and their applications: Advances in Applied Probability, v. 5, p. 439–468.
Ortiz, C. J., and C. V. Deutsch, 2002, Calculation of uncertainty in the variogram: Mathematical Geology, v. 34, p. 169–184.
Pardo-Igúzquiza, E., and P. Dowd, 2001, Variance–covariance matrix of the experimental variogram: Assessing variogram uncertainty: Mathematical Geology, v. 33, p. 397–420.
Pardo-Igúzquiza, E., and P. Dowd, 2004, Normality tests for spatially correlated data: Mathematical Geology, v. 36, p. 659–682.
Pauli, W., 1933, Handbuch der physik: Berlin, Springer, v. 24, pt. 1, 151 p.
Shannon, C. E., 1948, A mathematical theory of communication: The Bell System Technical Journal, v. 27, p. 379–423, 623–656.
Smith, H. F., 1938, An empirical law describing heterogeneity in the yields of agricultural crops: Journal of Agricultural Science, v. 28, p. 1–23.
Srivastava, R. M., 1997, Matheronian geostatistics: Where are we going?, *in* E. Y. Baafi and N. A. Schofield, eds., Geostatistics Wollongong '96: Dordrecht, Kluwer Academic Publishing, p. 53–68.
Stein, M., 1999, Interpolation of spatial data: Some theory for kriging: New York, Springer, 247 p.
Tavares, G., H. Lopes, S. Pesco and C. Poletto, 2001, Petbool: A software for stochastic modeling of geometric objects, *in* U. Bayer, H. Burger, and W. Skala, eds., IAMG 2002, Proceedings of the Annual Conference of the International Association for Mathematical Geology: Berlin, Terra Nostra, p. 203–208.
Van Niel, K. P., and S. W. Laffan, 2001, Caveat emptor: Random number generators in geospatial analysis, *in* D. V. Pullar, ed., Proceedings, 6th International Conference on Geocomputation: Queensland, University of Brisbane, 3 p.
Whitten, E. H. T., 2003, Mathematical geology in perspective: Has objective hypothesis testing become overlooked?: Mathematical Geology, v. 35, p. 1–8.
Zhang, R., A. Warrick, and D. E. Myers, 1990, Variance as a function of sample support size: Mathematical Geology, v. 22, p. 107–121.

Coburn, T. C., P. W. M. Corbett, A. Datta-Gupta, J. L. Jensen, M. Kelkar, D. S. Oliver, and C. D. White, 2006, A virtual conversation on the impact of geostatistics in petroleum, production, and reservoir engineering, *in* T. C. Coburn, J. M. Yarus, and R. L. Chambers, eds., Stochastic modeling and geostatistics: Principles, methods, and case studies, volume II: AAPG Computer Applications in Geology 5, p. 23–33.

Chapter 3

# A Virtual Conversation on the Impact of Geostatistics in Petroleum, Production, and Reservoir Engineering

**T. C. Coburn**
*Abilene Christian University*
*Abilene, Texas, U.S.A.*

**P. W. M. Corbett**
*Heriot-Watt University*
*Edinburgh, Scotland, United Kingdom*

**A. Datta-Gupta**
**J. L. Jensen**
*Texas A&M University*
*College Station, Texas, U.S.A.*

**M. Kelkar**
*University of Tulsa*
*Tulsa, Oklahoma, U.S.A.*

**D. S. Oliver**
*University of Oklahoma*
*Norman, Oklahoma, U.S.A.*

**C. D. White**
*Louisiana State University*
*Alexandria, Louisiana, U.S.A.*

## ABSTRACT

Throughout the years, geostatistics has greatly influenced the broad spectrum of petroleum and earth sciences. In a directed conversation, five experts from respected university petroleum engineering departments discuss the impact of geostatistics on petroleum, production, and reservoir engineering and debate the current state of the art of geostatistical methodology in these disciplines.

## INTRODUCTION

Geostatistics has been an increasingly important factor in the development of petroleum resources throughout the last 10 yr. The interface between geology, engineering, and geostatistics, and the interplay among the individuals who practice these disciplines, has become an essential ingredient to successful field development and optimization of oil and gas reserves. The increasing literature on applications of geostatistics to field development problems, reservoir simulation, and reservoir characterization is testimony to the degree to which geostatistics has penetrated more traditional engineering thinking and impacted corporate strategies and operations.

The authors recently conducted a virtual conversation in which they reflected on the function of geostatistics in petroleum, production, and reservoir engineering during the previous decade and looked


DOI:10.1306/1063804CA51306

ahead to what might be in store for the future. Tim Coburn, one of the editors of this volume, initiated the conversation and served as moderator and questioner. The comments and opinions that are expressed encompass a broad spectrum of ideas and points of view and represent a wealth of knowledge and experience within the petroleum engineering community.

## QUESTIONS AND ANSWERS

**Coburn:** Gentlemen, welcome to our meeting today via the exciting world of electronic media! I want to thank each of you for your willingness to participate in this unique conversation. I am looking forward to hearing what each of you has to say about the role and importance of geostatistics and its impact on petroleum reservoir development, particularly from the engineering standpoint.

What I would like to do today is ask several questions that I think will stimulate discussion and then give each of you an opportunity to respond. You might think of this as sort of an electronic panel discussion. Feel free to make whatever contribution you choose, and I will attempt to keep us all moving along in the same direction.

Let me ask you this first question. How would you characterize the contribution of geostatistics to the recent evolution of petroleum/production/reservoir engineering? What has geostatistics done for the engineering side of oil and gas production and development? Dean, would you mind starting us off?

**Oliver:** Sure. I'd be happy to. I think the major contribution of geostatistics to the evolution of reservoir engineering has been the provision for the possibility of evaluating heterogeneity in parts of the reservoir that are unobservable and whose properties are unknown. Prior to the development of geostatistics, folks sometimes worried about optimal layering in heterogeneous reservoirs; but the models were still typically "layer cake" and deterministic in nature. Geostatistics has altered the mindset that the best reservoir models are those with no features that are not demanded by the data. In doing so, it has provided a means for avoiding the bias in flow predictions inherent to assigning uniform properties where heterogeneity is known to exist. Another major contribution is the ability to evaluate reservoir behavior using the uncertainty expressed in a number of geologically plausible reservoir realizations (simulations, if you will), none of which is actually a true or exact representation.

**Jensen:** Let me add to that a bit if I may. Geostatistics provides a means to quantitatively analyze and model systematic geological and petrophysical aspects of the reservoir. It permits quantification of uncertainty in the unpredictable geological aspects and the impact of such uncertainty on reservoir performance.

**Datta-Gupta:** This is true. Geostatistics dramatically improved our ability to generate highly detailed geological models by integrating diverse data sources. Unfortunately, the result was a tremendous gap between geological modeling and flow simulation. Many of the recent developments in reservoir engineering have focused on addressing this gap. Some examples are the advancements in upscaling, streamline simulation, and history matching. Of course, improved reservoir characterization itself has contributed to better reservoir management and field development strategies, geostatistics being one of the important associated technologies. Thus, the impact of geostatistics extends well beyond geological modeling to the broader discipline of petroleum engineering in general.

**Kelkar:** I want to follow up on what Akhil has said and perhaps be even more explicit. I would say that the three most important contributions are the ability to (1) integrate various sources of geological, geophysical, and engineering data in a consistent manner; (2) provide history matching of production data by generating geologically defensible models; and (3) quantify uncertainties in future performance through history matching of multiple realizations.

**Coburn:** Mo, would you say these are actually "contributions of geostatistics," or are they advancements that have occurred in conjunction with the expansion of geostatistical theory, knowledge, and practice?

**Kelkar:** Well, the distinction you make is probably valid. We often attribute these developments to geostatistics, when in fact, changes and enhancements have occurred because bright people have put the ideas of geostatistics into practice. For example, on the idea of integration, I have visited many companies recently, and I would say that integrated reservoir description, of which geostatistical analysis and modeling is an important component, is being practiced fairly routinely.

**White:** Maybe, then, the most important contribution of geostatistics is that it has fostered discussion of uncertainty, knowledge, and integration among engineers, geoscientists, and managers. As Akhil and Mo have suggested, geostatisticians have notably helped initiate the recent increased focus on integration of uncertainty with the reservoir characterization workflow.

**Coburn:** Pat, would you agree with that?

**Corbett:** Yes. Geostatistics has been key to the development of integrated shared-earth models that we are familiar with today. The need for these models has been driven by a push to communicate uncertainty between disciplines more effectively. The requirement for data to feed these models has driven geologists to look much closer at reservoir architecture in a quantitative way, and this has improved their understanding of reservoirs.

**Coburn:** OK. If I can summarize your thoughts on this, the key points seem to be that geostatistics and geostatistical technology facilitates improved characterization of reservoir behavior and performance through the integration of data and information and an interdisciplinary approach to problem solving.

Now, let me be a bit more specific. From the collective standpoints of petroleum, production, and reservoir engineering, what do you think has been the single most important development in geostatistics throughout the last 10 yr? I'm expecting a wide variety of responses here! Akhil, why don't you go first this time.

**Datta-Gupta:** The increasing use of seismic data in routine reservoir characterization is certainly a major contribution of geostatistics to petroleum engineering throughout the past 10 yr. I think many of the recent developments related to seismic data integration have been readily adopted by the industry for their versatility and computational efficiency.

**Coburn:** Chris, how about you?

**White:** I would say the development of conditional categorical methods has been significant. These methods allow reservoir-modeling teams to create images that display the high contrasts expected in an environment with episodic depositional processes.

**Corbett:** I'll chime in here. The making available of the GSLIB library from Stanford University and the availability of commercial software that allowed both pixel- and object based-modeling, I see as key events. More recently, the development of multipoint statistical techniques is another step forward, but these have yet to filter down to everyday usage.

**Oliver:** For reservoir engineers, the ultimate goal of reservoir characterization is usually to develop a reservoir-simulation model (or an ensemble of models) capable of predicting future reservoir behavior under a variety of production scenarios. These models are, of course, most useful for prediction if they are capable of reproducing current and past performance. Therefore, it seems to me that the most important development from the reservoir engineering standpoint has been the ability to generate realizations of the reservoir property fields that are conditional to production data. The computational resources necessary to achieve this were simply not available 10 yr ago. Now, we have both the algorithms and the computing power we need, with more of both on the way.

**Kelkar:** I would consider the most important development in the last 10 yr to be routine use of sequential simulation procedures. Many versatile techniques used today are really offshoots of sequential simulation. An equally important development, in a negative sort of way, is the demise of, or reduced emphasis on, fractal-based approaches and simulated annealing. Rarely do I have see people using these techniques.

**Jensen:** I would broaden this discussion just a bit. Indeed, we have all these wonderful new tools at our disposal; but what really seems to have been important is the development of a pool of experience to guide the selection of methods we use to analyze and model particular geological characteristics.

**Coburn:** Well, I certainly got what I asked for—lots of diversity! I really like what Jerry has said here at the end. The implication that we are learning more about how to effectively choose and apply geostatistical methodology seems to be a key point. I have a gut feeling that, 10 yr ago, we were taking more of a shotgun approach to geostatistics, and I am glad to hear someone suggest that our collective approach has matured.

**Kelkar:** I personally believe that geostatistics has become a mature technology, and that the basic analytical framework is already well established. What we are largely doing today is modifying and enhancing existing techniques in ways that we can apply them to new or different types of data.

**Coburn:** That's a nice lead in to my next question. What current areas of geostatistical research and stochastic modeling do you think hold the most promise for further enhancements in petroleum/production/reservoir engineering and reservoir characterization? I realize I am asking you to engage in a bit of "crystal ball gazing," but I'd really like to hear your ideas on this. Again, we're probably all going to have a somewhat different perspective. Jerry, what are your thoughts?

**Jensen:** I actually have three ideas to suggest: (1) extracting more information from poorly sampled material, including guideline numbers and relationship of sampling to uncertainty of predictions; (2) more geologically orientated data analysis and modeling, including wavelet analysis and modeling; and (3) integration of measurements. I'm not sure what order of importance I would put on these, but they are certainly at the top of my list.

**Datta-Gupta:** Reconciling high-resolution geostatistical models to dynamic data in a timely manner is definitely an outstanding challenge to the industry. I think the full promise of geostatistics in terms of uncertainty quantification/risk assessment will not be realized until we come up with practical approaches to production data integration. This is certainly the most active research area at present, and the industry has seen numerous developments throughout the last 10 yr. However, we are yet to come up with a method for routine use by practicing engineers and geoscientists.

**Kelkar:** Following up on what Akhil has suggested, I also believe the most important future developments will involve the full integration of production data into geostatistical models and learning how to better quantify uncertainty in future reservoir performance. As I said before, the framework of geostatistical methodology is pretty well established; and for all practical purposes, static data integration is a fairly straightforward process. Some scale issues still remain to be solved.

**White:** From a more basic geologic perspective, stochastic models for fractures, shales, and diagenetic features are relatively undeveloped and poorly integrated with the reservoir-modeling workflow. Recent work in multipoint statistics is promising.

**Oliver:** Since Chris has taken us back to geology, I think that we have not yet made full use of facies models for rock types in reservoir-simulation models, and certainly not in history matching. It is somewhat complex to first distribute facies and then distribute permeability and porosity within facies, especially if it is necessary for the resulting model to honor complex flow data; but it allows the creation of far more realistic permeability and porosity distributions and presumably more realistic estimates of uncertainty than those that are obtained without explicit consideration of facies.

**White:** Tim, if you don't mind, I want to step back in. Dean's comment has inspired me to make another point or two.

**Coburn:** Sure, go ahead.

**White:** With all due respect to what Akhil has said, I think it's also important to recognize what we don't and can't know, and we must avoid confounding sources of variability, for example, vertical and areal heterogeneity.

I recall the disdain justifiably heaped on the use of relative permeability models to match reservoir behavior in the early days of the Stanford Center for Reservoir Forecasting (ca. 1985). Some geostatistical modelers have simply substituted the variogram range and random number seeds "knobs" for the former relative permeability "knob," and in many cases, little insight has been gained. In my opinion, we commonly overstate property variability (e.g., using core- or log-scale variance incorrectly for grid block-scale properties) and correlation (typical variogram ranges used in reservoir models are much longer than any ever observed in outcrop data sets, for example). We are artificially injecting the wrong sorts of variability into reservoir models using the wrong parameters. This may be done as an ad-hoc procedure, or it may occur in history matching. I would argue that our ability to create large, complex models is ahead of our ability to calibrate or understand those models. We need hierarchical methods to assess variability across model spaces. The necessary changes may include changes in the mathematical models used for porous media flow.

Further, sensitivity, uncertainty, and significance are difficult to visualize across a large suite of realizations. How does one discern which factor is generally the most important, and how does one examine how this importance varies over the range of the uncertain factors? This problem will take both ingenuity and software development resources to solve.

**Oliver:** I also want to make one more point. Another extremely promising technology that is just beginning to be applied to reservoir characterization problems is data assimilation. The main idea is that geostatistical reservoir models should be continuously updated as data are acquired—conditional simulation is done in real time. The driver for this area of research is the recent increase of the number of wells with permanent sensors and nearly continuous data streams. The potential advantages to this approach are obvious.

**Coburn:** Pat, you get the last shot at this. Anything to add?

**Corbett:** Yes. I would offer these two additional ideas that haven't been mentioned yet. The use of training images allows the most complex of geological structures to be reproduced, and these will make it easier for geoscientists to build representative models. The use of geostatistical tools with seismic data for time-lapse studies is another growth area.

**Coburn:** Boy, this is a pretty impressive list. It looks like there's enough to keep us all busy for quite some time to come. Let me ask you this question—and Chris hinted at it earlier—do you think that geostatistical science is perhaps too far ahead of actual practice? If so, what needs to be done to change the situation?

**Oliver:** My personal answer would be "no," particularly if we specifically associate geostatistics with reservoir characterization. I don't think that the science of reservoir characterization is too far ahead of actual practice at all. Science will, and should be, somewhat ahead. There are numerous examples of algorithms and methodologies for which the theory has existed for some time, but the limitations of computer resources kept the methods from being applied. A good example would be the application of Markov chain Monte Carlo (MCMC) methods to conditional simulation. The theory and science of MCMC has been available for some time; but only in the past few years has the method been used in practice (e.g., the Production Forecasting with Uncertainty Quantification, or PUNQ, project). Because reservoir characterization is an applied (as opposed to a basic) science, it is important, however, that theory be grounded in application to real problems, and that the theoreticians understand the real problems.

**Jensen:** I guess I would have to agree with Dean on this to a certain extent. The situation is certainly not as problematic as it used to be. More case studies, using both outcrop and subsurface examples, have been conducted and reported. Some interesting analyses have emerged recently. For example, Chris and his coauthors (White et al., 2003) showed that the flow performance of carefully mapped outcrops containing carbonate concretions and their geostatistically simulated counterparts can be similar. This lends confidence in the ability of geostatistical models to reproduce these features. On the other hand, studies such as the one by Jim Jennings et al. (2000) show the importance that subtle aspects, which could be missed in the geostatistical simulations, may have on displacements.

**White:** It is not a matter of being ahead or behind but a matter of being insensitive to applications and significance. Geostatisticians often fail to consider the recovery implications of their models sufficiently. A model is relevant only if it significantly changes flow predictions and affects project economics. Too many geostatistical studies end with the pseudoconclusion: "It looks geologically reasonable." What does that mean, and how can it be tested? Geostatistics should make better use of flow modeling, decision analysis, and hypothesis testing.

**Datta-Gupta:** I think geostatistics has been unique in the sense that the petroleum industry has been very quick to adopt the technology whenever it offered a niche. If anything, the reservoir engineers were a bit too fast to adopt the technology compared to geoscientists, and this led to its misuse sometimes.

**Corbett:** In some parts of the world, production data rule the reservoir monitoring domain, and geostatistics doesn't get a look in. There are other asset teams being driven to provide deterministic models in preference to geostatistical models. I think the management response to the trend is to use geostatistics to build models that are too complex, too big, and too difficult to update. We need to get smarter about building the appropriate complexity of model for the business decision needed. That way, management will really appreciate the value and not see modeling as a drain on time and resources.

**Kelkar:** Unfortunately, geostatistics is really not for the average user. It really requires a lot of practice and great care to use. To apply geostatistical principles correctly, you need to know the methodology and limitations, you need to have good understanding of the data available from various disciplines, you need to have the ability to customize to a specific situation, and you need to have software that you are familiar with. To achieve all of this is very difficult. The only way this situation can change is that if we can create a position of "Integrator" in oil companies. These integrators are not necessarily experts in any particular area. Instead, they rely on the expertise of others. Their job is to collect information from various disciplines, understand the uncertainties in various data sources, develop a workflow to accommodate various data sets and uncertainties, maintain sufficient computer skills to customize this workflow, and apply the necessary software expertise to generate multiple reservoir descriptions accounting for uncertainties in various data sources. These images should be verified with each discipline expert to ensure that they are consistent with the understanding of the reservoir.

**Coburn:** My sense is that we are not in agreement on this issue, but it seems to me to be an important one. While I concur with Dean that science needs to lead the way, I think that geostatistics and stochastic modeling have gotten a bit of a bad rap in some circles—particularly, as Pat suggests, in management circles. The complaints that I've personally heard have to do with the technology and the resulting models being too esoteric to be understood within the overall corporate context and the results not being well enough grounded in conventional earth science. Part of this perception might simply be a backlash against mathematical complexity, terminology, and notation with which more traditional geoscientists and managers are unfamiliar and uncomfortable. However, it might also have to do with the relevance factor to which Chris has alluded. I, too, do not have a definitive answer to the question; but I think it is one that deserves a lot more thought as the discipline matures.

It seems to me that this discussion naturally flows to questions about the quality of our educational and training programs. So, let me ask you this. How are we doing in educating students, seasoned engineers, and managers about the potential of geostatistics? Are we emphasizing too much or too little theory? To many or too few applications? Pat, why don't you go first?

**Corbett:** As an educator in this area, my perception is that we are very much engaged in teaching this subject within reservoir evaluation and management. However, we struggle to find space for geostatistics in the syllabi of traditional petroleum engineering and geoscience courses that are already pretty full. It's hard to add this additional technical material, along with information on business awareness, team skills, environment, etc., and still turn out students that are well grounded in the basics. Geostatistical techniques take time to learn and use properly; and ultimately, reservoir specialists who are dedicated to the technology are required for its effective application.

**Coburn:** Are we talking about undergraduate or graduate education?

**Datta-Gupta:** In many of the petroleum engineering schools, geostatistics is now part of the undergraduate curriculum. This is an indication of how mainstream the technology has become. It is no longer a toolbox for the specialists.

**Kelkar:** Actually, my experience has been different. I would say that, in most universities, courses on geostatistics are offered at the graduate level.

**Coburn:** OK. It looks like we've got sort of a two-pronged approach with regard to formal educational programs. What about other kinds of training programs, and what about our effectiveness in delivering it? Does the information stick with the students?

**Kelkar:** Frankly, our ability to truly educate students in this subject is limited at this point, particularly without structured hands-on experience. For seasoned engineers, it is very difficult to teach them in 1 week so that they can become very efficient. I think that 90% of the people who take a short course never use geostatistics. This is not a very high success ratio. The reason is without practice and familiarity with software, it is difficult to become proficient in the subject.

**Coburn:** Let's go back to the question about the proper mix between theory and applications.

**Oliver:** I believe that it is vitally important for students to have a solid understanding of the theory as well as the ability to apply it. For many students, much of their understanding of geostatistics comes from application and experimentation, but the ability to diagnose incorrect results and to understand the limitations of an algorithm is usually a result of understanding the theory. For managers, theory is probably of less importance and probably more difficult to communicate. In their case, applications are the appropriate vehicle for communication.

**Kelkar:** I, too, believe that some theoretical background is important in understanding the limitations and strengths of the methods.

**Datta-Gupta:** One could always argue that we are not emphasizing the theoretical background enough; however, this should not deter the practicing engineers and geoscientists. Sometimes, the best way to learn is from failure.

**White:** I would say that, from undergraduates to managers, we are often too keen to get to the applications, and we are left without sufficient theoretical grounding. Too often, the emphasis is on software rather than on models and data.

In my opinion, engineering and geoscience curricula should include courses devoted to conventional statistics as well as geostatistics. The Society of Petroleum Engineers has formulated guidelines to encourage this, and the guidelines are part of the United States engineering accreditation criteria.

To motivate management, we must ensure that our modeling processes integrate financial and infrastructure factors as well as geologic and geophysical features. We run too many megamodels and conduct too little analysis.

**Jensen:** I'm not so sure this is a black-and-white issue about theory vs. applications. Just like well-test interpretation, log analysis, or stratigraphic correlation, geostatistics is a specialty area requiring people with specific skills in order for it to be productively applied, as was suggested earlier. However, asset teams need all members to be acquainted with the methods, abilities, and limitations of geostatistics. Personally, I think many students simply lack the necessary multidisciplinary knowledge to understand the limitations of geostatistical procedures. Managers are becoming more comfortable with geostatistical-based evaluations, but we're certainly not "home and dry" yet.

**Coburn:** OK. With these thoughts in mind, how do we more effectively move theory into practice?

**Oliver:** One of the most effective ways to move theory into practice is through software/code development and distribution. The Levenberg-Marquardt minimization algorithm provides a good example from another discipline. The original algorithm was developed by Kenneth Levenberg and published in *Quarterly of Applied Mathematics* in 1944. It remained relatively unknown and unused until 1963 when Donald Marquardt published a paper that included a reference to a Fortran code implementing the algorithm. I suspect that there are many good and potentially useful geostatistical algorithms that are underutilized because they have not been made available in easily usable software packages.

**Corbett:** Industry can encourage more universities to get involved in this subject by providing data sets and software. We have been fortunate to be well supplied with both and, therefore, able to develop realistic industry projects for the students, which they enjoy. The skills they acquire are useful in helping them find employment in the industry.

**White:** Actually, I think the transfer from theory to practice has been pretty effective (with a great debt to Clayton Deutsch, André Journel, and their collaborators for moving GSLIB into the public domain). The uptake of the theory has been fairly rapid—if anything, too fast rather than too slow. For example, the applications of geostatistical techniques to huge geocellular models have been undertaken without sufficient theoretical understanding of basic modeling issues such as equiprobability, uncertainty in spatial correlation, and sampling.

**Datta-Gupta:** I agree with Chris. Within the petroleum industry, we have done pretty well in moving geostatistical theory into practice. As I mentioned before, it is sometimes healthy to have a gap between theoretical developments and practice to allow the industry to assimilate the technology and understand its limitations.

**Jensen:** I also want to follow up on something that Chris said. We need to better understand the geological contexts and the effects of our sampling strategies when we use geostatistics. This is why case studies are so important—the range of geological and flow scenarios is very large.

**Kelkar:** Ultimately, the need to integrate a particular data set in a specific project or problem is the primary driver in moving theory into practice. In this regard, I would like to mention two people who have made important contributions in this area: Mohan Srivastava, who developed the concept of probability kriging, which has many interesting applications; and Phillippe Doyen, who developed the concept of Bayesian kriging, which really helped the integration of seismic data in an efficient manner. I don't think we should force theory into practice. Instead, we should let practice draw in the best theories developed.

**Coburn:** I briefly mentioned something earlier that I now want to come back to. The concepts of geostatistics and stochastic modeling can be difficult to communicate, particularly to those without strong backgrounds or recent experience in mathematics and statistics. Language, terminology, and notation can be significant barriers. How do we overcome this situation and improve communication of scientific/technological developments?

**Kelkar:** Tim, I'm not sure that I agree with your premise. I believe that if you can explain the concepts in a simple enough manner, anyone should be able to understand them without the advanced knowledge of statistics and mathematics. We are not asking the end user to develop new theories; we are simply asking them to understand which method is appropriate under what conditions.

**Corbett:** Mo is right. One can't dumb down every subject to the lowest common denominator—if people wish to work in this area, they should have some geoscience, engineering, and geostatistical knowledge. Language, terminology, and notation can be significant barriers to petroleum engineering, log analysis, geology, and geophysics... so this is not a new problem for the industry! Proper teaching and allowing time for adsorption is the solution.

**Datta-Gupta:** Well, I'm not so sure. I think that language and notation really can be problem, but that it can be overcome. From my experience in teaching geostatistics to the students at our university and also in industry seminars, I have found it extremely effective to combine theory and practice at the same time. Otherwise, one can very easily be overwhelmed by the theory and jargon, and can lose interest in the subject. Some of the public domain software, such as GEOEAS, TOOLBOX, and GSLIB, have made enormous contributions to the community in this respect. I just wish that they could be updated to incorporate some of the relatively recent and commonly used algorithms such as sequential simulation/cosimulation and Bayesian updating. An updated and Windows-based version of GEOEAS and TOOLBOX are very much in order.

**White:** I would say that the communications problem is eased by use of well-balanced, long-running interdisciplinary reservoir-modeling teams. My experience in industry convinced me that sustained teamwork can pay valuable dividends in improved reservoir modeling. Team members can educate one another and develop a shared language for modeling. However, such long-term commitments are difficult to maintain in the volatile exploration and production business.

**Oliver:** Chris has a good point. Cross- or multidisciplinary training will almost certainly be the key to communication. The recent development of academic programs in reservoir characterization, especially those that integrate engineering and the geosciences, might be most effective. In some cases, an intermediary between the theoretician and the practitioners is needed. One major oil company has been particularly successful in communicating the essence of geostatistics to reservoir engineers and geoscientists without requiring either group to delve into the mathematics. The intermediary in their company is usually provided by a technical support group that provides software and advice.

**Coburn:** Jerry, what's your take on this?

**Jensen:** It's an interesting question, but I'm not sure I can add much to what has already been said. Ultimately, we motivate people by showing them improvements in productivity and value. If we're not doing a good job communicating—and I'm not totally convinced that that is the case—then perhaps it has more to do with our need/ability to demonstrate value than with our need/ability to explain methodological details.

**Coburn:** If I may, let me turn our attention to a different question. Recently, there has been an explosion of literature concerning the use of geostatistics and stochastic modeling in the earth sciences. Is this an indication of a significant scientific phenomenon, or are we simply experiencing a bandwagon effect? What's really driving all this apparent interest?

**Datta-Gupta:** I think it is a combination of genuine interest within the scientific community and also a bandwagon effect. Traditional variogram-based geostatistics has become a very mature technology, and many of the new developments in this area are probably going to be incremental in nature. On the other hand, as we have already suggested, there is increasing interest among the scientific community in looking beyond two-point statistics to generate realistic geologic models that can be conditioned to dynamic data. Clearly, the interest is driven by the past success of geostatistics and its future potential.

**White:** My perspective on this is slightly different from Akhil's. I'd say the explosion in interest is fueled more by the emergence of new algorithms and burgeoning computing power. Sequential Gaussian simulation is incredibly efficient compared to earlier methods, and storing and manipulating geomodels with approximately $10^7$ cells would have been technologically impossible a few years ago. Improved algorithms and computations allow application of new methods to explore uncertainty, such as adjoint methods and Markov chain Monte Carlo techniques. Combined with a realization that we are failing to assess risk adequately, the enthusiasm for these methods makes sense. Of course, these methods can be misused, and some geostatistical studies attain verisimilitude but provide no real insight.

**Oliver:** Chris is probably right about his reference to computing. I believe that the use of geostatistics in reservoir engineering was an idea that had to wait for advances in computer hardware and software. The usefulness for geostatistics in reservoir engineering has increased greatly as the size of reservoir-simulation models has increased. There was little incentive for using geostatistical methods on the small reservoir-simulation models of 20 yr ago.

**Jensen:** There is another important factor we don't want to forget. There is often a large economic interest in the results of geological and petrophysical models. Risk minimization and management requires having to examine various scenarios and their economic impacts. This need, coupled with a slow but steady drift towards quantification of geological and petrophysical phenomena, are placing geostatistics in a more important role.

**Corbett:** I'm not sure I'd call the phenomenon we're observing a bandwagon effect—perhaps we are simply in a revolution in quantitative geoscience.

**Kelkar:** Actually, I believe that the number of papers in this area has been slowly declining throughout the last 10 yr, indicating a maturation of technology as Akhil suggests. On the other hand, I do not believe that geostatistics is accepted as a standard procedure in many companies, including some of the major ones. Part of the reason is that, in the early stages, geostatistics promised too much, and delivered too little. For example, very few people realized that generating multiple realizations using the same variogram and same conditioning data cannot really capture true uncertainty in a reservoir model. Things have changed throughout the last decade, and we have gained a lot of experience in the correct application of geostatistical principles. However, we really need to conduct some type of cost-benefit analysis in order to emphasize the value of geostatistics. I think maybe we alluded to this earlier. The same point applies to technologies like 3-D seismic data acquisition. However, I have not come across any published examples in which this type of evaluation has been performed.

**Coburn:** OK. Now I'm going to ask you to be real prognosticators! If you had a crystal ball, what would you say is the future of geostatistics, particularly as it impacts petroleum/production/reservoir engineering? Where are we headed in the next 10 yr?

**Kelkar:** I believe that we will continue to see increasing use of geostatistics in reservoir description. With better software and better understanding, there will be more desire to use geostatistics. Ultimately, if the users cannot demonstrate the value of geostatistics through improved recovery, a successful infill well-drilling program, improved exploration, or some other means, the technology will die. Based on my own personal experience, I firmly believe that integrated reservoir description that is geostatistically based does save time in the history-matching process. In addition, the generated reservoir description results in a better reservoir management tool for predicting future performance. This type of experience has to be almost universal to sustain the technology.

**Datta-Gupta:** Let me add this to what Mo has said. I think that throughout the next 10 yr, we will see the gradual merger of many aspects of reservoir characterization and management/monitoring. To some extent, this is happening today through the use of time-lapse seismic information. In addition, we will be faced with an explosion of data from "smart wells" and permanent downhole sensors. Updating geologic models using these data will pose a significant challenge to the industry. Thus, dynamic data integration will continue to be an actively researched area throughout the next decade, requiring redevelopment or retooling of some of our methodologies.

**Coburn:** Jerry or Dean, how would address this question? What do you see on the horizon?

**Jensen:** Although from my perspective, the future looks much the same as recent past, here are three specific ideas that come to mind. First, faster flow simulation will enable us to take better advantage of the range of scenarios geostatistics provides. Because we'll be able to use more realizations in flow simulations, the integrity of the geostatistical product will be under greater scrutiny. For example, we might be asked to address the representativeness of extremes. Second, guidelines for data acquisition and integration will continue to evolve. Some techniques may be developed which require fewer data or which use the data more fully to provide more reliable interpretations and models. Third, wavelet-based analysis and simulation methods could become more important.

**Oliver:** Well, I suppose we've all got our favorite ideas, so here are mine. I believe that we will see two trends. The first one takes us back to geology. I believe we will see expanded use of more complex geostatistical/stochastic models that come closer to describing the connectivity and sequencing of facies resulting from erosion, deposition, faulting, diagenesis, etc. It is possible that this realism/complexity might result from a merging of geologic process modeling with stochastic methods for conditional simulation. The second trend will be towards better assessment of uncertainty in reservoir performance through the inclusion of more of the sources of uncertainty, not only those that are included in current software packages.

**Coburn:** Chris, how about you?

**White:** My response would be a more holistic one. It seems to me that the future of geostatistics in the petroleum industry will be determined by our ability to integrate probabilistic reservoir models with the asset development workflow. The integration will rely on better understanding of statistical properties such as correlation and conditional probability, visualization, and integration of static reservoir models with geophysical inversion and flow modeling. The ultimate goal is an easily updated suite of dynamic reservoir models that are conditional to inaccurately known core, log, seismic, and production data—Dean referred to such real-time updating earlier. The alternative models should consider the errors in all measurements and should consider metalevel uncertainties such as stratigraphic interpretation. The model should be integrated with algorithms for uncertainty assessment, flow simulation, history matching, optimization, and forecasting—including finance. Such developments will require a radical rethinking of our present workflow.

**Coburn:** And Pat...how would you respond?

**Corbett:** I'm not sure I can add anything to these excellent comments by my colleagues. This might sound somewhat simplistic, but I think the future of geostatistics is that we will use it everyday without thinking about it. Will that happen within 10 yr? I don't know; but I don't see us going back to a time when we didn't use geostatistics. I only see us moving forward.

**Coburn:** As Pat has suggested, these are all excellent points; and no doubt the foundational work necessary to bring them to fruition is already underway. Hopefully, this conversation will help to focus our collective efforts so that our future accomplishments can be maximized and the overall objectives of the discipline can be advanced. Before we close, are there any other comments you would like to make regarding the interface between geostatistics, stochastic modeling, petroleum/production/reservoir engineering, and reservoir characterization? Mo?

**Kelkar:** I think I've already said enough, so I will defer to my colleagues.

**Jensen:** Tim, I have a point I want to make and it has to do with sampling. The nonlinear interactions of fluids and geological/petrophysical events make it difficult to "transfer" results of case studies to new settings. Thus, sampling guidelines (i.e., where, how many, and what type) derived in one situation may not be readily applicable in another situation. For example, a decrease of mobility ratio can make the detailed assessment needed in a reservoir unnecessary. Right now, there seems to be no way around having inefficient sampling strategies. That is, either we sample poorly, evaluate, and sample more, or we oversample and risk wasting resources.

**Coburn:** It's interesting you should bring this up. Sampling is near and dear to my heart. It is fundamental to all of what we do and all the inferences we make. As you know, one of the original motivations of geostatistics was to help optimize the physical sampling process. It's an idea that I believe that the discipline of geostatistics has lost sight of and one that we can perhaps collectively encourage our colleagues to focus more effort on. Anybody else have closing comments or a particular point that you want to reemphasize?

**Oliver:** Since you and Jerry have brought up the idea of sampling, there's something else that I think is elemental. It has to do with the data values themselves and their relationship to "the model." I sometimes think that users deify "the model;" and in their quest to find the right one, they forget to adequately explore and portray the data. It's like seeing the forest but missing the trees. The use of geostatistics will not decrease the need for data in reservoir engineering; and model results are not necessarily the same as data. Further, in the future, it is likely that new measurements will be available, or that measurements will be made at scales that were not previously possible. It is critical to understand the problems of scale and the limitations of data as we attempt to assimilate measurements into reservoir model building in a realistic way, and it is essential to remember that models, though they may be good representations, are not truth and reality.

**Corbett:** I, too, feel we need to focus on using the tool appropriately—not introducing so much speculative data that we fool ourselves into thinking the model is right. The main objective of geostatistical methods is to communicate uncertainty—nothing more! Pretty pictures are nice to have—but they don't necessarily give a better answer.

**Coburn:** Chris, you get the last word.

**White:** Well, I suppose I would sum things up like this. Petrologists, sedimentologists, stratigraphers, geophysicists, petrophysicists, drilling engineers, production engineers, facility engineers, reservoir engineers, asset managers, and economists are all concerned with performance prediction, and therefore, all have a stake in stochastic reservoir modeling. However, it is the reservoir engineers that work at the nexus of these disciplines, and they will probably continue to be important users, developers, and coordinators of geostatistical modeling in the E&P industry. I would encourage all reservoir engineers to familiarize themselves with geostatistical technology because I see the methods and processes as critical components in a more unified and integrated approach to modern reservoir development and exploitation.

**Coburn:** Gentlemen, thank you so much for your time and your willingness to contribute to this important discussion. I look forward to our continued association in the future.

## REFERENCES CITED

Jennings Jr., J. W., S. C. Ruppell, and W. B. Ward, 2000, Geostatistical analysis of permeability data and modeling of fluid-flow effects in carbonate outcrops: SPE Reservoir Evaluation and Engineering, v. 3, p. 292–303.

White, C. D., D. Novakovic, S. P. Dutton, and B. J. Willis, 2003, A geostatistical model for calcite concretions in sandstone: Mathematical Geology, v. 35, p. 549–575.

## *Section II*

# Principles

The practice of geostatistics would not be where it is today, particularly in petroleum geology, without an ongoing commitment to theoretical research. It is fundamental theoretical work that enhances fundamental principles, produces new concepts and understanding, and leads to improved methodologies and computational strategies that can then be tested in practice.

This section includes seven chapters that address relatively new theoretical developments that have originated since the first volume of *Stochastic Modeling and Geostatistics* was published. The topics encompass a wide variety of geostatistical ideas, and the authorship has a genuinely international flavor. Readers should find much of the work referred to in this section to be quite interesting, yet fairly technical in nature.

In the first chapter, collaborators Amilcar Soares, José Almeida, and Luis Guerreiro from Lisbon, Portugal, present direct sequential simulation and cosimulation as alternatives to the well-known sequential Gaussian and sequential indicator simulation techniques. The authors discuss the methodologies in some detail, and then provide examples highlighting the advantages of the techniques. Next, Alexander Gribov and Konstantin Krivoruchko at Environmental Systems Research Institute (ESRI), along with Jay Ver Hoef at the Alaska Department of Fish and Game, discuss their work on new methods for modeling of the semivariogram. Many of these results have found their way into ESRI's products for modeling and analyzing spatial data from a geographic information systems (GIS) approach. In the third chapter in this section, Pierre Goovaerts presents his work on the multiple sources of uncertainty exhibited by petroleum reservoirs, specifically from the petrophysical perspective, and discusses their effect on the selection of a simulation approach. Readers who are routinely faced with the decision to choose among the various methods will want to be aware of Goovaerts' work and his ideas about the "spaces of uncertainty." This section also includes another chapter by Krivoruchko, Gribov, and Ver Hoef. In this chapter these authors provide the theoretical work for an alternative kriging technique, referred to as "new value kriging," which provides a different treatment of the nugget effect. "New value kriging" is shown to be equivalent to filtered kriging, and three very different applications of its use are provided. In the fifth chapter in the section, Jef Caers, Burc Arpat, and Cristina Arroyo Garcia, all from the Stanford Center for Reservoir Forecasting (SCRF), describe new approaches for identifying geobodies in seismic amplitude data based on automated pattern recognition. Next, Alain Galli, Gaëlle Le Loc'h, and Francois Geffroy, all from the Centre de Géostatistique at Ecole des Mines de Paris, and Rémi Eschard, from Institute Francais du Petrole (IFP) present the truncated pluri-gaussian (TPG) method for modeling geology. In recent years, TPG has been widely promoted within the French geostatistics community as a superior modeling approach. Finally, Michael Pyrcz (now at ChevronTexaco), and his collaborators Emanuel Gringarten (Earth Decision Sciences in Houston), Peter Frykman (Geological Survey of Denmark and Greenland), and Clayton Deutsch (University of Alberta) discuss their work on the effects of uncertainty in model inputs and the impact such effects have on the modeling outcome. Pyrcz et al. present a recommended best practices workflow for assembling inputs to geostatistical models that should facilitate more realistic and representative results in practice.

Soares, A., J. A. Almeida, and L. Guerreiro, 2006, Incorporating secondary information using direct sequential cosimulation, *in* T. C. Coburn, J. M. Yarus, and R. L. Chambers, eds., Stochastic modeling and geostatistics: Principles, methods, and case studies, volume II: AAPG Computer Applications in Geology 5, p. 35–43.

*Chapter 4*

# Incorporating Secondary Information Using Direct Sequential Cosimulation

**A. Soares**
*Instituto Superior Técnico*
*Lisbon, Portugal*

**J. A. Almeida**
*Instituto Superior Técnico*
*Lisbon, Portugal*

**L. Guerreiro**
*Partex Oil and Gas*
*Lisbon, Portugal*

## ABSTRACT

Direct sequential simulation is a new approach to stochastic simulation that does not require any prior transformation of the original variable or any posterior transformation of simulated values. To simulate continuous variables, the algorithm uses the estimated local mean and variance to sample from a global cumulative distribution function. An advantage of this approach over sequential Gaussian simulation or sequential indicator simulation is that it accommodates the joint simulation of several variables under the direct simulation principle; that is, coregionalization modeling is performed with the original variables. The direct joint simulation methodology is presented in this chapter, and a representative set of examples is provided to illustrate the potential of the method to incorporate secondary information in the characterization of oil reservoirs.

## INTRODUCTION

Stochastic sequential simulation is a family of stochastic models that aims at reproducing the spatial distribution and uncertainty of the main properties of petroleum reservoirs. Although they use the same sequential procedure, the different versions of sequential simulation require transformations of original variables and different approaches to estimate local distribution functions.

Direct sequential simulation (DSSIM), which requires no transformation of the original variables, has been widely used to characterize the spatial distribution of categorical attributes such as lithotypes (Soares, 1998). For continuous variables such as grades or porosity, the traditional approaches are sequential indicator simulation (SIS) and sequential Gaussian simulation (SGS), which do require such transformations. The original variables must be transformed into a set of indicator variables or a standard Gaussian


DOI:10.1306/1063805CA53225

variable. Various authors (e.g., Goovaerts, 1997) have reported the limitations of such approaches. Journel (1994) demonstrated that direct simulation of a continuous variable, without any prior transformation, can succeed in reproducing a covariance model, provided the simulated values are drawn from local distributions centered at the simple kriging estimates with a variance corresponding to the simple kriging estimation variance. Although the spatial covariance of the original variable is reproduced, the histogram is not, which is contrary to one of the main requirements for any simulation algorithm. This situation constitutes the most serious limitation to the practical application of the DSSIM approach.

J. Caers (1999, personal communication) resurrected the idea and proposed the use of postprocessing to transform the resulting pseudosimulated values into another set of values. This approach results in an approximate reproduction of the histogram of the original variable and preserves the data exactly. However, in some situations, a posterior transformation of this type (Journel and Xu, 1994) can eliminate the reproduction of the semivariogram model. Caers (2000) subsequently proposed a direct indicator simulation that incorporates a set of linear constraints after kriging, thereby avoiding the need for posterior transformation.

Another new and much simpler approach for the DSSIM is presented by Soares (2001). This simple algorithm, which consists of resampling the global cumulative distribution function, using the simple kriging and estimation variance, succeeds in reproducing, very satisfactorily, the semivariogram and histogram of a continuous variable.

The purpose of this chapter is to illustrate Soares' (2001) direct cosimulation procedure without requiring any transformation of the original variables through a set of examples of petroleum applications.

## DIRECT SEQUENTIAL SIMULATION

The sequential simulation algorithm for modeling a continuous variable $Z(x)$ follows the classical methodological sequence as follows:

1) Randomly choose the spatial location of a node $x_u$ in a regular grid of nodes to be simulated.
2) Estimate the local cumulative distribution function at $x_u$, conditioned to the original data $z(x_\alpha)$ and the previous simulated values $z^s(x_i)$. Estimation of the local cumulative distribution function (CDF) is commonly performed with the indicator formalism (SIS) or with the multi-Gaussian approach (SGS), both of which require the transformation of the original variable, as previously noted.
3) Draw a value $z^s(x_i)$ from the estimated local CDF.
4) Return to step $i$ until all nodes have been visited by the random path.

The simulated values $z^s(x)$ should reproduce the global CDF $F_Z(z) = \text{prob}\{Z(x) < z\}$ and the stationary semivariogram $\gamma(h)$ of $Z(x)$.

The principle of the DSSIM can be summarized as follows: If the local CDFs are centered at the simple kriging estimate, $z(x_u)^* - m = \sum_\alpha \lambda_\alpha\ (x_u)(z(x_\alpha) - m)$, with a conditional variance identified as the simple kriging variance $\sigma^2_{\text{sk}}(x_u)$, the spatial covariance model or semivariogram is reproduced in the final simulated maps. The problem is that this simulation approach does not reproduce the histograms of the original variables (the local CDF cannot be fully characterized by only the local mean and variance).

The idea of DSSIM (Soares, 2001) is to use the estimated local mean and variance, not to define the local CDF but to sample the constant global CDF $F_Z(z)$. Intervals of $z$ are chosen from $F_Z(z)$, defining a new $F'_Z(z)$, and then the simulated values $z^s(x_u)$ are sampled from them. These intervals, representing the range that is dependent on the simple kriging estimation variance $\sigma^2_{\text{sk}}(x_u)$, are centered at the simple kriging estimate $z(x_u)^*$. Simple kriging is performed with a unitary sill semivariogram and the simple kriging variance range. Hence, the sampling interval range can go from 0, when the point $x_u$ is estimated at a sample location $x_\alpha$ (i.e., $z(x_u)^* = z(x_\alpha)$ and $\sigma^2_{\text{sk}}(x_u) = 0$), to 1, corresponding to the maximum local variance. Note that $\sigma^2_{\text{sk}}(x_u) = 0$ means that the range interval is null, and the simulated value is equal to the sample value $z^s(x_u) = z(x_u)^* = z(x_\alpha)$, and $\sigma^2_{\text{sk}}(x_u) = 1$ means that the simulated value can be drawn from the entire range of values of the global CDF.

The simplest way to accomplish this resampling stems from the definition of the range of the global CDF assuming a Gaussian distribution (see Figure 1). Consider a Gaussian variable $y(x)$ obtained from the original $z(x)$ values by a normal score transformation $\varphi$:

$$y(x) = \varphi(z(x)) \tag{1}$$

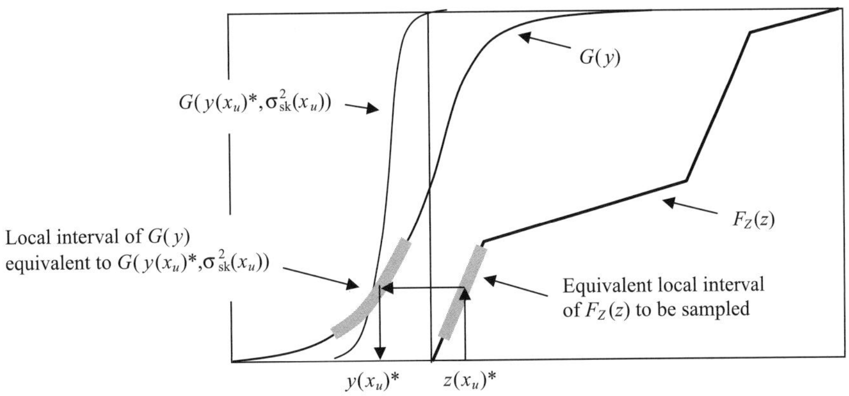


**Figure 1. Sampling of the global distribution $F_Z(z)$ by intervals defined by the local mean and variance of $z(x_u)$. The value $y(x_u)^*$ corresponds to the local estimate $z(x_u)^*$. The interval of $F_Z(z)$, from which is drawn the simulated value $z^s(x_u)$, is defined by $G(y(x_u)^*,\sigma^2_{sk}(x_u))$.**

with $G(y(x)) = F_Z(z(x))$. A simulated Gaussian value is drawn from a Gaussian CDF with a local mean $y(x_u)^* = \varphi(z(x_u)^*)$ and a local variance $\sigma^2_{sk}(x_u)$. To accomplish this, a value $p$ is selected from a uniform distribution $U(0,1)$. Then, a value $y^s$ from $G(y(x_u)^*, \sigma^2_{sk}(x_u))$ is generated:

$$y^s = G^{-1}(y(x_u)^*, \sigma^2_{sk}(x_u), p) \tag{2}$$

Finally, a simulated value $z^s(x_u)$ is obtained by the inverse transformation $\varphi^{-1}$:

$$z^s(x_u) = \varphi^{-1}(y^s) \tag{3}$$

This means that $z^s(x_u)$ is sampled from intervals of $F_Z(z)$ defined by the local estimates $z(x_u)^*$ and $\sigma^2_{sk}(x_u)$.

It is important to note that the Gaussian transformation is used solely for sampling local intervals from the distribution $F_Z(z)$. It does not have any function in the estimation of the local CDF; hence, the distribution of the transformed values is not assumed to be Gaussian. The entire sequential procedure is performed with the original variable $Z(x)$.

Because a nonlinear function, $\varphi$, is used to obtain the local intervals of the global CDF, the expectation of $z^s(x_u)$ may not be equal to $z(x_u)^*$. In other words, $E\{y^s\} = y(x_u)^* = \varphi(z(x_u)^*)$, but $E\{z^s(x_u)\} \neq z(x_u)^*$.

This theoretical limitation has not been shown to have any practical impact in case studies, and any potential impact has to be balanced against the simplicity of the algorithm. However, in situations where the histogram of $z(x)$ is based on a small amount of data, a bias can occur, especially in low-frequency classes. The bias arises from the disparity between a Gaussian CDF centered at $y^*(x_u)$ and a CDF that is not centered at $z^*(x_u)$. A correction for local bias is proposed in Soares (2001).

Summarizing, the DSSIM can be described by the following steps:

1) Calculate the normal score transform of $z(x)$ (for example, with a graphical method, Goovaerts, 1997).
2) Define a random path over the entire grid of nodes $x_u$, $u = 1, \ldots, Ns$, to be simulated.
3) Estimate the local mean and variance of $z(x_u)$ identified, respectively, with the simple kriging estimate $z(x_u)^*$ and estimation variance $\sigma^2_{sk}(x_u)$ conditioned to the experimental data $z^s(x_i)$ and previous simulated values $z^s(x_i)$.
4) Define the interval of $F_Z(z)$ to be sampled using the Gaussian CDF: $G(y(x_u)^*, \sigma^2_{sk}(x_u))$, where $y(x_u)^* = \varphi(z(x_u)^*)$.
5) Draw a value $z^s(x_u)$ from the CDF $F_Z(z)$.
6) Select a value $p$ from a uniform distribution $U(0,1)$.
7) Generate a value $y^s$ from $G(y(x_u)^*, \sigma^2_{sk}(x_u))$; $y^s = G^{-1}(y(x_u)^*, \sigma^2_{sk}(x_u), p)$.
8) Return the simulated value $z^s(x_u) = \varphi^{-1}(y^s)$.
9) Loop until all $N$ nodes have been visited and simulated.

## DIRECT SEQUENTIAL COSIMULATION

One of the main advantages of the proposed algorithm over traditional SIS and SGS is that it accommodates joint simulation of original variables. In this context, instead of simultaneously simulating a set of variables, each variable is simulated in turn, conditioned to the previously simulated variable (Gomez-Hernandez and Journel, 1993; Goovaerts, 1997). Consider the case of just two variables, $Z_1(x)$ and $Z_2(x)$. Choosing one variable, say $Z_1(x)$, as the most important or with more apparent spatial continuity (Almeida and Journel, 1994), the joint simulation algorithm proceeds as follows. $Z_1(x)$ is simulated first using the DSSIM algorithm described above. Then, the same algorithm is applied to simulate $Z_2(x)$, assuming the previously simulated $Z_1(x)$ as the abundant (known at every node) secondary variable. Collocated simple cokriging is used to calculate $z_2(x_u)^*$ and to estimate $\sigma^2_{\text{sk}}(x_u)$ conditioned to neighborhood data $z_2(x_\alpha)$ and the collocated datum $z_1(x_u)$ (Goovaerts, 1997). This process is described by

$$z_2(x_u)^*_{\text{sck}} = \sum_{\alpha=1}^{N} \lambda_\alpha(x_u)[z_2(x_\alpha) - m_2] + \lambda_\beta(x_u)[z_1(x_u) - m_1] + m_2 \quad (4)$$

where $m_1$ and $m_2$ are the means of $Z_1(x)$ and $Z_2(x)$, respectively. As in the case of DSSIM, $Z_2(x)^*$ is estimated with unitary sill semivariograms of $Z_1(x)$ and $Z_2(x)$.

Next, apply the transform $y(x_u)^* = \varphi_2(z_2(x_u)^*)$, with $\varphi_2$ as the normal score of the $Z_2(x)$ variable, and select a value $p$ from a uniform distribution $U(0,1)$. Then, generate a value $y^s$ from $G(y_2(x_u)^*, \sigma^2_{\text{sk}}(x_u))$, such that $y^s = G^{-1}(y_2(x_u)^*, \sigma^2_{\text{sk}}(x_u), p)$. Return the simulated value $z^s_2(x_u) = \varphi^{-1}_2(y^s)$, and repeat the process until all nodes are simulated.

This methodology can be applied to the joint simulation of a more extended set of variables.

## EXAMPLES OF SEQUENTIAL SIMULATION AND COSIMULATION

### Example 1: Direct Sequential Simulation

Figure 2 shows the spatial location of a set of 50 samples. The histogram of the data is bimodal, which is the target of the sequential direct simulation procedure. Figure 3 shows the results of the application of the DSSIM algorithm and indicates that the histogram and semivariogram have been satisfactorily reproduced. The range of the semivariogram is equal to 20 m (66 ft) (1/5 of the side of area $A$).

### Example 2: Direct Sequential Cosimulation of Acoustic Impedance

The direct sequential cosimulation was performed with a map of inverse acoustic impedance, as the secondary variable, known at every node of the grid. The primary variable to be cosimulated is the acoustic impedance known at the well logs, which is the product of density and seismic velocity. Both variables are at the same scale.

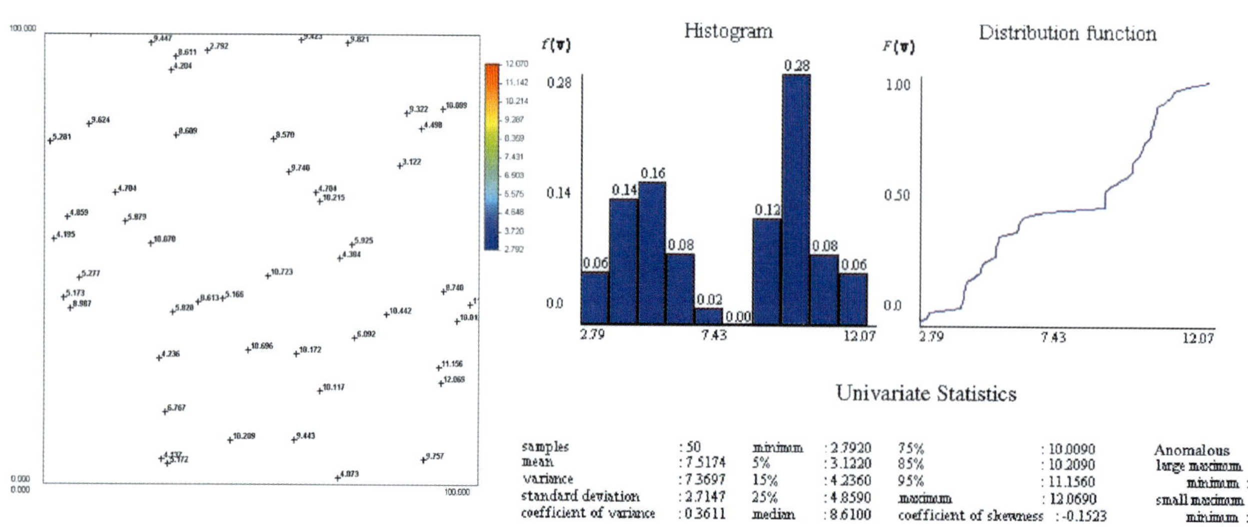


**Figure 2. Spatial location of the set of 50 experimental samples with the histogram (bimodal), which is the target of the simulation procedure.**

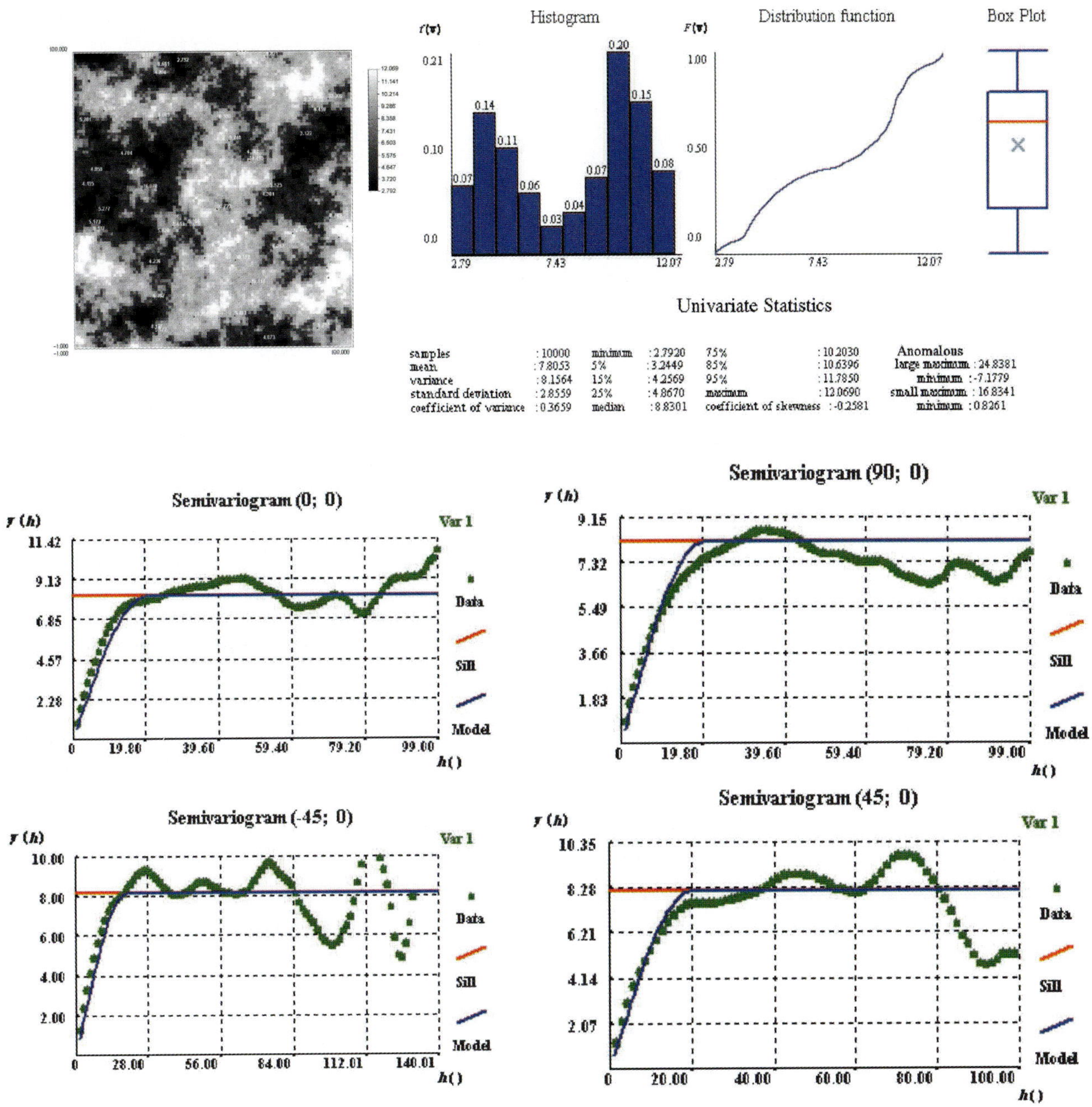


**Figure 3. One realization of DSSIM. The histogram and semivariogram range, 20 m (66 ft) (1/5 of the side of area *A*), are reproduced.**

For sake of simplicity, the colocated cokriging was applied with the Markov-type approximation (Goovaerts, 1997); that is, only the primary variable semivariogram and the correlation coefficient between primary and secondary variable are required. In terms of correlograms, the cross-correlogram $\rho_{12}(h)$ is obtained by the product of the correlation coefficient $\rho_{12}(0)$ and the correlogram of variable $Z_1(x)$: $\rho_{12}(h) \cong \rho_{12}(0) \cdot \rho_1(h)$. This approximation can be checked and validated because the experimental values of $\rho_{12}(h)$ can be calculated. For illustration purposes, two correlation coefficients were considered: $r = 0.95$ and $r = 0.6$.

Figure 4 illustrates the results of this approach, indicating a quite good match of the targets of the joint simulation. Figure 4a shows the image of the secondary variable TDROV cube. Figure 4b and c show two simulated realizations with $r = 0.60$ and

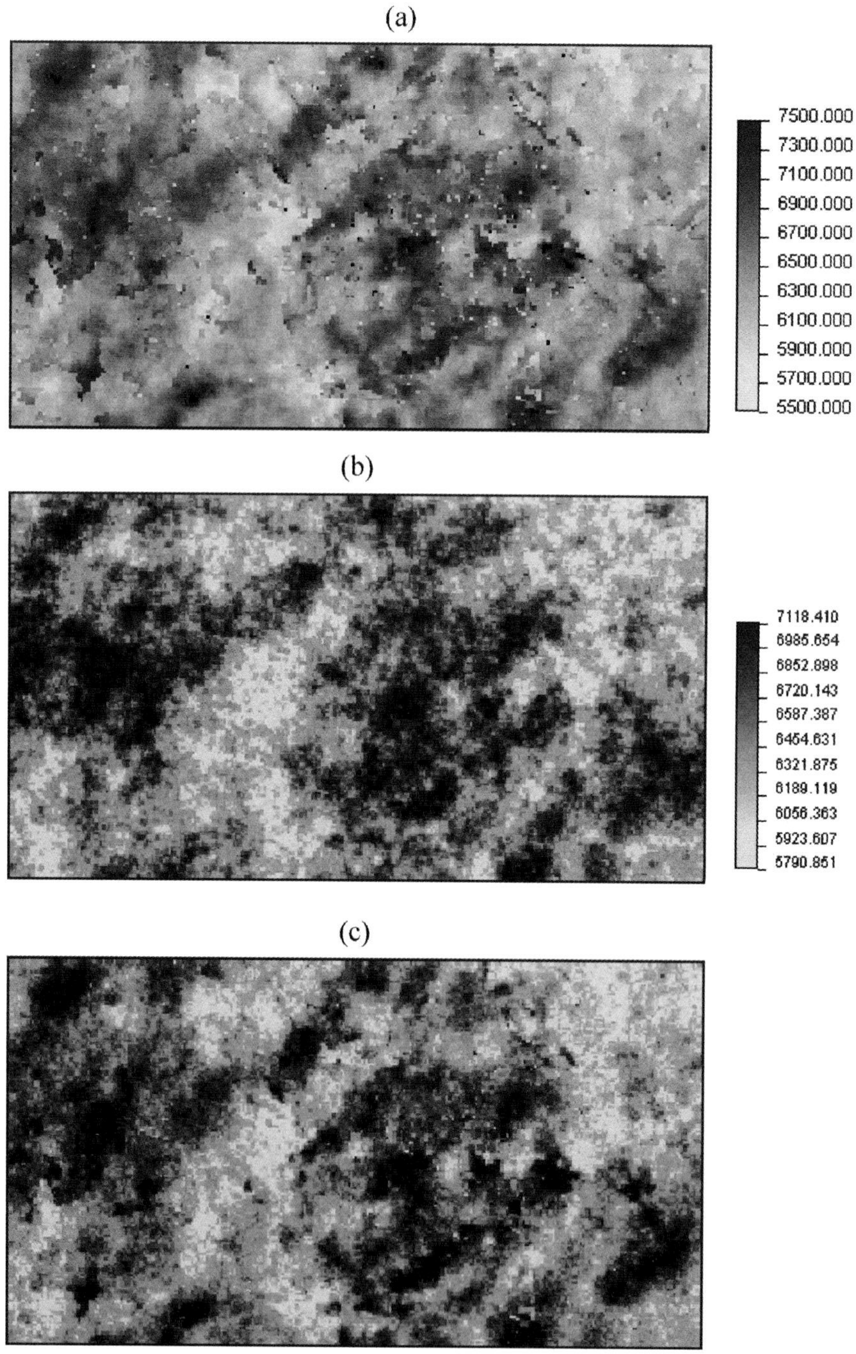


**Figure 4.** (a) Image of the secondary variable, the TDROV cube; (b) simulated realization with $r$ = 0.6; (c) simulated realization with $r$ = 0.95.

$r$ = 0.95, respectively. As expected, with $r$ = 0.95, a quite perfect match exists between the secondary variable (TDROV) and the cosimulated realization of acoustic impedance, except where the well logs did not agree with the TDROV value at the well location. With $r$ = 0.6, the cosimulated image of acoustic impedance presents some differences regarding the secondary image, which are coherent with the correlation coefficient and the local mismatch between primary and secondary variables at well locations.

## Example 3: Cosimulation of a Synthetic Cube of Impedance

In this example taken from E. Schuhler and J. P. Diet (2000, personal communication), the aim is to simulate a synthetic cube of values of acoustic impedance in a reservoir represented by four rock types. The secondary variable was a preexisting simulation of porosity per rock type obtained through the DSSIM algorithm.

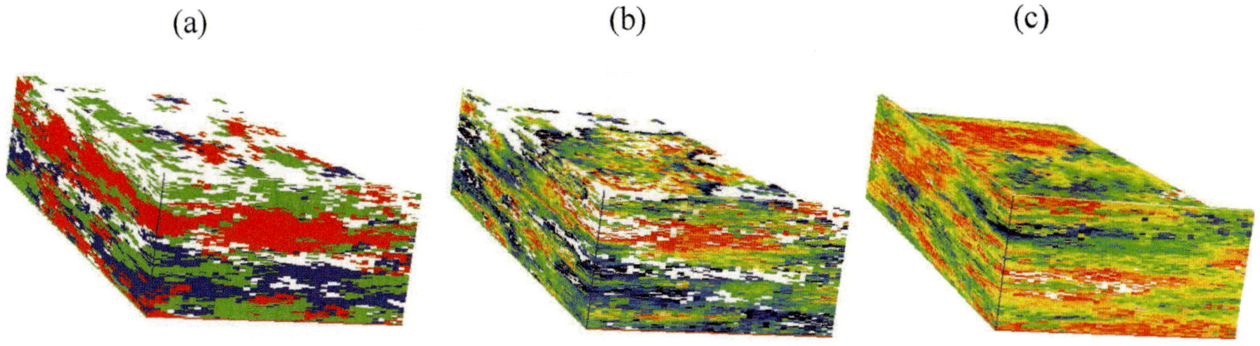


**Figure 5.** **(a) Image of rock types; (b) image of porosity simulated with DSSIM; (c) impedance values obtained by direct cosimulation.**

The relationship between porosity and acoustic impedance was calculated for each rock type showing the following correlation coefficients: −0.80 for type 1, −0.70 for type 2, −0.72 for type 3, and 0 for type 4, reflecting no influence of the secondary variable in this rock type. Hence, a map of local correlation coefficients was defined according the spatial distribution of rock types.

In the direct sequential cosimulation procedure, the porosity and acoustic impedance cross-correlograms are calculated according to the local values of correlation coefficients, assuming the Markov-type approximation mentioned above. Figure 5 shows the three-dimensional cube of simulated rock types together with porosity and the simulated acoustic impedance.

### Example 4: Reproducing Local Trends with Direct Sequential Cosimulation

In the previous example, the direct sequential cosimulation is performed using different correlation coefficients for each rock type. However, local trends representing continuous variation of correlation can also be accommodated in the proposed methodology.

Consider the case of example 2, where acoustic impedance is cosimulated having the TDROV map as a secondary variable. In the more realistic situation where the TDROV map, obtained by seismic inversion, reflects zones with good and noisy seismicity, the relation with log acoustic impedance depends on the spatial distribution of these good and noisy seismic zones. Local correlation coefficients are calculated at the well-log locations and interpolated for the entire area. Figure 6a shows an example of the correlation map in one layer.

Based on a secondary information image (Figure 4a), three different realizations were simulated (Figure 6b–d). It is worth noting that the spatial pattern of the central part of the secondary image, corresponding to the high correlation coefficients, is maintained in all three realizations, whereas the influence of the secondary image at the borders is negligible. The semivariograms of these realizations are reproduced in Figure 7, showing a good match with the proposed theoretical models.

## CONCLUSIONS

The methodology of DSSIM and cosimulation presented here uses the original variables without requiring any prior or posterior transformation.

Direct sequential simulation and cosimulation are as simple to implement as the multi-Gaussian approach, but with the great advantage of not requiring the calculation of Gaussian-transformed semivariograms and cross-semivariograms. Because the proposed algorithm deals with original variables, corrections for local means can be easily implemented to reproduce nonstationary patterns or trends in the primary variable or to better reproduce the extreme classes of highly skewed histograms.

This chapter illustrates that, among the class of sequential simulation algorithms, DSSIM and cosimulation are very appropriate and advantageous techniques with which to simulate and jointly simulate continuous variables.

## ACKNOWLEDGMENTS

This research was supported by grants of the Compagnie Generale de Geophysique. We acknowledge the constructive reviews of Jean Paul Diet.

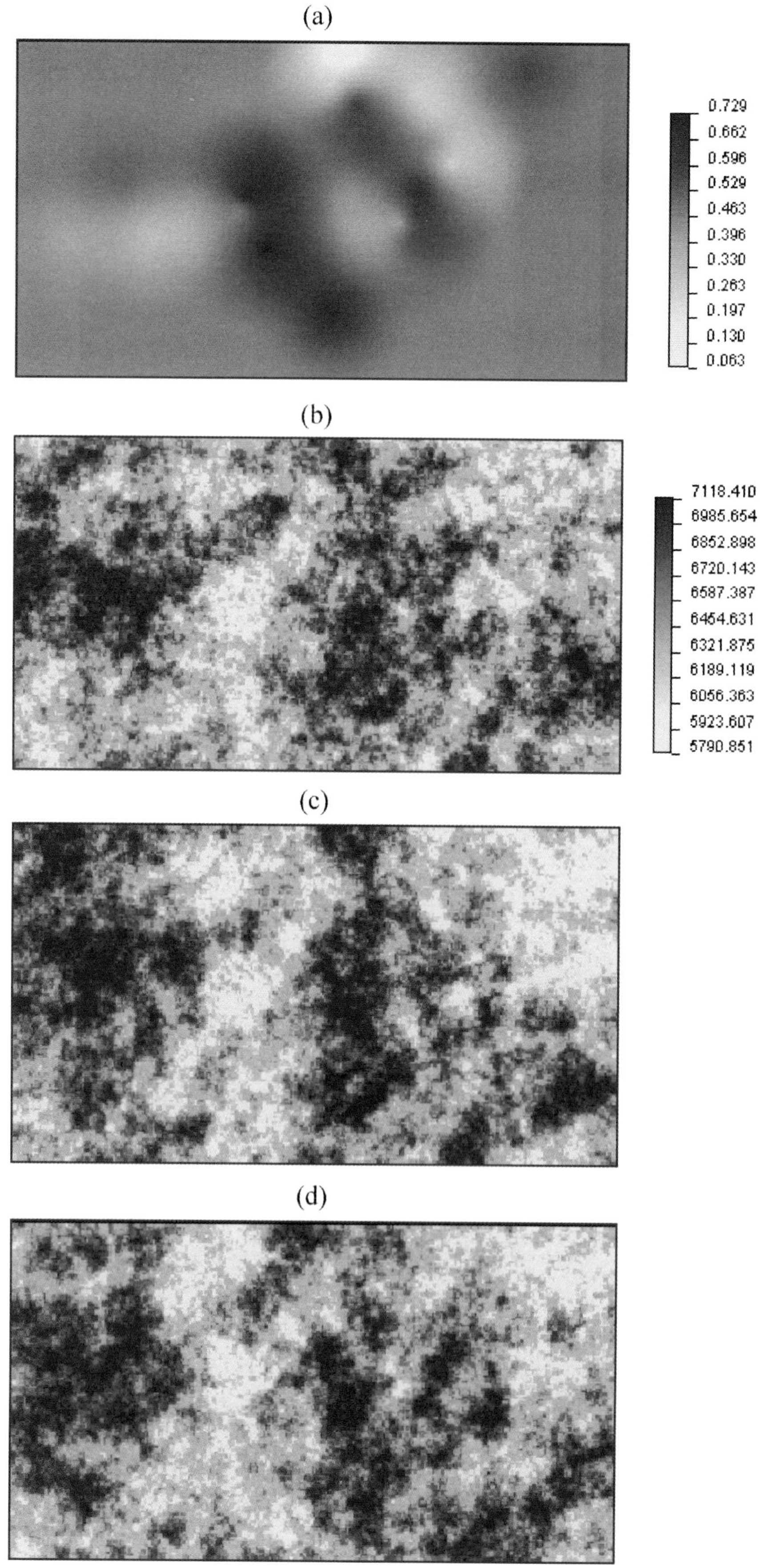


Figure 6. (a) Correlation coefficient map in one layer; (b–d) examples of three realizations of cosimulation.

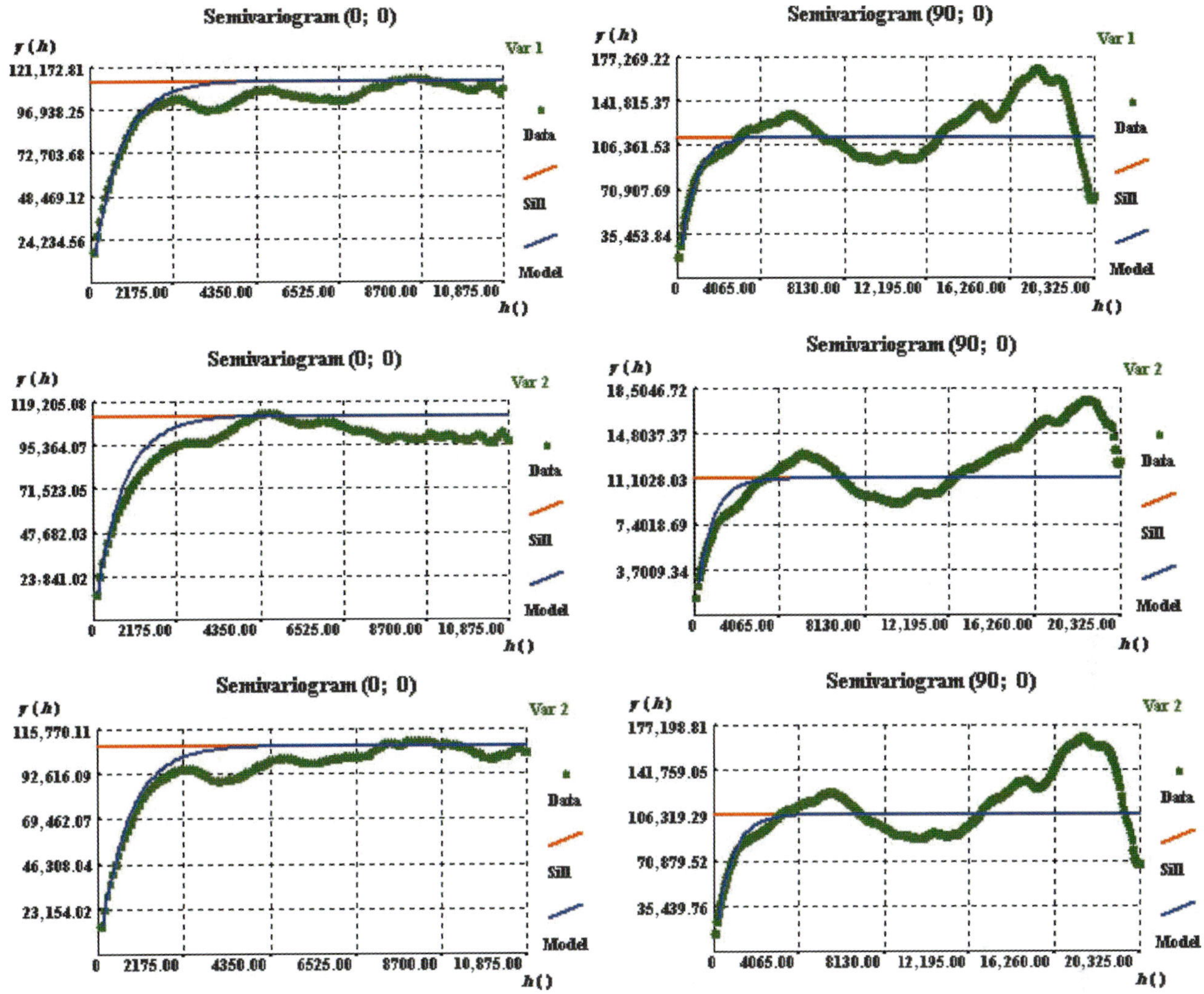


**Figure 7.** Semivariograms of the three realizations of cosimulation with local models of coregionalization: north–south direction is on the left, and east–west direction is on the right.

## REFERENCES CITED


Almeida, A., and A. G. Journel, 1994, Joint simulation of multiple variables with a Markov-type coregionalization model: Mathematical Geology, v. 26, p. 565–588.

Caers, J., 2000, Adding local accuracy to direct sequential simulation: Mathematical Geology, v. 32, no. 7, p. 815–850.

Gomez-Hernandez, J. J., and A. G. Journel, 1993, Joint sequential simulation of multi-Gaussian fields, *in* A. Soares, ed., Geostatistics Troia '92: Dordrecht, Kluwer Academic Publishers, p. 85–94.

Goovaerts, P., 1997, Geostatistics for natural resources characterization: New York, Oxford University Press, 483 p.

Journel, A. G., 1994, Modeling uncertainty: Some conceptual thoughts, *in* R. Dimitrakopoulos, ed., Geostatistics for the next century: Dordrecht, Kluwer Academic Publishers, p. 30–43.

Journel, A. G., and W. Xu, 1994, Posterior identification of histograms conditional to local data: Mathematical Geology, v. 22, p. 323–359.

Soares, A., 1998, Sequential indicator simulation with correction for local probabilities: Mathematical Geology, v. 30, p. 761–765.

Soares, A., 2001, Direct sequential simulation and cosimulation: Mathematical Geology, v. 33, p. 911–926.

Gribov, A., K. Krivoruchko, and J. M. Ver Hoef, 2006, Modeling the semivariogram: New approach, methods comparison, and simulation study, *in* T. C. Coburn, J. M. Yarus, and R. L. Chambers, eds., Stochastic modeling and geostatistics: Principles, methods, and case studies, volume II: AAPG Computer Applications in Geology 5, p. 45–57.

*Chapter 5*

# Modeling the Semivariogram: New Approach, Methods Comparison, and Simulation Study

**A. Gribov**
**K. Krivoruchko**
*Environmental Systems Research Institute*
*Redlands, California, U.S.A.*

**J. M. Ver Hoef**
*Alaska Department of Fish and Game*
*Fairbanks, Alaska, U.S.A.*

## ABSTRACT

This chapter proposes some new methods for computing empirical semivariograms and covariances and for fitting semivariogram and covariance models to empirical data. Grid-based empirical semivariograms and covariances are described, in which the grid values are smoothed using triangular kernels. A model-fitting procedure using modified iterative weighted least squares is presented. This algorithm is shown to be reliable for a wide range of data types and conditions, and its implementation in commercial software is discussed. Comparisons to restricted maximum likelihood estimation are also discussed.

## INTRODUCTION

Investigation of the structure functions known as the covariance and the semivariogram has a long history. Geostatistical textbooks commonly present structural analysis and spatial interpolation in the context of geological investigations. However, both the semivariogram and the related estimation procedure known as kriging were used in other fields such as meteorology years before they were popularized by geoscientists. Early in the 20th century, Keller and Friedmann (1925) introduced a generalized form of the correlation function for hydrodynamic applications for which they assumed that moments of order greater than 2 could be ignored when characterizing average motion. Essentially, they computed the covariance in four dimensions: three-dimensional space plus time. Later, Kolmogorov (1941) presented a simplification of the theory for locally homogeneous and isotropic turbulence in two dimensions in which he showed that a velocity structure function (semivariogram) is proportional to two-thirds the power of the distance in a range for which the assumptions of local stationarity are true. After Kolmogorov's publications on structural analysis, several case studies on structure functions appeared, mostly in the former Soviet Union. Later, Gandin (1959) presented an optimal spatial interpolation procedure in two dimensions


DOI:10.1306/1063806CA53228

(kriging) that the meteorological community embraced, using semivariogram construction as a necessary step in the process. Power functions, power exponential functions, and J- and K-Bessel functions were commonly used to model the semivariogram. Soviet meteorological monitoring networks in the 1950s and early 1960s included several dozen stations, and Gandin et al. commonly used all pairs of points to estimate the parameters of the semivariogram and covariance models (Gandin, 1963). Gandin mentioned that he used a least-squares algorithm to fit the parameters of the model but did not provide any details.

Today, several reliable methods are available to estimate or fit semivariogram and (cross-)covariance models. This chapter details the commercial software implementation of one of the most popular methods: least-squares model fitting. One particular advantage of least squares is that it produces a better visual fit of the empirical semivariogram than likelihood methods. This is true simply because least squares produces the best-fitting trend through the empirical semivariogram, whereas likelihood methods more closely honor individual points on the graph.

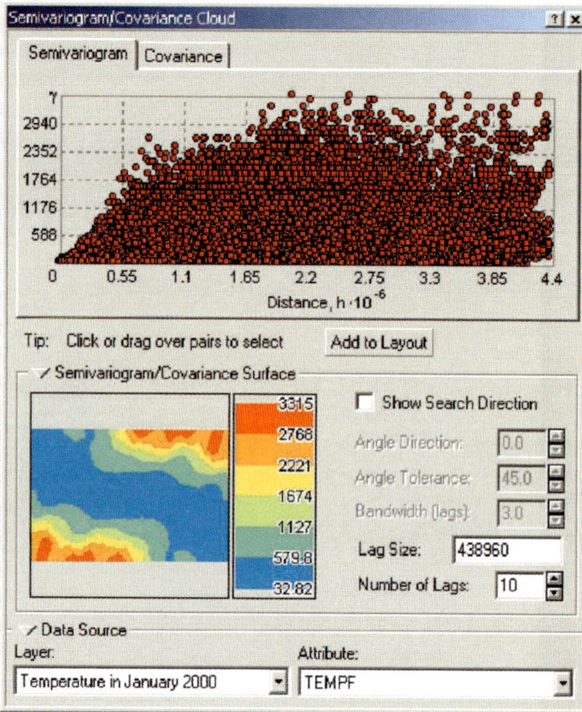


**Figure 1. Computer screen shot showing an empirical semivariogram cloud. The $y$-axis of each red dot represents the empirical semivariogram of a pair of locations.**

## EMPIRICAL SEMIVARIOGRAM ESTIMATION

Let $z(\mathbf{s}_p)$ be a value observed at the $p$th location $\mathbf{s}_p$, where $\mathbf{s}_p = (x_p, y_p)$ is the vector containing the $x$- and $y$-spatial coordinates. Define the lag, $\mathbf{h} = \mathbf{s}_2 - \mathbf{s}_1$, as the vector from point $\mathbf{s}_1$ to point $\mathbf{s}_2$. The semivariogram cloud is defined as half the squared differences, $\gamma_{pq} = 0.5 \cdot [z(\mathbf{s}_p) - z(\mathbf{s}_q)]^2$ for all possible pairwise lags $\{(\mathbf{s}_p, \mathbf{s}_q); p,q = 1, 2, \ldots, \mathbf{n}\}$, plotted as a function of the distance $\|\mathbf{h}\| = \|\mathbf{s}_p - \mathbf{s}_q\| = [(x_p - x_q)^2 + (y_p - y_q)^2]^{1/2}$. Typical applications deal with hundreds and sometimes thousands of data locations, and it is commonly difficult to see any pattern because so many pairwise lags exist (Figure 1).

Not only is it difficult to see any pattern, but fitting a semivariogram model to so many values is difficult even for modern computers. Instead, the squared differences $\gamma_{ij}$ are commonly binned into those distances and directions that are similar. Thus, the following empirical semivariogram formula

$$\hat{\gamma}(\mathbf{h}_k) \equiv \frac{1}{2|N(\mathbf{h})|} \sum_{N(\mathbf{h})} [z(\mathbf{s}_p) - z(\mathbf{s}_q)]^2$$

is commonly used, where $z(\cdot)$ are observed values, $N(\mathbf{h})$ is the set of pairs of values at $\mathbf{s}_p$ and $\mathbf{s}_q$ that have a similar lag $\mathbf{h} \in T(\mathbf{h}_k)$, with $T(\mathbf{h}_k)$ defined to be a tolerance region around $\mathbf{h}_k$ and $|N(\mathbf{h})|$ defined to be the number of distance pairs in the set $N(\mathbf{h})$. The question remains how to best bin the data, which is equivalent to choosing the sets $N(\mathbf{h})$.

Most commonly, the binning is computed on a tolerance region $T(\mathbf{h}_k)$ representing a radial sector. Figure 2 shows the method most commonly used in software packages, including Geostatistical Software Library (Deutsch and Journel, 1998), Splus (Kaluzny et al., 1998), and SAS (SAS Institute, Inc., 1996).

A rule of thumb has been to set the maximum lag for binning to be half the maximum distance between pairs in the data set and choose the number of bins so that there are more than 30 pairs per bin (Journel and Huijbregts, 1978). However, this choice seems to be based on tradition alone, and scant theoretical evidence exists to recommend its use over other schemes.

Instead of the radial sector method, the procedure described here employs a binning approach based on tolerance regions $T(\mathbf{h}_k)$ that are rectangles distributed uniformly on a grid (Figure 3). After averaging

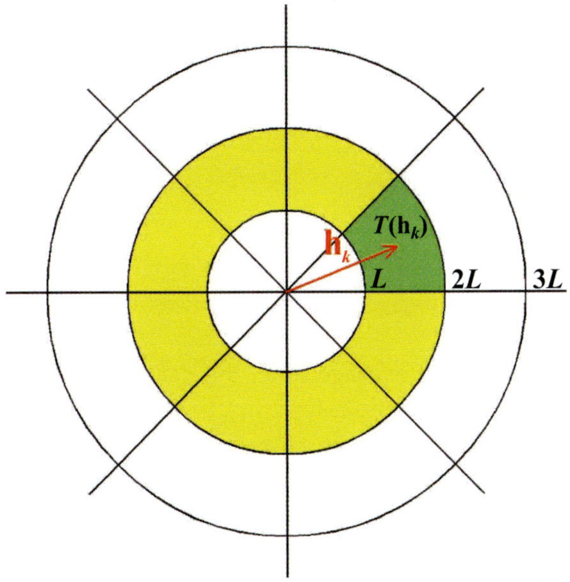


**Figure 2.** **Radial sector tolerance regions for binning empirical semivariograms. The red vector $\mathbf{h}_k$ is the center of the bin, which has a radial tolerance region $T(\mathbf{h}_k)$ shown in green. Any vector that falls within the green area is assigned to the bin for $\mathbf{h}_k$.**

the values in each bin, average values are plotted for each $\mathbf{h}_k$. This is done in two ways: constructing a semivariogram graph by plotting the average bin values versus distance and creating a semivariogram surface by plotting the average bin values directly on their grid, preserving the spatial orientation of the lags and using colors to indicate the lag values. The semivariogram surface uses cool colors like blue and green for low values and warm colors like red and yellow for high values. The color scale next to the semivariogram graph links it to both the graph and the semivariogram surface; that is, each point in the graph corresponds to a cell on the surface (Figure 4).

The grid method is more understandable because the semivariogram surface and semivariogram cloud are linked in an obvious manner (Figure 4). The grid method also helps to ensure that the number of pairs in each cell is approximately the same, which is advantageous for visually assessing the fit of the model.

If the data are uniformly distributed on rectangular domain having dimensions $A$ by $B$, the distance between points has the following distribution:

$$f(\Delta x, \Delta y) = \frac{(A - |\Delta x|) \cdot (B - |\Delta y|)}{A^2 \cdot B^2} = \frac{\left(1 - \frac{|\Delta x|}{A}\right) \cdot \left(1 - \frac{|\Delta y|}{B}\right)}{A \cdot B}$$

where $|\Delta x| \leq A$ and $|\Delta y| \leq B$. The most important pairs are separated by relatively short distances, $|\Delta x| << A$ and $|\Delta y| << B$, and their distance distribution is close to the uniform distribution. Hence, the number of pairs in a rectangular averaging scheme is approximately constant for small lags. In a sector averaging scheme, the number of pairs can differ significantly from sector to sector. Note that it is not necessary to explicitly define the lag interval.

A three-dimensional view of the semivariogram surface is presented in Figure 5 for a very large simulated data set with an anisotropic structure. The sector and grid methods are presented for a 90° sector for tolerance regions $T(\mathbf{h}_k)$ of fixed size (hereafter called fixed lag sizes) and for tolerance regions that increase logarithmically in size (hereafter called logarithmic lag sizes). Note that for small fixed lags, both the sector and grid methods may have bins that are too large near the origin, and hence, in this case, the logarithmic lag size may be more effective at describing variability in the data.

The requirement to have a relatively large number of points in each lag interval, say 30 or more for each bin, can be overcome using the algorithm proposed below. In addition, fewer samples per bin can be used if smoothing has been performed (e.g., by relying on values in neighboring bins).

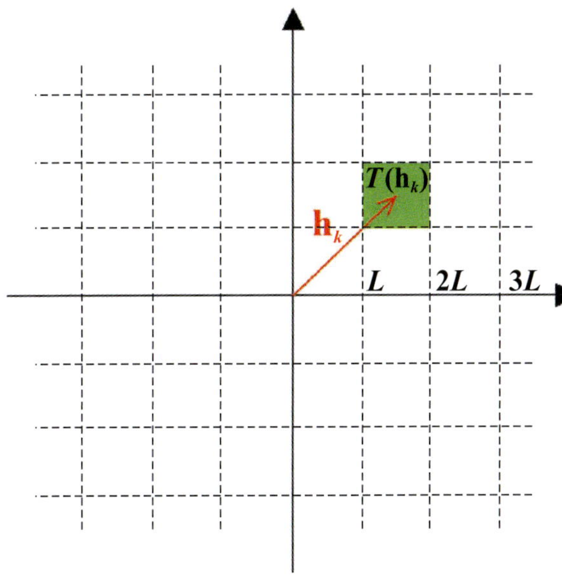


**Figure 3.** **Grid tolerance regions for binning empirical semivariograms. The red vector $\mathbf{h}_k$ is the center of the bin, which has a radial tolerance region $T(\mathbf{h}_k)$ shown in green. Any vector that falls within the green area is assigned to the bin for $\mathbf{h}_k$.**

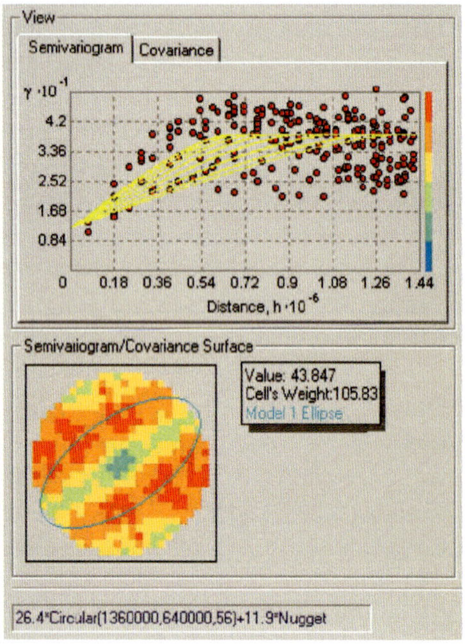


**Figure 4. Computer screen shot showing a semivariogram cloud and an anisotropic model based on the semivariogram surface. Yellow lines represent models in different directions. Ellipse of anisotropy is displayed over the semivariogram surface (blue).**

The algorithm proceeds as follows. A triangular kernel is used to assign weighted semivariogram values to each cell, depending on how close they are to the center of the cell (Figure 6). For example, the weight for the cell in Figure 6 containing the blue dot can be taken as the product of the two marginal profiles. The weight $w_k(\mathbf{h})$ for vector $\mathbf{h}$ for the $k$th bin with center $\mathbf{h}_k$ is

$$w_k(\mathbf{h}) = \left(1 - \frac{|x_k - x|}{L}\right)\left(1 - \frac{|y_k - y|}{L}\right) \times I(|x_k - x| \leq L)I(|y_k - y| \leq L)$$

where $L$ is the length of the side of a square bin, $x_k$ is the $x$-coordinate of $\mathbf{h}_k$, $y_k$ is the $y$-coordinate of $\mathbf{h}_k$, $x$ is the $x$-coordinate of $\mathbf{h}$, $y$ is the $y$-coordinate of $\mathbf{h}$, and $I(\bullet)$ is the indicator function that is equal to 1 if its argument is true and zero otherwise. Modification of this method can be used for other types of averaging presented in Figure 5.

Commonly, a table of binned semivariogram values shows the number of pairs for that bin. Because smoothing with weights is used, an equivalent presentation for the proposed method is to present the sum of the weights used in each bin.

## EMPIRICAL COVARIANCE AND CROSS-COVARIANCE ESTIMATION

It is also possible to estimate the empirical covariance function

$$\hat{C}(\mathbf{h}_k) \equiv \frac{1}{|N(\mathbf{h})|} \sum_{N(\mathbf{h})} [z(\mathbf{s}_p) - \bar{z}][z(\mathbf{s}_q) - \bar{z}]$$

where $\bar{z}$ is the average of all the data, and all other notation is the same as for $\hat{\gamma}(\mathbf{h}_k)$. In comparison, $\hat{\gamma}(\mathbf{h}_k)$ is an unbiased estimate of the semivariogram, whereas $\hat{C}(\mathbf{h}_k)$ is a biased estimate of the population covariance. Thus, $\hat{\gamma}(\mathbf{h}_k)$ is generally preferred.

Choosing a structure function for modeling the correlation between variables is more complicated (Ver Hoef and Cressie, 1993). The traditional cross-semivariogram is defined as

$$\tau_{ij}(\mathbf{h}) \equiv \frac{1}{2} E[z_i(\mathbf{s}) - z_i(\mathbf{s} + \mathbf{h})][z_j(\mathbf{s}) - z_j(\mathbf{s} + \mathbf{h})]$$

which has, as an estimator,

$$\hat{\tau}_{ij}(\mathbf{h}_k) \equiv \frac{1}{2|N(\mathbf{h})|} \sum_{N(\mathbf{h})} [z_i(\mathbf{s}_p) - z_i(\mathbf{s}_q)][z_j(\mathbf{s}_p) - z_j(\mathbf{s}_q)]$$

for the $i$th and $j$th variable types. The notation follows that of $\hat{\gamma}(\mathbf{h}_k)$ and $\hat{C}(\mathbf{h}_k)$. The cross-semivariogram, $\tau_{ij}(\mathbf{h})$, has two problems. First, it does not accommodate asymmetry; that is, the cokriging equations are only obtained by assuming that $C_{ij}(\mathbf{h}) = C_{ij}(-\mathbf{h})$. Second, both variables must be measured at the same location in order for that location to be used in the empirical estimator.

Alternatively, the cross-semivariogram can be defined as

$$\gamma_{ij}(\mathbf{h}) \equiv \frac{1}{2} \mathrm{var}[z_{\tilde{i}}(\mathbf{s}) - z_{\tilde{j}}(\mathbf{s} + \mathbf{h})]$$

which has an estimator

$$\hat{\gamma}_{ij}(\mathbf{h}_k) \equiv \frac{1}{2|N(\mathbf{h})|} \sum_{N(\mathbf{h})} [\tilde{z}_i(\mathbf{s}_p) - \tilde{z}_j(\mathbf{s}_q)]^2$$

where $\tilde{z}_i(\mathbf{s}_p)$ is standardized by subtracting off the mean and dividing by the standard deviation. The

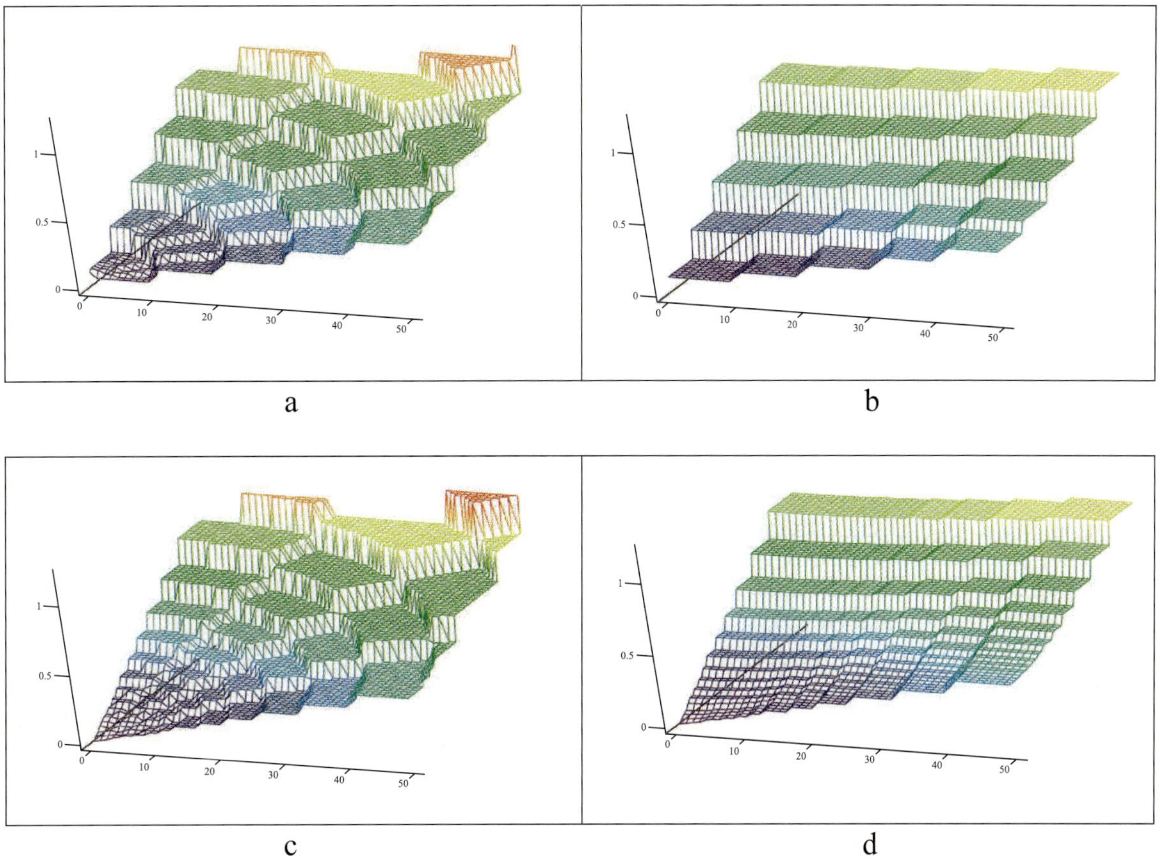


**Figure 5. Empirical semivariograms: (a) sector method with fixed lag size, (b) grid method with fixed lag size, (c) sector method with logarithmic lag size, and (d) grid method with logarithmic lag size.**

cross-semivariogram, $\gamma_{ij}(\mathbf{h})$, is asymmetric and allows the cokriging equations to be obtained more generally than $\tau_{ij}(\mathbf{h})$. However, because of standardization,

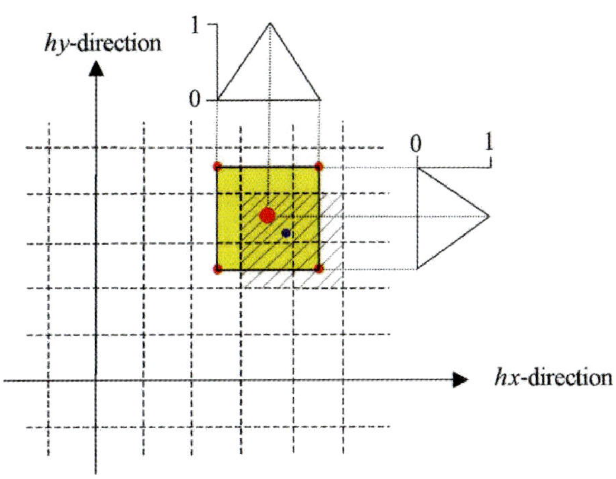


**Figure 6. Smoothing the empirical semivariogram. The yellow region is the area in which lags have some positive weighting for the bin with the large red circle. The lag indicated by the blue dot will have an influence in all bins with diagonal hatching.**

$\hat{\gamma}_{ij}(\mathbf{h}_k)$ is no longer unbiased, so there appears to be no advantage in using $\gamma_{ij}(\mathbf{h})$ over the cross-covariance. The first complete explanation of cokriging (Gandin, 1963) used the cross-covariance only. Hence, the present work concentrates on the cross-covariance, which is estimated by

$$\hat{C}_{ij}(\mathbf{h}_k) \equiv \frac{1}{|N(\mathbf{h})|}\sum_{N(\mathbf{h})}[z_i(\mathbf{s}_p)-\overline{z}_i][z_j(\mathbf{s}_q)-\overline{z}_j]$$

A spatial shift is an example of asymmetry, as illustrated in Figure 7. A shift that is present when the highest cross-covariance occurs for two variables at different locations represents the distance and direction where the calculated cross-covariance is at its maximum value. For example, in Figure 7, variable 1 at some location will have the highest correlation with variable 2 at a location to the west. Shifted cross-correlation is common in environmental applications (Krivoruchko, 2002).

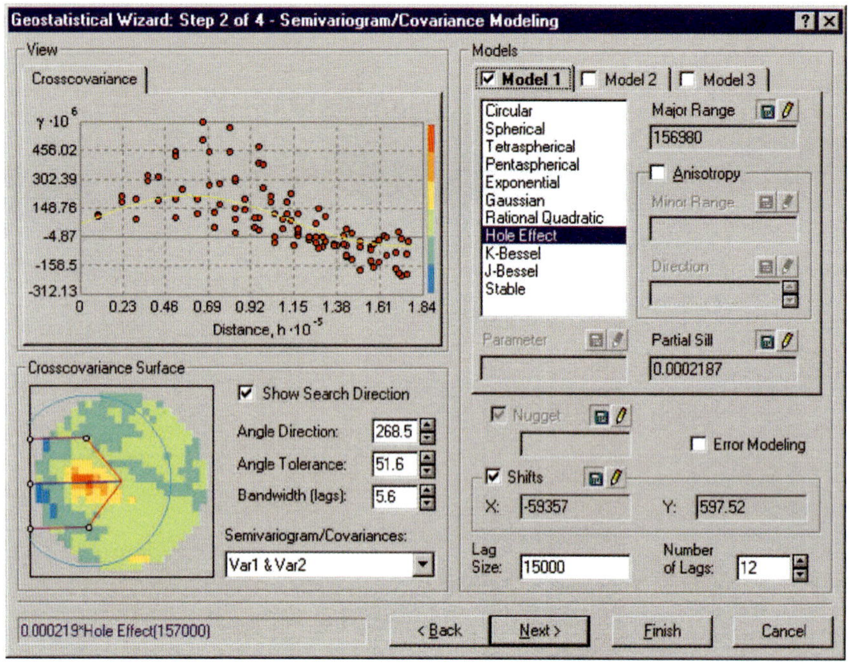


Figure 7. Computer screen shot showing how cross-covariance modeling is accomplished using the shift parameter. For the cross-covariance surface, the highest cross-covariance (red and orange colors) is shifted toward the west.

## FITTING SEMIVARIOGRAM AND COVARIANCE MODELS USING A MODIFIED ITERATIVE WEIGHTED LEAST-SQUARES ALGORITHM

After calculating the empirical semivariogram, the next step is to fit a theoretical model (e.g., spherical or exponential) to the values. Three main approaches exist for estimating the parameters of the semivariogram model: visual, (weighted) least-squares, and likelihood methods. In practice, a true semivariogram is almost never known, and consequently, no method exists to determine it exactly. Before modern computers and software became available, semivariograms were commonly fitted visually. More recently, statisticians have commonly used variants of maximum likelihood estimation to fit semivariograms. Researchers in disciplines such as meteorology, geology, and environmental sciences more commonly use variants of a least-squares approach. All three techniques provide useful results, especially if the user has experience in spatial data processing. Several good comparisons of these approaches have been published, including those of Cressie (1993) and Zimmerman and Zimmerman (1991). A modification of the weighted least-squares (WLS) approach (Cressie, 1985) is described below. As noted previously, one important reason to prefer WLS in commercial software implementations is that the results can be easily visualized.

The first step in the modified WLS approach is to select a reasonable lag size for the empirical semivariogram averaging by fitting a preliminary semivariogram using the sector method with logarithmic lags (stage 1 of the algorithm) and then using the range of the fitted model to define the lag used in the second stage, when the final semivariogram model is fitted using a grid method with constant lag size. Before detailing the algorithm, it is instructive to consider an extreme situation where the method gives very good results. Figure 8 presents a simulated data set based on a spherical covariance model further described below. The data locations consist of two clusters separated by a relatively long distance. An automatically derived semivariogram is shown, which looks very reasonable when compared to the true semivariogram given in Figure 9. A screenshot produced with ArcInfo 8.2 (geographic information software developed and marketed by the Environmental Systems Research Institute [ESRI]) shows a map produced using ordinary kriging with the default semivariogram.

### Stage 1

The algorithm first scales each data set, $\tilde{z}_j(\mathbf{s}_i) = \left(z_j(\mathbf{s}_i) - \bar{z}_j\right)/\sigma_j$, where $\sigma_j$ is the sample standard deviation for the $j$th variable type. Stage 1 begins by assuming an isotropic model. The empirical semivariogram (or covariance) is computed on the scaled data $\tilde{z}_j(\mathbf{s}_i)$, using the sector method with logarithmic

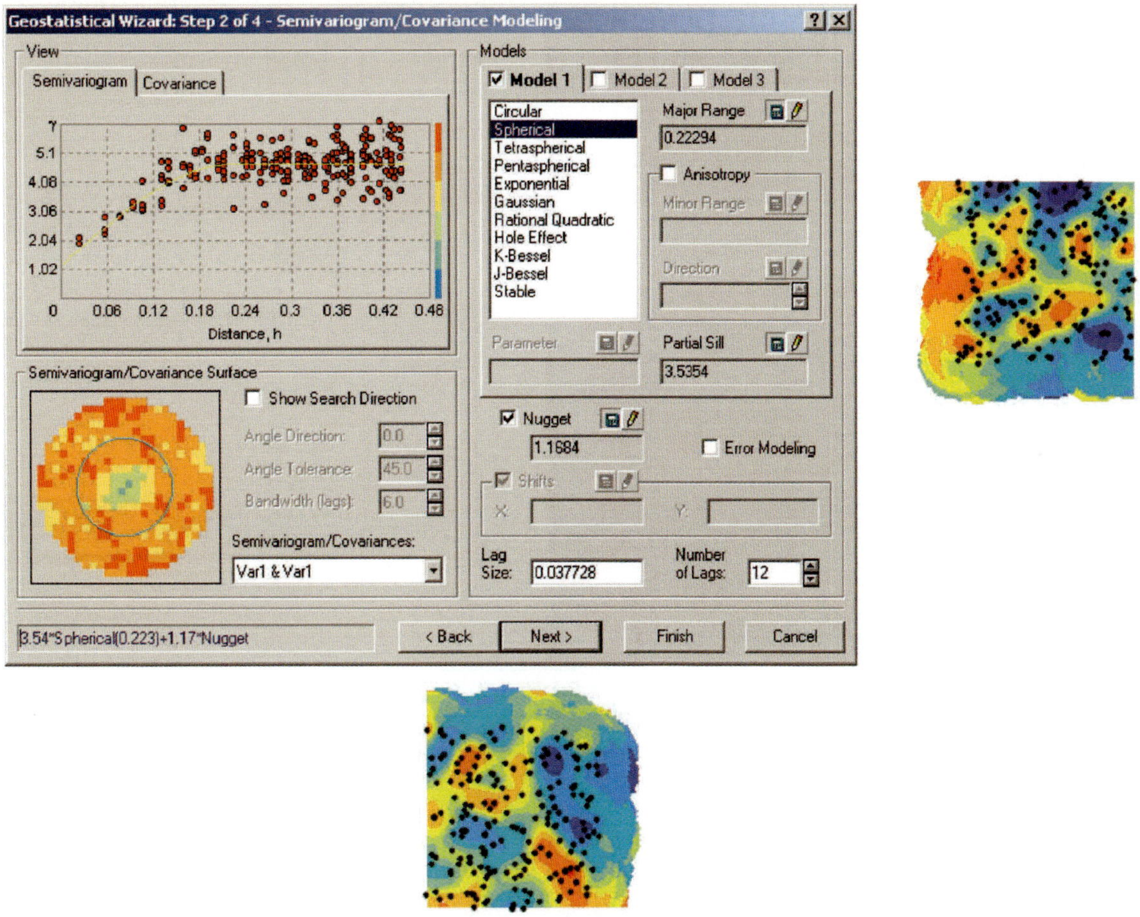


**Figure 8. Computer screen shot showing the estimated semivariogram obtained using an automatic model-fitting algorithm (default). The data locations are black dots in two clusters separated by a relatively long distance. The predicted surface using kriging with the estimated semivariogram is also given. For the kriging surface, the cool colors (blue and green) represent low values, and the warm colors (red and orange) represent high values.**

lag size (Figure 5c) across a large range of lag classes that progress in a geometric series. The lag classes are formed from intervals $[d^{k-1/2}, d^{k+1/2})$, where the empirical parameter $d$ equals 1.25, and $k$ ranges from large negative to large positive integers. Part of the series can be seen in Figure 5c. The center of each lag class is taken to be $(d^{k-1/2} + d^{k+1/2})/2$. Only lag classes that have data are used.

The empirical (cross-)covariance can be given by $\hat{C}_{ij}(\mathbf{h}_k)$, where $i$ indicates the $i$th variable, $j$ indicates the $j$th variable type, and $k$ indicates the $k$th lag class. The first iteration of covariance parameter estimates is obtained by minimizing

$$\sum_{i=1}^{T} \sum_{j=i}^{T} \sum_{k=1}^{n_{ij}} w_{ij}(\mathbf{h}_k)\left(\tilde{C}_{ij}(\mathbf{h}_k; \boldsymbol{\theta}) - \hat{C}_{ij}(\mathbf{h}_k)\right)^2 \quad (1)$$

where $\boldsymbol{\theta}$ is the vector of parameters for the $i,j$th covariance function, and

$$w_{ij}(\mathbf{h}_k) = N_{ij}(\mathbf{h}_k) / \sum_{m=1}^{n_{ij}} N_{ij}(\mathbf{h}_m) \quad (2)$$

where $N_{ij}(\mathbf{h}_k)$ is the number of pairs in the empirical (cross-)covariance function for variables $i$ and $j$ in lag class $k$. Let this estimate be specified as $\boldsymbol{\theta}^{(1)}$. In the next iteration, a modification of the WLS algorithm given by Cressie (1985) is used to minimize equation 1 again. In this case,

$$\varpi_{ij}(\mathbf{h}_k; \boldsymbol{\theta}^{(1)}) = \frac{N_{ij}(\mathbf{h}_k)}{\tilde{C}_{ii}(0; \boldsymbol{\theta}^{(1)})\tilde{C}_{jj}(0; \boldsymbol{\theta}^{(1)}) + \tilde{C}_{ij}^2(\mathbf{h}_k; \boldsymbol{\theta}^{(1)})} \quad (3)$$

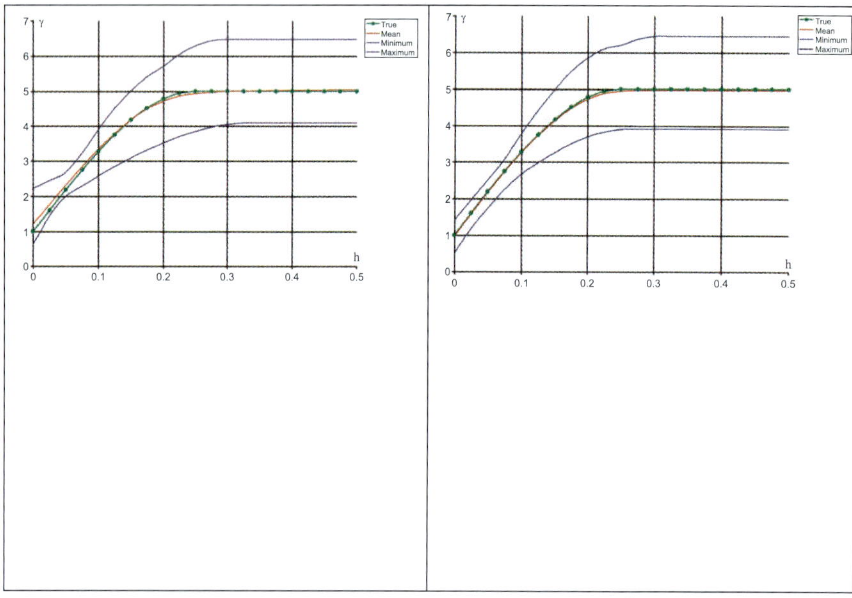


| | $a$ | $b$ | MISE |
|---|---|---|---|
| REML | 0 | 0.1 | 0.981946 |
| WLS | 0 | 0.1 | 1.764688 |
| REML | 0.1 | 0.2 | 4.407449 |
| WLS | 0.1 | 0.2 | 5.96006 |
| REML | 0.2 | 0.3 | 8.224324 |
| WLS | 0.2 | 0.3 | 7.892238 |
| REML | 0.3 | 0.5 | 9.090794 |
| WLS | 0.3 | 0.5 | 6.945768 |

a b c

**Figure 9. Comparison of true and fitted semivariogram models. The green line is the true semivariogram, the red line is the average estimated semivariogram across the 22 simulations, and the blue lines are the minimum and maximum semivariogram values from the simulations. Shown in the figure are results from (a) modified WLS and (b) REML, as well as (c) a quantitative comparison for different ranges.**

These weights are normalized so that each (cross-) covariance gets equal weight, that is

$$w_{ij}(\mathbf{h}_k) = \varpi_{ij}(\mathbf{h}_k; \boldsymbol{\theta}^{(1)}) / \sum_{m=1}^{n_{ij}} \varpi_{ij}(\mathbf{h}_m; \boldsymbol{\theta}^{(1)}) \qquad (4)$$

This estimate is specified as $\boldsymbol{\theta}^{(2)}$. If a semivariogram is used instead of the covariance, the estimate $\boldsymbol{\theta}^{(1)}$ is given by minimizing

$$\sum_{i=1}^{T} \left( \begin{array}{c} \sum\limits_{k=1}^{n_{ii}} w_{ii}(\mathbf{h}_k)(\tilde{\gamma}_{ii}(\mathbf{h}_k; \boldsymbol{\theta}) - \hat{\gamma}_{ii}(\mathbf{h}_k))^2 + \\ \sum\limits_{j=i+1}^{T} \sum\limits_{k=1}^{n_{ij}} w_{ij}(\mathbf{h}_k)\left(\tilde{C}_{ij}(\mathbf{h}_k; \boldsymbol{\theta}) - \hat{C}_{ij}(\mathbf{h}_k)\right)^2 \end{array} \right) \qquad (5)$$

where $w_{ii}(\mathbf{h}_k)$ is given by equation 2. $\boldsymbol{\theta}^{(2)}$ is then obtained from equation 5, with weights determined by equation 4, except that

$$\varpi_{ii}(\mathbf{h}_k; \boldsymbol{\theta}^{(1)}) = \frac{N_{ii}(\mathbf{h}_k)}{\tilde{\gamma}_{ii}^2(\mathbf{h}_k; \boldsymbol{\theta}^{(1)})}$$

Determining the estimates $\boldsymbol{\theta}^{(1)}$ and $\boldsymbol{\theta}^{(2)}$ are two steps in an iteratively reweighted least-squares algorithm. It is possible to use additional iterations as described in Shapiro and Botha (1991). $\boldsymbol{\theta}^{(2)}$ provides a value for the range of autocorrelation that is used to obtain a default lag size for the grid method in estimating the empirical semivariogram or covariance (see stage 2 below). Based on the authors' experience, a dozen lags is commonly enough to estimate the parameters of the model using the grid method, so the default lag size for the grid method for the next stage of the algorithm is taken to be $(2 \times \text{range})/12$.

The main difference between this approach and that of Cressie (1985) is that the weights do not change during optimization. One of the consequences of this modification is that the algorithm is not sensitive to the number of lags in the grid cells. In fact, it is possible to use even one pair in the cell; its weight will be simply less than the weights of more populated cells.

A similar approach for semivariogram calculation is discussed by Shapiro and Botha (1991) for an isotropic process using one variable. Here, however, anisotropic structure functions (covariance, semivariogram, and cross-covariance) are estimated for transformed variables in two steps, using logarithmic lag scales and then regular lag scales, with additional kernel smoothing of empirical structure functions.

## Stage 2

Stage 2 essentially repeats stage 1, using an empirical semivariogram or (cross-)covariance for the scaled data, $\tilde{z}_j(\mathbf{s}_i)$, that employs the grid method (Figure 5b), where the default lag size is obtained from the range estimate in stage 1. It also allows for anisotropy and linear combinations of (cross-)covariance or semivariogram models for each data set, such as

$$\tilde{C}_{ij}(\mathbf{h};\boldsymbol{\theta}) = \sum_{u=1}^{M} B_u(i,j)\rho_u(\mathbf{h};\boldsymbol{\varphi}_u)$$

Here, $B_u(i,j)$ is the $i,j$th component of $\mathbf{B}_u$, a $T \times T$ positive-definite matrix, where $T$ is the number of variables; $\boldsymbol{\theta}$ contains the parameters $B_u(i,j)$ and $\boldsymbol{\varphi}_u$ for all $u$; $M$ is the number of different (cross-)covariance models used in a linear combination; and the function $\rho_u(\mathbf{h};\boldsymbol{\varphi}_u)$ is a normalized covariance model with $\rho_u(h = 0;\boldsymbol{\varphi}_u) = 1$. The third iteration of parameter estimates, $\boldsymbol{\theta}^{(3)}$, is obtained by minimizing equation 1 with weights given in equation 2 using the empirical covariance and the grid method. $\boldsymbol{\theta}^{(4)}$ is obtained by minimizing equation 1 with weights from equations 3 and 4 using the grid method and the empirical covariance. These formulas are modified when using semivariograms, as shown in stage 1 (equation 5).

At this point, it is necessary to change back to the original scale. The final (cross-)covariance model is

$$C_{ij}(\mathbf{h}) = \sigma_i\sigma_j\tilde{C}_{ij}(\mathbf{h};\boldsymbol{\theta}^{(4)})$$

or, in the case of the semivariogram,

$$\gamma_{ii}(\mathbf{h}) = \sigma_i^2\tilde{\gamma}_{ii}(\mathbf{h};\boldsymbol{\theta}^{(4)})$$

If the user changes any of the default parameters of the semivariogram and covariance models (lag size, range, nugget, and partial sill), then estimates are recalculated beginning at stage 2.

In addition to the WLS, several other good methods of fitting spatial semivariogram models exist, including maximum likelihood, restricted maximum likelihood (REML), and generalized least squares. For a review, see Cressie (1993) and Genton (2001). The next section compares the proposed modified WLS approach and REML.

If the covariance matrix is written as $\boldsymbol{\Sigma}(\boldsymbol{\theta})$, which depends on spatial parameters $\boldsymbol{\theta}$, the REML equation to be minimized for $\boldsymbol{\theta}$ is

$$L(\boldsymbol{\theta}) = (\mathbf{z} - \mathbf{X}\hat{\boldsymbol{\beta}})'\boldsymbol{\Sigma}(\boldsymbol{\theta})^{-1}(\mathbf{z} - \mathbf{X}\hat{\boldsymbol{\beta}}) + \log(|\boldsymbol{\Sigma}(\boldsymbol{\theta})|) + \log(|\mathbf{X}'\boldsymbol{\Sigma}(\boldsymbol{\theta})^{-1}\mathbf{X}|) + (n - p)\log(2\pi) \qquad (6)$$

Equation 6 can be minimized by any of the computer-intensive minimization methods such as function FMINCON in MATLAB (software developed and marketed by MathWorks, Inc.).

## A SIMULATION EXAMPLE

Suppose data are simulated using the spherical semivariogram model

$$\gamma(\mathbf{h};\boldsymbol{\theta}) = \theta_n + \begin{cases} \theta_s\left[\frac{3}{2}\frac{\|\mathbf{h}\|}{\theta_r} - \frac{1}{2}\left(\frac{\|\mathbf{h}\|}{\theta_r}\right)^3\right] & \text{for} \quad \|\mathbf{h}\| < \theta_r \\ \theta_s & \text{for} \quad \theta_r \le \|\mathbf{h}\| \end{cases}$$

where $\theta_n \ge 0$ is the nugget, $\theta_s \ge 0$ is the partial sill, and $\theta_r \ge 0$ is the range. For each of the simulations, $\theta_s = 4$, $\theta_r = 0.25$, and $\theta_n = 1$. For semivariogram models with sills, it is easy to go back and forth between semivariogram and covariance functions obtained using the relations

$$\gamma(\mathbf{h}) = C(\mathbf{0}) - \mathbf{C}(\mathbf{h}) \quad \text{and} \quad \mathbf{C}(\mathbf{h}) = \gamma(\infty) - \gamma(\mathbf{h})$$

and this approach is employed here. After converting the spherical semivariogram to a covariance function, the Cholesky decomposition method is applied (Cressie, 1993) to simulate 200 values from a Gaussian distribution in the region $0 \le x \le 1$ and $0 \le y \le 1$. Twenty-two such data sets of size 200 are simulated. Then, the grid method is used to compute the empirical semivariogram for each data set, and a spherical semivariogram model is subsequently fit to the empirical semivariogram values using modified WLS as described above. Additionally, REML is used to fit a spherical model to the empirical covariance, and the result is subsequently converted to a semivariogram.

**Table 1. Estimation of the semivariogram parameters by REML and modified WLS.**

| | REML | | | WLS | | |
|---|---|---|---|---|---|---|
| *Quantity* | *Nugget* | *Sill* | *Range* | *Nugget* | *Sill* | *Range* |
| Average value | 0.973295 | 4.000645 | 0.250886 | 1.199273 | 3.839545 | 0.269727 |
| $(1/22)\sum_{i=1}^{22}(\hat{\theta}_{r,i}-\theta_r)^2$ | 0.051663 | 0.513559 | 0.000717 | 0.161678 | 0.583832 | 0.004794 |

The true semivariogram model is shown as the green curve in Figure 9. The average fitted semivariogram is shown as the red curve in Figure 9, with lower and upper bounds given by the blue curves. These curves are calculated using the following formulas:

$$\gamma_{\text{minimum}}(h) = \min(\gamma_i(h)),\ i = 1\ldots22$$
$$\gamma_{\text{maximum}}(h) = \max(\gamma_i(h)),\ i = 1\ldots22,\ \text{and}$$
$$\gamma_{\text{average}}(h) = \text{average}(\gamma_i(h)),\ i = 1\ldots22$$

Figure 9 shows that, on average, both algorithms for fitting the semivariogram worked quite well. However, REML estimates all the semivariogram parameters ($\theta_r$, $\theta_n$, and $\theta_s$) more precisely, as indicated in the Table 1.

Further, the difference between true and estimated models near the origin is smaller for REML. To quantify this, the mean integrated squared error (MISE) was used:

$$\text{MISE}(a,b) = \frac{1}{22(b-a)}\sum_{i=1}^{22}\int_a^b [\hat{\gamma}_i(h)-\gamma(h)]^2 dh$$

An approximation is given by

$$\text{MISE}(a,b) \cong \frac{1}{22|B|}\sum_{i=1}^{22}\sum\nolimits_{j\in B}[\hat{\gamma}_i(h_j)-\gamma(h_j)]^2$$

where $H = \{h_j{:}j/c\}$ for $j = 1, 2, \ldots$ and some value $c$, and $B \subset H$, where $B = \{h_j{:}a \le h_j \le b\}$, and $|B|$ is the number of elements in $B$. The larger the value $c$, the closer the approximation becomes.

It is interesting that MISE is different for different ranges. For example, MISE(0,0.1) and MISE(0.1,0.2) are smaller for REML, but MISE(0.3,0.5) is smaller for modified WLS (Figure 9c). Thus, REML yields better estimates of the semivariogram parameters for the most important short lags.

Figure 10 presents the parameters estimated for the spherical semivariogram model on each of the 22 simulated data sets using modified WLS. The parameters are scattered around the true values given in magenta. The figure also suggests that the parameters are correlated (e.g., the estimate of the nugget effect is commonly low when the partial sill is high). Table 2 contains the estimated pairwise correlations under both the REML and modified WLS model-fitting approaches.

Estimates of the correlation between semivariogram parameters obtained using the modified WLS model-fitting approach are larger because the range is estimated first, and range influences the determination of the other semivariogram parameters. However, this is theoretically undesirable, and so the REML algorithm is considered to have performed better.

## SMOOTH EMPIRICAL SEMIVARIOGRAM, COVARIANCE, AND CROSS-COVARIANCE SURFACES

Having access to a smooth semivariogram and covariance surface is useful in exploratory data analysis when investigating spatial auto- and cross-correlation. An algorithm for obtaining a smooth semivariogram surface is presented below.

The sector method with logarithmic lag size (Figure 5c) is used for averaging pairs of data values. Estimation at the point $\vec{v}$ of the semivariogram surface is accomplished using the following formula:

$$\bar{\gamma}(\vec{v}) = \frac{\sum_{i=1}^{N} n_i \cdot e^{-\Theta\cdot\rho(\vec{v},\vec{c}_i)} \cdot \hat{\gamma}_i}{\sum_{i=1}^{N} n_i \cdot e^{-\Theta\cdot\rho(\vec{v},\vec{c}_i)}}$$

where $N$ is the number of sectors, $\hat{\gamma}_i$ is the average semivariogram value in the sector $i \in \overline{1,N}$, $n_i$ is the number of pairs in sector $i \in \overline{1,N}$, and $\vec{c}_i$ are

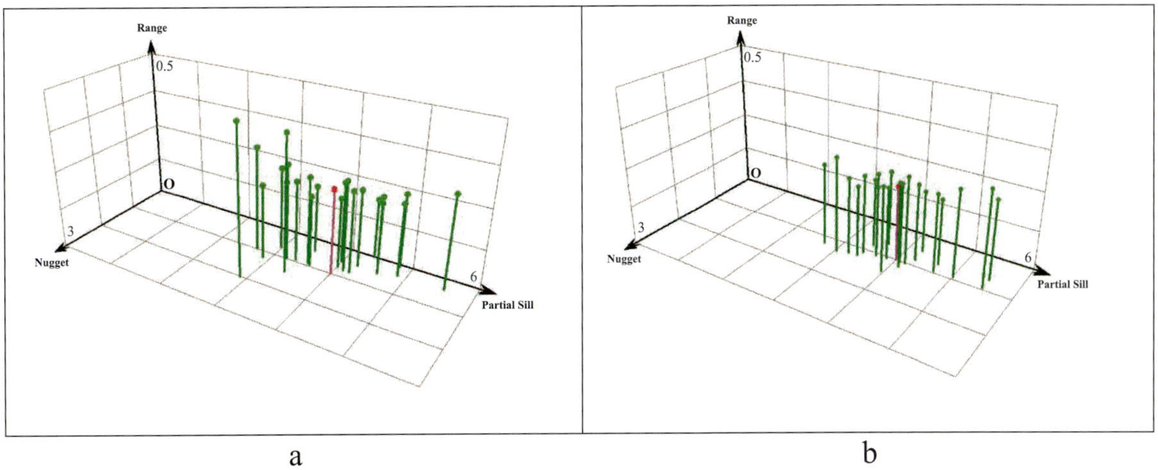


**Figure 10. Three-dimensional scatter diagram of 22 sets of parameter estimates of the spherical semivariogram model. Each estimate of the nugget, range, and partial sill is given in green, and the true value is given in magenta. Results are shown for (a) WLS and (b) REML.**

the coordinates of the center of sector $i \in \overline{1,N}$. Parameter $\Theta$ depends on data and data locations: $\Theta = \sqrt{\frac{8 \cdot \pi \sqrt{\sum_{i=1}^{N} n_i}}{S}}$, where $S$ is an area of the region under investigation. Figure 11 presents examples of empirical and smooth semivariogram surfaces.

Using the smooth semivariogram and covariance surface defined above for analyses other than data exploration requires some further improvements to surface calculation. For example, $\rho(\vec{a}, \vec{b})$ can be based on a non-Euclidean distance, considering the anisotropy of the covariance surface, where the parameter $\Theta$ depends on its position on the covariance surface.

## DISCUSSION AND CONCLUSION

This chapter describes grid-based empirical semivariograms and covariances, along with model fitting using modified WLS. These procedures have been implemented in commercial software packages (e.g., the Geostatistical Analyst extension to ArcGIS 8.1, a product developed and marketed by ESRI; Johnston et al., 2001). For data sets with more than 100 data points and with significant spatial dependence, the algorithm gives very good defaults, which allows users to easily update parameters using cross-validation.

However, situations exist in which modified WLS does not perform well. This is particularly true for situations in which the data set is small, a large amount of measurement error exists, and the data include clusters of high or low values (i.e., they are nonstationary). In such cases, the lag and, consequently, the range are commonly overestimated, and the default algorithm produces oversmoothed surfaces. However, the user can control this situation by selecting different lag sizes with graphical tools. Cross-validation is also an important diagnostic in these instances. If the lag size is selected properly (commonly, it is obvious how to manually improve

**Table 2. Estimates of correlation between the estimated parameters of the spherical semivariogram model for 22 simulated data sets.**

| | *REML* | | | *WLS* | | |
|---|---|---|---|---|---|---|
| | *Nugget* | *Sill* | *Range* | *Nugget* | *Sill* | *Range* |
| Nugget | | −0.470 | 0.488 | | −0.703 | 0.789 |
| Sill | −0.470 | | 0.139 | −0.703 | | −0.384 |
| Range | 0.488 | 0.139 | | 0.789 | −0.384 | |

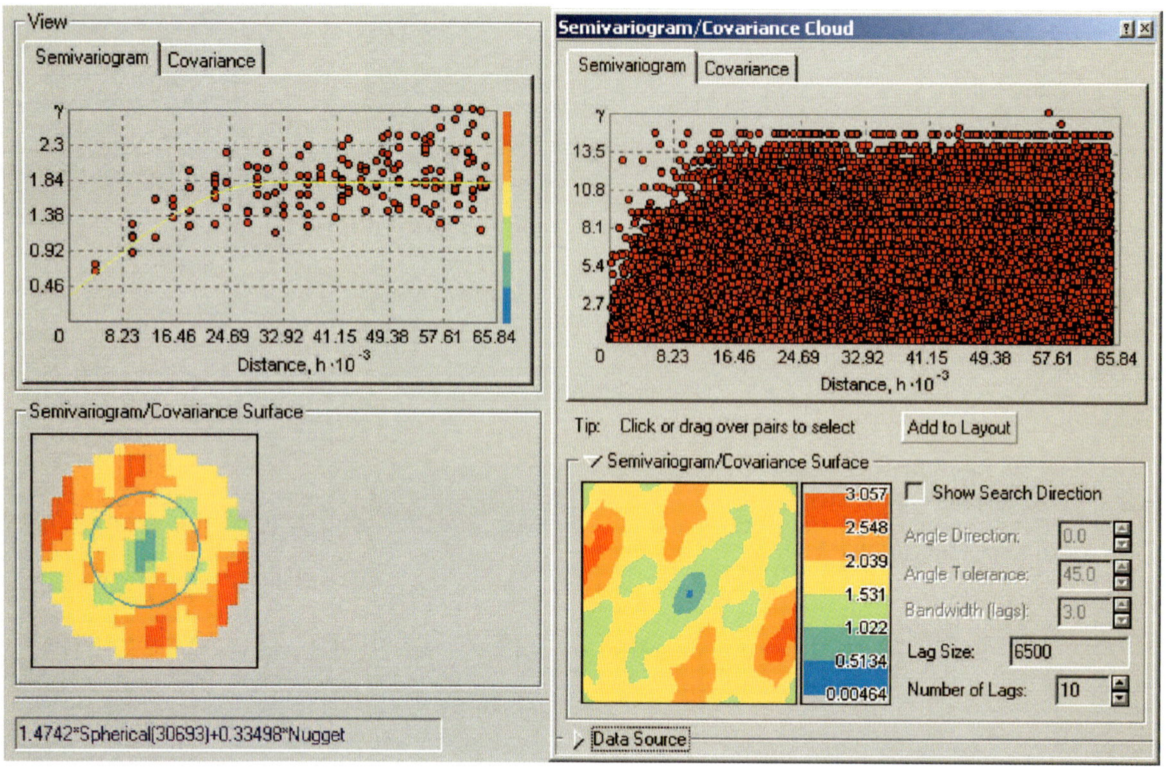


a b

**Figure 11. Computer screen shot of an empirical semivariogram cloud and surface using (a) the grid method and (b) all semivariogram pairs and a smooth semivariogram surface using the same data set.**

default lag estimation), then other model parameters are generally well estimated.

For relatively small data sets, REML is almost always better, and this algorithm should be used as an alternative. The only serious drawback to this technique is that it requires much more computing time than modified WLS.

## REFERENCES CITED

Cressie, N., 1985, Fitting variogram models by weighted least squares: Mathematical Geology, v. 17, p. 653–702.

Cressie, N., 1993, Statistics for spatial data: New York, John Wiley & Sons, 900 p.

Deutsch, C. V., and A. G. Journel, 1998, GSLIB—Geostatistical software library and user's guide: New York, Oxford University Press, 369 p.

Gandin, L. S., 1959, The problem on optimal interpolation: Leningrad, Trudy Glavnaya geofizicheskaya observatoria, v. 99, p. 67–75.

Gandin, L. S., 1963, Objective analysis of meteorological fields: Leningrad, Gidrometeorologicheskoe Izdatel'stvo, 286 p. (translated by Israel Program for Scientific Translations, Jerusalem, 1965, 242 p.).

Genton, M. G., 2001, Robustness problems in the analysis of spatial data, *in* M. Moore, ed., Spatial statistics: Methodological aspects and applications: Springer Lecture Notes in Statistics, v. 159, p. 21–37.

Johnston, K., J. Ver Hoef, K. Krivoruchko, and N. Lucas, 2001, Using ArcGIS geostatistical analyst: GIS by ESRI: ESRI Press, 300 p.

Journel, A. G., and C. J. Huijbregts, 1978, Mining geostatistics: London, Academic Press, 600 p.

Kaluzny, S. P., S. C. Vega, T. P. Cardoso, and A. A. Shelly, 1998, S+ Spatial Stats: User's manual for Windows and Unix: New York, Springer-Verlag, 384 p.

Keller, L., and A. Friedmann, 1925, Differentialeichungen fur die turbulente bewegungner kompressibelen flussigkeit: Proceedings of the 1st Congress for Applied Mathematics, Delft, p. 395–405.

Kolmogorov, A., 1941, Local turbulence structure in an incompressible viscous liquid for very high Reynolds numbers: Doklady Akademii nauk SSSR, v. 30, no. 4, p. 229–232.

Krivoruchko, K., 2002, Geostatistical analysis of California air quality: ESRI Technical Report: Available at www.esri.com/software/arcgis/arcgisxtensions/geostatistical/airqualityjgra.pdf (accessed July 2004).

SAS Institute Inc., 1996, SAS/STAT technical report: Spatial prediction using the SAS system: Cary, North Carolina, SAS Institute, Inc., 80 p.

Shapiro, A., and J. D. Botha, 1991, Variogram fitting with a general class of conditionally nonnegative definite functions: Computational Statistics and Data Analysis, v. 11, p. 87–96.

Ver Hoef, J. M., and N. Cressie, 1993, Multivariable spatial prediction: Mathematical Geology, v. 25, p. 219–240.

Zimmerman, D. L., and M. B. Zimmerman, 1991, A comparison of spatial semivariogram estimators and corresponding ordinary kriging predictors: Technometrics, v. 33, p. 77–91.

Goovaerts, P., 2006, Geostatistical modeling of the spaces of local, spatial, and response uncertainty for continuous petrophysical properties, *in* T. C. Coburn, J. M. Yarus, and R. L. Chambers, eds., Stochastic modeling and geostatistics: Principles, methods, and case studies, volume II: AAPG Computer Applications in Geology 5, p. 59–79.

*Chapter 6*

# Geostatistical Modeling of the Spaces of Local, Spatial, and Response Uncertainty for Continuous Petrophysical Properties

**P. Goovaerts**
*Biomedware, Inc.*
*Ann Arbor, Michigan, U.S.A.*

## ABSTRACT

In the characterization of petroleum reservoirs, three types of uncertainty typically arise: (1) the uncertainty about the value of a petrophysical attribute at an unsampled location (local uncertainty); (2) the joint uncertainty about attribute values at several locations taken together (spatial uncertainty); and (3) the uncertainty about production forecasts, such as time to recover a given proportion of the oil (response uncertainty). In each case, the probabilistic way to assess the uncertainty consists of determining the distribution or set of possible outcomes (e.g., local permeability value, permeability grid, or production parameters), which is referred to as the space of uncertainty.

This chapter reviews the major geostatistical algorithms available to model both local and spatial uncertainties of continuous attributes. Goodness criteria are introduced for each type of space of uncertainty, and the impacts of the following parameters are discussed: stochastic-simulation algorithm, number of realizations, and ergodic fluctuations. Conclusions are drawn on the relations between the different spaces of uncertainty. The discussion is illustrated using an exhaustive set of 102 $\times$ 102 permeability values.

## INTRODUCTION

Effective management of petroleum reservoirs commonly requires detailed three-dimensional (3-D) models of lithofacies, porosity, and permeability. Because spatial heterogeneity of petrophysical properties is commonly complicated and the available information is limited, it is impossible (or rather unrealistic) to construct a detailed deterministic model that represents the actual heterogeneity of the reservoir. Once this limitation is recognized, it becomes sensible to translate imperfect knowledge of reservoir properties into a probabilistic framework. In most situations, an equiprobable set of representations of the petrophysical properties of interest would be preferred over a single representation having dubious


DOI:10.1306/1063807CA53229

quality, so that a distribution of possible values for each property (such as time to reach different water cuts) could be established. The set of possible outcomes is commonly referred to as the space of uncertainty, and three types of spaces can be distinguished. Identifying or specifying the type of space depends on whether the focus is on the value(s) obtained by the property at a specific location (local uncertainty), jointly across several locations (spatial uncertainty), or on the results of propagating this uncertainty through transfer functions (response uncertainty).

Geostatistical simulation is increasingly used to build probabilistic models of petroleum reservoirs conditioned on a variety of information, like seismic survey data, values of core porosity or permeability, well-log responses, etc. The set of geostatistical realizations allows one to model the three types of uncertainty, and at first glance, it provides an answer to the quest for an objective (unquestionable, theoretically sound) assessment of uncertainty. This sense of confidence is quickly eroded during the implementation stage when the geologists, geophysicists, and reservoir engineers dare to open the Pandora's box of geostatistical-simulation algorithms. As emphasized by Srivastava (1994a, p. 3) in his overview of stochastic methods for reservoir characterization, "Choosing a specific approach can be a bewildering and daunting task for a newcomer to the field of stochastic methods." Further, in his article on the past, present, and future of petroleum geostatistics, Journel (1999, p. 16) lucidly stated "Geostatistics cannot provide a fully objective assessment of uncertainty since all probabilistic statements rely on a prior (multivariate) distribution model... When generating a set of stochastic realizations of a reservoir model, one necessarily relies on a model that goes well beyond the data available and is implicit to the series of algorithmic decisions taken." In fact, one must keep in mind that the generation of a set of realizations, and hence, the models of uncertainty and subsequent risk assessment, is influenced by a succession of (subjective) decisions made along the modeling process, in particular (1) the choice of a conceptual model for the reservoir, which relies on geologists' expertise; (2) the selection of a particular stochastic-simulation algorithm and number of realizations to explore the space of uncertainty; and (3) the inference of the parameters (e.g., histogram or semivariogram) of the random function model.

The impact of the simulation algorithm on uncertainty assessment has received increasing attention throughout the last few years and triggered the coinage of the term "algorithmically defined" space of uncertainty. For example, Deutsch and Journel (1992) compared the spaces of uncertainty generated by three simulation algorithms: sequential indicator simulation (SIS), sequential Gaussian simulation (SGS), and simulated annealing (SA). In their study, all algorithms used the same information (i.e., the normal-score semivariogram model), and the realizations were unconditional. Regardless of the response variable (time to recover 50% of the oil, times to achieve 5 and 95% water cut, etc.), the three algorithms generated comparable spaces, although the response distribution was slightly wider for SA. For the same three algorithms (SIS, SGS, and SA) plus p-field simulation, Goovaerts (1999) also found some overlap between spaces, but their size varied greatly depending on both the flow property and the simulation algorithm considered. Differences between the algorithms were more pronounced for long-term responses (e.g., 95% vs. 5% water cut) because the intrinsic properties of permeability maps had an increasing impact as the flow simulation proceeded.

The issue of the number of realizations required to explore the space of uncertainty has received much less attention. Following Deutsch (2001), that number depends on the parameters being quantified (i.e., average values would require fewer realizations than extreme percentiles of the distribution) and the precision that needs to be achieved for the uncertainty assessment. For properties such as effective permeability or time to reach different water cuts, Goovaerts (1999) showed that the impact of the number of realizations on the size of the space of uncertainty depends, to a large extent, on the simulation algorithm. In that study, the rate of expansion of the space weakened beyond 20 realizations; the attenuation was slower for short-term simulation results (5% water cut and 10% oil recovery), particularly for SA.

According to Wingle and Poeter (1993), stochastic-simulation techniques are incapable of accounting for all the uncertainty if only a single deterministic semivariogram model is used. A common approach is to use the ergodic fluctuations (i.e., deviations between model [target] and realizations) that are displayed by most realizations as an indirect way to integrate the uncertainty about these estimates. For example, Goovaerts (1999) showed that algorithms, such as SIS or p-field simulation, generate realizations that may depart significantly from a target frequency distribution or semivariogram. The benefit of reducing the ergodic fluctuations, as well as the impact of a misspecification of the semivariogram parameters, is still controversial. Several authors (Deutsch and Journel, 1992; Goovaerts, 1999) found that a reduction of ergodic fluctuations in the permeability histogram increases the precision (narrowness) of the response distribution generated by SIS and p-field simulation. A substantial gain in accuracy also occurred when the sample histogram was reasonably close to the reference histogram. However, algorithms such as SA that impose reproduction of the permeability histogram did not necessarily yield the narrowest distributions of response values, particularly in the case of flow simulation, because they are not controlled by a distribution of permeability values that is univariate.

This chapter provides a brief overview of geostatistical techniques available to model local and spatial uncertainty of continuous attributes and presents one more case study on the comparison of algorithmically defined spaces of uncertainty. Because the conclusions that can be drawn from any case study are always, to some extent, specific to the data analyzed, criteria are introduced to assess the goodness of uncertainty models and to guide the selection of the algorithm best suited to the problem at hand. Although the author recognizes that a first and critical step in any reservoir characterization is the modeling of its geological architecture (e.g., genetic units, sand bodies, lithofacies, or rock types), the modeling of the spatial distribution of categorical variables is beyond the scope of this chapter (e.g., see Deutsch, 2001). Discussion about geological-architecture modeling and fluid-displacement processes will thus be limited. Interested readers should refer to Chu et al. (1994) for a more complete application of the geostatistical approach to real reservoir data.

## MODELING THE LOCAL UNCERTAINTY

The cornerstone of any spatial uncertainty assessment is the modeling of the uncertainty prevailing at any single location $\mathbf{u}$, where $\mathbf{u}$ denotes a vector of coordinates in 3-D space. The probabilistic way to model the uncertainty about a continuous attribute $z$ at $\mathbf{u}$ consists of (1) viewing the unknown value $z(\mathbf{u})$ as the realization of a random variable $Z(\mathbf{u})$ and (2) deriving the conditional cumulative distribution function (CCDF) of $Z(\mathbf{u})$:

$$F(\mathbf{u}; z|(n)) = \text{Prob}\{Z(\mathbf{u}) \leq z|(n)\} \quad (1)$$

where the notation "$|(n)$" expresses conditioning to the local information, say, $n(\mathbf{u})$ neighboring permeability values, $z(\mathbf{u}_\alpha)$. The CCDF fully models the uncertainty at $\mathbf{u}$ because it gives the probability that the unknown is no greater than any given threshold $z$. This section introduces the two main classes of algorithms (parametric and nonparametric) available for determining CCDFs. The practical evaluation of properties of models of uncertainty, such as accuracy and precision, is then discussed.

### Parametric vs. Nonparametric Approach

Determination of CCDFs is straightforward if an analytical model defined by a few parameters can be adopted for such distributions. Multi-Gaussian kriging (Verly, 1986) capitalizes on the congenial properties of the multi-Gaussian model that ensures that the CCDF given by equation 1 is Gaussian, with mean and variance given by the simple kriging estimate and variance. Because the histogram of many petrophysical attributes is commonly asymmetric with a few very large values (positive skewness), a prior transform of observations is typically required, and the multi-Gaussian approach proceeds as follows.

1) The original $z$-data are first transformed into $y$-values with a standard normal histogram. Such a transform is referred to as normal score transform, and the $y$-values $y(\mathbf{u}_\alpha) = G^{-1}[F(z(\mathbf{u}_\alpha))]$ are called normal scores. The normal score transform can be seen as a correspondence table between equal $p$-quantiles $z_p$ and $y_p$ of the $z$ cumulative distribution function (CDF), $F(z)$, and the standard Gaussian CDF, $G(y)$. In other words, $z_p$ and $y_p$ correspond to the same cumulative probability $p$:

$$G(y_p) = G[G^{-1}(F(z_p))] = F(z_p) = p \; \forall \; p \in [0, 1]$$

2) The semivariogram of normal scores is then estimated and modeled, and the simple kriging (SK) estimate and variance are computed and combined to construct the CCDF as

$$\hat{G}(\mathbf{u}; y|(n)) = \text{Prob}\{Y(\mathbf{u}) \leq y|(n)\} = G\left(\frac{y - \hat{y}_{\text{SK}}(\mathbf{u})}{\hat{\sigma}_{\text{SK}}(\mathbf{u})}\right) \quad (2)$$

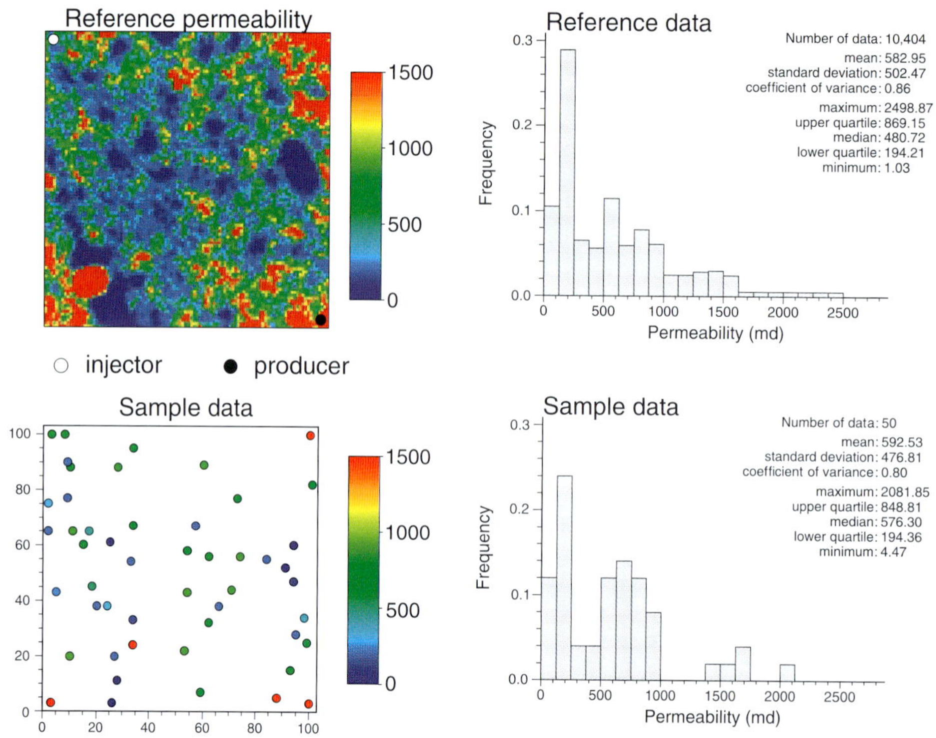


**Figure 1.** Reference permeability map with the injection and production pattern used for flow simulation (producer and injector are in opposite corners). The information available consists of 50 randomly drawn values and the sample histogram.

3) The CCDF of the original variable is retrieved as

$$
\begin{aligned}
\hat{F}(\mathbf{u}; z|(n)) &= \text{Prob}\{Z(\mathbf{u}) \le z|(n)\} \\
&= \text{Prob}\{Y(\mathbf{u}) \le y|(n)\} \\
&= \hat{G}(\mathbf{u}; \phi(z)|(n))
\end{aligned}
$$

under the condition that the transform function $\phi(.) = G^{-1}[F(.)]$ is monotonically increasing.

The 102 × 102 image of Figure 1 (left top graph) is considered to be the exhaustive distribution of permeability values in a two-dimensional (2-D) section of a reservoir. This data set is not synthetic but instead results from the scanning of a block of sandstone, and the resulting computed tomography values have been rescaled so that their histogram reproduces the positively skewed histogram of well-log–derived permeability values from a deep-water turbiditic reservoir (Srinivasan and Journel, 1998). Fifty locations were drawn at random and form the sample data set available to compute experimental semivariograms and reconstruct the reference image (Figure 1, bottom graphs). Figure 2 (right top graphs) shows multi-Gaussian-based CCDF models at $\mathbf{u}_1$ (small open circle) and $\mathbf{u}_2$ (large circle). At each location, the mean and variance of the CCDF in the normal space were identified with the simple kriging estimate and variance, and then the horizontal axis was rescaled through a normal score backtransform.

Unlike the Gaussian-based approach, nonparametric algorithms do not assume any particular shape or analytical expression for the conditional distributions. Instead, the value of the function $F(\mathbf{u};z|(n))$ is determined for a series of $K$ threshold values $z_k$ discretizing the range of variation of $z$:

$$
F(\mathbf{u}; z_k|(n)) = \text{Prob}\{Z(\mathbf{u}) \le z_k|(n)\} \quad k = 1, \ldots, K \tag{3}
$$

The nonparametric geostatistical estimation of CCDF values (Journel, 1983) is based on the interpretation of the conditional probability (see equation 1) as the conditional expectation of an indicator random variable $I(\mathbf{u};z_k)$ given the information $(n)$:

$$
F(\mathbf{u}; z_k|(n)) = E\{I(\mathbf{u}; z_k)|(n)\} \tag{4}
$$

with $I(\mathbf{u};z_k) = 1$ if $Z(\mathbf{u}) \le z_k$ and zero otherwise. Conditional cumulative distribution function values can thus be estimated by least-squares (kriging) interpolation of indicator transforms of data, which commonly requires the estimation and modeling of an indicator semivariogram for each threshold. The resolution of the discrete CCDF is then increased by interpolation within each class $(z_k, z_{k+1})$ and extrapolation beyond the two extreme threshold values $z_1$ and $z_K$.

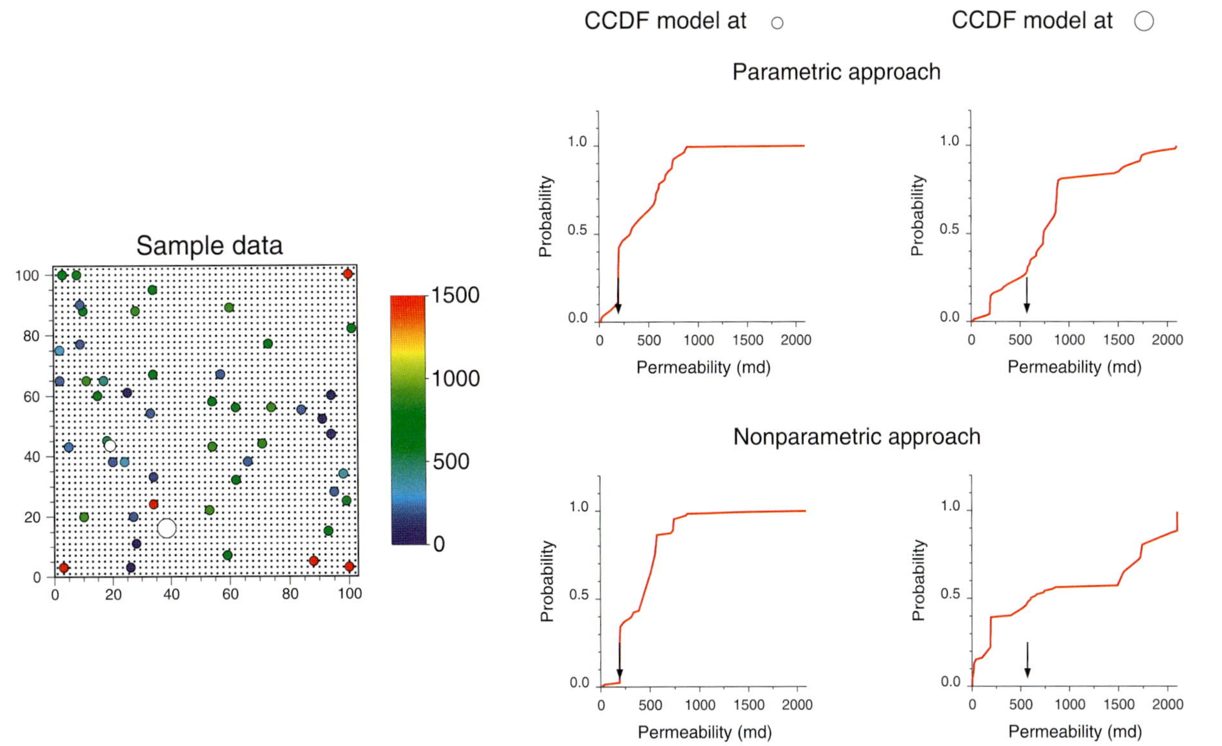


**Figure 2. Location map of 50 permeability values and CCDF models provided by parametric (multi-Gaussian) and nonparametric (indicator) algorithms at two grid nodes depicted by open circles. Arrows depict true permeability values at these two locations.**

For example, Figure 2 (bottom graphs) shows the model fitted to the nine CCDF values estimated at $\mathbf{u}_1$ and $\mathbf{u}_2$ using ordinary indicator kriging and models fitted to nine experimental indicator semivariograms.

The indicator approach appears much more demanding than the multi-Gaussian approach both in terms of semivariogram modeling and computer requirements. This additional complexity is balanced by the possibility of modeling spatial correlation patterns specific to different classes of attribute values through indicator semivariograms. As an aside, note that this does not replace the common approach, whereby the facies distribution is modeled first, and then the spatial distribution of continuous parameters is simulated in each facies. In particular, the connectivity of extreme values can be accounted for, whereas the Gaussian model does not allow for any significant spatial correlation of very large or very small values, a property known as the destructuration effect. A potential pitfall for the indicator approach is the necessity to interpolate or extrapolate the estimated probabilities to derive a continuous CCDF model. Characteristics of the CCDF such as the mean or variance may overly depend on the modeling of the upper and lower tails of the distribution (Goovaerts, 1997). A linear model is commonly adopted for interpolation in each class $(z_k,z_{k+1})$, whereas power or hyperbolic models are used for extrapolation beyond the two extreme threshold values $z_1$ and $z_K$ (Deutsch and Journel, 1998). The choice of these models is fully arbitrary, and some practitioners prefer to capitalize on the higher level of discretization of the CDF (i.e., the cumulative histogram) to improve the within-class resolution of the CCDF (Goovaerts, 1997). For example, the resolution of the discrete CCDFs of Figure 2 (bottom graphs) has been increased by performing a linear interpolation between the tabulated bounds provided by the sample histogram of Figure 1.

Comparison of CCDFs in Figure 2 indicates that both parametric and nonparametric approaches yield similar models at $\mathbf{u}_1$ (small circle). The steep CCDF means that the unknown permeability value is more likely smaller than 1000 md, and the proximity to observations explains the smaller uncertainty prevailing at $\mathbf{u}_1$ relatively to $\mathbf{u}_2$. Differences between uncertainty models are more pronounced at $\mathbf{u}_2$. The spread of the parametric CCDF model is mainly controlled by the kriging variance that is larger than at $\mathbf{u}_1$ because of the greater distance to neighboring observations. However, the large spread of the nonparametric CCDF model is caused by the wide range of surrounding permeability values, which leads to a bimodal probability distribution and a high probability to be either smaller than 250 or larger than 1500 md.

## How Good is the Model?

Before making any decision on the basis of uncertainty models, it is critical to evaluate how well the CCDFs capture the uncertainty about the unknown values. As for spatial interpolation, a reasonable approach is to compare geostatistical predictions with observations that have been either temporarily removed one at a time (cross-validation) or set aside for the whole analysis (jackknife). The major difficulty resides in the selection of performance criteria for CCDF modeling.

At any test location $\mathbf{u}$, a series of symmetric $p$-probability intervals can be constructed by identifying the lower and upper bounds with the $(1-p)/2$ and $(1+p)/2$ quantiles of $F(\mathbf{u};z|(n))$, respectively. For example, the 0.5 probability interval will be bounded by the lower and upper quartiles $[F^{-1}(\mathbf{u};0.25|(n)), F^{-1}(\mathbf{u};0.75|(n))]$. Correct modeling of local uncertainty would imply that the interval covers the actual $z$-value at $\mathbf{u}$ with 0.5 probability, or equivalently, 50% of the probability intervals constructed this way include the true value. If a set of $z$-measurements and independently derived CCDFs is available at $N$ locations $\mathbf{u}_j$, $\{[Z(\mathbf{u}_j), \hat{F}(\mathbf{u}_j;z|(n))], j=1,\ldots,N\}$, the fraction of the distribution represented by the $p$-probability interval is readily computed as

$$\bar{\xi}(p) = \frac{1}{N}\sum_{j=1}^{N}\xi(\mathbf{u}_j;p)\ \forall\ p \in [0,1] \tag{5}$$

with

$$\xi(\mathbf{u}_j;p) = \begin{cases} 1 \text{ if } \hat{F}^{-1}(\mathbf{u}_j;(1-p)/2) < z(\mathbf{u}_j) \le \hat{F}^{-1}(\mathbf{u}_j;(1+p)/2) \\ 0 \text{ otherwise} \end{cases} \tag{6}$$

The closeness of the estimated and theoretical fractions can be assessed using the goodness statistic (Deutsch, 1997) defined as

$$G = 1 - \int_0^1 [3a(p)-2][\bar{\xi}(p)-p]\mathrm{d}p \tag{7}$$

where the indicator function $a(p)$ is defined as:

$$a(p) = \begin{cases} 1 \text{ if } \bar{\xi}(p) \ge p \\ 0 \text{ otherwise} \end{cases} \tag{8}$$

Somewhat arbitrarily, twice more importance is given to deviations when $\bar{\xi}(p) < p$ (inaccurate case). In this case, the weight $|3a(p) - 2|$ is 2, whereas it takes a value of 1 for the accurate case; that is, the case where the fraction of the distribution covered by the $p$-probability interval is larger than expected.

The goodness statistic is complemented by the so-called accuracy plot that allows departures between estimated and expected fractions to be visualized as a function of the probability $p$. Figure 3 (left column) shows accuracy plots computed for both parametric and nonparametric CCDFs. In the first case (top graph), each of the $N = 50$ permeability values has been temporarily removed, with the CCDF being modeled with the remaining 49 values using multi-Gaussian or indicator kriging (cross-validation approach). Most of the points fall below the 45° line (i.e., $\bar{\xi}(p) < p$ for most $p$), which indicates that the probabilistic models are inaccurate. For example, the 0.5 probability intervals derived from the parametric approach only contain 40% of the true values. Whereas the nonparametric models yield smaller deviations between estimated and theoretical fractions for small $p$-values, the parametric models perform better for large $p$-values. On average, the differences between the two approaches, as assessed by the goodness statistic, $G$, are negligible: 0.902 (nonparametric) and 0.899 (parametric), which agrees with Papritz and Dubois (1999), who found that both approaches perform equally well for CCDF modeling. In practice, it is desirable for the conclusions inferred by cross-validation of the available observations to apply to all unsampled locations in the reservoir. The left bottom accuracy plot shows the results obtained at the $N$ = 10,354 locations that have not been sampled (jackknife approach). For both algorithms, the curves are less erratic because of the larger number of locations used in the comparison, yet the conclusions are essentially the same: inaccurate CCDF models, with slightly better performances for the nonparametric approach for small $p$-values, and overall similar values of the goodness statistic (0.909 and 0.877).

Besides the fact that the true value should be covered by the probability interval according to the

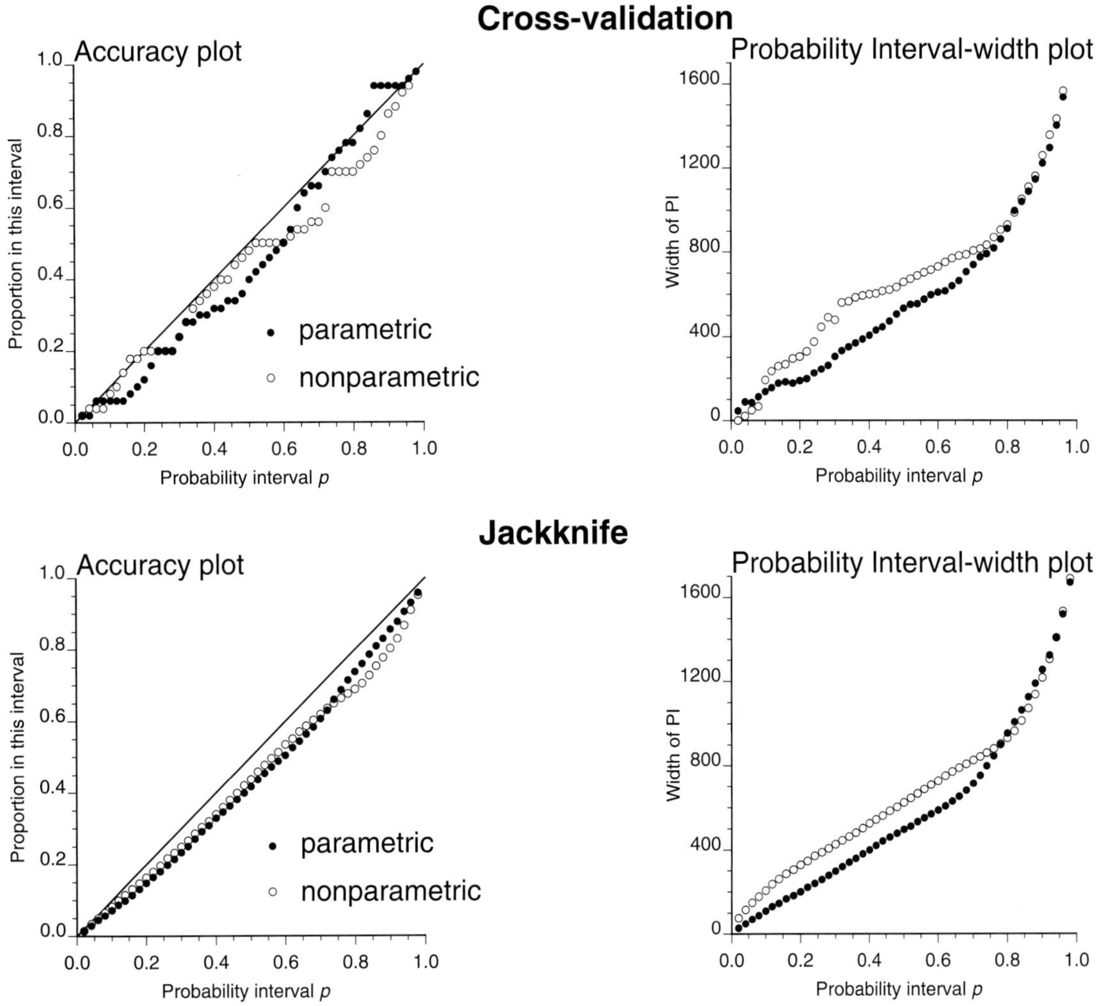


**Figure 3. Plots of the proportion of true values falling within probability intervals (accuracy plot) and the width of these intervals vs. the probability *p*. Parametric (multi-Gaussian) and nonparametric (indicator) algorithms are used to derive CCDF models at the data locations (cross-validation) or at the unsampled grid nodes (jackknife).**

expected probability $p$, it is desirable that this interval be as narrow as possible to reduce the uncertainty about that value. In other words, between two probabilistic models with similar goodness statistics, the one with the smallest spread (less uncertainty) would be preferred. Different measures of CCDF spread can be used (variance, interquartile range, entropy, etc.). Following Deutsch (1997), the variance of each of the $N = 50$ cross-validated CCDFs was computed, and their average, denoted by $U$, was $1.819 \times 10^{-5}$ and $1.688 \times 10^{-5}$ for the nonparametric and parametric models, respectively. The shortcoming of the statistic $U$ is that it provides information only about the global spread of the CCDF, ignoring whether the model of uncertainty is accurate. A more informative procedure consists of plotting the average width of the probability intervals that include the true values for a series of probability values $p$ (see Figure 3, right column). For a given probability $p$, the average width, $\overline{W}(p)$ is computed as

$$\overline{W}(p) = \frac{1}{N\overline{\xi}(p)} \sum_{j=1}^{N} \xi(\mathbf{u}_j; p) \times \left[\hat{F}^{-1}(\mathbf{u}_j; (1+p)/2) - \hat{F}^{-1}(\mathbf{u}_j; (1-p)/2)\right] \quad (9)$$

For both the cross-validation and jackknife procedures, the parametric intervals are narrower until $p = 0.8$, which may explain why the proportion of true values covered by these intervals is smaller than for nonparametric models. In summary, for this particular data set, the different performance criteria do not provide a clear separation of the two classes of algorithms.

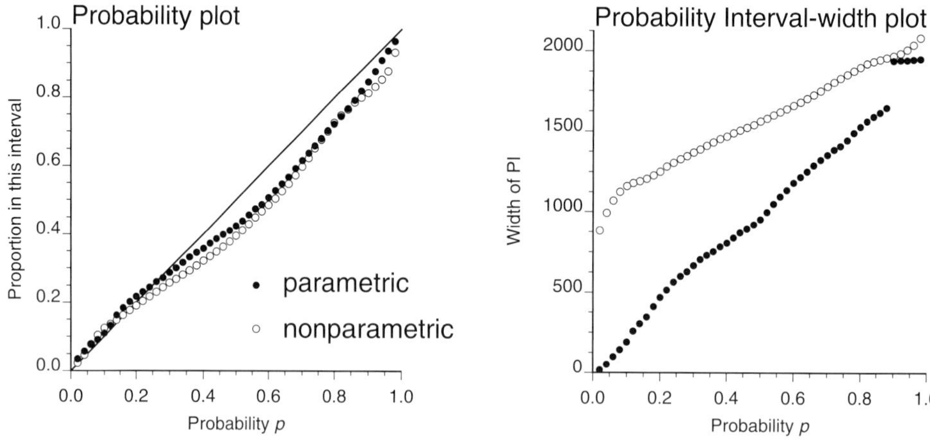


**Figure 4. Plots of the proportion of true values falling within probability intervals defined as [(1 − $p$)-quantile, $z_{max}$], and the width of these intervals vs. the probability $p$. Parametric (multi-Gaussian) and nonparametric (indicator) algorithms are used to derive CCDF models at the unsampled grid nodes (jackknife).**

For applications where the focus is on the probability of exceeding specific threshold values, median-centered probability intervals could be replaced by unbounded probability intervals; for example, the 0.5 probability interval is now defined as $[F^{-1}(\mathbf{u};0.5)|(n)), +\infty]$, or, in practice, $\lfloor \hat{F}^{-1}(\mathbf{u};0.5|(n)), z_{max} \rfloor$. A correct modeling of local uncertainty would entail that across the study area, 50% of the distribution exceeds the CCDF median. Fractions (see equation 5) are thus computed as

$$\bar{\xi}(p) = \frac{1}{N}\sum_{j=1}^{N} \xi(\mathbf{u}_j; p)\ \forall\ p \in [0,1] \tag{10}$$

with

$$\xi(\mathbf{u}_j; p) = \begin{cases} 1 \text{ if } \hat{F}^{-1}(\mathbf{u}_j; p') < z(\mathbf{u}_j) \\ 0 \text{ otherwise} \end{cases} \tag{11}$$

where $p' = 1 - p$. Particular types of accuracy plots, called exceedence probability plots, are created by plotting the estimated fractions vs. the expected ones (see Figure 4). Both parametric and nonparametric models yield similar fractions that are slightly higher than the expected ones for small probabilities $p$ corresponding to high permeability values, $p'$-quantiles, whereas the proportion of true values that exceed low permeability thresholds (high $p$-values) is smaller than expected. Differences between the indicator and multi-Gaussian approaches are much more pronounced for the width of probability intervals, which corresponds to the difference between the maximum $z$-value (2100 md) and the $p'$-quantile of the CCDF. These discrepancies essentially reflect differences in the modeling of CCDF upper tails. Narrow probability intervals indicate that the multi-Gaussian models allow a better detection of high values, whereas the more demanding indicator algorithm would be preferred for low values.

## MODELING THE SPATIAL UNCERTAINTY

In petroleum applications, the focus is on geological and physical processes, such as flow and solute transport, which are primarily controlled by the value, variability, and spatial distribution of petrophysical attributes. Conditional cumulative distribution functions that are measures of local (single-point) uncertainty thus need to be complemented by measures of multiple-point or spatial uncertainty, such as the probability of occurrence of a string of high or low permeability values that may represent a flow path or barrier. Mathematically speaking, the problem is to determine the joint probability that the $z$-values at $J$ locations $\mathbf{u}_j$ are no greater than a critical threshold $z_c$, which can be represented as

$$\begin{aligned} &F(\mathbf{u}_1, \ldots, \mathbf{u}_j; z_c|(n)) \\ &\quad = \text{Prob}\{Z(\mathbf{u}_j) \leq z_c,\ j = 1, \ldots, J|(n)\} \end{aligned} \tag{12}$$

Unless the $J$ random variables $Z(\mathbf{u}_j)$ are independent (a case of little interest), the joint probability given by equation 12 cannot be inferred from the set of $J$ CCDFs, $F(\mathbf{u}_j;z_c|(n))$. The joint probability may, however, be assessed numerically from a set of $L$ realizations of the spatial distribution of $z$-values over the $J$ locations $\mathbf{u}_j$, represented as

$$F(\mathbf{u}_1, \ldots, \mathbf{u}_j; z_c|(n)) \approx \frac{1}{L}\sum_{l=1}^{L}\sum_{j=1}^{J} i^{(l)}(\mathbf{u}_j; z_c) \tag{13}$$

where the indicator value $i^{(l)}(\mathbf{u}_j;z_c)$ is 1 if the simulated $z$-value at $\mathbf{u}_j$ does not exceed the threshold $z_c$ and zero otherwise. The most common geostatistical-simulation algorithms (sequential simulation, p-field simulation, and SA) are reviewed in this section before addressing the issue of the goodness of spatial uncertainty models.

## Stochastic Simulation Algorithms

Simulation of spatial phenomena can be accomplished using a growing variety of techniques that differ in the underlying random function model (multi-Gaussian or nonparametric), the amount and type of information that can be accounted for, and the computer requirements (Gotway and Rutherford, 1994; Myers, 1996; Srivastava, 1996; Deutsch and Journel, 1998). Among the algorithms that are the most commonly used today to simulate the spatial distribution of porosity or permeability in petroleum reservoirs, the following three all involve a sampling of CCDFs as described in the previous section.

Sequential simulation amounts to modeling the CCDF, $F(\mathbf{u}_j;z\ |(n))$ and then randomly sampling it at each of the $N$ grid nodes visited along a random sequence. To ensure the reproduction of the $z$-semivariogram model, each CCDF is made conditional not only to the original $n$ data but also to all values simulated at previously visited locations. Two major classes of sequential simulation algorithms can be distinguished, depending on whether the series of CCDFs are determined using the multi-Gaussian or the indicator formalism.

P-field simulation also requires the sampling of $N$ successive CCDFs, but unlike the sequential approach, all CCDFs are conditioned only to the original $n$ data. Reproduction of the $z$-semivariogram model is approximated by imposing an autocorrelation pattern on the probability values (probability field) used for sampling these CCDFs.

Simulated annealing (SA) approaches the creation of a stochastic image as an optimization problem without any reference to a random function model. The basic idea is to perturb an initial (seed) image gradually so as to match target constraints, such as reproduction of a histogram or semivariogram model. For purposes of the discussion present here, the creation and perturbation of initial images were performed by a random sampling of CCDFs that were conditioned only to the original $n$ data (Goovaerts, 2000). This implementation of SA is thus similar to a p-field approach, where reproduction of spatial patterns is controlled directly by the minimization of an objective function that measures the deviation from the target $z$-semivariogram model instead of through the generation of a probability field.

Figure 5 shows the first realization (out of 100) of the spatial distribution of permeability values generated by each simulation algorithm. For p-field and SA, CCDFs have been modeled using a nonparametric approach. A display of the first realizations suffices to highlight the significant impact of the simulation algorithm on the permeability maps. Note, for example, the strong clustering of high values for SIS and p-field realizations, whereas high permeability values are scattered over the SGS map. Visually, and perhaps subjectively, the SA realization seems to better reproduce the spatial features of the reference permeability map. In practice, the actual distribution of permeability values is unknown, and the problem is thus to be able to quantify the quality of each realization or, more generally, to assess how well the set of realizations depict the uncertainty about the spatial distribution of permeability values in the reservoir.

## How Good is the Model?

Let $\{z^{(l)}(\mathbf{u}_j), j = 1,\ldots, N\}$, $l = 1,\ldots, L$ be a set of $L$ realizations generated by a given algorithm, and let $\{z(\mathbf{u}_\alpha), \alpha = 1,\ldots, n\}$ be the set of true permeability values that are known at only a limited number of sampled locations $\mathbf{u}_\alpha$ (measurement error is assumed negligible). Two common issues are (1) how the quality (goodness, realism) of each individual realization can be defined and quantified and (2) what information can be retrieved from a set of $L$ realizations and how the goodness of such a model of uncertainty can be assessed.

### *Spatial Distribution of Permeability Values*

Disregarding for now the flow characteristics of a permeability map, the quality of a realization could be defined as its ability to match knowledge about the spatial distribution of permeability values a priori. This knowledge can take various forms, ranging from geologists' or petroleum engineers' expertise to information inferred directly from the permeability data, such as a global distribution (histogram) and pattern of spatial variability (semivariogram of the attribute values or of indicator transforms).

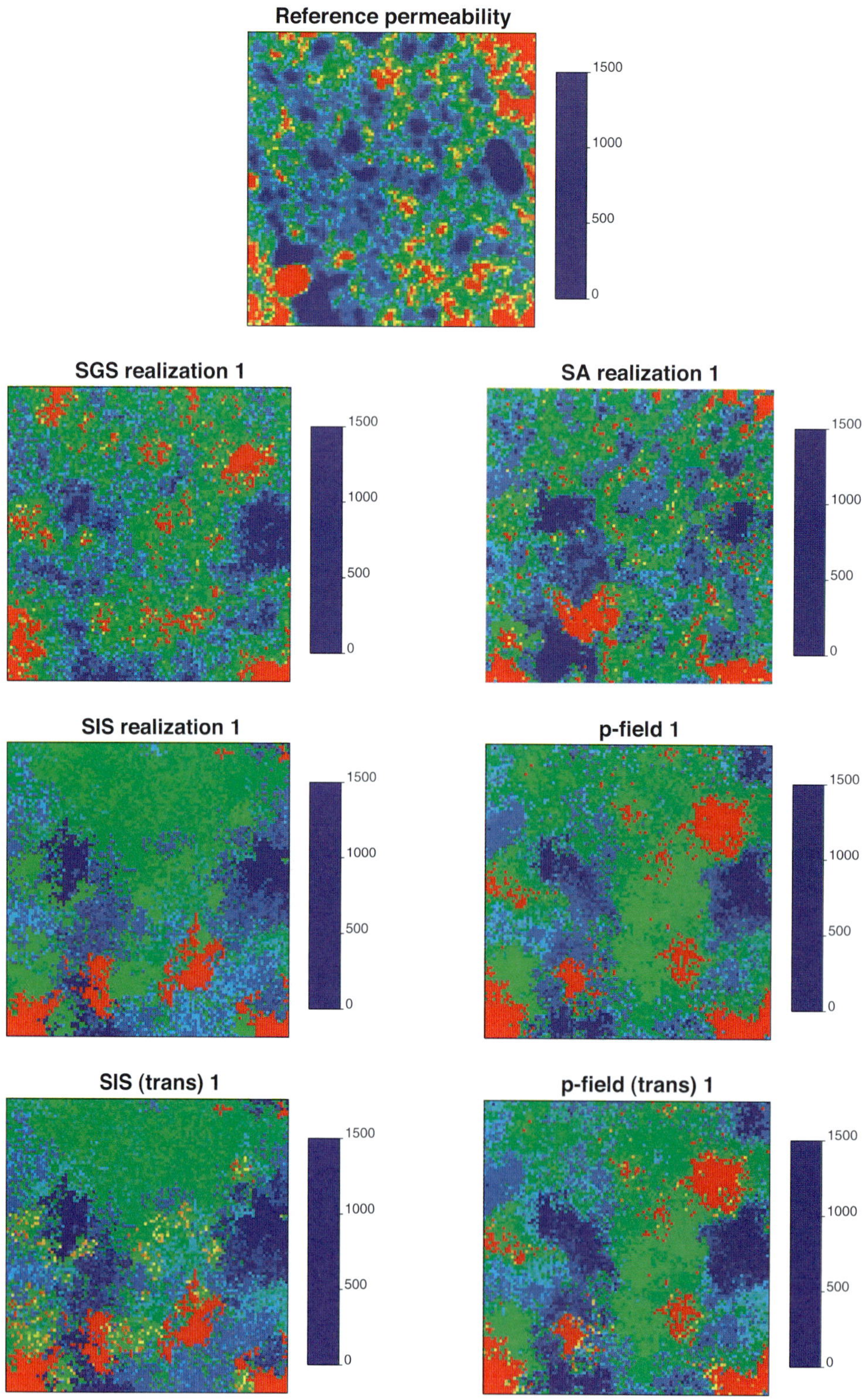


Figure 5. Reference permeability map and the first realization generated using four algorithms: sequential Gaussian simulation (SGS), simulated annealing (SA), sequential indicator simulation (SIS), and p-field simulation. Bottom maps show SIS and p-field realizations after postprocessing using a rank-preserving algorithm to reproduce the sample histogram.

Reproduction of a target histogram and semivariogram is a common way to assess the quality of a simulated map (Chilès and Delfiner, 1999). However, caution must be exercised, and several issues must be kept in mind. First, the magnitude of discrepancies between realization statistics and model parameters, which are referred to as ergodic fluctuations, depends on these and other factors:

1) Selection of the simulation algorithm. Unlike SA, which allows a direct control on the reproduction of target values, most algorithms reproduce

**Table 1. Average discrepancies between realization and target statistics (histogram and semivariogram model) or reference connectivity functions obtained for different stochastic simulation algorithms (results expressed as percentages of the largest value).***

| | Ergodic Fluctuations | | | | |
|---|---|---|---|---|---|
| *Algorithm* | *Histogram* | *Semivariogram* | *Connectivity* | *G* | *W* |
| SIS | 100 | 100 | 100 | 0.84 | 600 |
| SIS transformed | 40 | 62 | 95 | 0.88 | 613 |
| SGS | 65 | 75 | 75 | 0.86 | 550 |
| p-field | 89 | 87 | 99 | 0.88 | 659 |
| p-field transformed | 38 | 63 | 89 | 0.90 | 650 |
| SA | 1 | 25 | 66 | 0.75 | 568 |

*The last two columns give measures of accuracy and precision of the local distributions of 100 simulated permeability values.

model parameters only in expected value, that is, on average, over many realizations. Moreover, it is either the semivariogram of the normal scores (SGS) or indicator transforms (SIS) of attribute values that are considered in the simulation algorithm, and so reproduction of the semivariogram of attribute values should not be expected. Table 1 (first two columns) shows that, on average over 100 realizations, SA leads to the best reproduction of the histogram and semivariogram model, whereas ergodic fluctuations are the most important for SIS and p-field simulation, which confirms results reported in other comparative studies (Goovaerts, 1999).

2) Realization statistics. Realization statistics become increasingly similar and closer to the target values, if these were modeled from the same data.
3) Semivariogram parameters and the size of the simulation grid. Ergodic fluctuations of realization semivariograms are generally important when the range of the semivariogram model is large with respect to the size of the simulated area, particularly if the relative nugget effect is small.

Second, the choice of target values is commonly not trivial, particularly when data are scarce or nonrepresentative of the study area (e.g., preferential sampling of high values). For example, even for the well-sampled 2-D section of Figure 1, the sample histogram and semivariogram model deviate somewhat from the reference values (see Figure 6); hence, their strict reproduction may not be a reliable goodness criterion. A sloppy way to consider the uncertainty about parameter values is to let ergodic

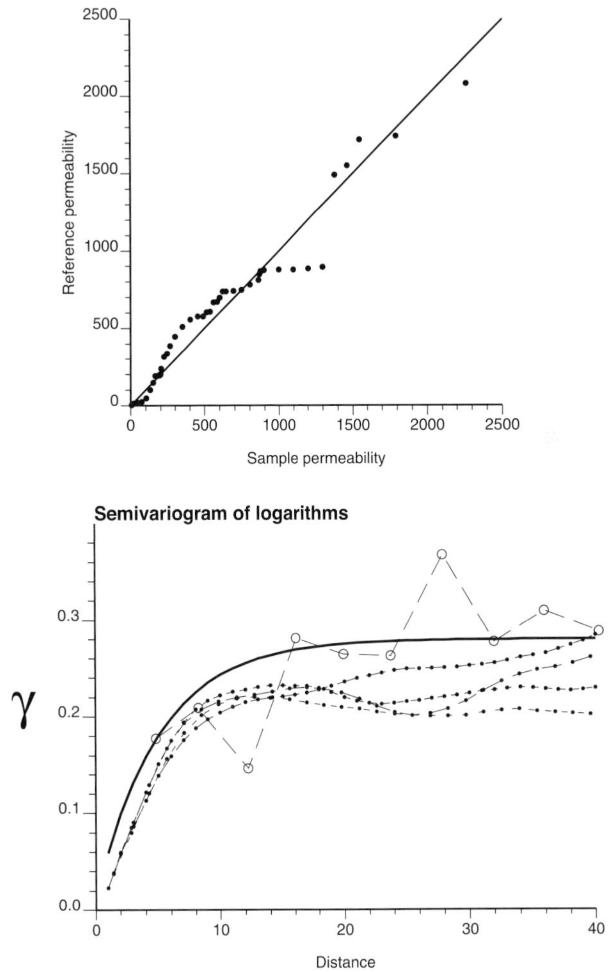


**Figure 6.** Q–Q plot of the distribution of 10,404 reference permeability data vs. that of the 50 sampled data of Figure 1. Bottom graph shows the sample semivariogram (open circles) computed from logarithmic transforms of 50 data with the model fitted; note the discrepancies with the directional semivariograms computed from the reference image (black dots).

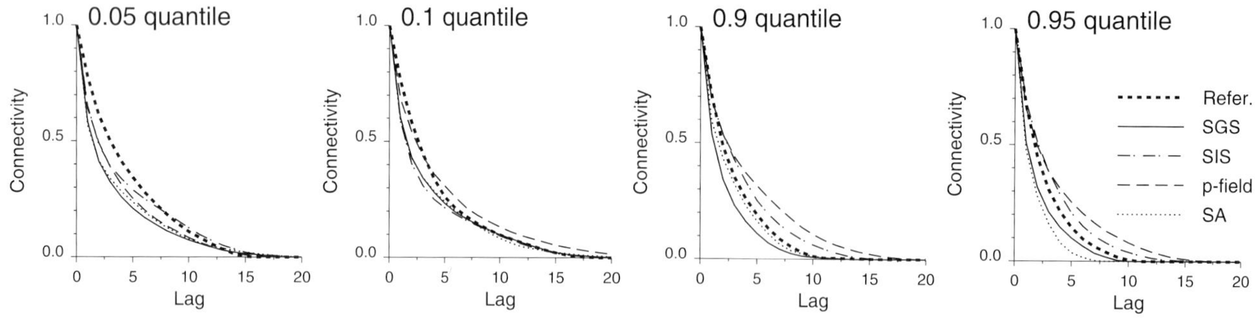


**Figure 7. Connectivity functions displaying the frequency of *J* (1–20) pixels in a row (east–west string) being valued less than the 0.05 or 0.1 quantile of the histogram of simulated values (connectivity of low values) or exceed the 0.9 or 0.95 quantile (connectivity of high values). The thick dashed line depicts the curve obtained from the reference image of Figure 1. All curves represent the average over 100 realizations and have been standardized to unit connectivity for zero-lag distance.**

fluctuations do the job; that is, view these fluctuations as the result of parameter uncertainty. A more rigorous approach (Journel, 1994a) consists of a formal randomizing of the semivariogram models and then accounting for such variation in the simulation algorithms. Wingle and Poeter (1993) proposed to bound the limits of model semivariograms through jackknifing and to use Latin hypercube sampling of the distribution of semivariogram parameters, combined with expert opinion, to select a limited number of semivariogram models that encompass the range of uncertainty.

In the author's opinion, the importance given to reproducing the target values should depend on (1) the reliability of the information available regarding the histogram and spatial pattern of attribute values and (2) the goal of the simulation. In some situations, the objective is to generate a single map that reproduces the pattern of spatial variability inferred from the data, which is seen as a better alternative to a single, smooth, estimated map. In this case, ergodic fluctuations are annoying, and realizations can be postprocessed using one of several methods, SA for example, to improve reproduction of both histogram and semivariogram or the rank-preserving transform algorithm developed by Journel and Xu (1994) for better reproduction of the histogram only. Figure 5 (bottom graphs) shows the results of the rank-preserving transform on the first SIS and p-field realizations. Results shown in Table 1 confirm the better reproduction of the target histogram, and indirectly of the semivariogram model, achieved by that transform. Note that, for SGS, reproduction of the target histogram is theoretically ensured by the backtransform of simulated normal scores as long as the distribution of simulated normal scores is standard Gaussian. Deviations from this target distribution in the normal space explain why, after backtransform, SGS realizations deviate more from the target histogram than transformed SIS and p-field realizations (see Table 1).

Beside features that the permeability maps should reproduce, there might be features that are unlikely to be observed in practice and that should not be displayed by realizations. A common example is the pattern of connectivity of extremely high or low permeability values. All the simulation algorithms will generate permeability maps that reproduce a histogram and semivariogram model to some extent, if only after postprocessing to reduce ergodic fluctuations. As demonstrated by several authors (Journel and Alabert, 1988; Wen and Kung, 1993), permeability fields with similar histograms and semivariograms can exhibit very different flow and transport behaviors because of connectivity properties that are implicitly imparted by the geostatistical algorithm used to generate them. A well-known example is the multi-Gaussian model that typically leads to realizations that show no significant correlation of extreme values (Gómez-Hernández and Wen, 1998) and, for a given covariance function, maximizes scattering of extreme values in space (destructuration effect). The *J*-order connectivity of low values can be computed as

$$\phi(J; z_c) = E\left\{\prod_{j=1}^{J} i(\mathbf{u} + (j-1)\mathbf{h}; z_c)\right\} \qquad (14)$$

where $i(\mathbf{u}; z_c) = 1$ if $z(\mathbf{u}) \leq z_c$. Figure 7 shows, for each simulation algorithm (average over 100 realizations) and as much as $J$ = 21, the connectivity of values

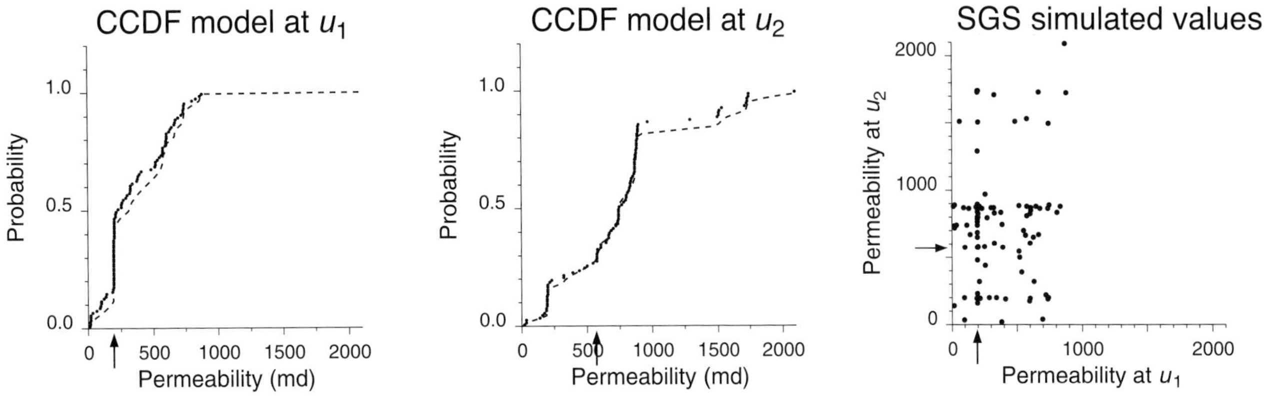


**Figure 8. Cumulative distributions of 100 permeability values (black dots) generated using SGS at the two grid nodes depicted in Figure 2. In each case, the dashed line represents the CCDF model provided by multi-Gaussian kriging, whereas the arrow depicts the true permeability value. The relationship between these two sets of 100 simulated values is displayed in the scattergram (right graph).**

below $q_{0.05}$ = 43.5 md and $q_{0.1}$ = 117.2 md. Connectivity of high values is assessed using expression 14, with $i(\mathbf{u};z_c) = 1$ if $z(\mathbf{u}) > z_c$; see Figure 7 for $q_{0.9}$ = 1340 md and $q_{0.95}$ = 1459 md. As expected, SGS realizations (solid line) tend to display smaller connectivity than the other simulated maps and the reference permeability map (thick dashed line). In practice, well data do not allow the inference of target connectivity functions because a series of strings of $J$-values are not often available, except for $J = 2$ in which case quantity 14 can be seen as a noncentered indicator covariance (Journel, 1989). However, outcrop data or geologic knowledge may indicate that a poor connectivity of high or low values is neither a desirable (conservative) nor realistic feature, and so these realizations should be discarded. Table 1 (third column) indicates that, on average over 100 realizations and four directions, the connectivity of SA realizations is the closest to the reference, whereas SGS yields surprisingly good results. It is worth mentioning that when flow and transport behaviors are mainly controlled by the facies distribution, then the impact of the algorithm used to simulate permeability values in each facies is reduced.

## *Spatial Uncertainty*

In most situations, multiple realizations are generated to depict the uncertainty about the spatial distribution of permeability values; hence, the set of realizations must be considered as a whole. Srivastava (1994b) has reviewed several ways to visualize the spatial uncertainty depicted by a set of $L$ realizations: probability maps, quantile maps, maps of spread, and even uncertainty animation through the creation and rapid display of realizations that are incrementally different using p-field simulation.

The set of realizations can be used to derive measures of uncertainty of increasing complexity as more grid nodes are involved in their computation. First, the local or location-specific uncertainty can be assessed by looking at the distribution of $L$ simulated values at any particular location $\mathbf{u}$. Figure 8 (left graphs) shows the cumulative distributions of 100 permeability values (black dots) generated using SGS at locations $\mathbf{u}_1$ and $\mathbf{u}_2$ of Figure 2. The dashed lines depict the CCDFs derived using multi-Gaussian kriging and displayed previously in Figure 2. As expected in theory for $L \to \infty$, the local distribution of simulated values matches that provided by multi-Gaussian kriging. Thus, no apparent gain is present from using stochastic simulation for modeling local uncertainty because (1) generating a set of realizations is computationally much more demanding than solving a single set of multi-Gaussian or indicator kriging systems; and (2) the goodness of local uncertainty models is not easily checked because a cross-validation approach would require the generation of $n \times L$ realizations, where $n$ is the number of permeability data that will be discarded one at a time before generating $L$ realizations.

One advantage of stochastic simulation over kriging, however, is its flexibility in modeling block CCDFs, which provide the probability for the average $z$-value over a block V centered at $\mathbf{u}$ to be no greater than any specific threshold:

$$F_V(\mathbf{u};z|(n)) = \text{Prob}\{Z_V(\mathbf{u}) \leq z|(n)\} \qquad (15)$$

Equation 15 is numerically approximated by averaging the point-simulated values within the block for each realization and computing the proportion of simulated block values below the threshold. Although such change of support (upscaling and aggregation) can also be conducted using either block multi-Gaussian kriging (e.g., see Chilès and Delfiner, 1999) or the volume-variance correction of the point CCDF at the center of the block (Isaaks and Srivastava, 1989), stochastic simulation allows for a nonparametric uncertainty assessment (i.e., no prior assumption is made about the shape of the local distribution of block values, a straightforward modeling of uncertainty for irregular blocks, and nonlinear upscaling). With respect to nonlinear upscaling, block kriging applies only to the linear average of attribute values over the block, whereas in stochastic simulation, the averaging functions (e.g., geometric or harmonic mean) can be applied directly to the simulated point values inside the block, which is particularly important for permeability values that do not average linearly in space.

As mentioned earlier, the benefit of stochastic simulation over kriging becomes more apparent as one seeks measures of multiple-point uncertainty. For example, the joint uncertainty about the permeability values at the locations $\mathbf{u}_1$ and $\mathbf{u}_2$ of Figure 2 can be assessed using the scattergram of the 100 pairs of values jointly generated using SGS (see Figure 8, right graph). However, the accuracy of such models is difficult to assess because often not enough data pairs exist to compute the proportions of true values that fall within the bivariate probability intervals. Such checks become virtually impossible as the number of grid nodes that are jointly considered increases.

In summary, stochastic simulation provides numerical assessments of multiple-point or spatial uncertainty, but the goodness of such models cannot be evaluated from the sample information alone.

For the example discussed here, knowledge of actual permeability values (reference image) allows one to assess the goodness of simulation-based models of local uncertainty using a jackknife approach. At the 10,354 unsampled locations, the actual permeability values have been compared to the distribution of 100 values generated by the different simulation algorithms. The goodness ($G$) and width ($W$) of accurate probability intervals were computed using expressions 7 and 9 (see Table 1, last two columns). According to the goodness statistics, the better reproduction of spatial statistics, like semivariogram and connectivity functions, by SA realizations does not lead to better CCDF models. This indicates the necessity for a tradeoff between local and global features of simulated maps and leads to the observation that one algorithm may not be best for different objectives. Sequential and p-field simulations yield greater $G$ values that are slightly improved by the rank-order transform. In view of the second statistic, $W$, one would prefer SGS because the probability intervals are narrower than for other algorithms.

## MODELING THE UNCERTAINTY ABOUT RESERVOIR PERFORMANCE

Representations of the spatial distribution of petrophysical attribute values are typically used as inputs to other systems having various complexities, such as pressure solvers or flow simulators. The uncertainty about the input attribute values will propagate through these transfer functions, leading to uncertain response values, such as effective permeability or time to reach a given water cut. Such propagation of uncertainty and the goodness of the resulting probabilistic models for response values are discussed in this section.

### Propagation of Uncertainty

Depending on the nature of the transfer function, various approaches can be used to assess the response uncertainty.

A transfer function is said to be local if it is applied to a single location $\mathbf{u}$ or block $V(\mathbf{u})$ at a time. The propagation of uncertainty from $z$- to $y$-values can be approached numerically (Monte Carlo simulation): the CCDF of $z$ at $\mathbf{u}$ is randomly sampled many times, yielding a set of simulated $z$-values $\{z^{(l)}(\mathbf{u}), l = 1,\ldots, L\}$, which are then fed into the local transfer functions to generate a set of simulated $y$-values $\{y^{(l)}(\mathbf{u}), l = 1,\ldots, L\}$. Random sampling of CCDFs can be very time consuming, and a more efficient strategy is Latin hypercube sampling, which consists of dividing each distribution into $N$ equiprobable classes (100, for example) and randomly sampling these classes without replacement to generate a set of $N$ outcomes (stratified random sampling). In this way, the input distributions are represented in their entirety, requiring a much smaller sample than the Monte Carlo approach for a given degree of precision (McKay et al., 1979; Helton, 1997).

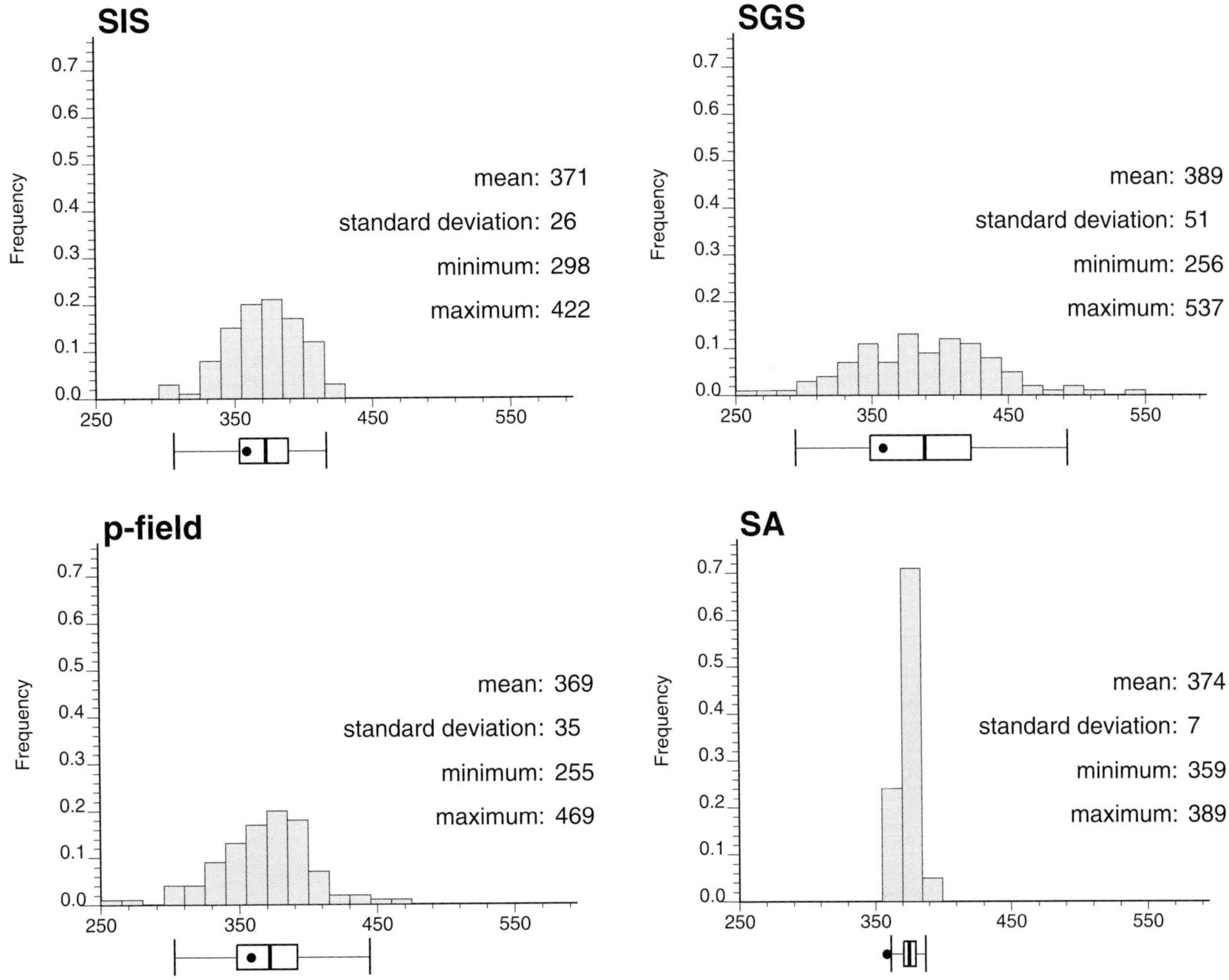


**Figure 9. Histograms of the east–west effective permeability assessed from 100 realizations generated by each simulation algorithm. The black dot in the box plot below each histogram is the true value obtained from the reference image of Figure 1. Five vertical lines are the 0.025 quantile, lower quartile, median, upper quartile, and 0.975 quantile of the distribution.**

Stochastic spatial simulation is commonly not recommended for local transfer functions because a Latin hypercube sampling of the CCDF provides similar results with much less central processing unit (CPU) cost. An exception is when the transfer functions are available only for supports much larger than the measurement support, which requires a prior simulation-based derivation of the block CCDF.

In the petroleum industry, transfer functions are typically global in that they involve attribute values at many grid nodes considered simultaneously or sequentially. The only approach is to generate many realizations of the spatial distribution of petrophysical attributes that are then fed into the transfer function to yield a distribution of response values. In this study, two transfer functions were applied to the set of permeability maps generated by each simulation algorithm: (1) the pressure solver flowsim (Deutsch and Journel, 1992) was used to compute the effective permeability in both the east–west and north–south directions; and (2) a waterflood simulator (Eclipse, 1991) was applied to each realization using the injection-production pattern in Figure 1. The fractional flow of oil vs. time was recorded, and four values were retrieved: the times to reach 5 and 95% water cut and the times to recover 10 and 50% of the oil.

The space of uncertainty is commonly represented by the histogram of response values for each variable taken separately, that is, six histograms in this case: two simulation responses (effective permeability) plus four displacement responses (times for recovery and water cut). For example, Figure 9 shows the histograms for the east–west effective permeability. Because of the usual correlation between

Figure 10. Scatter of the 400 realizations in the plane defined by the first two rotated principal components of the correlation matrix of six flow properties of Table 2 (multivariate space of response uncertainty). Dashed lines depict the true values for each principal component.

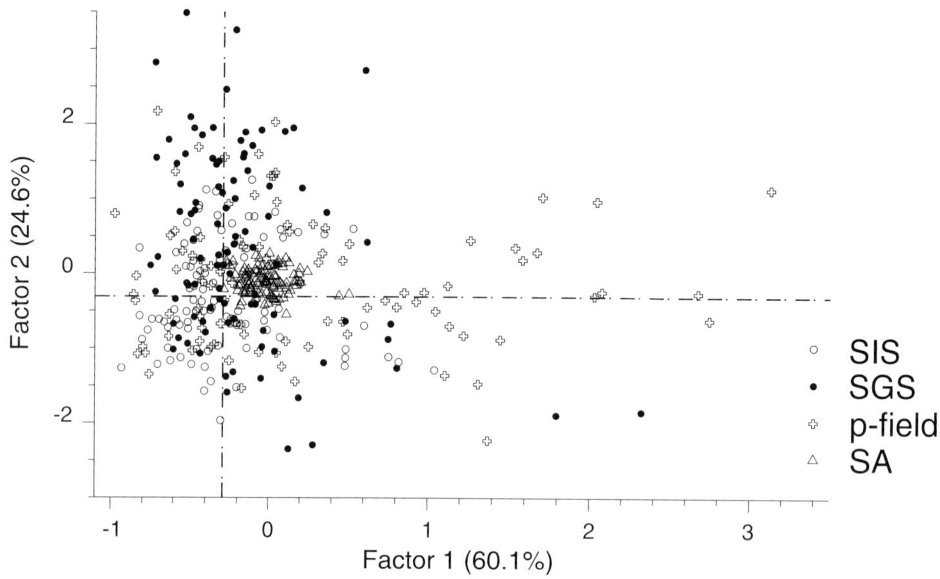


response variables (e.g., negative correlation between effective permeability and the time to reach a given water cut or to recover a given proportion of the oil), Goovaerts (1999) proposed to look at the scatter of realizations in the six-dimensional space spanned by the six flow characteristics, that is, to characterize the multivariate space of uncertainty. Visualization of this space is facilitated by the display of the realizations in a subspace created using principal component analysis. The basic idea of principal component analysis (Davis, 1986) is to create new orthogonal variables, the principal components, as linear combinations of the original variables (i.e., the six response variables). The first few components generally account for most of the variance and so are the most informative. Figure 10 shows the scatter of the 400 realizations in the plane of the first two principal components that explain 84.7% of the total variance. These components have been rotated (varimax rotation) to achieve a simple structure that is easier to interpret; that is, each response variable correlates mainly with one of the two principal components (see Table 2). The first factor essentially reflects flow-simulation (i.e., displacement) results, whereas the second factor is a combination of east–west and north–south effective permeability values.

Figures 9 and 10 clearly show that the four simulation algorithms generate spaces of uncertainty that, despite some overlap, greatly differ: p-field space encompasses most of SIS space that itself includes SA space. Sequential Gaussian simulation space is wider along the vertical axis (second factor), which reflects a larger variation of effective permeability values (recall Figure 9). The key issue is then the assessment of the quality of each space (model) of uncertainty.

## How Good is the Model?

Like other models of uncertainty, two desirable features are the accuracy and precision of the response distribution. In other words, one would like the histogram of response values to encompass the true value while being as narrow as possible to reduce the uncertainty about that value.

In many situations, flow rate history and well-test–derived effective permeability values are unavailable; hence, the accuracy of the response distribution cannot be checked. However, the precision is directly measured by the spread of the response distribution that differs greatly among algorithms (see Figure 9; Table 3). Simulated annealing systematically yields

**Table 2. Correlation coefficients (i.e., factor loadings) between the six response variables and the first two rotated principal components or factors displayed in Figure 10.***

| *Response* | $\overline{K}^x_{\text{eff}}$ | $\overline{K}^y_{\text{eff}}$ | $Wc_1$ | $Wc_2$ | $Oil_1$ | $Oil_2$ |
|---|---|---|---|---|---|---|
| Factor 1 | −0.14 | −0.03 | 0.95 | 0.91 | 0.98 | 0.95 |
| Factor 2 | 0.83 | 0.86 | −0.08 | −0.19 | −0.06 | −0.05 |

*East–west effective permeability ($\overline{K}^x_{\text{eff}}$), north–south effective permeability ($\overline{K}^y_{\text{eff}}$), times to reach 5% ($Wc_1$) and 95% ($Wc_2$) water cut, and the times to recover 10% ($Oil_1$) and 50% ($Oil_2$) of the oil.

**Table 3. Standard deviation of the output distributions for the six response variables.***

| | Response | | | | | |
|---|---|---|---|---|---|---|
| *Algorithm* | $\overline{K}^x_{\text{eff}}$ | $\overline{K}^y_{\text{eff}}$ | $Wc_1$ | $Wc_2$ | $Oil_1$ | $Oil_2$ |
| SIS | 51.9 | 53.0 | 0.417 | 2.55 | 0.176 | 2.01 |
| SIS transformed | 25.9 | 25.6 | 0.344 | 2.01 | 0.133 | 1.54 |
| SGS | 50.9 | 44.0 | 0.472 | 2.80 | 0.194 | 2.04 |
| p-field | 65.0 | 63.2 | 1.604 | 6.14 | 0.798 | 6.98 |
| p-field transformed | 34.5 | 33.7 | 1.634 | 5.77 | 0.819 | 6.89 |
| SA | 6.62 | 5.90 | 0.134 | 0.77 | 0.045 | 0.55 |

*East–west effective permeability ($\overline{K}^x_{\text{eff}}$), north–south effective permeability ($\overline{K}^y_{\text{eff}}$), times to reach 5% ($Wc_1$) and 95% ($Wc_2$) water cut, and the times to recover 10% ($Oil_1$) and 50% ($Oil_2$) of the oil.

the narrowest response distribution, leading to a very compact multivariate space of uncertainty in Figure 10. The widest distributions are observed for p-field and SIS, which confirms previous results (Goovaerts, 1999). Rank-order transform allows a substantial reduction of the spread of effective permeability distributions, whereas the impact is much less pronounced for flow responses, in particular for p-field simulation.

Selection of a response distribution on the basis of its spread is hazardous in the absence of information about its accuracy. Depending on the objective pursued (identification of extreme scenarios or of the most likely response) and the number of realizations that can be generated and processed through the flow simulator, one may prefer wide distributions to identify pessimistic and optimistic scenarios (which means that the focus is on the edges of the space of uncertainty) or narrow distributions if the most likely response needs to be inferred from a few realizations. As a consequence, the same simulation algorithm should not be used for all purposes. For example, Table 4 gives for each response variable the average absolute prediction error (AAPE) and absolute prediction error of the average (APEA), computed as

$$\text{AAPE} = \frac{1}{100}\sum_{l=1}^{100} |y^{(l)} - y| \qquad (16)$$

$$\text{APEA} = \left| y - \frac{1}{100}\sum_{l=1}^{100} y^{(l)} \right| \qquad (17)$$

Except for the prediction of oil recovery, the average deviation between the predicted and true response value is much smaller for the narrow response distribution of SA realizations. The benefit of SA is substantially reduced when 100 realizations are generated, and the mean of the response distribution is used as a predictor (APEA). It is worth noting that the rank-order transformation does not systematically improve prediction performances, which contradicts previous results (Goovaerts, 1999)

**Table 4. Measures of the prediction performance for effective permeability, along with the water cut and oil recovery curves.***

| | AAPE | | | | APEA | | | |
|---|---|---|---|---|---|---|---|---|
| *Algorithm* | $\overline{K}^x_{\text{eff}}$ | $\overline{K}^y_{\text{eff}}$ | *WC* | *Oil* | $\overline{K}^x_{\text{eff}}$ | $\overline{K}^y_{\text{eff}}$ | *WC* | *Oil* |
| SIS | 49.6 | 43.7 | 8.62 | 3.59 | 32.6 | 5.74 | 0.448 | 0.515 |
| SIS transformed | 23.7 | 23.3 | 7.12 | 3.73 | 12.5 | 11.0 | 0.424 | 0.682 |
| SGS | 47.5 | 35.1 | 9.40 | 4.56 | 30.7 | 13.8 | 0.517 | 0.792 |
| p-field | 54.2 | 51.4 | 18.4 | 10.7 | 25.5 | 14.2 | 0.698 | 1.980 |
| p-field transformed | 28.3 | 25.7 | 17.1 | 10.7 | 10.3 | 0.53 | 0.783 | 2.090 |
| SA | 16.1 | 5.57 | 4.55 | 4.09 | 16.1 | 3.69 | 0.409 | 0.817 |

*Average absolute prediction error (AAPE) and absolute prediction error of the average (APEA). WC = water cut.

that were based on a sampling intensity of 5% instead of the 0.48% used here. As pointed out by the same author, when too few data are available to infer reliable histograms, one should be cautious in imposing the strict reproduction of a target histogram that might be far from the true one and so leads to worse predictions. Among simulation algorithms, SIS outperforms its Gaussian counterpart because, for all response variables, it yields narrower distributions and smaller prediction errors for either individual realizations (AAPE) or their averages (APEA).

The accuracy and precision of the four simulation algorithms over the range of response variables are summarized in Figure 10. The six reference response values were combined to derive the true values for the first two principal components (see vertical and horizontal dashed lines). True values are on the margins of the condensed multivariate space of uncertainty produced by SA (small accuracy). However, if one can afford the generation and processing of only one realization, SA greatly reduces the risk of falling far away from the true value relative to p-field simulation or SGS.

### Impact of the Number of Realizations

The characterization of the space of uncertainty is rendered difficult by the fact that only a limited number of realizations are commonly generated. A frequent and still open question relates to the number of realizations needed to sample and represent this space fairly. Looking at Figure 10, it becomes clear that some simulation algorithms, like SA, do not generate the same set of outcomes as other algorithms. One can safely expect that the same conclusions would apply if 1000 instead of 100 realizations were generated. Thus, it appears that the best way to explore the entire set of outcomes is to use several simulation algorithms and to combine the response distributions. An alternative is to preferentially use p-field simulation, which has the advantage of being fast while yielding a large spread of response values. If the generation of realizations is much less time consuming than their processing through flow simulators, a cost-efficient strategy would be to rank the generated set of realizations and to run the flow simulator on a subset of them, say the low-side, expected, and high-side realizations (Deutsch and Srinivasan, 1996; Deutsch, 2001).

For a given simulation algorithm and response variable, the impact of the number of realizations on accuracy and precision can be investigated using the procedure described in Goovaerts (1999). The idea is to create subsets of increasing size through the random sampling of an initial set of 100 response values. For example, each distribution of 100 effective permeability values in Figure 9 was randomly sampled to create 100 subsets of two values. The operation was repeated for increasing subset sizes: 3, 4,. . ., 100 values. Note that a value is included only once in a subset (sampling without replacement). This resampling creates 100 subsets for each of the 99 possible sizes, and the upper and lower bounds of the space of uncertainty for a given size are identified with the average minimum and maximum response values over the 100 subsets.

For each simulation algorithm and response variable, Figure 11 shows the evolution of the upper and lower bounds of the space of uncertainty for an increasing number of generated realizations. As expected, the extent of the space increases with the number of realizations but at different rates, depending on the response variable and simulation algorithm. Except for p-field simulation, the increase generally becomes small beyond 20 realizations. With too few realizations, the space of uncertainty may not contain the true value depicted by the horizontal dashed line (e.g., less than 10 SIS realizations for 50% oil recovery). This risk is very high for the narrow SA space. Regardless of the response variable, the SA space of uncertainty is included in the SIS space that is itself encompassed by the SGS and p-field spaces.

## CONCLUSIONS

In the oil industry, uncertainty is attached to every step of reservoir modeling, and such imperfect knowledge of the spatial distribution of rocks and petrophysical properties will impact flow simulation and recovery forecasting. This chapter describes various forms of uncertainty assessment for continuous properties, ranging from the modeling of location-specific probability distributions to the generation of a set of conditional realizations that are then fed into complex transfer functions to yield a probability distribution of response values. As for spatial interpolation, users are faced with the difficult choice of selecting an algorithm from an extensive geostatistical toolbox.

Results of this study substantiate the conclusion of Srivastava (1994a, p. 16): "There is no stochastic

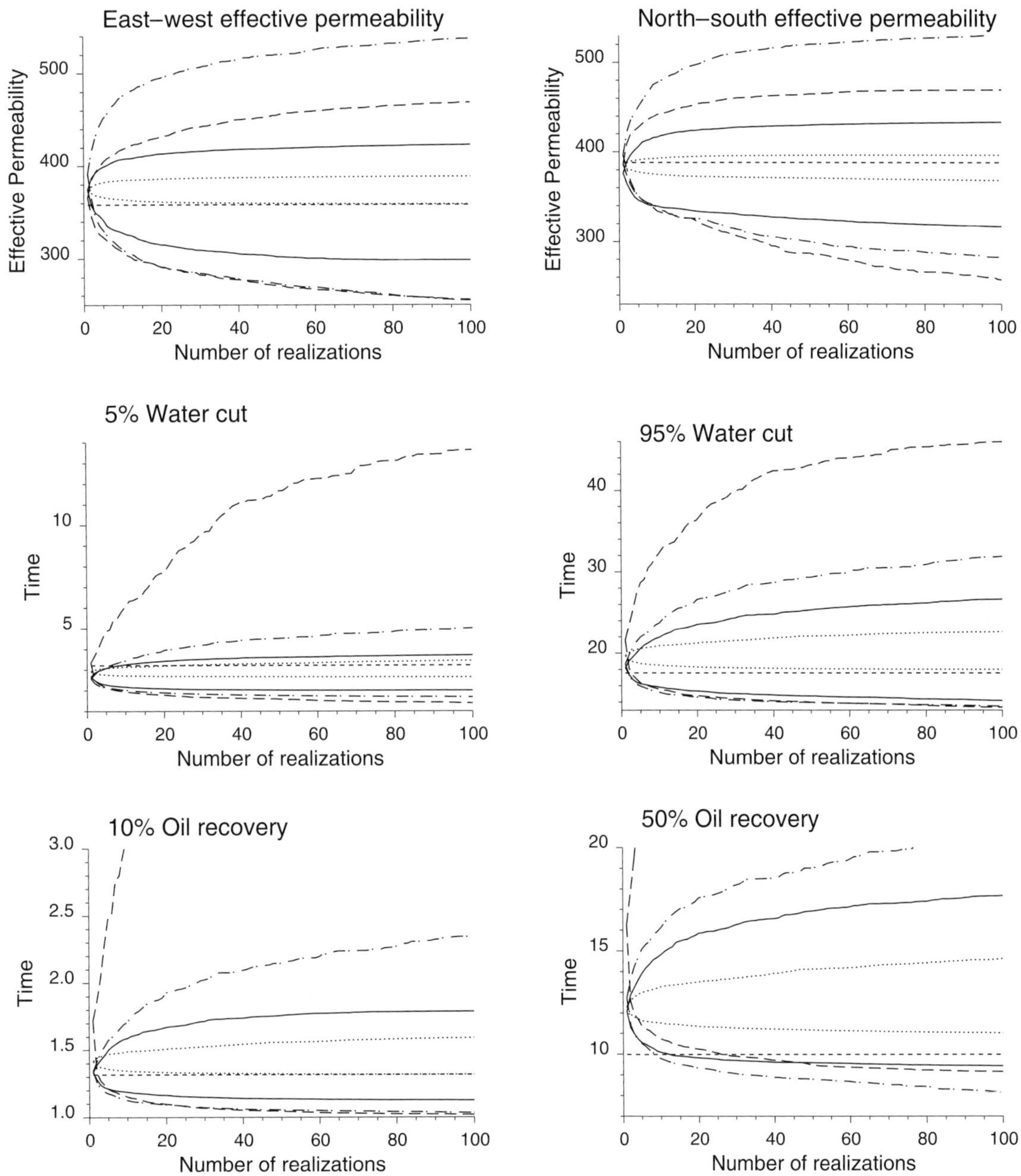


**Figure 11. Impact of the number of generated realizations on the upper and lower bounds of the space of uncertainty for the six flow properties. The horizontal dashed line denotes the true value derived from the reference image of Figure 1. It is best if the space is narrow and includes the true value. The algorithms are denoted by — (SIS transformed), -·-·- (SGS), - - - (p-field transformed), ··· (SA).**

modeling method that is universally best for all possible petroleum reservoir problems… the most successful case studies will be those that view the growing assortment of methods as a tool kit rather than as a set of competing methods." It also stresses the importance of goodness criteria to guide the choice of a specific algorithm.

Models of local uncertainty are the easiest to build and to evaluate; implementation of multi-Gaussian kriging and, to a lesser extent, indicator kriging are quite straightforward and do not require huge CPU time even for large reservoirs. Thus, one should be able to apply and to systematically employ both approaches and then use accuracy plots and associated statistics (goodness and width of the probability intervals) to select the technique that provides the best models for the data. An implicit assumption is that the conclusions inferred by cross-validation of the available observations hold true for all unsampled locations in the reservoir.

Unfortunately, most applications in the oil industry require models of spatial uncertainty that are much more expensive to build because they require the generation of multiple realizations of the spatial distribution of petrophysical attributes. Their evaluation is also hampered by the lack of spatial information available, and it is typically limited to checking the reproduction of a few target statistics, such as a histogram or semivariogram model, for each individual realization. According to these criteria, p-field and SIS are commonly outperformed by SGS and SA; however, a rank-preserving transformation can quickly and substantially reduce the magnitude of ergodic fluctuations. Connectivity functions are certainly more informative for flow and transport properties, but they cannot be compared to experimental values.

Modeling local or spatial uncertainty is typically a preliminary step toward the modeling of uncertainty about response values, such as effective permeability or times to reach different water cuts. Propagation of uncertainty through local transfer functions is quite straightforward, and the efficiency of Monte Carlo simulation is greatly enhanced by the use of stratified random sampling (Latin hypercube approach) that allows one to ensure that the full range of possible outcomes is generated for a given input distribution. The problem is much more complex for multiple-point or spatial transfer functions, such as flow simulators. For example, although the same set of CCDFs is sampled in both p-field simulation and SA, the resulting spaces of response uncertainty greatly differ, which illustrates the multitude of possible combinations. That number should increase with the size of the simulation grid; hence, the small 2-D example of this chapter likely understates the variability that could be expected for realistic 3-D problems.

In the absence of information regarding the expected response value, goodness criteria developed for local uncertainty models cannot be applied to response distributions. If one can afford the generation of many realizations, a good practice would be to use different simulation algorithms and to view the overlap between response distributions as the most likely. If only a few realizations can be generated and processed while histogram and semivariogram models are deemed reliable, SA allows a drastic reduction of ergodic fluctuations, and in this case study, it yields the smallest deviations from the reference responses. If the focus is on the detection of extreme scenarios, p-field simulation tends to generate the widest flow-response distributions.

Although facies modeling has not been covered in this chapter, it is a critical step in any reservoir characterization that might outweigh the impact of the continuous simulation algorithms compared here. Therefore, the choice of a conceptual model for the reservoir, which relies on the geologists' expertise, is commonly considered to be the key decision in the modeling process.

In summary, one must keep in mind that all predictions of reservoir performance are uncertain. Uncertainty assessments depend on the choice of the stochastic-simulation algorithm because it seems unlikely that the entire set of possible outcomes could be sampled by a single class of algorithms, regardless of the number of realizations generated. However, to quote Journel (1994b, p. 20), "Better that we create a subjective model of uncertainty than an illusion of certainty."

## REFERENCES CITED

Chilès, J.-P., and P. Delfiner, 1999, Geostatistics: Modeling spatial uncertainty: New York, John Wiley & Sons, 695 p.

Chu, J., W. Xu, and A. G. Journel, 1994, 3-D implementation of geostatistical analyses—The Amoco case study, *in* J. M. Yarus and R. L. Chambers, eds., Stochastic modeling and geostatistics: Principles, methods, and case studies: AAPG Computer Applications in Geology 3, p. 201–216.

Davis, J. C., 1986, Statistics and data analysis in geology, 2d ed.: New York, John Wiley & Sons, 646 p.

Deutsch, C. V., 1997, Direct assessment of local accuracy and precision, *in* E. Y. Baafi and N. A. Schofield, eds., Geostatistics Wollongong '96: Dordrecht, Kluwer Academic Publishers, p. 115–125.

Deutsch, C. V., 2001, Geostatistical reservoir modeling: New York, Oxford University Press, 376 p.

Deutsch, C. V., and A. G. Journel, 1992, Annealing techniques applied to the integration of geological and engineering data: Stanford University Special Report of the Stanford Center for Reservoir Forecasting.

Deutsch, C. V., and A. G. Journel, 1998, GSLIB: Geostatistical software library and user's guide, 2d ed.: New York, Oxford University Press, 369 p.

Deutsch, C. V., and S. Srinivasan, 1996, Improved reservoir management through ranking stochastic reservoir models: Society of Petroleum Engineers–Department of Energy Tenth Symposium on Improved Oil Recovery, Tulsa, Oklahoma, SPE Paper 35411, p. 105–113.

ECLIPSE 100 Reference Manual, 1991, Intera ECL Petroleum

Technologies, Highlands Farm, Greys Road, Henley-on-Thames, Oxfordshire, England.

Gómez-Hernández, J., and X.-H. Wen, 1998, To be or not to be multi-Gaussian? A reflection on stochastic hydrogeology: Advances in Water Resources, v. 21, p. 47–61.

Goovaerts, P., 1997, Geostatistics for natural resources evaluation: New York, Oxford University Press, 483 p.

Goovaerts, P., 1999, Impact of the simulation algorithm, magnitude of ergodic fluctuations and number of realizations on the spaces of uncertainty of flow properties: Stochastic Environmental Research and Risk Assessment, v. 13, p. 161–182.

Goovaerts, P., 2000, Estimation or simulation of soil properties? An optimization problem with conflicting criteria: Geoderma, v. 3–4, p. 165–186.

Gotway, C. A., and B. M. Rutherford, 1994, Stochastic simulation for imaging spatial uncertainty: Comparison and evaluation of available algorithms, *in* M. Armstrong and P. A. Dowd, eds., Geostatistical simulations: Dordrecht, Kluwer Academic Publishers, p. 1–21.

Helton, J. C., 1997, Uncertainty and sensitivity analysis in the presence of stochastic and subjective uncertainty: Journal of Statistical Computation and Simulation, v. 57, p. 3–76.

Isaaks, E. H., and R. M. Srivastava, 1989, An introduction to applied geostatistics: New York, Oxford University Press, 561 p.

Journel, A. G., 1983, Non-parametric estimation of spatial distributions: Mathematical Geology, v. 15, p. 445–468.

Journel, A. G., 1989, Fundamentals of geostatistics in five lessons: Washington, D.C., American Geophysical Union, 40 p.

Journel, A. G., 1994a, Resampling from stochastic simulations: Environmental and Ecological Statistics, v. 1, p. 63–84.

Journel, A. G., 1994b, Geostatistics and reservoir geology, *in* J. M. Yarus and R. L. Chambers, eds., Stochastic modeling and geostatistics: Principles, methods, and case studies: AAPG Computer Applications in Geology 3, p. 19–20.

Journel, A. G., 1999, Past, present and future of petroleum geostatistics: Stanford University Special Report of the Stanford Center for Reservoir Forecasting.

Journel, A. G., and F. G. Alabert, 1988, Focusing on spatial connectivity of extreme valued attributes: Stochastic indicator models of reservoir heterogeneities, SPE Paper 18324, p. 621–632.

Journel, A. G., and W. Xu, 1994, Posterior identification of histograms conditional to local data: Mathematical Geology, v. 26, p. 323–359.

McKay, M. D., R. J. Beckman, and W. J. Conover, 1979, A comparison of three methods for selecting values of input variables in the analysis of output from a computer code: Technometrics, v. 21, p. 239–245.

Myers, D. E., 1996, Choosing and using simulation algorithms, *in* H. T. Mowrer, R. L. Czaplewski, and R. H. Hamre, eds., Spatial accuracy assessment in natural resources and environmental sciences: Second International Symposium, U.S. Department of Agriculture, Forest Service, General Technical Report RM-GTR-277, p. 23–29.

Papritz, A., and J.-P. Dubois, 1999, Mapping heavy metals in soil by (non-)linear kriging: An empirical validation, *in* J. Gómez-Hernández, A. Soares, and R. Froidevaux, eds., GeoENV II: Geostatistics for Environmental Applications, Proceedings of the Second European Conference on Geostatistics for Environmental Applications: Dordrecht, Kluwer Academic Publishers, p. 429–440.

Srinivasan, S., and A. G. Journel, 1998, Direct simulation of permeability field conditioned to well test data: Stanford University Special Report of the Stanford Center for Reservoir Forecasting.

Srivastava, R. M., 1994a, An overview of stochastic methods for reservoir characterization, *in* J. M. Yarus and R. L. Chambers, eds., Stochastic modeling and geostatistics: Principles, methods, and case studies: AAPG Computer Applications in Geology 3, p. 3–16.

Srivastava, R. M., 1994b, The visualization of spatial uncertainty, *in* J. M. Yarus and R. L. Chambers, eds., Stochastic modeling and geostatistics: Principles, methods, and case studies: AAPG Computer Applications in Geology 3, p. 339–345.

Srivastava, R. M., 1996, An overview of stochastic spatial simulation, *in* H. T. Mowrer, R. L. Czaplewski, and R. H. Hamre, eds., Spatial accuracy assessment in natural resources and environmental sciences: Second International Symposium, U.S. Department of Agriculture, Forest Service, General Technical Report RM-GTR-277, p. 13–22.

Verly, G., 1986, MultiGaussian kriging— A complete case study, *in* R. V. Ramani, ed., Proceedings of the 19th International Application of Computers and Operations Research in the Mineral Industry (APCOM) Symposium: Society of Mining Engineers, Littleton, Colorado, p. 283–298.

Wen, X.-H., and C.-S. Kung, 1993, Stochastic simulation of solute transport in heterogeneous formations: A comparison of parametric and nonparametric geostatistical approaches: Ground Water, v. 31, p. 953–965.

Wingle, W. L., and E. P. Poeter, 1993, Uncertainty associated with semivariograms used for site simulation: Ground Water, v. 31, p. 725–734.

Krivoruchko, K., A. Gribov, and J. M. Ver Hoef, 2006, A new method for handling the nugget effect in kriging, *in* T. C. Coburn, J. M. Yarus, and R. L. Chambers, eds., Stochastic modeling and geostatistics: Principles, methods, and case studies, volume II: AAPG Computer Applications in Geology 5, p. 81–89.

*Chapter 7*

# A New Method for Handling the Nugget Effect in Kriging

**K. Krivoruchko**
**A. Gribov**
*Environmental Systems Research Institute*
*Redlands, California, U.S.A.*

**J. M. Ver Hoef**
*Alaska Department of Fish and Game*
*Fairbanks, Alaska, U.S.A.*

## ABSTRACT

This chapter discusses methods for estimating the nugget in semivariogram models. Commonly used exact and filtered kriging methods are compared with an alternative method, which predicts a new value at the sampled location. Using the alternative method, estimation at a location where data have been collected involves predicting the smooth underlying value plus a new observation from measurement error. This is exactly what is necessary for validation and cross-validation diagnostics. Three examples of using new value kriging are presented that involve comparison of simulated results, porosity estimation for the North Cowden unit in west Texas, and analysis of radiocesium soil-contamination data collected in Belarus after the Chernobyl accident.

## INTRODUCTION

Kriging is a spatial interpolation and estimation method first used in meteorology, subsequently rediscovered and used in mining applications, and later widely applied in geology, environmental sciences, agriculture, and other disciplines. It operationally and theoretically depends on models of spatial autocorrelation, which can be formulated in terms of covariance or semivariogram functions. In either formulation, the models of spatial autocorrelation commonly exhibit similar characteristics, which are called the nugget, sill, and range. These are also referred to as the parameters of the semivariogram. The sill represents the total variation in the data, and the range is the distance between sampling locations at which autocorrelation vanishes or nearly vanishes. The nugget effect refers to the situation in which the difference between measurements taken at sampling locations that are close together is not zero. From a graphical perspective, this results in a discontinuity at or near the origin on a semivariogram plot. Generally speaking, the nugget effect is unlikely to be a physical reality, because for many physical processes,


DOI:10.1306/1063808CA53227

observations taken indistinguishably close together should converge in value, creating a smooth response. The primary reasons why discontinuities occur near the origin for semivariogram and covariance models are the presence of measurement error or variation at scales too fine to detect (microscale variation).

In geological applications, kriging is commonly associated with exact interpolation. Exact interpolation means that when semivariogram and covariance models have a nugget effect, there will potentially be a discontinuity at prediction locations where data have been collected. In other words, the kriging predictions will change gradually and relatively smoothly in space until they get to a location where data have been collected, and then there will be a jump in the prediction to the exact value that is measured. Because of this jump in predictions to the exact value, there will also be a discontinuity in the standard error of the predicted value, which becomes zero at measured locations. When multiple values are obtained at a single location and these values are different (a common situation in earth science applications), exact kriging cannot be used.

Variations of kriging can produce noiseless (or filtered) predictions (Gandin, 1959). Interpolation based on filtered kriging produces smoother maps without the jumps. A consequence of filtering is that the standard error of prediction is smaller because measurement error is not included in the nugget effect.

This chapter presents an alternative kriging method called "new value kriging," which predicts a new value at locations where data have been observed. This method causes no discontinuities in predictions or in their standard errors, and the standard error is equivalent to that of exact kriging. Cross-validation is an area of interest for which new value predictions can be highly useful. Comparisons of three prediction methods (exact, filtered, and new value) are presented here. The chapter also details the theory and provides examples using real geological and radioecological data to illustrate the differences between the methods for covariance models having nugget effects. New value kriging, as well as exact and filtered kriging, has been implemented in commercial software (e.g., MapStudio [Maignan and Krivoruchko, 1997; Krivoruchko et al., 1997] and Geostatistical Analyst, an extension to ArcInfo 8.1 that is developed and marketed by Environmental Systems Research Institute [Johnston et al., 2001]).

## KRIGING MODELS

This section presents the assumptions and theory for ordinary kriging using covariances, although the same concepts can be applied to semivariograms and to other types of linear kriging, including simple kriging and universal kriging. To begin, assume the data are realizations of a spatially autocorrelated process with independent random errors represented by

$$Z_t(\mathbf{s}) = \mu(\mathbf{s}) + Y(\mathbf{s}) + \eta(\mathbf{s}) + \varepsilon_t(\mathbf{s})$$

where $Z_t(\mathbf{s})$ denotes the $t$ − th realization at location $\mathbf{s}$. Let $n_i$ be the number of measurements at location $\mathbf{s}_i$. Commonly, $n_i = 1$, and if $n_i > 1$, $Z_t(\mathbf{s})$ comprises a measurement error model (see Fuller, 1987, for a detailed presentation on measurement error models). The following additional assumptions are required: $\mu(\mathbf{s}) = \mu$ is the unknown mean value; $Y(\mathbf{s})$ is a smooth, second-order stationary process whose range of autocorrelation is detectable with an empirical semivariogram or covariance; $E(Y(\mathbf{s})) = 0$; $C_y(\mathbf{h}) = \mathrm{Cov}(Y(\mathbf{s}),Y(\mathbf{s} + \mathbf{h}))$; $\eta(\mathbf{s})$ is a smooth, second-order stationary process whose semivariogram range is so close to zero that it is shorter than all practical distances between data and prediction locations; $E(\eta(\mathbf{s})) = 0$; $C_\eta(\mathbf{h}) = \mathrm{Cov}(\eta(\mathbf{s}),\eta(\mathbf{s} + \mathbf{h}))$ with $C_\eta(\infty) = 0$ (no nugget effect); $\varepsilon_t(\mathbf{s})$ is a white noise process composed of measurement errors; $E(\varepsilon_t(\mathbf{s})) = 0$ for all $\mathbf{s}$ and $t$; $\mathrm{Cov}(\varepsilon_t(\mathbf{s}),\varepsilon_u(\mathbf{s} + \mathbf{h})) = \sigma^2$, where $t$ and $u$ are different observations in the same datum location if $\mathbf{h} = \mathbf{0}$ and $t = u$ (otherwise, it is 0); and $Y(\bullet)$, $\eta(\bullet)$, and $\varepsilon(\bullet)$ are independent of each other. The formulas for situations when processes $Y(\bullet)$, $\eta(\bullet)$, and $\varepsilon(\bullet)$ are correlated are given by Gandin (1959), although it is unclear how to model such correlations. It must also be assumed that the nugget effect $\delta$ is composed of two parts— microscale variation and measurement error— such that $\delta = C_\eta(\mathbf{0}) + \sigma^2$. From this model, it can be deduced that

$$\mathrm{Cov}(Z_t(\mathbf{s}), Z_u(\mathbf{s} + \mathbf{h})) = \begin{cases} C_y(\mathbf{h}) + C_\eta(\mathbf{h}) & \text{if } \mathbf{h} \neq \mathbf{0} \text{ or } t \neq u \\ C_y(\mathbf{0}) + C_\eta(\mathbf{0}) + \sigma^2 & \text{if } \mathbf{h} = \mathbf{0} \text{ and } t = u \end{cases}$$

From these assumptions, different predictions and standard errors can be derived for exact prediction, filtered prediction, and new value prediction.

## Exact Prediction

Exact prediction is the usual form of ordinary kriging given in all geostatistical textbooks. It is obtained for the model $Z(\mathbf{s}) = \mu + Y(\mathbf{s}) + \eta(\mathbf{s})$, with $\sigma^2 = 0$. The quantity $Z(\mathbf{s}_0)$ at location $\mathbf{s}_0$ is estimated using the linear predictor $\hat{Z}(\mathbf{s}_0) = \lambda'\mathbf{z}$, minimizing $E(Z(\mathbf{s}_0) - \lambda'\mathbf{z})^2$, where $\mathbf{z}$ is a vector of the observed data, and $\lambda$ is a vector of the kriging weights. Coincident measurements are identical, and each new measurement is exactly the same as previous ones at the same location. An unbiasedness condition, $E(Z(\mathbf{s}_0) - \lambda'\mathbf{z}) = 0$, leads to the constraint $\lambda'\mathbf{1} = 1$. Then, the ordinary kriging equations are expressed as

$$\begin{pmatrix} \boldsymbol{\Sigma}_\mathbf{z} & \mathbf{1} \\ \mathbf{1}' & 0 \end{pmatrix} \begin{pmatrix} \boldsymbol{\lambda} \\ m \end{pmatrix} = \begin{pmatrix} \mathbf{c}_\mathbf{z} \\ 1 \end{pmatrix}$$

where $m$ is the Lagrange multiplier, $\boldsymbol{\Sigma}_\mathbf{z}$ is the empirical covariance matrix, and $\mathbf{c}_\mathbf{z} = \mathrm{Cov}(\mathbf{z}, Z(\mathbf{s}_0))$. The weights $\lambda$ for the ordinary kriging predictor are found by solving

$$\lambda = \Sigma_\mathbf{z}^{-1}(\mathbf{c}_\mathbf{z} - \mathbf{1}m) \tag{1}$$

where $m = (\mathbf{1}'\Sigma_\mathbf{z}^{-1}\mathbf{c}_\mathbf{z} - 1)/((\mathbf{1}'\Sigma_\mathbf{z}^{-1}\mathbf{1})$. Substituting to get the mean squared prediction errors,

$$\begin{aligned} E(Z(\mathbf{s}_0) - \lambda'\mathbf{z})^2 &= C_y(\mathbf{0}) + C_\eta(\mathbf{0}) + \lambda'\boldsymbol{\Sigma}_\mathbf{z}\lambda - 2\lambda'\mathbf{c}_\mathbf{z} \\ &= C_y(\mathbf{0}) + C_\eta(\mathbf{0}) - \lambda'\mathbf{c}_\mathbf{z} - m \end{aligned}$$

so that the prediction standard errors are given by

$$\hat{\sigma}_Z(\mathbf{s}_0) = \sqrt{C_y(\mathbf{0}) + C_\eta(\mathbf{0}) - \boldsymbol{\lambda}'\mathbf{c}_\mathbf{z} - m} \tag{2}$$

When $\mathbf{s}_0 = \mathbf{s}_i$ for $z(\mathbf{s}_i)$, the prediction $\hat{Z}(\mathbf{s}_0) = z(\mathbf{s}_i)$, and the prediction standard error is zero.

## Filtered Prediction

A situation with absolutely precise measurements is rather artificial; measurement error always exists. The sources of measurement uncertainties are not just errors of measurement device but also positional errors (uncertainty in data coordinates), censored data error, and error because of data integration (the so-called change of support problem), among others.

Filtered prediction, discussed in detail by Gandin (1959, 1963), is represented by the model $Z(\mathbf{s}) = \mu + Y(\mathbf{s}) + \eta(\mathbf{s}) + \varepsilon(\mathbf{s})$, with $\sigma^2 \neq 0$. In this case, the objective is to predict the filtered (noiseless) quantity $S(\mathbf{s}_0) = \mu + Y(\mathbf{s}_0) + \eta(\mathbf{s}_0)$ at location $\mathbf{s}_0$. If no measurement error occurs, $S(\mathbf{s}_0) = Z(\mathbf{s}_0)$. Otherwise, ordinary kriging in the presence of measurement error is employed using the linear predictor $\hat{S}(\mathbf{s}_0) = \lambda'\mathbf{z}$. The process minimizes $E(S(\mathbf{s}_0) - \lambda'\mathbf{z})^2$, where $\mathbf{z}$ is a vector of the observed data, and $\lambda$ is a vector of kriging weights. The unbiasedness condition, $E(S(\mathbf{s}_0) - \lambda'\mathbf{z}) = 0$, leads to the constraint $\lambda'\mathbf{1} = 1$. The kriging equations are obtained as follows:

$$\begin{pmatrix} \boldsymbol{\Sigma}_\mathbf{z} & \mathbf{1} \\ \mathbf{1}' & 0 \end{pmatrix} \begin{pmatrix} \boldsymbol{\lambda} \\ m \end{pmatrix} = \begin{pmatrix} \mathbf{c}_\mathbf{s} \\ 1 \end{pmatrix}$$

where $m$ is the Lagrange multiplier; $\boldsymbol{\Sigma}_\mathbf{z}$ is the empirical covariance matrix; and $\mathbf{c}_\mathbf{s} = \mathrm{Cov}(\mathbf{z}, S(\mathbf{s}_0)) = \mathrm{Cov}(\mathbf{z}, Y(\mathbf{s}_0) + \eta(\mathbf{s}_0))$. Because the range of $\eta(\bullet)$ is assumed to be close to 0, $\mathrm{Cov}(\mathbf{z}, \eta(\mathbf{s}_0)) = \mathbf{0}$ for all practical distances. An exception arises when $\mathbf{s}_0 = \mathbf{s}_i$, where $\mathbf{s}_i \in A$, $A$ being the set of spatial locations for observed data. In this case, $\mathrm{Cov}(Z(\mathbf{s}_i), \eta(\mathbf{s}_i)) = C_\eta(\mathbf{0})$, which must be estimated. The total nugget effect is composed of two parts, $\delta = C_\eta(\mathbf{0}) + \sigma^2$. With an independent estimate of $\sigma^2$, $C_\eta(\mathbf{0}) = \delta - \sigma^2$. This is equivalent to specifying a portion $\pi$ of the nugget effect to be measurement error and a portion to be microscale variation. For $0 \leq \pi \leq 1$, set $\sigma^2 = \pi\delta$ and $C_\eta(\mathbf{0}) = (1 - \pi)\delta$.

If multiple observations occur per location, then measurement error can be estimated as

$$\hat{\sigma}^2_{ME} = \frac{\sum_{\mathbf{s}_i \in D} \sum_{j=1}^{n_{\mathbf{s}_i}} (Z_j(\mathbf{s}_i) - \overline{Z}(\mathbf{s}_i))^2}{N - n_D}$$

where $D$ is the set of all data locations that have more than one measurement, $Z_j(\mathbf{s}_i)$ is the $j$th measurement at location $\mathbf{s}_i$, $\overline{Z}(\mathbf{s}_i)$ is the mean value at location $\mathbf{s}_i$, $n_i$ is the number of observations at location $\mathbf{s}_i \in D$, $N = \Sigma_i n_i$ for all $\mathbf{s}_i$ in $D$, and $n_D$ is the number of spatial locations in $D$.

To make a filtered prediction, $\sigma^2$ and $C_\eta(\mathbf{0})$ must be specified. If the entire nugget effect is microscale variation (no measurement error), then the solution to the kriging equations yields exact kriging. Otherwise, solving for $\lambda$ produces the kriging weights for the measurement errors model given by

$$\lambda = \Sigma_\mathbf{z}^{-1}(\mathbf{c}_\mathbf{s} - \mathbf{1}m) \tag{3}$$

where $m = (\mathbf{1}'\Sigma_{\mathbf{z}}^{-1}\mathbf{c}_{\mathbf{s}} - 1)/(\mathbf{1}'\Sigma_{\mathbf{z}}^{-1}\mathbf{1})$. Equation 3 can be compared to equation 1. Substituting to get the mean squared prediction errors,

$$E(S(\mathbf{s}_0) - \lambda'\mathbf{z})^2 = C_y(\mathbf{0}) + C_\eta(\mathbf{0}) - \lambda'\mathbf{c}_{\mathbf{s}} - m$$
$$= C_y(\mathbf{0}) + (1 - \pi)\delta - \lambda'\mathbf{c}_{\mathbf{s}} - m$$

so the prediction standard errors are

$$\hat{\sigma}_S(\mathbf{s}_0) = \sqrt{C_y(\mathbf{0}) + (1 - \pi)\delta - \boldsymbol{\lambda}'\mathbf{c}_{\mathbf{s}} - m} \qquad (4)$$

Equation 4 can be compared to equation 2.

## New Value Prediction

In the case of new value prediction, the estimation of a new value is obtained for the linear predictor $\hat{Z}_u(\mathbf{s}_0) = \lambda'\mathbf{z}$ by minimizing $E(Z_u(\mathbf{s}_0) - \lambda'\mathbf{z})^2$. We assume that if $\mathbf{s}_0 = \mathbf{s}_i \in D$, then $u > n_i$. Proceeding as before, the kriging equations are obtained using

$$\begin{pmatrix} \boldsymbol{\Sigma}_{\mathbf{z}} & \mathbf{1} \\ \mathbf{1}' & \mathbf{0} \end{pmatrix} \begin{pmatrix} \boldsymbol{\lambda} \\ m \end{pmatrix} = \begin{pmatrix} \mathbf{c}_{\mathbf{z}} \\ 1 \end{pmatrix}$$

where $m$ is the Lagrange multiplier, $\Sigma_{\mathbf{z}}$ is the empirical covariance matrix, and $\mathbf{c}_{\mathbf{z}} = \mathrm{Cov}(\mathbf{z}, Z_u(\mathbf{s}_0)) = \mathrm{Cov}(\mathbf{z}, Y(\mathbf{s}_0) + \eta(\mathbf{s}_0) + \varepsilon_u(\mathbf{s}_0))$. Solving for $\lambda$,

$$\lambda = \Sigma_{\mathbf{z}}^{-1}(\mathbf{c}_{\mathbf{z}} - \mathbf{1}m) \qquad (5)$$

where $m = (\mathbf{1}'\Sigma_{\mathbf{z}}^{-1}\mathbf{c}_{\mathbf{z}} - 1)/(\mathbf{1}'\Sigma_{\mathbf{z}}^{-1}\mathbf{1})$. Equation 5 can be compared to equations 1 and 3.

Note that when $\mathbf{s}_0 = \mathbf{s}_i \in D$, then the prediction $\hat{Z}_u(\mathbf{s}_i)$ is commonly not equal to one of the observed values $z_t(\mathbf{s}_i)$, where $t \le n_i$ and $u > n_i$. Instead, it is identical to the noiseless prediction $S(\mathbf{s}_0)$. Substituting to get the mean squared prediction errors,

$$E(Z_u(\mathbf{s}_0) - \lambda'\mathbf{z})^2 = C_y(\mathbf{0}) + C_\eta(\mathbf{0}) + \sigma^2 - \lambda'\mathbf{c}_{\mathbf{z}} - m$$
$$= C_y(\mathbf{0}) + \delta - \lambda'\mathbf{c}_{\mathbf{z}} - m$$

so the prediction standard error is given by

$$\hat{\sigma}_Z(\mathbf{s}_0) = \sqrt{C_y(\mathbf{0}) + \delta - \boldsymbol{\lambda}'\mathbf{c}_{\mathbf{z}} - m} \qquad (6)$$

which can be compared to equations 2 and 4. When $\mathbf{s}_0 = \mathbf{s}_i$ for one of the observed data locations $\mathbf{s}_i \in D$, then the prediction standard error will not be zero.

Cross-validation is a particularly important motivation for new value kriging prediction. Cross-validation is the process of removing each datum, one at a time, predicting the value of the removed datum and then comparing the measured and the predicted values. One such means of comparison is the root mean squared prediction error, which is the average of the squared differences between the measured and predicted values. For cross-validation, it is necessary to consider the prediction $Z_u(\mathbf{s}_0)$ with a measurement error instead of predicting the noiseless version of data $S(\mathbf{s}_0)$ in order for the prediction standard errors to reflect the root mean squared prediction error from cross-validation.

## EXAMPLES

Three examples are presented to further elucidate the theory described above. First, to compare the different methods, a series of displays like those presented in Chiles and Delfiner (1999) are shown in Figure 1. Predictions are to be made from two observed values in one dimension, located at positions $x_1 = 1$ and $x_2 = 2$. The observed data are $Z_1(x_1) = 1$ and $Z_1(x_2) = 1$, with $n_1 = n_2 = 1$. Prediction proceeds from 0 to 3 along the $x$-axis in increments of 0.1, using an exponential covariance model $C_y(h) = \theta_s \exp(-h/\theta_r) + \sigma^2 I(h = 0)$, where $\theta_s$ is the partial sill, $\theta_r$ is the range parameter, $\sigma^2$ is the nugget parameter, and $I(\bullet)$ is an indicator function.

For Figure 1a and b, $\theta_s = 1$, $\theta_r = 2$, and $\sigma^2 = 0$. No differences in predictions (Figure 1a) or prediction standard error (Figure 1b) exist when using exact kriging, filtered kriging, or new value kriging. For Figure 1c and d, $\theta_s = 0.5$, $\theta_r = 2$, and $\sigma^2 = 0.5$, where the nugget effect is assumed to be pure measurement error (or $\pi = 1$). In this case, there are no differences in predictions among the methods at locations where data have not been observed. However, at locations where data have been observed ($x_1 = 1$ and $x_2 = 2$), exact kriging results in a discontinuity for $Z(x_1)$ or $Z(x_2)$. However, new value kriging yields the same results as filtered kriging (Figure 1c). The prediction standard error is the same for exact kriging and new value kriging, except when predicting at a location where data have been observed (Figure 1d). In this case, new value kriging yields a smooth standard error surface, whereas exact kriging produces a discontinuity with a jump to zero. Filtered kriging produces uniformly lower standard error than exact kriging or new value

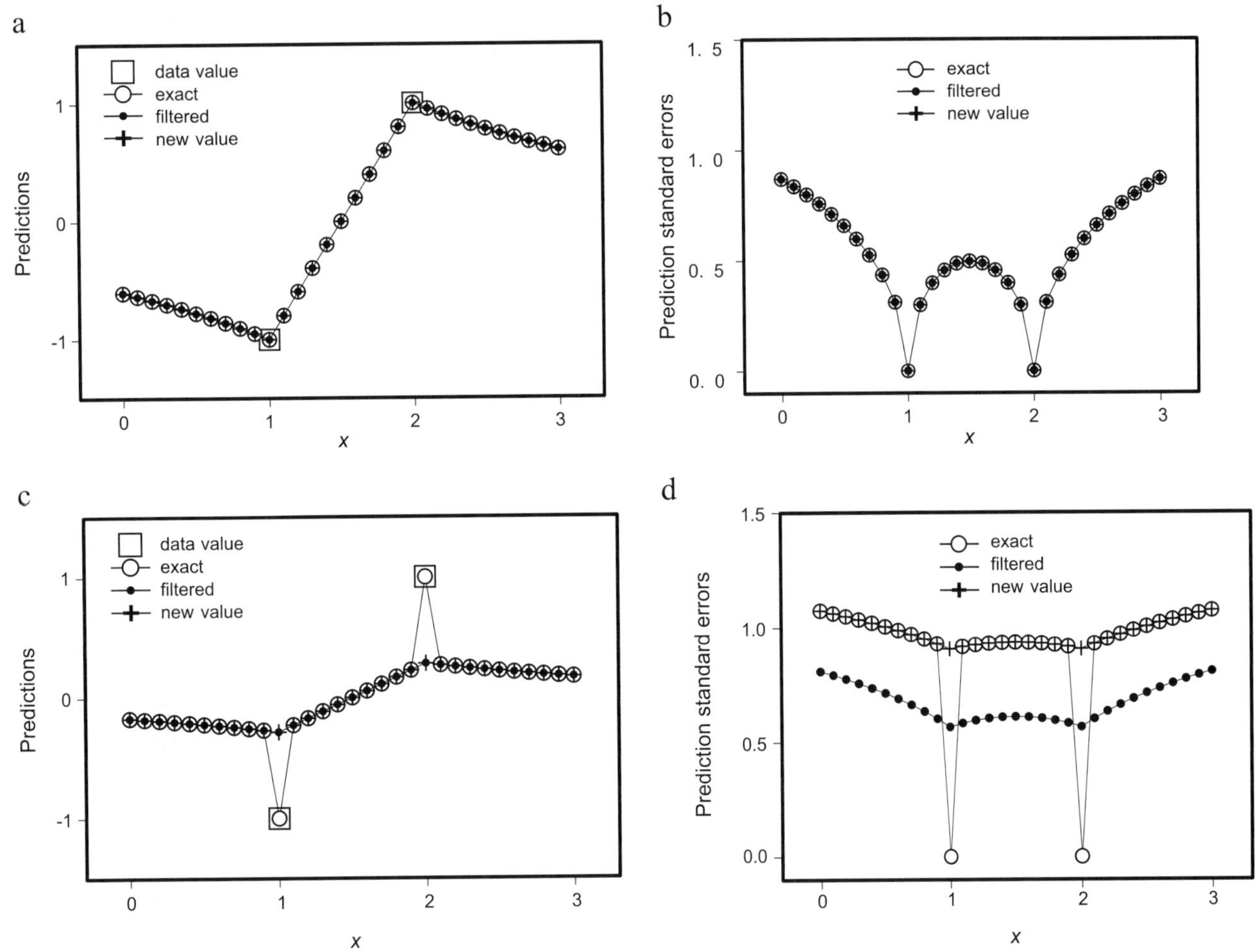


**Figure 1. Comparison of predictions and prediction standard errors for exact, smooth, and new value kriging: (a) predictions using an exponential model with no nugget effect, a partial sill of 1, and a range of 2; (b) prediction standard errors for the data and model shown in (a); (c) predictions using an exponential model with a nugget effect of 0.5, a partial sill of 0.5, and a range of 2; and (d) prediction standard errors for the data and model shown in (c).**

kriging by an amount equal to the measurement error part of the nugget effect, except at locations where data have been observed. At such locations, exact kriging yields a standard error of zero (Figure 1d).

The second example illustrates the difference between exact kriging, filtered kriging, and new value kriging using a semivariogram model that has a nugget effect for porosity data from the North Cowden unit in the Permian basin of west Texas (see a detailed description of the data in Chambers et al., 1994). Figure 2a shows a map created by exact kriging, applying a contouring algorithm to a regular grid of predicted values and to the locations where data are actually recorded. Because the data locations are included in the prediction set, the contouring algorithm causes small-scale bumps in the surface. Theoretically, the discontinuity occurs only at the data location, but the contouring algorithm interpolates from the nearest points on the grid. In comparison, no discontinuities are produced when using filtered or new value kriging, so the resulting map does not have the bumps (Figure 2b).

The last example involves an analysis of soil contamination data. The data set consists of radiocesium ($^{137}Cs$) measurements obtained in 1992 in Belarus, 6 yr after the Chernobyl accident (Krivoruchko, 1997) (see Figure 3). According to regulations, the radiation dose in unrestricted areas is limited to no more than 1 milliSieverts (mSv)/yr. It has been suggested that a linear correlation exists between soil contamination and human dose loads. According to Konoplya and Rolevich (1996), permanent living in a territory with $^{137}Cs$ soil contamination of 15 curie (Ci)/km$^2$ is equivalent to receiving an annual dose

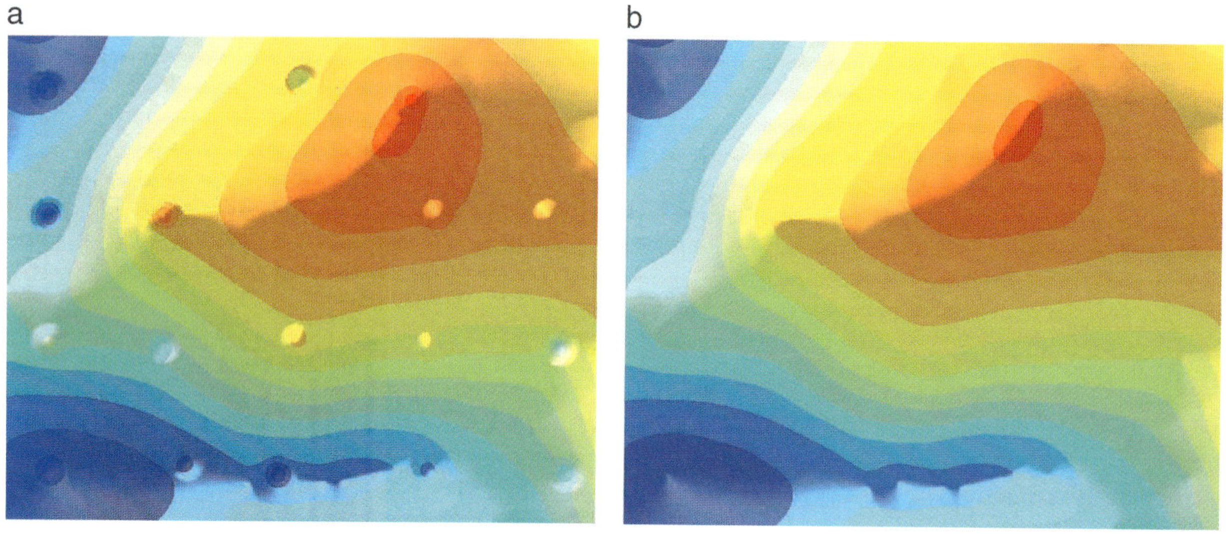


**Figure 2.** A contour map of estimated porosity in the North Cowden unit based on ordinary kriging: (a) exact kriging; (b) filtered and new value kriging.

of 5 mSv. People living in such territories should be evacuated, and countermeasures must be provided.

Figure 4a presents $^{137}Cs$ soil contamination data in curie per square kilometer in some of the settlements in the northeastern part of the Gomel province, Belarus (see Figure 3). Isolines for 10, 15, 20, and 25 Ci/km$^2$ are shown. The prediction map has been created using ordinary kriging with the J-Bessel semivariogram (Chiles and Delfiner, 1999) shown in Figure 4b. Two adjacent measurement locations separated by 1 mi (1.6 km) are circled. These locations have radiocesium values of 14.56 and 15.45 Ci/km$^2$, respectively. It is known that the error of $^{137}Cs$ soil measurements is about 15–30%. Hence, measurements at the circled locations are approximately the same, and it would be difficult to conclude that either location is safe or unsafe. Using the 15 Ci/km$^2$ value as a safety threshold, only the location with 14.56 Ci/km$^2$ can be considered safe. Because the data are subject to measurement error, one possible solution to improve

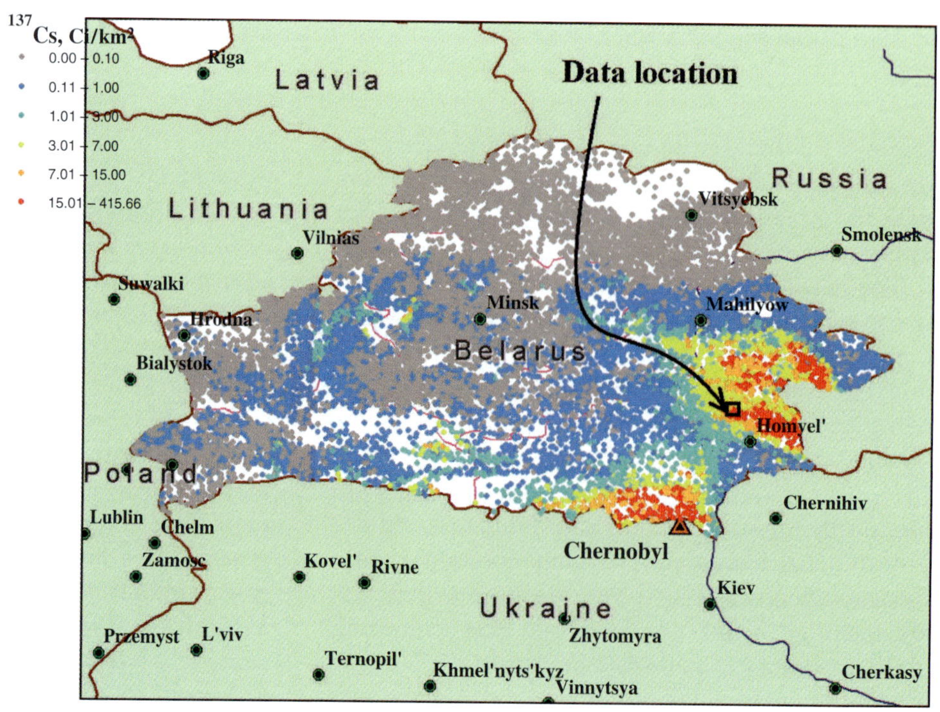


**Figure 3.** A map of Belarus indicating the location of Chernobyl. Symbols show measurements of $^{137}Cs$ soil contamination in 1992. The area from which measurements were taken to create the map in Figure 4 is highlighted.

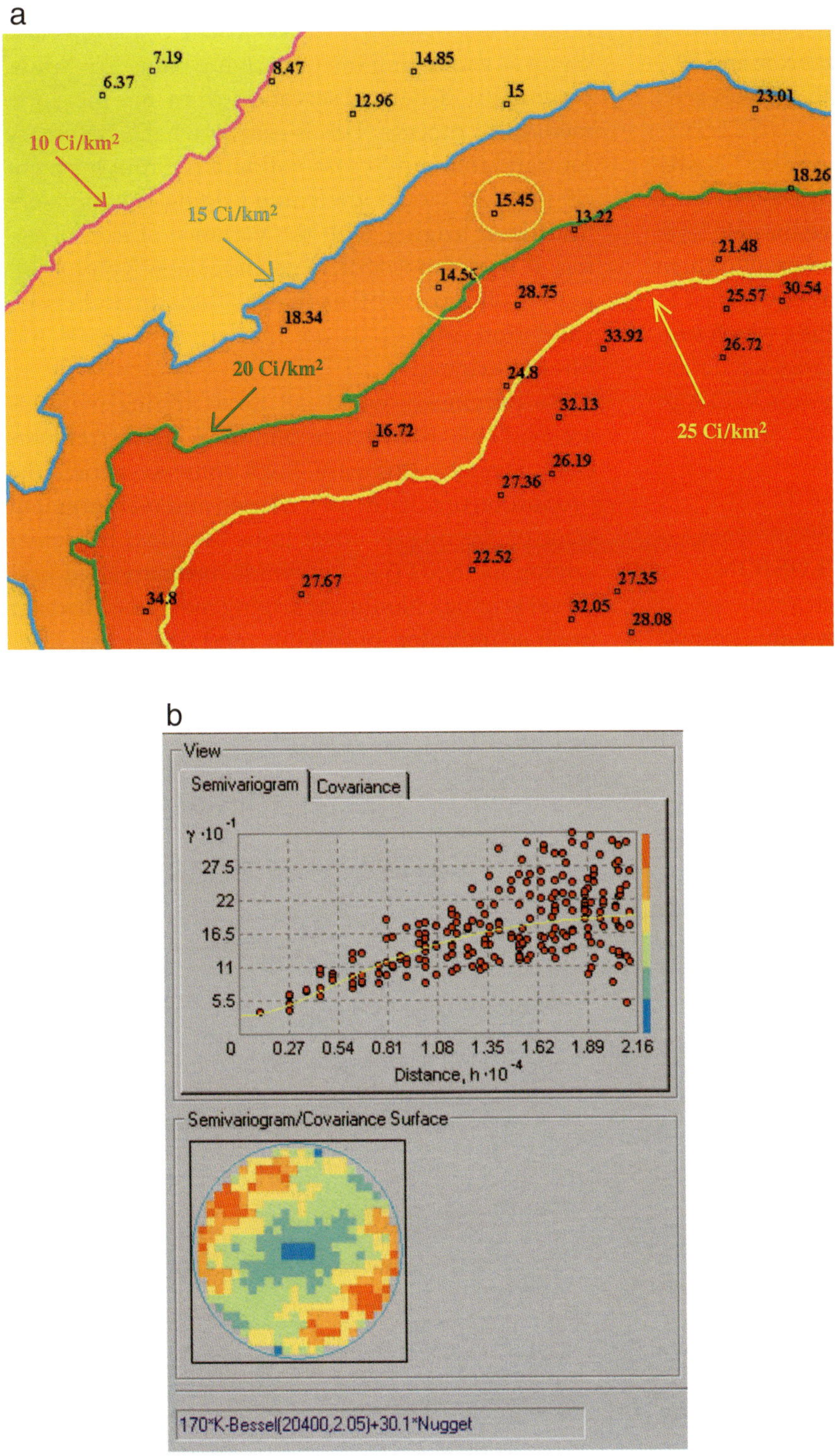


Figure 4. $^{137}$Cs soil contamination in the northeastern part of Gomel province, Belarus. (a) Isolines of 15, 20, and 25 Ci/km$^2$. Circles highlight two selected measurement points with the soil values of $^{137}$Cs given as 14.56 and 15.45 Ci/km$^2$. (b) A fitted semivariogram model and a semivariogram and covariance surface plot as captured in a screen shot produced with commercial software.

decisions about soil contamination is to use filtered kriging. Notice that the nugget is estimated to be about 30, and the square root of 30 is about 5.5, which is approximately 30% of the observed values of about 15 and about 15% of the observed values of about 30. With measurement error between 15 and 30% of recorded values, the measurement error proportion of the nugget effect is between 50 and 100%.

Taking the nugget effect as 100% measurement error, the location with the observed value 14.56 Ci/km$^2$ is predicted to be 19.53 Ci/km$^2$, and the location with

15.45 Ci/km$^2$ is predicted to be 18.31 Ci/km$^2$. Taking the nugget effect as 50% measurement error and 50% microscale variation, the location with the observed value of 14.56 Ci/km$^2$ is predicted to be 17.05 Ci/km$^2$, and the location with value of 15.45 Ci/km$^2$ is predicted to be 16.88 Ci/km$^2$. From a decision-making point of view, both locations are above the safe level of 15 Ci/km$^2$, and people living in and around them should probably be evacuated. This is in contrast to decision making based on the raw values, which suggests that one of the locations is safe. Cross-validation using new value kriging prediction yields values of 20.3 and 18.95 Ci/km$^2$, respectively.

## DISCUSSION AND CONCLUSION

Comparing exact, filtered, and new value kriging requires some discussion of the nugget effect, which is really more of a mathematical construction than an actual physical feature. In a practical sense, it gives the kriging equations stability. Without a nugget effect, inverting the kriging matrices may lead to computational round-off error. The nugget effect also ensures robustness. The standard error of prediction depends on the behavior of the semivariogram at the origin, and kriging practitioners know that without a nugget effect, the prediction standard error is commonly underestimated.

As noted above, there can be two components of the nugget effect; measurement error and microscale variation. In the case of microstructure variation, errors are correlated at small distances. For physical reasons, this distance is small but not zero. Processes characterized by microscale variation can be modeled using two semivariogram models, one of them having a small range (e.g., see Journel and Huijbregts, 1978). Microscale variation is spatial autocorrelation at a scale smaller than the closest data locations. Measurement error is variability caused by uncertainties in datum collection, including imprecise measurement device, local integration, coordinates uncertainties, and rounding off.

New value kriging assumes that when the nugget effect includes measurement error, the smooth underlying process $Y(\bullet)$ is unchanged, and potentially, many observations are possible for each location. Measurement error associated with the nugget effect is equivalent to the variation among multiple potential observations at a specific location. Prediction at a location $\mathbf{s}_0$ (where data have been collected) involves estimating the underlying value $Y(\mathbf{s}_0)$ plus, potentially, a new observation from the measurement error process. This is exactly what is necessary for validation and cross-validation. From a practical point of view, new value kriging is equivalent to filtered kriging with the variance of exact kriging. It is also possible to consider new value kriging for situations other than validation and cross-validation, that is, as an alternative to exact and filtered kriging.

As previously noted, all three kriging methods are available as functions and modules in commercially available software. Such software allows the proportion of microscale variation and measurement error of the nugget effect to be specified for linear kriging methods (i.e., for simple, ordinary, and universal kriging). Nonlinear kriging methods, however (such as indicator, probability, and disjunctive kriging), are exact interpolators because they require zero measurement error by definition, a situation that must be considered before selecting a kriging methodology.

## REFERENCES CITED

Chambers, R., M. Zinger, and M. Kelly, 1994, Constraining geostatistical reservoir descriptions with 3-D seismic data to reduce uncertainty, *in* J. Yarus and R. Chambers, eds., Stochastic modeling and geostatistics: Principles, methods, and case studies: AAPG Computer Applications in Geology 3, p. 143–157.

Chiles, J., and P. Delfiner, 1999, Geostatistics: Modeling spatial uncertainty: New York, John Wiley & Sons, 695 p.

Fuller, W. A., 1987, Measurement error models: New York, John Wiley & Sons, 440 p.

Gandin, L. S., 1959, The problem on optimal interpolation: Trudy Glavnaya geofizidneskaya observatoria, v. 99, p. 67–75.

Gandin, L. S., 1963, Objective analysis of meteorological fields: Leningrad, Gidrometeorologicheskoe Izdatel'stvo, 286 p. (translated by Israel Program for Scientific Translations, Jerusalem, 1965, 242 p.)

Johnston, K., J. Ver Hoef, K. Krivoruchko, and N. Lucas, 2001, Using ArcGIS geostatistical analyst: GIS by ESRI: Environmental Systems Research Institute, 300 p.

Journel, A. G., and C. J. Huijbregts, 1978, Mining geostatistics: London, Academic Press, 600 p.

Konoplya, E. F., and I. V. Rolevich, eds., 1996, The Chernobyl catastrophe consequences and their overcoming in the Republic of Belarus: Ministry for Emergencies and Population Protection from the Chernobyl NPP Catastrophe Consequences, Academy of Sciences of Belarus, 88 p.

Krivoruchko, K., 1997, Geostatistical picturing of Chernobyl fallout and estimation of cancer risk among Belarus population, *in* S. Hodgson and M. Rumor, eds., Geographical information '97: Third Joint European Conference and Exhibition on Geographical Information: Vienna, IOS Press, p. 676–685.

Krivoruchko, K., A. Gribov, I. Figurin, S. Karebo, D. Pavlushko, D. Remesov, and A. Zhigimont, 1997, MapStudio: The specialized GIS integrating possibilities of geostatistics, *in* S. Hodgson and M. Rumor, eds., Geographical information '97: Third Joint European Conference and Exhibition on Geographical Information: Vienna, IOS Press, p. 187–196.

Maignan, M., and K. Krivoruchko, 1997, Integration of GIS and geostatistics: A software and a case study, *in* L. Ottoson, ed., Proceedings of the 18th ICA/ACI International Cartographic Conference: Stockholm, International Cartographic Association, p. 925–932.

Caers, J., B. G. Arpat, and C. A. Garcia, 2006, Feature-based probabilistic interpretation of geobodies from seismic amplitudes, *in* T. C. Coburn, J. M. Yarus, and R. L. Chambers, eds., Stochastic modeling and geostatistics: Principles, methods, and case studies, volume II: AAPG Computer Applications in Geology 5, p. 91–107.

*Chapter 8*

# Feature-based Probabilistic Interpretation of Geobodies from Seismic Amplitudes

**J. Caers**
**B. G. Arpat**[1]
**C. A. Garcia**[2]
*Stanford University*
*Stanford, California, U.S.A.*

## ABSTRACT

Manual interpretation of large volumes of seismic data is a tedious and time-consuming process. Seismic interpretation relies on extensive expert knowledge of geological rules, as well as strong geophysical interpretation skills. Moreover, in modern approaches to reservoir characterization, a single, deterministic reservoir model interpreted from seismic information is commonly of less interest than the multiple potential interpretations that can arise from the data. Hence, an understanding of the uncertainty associated with seismic interpretation and its quantification are extremely important to reservoir management and decision analysis. In this chapter, several new tools are presented, most of them based on statistical pattern recognition, that can aid the interpreter in constructing a seismic-based reservoir model and provide some uncertainty quantification. Two groups of methods based on unsupervised and supervised pattern detection are discussed. A new geostatistical approach termed feature-based geostatistics is introduced, the aim of which is to accurately reproduce facies shapes. All methods are validated using an interpreted seismic data set representing channel facies from a turbidite sequence in Gabon, west Africa.

## INTRODUCTION

Seismic data bring an important constraint to geological variability in geostatistical reservoir characterization. Manipulation of seismic and rock physics data requires skills in both seismic data processing and interpretation. Such interpretation is a tedious, time-consuming task and is based on certain geological and geophysical rules emanating from years of experience. However, the challenge is not to replace traditional seismic interpretation by an automatic tool, but to design automated devices, requiring little

[1]*Present address*: Earth Decision Sciences, Houston, Texas, U.S.A.
[2]*Present address*: Shell Oil Company, New Orleans, Louisiana, U.S.A.


DOI:10.1306/1063809CA53032

expertise in their use, that can aid and complement manual expert interpretations.

A major advantage of automated seismic interpretation is that it allows uncertainty to be quantified, a task that is essentially impossible to accomplish with manual interpretation. Having several experts repeat the same interpretation would allow uncertainty quantification, but this is simply impractical and too costly. In addition, expert interpretations of the same reservoir are commonly inconsistent, which means that a single seismic feature could be manually interpreted as different geological events. Moreover, the seismic data itself leave a large amount of ambiguity about facies types. Local variation in fluid saturation, layer geometries, and large variabilities in petrophysical properties can result in large ambiguities or uncertainties about the presence of geobodies. However, uncertainty quantification with an automated device is possible because learning from expert knowledge is a typical exercise in statistical artificial intelligence.

The first part of this chapter deals mainly with the seismic data itself. A tool is presented that can automatically recognize distinct patterns from seismic amplitudes. A local three-dimensional (3-D) window is defined within which the seismic data are sampled. Each sample represents a local seismic feature. This large data set of seismic features is then clustered into seismic clusters using an unsupervised clustering technique. Next, the clusters are calibrated to facies observations obtained from wells. The methodology is virtually entirely automatic: the clustering technique does not require the number of clusters to be known a priori, yet human experts can intervene and decide on which seismic clusters to retain for further interpretation.

The second part of the chapter concentrates on both seismic and facies geometries. A feature-based probabilistic approach to classifying facies geometries from seismic data is introduced, based on partial expert interpretations. The expert manually interprets a small volume of seismic data by relating facies features to seismic patterns. A neural network structure inspired by the human vision system attempts to mimic this process. Once the neural net is trained, it can be applied to parts of the reservoir that have not yet been interpreted to obtain fast interpretations of geobodies from seismic data without relying on time-consuming expert work. A crucial part of the automated methodology is the reproduction of channel shape morphologies. For this purpose, a new concept in geostatistics termed feature-based geostatistics is introduced.

Both approaches are demonstrated on a seismic-amplitude data set from a turbidite sequence in Gabon, west Africa.

## UNSUPERVISED PATTERN RECOGNITION

Typical to seismic interpretation is the qualitative nature and art of detecting geological patterns in the seismic data. In this section, a quantitative interpretation of seismic data is developed. A statistical approach is proposed that can automatically recognize seismic patterns from amplitude data and then relate them to facies variations. The methodology can be used in a stand-alone form or can guide the geologist and geophysicist in the interpretation.

### Methodology

There are two general approaches to this pattern recognition problem: supervised and unsupervised. The supervised approach uses well data and seismic data to directly build a relation between facies and lithology and seismic data. In the traditional colocated approach in geostatistics (e.g., Deutsch and Journel, 1998), this amounts to modeling the conditional probability of a facies indicator,

$$I(\mathbf{u}, s_k) = \begin{cases} 1 & \text{if facies } s_k \text{ occurs at location } \mathbf{u} \\ 0 & \text{else} \end{cases}$$

with $k = 1, \ldots, K$, where $s_k$ are the facies categories and $k$ is the number of facies, given the colocated seismic datum $s(\mathbf{u})$. This conditional probability is mathematically denoted as $\Pr\{I(\mathbf{u}, s_k) = 1 \mid s(\mathbf{u})\}$. In the unsupervised approach, one attempts to first detect certain patterns or features in the seismic data without relying on any well logs for guidance.

To capture such patterns from seismic data, a window approach (originally described by Payrazyan, 1998; see also Caers and Ma, 2002) is proposed. The window $W(\mathbf{u})$ is formed by a series of neighbor locations that include a central location $\mathbf{u}$. It captures information about the local spatial pattern of seismic amplitudes and is termed a seismic feature. In the window approach, one aims at modeling the conditional probability of all facies categories given a window of seismic data $s(\mathbf{u})$. Figure 1 provides an example of such a window that is strictly vertical; yet any 3-D window geometry can be used.

The approach taken here is an unsupervised one consisting of two steps: an unsupervised clustering of seismic features in distinct clusters (this step does not involve any well data) and calibration of the clusters against well data to determine the conditional probabilities. These steps are described more rigorously as follows.

### *Step 1*

Using a predefined template or window, the seismic-amplitude data $s(\mathbf{u})$, $\mathbf{u} \in A$, where $A$ is the study area, are scanned to obtain realizations of the random vector $S(\mathbf{u}) = \{S(\mathbf{u}), S(\mathbf{u}+\mathbf{h}_1), \ldots, S(\mathbf{u}+\mathbf{h}_{n_t})\}$, where $\{\mathbf{h}_\beta, \beta = 1, \ldots, n_t\}$ is a vector defining the template geometry. Using an unsupervised clustering method, the samples $s(\mathbf{u}) = \{s(\mathbf{u}), s(\mathbf{u}+\mathbf{h}_1), \ldots, s(\mathbf{u}+\mathbf{h}_{n_t})\}$, $\mathbf{u} \in A$, obtained by scanning the seismic data are clustered into an unknown number of clusters $t_m$, $m = 1, \ldots, M$. $M$ itself is an unknown quantity. Once each seismic feature is clustered into a specific cluster, the cluster label $m$ is assigned to the original location $\mathbf{u}$ of the seismic feature $s(\mathbf{u})$. The result of such clustering is a 3-D cube of cluster labels, mathematically denoted as

$$J(\mathbf{u}, t_m) = \begin{cases} 1 & \text{if feature } s(\mathbf{u}) \text{ belongs to cluster } t_m \\ 0 & \text{else} \end{cases}$$

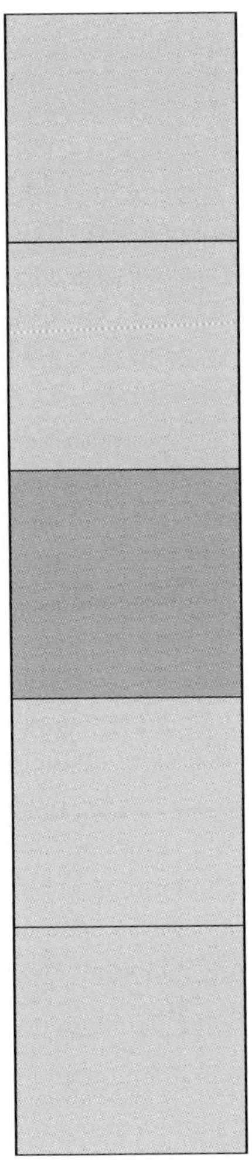

**Figure 1. A simple vertical template.**

### *Step 2*

The interpreted sequence of facies from well logs is tied to the 3-D cube of cluster labels; hence, a frequency table of facies occurrences per cluster label can be generated. The entries in this frequency table are the desired facies probabilities given seismic data.

There are three main advantages to this two-step approach:

1) Seismic features are clustered without being influenced by well data.
2) Some clusters of seismic features have a direct geological significance, whereas others may be attributed to recording artifacts or noise. Therefore, the clustering may serve as a filter of the seismic information. Experts can use these classes as a tool to guide further manual interpretation.
3) The clustering method proposed does not rely on a priori knowledge of $M$, the number of classes. $M$ is determined from the data itself.

## Unsupervised Clustering

Clustering aims at grouping a set of observed seismic features into separate clusters. Each feature is described by a set of $n_t + 1$ amplitudes. In the case of supervised clustering, each seismic feature would have a cluster label already attached; hence, the aim of supervised clustering is to find a classification system or model that is able to classify future seismic features that do not have class labels. In such a case, the set of seismic features' measurements and labels serves as a training set of information from which a classification device can learn on. In statistical terms, a classification system is a regression model because it builds a relationship (most likely nonlinear) between amplitudes in a window and a cluster label. In the application phase, amplitudes of new features are provided, and the regression model is used to determine the missing cluster labels. Supervised classification is well known, and many techniques have been developed (see Ripley, 1996, for a general overview).

The problem becomes much more complex in unsupervised classification, where cluster labels are not available. First, the feature database now has to be mined to find clusters in the data. Second, the number of classes is not known a priori and needs to be determined from the database itself. This mining process is not likely to be trivial and will actually depend on certain criteria or strategies by which the search process is conducted. Several existing unsupervised clustering methods (such as the widely popular $k$ means and mixture distribution models; see Ripley, 1996) rely on the knowledge of the existing number of clusters. However, in many real cases (such as the geobody detection problem), the number of existing clusters is unknown a priori, and any wrong assumption might lead to incorrect geobody determination.

An unsupervised clustering routine is used here that does not rely on a priori knowledge of the number of classes or clusters. As mentioned above, some measure of similarity between measurements needs to be defined to produce clusters that contain similar seismic features. The so-called minimum message length (MML) clustering method (Wallace and Boulton, 1986; Wallace and Dowe, 1994) relies on principles of information theory to measure a goodness of a clustering, which is found to be well suited for the geobody detection problem.

Clustering is essentially a dimensionality reduction problem, reducing the number of data $(n_t + 1) \times N$, with $N$ the number of features, into a set of discrete classes $M$, $M << (n_t + 1) \times N$. In the geobody detection problem, one essentially reduces the dimension of order $10 \times 10^6$ $((n_t + 1) \times N)$ into one of dimension 100 ($M$). This allows the seismic interpreter to make more efficient use of large data sets.

## Methodology of the MML Clustering

The MML methodology relies on the well-known Shannon principle for message coding. To optimally code (in terms of smallest number of bytes) a simple message, such as a sentence, one should attach the shortest code to the most frequent words in a sentence. A similar principle can be envisioned in clustering seismic features: if the entire database of seismic features sampled from a seismic cube is considered as a message, then to present that message in its shortest form, one should attach the shortest (cluster) name to the cluster of the most commonly appearing features. At the same time, the variability or spread of features in a single cluster should not be too large because it would require a cost to code (explain) that variability. Minimum message length therefore attempts to transfer the entire database of seismic features into a new database containing the following elements:

1) the cluster label to which each seismic feature belongs
2) some summary of the average properties of each cluster (mean, standard deviation)
3) the deviation of each measurement from the average properties

Each of these elements is bit-coded and can be attributed to a cost calculated in bytes. Using an optimization method, MML searches for the optimal clustering of features such that the number of bytes used to represent this new database is as small as possible. To understand this process better, consider two extremes: (1) all seismic features are clustered into one single cluster, and (2) there are as many clusters as seismic features. In the first case, the cost of labeling clusters is small, but the spread in the single cluster is extremely large, hence requiring a large coding cost; in the second case, the variability in each cluster is zero, but the cost, in terms of bytes, to represent each cluster label is large. Minimum message length searches for a balance between these two extremes.

The application of information-based clustering to the problem of geobody detection is motivated as follows: if a clear geological body exists, then the seismic response is likely to be similar over that entire geobody. Hence, local seismic-amplitude measurements will tend to group into a single cluster. This means that the entire cluster of seismic-amplitude information can be largely compressed into information provided by the cluster label and the average properties of amplitude features in a cluster. The aim of MML clustering is to represent the amount of information into the briefest possible representation of all amplitude measurements.

## Application

Seismic data obtained from the lower Miocene age Baliste–Crecerelle canyon fill deposits off the African coast in Gabon are used to illustrate the proposed interpretation tool. The complete canyon fill deposits extend for 120 km (75 mi) basinward from the paleoshelf edge with a width of as much as 10 km (6 mi) and a thickness of 500 m (1600 ft). Wonham et al. (2000) investigated the sedimentary architecture of the canyon fill using high-resolution seismic stratigraphy, seismic attribute analysis, and borehole data calibrated to seismic data. The gOcad software (Mallet, 2002) has been used to integrate and interpret these data to develop a model of stratigraphic architecture from the scale of the canyon (4 km; 2.5 mi) to the scale of individual channels (200 m; 660 ft).

Individual channels are located through body-check picks of high-amplitude reflectors. More than 100 individual channel morphologies have been manually created and discretized on a regular stratigraphic grid. A large seismic cube (more than 40 million cells) and corresponding interpreted channel facies interpretations are available.

The clustering approach is applied to the entire seismic data cube to test its capabilities against the expert interpretation. Only results of selected but representative channel bodies are shown. Figure 2 shows the expert interpretation: a single channel is apparent, running from east to west. The seismic data are not shown.

The five-pixel vertical template in Figure 1 is used to scan the seismic-amplitude data set. Using this window, a set of seismic features $\mathbf{s}(\mathbf{u})$ is compiled by scanning the entire 3-D cube. The MML algorithm clusters these features into 27 clusters. Recall that the number 27 is unknown a priori and obtained as a result of the MML clustering. The cluster label of each feature $\mathbf{s}(\mathbf{u})$ is mapped at the location $\mathbf{u}$, generating the cluster label cube shown in Figure 3. A horizontal body can be distinguished in Figure 3b and f corresponding mainly to cluster label 26. This feature is present in most horizontal slices although represented by different cluster labels. In fact, in Figure 3a, c, and e, cluster labels 3–5 correspond to the same channel feature. Comparing the results to the interpreted reservoir horizontal slices in Figure 2, the geobody obtained from clustering corresponds well to the interpretation model. Next, the cluster labels

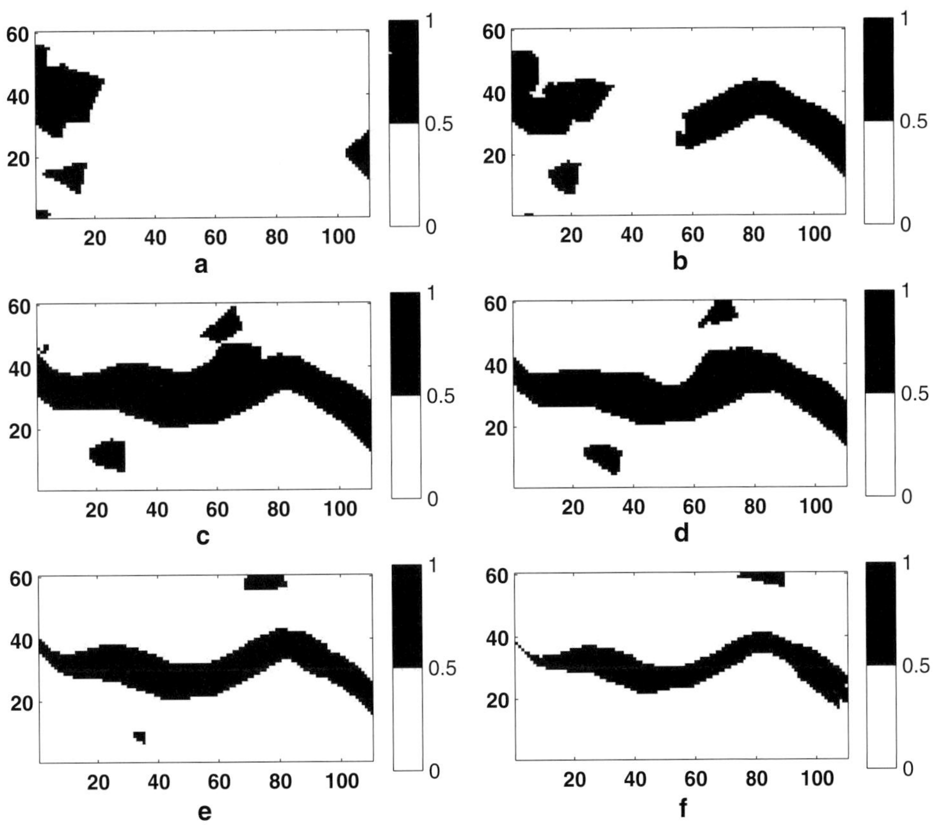


**Figure 2. Horizontal slices of a selected region of the expert seismic interpretation. (a) is the top slice, (f) is the bottom slice.**

can be calibrated with well data (step 2 in the algorithm) to obtain a probability of channel facies for each grid cell in the reservoir (results not shown).

A second example is shown in Figure 4, showing two slices from a 3-D manually interpreted seismic cube next to the results of the MML clustering. In this case, cluster numbers 3–5 appear to be indicating a channel.

## FEATURE-BASED INTERPRETATION OF SEISMIC DATA

### Motivation

In the previous section, a pixel-based method for detecting patterns from seismic data is presented. Although the method provides satisfactory results locating channel objects, the shape of these channels is poorly reproduced. For example, compare the object-based manual interpretation with the pixel-based classification in Figure 4.

In this section, a novel neural network approach is combined with a feature-based geostatistical-simulation method to overcome the poor-shape reproduction. The proposed method attempts to establish an explicit relationship between seismic patterns and facies geometries instead of pixels. The automated interpretation methodology consists of two phases: a learning phase and a reproduction and generalization phase. In the learning phase, the aim is to understand the relationship between facies geometry and local seismic patterns. In the reproduction phase, the trained neural network is applied to uninterpreted areas of the seismic cube to determine channel probability patterns from seismic data. Finally, a feature-based geostatistical-simulation technique conditioned to the channel probability patterns is used to reproduce channel shapes accurately.

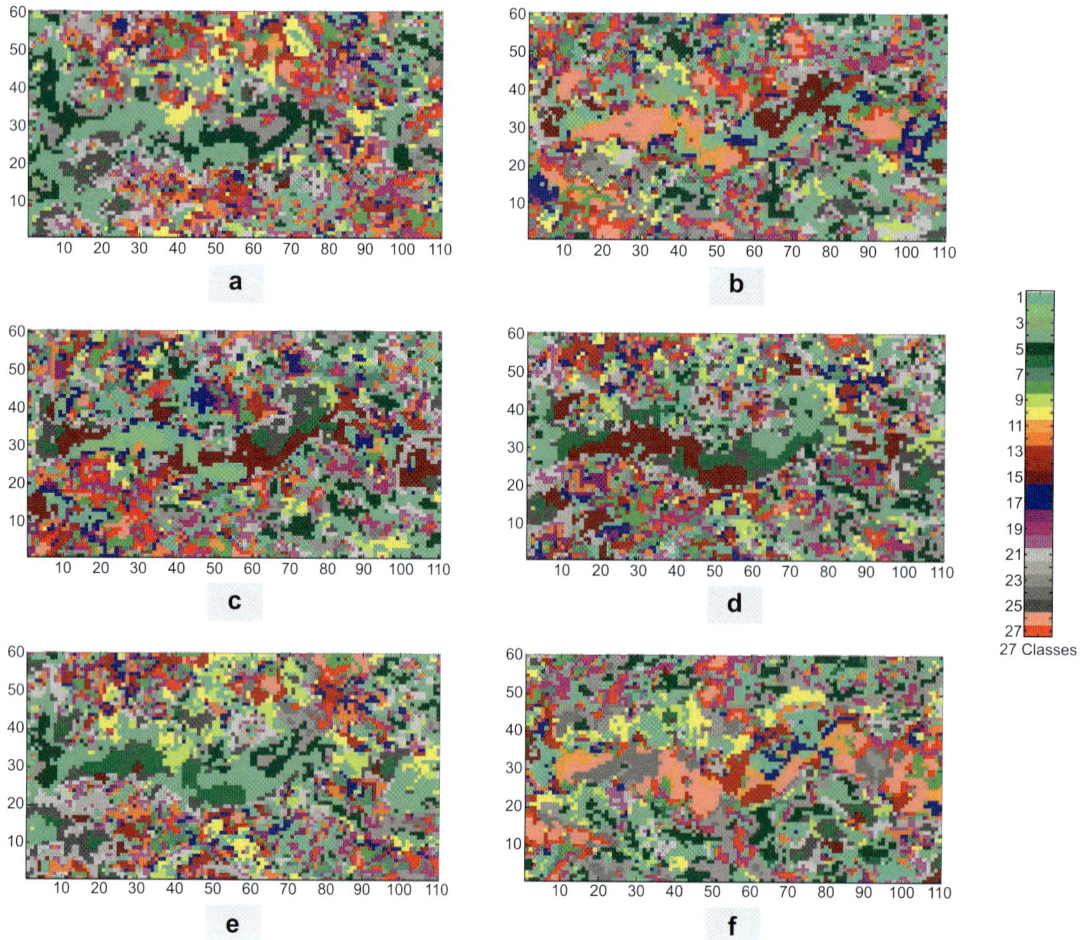


Figure 3. Cluster labels from the MML clustering (label values have no physical meaning).

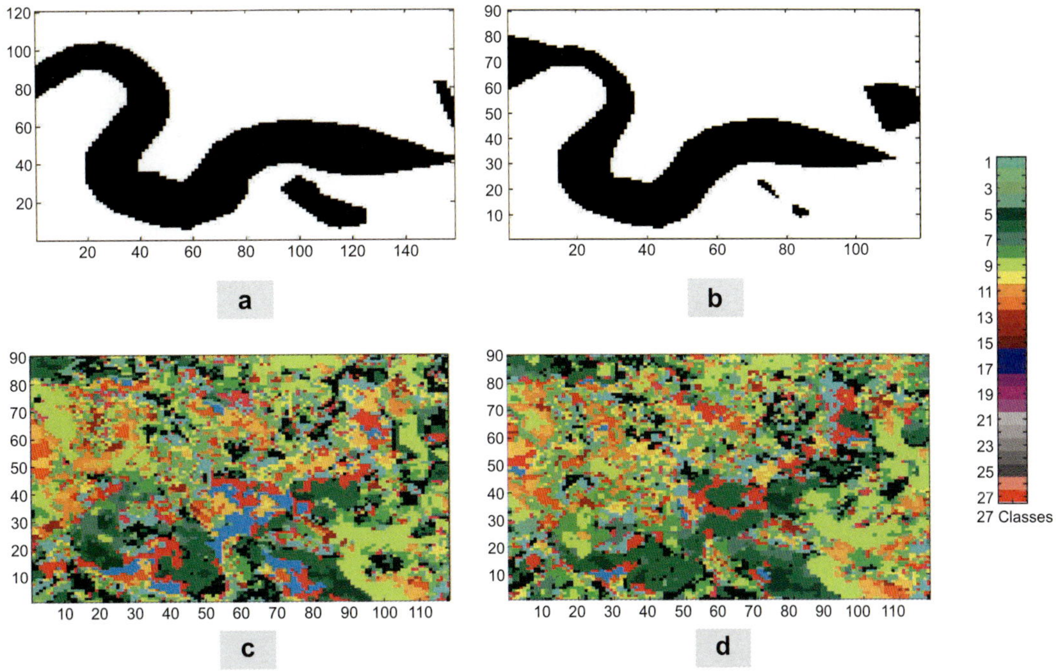


**Figure 4. (a, b) Expert interpretation; (c, d) cluster labels from the MML clustering.**

To model the seismic-to-facies relationship during the learning phase, a training data set is required that consists of a 3-D cube of seismic data and the corresponding expert interpretation. Such data may be obtained from partial seismic interpretations or from partial interpretations near the wells based on well logs.

In geostatistics, the traditional approach is to build the reservoir one pixel at a time (pixel-based geostatistics) or one object at a time (object-based modeling). However, the human expert is not likely to build a channel or a shale body one pixel at a time or one body at a time. Experts look for patterns in the seismic data and relate them to geobody features.

In this terminology, a piece of a large facies body is termed a facies feature. For example, an oxbow can be a feature of a larger channel complex. One can think of features as the pieces of a jigsaw puzzle and feature-based simulation as an attempt to solve this puzzle consistent with the given channel probabilities.

## IMPLEMENTATION OF THE NEURAL NETWORK

The neural network accepts a seismic window as its input. The information in seismic data is mainly coded in the vertical direction. Thus, an input window with a much higher aspect ratio in the vertical direction is chosen. The output of the neural network is a pattern of channel probabilities. Typical dimensions for these windows are $5 \times 5 \times 9$ for the input (seismic data) and $3 \times 3 \times 1$ for the output (facies), as shown in Figure 5. The seismic window is 3-D and has a large vertical component, whereas the channel is flat, although in general, any type of window can be selected.

The connections of the neural network loosely follow the human eye neural wiring architecture (Ganong, 1997). Traditionally, back-propagation networks are designed to be fully connected with one or two hidden layers regardless of the task at hand (Haykin, 1999). By borrowing concepts from the human eye, architecture specific for this application is favored over a more traditional, fully connected architecture. The two main concepts adopted from vision science are overlapping receptive fields and hierarchical processing.

In neural network jargon, the term "receptive field" refers to the area of an image or scene that is processed by a single neuron (Palmer, 1999). For the seismic interpretation, the receptive field coincides with a seismic window. The hidden neuron receives input only from its corresponding neighborhood and not from all available input nodes. In other words, the neuron is partially connected to the input layer

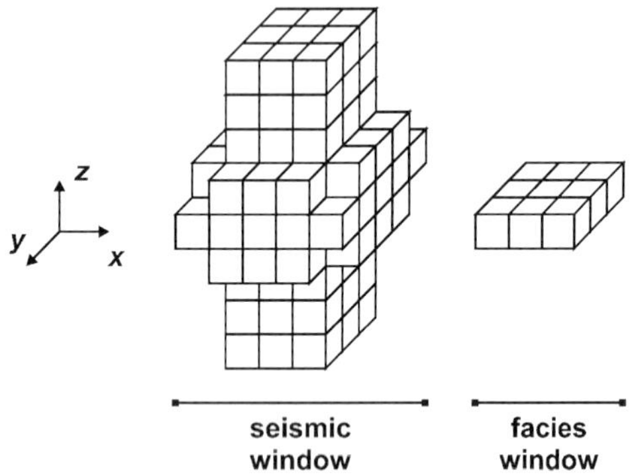


**Figure 5.** **(Left) The template used to scan the seismic cube, and (right) the template used to scan the expert interpretation (channels). The information obtained from seismic data is mainly coded in the vertical direction. Thus, an input window with a much higher aspect ratio in the vertical direction is chosen.**

(see Figure 6a). The human brain introduces redundancy to this structure through neurons with partially overlapping receptive fields. In case a neuron fails, combining information from other neurons can still retrieve any lost information.

Another concept borrowed from hierarchical processing is a cascading set of layers of hidden neurons with an overlapping receptive field principle applied to each layer (Palmer, 1999). Consider the example in Figure 6b. Here, the hidden neuron connects to a total of four neurons in the previous layer (its immediate receptive field), and each of those neurons connect to their respective overlapping $3 \times 3$ receptive fields in the input layer. The result of this structure is an effective receptive field of $5 \times 5$ for the hidden neuron of interest. This $5 \times 5$ area is said to be processed hierarchically by the four neurons in the first hidden layer and the hidden neuron in the second hidden layer.

The combination of overlapping receptive fields and hierarchical processing poses two main advantages over traditional fully connected networks: (1) by introducing structure to the neural network architecture, the neural network is allowed to detect patterns by looking at small-scale local changes (receptive fields) and by considering the global relations of these local changes to each other (hierarchical

**Figure 6.** **(a) The receptive field of a neuron. Instead of fully connecting the neurons to all inputs, a local area is chosen for the neuron to process. (b) A network with two hidden layers, each layer employing the receptive field concept. The highlighted $5 \times 5$ area corresponds to the effective receptive field of a second hidden layer neuron.**

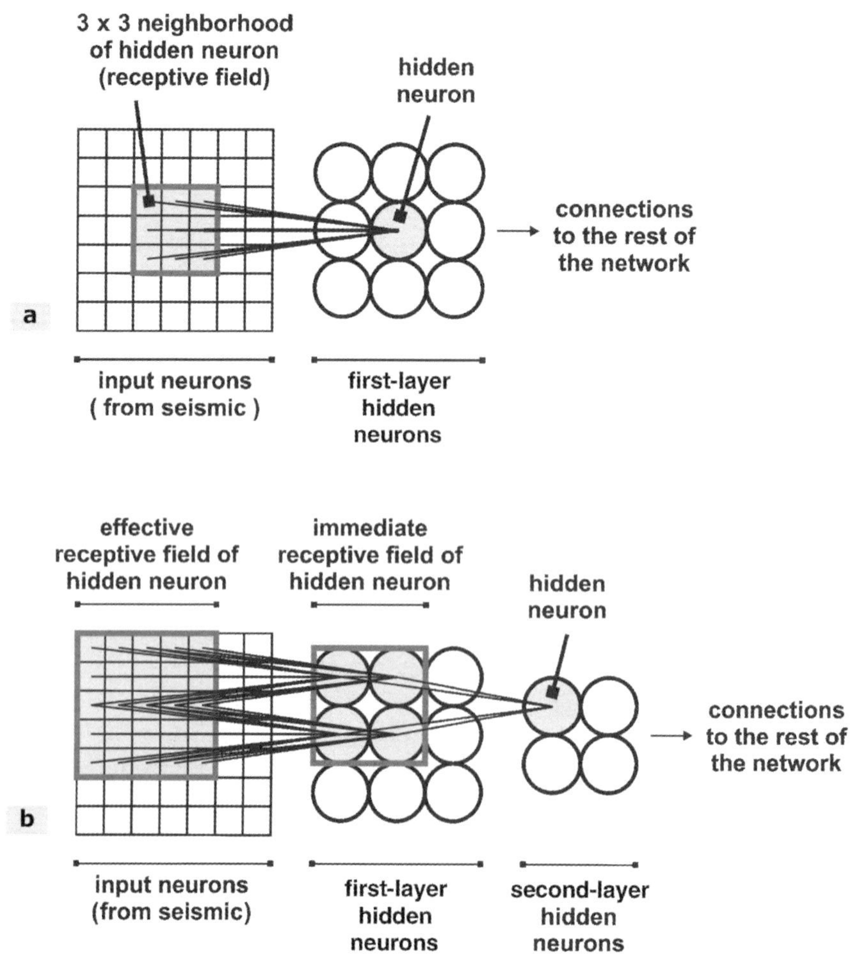

processing); and (2) although the structure uses a relatively large number of neurons, the computational performance is not degraded because the connections between these neurons are selective. Typically, the processing power lost because of the missing connections is compensated for by introducing an extra, fully connected layer right before the output (Reed and Marks, 1999).

The application-specific architecture of the neural network is critical to the success of the proposed seismic interpretation methodology. The relation between seismic-amplitude data and the facies occurrence is not a direct relation. A particular seismic-amplitude datum does not contribute much to its colocated facies interpretation. Instead, facies are determined on the basis of change in neighboring amplitude reflections. The receptive field concept allows the network to capture the small-scale local changes in the seismic data, and the hierarchical processing relates these small-scale changes to the overall large-scale change in the entire seismic window. Traditional, fully connected networks commonly lack the capability to capture the nature of seismic-amplitude patterns.

Because of the application-specific nature of the neural network, the final architecture depends on the size and shape of both the seismic template and the channel template. In this chapter, the templates shown in Figure 5 are used. Use of such templates provides a network architecture shown in Figure 7.

This partially connected neural network structure is trained using the classical online back-propagation learning algorithm. A presentation of the back-propagation algorithm is beyond the scope of this chapter. The reader is referred to Haykin (1999).

During the training phase, the content of the seismic window is given to the neural network as the input and its corresponding expert interpretation as the desired output. These expert interpretations are binary in nature for this application: channel vs. no channel. Because interpretations are never certain, a probabilistic neural network modeling the probability of observing certain facies from seismic data is required. This network should be capable of producing channel probabilities even when trained only by binary channel data. Caers and Ma (2002) suggest one such neural network structure for pixel-based estimations. Their methodology is adapted to the problem at hand and is reviewed briefly in the next section.

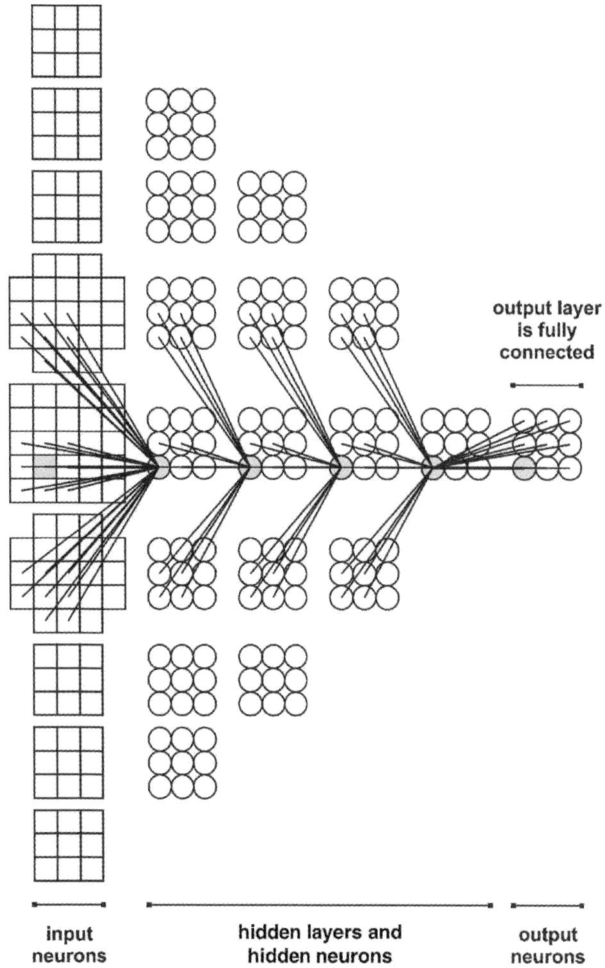


**Figure 7. The neural network structure used in this chapter. The vertical information content of seismic data is high, and therefore, it is emphasized in the structure by introducing more hierarchical levels along the Z-axis.**

## A PROBABILISTIC NEURAL NETWORK METHODOLOGY

Traditionally, neural networks are used for modeling a complex, nonlinear interaction between a set of variables. Commonly, neural network aims at predicting a set of target variables $Y$ (e.g., petrophysical properties) from a set of input variables $X$ (e.g., well-log measurements). The neural network models a specific nonlinear function $g(X)$ between these two types of variables. The neural network above is an example of a regression-type neural network because it models $E[Y \mid X]$ (conditional expectations). The traditional framework, however, lacks the commonly important capability of modeling the uncertainty about these predictions (i.e., $P(Y \mid X)$, the conditional probabilities).

Caers and Ma (2002) recently published a general methodology for turning deterministic neural networks into probabilistic neural networks, the output of such being the conditional probability $P(Y \mid X)$. Seismic data are known to potentially leave a large degree of uncertainty about the presence of reservoir facies (smoothing and tuning effects). Uncertainty quantification is needed, because such ambiguities commonly cannot be resolved. In this chapter, the approach of Caers and Ma (2002) is applied to the neural network architecture presented in the previous section.

## RESULTS OF THE NEURAL NETWORK APPLICATION

The neural network structure above is applied to the reservoir studied in the previous part of this chapter (Figure 4). Figure 8 shows the training results; that is, the neural network is trained using the expert interpretation in Figure 8a and the seismic data (not shown). The templates of Figure 5 are used. To check the performance, the trained neural network is applied to the same training set. Figure 8b shows the network prediction, and Figure 8c is the overlaid comparison of the network prediction with an actual expert interpretation. Figure 8c confirms that the neural network manages to predict the location of the channels well, at least in conformance with the expert interpretation. However, the results lack the crisp shape reproduction desired.

Figures 9 and 10 contain the test results for two test areas. Test results are obtained by applying the trained neural network to other parts of the seismic cube where no expert interpretation is available. An independent expert interpretation is also available in the test area that can be compared with the neural network result.

Figure 10 is a typical test output that one would expect from a neural network predictor. The result is not as good as the training area result, but the channel locations are still easily identifiable. The more interesting result is that of Figure 9. The neural network seems to predict certain channel-like structures near the top-left corner of the grid, although the expert interpretation does not contain any channels

Figure 8. Neural network results for the training area.

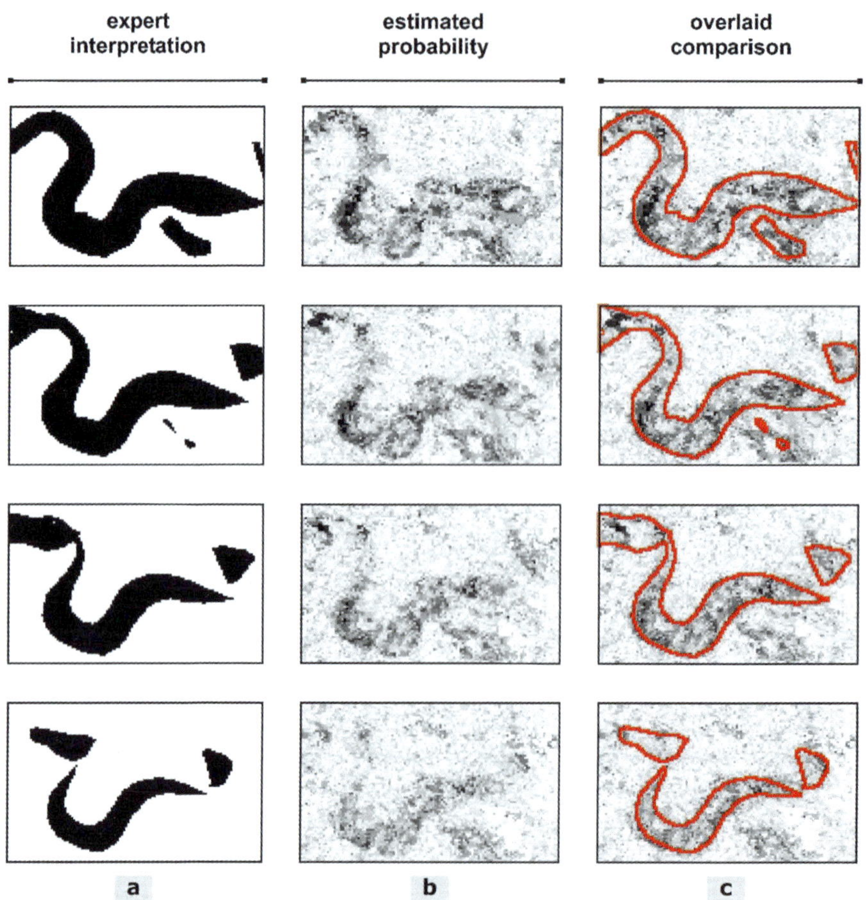

in that area. In this region, a salt dome is known to exist, yet the neural network cannot distinguish between seismic reflections belonging to the salt dome and those belonging to the channel body. It is hard to avoid such misinterpretations because the original neural network is never trained on a salt dome seismic response. Next, a method is developed to overcome such problems.

## A FEATURE-BASED GEOSTATISTICAL-SIMULATION ALGORITHM

The neural network estimator is essentially a stand-alone module that can be used to estimate facies channel probabilities. As shown in Figures 8–10, it is not sufficient to accurately identify facies features and geometries. Although the neural network explicitly considers the channel shapes using a facies template (right side of Figure 5), the results lack the crispness of channel objects. The neural network is applied independently on each window of seismic amplitudes, not considering the spatial continuity of the channel objects.

In this section, a novel geostatistical approach that restores the channel object geometry is presented. This step uses the channel probabilities of Figures 8–10 as conditioning soft data. A feature-based conditional simulation method is used for this purpose.

Feature-based geostatistics deals with patterns instead of single pixels (pixel-based geostatistics) or entire objects (object-based modeling). Versatile feature-based simulation algorithms (Arpat et al., 2002), as well as various approaches to the use of features that are not simulation based (Arpat et al., 2001), have been published by the authors before. In this chapter, a new, relatively simple feature-based simulation method is developed to process the channel probabilities estimated by the neural network estimated. The methodology is illustrated on a simple example and then applied to the case study.

## A SIMPLE EXAMPLE

Consider the simple $10 \times 10$ expert interpretation in Figure 11a depicting a small channel. This image contains 35% channel facies. Assume that a channel

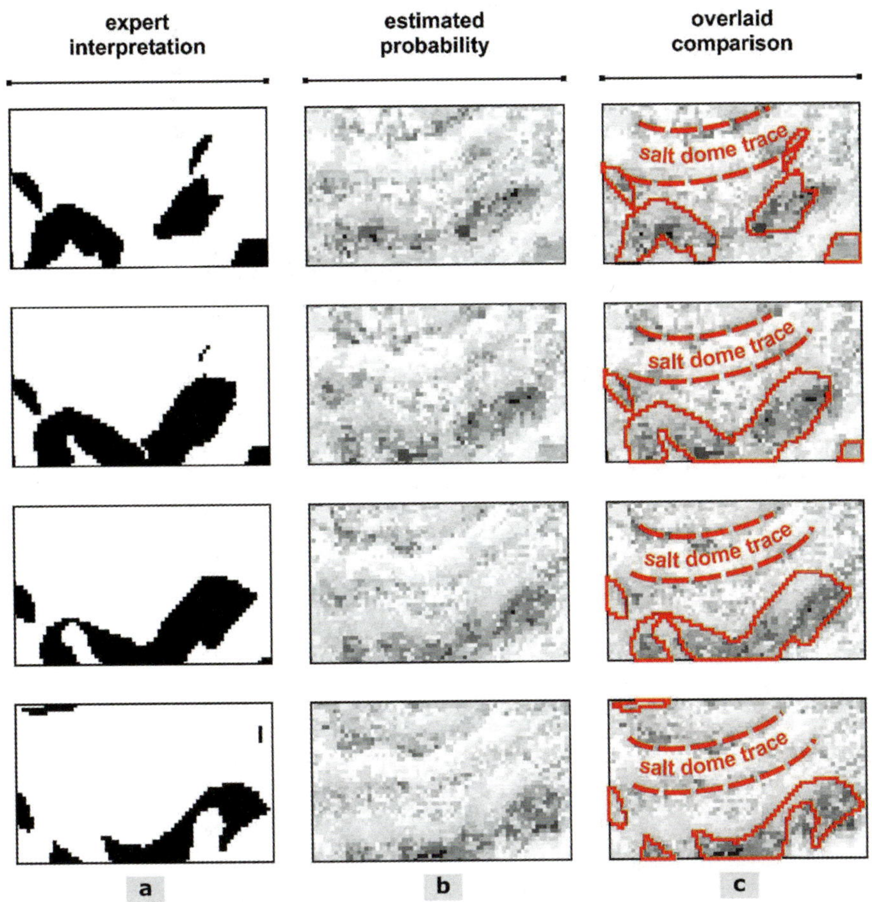


**Figure 9. Neural network results for a test area.**

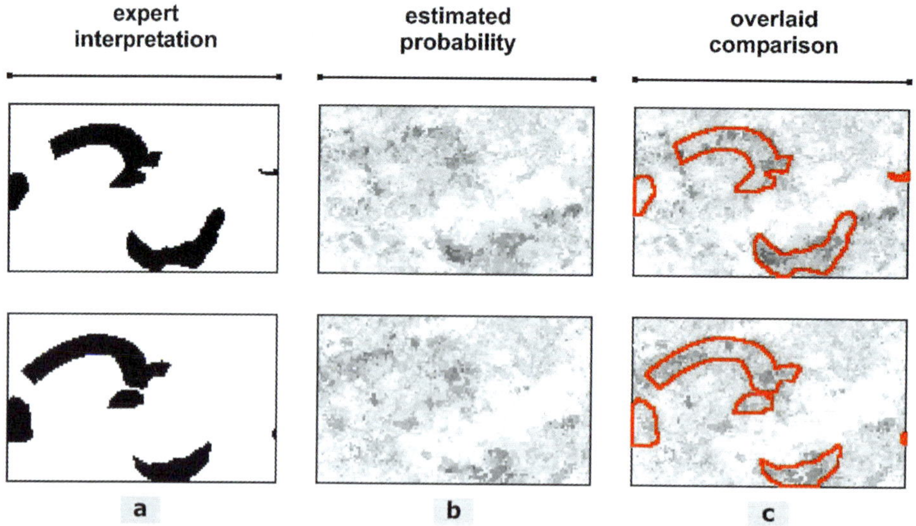


Figure 10. Neural network results for another test area.

probability map derived from seismic data using a neural network as above is also given (Figure 11b). Note that this map lacks the crispness of an actual channel body.

## Step 1: Extracting Features

The algorithm starts by extracting features from the expert interpretation. The feature template is placed on this image and scanned in a sliding fashion. Figure 11c illustrates the scanning process. All features extracted from expert interpretation are given in Figure 11d.

## Step 2: Calculating Skins

Once the features are extracted, the training image is rescanned to calculate the so-called skins belonging to each feature. Feature skins constitute pixels in the immediate neighborhood of a feature. These skins serve to quantify the direction of continuity and connectivity associated with each feature. For example, in Figure 12a, the $3 \times 3$ feature has a skin of two pixels in each direction. The pixel values of the skin express the probability of channel occurrence for any feature in the feature set. The skin of a feature is calculated by averaging the neighboring pixels of all occurrences of that feature in the expert interpretation. Figure 12b illustrates this process. If one considers features as puzzle pieces, then the skin corresponds to the way the puzzle pieces interconnect with each other.

## Step 3: Simulation

A random path is determined. The leftmost map of Figure 13a shows the first node visited in the grid. At this node, feature assignment is constrained by the channel probabilities obtained from seismic data (Figure 11b). The assigned feature needs to be selected from the existing feature set. This selection will retain the feature that is most similar to the colocated channel probability pattern of Figure 11b. The Tanimoto distance function is used as the similarity metric:

$$\text{similarity} = \frac{\sum_{i=1}^{n} |a_i - b_i|}{\sum_{i=1}^{n} (a_i + b_i) - \sum_{i=1}^{n} |a_i - b_i|} \qquad (1)$$

where $n$ is the total number of pixels in the feature body and skin ($7 \times 7$ in this case), $a_i$ are the pixel values of the probability pattern, and $b_i$ are the pixel values of the candidate feature. A sample calculation for the Tanimoto metric is illustrated in Figure 14. Other distance measures could be used, such as Euclidean or Manhattan distance (Duda et al., 2001). For the particular application to channel reservoirs, a sensitivity study has found that the Tanimoto distance function promotes the channel continuity best.

Figure 13a shows the selected feature for the first visited node. Once this most similar node is determined, corresponding skin values are pasted into the simulation grid, replacing the original channel

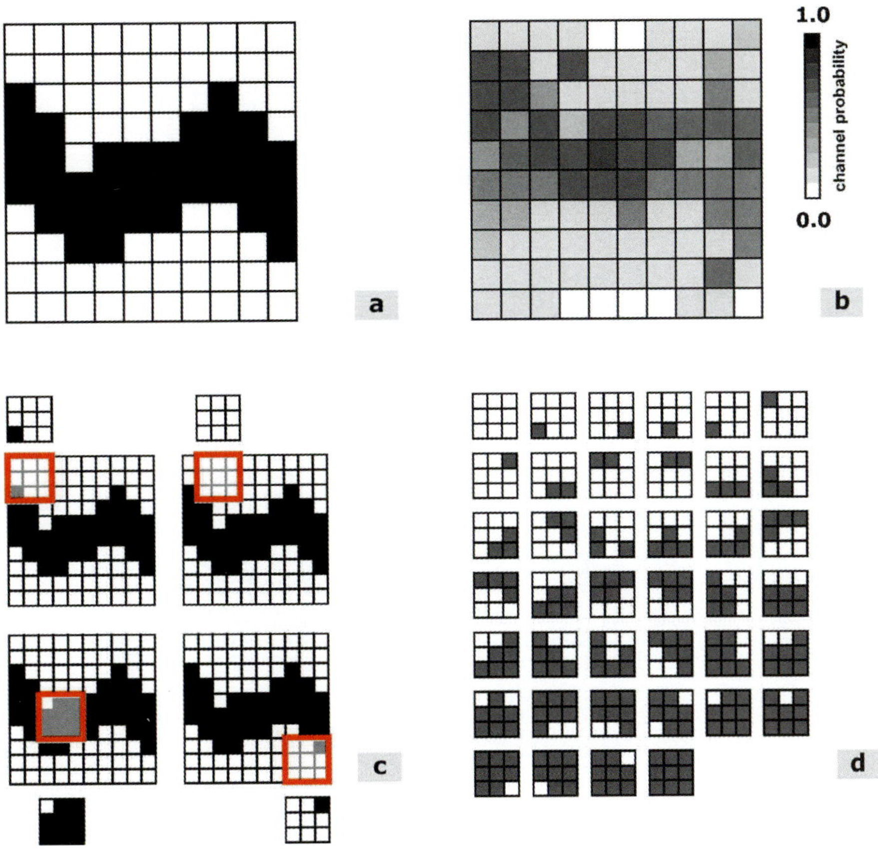


Figure 11. (a) A 10 × 10 expert interpretation used for the simple example. (b) The channel probability from seismic data used for the simple example. For demonstration purposes, it is assumed that this map is obtained from a neural network structure as explained in the text. (c) Various steps of feature extraction from expert interpretation. (d) All features extracted from the expert interpretation.

probabilities from seismic data as shown in the last column of Figure 13a.

Next, another node along the random path is selected, for example, the one right next to the first visited node. Figure 13b shows the location of this node. Again, the most similar feature is selected, considering the channel probabilities from seismic data and any previously determined features (sequential simulation). Figure 13b shows the selected feature and the updated probability map. Figure 13c shows some intermediate steps during the simulation. The final result, after all nodes have been simulated, is shown in the last column of Figure 13c. If another realization is required, then the random path needs to be changed.

## RESULTS

The proposed feature-based simulation method is applied to the channel probabilities obtained using the neural network (Figure 8). Figure 15 shows the results for the training area. The feature-based

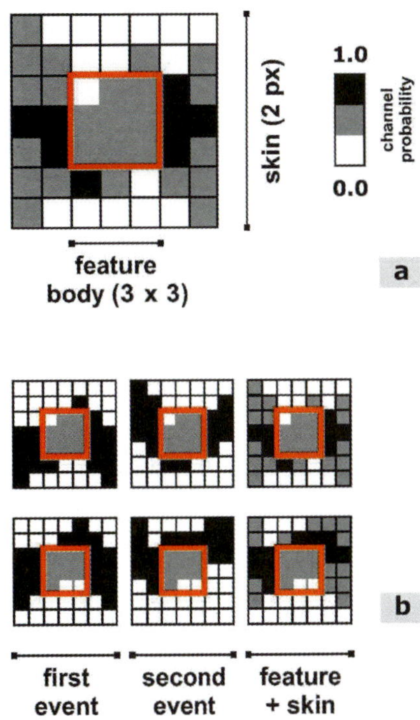


Figure 12. (a) A 3 × 3 feature with two units of skin. (b) Two samples showing how the skin of a feature is calculated. Simply, all occurrences (twice only in the above examples) are added up and then divided by the number of occurrences.

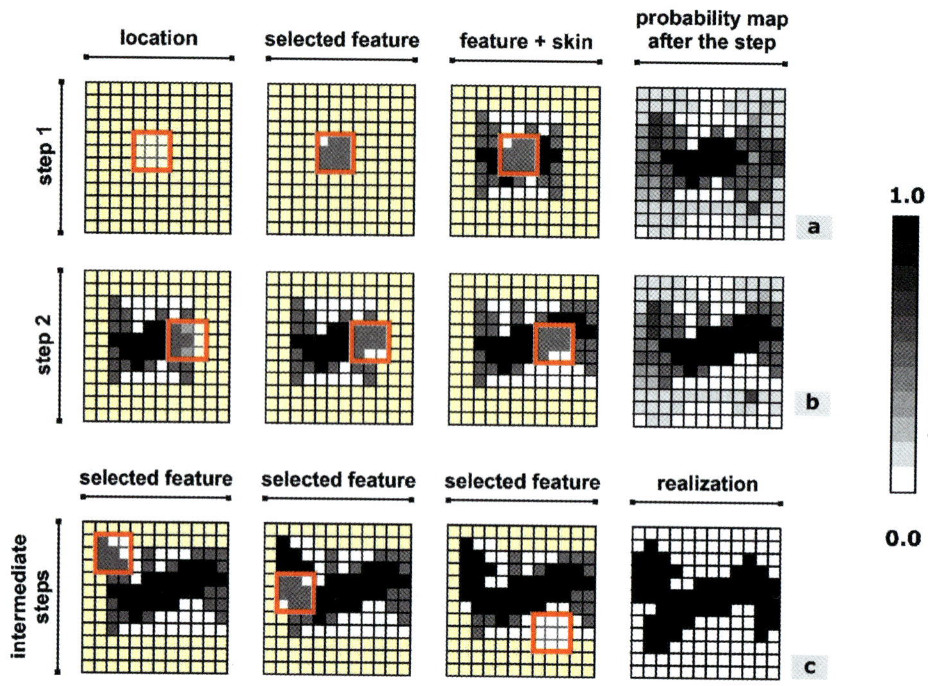


Figure 13. The steps of the feature-based simulation. In (a) and (b), the location of the node dictated by the random path is shown, followed by the most similar feature selected, followed by the final pasting of the skin values. (c) An illustration of the intermediate steps and the final realization.

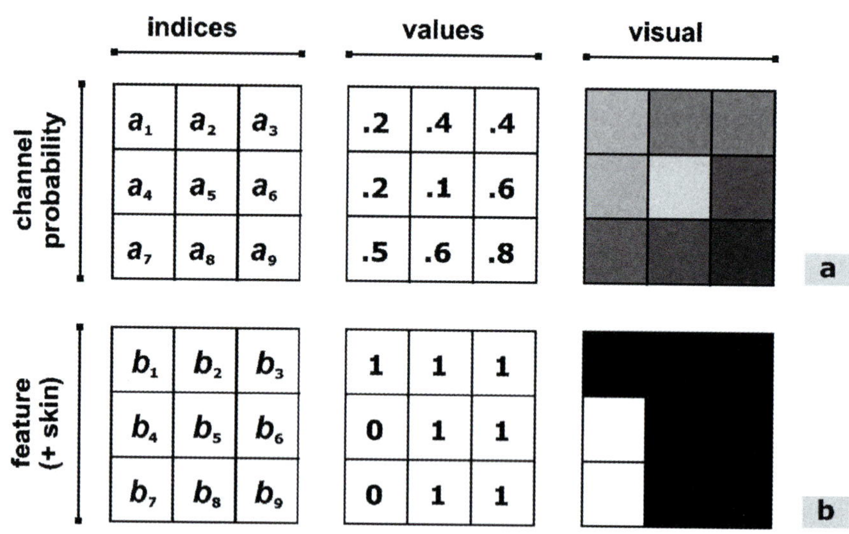


$$\text{similarity} = \frac{\sum_{1=1}^{n} |a_i - b_i|}{\sum_{1=1}^{n} (a_i + b_i) - \sum_{1=1}^{n} |a_i - b_i|}$$

$$\text{similarity} = \frac{|0.2 - 1| + |0.4 - 1| + \ldots + |0.8 - 1|}{[(0.2 + 1) + (0.4 + 1) + \ldots + (0.8 + 1)] - 4.6}$$

$$\text{similarity} = 0.742 \quad (0: \text{most similar} - 1: \text{least similar})$$

c

Figure 14. Sample Tanimoto distance calculation.

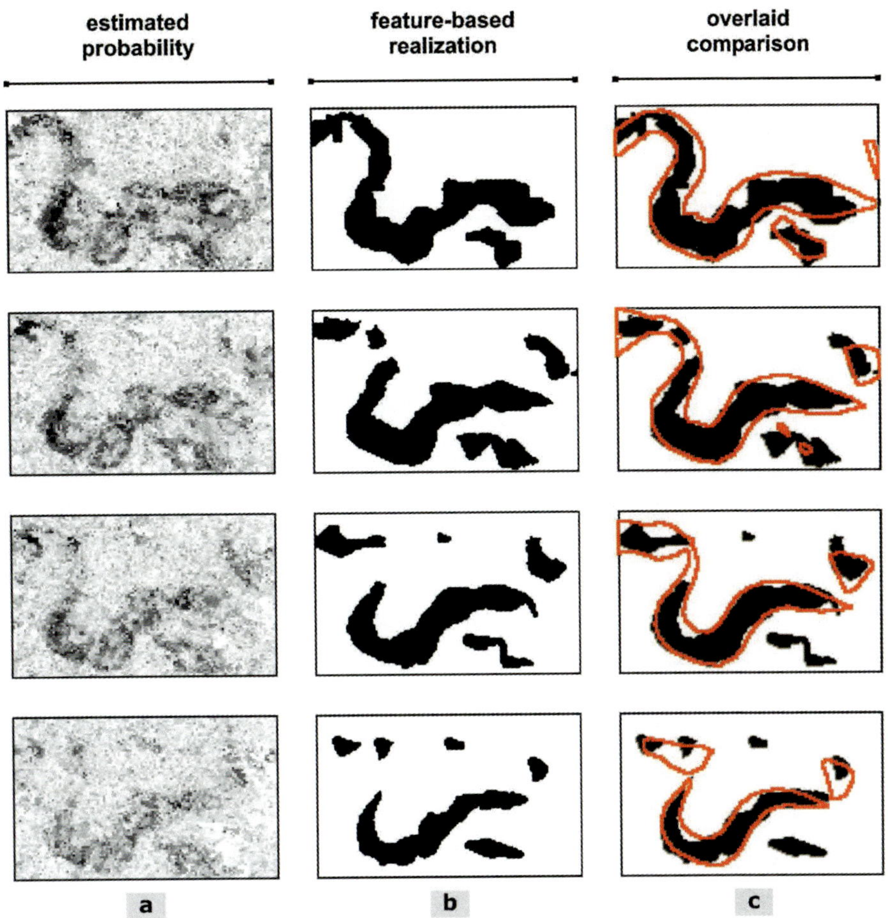


**Figure 15. Final results for the training area.**

simulation manages to convert the channel probability patterns into actual channel shapes successfully. Notice that wherever there is a discontinuity in the probabilities, the simulation algorithm fills in the channel and promotes continuity.

Figure 16 presents the results for the test area containing the salt dome. This problematic test area produces results without the salt dome probabilities converted into actual channels. This is possible because the salt dome probabilities do not correspond to a channel pattern found in the expert interpretation. The feature-based algorithm naturally recognizes the difference between the channel shapes and the salt dome response.

## CONCLUSION

Building a single or multiple facies model from seismic data should rely on the existing multiple-point relation between seismic attribute data and the continuity of the underlying facies. Expert interpretations are explicit representations of the multiple-point relationship. Such interpretations are mappings of seismic features (consisting of multiple pixels or voxels) into channel shapes or object pieces.

In this chapter, an automated interpretation tool is presented that capitalizes on this notion. Building on ideas from human vision science, a neural network approach is developed that maps seismic features into facies features. However, because neural network approaches commonly neglect spatial continuity, an additional geostatistical step is warranted.

The geostatistical-simulation algorithm developed is also somewhat nontraditional in the sense that features (parts of objects) are simulated instead of pixels and objects. The feature-based algorithm attempts to build larger objects by connecting smaller object pieces constrained to the presence of facies detected in the seismic data by the neural network.

Figure 16. Final results for a test area.

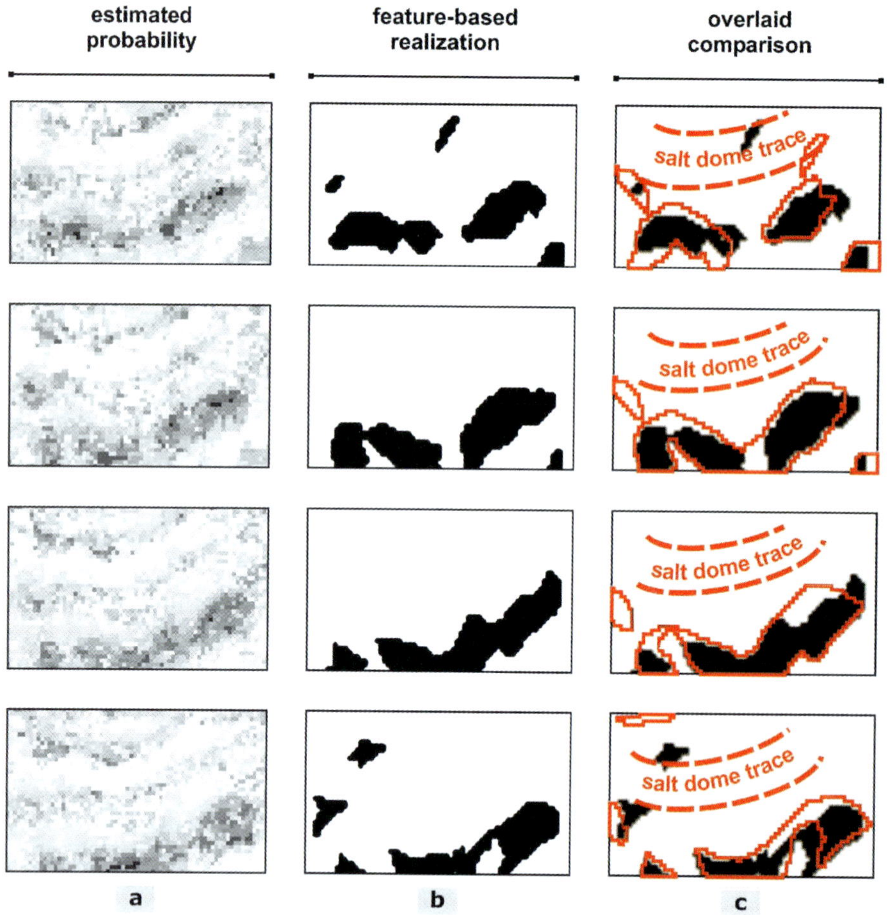


## ACKNOWLEDGMENTS

The authors appreciate the opportunity provided by TotalFinaElf to work on the submarine-channel data set. Comments on previous parts of this work by Andre Haas are also appreciated.

## REFERENCES CITED

Arpat, B., J. Caers, and A. Haas, 2001, Characterization of west Africa submarine channel reservoirs: A neural network based approach to integration of seismic data: Society of Petroleum Engineers Annual Technical Conference and Exhibition, New Orleans, September 2001, SPE Paper 71345, 16 p.

Arpat, B., J. Caers, and S. Strebelle, 2002, Feature-based geostatistics: An application to a submarine channel reservoir: Society of Petroleum Engineers Annual Technical Conference and Exhibition, San Antonio, September 2002, SPE Paper 77426, 16 p.

Caers, J., and X. Ma, 2002, Modeling conditional distribution of facies from seismic using neural nets: Mathematical Geology, v. 34, no. 2, p. 139–163.

Deutsch, C., and A. Journel, 1998, GSLIB— Geostatistical software library and user's guide, 2d ed.: New York, Oxford University Press, 369 p.

Duda, O. R., P. E. Hart, and D. G. Stork, 2001, Pattern classification, 2d ed.: New York, John Wiley & Sons, 186 p.

Ganong, W. F., 1997, Review of medical physiology: Norwalk, Connecticut, Appleton and Lange, 673 p.

Haykin, S., 1999, Neural networks: A comprehensive foundation: Upper Saddle River, New Jersey, Prentice Hall, 842 p.

Mallet, J.-L., 2002, Geomodeling: New York, New York, Oxford University Press, 590 p.

Palmer, S., 1999, Vision science: Photons to phenomenology: Massachussets, MIT Press, 253 p.

Payrazyan, K., 1998, Modelisation 3D des reservoir petroleirs par l'integration des donnes sismiques et geologiques: approaches quantitatives multivariables: Ph.D. dissertation, Ecole Nationale Superieure de Geology de Nancy, Nancy, France, 192 p.

Reed, R. D., and R. J. Marks II, 1999, Neural smithing:

Supervised learning in feed— Foreword artificial neural networks: Cambridge, MIT Press, 152 p.
Ripley, B. D., 1996, Pattern recognition and neural networks: Cambridge, UK, Cambridge University Press, 403 p.
Wallace, C., and D. Boulton, 1986, An information measure for classification: Computer Journal, v. 11, p. 185–195.
Wallace, C., and D. Dowe, 1994, Intrinsic classification by MML— The SNOB program, *in* Proceedings of the 7th Australian Joint Conference on Artificial Intelligence, p. 37–44.
Wonham, J. P., S. Jayr, R. Mougamba, and P. Chuilon, 2000, 3D sedimentary evolution of a canyon fill from the Mandorove Formation, offshore Gabon: Marine and Petroleum Geology, v. 17, p. 175–197.

Galli, A., G. Le Loc'h, F. Geffroy, and R. Eschard, 2006, An application of the truncated pluri-gaussian method for modeling geology, *in* T. C. Coburn, J. M. Yarus, and R. L. Chambers, eds., Stochastic modeling and geostatistics: Principles, methods, and case studies, volume II: AAPG Computer Applications in Geology 5, p. 109–122.

*Chapter 9*

# An Application of the Truncated Pluri-gaussian Method for Modeling Geology

**A. Galli**
**G. Le Loc'h**
**F. Geffroy**
*Centre de Geostatistique*
*Ecole des Mines de Paris*
*Paris, France*

**R. Eschard**
*Institute Français du Petrole*
*Rueil Malmaison, France*

## ABSTRACT

The truncated pluri-Gaussian approach is a powerful method for modeling geology. Its main strength is its flexibility in representing complex lateral and vertical facies transitions with different anisotropies. In addition, it is easy to condition simulations to data points. This makes it an excellent method for modeling reservoirs with a complex architecture, such as carbonate bioconstructions or reservoirs affected by diagenesis.

This chapter presents a method for obtaining a tractable and mathematically consistent model for lithotype semivariograms and cross-semivariograms in complex cases. The method is illustrated using an example involving algal bioconstructions in the outcrops of the Paradox basin (Utah). Complex facies transitions, both vertically and laterally between the mound and intermound facies, together with the complex geometry of the algal mounds, make it virtually impossible to simulate these sorts of deposits using object-based models or classical pixel-based methods. The truncated pluri-Gaussian model is introduced to handle these complex facies transitions using the concept of a lithotype rule.

Such a rule, when expressed diagrammatically, can be a valuable tool for synthesizing geological information, and it can serve as one of the key inputs into the stochastic model. As illustrated here, combining the rule with proportion curves is a very effective way for analyzing and modeling geology in terms of facies sequences, even in complex depositional environments.


DOI:10.1306/1063810CA53226

## INTRODUCTION

Stochastic models are widely used to simulate geology and petrophysical properties in petroleum reservoirs. Several modeling approaches are available, and the selection of an appropriate technique depends on the context in question. No matter which method is used, it should respect available well data, reproduce trends and directionality among facies (e.g., anisotropy), respect the experimental cross-correlations and the transition probabilities among facies, and guarantee the consistency of models used for the semivariograms and cross-semivariograms.

Two broad categories of modeling techniques exist: those that are object based and those that are pixel based. Both categories accommodate the above four requirements but in different and important ways. Consequently, it is both interesting and instructive to consider these differences, particularly as they relate to the ideas of stochastic simulation applied to facies modeling, which is the primary focus of this chapter.

The remainder of this section briefly contrasts object- and pixel-based methods and provides the rationale for using truncated Gaussian (TG) and truncated pluri-Gaussian (TPG) random functions. The next section discusses some basic properties of indicator variables. Subsequent sections describe the geological setting of a particular motivating example from the Paradox basin, along with the development and application of a modeling approach for this area based on the use of TPG random functions.

### Object-based Modeling Techniques

In light of the four requirements noted above, it is useful to first consider object-based techniques because of their popularity. With regard to anisotropy, trends and directionality are treated quite naturally in object-based methods, and inconsistencies between models for semivariograms and cross-semivariograms do not arise. In terms of respecting well data, however, conditioning is difficult when many wells are present or when the objects are large. Constraints from seismic and/or production data make this even more challenging. Furthermore, it seems to be difficult for object-based methods to reproduce semivariograms in this case and also when transitions among facies are involved. Semivariograms clearly provide valuable information in the vertical direction, even when only a few wells are available, and in the horizontal direction, when either many wells or seismic information is available. Consequently, the ability to appropriately treat them is important. One point of contention is that, commonly, the need to reproduce the experimental cross-correlation is not even addressed, which may in part be because of the users' unfamiliarity with computing simple and cross-semivariograms.

### Pixel-based Modeling Techniques

In contrast to object-based methods, conditioning simulations to many wells is generally not too difficult with pixel-based methods, and it is easier to integrate seismic and/or production data. However, some methods are plagued with an inability to reproduce cross-correlations. Stochastic indicator simulation proposed by Journel and Isaaks (1984) is one such method. This is why simulated annealing is sometimes used to postprocess resulting images (e.g., see Goovaerts, 1996). Most of the first-generation methods, except for those based on Markov random fields, fail to reproduce the transition properties; but even Markov random field methods are difficult to use because their parameters are not directly related to the properties of interest. Unfortunately, it is not uniformly understood that consistency between semivariograms and cross-semivariograms is not guaranteed if they are fitted separately. It is also not well known that specific semivariogram models should be used for fitting indicator semivariograms. These considerations and the desire to honor the observed cross-correlation are the main motivations that led Matheron et al. (1987) to model facies with a TG random function. The TG model is one of the simplest, complete models for indicators. Furthermore, it makes simulations easy to perform (see Rivoirard, 1993, for a discussion on indicator modeling).

Although the TG random function naturally honors a sequential order within lithofacies (Matheron et al., 1987; Ravenne et al., 1991) and can be extended to some other types of transitions, it cannot be used in the general case. One limitation is that the anisotropy has to be the same for all the lithofacies. The TPG random function is a generalization of the TG random function, which has the same advantage of flexibility and mathematical consistency and overcomes the limitations associated with transitions and anisotropies. One of the key steps in the pluri-Gaussian methodology is defining the lithotype rule.

This rule is represented by a diagram that synthesizes contacts between facies. Geological knowledge can be used to help define the rule, and geostatistical tools can be applied to quantify the extent of the transitions.

Before describing the TPG methodology in further detail, it is instructive to review the properties of indicator functions and indicator semivariograms. Proportion curves and three-dimensional (3-D) proportion matrices are also discussed. To enhance the presentation of these ideas, reference is made to an outcrop observed in the Paradox basin, which is described in more detail later in a succeeding section.

## BASIC PROPERTIES OF INDICATORS

Indicator random functions are very special random functions. They take only two values, 0 or 1. They are used, in particular, when working with categorical data such as lithofacies. In such a case, an indicator is defined for each facies $F_i$ by

$$1_{F_i}(x) = \begin{cases} 1 & \text{if } \ x \in F_i \\ 0 & \text{if } \ x \notin F_i \end{cases}$$

More generally, given a random set $F$ (that is, a set generated in a probabilistic way), its indicator is the function whose value is 1 on $F$ and 0 elsewhere. Let $\overline{F}$ be the complement of $F$. Then, the following elementary properties hold:

1) $1_{F \cup \overline{F}}(x) = 1_F(x) + 1_{\overline{F}}(x) = 1$.
2) If $p_F(x) = E(1_F(x))$, $p_F$ is called the proportion of $F$ and $0 \leq p_F(x) \leq 1$.
3) $\mathrm{Var}(1_F(x)) = p_F(x)(1 - p_F(x))$.
4) $\sigma_F(x,y) = E(1_F(x)1_F(y)) = \mathrm{Prob}(x \in F \text{ and } y \in F)$. $\sigma_F(x,y)$ is called the noncentered covariance.
5) The indicator semivariogram of $F$, $\gamma_F(x,y)$, always lies between 0 and 0.5.

These elementary properties are the ones that make indicator random functions so special. Property 3 has interesting implications in terms of nonstationarity. Nonstationarity in the mean is equivalent to nonstationarity in the variance for indicators. From Property 5, we see that power semivariograms of the form $\gamma(h) = h^\alpha$ are forbidden for indicators (because $h^\alpha$ is not bounded).

Using the previous properties, it is easy to prove the following theorems.

### Theorem 1

If $F$ is a stationary random set and $C_F$ is its covariance, $C_{F\overline{F}}$ is its cross-covariance, and $\gamma_F$ is its semivariogram, then

$$\gamma_F(h) = \gamma_{\overline{F}}(h) \quad \begin{matrix} \gamma_F(x) = \gamma_{\overline{F}}(x) \\ C_F(h) = C_{\overline{F}}(h) = -C_{F\overline{F}}(h) = -C_{\overline{F}F}(h) \end{matrix}$$

### Theorem 2

If $A$, $B$, and $C$ are three facies whose indicators are second-order stationary, and if $A \cup B \cup C$ fills the entire space, then

$$\begin{aligned} C_A &= -C_{AB} - C_{AC} \\ C_B &= -C_{BA} - C_{BC} \\ C_C &= -C_{CA} - C_{CB} \\ C_{AB} - C_{BA} &= C_{BC} - C_{CB} = C_{AC} - C_{CA} \end{aligned}$$

Galli et al. (1994) and Armstrong (1992) provide the respective proofs.

Up to this point, the discussion has focused on consistency relations when working with different indicators, but there are also strong requirements for a covariance to be the covariance of an indicator (e.g., more than just being a positive definite function; see Matheron, 1989; Armstrong, 1992). In particular, near the origin, indicator semivariograms must have a behavior in $h^\alpha$ with $\alpha \leq 1$. Consequently, models of semivariograms with a parabolic behavior close to the origin, such as the Gaussian model or the cardinal sine model, are forbidden for indicators, as previously suggested.

### Anisotropy

Anisotropy is a simple and intuitive concept. An object is said to be anisotropic if it behaves differently in different directions. A simple example in geology is a channel that is more continuous along its principal direction than along the orthogonal direction. As a simple example, consider channels to be parallelepipeds with constant length and width.

Figure 1 shows a diagram corresponding to this model where the main direction is east–west.

Figure 2 shows the semivariograms of these channels along the east–west and north–south directions. The curves are consistent with intuition about anisotropy. However, because of Theorem 1, the background semivariogram has to be the same as that of the channels; that is, the background has the same anisotropy as the channels. This result is somewhat disturbing, until it is realized that the background is linked with the channels. Being in the background is different from being in a channel. So, the intuitive feeling about the isotropy of the background derives from local analysis based on visual inspection, not from a global view. Basically, this is the meaning of Theorem 1.

Squares shown in Figure 3 were eroded by the channels shown in Figure 1, as shown in Figure 4. Now there are three facies—the background, small squares, and channels.

Figure 5 shows the simple and cross-semivariograms of these three facies. Surprisingly, both the background and the small squares have anisotropic semivariograms, as do the channels. This is a part of the meaning of Theorem 2: to a smaller or greater extent, anisotropy is observed in the indicator semivariogram for a facies as well as in each simple semivariogram.

The other meaning of Theorem 2 is understood when looking at the cross-semivariograms. They have some common features with the simple semivariograms. This indicates that all semivariograms (simple and cross) have to be fitted simultaneously to respect the constraints, and furthermore, that anisotropy has to be present on all the simple and cross-semivariograms. Theorem 2 is a more precise way of stating the relationship that should be followed when fitting the model. Consequently, to respect anisotropy during simulation, a model must be used that honors the way the different anisotropies interact.

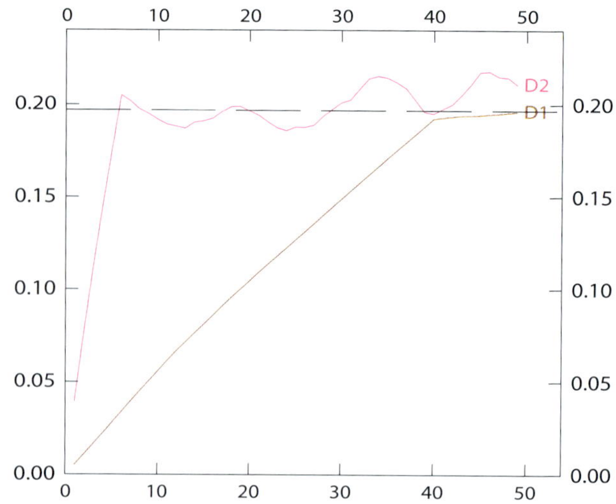

**Figure 2. Semivariograms east–west and north–south of the anisotropic objects.**

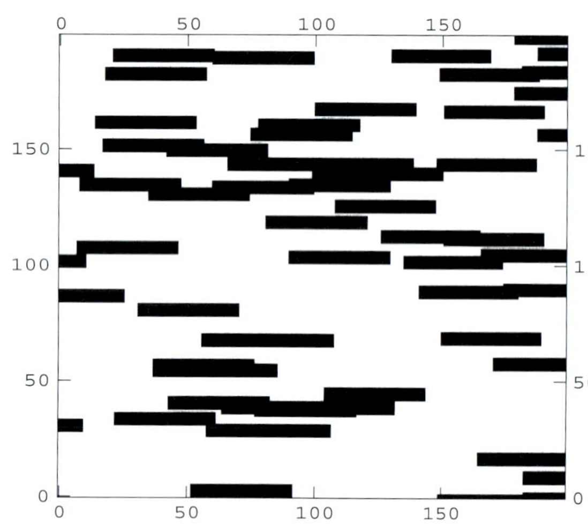

**Figure 1. Anisotropic objects.**

## Transitions among Facies

Transitions describe contacts between facies. More precisely, they are quantified as probabilities of being one facies or another. Transitions were intensively studied about 20 yr ago in the context of using Markov chains to model geology (Allègre, 1964; Krumbein, 1967). However, transitions can vary with the direction (horizontal vs. vertical) and orientation (upward or downward). Commonly, a vertical nonstationarity is present because of the cyclicity of sedimentation, such as low-order cycles being superimposed on higher order cycles that vary rapidly in a small amount of time (i.e., small variation in depth). This makes it more difficult to characterize the probabilities because they change vertically.

# A MOTIVATING EXAMPLE

To better demonstrate the principles of TPG methodology and to explain its potential, the following example is presented.

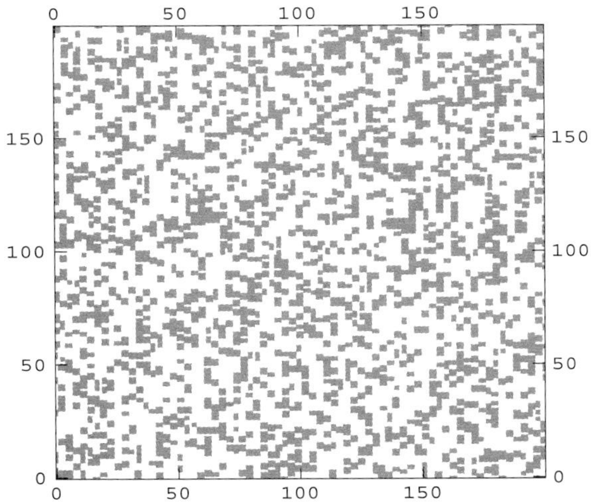


**Figure 3. Isotropic objects.**

## Geological Setting

The Paradox basin, which extends through southwest Colorado and in southern Utah, is an asymmetric basin that was probably created in a strike-slip tectonic context (Baars and Stevenson, 1981). During the Pennsylvanian, the basin paleogeography was characterized by thick evaporites that interfingered southward with thinner, shallow-marine carbonated platform sediments (the Hermosa Group). These carbonate-platform deposits are now exposed in the canyon of the San Juan River (New Mexico), where they are characterized by a tabular organization at the outcrop scale caused by the vertical stacking of several depositional cycles. The fourth-order sequences have been interpreted as the result of high-frequency, relative sea level variations, probably of climatic origin (Goldhammer et al., 1991). Each sequence corresponds to a transgressive-regressive cycle, commonly bounded by exposure surfaces. Significant algal buildups have been identified in some of the sequences. In outcrops, phylloid algal mounds can be observed in the Akhah and lower Ismay formations, whereas in the subsurface, algal mounds of the Desert Creek Formation constitute the main hydrocarbon reservoir of the Aneth field area (Chidsey et al., 1996, Montgomery et al., 1999).

## Sedimentology and Stratigraphic Architecture of the Lower Ismay Formation in Outcrops

The present work focuses on the phylloid algal mounds of the lower Ismay Formation, which are beautifully exposed along the San Juan River canyon walls in the Eight Foot Rapid area (Figure 6). The sedimentary facies and the sequential organization of the mounds have been studied in detail by Grammer et al. (1996). Geostatistical modeling of the area has been conducted by Van Buchem et al. (2000). These authors identify and describe 12 sedimentary facies in detail, which are here regrouped into nine facies for simulation purposes. The main sedimentological characteristics of these nine facies are summarized in Table 1.

The black laminated mudstones (facies 1) were deposited during periods of relative sea level rise, which first induced a rapid flooding of the platform, and then the development of an open carbonate-platform setting (facies 2 and 3). During the period of relative sea level highstand, algal mounds started to aggrade, incipient mounds (facies 4) being followed by bioherm construction (facies 5 and 6). Intermound troughs were progressively filled by apron debris as mounds grew and by open-platform carbonates (facies 7). Mound tops were capped by high-energy bioclastic beds deposited close to sea level. Finally, the sand brought into the basin by Eolian winds during periods of relative sea level lowstand was reworked around the mounds by tidal and wave dynamics (facies 8). The sequence ended with the emersion of the platform and the development of soil horizons (facies 9).

Two sedimentary cycles have been identified in the lower Ismay Formation (Van Buchem et al., 2000). The first cycle starts with a marine transgression and the deposition of the black laminated mudstones (Gothic Shales Member). It ends with the platform exondation, which is marked by an

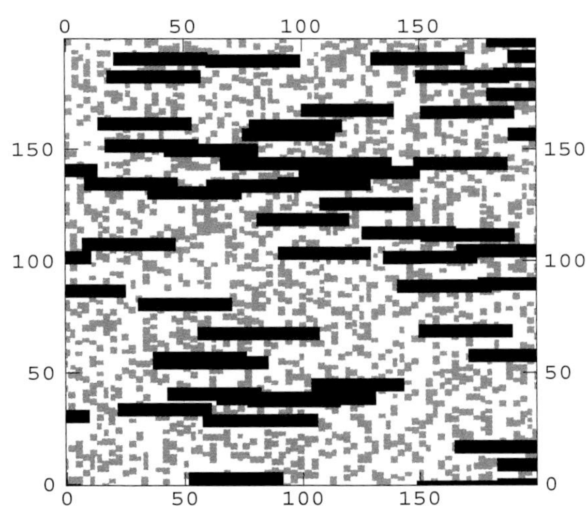


**Figure 4. Erosion of Figure 3 by Figure 1.**

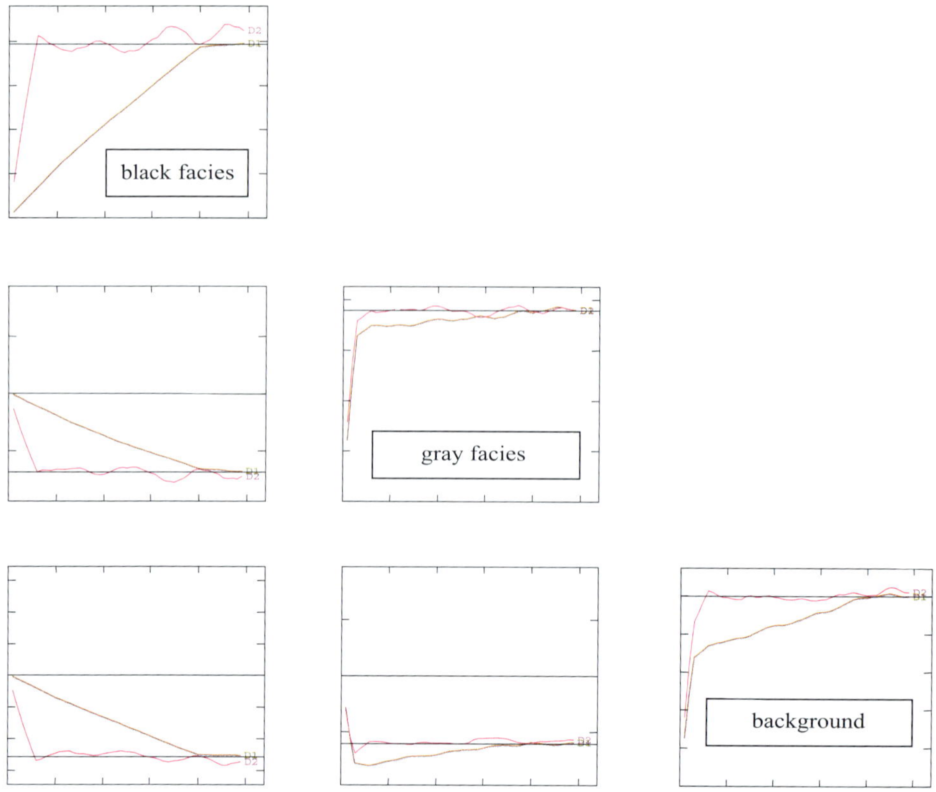


**Figure 5.** Semivariograms and cross-semivariograms of the facies shown in Figure 4.

important soil horizon on top of an extensive sandstone bed. Algal mounds were well developed in this cycle. The second cycle is marked by another transgressive event and new development of open-marine platform carbonates. Algal mounds have only low-relief morphology in this interval (Horn

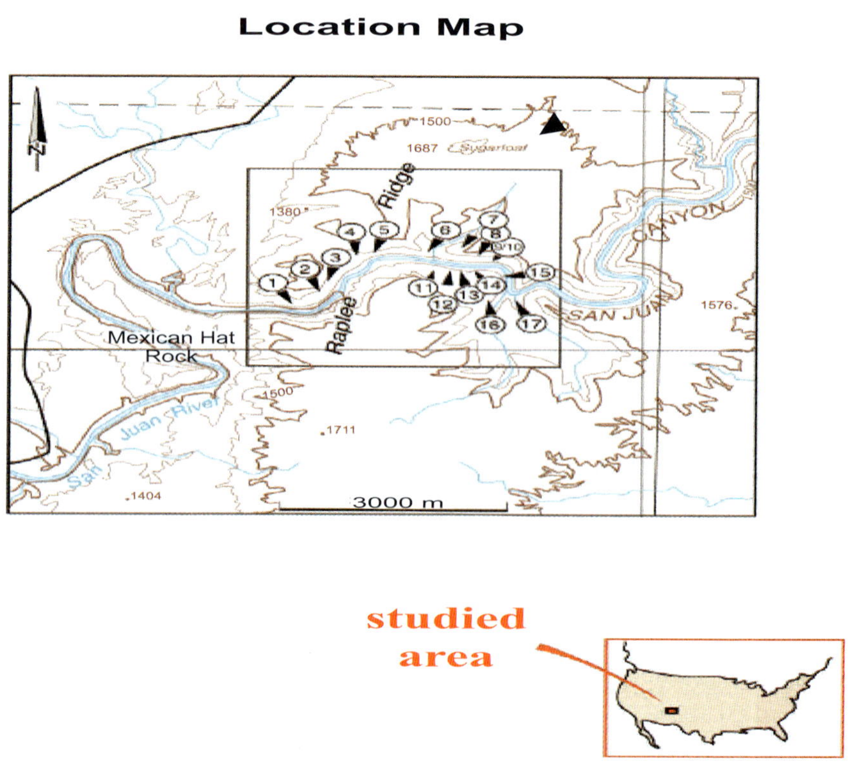


**Figure 6.** Location map of the sedimentological section in the Eight Foot Rapid area, along the San Juan River canyon.

Table 1. Description of facies.

| *Facies Name* | *Number* | *Description* |
|---|---|---|
| Black laminated mudstones | 1 | Black organic-rich fissible mudstones, with millimetric silty lamination. Deposited in anoxic, shallow-marine environments during transgressions. |
| Sponge mudstone facies | 2 | Argillaceous spiculitic mudstones, with chert nodules, deposited in shallow low-energy subtidal environments. |
| Intermediate facies | 3 | Laminated bioturbated silty mudstones to skeletal wackestones and packstones, with wavy bedding and a rich fauna, deposited in a low energy restricted to open-platform environment. |
| Incipient algal mound | 4 | Phylloid algal mudstones to wackestones in a muddy matrix. Phylloid algae are disseminated in a muddy matrix found in a moderate-energy, open-platform setting. The facies is not bioconstructed. |
| Algal mound, initial stage | 5 | Phylloid bafflestones and algal wackestones to packstones, forming mound-shaped bioherms with a flat base. Maximum thickness of individual mounds reaches 15 m (50 ft) in outcrops. These bioherms are constructed in a moderate-energy, open-platform setting. Two phases of bioconstruction are evidenced in the outcrops (initial and final stages; see below). |
| Algal mound, final stage | 6 | This facies is identical to facies 5 and corresponds to the final stage of bioconstruction of the algal mounds. Emersion surfaces have been identified in the topmost part of the mounds. |
| Skeletal bioclastic facies | 7 | Sandy crinoidal-rich wackestones, packestones, and grainstones, with sigmoidal cross-bedding, interpreted as moderate- to high-energy platform shoals. Wackestones and packstones mostly fill the intermound troughs, whereas oolitic grainstones form high-energy shoals capping the mounds. |
| Quartz sandstone facies | 8 | Siltstones to very fine-grained quartz sandstones with skeletal fragments and lithoclasts, with hummocky or sigmoidal cross-bedding. Sands and silts probably have an Eolian origin, having been transported by winds to the shelf and then reworked by a marine dynamic. |
| Nonskeletal facies | 9 | Massive mudstones to wackestones, deposited in a restricted marine setting. This facies is commonly affected by karstification and pedogenesis. |

Point Member). A well-developed karstified surface marks a significant emersion at the end of the cycle.

### Phylloid Algal Mound Geometry

In the outcrops, mound height can reach 15 m (50 ft), for a lateral extension ranging between 200 and 700 m (660 and 2300 ft). Mounds also have relatively symmetrical circular shapes, although outcrops do not allow detailed 3-D reconstructions of their internal architecture. The complexity of the algal mound facies architecture from the point of view of shapes and transitions makes this outcrop a challenge for stochastic modeling.

### Database Compilation

Outcrops in the Eight Foot Rapid area make it possible to reconstruct the lower Ismay stratigraphic architecture. The model is nearly 3-D because of the bends in the San Juan River and the occurrence of side canyons. A continuous photomosaic of the canyon sides has been constructed and then interpreted in terms of sedimentary facies (Figure 7) with the help of 17 detailed vertical sedimentological sections. This 6-km (3.7-mi)-long transect has been digitized, and pseudowells have subsequently been extracted from the transect and compiled into a working database.

## DEVELOPMENT AND APPLICATION OF THE TRUNCATED PLURI-GAUSSIAN

As noted above, the TG methodology was first introduced by Galli et al. (1994). The basic idea is to define facies by truncating several Gaussian random functions. Using more than one Gaussian random function provides greater flexibility to handle complex transitions and anisotropy among facies.

The general approach is to start from a partition $D_i$, $i = 1, \ldots, n$, of the $p$-dimensional space (i.e., $D_i \cap D_k = \emptyset$ and $\cup D_i = \mathcal{R}^p$) and a vector of $p$ Gaussian

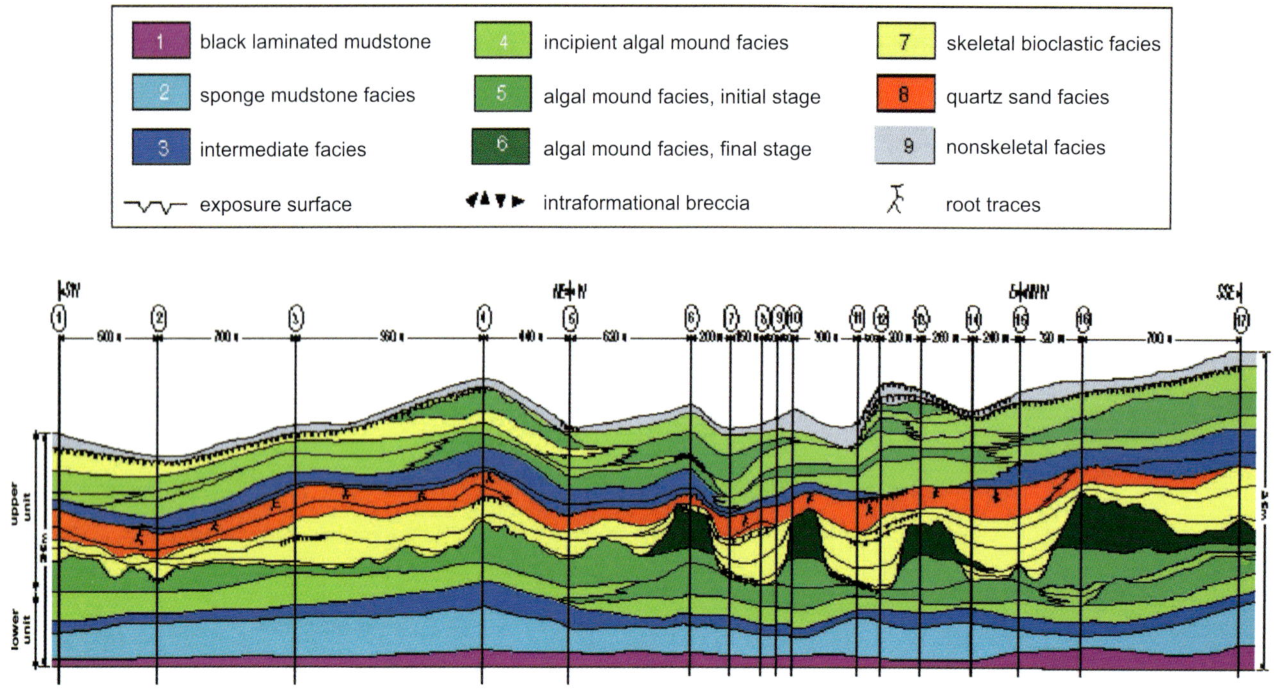


**Figure 7. Transect of the lower Ismay member interpreted in terms of sedimentary facies. The bioherms (in green) show at least two phases of bioconstruction that contrast with the tabular organization of the platform deposits below and above.**

random functions, $Y(x) = (Y_1(x), Y_2(x),\ldots, Yp(x))$, where $x$ is the location in 3-D space. By definition, facies $F_i$ consists of the points $x$ of the 3-D space for which $Y(x) \in D_i$. This truncation rule is called the lithotype rule because it synthesizes transitions between facies. If there are two Gaussian random functions, the lithotype rule can be visualized as a two-dimensional (2-D) diagram. Figure 8 shows a simple example where the partition is defined in terms of rectangles (because this simplifies the computations) and where the two Gaussian random functions are uncorrelated. Even with these simplifications, the results are still realistic. In fact, they correspond to erosion that could have been obtained in two steps using the regular TG approach. Note also that the anisotropy exhibited by the Gaussian random functions is intuitively expected on the facies, although as shown previously, the anisotropy indicated on the indicator semivariograms would be far more complicated. To this extent, TPG is more natural: knowing the anisotropy of facies, it is possible to directly choose the same anisotropy on the semivariograms of the Gaussian random function. The TPG is far more general than the previous example suggests. In some cases, it turns out to be instructive to use more than two Gaussian random functions. In addition, rectangles can be replaced by polygons that could give even more complex transitions.

Furthermore, Gaussian random functions that are correlated could be used. In such a case, one way to obtain a consistent model for the semivariograms and cross-semivariograms is to use the linear model for coregionalization (see Wackernagel, 1998). Another possibility is to use Gaussian random functions linked by linear transforms. Armstrong and Galli (1999) present a case of Gaussian random functions linked by partial differential equations. In Roth et al. (1998) a mining example is presented.

## Conditional Simulations

To model the facies, the Gaussian random functions are first simulated, and then the lithotype rule is applied. This means that, at each observed location, the values of the Gaussian random functions have to belong to the set $D$ corresponding to the lithofacies. In other words, for facies $F_i$ at point $x_0$ in a well, the values at $x_0$ have to be in the set $D_i$. To honor this information, Gibbs sampling is used to reconstitute consistent values at the well locations (Le Loc'h and Galli, 1997; Armstrong et al., 2003). Once the values at observed locations are obtained, classical Gaussian

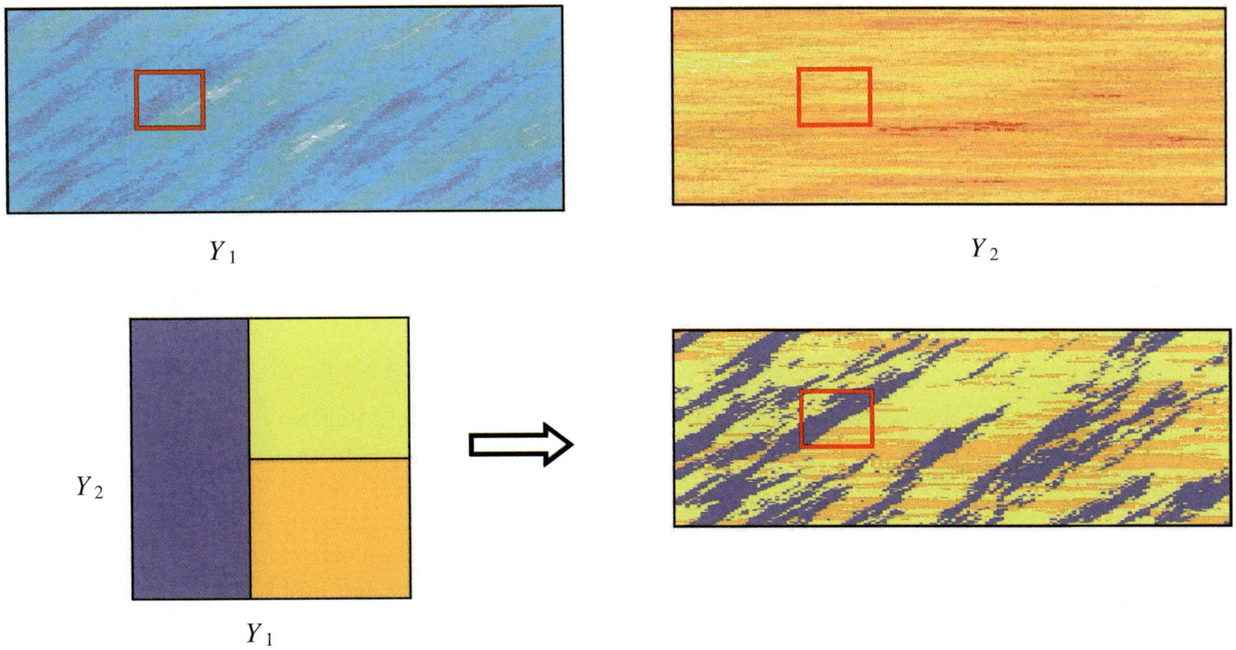


**Figure 8. A simple example showing how to construct three facies (bottom right) with different anisotropies from two Gaussian distributions (top) and a lithotype rule (bottom left). The red square highlights the way the lithotype rule is used.**

conditional simulation is performed (see Lantuejoul, 1994), where the conditioning part has to be accomplished by cokriging involving all the Gaussian random functions considered.

## Indicator Semivariograms

As is the case for the TG approach, there is a well-known correspondence between simple and cross-semivariograms of indicators and simple and cross-semivariograms of a Gaussian random function. The relationship is even simpler when written in terms of noncentered indicator covariances, $\sigma_{FiFk}$, between facies $F_i$ and facies $F_k$. As previously noted, this noncentered indicator covariance (property 4) is a bivariate probability. Because of the definition of the indicators in the TPG approach, this becomes

$$\sigma_{F_iF_k}(x,y) = P(Y(x) \in D_i \quad \text{and} \quad Y(y) \in D_k) \qquad (1)$$

The right-hand side of equation 1 represents a multi-Gaussian probability, so it can be written as an integral of the Gaussian density over $D_i \times D_k$ (that is, an integral in $2p$ dimensions). Because this density depends only on the covariance matrix between $Y(x)$ and $Y(y)$, the required relationship between the indicator (cross-)covariances and the Gaussian random function covariances and cross-covariances is obtained. In practice, it is used in the reverse way: the experimental semivariograms and cross-semivariograms of the indicators are computed, then the semivariograms and cross-semivariograms of the Gaussian random functions are selected, and the resulting fit is obtained using equation 1. Note that the previous formula is also valid in the nonstationary case. For more details on these relationships and their use in fitting indicator semivariograms, see Galli et al. (1994), Le Loc'h and Galli (1997), and Armstrong et al. (2003).

## Lithotype Rule

It is now necessary to choose a partition of the space $\mathcal{R}^P$ ($\mathcal{R}^2$ in the present case). The simplest way is to use rectangles. Information on lithotype transitions is used to determine where to place these rectangles in the space. In the present example, the unit has been split into upper and lower units, with one lithotype rule for each of them. For the lower unit, which is simpler (Figure 9b), only the first Gaussian random function is used for defining the facies. This corresponds to the regular TG approach. The upper unit is more complex. For application of the lithotype rule illustrated in Figure 9a, the three green facies and the yellow one are defined by the union of two rectangles. This is merely for convenience. It avoids having to split this unit into two subunits or to define different facies that would need to be regrouped after simulation. The lithotype rule shows which facies are

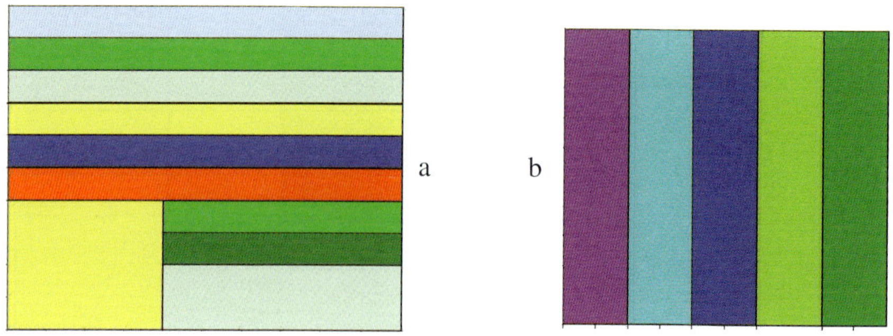

**Figure 9. (a) Lithotype rule for the upper unit in the lower Ismay Formation. (b) Lithotype rule for the lower unit in the lower Ismay Formation. See Figure 11 for the color code.**

in contact. For example, the red facies can touch (or have transitions with) the dark blue, the yellow, and the medium green ones. The red facies separates the dark blue one and the green one below, and it also separates the two yellow facies and green one. This pattern corresponds to what is seen on the outcrop. Looking carefully at the lithotype rule, it is apparent that the facies starting from the red to the light blue have properties determined only by the first Gaussian random function; but below the red facies, the yellow one and the three green ones have properties determined by both Gaussian random functions. This means that the spatial characteristics of the green and the yellow facies will be different according to the rectangle from which they originate.

When the lithotype rule has been decided, the precise boundaries of all the rectangles have to be computed. For a fixed partition, these boundaries depend on the proportions of each lithofacies. By definition, $p_{Fi}(x)$, the proportion of facies $F_i$ around location $x$, is given by

$$p_{F_i}(x) = P(Y(x) \in D_i) \tag{2}$$

So knowing the experimental proportion of each facies (and the lithotype rule), finding the location of the boundaries of the $D_i$ is what remains.

As an aside, it is worthwhile to emphasize that these formulas are valid even if the proportions are nonstationary (as previously noted, this also means that the indicators are nonstationary). This nonstationarity can be complex, involving variations in the proportions in the whole space, or can be simple, in which case they depend only on the elevation (after flattening according to a reference level). The latter situation is always present simply because of the cyclic character of sedimentation.

It is clear from equation 2 that the location of the boundaries depends on the correlation between the two Gaussian random functions, as well as on the proportions. This is true because equation 2 involves the $p$-dimensional integral of the density of the vector $Y(x)$. In the present example, it is only a 2-D integral involving the bivariate density of the two Gaussian random functions at the same point. This density involves the covariance matrix of $Y_1(x)$ and $Y_2(x)$, which has diagonal terms of 1 and off-diagonals terms that are just the correlation between $Y_1(x)$ and $Y_2(x)$.

## Vertical Proportion Curves and 3-D Matrices of Proportions

In most cases, a reservoir comprises different genetic units. Each of them corresponds to a parasequence of high order. Because of the relative change of position compared to the source of material, the amount of each facies varies in time for a given areal position. In a chronostratigraphical system (with changing coordinates), time is roughly equivalent to depth or thickness, so a marked evolution in the proportion of each facies is expected with depth. Mathematically speaking, this implies that the proportion $p_{Fi}$ of a given facies $F_i$ is a function of $z$.

The proportion is calculated, then, level by level, by computing the number of wells showing a particular facies as a proportion of the number of wells that penetrate that level. It is important to have an overview of this vertical variation. This is why vertical proportion curves have been designed (Matheron et al., 1987; Volpi et al., 1997).

Having chosen an order for the lithotypes, a vertical proportion curve is the representation of the accumulated proportions of the previous facies vs. depth. At a given depth, the space between curve $i$ and $i + 1$ is just the proportion of facies $i + 1$ at that depth. This is illustrated in Figure 9 for the two units from the Paradox basin. In the vertical proportion curve of the upper unit (Figure 9a), the dark green facies and the yellow one are present before and after the red one.

Note that an interaction exists between the vertical proportion curve and the lithotype rule. The vertical proportion curve also controls the transitions. If the proportion of a facies is small at a given level, the transition probability between this facies and another is generally also small, so that the lithotype rule is adjusted for each level. Figure 8 illustrates the situation based on the general rules, along with the possible transitions, but for each level that will be modulated at each level according the vertical proportion curve.

In some cases where the situation is more complex or where more information is available from seismic data, it might be useful to consider more than one proportion curve for the genetic unit. In addition, proportion curves can be computed for different areas, or alternatively, vertical proportion curves can be computed on the nodes of a coarse grid regrouping several simulation nodes. This gives a 3-D matrix of proportions (Beucher et al., 1993).

## MODELING RESERVOIR ARCHITECTURE WITH PLURI-GAUSSIAN RANDOM FUNCTIONS: THE PARADOX BASIN EXAMPLE

As noted previously, geological complexity can be difficult to reproduce in simulations using conventional object- or pixel-based algorithms. The facies architecture of the algal mounds in the example from the Paradox basin is a clear example. In such situations, the facies have a complex sequential organization that varies both vertically and laterally. Spatial relationships among facies depend on the stratigraphic level. In the Paradox basin example, the lower half of the first cycle has a tabular geometry with a strict vertical ordering among the facies. This layer-cake geometry contrasts with the mound and trough architecture of the bioherms, in which intermound facies laterally interfinger with the mounds. In addition, in this example, early diagenetic processes may affect some of the facies differentially, making the ordering more complex. Furthermore, the mounds have a complex geometry. Their shape is irregular, which influences the facies distribution within and around each mound. The system is characterized by concentric facies belts in the mounds and draping geometries for the intermound facies and capping beds.

The main mound complexes are located preferentially in the easternmost part of the regional transect. Van Buchem et al. (2000) have simulated the algal mounds at a regional scale with a pixel-based model, and in doing so, they have consequently insisted on a nonstationary configuration of the mound distribution. However, this observation is less crucial at the scale of the present study, in which mound distribution can be considered stationary along the 6-km (3.7-mi)-long transect.

Lithotype rules are a very convenient tool for reproducing the complex facies organization of the platform deposits and of the bioherms. For simulation purposes, the series has been divided into two units. This layering is based on the facies architecture and does not necessarily correspond to the sedimentary cycles identified by Van Buchem et al. (2000). The lower unit comprises the layer-cake platform facies, from the organic-rich mudstone at the base to the tabular phylloid layer. Its lithotype rule shows adjoining facies bands (facies 1–5, Figure 9b), completed by a marked vertical organization in layers identified by the vertical proportion curve (Figure 10b). This will prevent interfingering between facies (e.g., facies 1 and 5), which is in agreement with the facies sequence. The upper unit comprises the bioherms, the intermound facies, the capping and the sandstone beds, and the upper part of the series where mounds have low relief. Its lithotype rule (Figure 9a) is more complex.

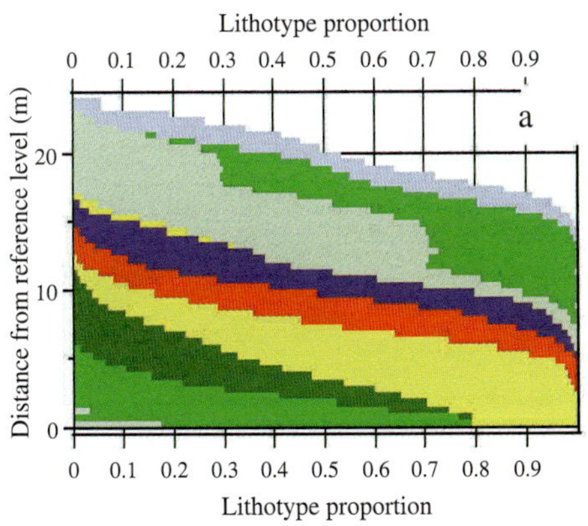


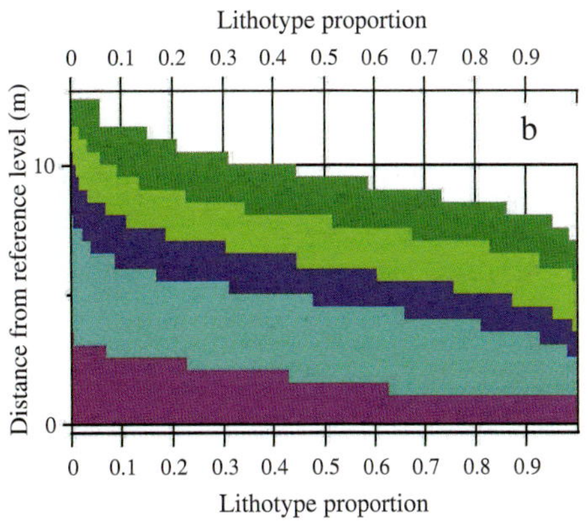


**Figure 10.** (a) Vertical proportion curves for the upper unit in the lower Ismay Formation. (b) Vertical proportion curves for the lower unit in the lower Ismay Formation.

**Figure 11.** Vertical section of simulation showing the tabular platform deposits (lower unit) and the algal mounds (upper unit) in the lower Ismay Formation.

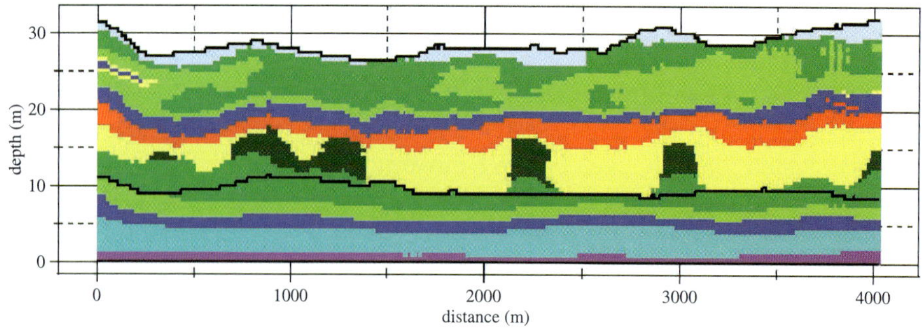


Intermound facies 7 can be in contact with the mound facies 5 and 6 and also with the draping sandstone beds (facies 8). The more tabular organization of the upper part of the series is respected by the different facies bands in the upper part of the lithotype rule. The corresponding vertical proportion curve (Figure 10a) also reflects this more complex geometry.

As previously noted, outcrop observations suggest the existence of large-scale lateral nonstationarity, but this is difficult to check or to quantify with the available information. Because the aim of the present work is to concentrate on the complex geometry of the mounds, the possible nonstationarity is ignored. The resulting layout of the 17 vertical sections (Figure 7) makes it difficult to calculate meaningful horizontal semivariograms for a stationary model. For these reasons, the semivariogram models are chosen to honor the spatial continuity of the facies that can be seen in the outcrop. After a few trials, a model consisting of two Gaussian random functions, each based on a Gaussian semivariogram, can be shown to be adequate. The first semivariogram has an isotropic horizontal range of 400 m (1312 ft) and a vertical range of 10 m (33 ft), whereas the second one has an isotropic horizontal range of 2000 m (6561 ft) and a vertical range of 7 m (23 ft). Furthermore, a correlation coefficient of −0.5 between the two Gaussian random functions has been chosen to accentuate the draping of the red facies over the mounds. With these models and the lithotype rule, the yellow and green lithofacies below the red facies in the lithotype rule are expected to be less horizontally continuous than the same colors above it. For the lower unit only, a Gaussian random function based on a Gaussian semivariogram is used. It has an isotropic horizontal range of 2000 m (6561 ft) and a vertical range of 7 m (23 ft).

## Simulation Results

Key features of these carbonate-platform deposits can be reproduced very realistically, as illustrated in Figures 11–13. The cross section (Figure 11) shows the distribution of mound and intermound facies, with the overlying facies draped over them.

**Figure 12.** Horizontal section of a simulation of the upper unit in the lower Ismay Formation.

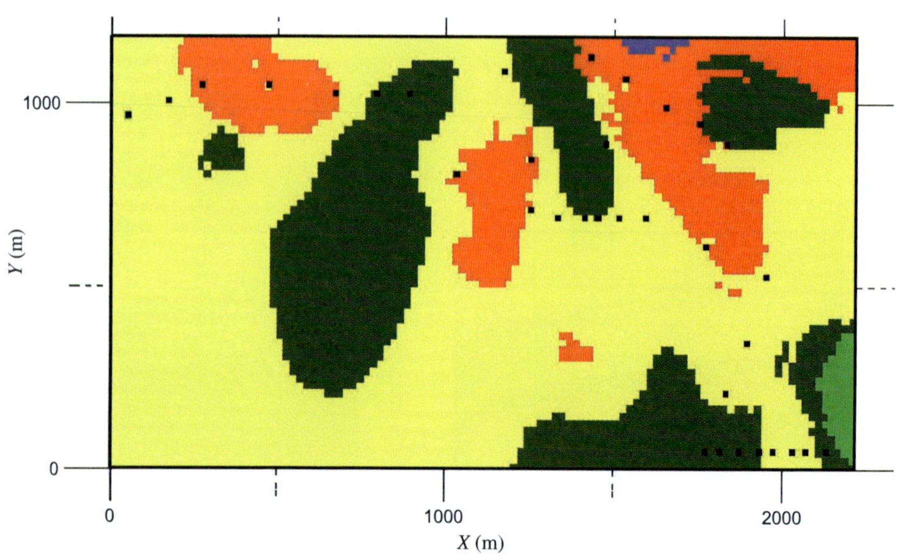

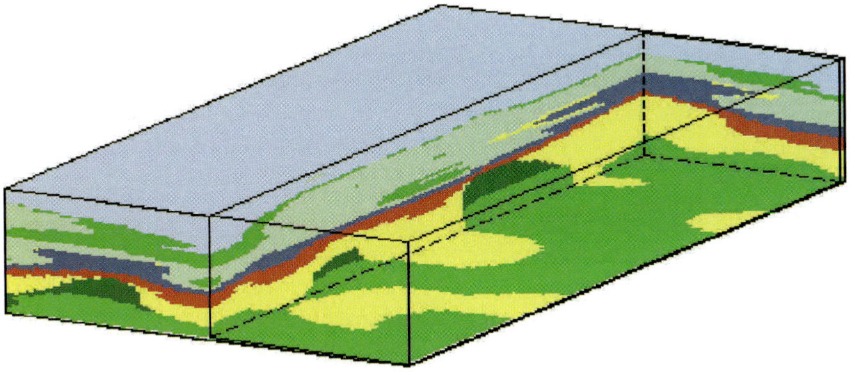

**Figure 13. Block diagram of the simulation of the upper unit showing a vertical section.**

Below and above the mounds, the layer-cake geometry of the carbonate-platform deposits is apparent. In plan view (Figure 12), the mounds show an irregular, rounded shape with concentric facies belts, which is precisely what is expected from a sedimentological point of view. Figure 13 shows these features in an excavated block diagram.

## CONCLUSION

Although stochastic models are widely used for simulating geology in the petroleum industry, few methods can honor the transitions between facies together with nonstationarity because of the cyclic aspect of sedimentation. This chapter illustrates how the TPG approach can be used to successfully accomplish this result. One feature of the method is its simple and intuitive way of representing transitions. Furthermore, it is easy to incorporate anisotropy. The directions are chosen intuitively, for example, following the direction of channels. This contrasts with the complexity of modeling indicator semivariograms directly, where part of the anisotropy has to be incorporated in each of the semivariograms. One of the strong points of the TPG approach is that it provides a very simple way to consistently model all of the indicator semivariograms. Another feature is that the truncation rule expressed simply as rectangles is easy to interpret in terms of geology (transitions between facies). Figures 11–13 show how well the method simulates complex real situations like the carbonate-platform deposits from the Paradox basin.

## REFERENCES CITED

Allègre, C., 1964, Vers une logique mathématique des séries sédimentaires: Bulletin de la Société Géologique de France, Série 7, v. 6, p. 214–218.

Armstrong, M., 1992, Positive definiteness is not enough: Mathematical Geology, v. 24, p. 135–143.

Armstrong, M., and A. Galli, 1999, Derivative based plurigaussian simulations, *in* S. J. Lippard, A. Naess, and R. Sinding-Larsen, eds., Proceedings of IAMG '99: The 5th Annual Conference of the International Association for Mathematical Geology, Tapir, p. 591–596.

Armstrong, M., A. Galli, G. Le Loc'h, F. Geffroy, and R. Eschard, 2003, Plurigaussian simulations in geosciences: New York, Springer-Verlag, 149 p.

Baars, D. L., and G. M. Stevenson, 1981, Tectonic evolution of the Paradox basin, Utah and Colorado, *in* D. L. Wiegand, ed., Geology of the Paradox basin: Rocky Mountain Association of Geologists field conference guidebook, p. 21–23.

Beucher, H., A. Galli., G. Le Loc'h, and C. Ravenne, 1993, Including a regional trend in reservoir modeling using the truncated Gaussian, *in* A. Soares, ed., Geostatistics Troia '92: Dordrecht, Kluwer Academic Publishers, p. 555–566.

Chidsey Jr., T. C., D. E. Eby, and D. M. Lorenz, 1996, Geological and reservoir characterization of small shallow-shelf fields, southern Paradox basin, Utah, *in* A. C. Huffman Jr., W. R. Lund, and L. H. Godwin, eds., Geology and resources of the Paradox basin: Utah Geological Association Publication, v. 25, p. 39–56.

Galli, A., H. Beucher, G. Le Loc'h, and B. Doligez, 1994, The pros and cons of the truncated Gaussian method, *in* M. Armstrong and P. Dowd, eds., Geostatistical simulation (Proceedings of the Geostatistical Simulation Workshop): Dordrecht, Kluwer Academic Publishers, p. 217–233.

Goldhammer, R. K., E. J. Oswald, and P. A. Dunn, 1991, The hierarchy of stratigraphic forcing: An example of middle Pennsylvanian shelf carbonates of the Paradox basin, *in* E. K. Franseen, W. L. Watney, C. G. St. C. Kendall, and W. Ross, eds., Sedimentary modeling: Computer simulations and methods for improved parameter definition: Kansas Geological Survey Bulletin, v. 233, p. 235–266.

Goovaerts, P., 1996, Stochastic simulation of categorical variable using a classification algorithm and simulated annealing: Mathematical Geology, v. 28, p. 909–922.

Grammer, G. M., G. P. Eberli, F. S. Van Buchem, G. M. Stevenson, and P. Homewood, 1996, Application of high-resolution sequence stratigraphy to evaluate lateral variability in outcrops and subsurface—Desert Creek and Ismay intervals, Paradox basin, *in* M. W. Longman and M. D. Sonnenfeld, eds., Paleozoic of the Rocky Mountain region: Rocky Mountain Section SEPM, p. 235–266.

Journel, A. G., and E. H. Isaaks, 1984, Conditional indicator simulation—Application to a Saskatchewan uranium deposit: Mathematical Geology, v. 16, p. 685–718.

Krumbein, W. C., 1967, Fortran IV computer program for Markov chain experiment in geology: Kansas Geological Survey Computer Contribution 13, 38 p.

Lantuejoul, C., 1994, Non-conditional simulation of stationary isotropic multigaussian random functions, *in* M. Armstrong and P. Dowd, eds., Geostatistical simulation (Proceedings of the Geostatistical Simulation Workshop): Dordrecht, Kluwer Academic Publishers, p. 147–177.

Le Loc'h, G., and A. Galli, 1997, Truncated plurigaussian method: Theoretical and practical points of view, *in* E. Y. Baafi and N. A. Schofield, eds., Geostatistics Wollongong '96: Dordrecht, Kluwer Academic Publishers, p. 211–222.

Matheron, G., 1989, The internal consistency of models in geostatistics, *in* M. Armstrong, ed., Geostatistics: Proceedings of the Third International Geostatistics Congress: Dordrecht, Kluwer Academic Publishers, p. 21–38.

Matheron, G., H. Beucher, C. de Fouquet, A. Galli, D. Guerillot, and C. Ravenne, 1987, Conditional simulation of the geometry of fluvio-deltaic reservoirs, *in* Proceedings of the Society of Petroleum Engineers Annual Technical Conference and Exhibition, Dallas, Texas, September 27–30, Reservoir Engineering, SPE Paper 16753, p. 591–599.

Montgomery, S. L., T. C. Chidsey Jr., D. E. Eby, D. M. Lorenz, and W. E. Culham, 1999, Pennsylvanian carbonate buildups, Paradox basin: Increasing reserves in heterogeneous shallow-shelf reservoirs: AAPG Bulletin, v. 83, p. 193–210.

Ravenne, C., A. Galli, H. Beucher, R. Eschard, and D. Guerillot, 1991, Outcrop studies and geostatistical modeling of a Middle Jurassic Brent analogue, *in* G. Imarisio, M. Frias, and J. M. Bemtgen eds., Proceedings of the European Oil and Gas Conference, A Multidisciplinary Approach in Exploration and Production R&D: London, Graham and Trotman, p. 497–520.

Rivoirard, J., 1993, Relation between the indicators related to a regionalized variable, *in* A. Soares, ed., Geostatistics Troia '92: Dordrecht, Kluwer Academic Publishers, p. 273–284.

Roth, C., M. Armstrong, A. Galli, and G. Le Loc'h, 1998, Using plurigaussian simulations to reproduce lithofacies with contrasting anisotropies, *in* Applications of Computers and Operations Research in the Mineral Industry '98, Computer applications in the mineral industries: London, Institution of Mining and Metallurgy, p. 201–213.

Van Buchem, F. S., B. Doligez, R. Eschard, O. Lerat, M. Grammer, and C. Ravenne, 2000, Stratigraphic architecture and stochastic simulation of a mixed carbonate/siliciclastic platform (Upper Carboniferous, Paradox basin, U.S.A.), *in* P. Homewood and G. P. Eberli, eds., Genetic stratigraphy on the exploration and production scales: Case studies from the Upper Devonian of Alberta and the Pennsylvanian of the Paradox basin: Elf EP Editions Memoire 24, p. 109–130.

Volpi, B., A. Galli, and C. Ravenne, 1997, Vertical proportion curves: A qualitative and quantitative tool for reservoir characterization, *in* Memorias del I Congreso Latinoamericano de Sedimentologia, Sociedad Venezolana de Geólogos Tomo 1 (November), p. 351–358.

Wackernagel, H., 1998, Multivariate geostatistics, 2d ed.: New York, Springer-Verlag, 291 p.

Pyrcz, M. J., E. Gringarten, P. Frykman, and C. V. Deutsch, 2006, Representative input parameters for geostatistical simulation, *in* T. C. Coburn, J. M. Yarus, and R. L. Chambers, eds., Stochastic modeling and geostatistics: Principles, methods, and case studies, volume II: AAPG Computer Applications in Geology 5, p. 123–137.

*Chapter 10*

# Representative Input Parameters for Geostatistical Simulation

**M. J. Pyrcz**
*Chevron Energy Technology Company*
*Houston, Texas, U.S.A.*

**E. Gringarten**
*Earth Decision*
*Houston, Texas, U.S.A.*

**P. Frykman**
*Geological Survey of Denmark and Greenland*
*Copenhagen, Denmark*

**C. V. Deutsch**
*University of Alberta*
*Edmonton, Alberta, Canada*

## ABSTRACT

Geostatistical-simulation techniques are increasingly being used to create heterogeneous realizations for flow modeling and to assess uncertainty in hydrocarbon resources and reserves. These geostatistical-simulation techniques reproduce the input statistics within ergodic fluctuations. The input statistics representing various model parameters must be computed from data that are representative of the entire domain being modeled. Geostatistical simulation does not accommodate a lack of representativeness in the data. Moreover, the extent to which the input statistics are reproduced depends almost exclusively on the size of the modeling domain relative to the range of spatial correlation; fluctuations in realizations of the full reservoir model do not depend entirely on the uncertainty of the input statistics. It is necessary to explicitly incorporate the uncertainty of the input statistics because they have a much larger and more realistic impact on the uncertainty of the full reservoir model than stochastic fluctuations. The best practices for determining representative input values of model parameters and quantification of their uncertainty are presented in this chapter.


DOI:10.1306/1063811CA53230

## INTRODUCTION

The combinatorial space of uncertainty in reservoir modeling is inconceivably vast. Local well data measure less than one trillionth of a typical reservoir volume. Seismic data provide excellent coverage, but its scale of measurement is very large, and it must be calibrated to reservoir properties using limited well data. Reservoir models are commonly not adequately constrained by local well data. The choice of a geostatistical modeling approach and the values specified for the associated input model parameters provide the most significant constraints on the modeling results. It is essential that the input statistics be representative of the entire domain being modeled.

Histograms and other statistical characteristics must be assembled from the local well data and commonly supplemented by outcrop studies and analog field data. However, wells are not drilled to be statistically representative, and core data are not always taken uniformly from good- and poor-quality reservoir rock. The geostatistical modeler must take great care to assemble values of the model parameters that are derived from representative data.

### Impact of the Histogram

To demonstrate the importance of adequately specifying the input histogram of a reservoir property such as porosity, consider the four wells shown in Figure 1. These wells are located in the North Cowden field in west Texas, where 62 wells have been drilled (the data were kindly provided by Amoco Oil Company, now part of British Petroleum, as an industry training set for the development of geostatistical methodologies). Porosity is given in units of percent averaged over the vertical thickness of the reservoir. The histogram on the right side of Figure 1 characterizes the vertical average of porosity for the 62 wells. The four wells have been randomly chosen to provide realistic local conditioning for the geostatistical realizations of porosity.

Note that the observed porosity distribution appears to be bimodal (see right of Figure 1). This indicates the presence of two distinct facies types: dolomite and siltstone. In practice, the two facies should be treated independently; however, for the purpose of the present discussion this does not need to be done.

The mean of the vertical average porosity (henceforth referred to as average porosity) of the 62 wells is 8.4%. Two realizations constructed with sequential Gaussian simulation using the reference (base case) histogram and the four wells as conditioning data are shown in the center of Figure 2. An isotropic semivariogram is used to fit the experimental normal-scores semivariogram of the set of 62 wells based on a spherical model with a range of 3000 ft (914 m). The histogram in the top center is the observed base case histogram of average porosity. The histogram in the bottom center characterizes the resulting distribution of pore volume over 101 realizations. The units of pore volume are millions of cubic feet, assuming an average reservoir thickness of 10 ft (3 m). Note the uncertainty in pore volume caused by fluctuations among the realizations.

Considerable uncertainty exists with regard to the true distribution of average porosity. Consider three estimates of the distribution of average porosity, with associated means of 7.5, 8.4, and 10%. These estimated distributions are shown on the top of Figure 2. Each of these estimated distributions were applied as reference distributions to calculate 101 realizations with sequential Gaussian simulation (two realizations of each case are shown in Figure 2). Although all 101 realizations are constrained to the

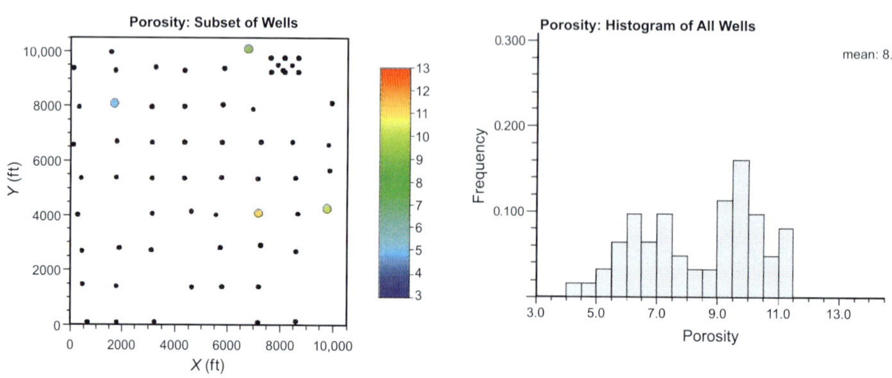


**Figure 1. The locations of all 62 wells in North Cowden field and four randomly selected wells (colored dots), along with the observed distribution of average vertical porosity (in percent) for all wells.**

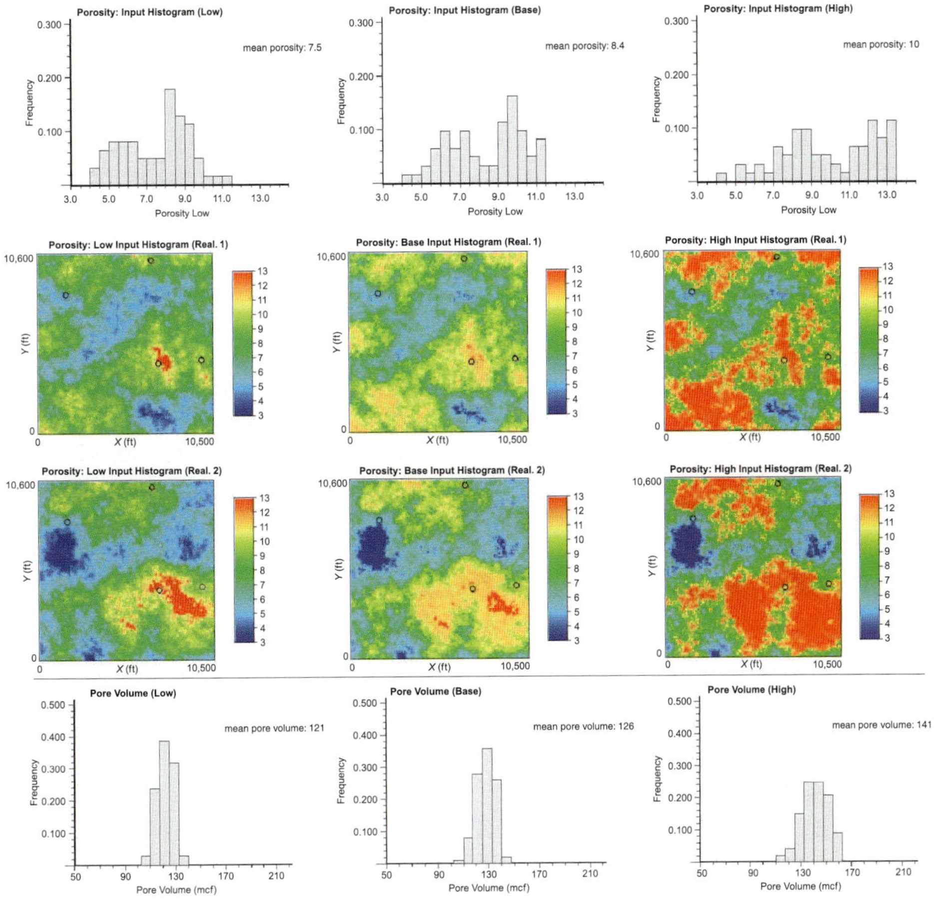


**Figure 2. Three input distributions (low, base, and high cases) of average vertical porosity, two example realizations of a geostatistical model of average vertical porosity, generated with sequential Gaussian simulation conditioned to the four sampled wells, based on each distribution, and the resulting distributions of pore volume (million cubic feet assuming an average reservoir thickness of 10 ft [3 m]) compiled from 101 realizations generated for each case.**

data in the four wells shown in Figure 1, the estimated porosity distribution has a first-order impact on the estimate of pore volume. A low case is shown on the left of Figure 2. This case represents an 11% reduction in mean average porosity, which translates to a 5% reduction in mean pore volume compared to the base case (the histogram of pore volume resulting from the 101 realizations is shown in the lower left of Figure 2). The right side of Figure 2 shows the high case based on a mean average porosity of 10%. The histogram of pore volume resulting from 101 realizations is shown in the lower right of Figure 2. The small number of wells and the relatively large range of correlation dampen the effect of the change in mean porosity (20%); the mean pore volume increases by 10% compared to the base case.

This example demonstrates the effect that estimates of model parameters can have on the ultimate estimates of reservoir performance or capacity, such as total pore volume or in-place reserves. In this example, the impact of the observed or empirical histogram of average porosity would have been even more dramatic if the correlation range had been shorter; in geostatistical simulation, the entire emphasis is placed on the histogram beyond the range of correlation among the wells.

The performance of geostatistical simulation models is directly controlled by the input statistics. Imprecision in these values is directly imputed to the simulation results, and decisions based on such realizations may be ill informed. Determining the best estimates of model parameters is of first-order importance.

Geostatistical simulation algorithms reproduce the inputs in ergodic statistical fluctuations. These statistical fluctuations depend almost entirely on the size of the study area relative to the range of correlation; they do not depend on the uncertainty in the inputs. Such uncertainty must be explicitly treated outside the simulation algorithm because the inputs themselves can be statistics computed for purposes of estimating model or algorithm parameters. These concepts are illustrated below.

## Effect of the Domain Size

The effect of domain size on statistical fluctuations in the input histogram is demonstrated with the same four wells shown in Figure 1. Three different domain sizes are considered: area 1 (8000 × 7000), area 2 (16,000 × 15,000), and area 3 (24,000 × 23,000) (all units are in feet). The locations of the four wells and the nested domains are shown in Figure 3. The semivariogram model was held constant as a single, isotropic spherical structure with a range of 3000 ft (914 m) and with no nugget effect. The reference (base-case) distribution was considered as the input histogram for all cases. Thus, all parameters are held constant except for the size of the modeling domain.

The histograms for three example realizations are shown for each domain size, with histograms from area 1 in the left column, area 2 in the center column, and area 3 in the right column of Figure 4.

To illustrate the statistical fluctuations between realizations, the mean porosity of each realization was calculated, and the histograms of the means are shown for 101 realizations for each domain. The means associated with the smallest domain size, area 1, have a standard deviation of 0.69, whereas those for area 2 have a standard deviation of 0.41. This represents a 20% decrease in variability in the mean caused by increasing the domain by a factor of 4. The means associated with area 3 have a standard deviation of 0.35, which represents a 50% decrease in variability in the mean caused by increasing the domain size by a factor of 8. The magnitude of variability in mean porosity between realizations decreases as the domain size increases. This is caused by volume variance relations and is independent of the reliability of the input statistics.

The need to explicitly account for uncertainty in input statistics applies to both categorical and continuous variables. All stochastic facies modeling techniques require the values of input facies proportions and spatial parameters such as semivariograms or size distributions to be specified. These values are almost exactly reproduced regardless of how well they represent the underlying truth. Moreover, the accuracy with which they are reproduced depends on ergodic fluctuations.

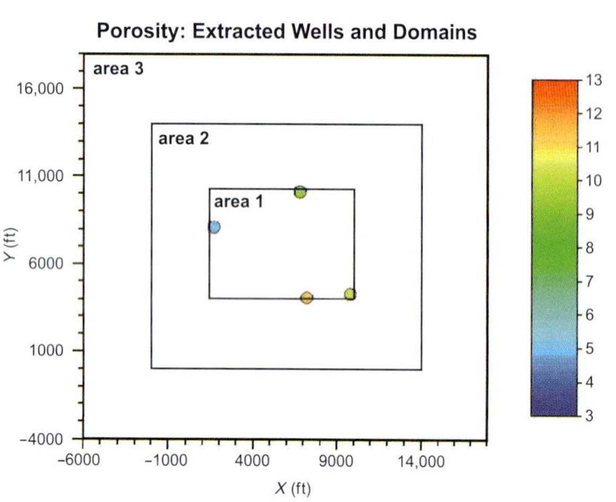


**Figure 3. The four sampled wells relative to three nested domains: area 1, area 2, and area 3. The color bar indicates average vertical porosity.**

Petrophysical properties are modeled in facies associated with a structural framework. The histogram and semivariogram for continuous variables, such as porosity and permeability, are subject to the same issues of representativeness and ergodicity.

This chapter is divided into two parts: the first part addresses the need to derive input statistics such as the histogram and semivariogram from representative data, whereas the second part addresses the characterization of uncertainty in input statistic and the transfer of that uncertainty to realizations of the full reservoir model.

## INPUT STATISTICS AND THE REPRESENTATIVENESS OF SAMPLE DATA

Commonly, too little attention is paid to the representativeness of the sample data and its impact on the calculation of input statistics. Many software packages provide only cursory tools to permit the modeler to evaluate data quality, establish local variations in model parameters, and assess uncertainty in input statistics. For example, the input proportions of facies are absolutely critical for both static resource assessment and production forecasts. However, modelers commonly fix them as the naïve proportions from well data without consideration of vertical and areal trends or the possibility that the existing wells are targeted at high net-to-gross areas of the reservoir.

The reservoir or model parameters considered here include facies proportions, histograms (or at least the mean) of continuous variables, correlation coefficients with secondary data sources, size distributions for object-based modeling, or semivariograms for cell-based facies modeling and the assignment of porosity and permeability. Conventional statistical methods used to estimate such parameters inherently assume they are stationary. The concept of stationarity is described below, and ways to ensure data representativeness are discussed.

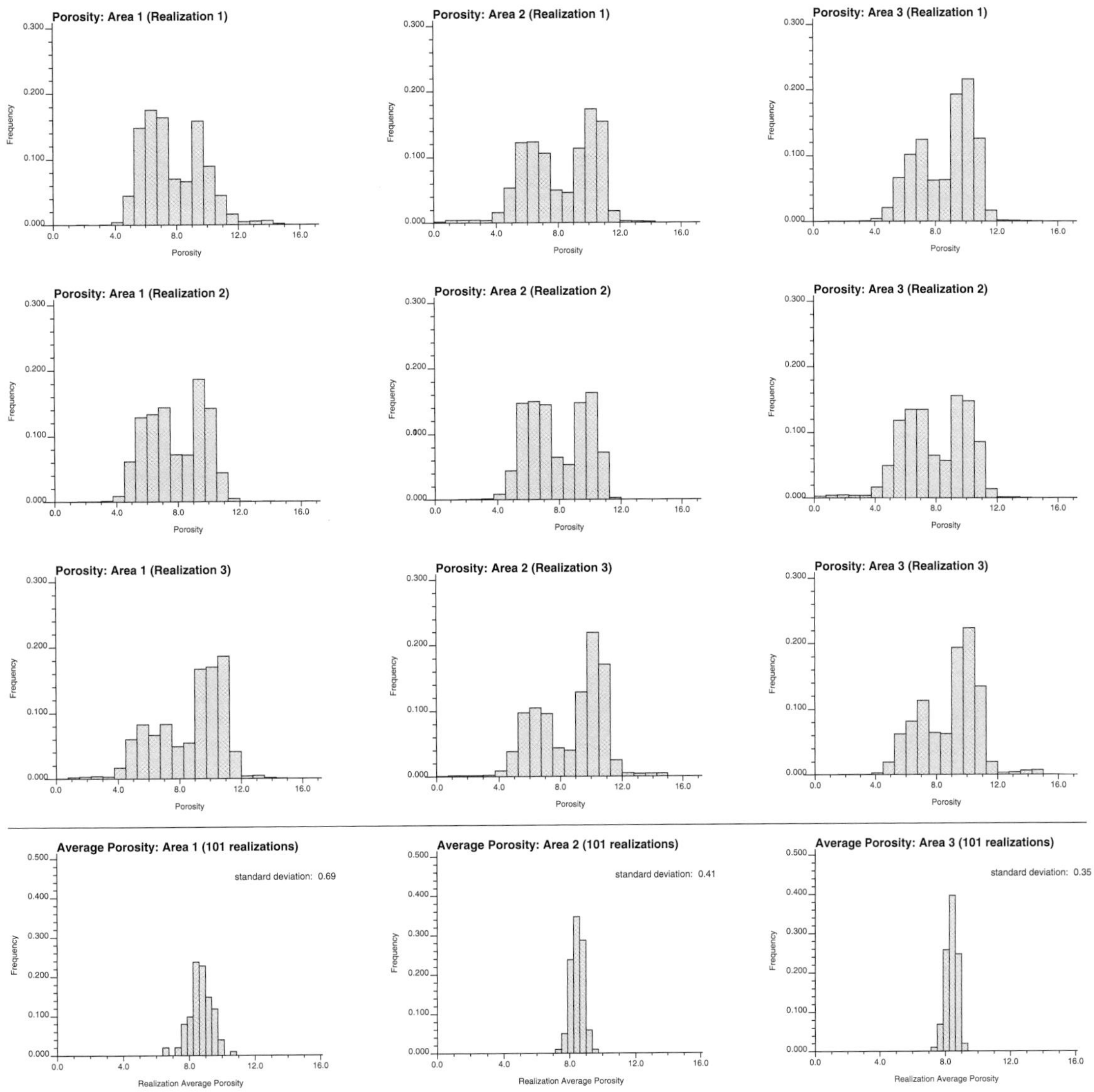


**Figure 4. The empirical distributions of average vertical porosity based on three realizations generated with sequential Gaussian simulation for each of the three domains. The distributions of the mean values from 101 realizations over each of the three domains are shown at the bottom.**

## STATIONARITY

Stationarity is a property of a geostatistical model in which the expected value of a particular statistic $S$ is independent of location:

$$E\{S(\mathbf{u})\} = s, \text{ for all } \mathbf{u} \text{ in area } A \qquad (1)$$

where $\mathbf{u}$ is a location vector, the statistic $S$ could be the mean, and $A$ is the chosen area of interest such as a particular facies. All statistical and geostatistical procedures assume stationarity at the time of modeling. Geological variables, however, commonly exhibit areal and vertical variations that invalidate this statistical assumption. Stationarity can be relaxed by working with residuals from a locally varying (or nonstationary) mean value or using the geological trend as some form of secondary conditioning data.

Practitioners have become quite creative in working around stationarity in geostatistical modeling. It is common to use locally varying proportion or mean models for facies and for continuous properties like porosity and permeability. The variance may be gridded and considered locally variable; the directions of anisotropy and other continuity parameters such as the semivariogram range or relative nugget effect may be gridded and used for local control; and the correlation between multiple variables may be made locally varying to account for data quality and geological variability.

A typical method for constructing a local varying mean or trend model is demonstrated for North Cowden field (see Figure 5). The trend in the horizontal direction may be calculated as a smooth map of the vertically averaged porosity data. The vertical trend may be calculated by a smooth representation of vertical porosity averaged over horizontal bins. Care must be taken to account for stratigraphic correlation between wells. The $z$-coordinate should be transformed to conform to the stratigraphic correlation prior to calculating the vertical trend.

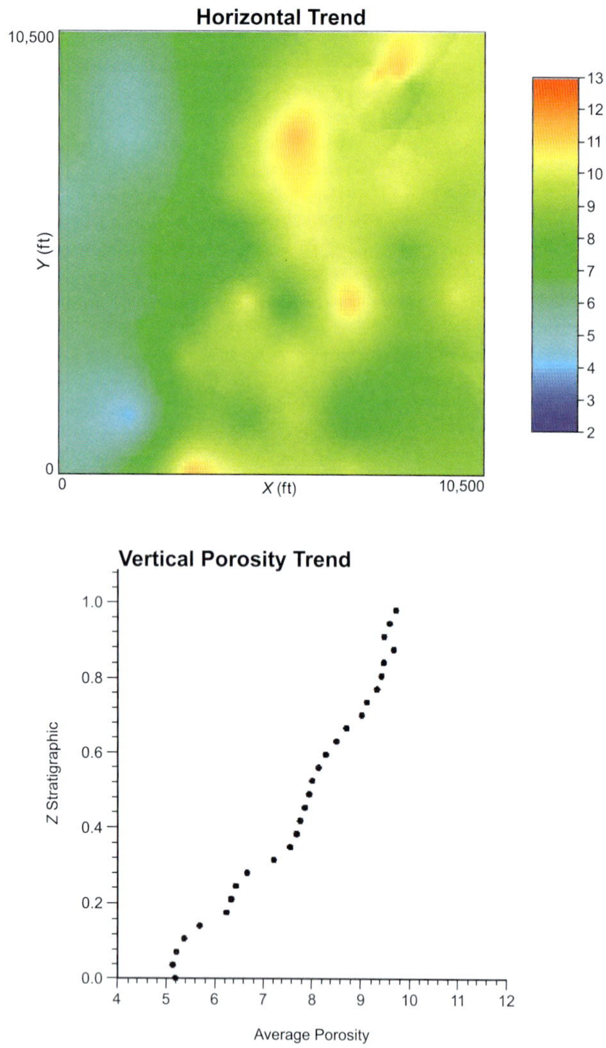


**Figure 5. An example horizontal and vertical trend calculated from all 62 wells. Average porosity is given in percent. Top: the trend in the horizontal may be calculated by a smooth fit of the vertically averaged data. Bottom: the vertical trend may be calculated using a smooth representation of porosity averaged over horizontal bins. Note the use of a stratigraphic $z$-coordinate 0.0–1.0, indicating a proportional correlation style.**

Horizontal and vertical trends may be merged into a three-dimensional trend model. By assuming conditional independence, the following relation may be applied that ensures that the global mean, as well as vertical and horizontal trends, are reproduced in the trend model:

$$\overline{\phi}(x, y, z) = \overline{\phi}(x, y) \cdot \frac{\overline{\phi}(z)}{\overline{\phi}} \tag{2}$$

where $x$, $y$, and $z$ are horizontal and vertical coordinates; and $\phi$ (with no parenthetical argument) is the overall average of vertical porosity. Despite creativity in mapping the values of locally varying parameters, the inherent property of stationarity is invoked in the construction of geostatistical models, and the input statistics will be reproduced. Hence, it is essential to assemble representative data from which input statistics can be computed. Methods for improving the representativeness of the sample histogram include declustering and debiasing.

## DECLUSTERING

Declustering is well documented and widely applied (Isaaks and Srivastava, 1989; Deutsch and Journel, 1997; Goovaerts, 1997; Deutsch, 2002). Common declustering methods include cell and polygonal declustering. Declustering methods rely on the weighting of the sample data to account for spatial representativeness.

All declustering techniques assume that the entire range of the true distribution has been sampled; that is, the presence of both low and high values is known and is sampled in varying degrees. Declustering is not effective when the entire range of values has not been sampled. The weight or influence of each sample value can be adjusted, but the potential effect of unsampled values is generally ignored.

The effect of weighting data values is demonstrated in Figure 6. Note that only the height of the bars in the histogram changes as the data weights are adjusted.

The cell declustering technique is the most common approach. It is insensitive to the location of the boundary and is simple to apply in three dimensions. For these reasons, it is seen as more robust than polygonal declustering. The essential idea of cell declustering is to assign a weight, $w_i$, to each data value that is inversely proportional to

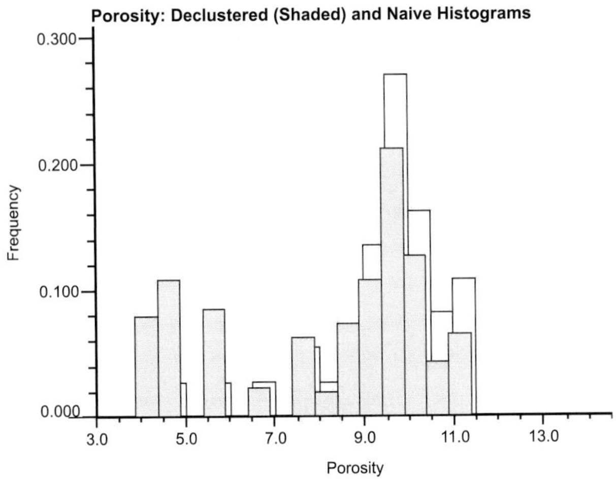


**Figure 6. The effect of weighting data values. The data values remain the same, but their relative weights are modified to consider spatial representativeness. Note that the naïve histogram is slightly offset for visibility.**

the product of the number of occupied cells ($L_o$) and the number of data in the same cell as datum $i$, $n_c(i)$:

$$w_i = \frac{1}{L_o - n_c(i)} \quad i = 1, \ldots, n \tag{3}$$

Clearly, the weights assigned by cell declustering depend on the cell size. If the cell size is set as very small, then every sample occupies its own cell ($n_c(i)$ is 1 for all data) and the result is equal weighting or the naïve sample distribution. If the cell size is very large, then all samples reside in the same cell ($n_c(i)$ is $n$ for all data), and the result is once again equal weighting. The spacing between the widely spaced data or the spacing of an underlying regular grid is suitable for cell declustering.

When it is difficult to make a choice, a common procedure is to assign a cell size that maximizes or minimizes the declustered mean (the declustered mean is maximized if the data are clustered in low-valued areas, and it is minimized if the data are clustered in high-valued areas). This procedure is applied when the sample values are clearly clustered in a low or high range. Automatically assigning the minimizing or maximizing cell size may lead to less representative results than simply using the original distribution.

To illustrate the approach, a subset of 37 wells was selected from the 62 wells in the North Cowden field. The selection was performed so that the high-porosity area in the top right of the domain is over-represented, and the remaining low-porosity zone is underrepresented (see the top left of Figure 7). The clustered histogram is shown in the top right of Figure 7 with a mean porosity of 9.4% that is 12% higher than the reference mean porosity of 8.4%. The declustering cell size was chosen such that the declustered mean was minimized, as shown in the bottom right of Figure 7. The application of the declustering weights resulted in a histogram with a mean porosity of 8.4% (see bottom right of Figure 7).

Polygonal declustering is also commonly applied in a variety of scientific disciplines for the purpose of correcting clustered data. The method is flexible and straightforward. The polygonal declustering technique is based on the construction of polygons of influence or Voronoi polygons for each of the sample values. The weight applied to each value is proportional to the area of its polygon of influence.

The weight assigned to edge values is very sensitive to the boundary location. If the boundary is located far from the data, then the edge values will receive a large weight, because the area of their polygons of influence increases. In general, this sensitivity to the boundary is perceived as a weakness of the approach, although polygonal declustering is well suited to two-dimensional settings with well-defined boundaries or to declustering within facies.

To illustrate this approach, polygonal declustering was applied to the clustered subset of wells described above. The well locations and the associated Voronoi polygons are shown on the left of Figure 8. The resulting polygonal declustering weights were applied to the clustered histogram (see top right of Figure 8). The weighted histogram (see right of Figure 8) has a mean porosity of 8.3% compared to the reference of 8.4% and the clustered mean of 9.4%.

In both examples described above, declustering performed well because the entire underlying distribution was sampled, even if not in a representative manner. If the entire distribution had not been sampled (a condition referred to as spatial bias), trend modeling or debiasing would need to have been considered.

## TREND MODELING AND DEBIASING

There may be evidence of geological trends even if there are inadequate data to apply conventional declustering techniques. This information may indicate that the sample data are spatially biased. In this case, declustering weights are not able to correct

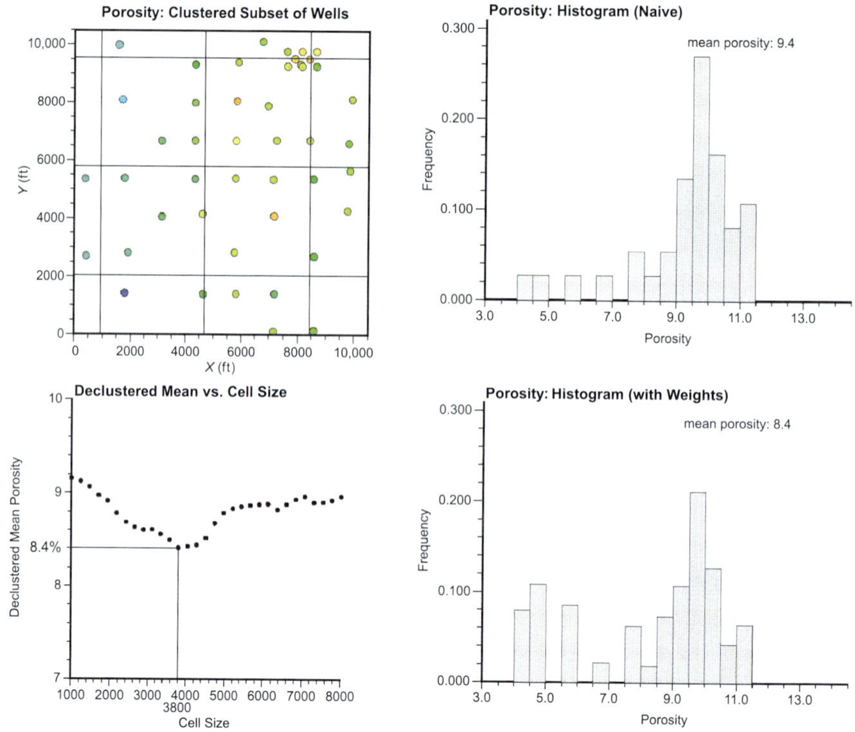


Figure 7. An example of cell declustering applied to the histogram of average porosity from a clustered subset of 37 wells in the North Cowden field. Top left: the clustered subset of wells with the declustering cell size illustrated. Top right: the naïve, equally weighted, histogram. Bottom left: the declustered mean vs. cell size. Bottom right: the weighted histogram. Average porosity is given in percent.

the distribution for representativeness. Two related methods (trend modeling and debiasing) can be used to make the adjustment, both of which rely on soft or secondary data.

In the presence of a clear and persistent trend, trend modeling may be applied to ensure that the correct distribution is reproduced. Trend modeling is well established (Goovaerts, 1997; Deutsch, 2002). The steps are as follows: (1) remove the trend a priori; (2) stochastically model residuals; and (3) replace the trend a posteriori. It is necessary to make a decision on the scale of the trend model. A large-scale trend captures course features and leaves the remaining variability to be modeled as residual, whereas a small-scale trend also captures fine features and leaves a smaller residual component. The results, by construction, reproduce the trend. The results of applying trend modeling to vertical porosity in the wells from North Cowden field are illustrated in Figure 5.

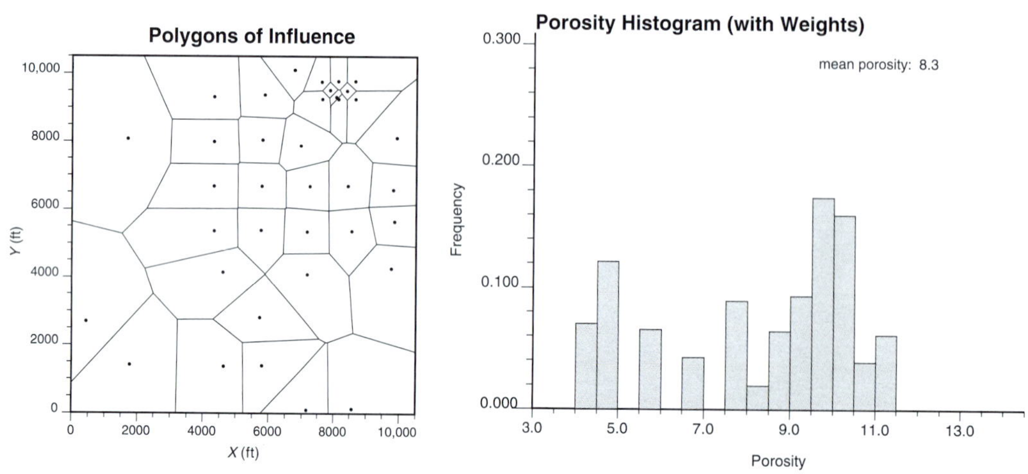


Figure 8. An example of polygonal declustering applied to the histogram of average porosity from a clustered subset of 37 wells in the North Cowden field. Left: the polygons of influence used to assign data weights. Right: the weighted histogram. Average porosity is given in percent.

Both advantages and disadvantages exist to this technique. One advantage is that the simulation step may be simplified because cosimulation is not necessarily required to integrate the soft information. Further, the use of trend modeling has a direct impact on the level of uncertainty inherent in the full reservoir model. The trend modeling procedure entails a decomposition of each reservoir property of interest into deterministic and stochastic components that leads to a reduction in overall model uncertainty. The major disadvantage of trend modeling is that data constraints are not intrinsically honored. For example, a porosity model may have negative porosity values after the addition of trend and residual components.

Although this technique corrects the trend features, no direct control over the resulting histogram is present. Introducing a trend with collocated cokriging or Gaussian simulation and a local variable mean model amounts to changing the locations of relative high and low values but not the actual histogram because the data are transformed. Hence, care should be taken in the trend-modeling phase to insure that the mean of the residuals is close to 0.0 and the correlation between the trend components and residuals is close to 0.0.

Another technique is to use soft (secondary) data that are representative of the entire area of interest, along with an understanding of the relationship between the primary and soft (secondary) data to adjust the primary distribution (Frykman and Deutsch, 1998). This adjusted distribution is used as a reference for the subsequent simulation of the primary variable. The underlying relationship between the primary and secondary data may be inferred from geologic information or other analog data, but it commonly may not be observed directly. Observable or not, such a relationship between the soft (secondary) and primary data, expressed as the bivariate distribution $\hat{f}_{x,y}(x, y)$, must be established for debiasing.

A variety of techniques can be used to establish this bivariate distribution. The simplest and most flexible approach is to empirically construct data pairs that describe the relationship without regard to some form of calibration or prediction process. For each pair, a weight is then assigned to the primary data value based on the secondary data distribution.

Another method is to calculate a series of conditional distributions of the primary data given the secondary data, $\mathbf{f}_{\text{primary}|\text{secondary}}$, over the range of the collocated secondary values. This relationship can be extrapolated over the range of all secondary data by a bivariate trend. The primary distribution is then constructed by scaling the binned bivariate relationship with respect to the secondary distribution. This is a discrete approximation to the primary distribution as expressed in equation 4.

$$f_y(y) = \int_x f_x(x) \cdot f_{y|x}(y|x)\mathrm{d}x \qquad (4)$$

The debiasing method explicitly adjusts the global distribution and retains consistency by employing the secondary data as collocated data in the simulation. This results in direct control over the shape of the histogram, through a reference distribution, and indirect control over trend reproduction through the secondary data. The method has been successfully applied in a recent reservoir modeling case study (Vejbæk and Kristensen, 2000).

To illustrate the debiasing technique, a subset of 38 wells was chosen from the 62 wells in North Cowden field such that the low-porosity regions were not sampled (see the top left of Figure 9). Available seismic information was used as a representative data source to aid in inferring the entire porosity distribution (see top right of Figure 9). A potential bivariate relationship is indicated on the scatter plot of the porosity from the 38 spatially biased wells and the collocated seismic attribute (see center of Figure 9). The debiased porosity distribution is established by applying equation 4. The original biased distribution and the resulting debiased distribution are shown on the bottom left and right of Figure 9, respectively.

Trend modeling and debiasing yield different results for integration into the full reservoir model. With trend modeling, the geostatistical simulation is augmented by information concerning the spatial behavior of the primary variable. Debiasing, however, relies on information concerning a more representative collection of secondary data and its relationship to the primary data. Data quality and sufficiency is key to the successful application of both methods, and the use of both methods has an impact on the uncertainty associated with the full reservoir model.

## PARAMETERS FOR OBJECT-BASED MODELING

Facies models constructed with object-based techniques reflect the well-defined parametric shapes used as input. These shapes include channels, levees,

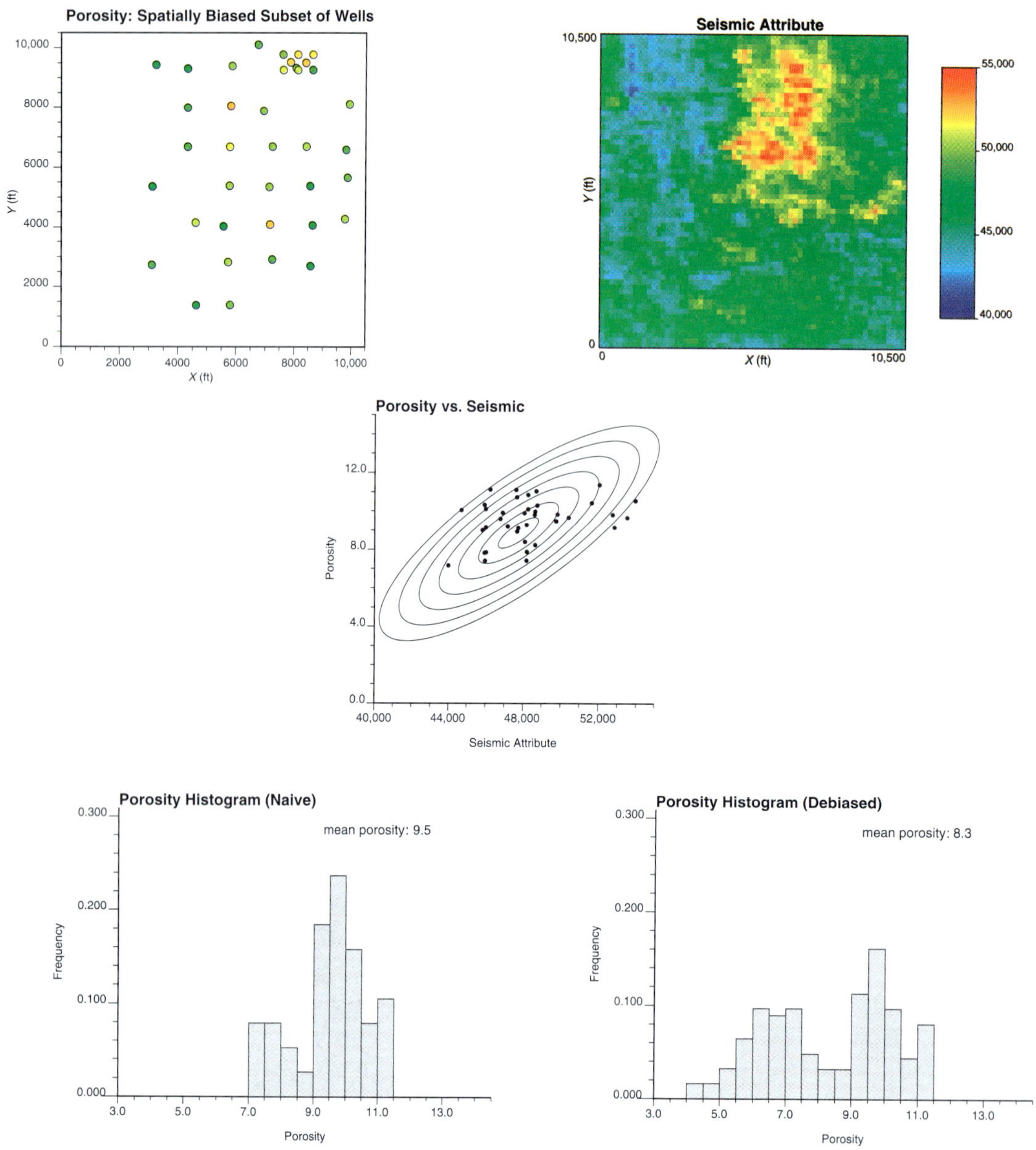


**Figure 9. An example of debiasing applied to a spatially biased subset of 38 wells in the North Cowden field. Top left: the spatially biased subset of wells. Top right: the representative secondary soft data (seismic attribute). Center: the scatter plot of the average porosity data and the collocated seismic data illustrating the bivariate relationship. Bottom left: the naïve histogram of the spatially biased subset. Bottom right: the debiased histogram based on the seismic distribution and the bivariate relationship. Average porosity is given in percent.**

crevasse splays, ellipsoidal concretions or remnant shales, barchan dunes, beach sand bars, submarine fans, and so on. Provided the parametric shape is relevant to the reservoir, object-based models are very appealing. Semivariogram-based facies models cannot represent shapes. However, they may be appropriate in settings where there are no clear shapes, as in the case of many carbonate reservoirs or diagenetically controlled facies.

The input shapes are commonly based on a conceptual geologic model and are not directly observed with reservoir data. There are notable exceptions where channel forms are sometimes observed in seismic reflections and wells intersect geologically well-defined rock types. Little can be done to validate the representativeness of estimates of model parameters such as these derived from a conceptual model. Therefore, it is important that the full range of shapes and sizes be considered.

The extent of various reservoir components must be evaluated in conjunction with the estimated shapes to arrive at unbiased size estimates. Consider the problem of inferring channel thickness from observed intersections in wells. Because the maximum channel thickness is the default value in most channel modeling software, and because channels are then determined on the basis of some postulated cross section, the intersections in wells may be thinner than anticipated. They intersect the margin of an abandoned channel, or the channel may have been eroded. Of course, the channels may also have amalgamated, leading to the possibility of choosing an erroneously large thickness. Consequently, all available information and expert judgment must be employed to estimate appropriate input distributions.

## ANALOG DATA FOR THE SEMIVARIOGRAM

The semivariogram and other spatial parameters are commonly difficult to estimate because the wells are too sparse or widely spaced. This does not detract from the importance of geostatistics. On the contrary, it makes the methodological choices even more important. Every numerical model has implicit (hidden from the modeler and beyond their control) or explicit spatial statistical controls that must be considered.

A reliable horizontal semivariogram is particularly difficult to establish because the experimental horizontal semivariograms are commonly too noisy to interpret. However, the goal is to describe and represent the underlying phenomenon as accurately as possible and not necessarily to obtain the best possible fit of the semivariogram. To do so, secondary information in the form of horizontal well data, seismic data, conceptual geological models, and analog data must be considered; and expert judgment must be used to integrate global information from analog data with sparse local data. In all cases, a systematic approach to the semivariogram interpretation is required (Gringarten and Deutsch, 2001).

In the absence of sufficient horizontal data, a horizontal semivariogram may be inferred by (1) determining the fraction of the variance that is explained by zonal anisotropy (i.e., stratification that leads to persistent positive correlation in the horizontal direction) and then (2) establishing the horizontal-to-vertical anisotropy ratio based on secondary data. The inferred horizontal semivariogram consists of the zonal anisotropy (step 1) and the scaled vertical semivariogram. Deutsch (2002) has published a table of typical anisotropy ratios that may aid in establishing the ratio of horizontal to vertical anisotropy.

Figure 10 illustrates the relationship between these two types of anisotropy. Both the zonal anisotropy contribution and the horizontal-to-vertical anisotropy ratio have considerable uncertainty, which should be investigated via a sensitivity study and/or geostatistical simulation (Deutsch, 2002).

## SIZE SCALING

The input model parameters must also be consistent with the support size of the model. Sample data are not often, if ever, available at the support size of the model; therefore, the input values must be explicitly adjusted to reflect the support size prior to geostatistical simulation. Variances and semivariograms change with respect to well-understood scaling laws (Frykman and Deutsch, 2002). Figure 11 illustrates how a change in support size affects the histogram, variance, and semivariogram of porosity.

A change in the variance with respect to support size is characterized by a difference in semivariograms given by equation 5.

$$D^2(v, V) = \bar{\gamma}(V, V) - \bar{\gamma}(v, v) \tag{5}$$

This equation states that the variance of volumes $v$ within the larger volume $V$ is the average point-scale semivariogram within the larger volume minus the average point scale semivariogram within the smaller volume.

The required correction in variance is applied to the input histogram prior to geostatistical simulation

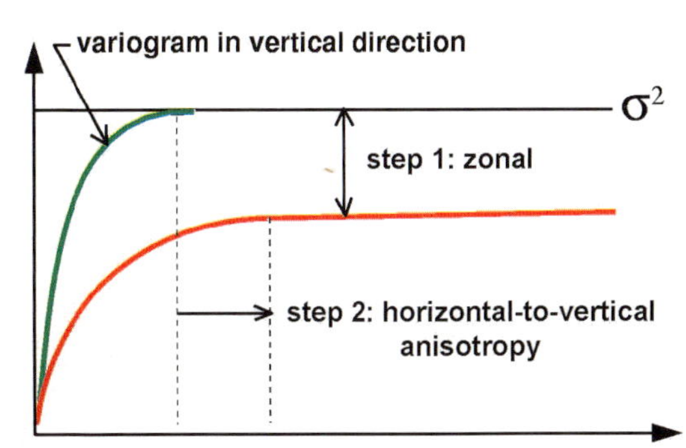


**Figure 10. A schematic illustration of the zonal anisotropy contribution and the horizontal-to-vertical anisotropy parameters required to infer a horizontal semivariogram in the absence of reliable horizontal experimental semivariograms (adapted from Deutsch, 2002).**

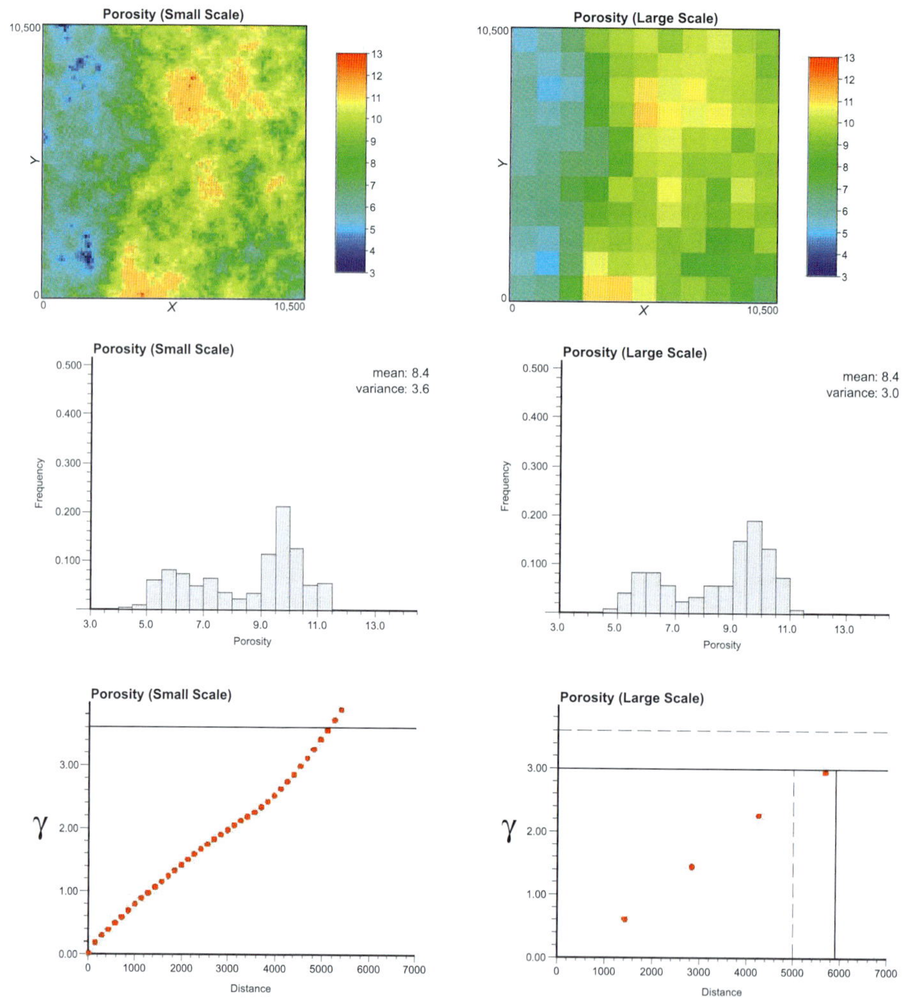


Figure 11. An example of the effect of change in support size on the variance and the semivariogram. Note the decrease in variance and semivariogram sill and increase in semivariogram range with increase in support size. Porosity is given in percent.

by techniques such as affine and lognormal correction. The nugget effect, variance contribution, and range are corrected for change in support size, respectively, as follows:

$$C_V^o = C_v^o \cdot \frac{|v|}{|V|} \tag{6}$$

$$C_V^i = C_v^i \frac{1 - \overline{\gamma}(V, V)}{1 - \overline{\gamma}(v, v)} \tag{7}$$

$$a_V^i = a_v^i + [|V| - |v|] \tag{8}$$

In the above equations, the large $V$ represents the block scale, and the small $v$ represents the data scale. For example, in equation 6, $|V|$ is the physical size in, say, cubic meters; and in equation 8, $|V|$ is the size of the domain in a particular direction. The range values in equation 8 are different in different directions, and they are affected by the original small-scale ranges as well as the geometry of the blocks under consideration.

## UNCERTAINTY IN HARD AND SOFT DATA

Log-derived facies and porosity values are interpretations of (error-prone) wire-line measurements. Uncertainty thus derives from both the gathering and the interpretation process. More and more, practitioners are realizing that well data, generally considered hard in geostatistical simulation, have a certain degree of softness that must be considered in the modeling exercise. Similarly, secondary data sources (geological maps and seismically derived attributes) that are considered as soft data limit stochastic simulation by removing a great deal of spatial variability from one realization to the next. It is important from a modeling point of view to account for variability in secondary data by considering multiple net-to-gross maps, multiple seismic attributes, or multiple representations of other such reservoir facets.

## UNCERTAINTY IN THE INPUT PARAMETERS

Modern geostatistical reservoir modeling consists of generating realizations of all critical spatial parameters. The spatial parameters may include the top structure, the thickness, the facies, porosity, and permeability. These parameters are commonly modeled hierarchically; for example, facies identification depends on structure, porosity depends on the identification of facies, and permeability depends on porosity. The number of realizations depends on the goals and the available computing and professional resources. Typically, 10–200 realizations are generated. A common approach is to produce several realizations independently and to avoid creating multiple facies models within a specified structural framework, multiple porosity models for a given facies model, and so on. The resulting set of $L$ realizations could be denoted as

$$\{\text{top}_l(x,y),\ \text{thk}_l(x,y),\ \text{facies}_l(x,y),\ \text{por}_l(x,y,z),\ \text{perm}_l(x,y,z)\},\quad l=1,\ldots L \tag{9}$$

where $x$, $y$, and $z$ represent the areal and stratigraphic coordinates. Different deterministic and geostatistical simulation techniques may be used to construct the realizations (estimates) of each parameter. In addition, the realizations (estimates) generated for each parameter must encompass the range of possible input model parameters. For example, the net sand proportion may initially be set to 0.6; whereas it could very well fall between 0.5 and 0.7 because of limited well data. Nonetheless, whereas the different facies realizations may reflect a range of results, the practical difference between them is commonly quite small. As previously suggested, the ergodic fluctuations between realizations depend more on the size of the domain relative to the range of correlation than on the actual uncertainty in the parameter.

The operational flow of the simulation process proceeds as follows. An empirical or theoretical distribution is established for the estimates of each input model parameter, or an empirical histogram is produced. A value is selected at random from this distribution to be used in each realization. For example, the sand proportion could be modeled as a triangular distribution with a minimum of 0.5, a mode of 0.6, and a maximum of 0.7. A new target proportion would be drawn from this distribution for each realization. This basic idea, which is analogous to the Monte Carlo simulation, has been around for many years (Haldorsen and Damsleth, 1990) and is implemented in various software packages used for geostatistical modeling.

Note that the distribution of the estimates of each parameter must be established or postulated on the basis of concomitant information. Ideally, the choice derives from expert knowledge. For example, the project geologist chooses the range of object sizes; the geophysicist selects the range of correlation between acoustic impedance and porosity; the engineer determines the uncertainty in an interpreted well test $k \cdot h$; and the geostatistician decides the minimum, most likely, and maximum semivariogram range. In the absence of expert knowledge, some quantitative tools help establish the uncertainty in certain parameters. The bootstrap methodology (Efron, 1979) is one such approach.

## THE BOOTSTRAP APPROACH

The bootstrap approach can be used in certain cases to assess the uncertainty in an input parameter. Given $n$ data values from which a statistic $s$ is calculated, additional sets of $n$ data values are produced by sampling the original $n$ values with replacement (e.g., using a Monte Carlo sampling process). The statistic $s$ is calculated for each new set of $n$ values, with the process being repeated a large number of times (on the order of 1000). The values of the statistic $s$ computed for each of the new data sets are denoted $s'$ to distinguish them from the original value of $s$. The empirical distribution of $s'$ is then used to approximate the unknown distribution of $s$. Efron (1979) discusses the reasonableness of using such an approximation.

Consider the crossplot of vertical average porosity and seismic energy for the wells in North Cowden field shown in Figure 12. The correlation coefficient is calculated to be 0.62. Being a sample statistic, this value has uncertainty associated with it. Clearly, the uncertainty would be greater if there were only three wells, and the uncertainty would be less with more wells. The histogram on the right of Figure 12 is an empirical distribution of the correlation coefficient resulting from the bootstrap of pairs that can subsequently be carried through successive modeling steps.

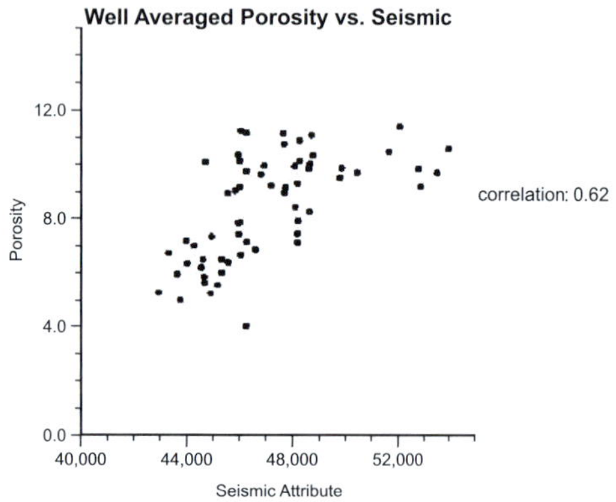


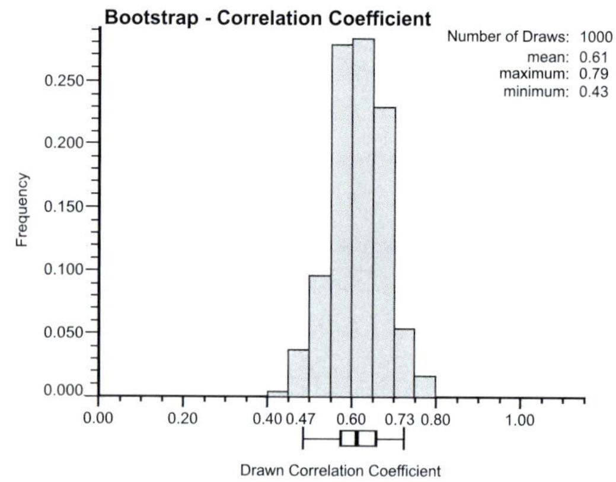


**Figure 12. An example of the bootstrap applied to the correlation coefficient between average porosity and a seismic attribute for all 62 wells in North Cowden field. Left: the scatter plot between porosity and the seismic attribute. Right: the distribution of correlation coefficients from 1000 bootstrap iterations.**

An estimate of the correlation coefficient is simply selected at random from this distribution each time a geostatistical realization (estimate) of average porosity is produced.

Two important assumptions are implicit to the use of the bootstrap: (1) the data are independent, and (2) the data histogram is representative. The issue of data representativeness has been previously addressed, and the steps proposed there should be undertaken prior to implementing the bootstrap procedure. The assumption of data independence is much more difficult to resolve. Although wells in fields are commonly widely spaced making the independence assumption seem reasonable, core or well-log data from within a single well is closely spaced, and the values of reservoir properties they represent certainly may not be independent.

## UNCERTAINTY IN THE SEMIVARIOGRAM

The semivariogram has a direct influence on the spatial arrangement of the petrophysical properties in a geostatistical reservoir model, but it does not directly affect the static resource. The level of model spatial variability, as modeled by the semivariogram, indirectly affects sweep efficiency and recovery. The semivariogram also affects the nature of ergodic fluctuations; for example, a larger range of correlation implies greater ergodic fluctuations. For these reasons, it is necessary to account for the uncertainty in the semivariogram and other spatial controls.

Uncertainty in the semivariogram may be assessed by expert judgment or by comparison to analogs representing similar depositional settings. The uncertainty in the semivariogram may also be estimated directly from the data as demonstrated by Ortiz and Deutsch (2002). For example, consider the experimental standardized normal-score semivariogram of average porosity in the 62 North Cowden wells (Figure 13). The experimental semivariogram is very noisy because of limited data, and so the choice of a final model (curve) will not be very precise. To address this situation, it is possible to construct a distribution of semivariogram ranges, as shown in Figure 13, so that for each realization, a semivariogram range is drawn at random from this distribution. This procedure transfers the uncertainty about the form of the semivariogram model (curve) to the full reservoir model.

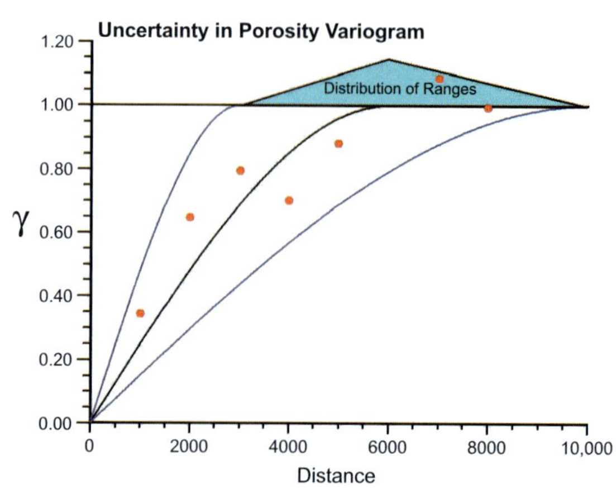


**Figure 13. Uncertainty in the semivariogram. The experimental semivariogram (red dots) and three possible semivariogram models (curves) that encompass a possible distribution of the semivariogram range based on the experimental semivariogram.**

**Table 1. Identification of reservoir properties in multiple realizations of a full reservoir model.**

| *Realization* | *Structure* | *Facies* | *Porosity* | *Permeability* |
|---|---|---|---|---|
| $l = 1$ | Structure 1 | Facies 1 | $\phi_1$ | $k_1$ |
| $l = 2$ | Structure 2 | Facies 2 | $\phi_2$ | $k_2$ |
| ... | ... | ... | ... | ... |
| $l = L$ | Structure $L$ | Facies $L$ | $\phi_L$ | $k_L$ |

## BEST PRACTICES

The following steps constitute the recommended workflow for the assembly of inputs to stochastic models:

1) Review all data.
2) Assemble the most representative data.
3) Compute appropriate statistics with which to estimate all reservoir parameters.
4) Generate realizations of the full reservoir model.

For the purpose of generating realizations of the full reservoir model that express the uncertainty in the input statistics, it is recommended that multiple realizations of each property be produced, and that these realizations be linked as illustrated in Table 1 and described in equation 9. The important idea is that any realization of one reservoir property (e.g., porosity) may be dependent on the preceding realization of another (e.g., facies). Sensitivity analysis should be undertaken to evaluate the overall impact of each parameter. A sensitivity study can be performed by varying a single parameter while holding the others constant. This identifies key controlling parameters for which uncertainty in the input values must be refined to minimize uncertainty in the full reservoir model.

## SUMMARY

The assembly of representative data is extraordinarily important to reservoir modeling; the input statistics will be reproduced regardless of how representative they are of the underlying distributions; sound geological interpretation of major subsurface features and trends is a key step; they will not be reproduced by chance in the resulting geostatistical realizations. Interpretive and analog information must be incorporated to ensure use of the most accurate and precise statistics, such as the histograms and semivariograms.

Uncertainty in the input statistics must be explicitly integrated into the full geostatistical reservoir model. Variation between realizations of this model is a function of the ratio of domain size to the range of correlation and not of the level of certainty in the input statistics. Uncertainty in the input statistics must be quantified by expert judgment and, where possible, by use of tools such as the bootstrap. This uncertainty may be imparted to the full reservoir model by varying the estimates of the model parameters according to their sampling distributions.

## REFERENCES CITED


Deutsch, C. V., 2002, Geostatistical reservoir modeling: New York, Oxford University Press, 376 p.

Deutsch, C. V., and A. G. Journel, 1997, GSLIB—Geostatistical software library and user's guide: New York, Oxford University Press, 369 p.

Efron, B., 1979, Bootstrap methods: Another look at the jackknife: The Annals of Statistics, v. 7, p. 1–26.

Frykman, P., and C. V. Deutsch, 1998, Model-based declustering for deriving representative distributions prior to conditional simulations: Second Institute of Mathematics and its Applications Conference on Modelling Permeable Rocks, March 23–25, Churchill College, Cambridge, Essex, United Kingdom, Institute of Mathematics and its Application, 1 p.

Frykman, P., and C. V. Deutsch, 2002, Practical application of the geostatistical scaling laws for data integration: Petrophysics, v. 43, no. 3, p. 153–171.

Goovaerts, P., 1997, Geostatistics for natural resources evaluation: New York, Oxford University Press, 487 p.

Gringarten, E., and C. V. Deutsch, 2001, Variogram interpretation and modeling: Mathematical Geology, v. 33, p. 507–534.

Haldorsen, H. H., and E. Damsleth, 1990, Stochastic modeling: Journal of Petroleum Technology, v. 42, p. 404–412.

Isaaks, E. H., and R. M. Srivastava, 1989, An introduction to applied geostatistics: New York, Oxford University Press, 561 p.

Ortiz C. J., and C. V. Deutsch, 2002, Calculation of the uncertainty in the variogram: Mathematical Geology, v. 34, p. 169–183.

Vejbæk, O. V., and L. Kristensen, 2000, Downflank hydrocarbon potential identified using seismic inversion and geostatistics: Upper Maastrichtian reservoir unit, Dan field, Danish Central Graben: Petroleum Geoscience, v. 6, p. 1–13.

## *Section III*

# Methods and Applications

On some level, concepts, theories, and ideas must eventually find their way into practice in order to be truly useful to society. Theoretical developments do have beauty and intrinsic value of their own; but it is the foundation and structure they bring to problem solving situations that gives them their greatest credence and viability.

The previous section of the book focused specifically on theoretical developments in geostatistics and stochastic modeling that have emerged throughout the past decade. This section makes the natural transition to methodological approaches and applications. Some of the methods presented in these chapters are new, deriving from recently shaped theoretical constructs, while others represent applications of previously established approaches to new situations. Of course, there is always a fine line between theory and application, but the material contained in this section is clearly methods-oriented, representing creative solutions to some challenging earth science problems.

Sebastian Strebelle at ChevronTexaco initiates the section with a chapter describing the use of sequential simulation to develop reservoir models that incorporate complex geological and structural information provided in the form of prior training images. Strebelle's approach is based on multi-point statistics and combines features of both pixel-based algorithms and object-based techniques. Referred to as the snesim algorithm, this simulation method appears to be both straightforward and relatively fast. In the second chapter, the emphasis turns to groundwater modeling. Van Kelly and Srikanta Mishra, INTERA Inc., present a computationally efficient approach to evaluate and rank geostatistical realizations of groundwater models. The authors demonstrate the effectiveness of their methodology, which is based on a surrogate traveltime measure, using a field example from the Waste Isolation Pilot Plant (WIPP) in New Mexico. Next, Magali Lecour, Pierre Thore, and Richard Cognot discuss a method for incorporating the uncertainty associated with geological structure into reservoir models using the p-field simulation technique. The principal motivation for this work is to improve the precision of reservoir volumetric calculations. Sean McKenna and Paul Reeves from Sandia National Laboratory discuss another geostatistical modeling application in the groundwater arena. McKenna and Reeves present their development of the fractured continuum approach to the modeling of fracture permeability that incorporates the advantages of discrete fracture network models and the processing speed of effective continuum representations. In collaboration with Harry Debeye at Jason Geosystems, Carlos Torres-Verdin and Alvaro Grijalba-Cuenca at the University of Texas describe the merits of geostatistical inversion of three-dimensional seismic data for reservoir delineation relative to more conventional geostatistical simulation practices in the next chapter. Advantages and disadvantages of both techniques are discussed. Geostatistical inversion, in particular, provides a quantitative way to integrate the high vertical resolution of well logs with the dense areal coverage of prestack three-dimensional seismic data. Yet it may be computationally prohibitive when applied to large seismic and well log data sets. Geostatistical inversion of seismic data is also the theme of the last chapter in this section, which is authored by Paul van Riel and his colleagues at Fugro-Jason. These authors suggest that geostatistical inversion results in improved control on the spatial distribution of reservoir properties; but that, as suggested in the previous chapter, this advantage comes at the cost of computational performance.

Strebelle, S. B., 2006, Sequential simulation for modeling geological structures from training images, *in* T. C. Coburn, J. M. Yarus, and R. L. Chambers, eds., Stochastic modeling and geostatistics: Principles, methods, and case studies, volume II: AAPG Computer Applications in Geology 5, p. 139–149.

*Chapter 11*

# Sequential Simulation for Modeling Geological Structures from Training Images

**S. B. Strebelle**
*Chevron*
*San Ramon, California, U.S.A.*

## ABSTRACT

Two geostatistical approaches are traditionally used to build numerical models of facies spatial distributions: the variogram-based approach and the object-based approach. Variogram-based techniques aim at generating simulated realizations that honor the sample data and reproduce a given semivariogram that models the two-point spatial correlation of the facies. However, because the semivariogram is only a measure of linear continuity, variogram-based algorithms give poor representations of curvilinear or geometrically complex actual facies geometries. In contrast, object-based techniques allow modeling crisp geometries, but the conditioning on sample data requires iterative trial-and-error corrections that can be time consuming, particularly when the data are dense relative to the average object size. This chapter presents an alternative approach that combines the easy conditioning of pixel-based algorithms with the ability to reproduce shapes of object-based techniques, without being too time and memory demanding.

In this new approach, the geological structures believed to be present in the subsurface are characterized by multiple-point statistics, which express joint variability or joint continuity at many more than two locations at a time. Multiple-point statistics cannot be inferred from typically sparse sample data, but they can be read from training images depicting the expected patterns of geological heterogeneity. Training images are simply graphical representations of a prior geological or structural concept; they need not carry any locally accurate information about the field to be modeled. The multiple-point patterns borrowed from the training image(s) are exported to the model, where they are anchored to the actual subsurface data, both hard and soft, using a pixel-based sequential simulation algorithm.

This multiple-point statistics simulation algorithm is tested on the modeling of a fluvial hydrocarbon reservoir where flow is controlled by meandering sand channels. The simulated numerical model reproduces channel patterns and honors exactly all well data values at their locations. The methodology proposed appears to be easy to apply, general, and fast.


DOI:10.1306/1063812CA53231

## INTRODUCTION

The goal of most current stochastic reservoir-simulation algorithms is to reproduce statistics of spatial continuity inferred from the sample data solely. Because sample data are typically sparse, they can give, at best, only an estimate of the two-point correlation of the attribute to be simulated within the field. Hence, most traditional simulation techniques, such as sequential Gaussian simulation or its close cousin truncated Gaussian simulation (Deutsch and Journel, 1998), are limited to the reproduction of semivariogram and covariance models, which are two-point statistics. Identification of two-point statistics is, however, not sufficient to reproduce curvilinear and large-scale connected structures, e.g., meandering channels. The modeling of such structures calls for characterizing the spatial continuity at three or more locations at a time. In petroleum or groundwater applications, reproduction of multiple-point patterns of facies continuity is critical, not so much to produce geologically realistic looking models, but to provide accurate input numerical models for flow performance predictions (Caers et al., 1999).

This quest for more powerful simulation algorithms is motivated by the fact that structural information much beyond facies two-point correlation, calculated from the sample data, is actually available. Geological interpretation from well logs, seismic data analysis, and comparison with field analogs do provide information about the actual reservoir architecture. Algorithms based on semivariograms and, more generally, any two-point statistics mostly ignore such valuable information.

One straightforward way to reproduce geological structures with crisp shapes consists of generating numerical objects with similar shapes, then distributing these objects over the reservoir model (Haldorsen and Damsleth, 1990; Dubrule, 1993; Deutsch and Wang, 1996; Viseur, 1999). However, because of the mathematical complexity of parameterizing object shapes, object-based techniques lack flexibility. In addition, multiple-type data conditioning remains a major challenge. Problems arise particularly when the sample data are dense relative to the average object size or when deterministic facies geobodies, which could have been interpreted directly from seismic data, for example, are integrated as hard data into the model. Conditioning to exhaustive soft data, typically three-dimensional (3-D) seismic data, is another limitation of object-based algorithms: most commonly, 3-D soft data are transformed into a two-dimensional (2-D) trend map, ignoring the information about the facies vertical distribution.

We propose in this chapter an alternative approach to reproduce specific facies geometries that uses multiple-point statistics, i.e., (cross-) correlation relating three or more locations at a time. To provide an analytical expression for multiple-point statistics, some terminology is now introduced.

Consider an attribute $S$, taking $K$ possible states $\{s_k, k = 1,\ldots, K\}$. $S$ can be a categorical property, for example, facies type, or a continuous value, such as porosity or permeability, with its interval of variability discretized into $K$ classes by $(K - 1)$ threshold values.

A data event $d_n$ of size $n$ centered at location $\mathbf{u}$ is constituted by

1) the $n$ data locations geometry defined by $n$ vectors $\{\mathbf{h}_\alpha, \alpha = 1,\ldots, n\}$
2) the $n$ data values $\{s(\mathbf{u} + \mathbf{h}_\alpha), \alpha = 1,\ldots, n\}$

A data template $\tau_n$ comprises only the previous data geometry. The data event $d_n$ is said to be associated with the geometric template $\tau_n$.

The categorical transform of the variable $S$ at location $\mathbf{u}$ is defined as

$$I(\mathbf{u};k) = \begin{cases} 1 & \text{if } S(\mathbf{u}) = s_k \\ 0 & \text{if not} \end{cases}$$

The multiple-point statistics used in this chapter actually refer to probabilities of occurrence of data events $d_n = \{S(\mathbf{u}_\alpha) = s_{k_\alpha}, \alpha = 1,\ldots,n\}$, i.e., probabilities that the $n$ values $s(\mathbf{u}_1),\ldots, s(\mathbf{u}_k)$ are jointly in the respective states $s_{k_1},\ldots,s_{k_n}$. The probability of occurrence of any such data event $d_n$ is also the expected value of the product of the $n$ corresponding indicator data:

$$\begin{aligned} \text{Prob}\{d_n\} &= \text{Prob}\{S(\mathbf{u}_\alpha) = s_{k_\alpha};\ \alpha = 1,\ldots,\ n\} \\ &= \text{E}\left[\prod_{\alpha=1}^{n} I(\mathbf{u}_\alpha;k_\alpha)\right] \end{aligned} \quad (1)$$

Note that for $n = 2$, equation 1 is the expression of the noncentered indicator covariance.

Multiple-point statistics of type 1 cannot be inferred from sparse sample data; their inference requires a densely and regularly sampled training image depicting the geometries believed to be present in the subsurface. Training images are not conditioned by any local field information; they simply reflect a prior geological or structural concept. Object-based algorithms freed of the constraint of data conditioning can be used to generate such images. Other possible sources of training images include photographs of outcrops processed with a computer-aided drawing (CAD) algorithm. For 3-D models, a 3-D training cube is required.

For any given data template $\tau_n = \{\mathbf{h}_\alpha, \alpha = 1, \ldots, n\}$ and any given training image $T$, the eroded training image $T_n$ is defined as the set of locations $\mathbf{u}$ such that all $n$ nodes of the data template $\tau_n$ centered at $\mathbf{u}$ are within the training image $T$. Under a prior decision of stationarity, the probability 1 of occurrence of the data event $d_n$ associated with $\tau_n$ can be inferred from the training image by counting the number $c(d_n)$ of replicates of $d_n$ found in the eroded training image $T_n$. A replicate should have the same geometric configuration and the same data values as $d_n$. If the size of the eroded training image $T_n$ is denoted by $N_n$, then the multiple-point statistics moment 1 can be identified by the following proportion:

$$\text{Prob}\{s(\mathbf{u}_\alpha) = s_{k_\alpha}; \alpha = 1, \ldots, n\} \approx \frac{c(d_n)}{N_n} \quad (2)$$

We propose in this chapter to capture the patterns of the training image through their multiple-point moments of type 2, then to export those multiple-point statistics to the model using a pixel-based sequential simulation algorithm.

## METHODOLOGY OUTLINE

The algorithm presented hereafter is based on the theory of extended normal equations (Journel and Alabert, 1989; Guardiano and Srivastava, 1993; Journel, 1993), which is now briefly reviewed.

### Extended Normal Equations

The key of any sequential simulation algorithm is to determine the local conditional probability distribution function (cpdf) at any unsampled location $\mathbf{u}$ and to evaluate the probability that the unknown attribute value $S(\mathbf{u})$ takes any of the $K$ possible states $s_k$, given $n$ conditioning data in the neighborhood of $\mathbf{u}$, denoted $s(\mathbf{u}_\alpha) = s_{k_\alpha}$, $\alpha = 1, \ldots, n$. That probability is denoted by $\text{Prob}\{s(\mathbf{u}) = s_k | d_n\}$, where $d_n$ is the data event constituted jointly by the $n$ data $s(\mathbf{u}_\alpha)$.

When using traditional semivariogram-based algorithms, the conditioning of $S(\mathbf{u})$ to the global data event $d_n$ is approximated by the sum of the $n$ conditionings of $S(\mathbf{u})$ to each datum $s(\mathbf{u}_\alpha)$ taken one at a time. The conditional probability is then estimated by the following simple indicator kriging expression:

$$\text{Prob}\{S(\mathbf{u}) = s_k | d_n\} \approx E[I(\mathbf{u}; k)] + \sum_{\alpha=1}^{n} \lambda_\alpha \times (i(\mathbf{u}_\alpha; k_\alpha) - E[I(\mathbf{u}_\alpha; k_\alpha)])$$

where $i(\mathbf{u}_\alpha; k_\alpha) = 1$ is the indicator value attached to the datum value $s_{k_\alpha}$ at location $\mathbf{u}_\alpha$, $E[I(\mathbf{u}_\alpha; k_\alpha)] = \text{Prob}\{S(\mathbf{u}_\alpha) = s_{k_\alpha}\} = p_{k_\alpha}$ is the prior probability of state $s_{k_\alpha}$; $E[I(\mathbf{u}; k)] = \text{Prob}\{S(\mathbf{u}) = s_k\} = p_k$ is the prior probability of state $s_k$.

Determination of the simple kriging weights, $\lambda_\alpha$, calls for solving a system of $n$ normal (kriging) equations. To ensure that this system has a unique solution, a permissible set of indicator semivariogram models commonly inferred from the sample two-point correlation moments is required.

If all multiple-point statistics are available, then the conditioning of $S(\mathbf{u})$ to the single global data event $d_n$ can be considered, and the exact conditional probability is now given by the simple kriging expression:

$$\text{Prob}\{S(\mathbf{u}) = s_k | d_n\} = E[I(\mathbf{u}; k)] + \lambda \left( \prod_{\alpha=1}^{n} i(\mathbf{u}_\alpha; k_\alpha) - E\left[ \prod_{\alpha=1}^{n} I(\mathbf{u}_\alpha; k_\alpha) \right] \right)$$

where $\prod_{\alpha=1}^{n} i(\mathbf{u}_\alpha; k_\alpha) = 1$ is the product of indicator values corresponding to the observed conditioning data event $d_n$, $E\left[\prod_{\alpha=1}^{n} I(\mathbf{u}_\alpha; k_\alpha)\right] = \text{Prob}\{S(\mathbf{u}_\alpha) = s_{k_\alpha}; \alpha = 1, \ldots, n\}$ is the probability of occurrence of $d_n$, and $E[I(\mathbf{u}; k)] = \text{Prob}\{S(\mathbf{u}) = s_k\} = p_k$ is the prior probability that the (unknown) state at $\mathbf{u}$ is $s_k$.

A single extended normal equation needs to be solved to calculate the single weight $\lambda$, which leads to the following expression for the conditional probability (Guardiano and Srivastava, 1993):

$$\text{Prob}\{S(\mathbf{u}) = s_k | d_n\} = \frac{\text{Prob}\{S(\mathbf{u}) = s_k \text{ and } S(\mathbf{u}_\alpha) = s_{k_\alpha}; \alpha = 1, \ldots, n\}}{\text{Prob}\{S(\mathbf{u}_\alpha) = s_{k_\alpha}; \alpha = 1, \ldots, n\}} \quad (3)$$

The limit case where all multiple-point statistics are available is thus straightforward in that it reduces to the very definition of the conditional probability. The multiple-point statistics moments of the numerator and denominator of relation 3 can hardly be inferred from sparse sample data; hence, the idea is to borrow them from one or more training images using relation 2 under a prior decision of stationarity, as follows:

1) $\text{Prob}\{S(\mathbf{u}_\alpha) = s_{k_\alpha}; \alpha = 1, \ldots, n\}$ can be inferred by counting the number $c(d_n)$ of replicates of the conditioning data event $d_n = \{S(\mathbf{u}_\alpha) = s_{k_\alpha}, \alpha = 1, \ldots, n\}$ found in the training image(s), and
2) $\text{Prob}\{S(\mathbf{u}) = s_k \text{ and } S(\mathbf{u}_\alpha) = s_{k_\alpha}; \alpha = 1, \ldots, n\}$ can be inferred by counting the number $c_k(d_n)$ of those replicates, among the $c(d_n)$ previous ones, associated to a central value $S(\mathbf{u})$ equal to $s_k$.

The conditional probability can then be identified by the following training proportion:

$$\text{Prob}\{S(\mathbf{u}) = s_k | S(\mathbf{u}_\alpha) = s_{k_\alpha}; \alpha = 1, \ldots, n\} = p(\mathbf{u}; s_k | d_n) \approx \frac{c_k(d_n)}{c(d_n)} \tag{4}$$

Note the simplicity of this approach, compared to semivariogram-based techniques. The cpdf is inferred directly and consistently from the training image; the intermediate step of modeling permissible functions to sample statistics, e.g., a covariance model, is completely shortcut. Further, the training proportions of equation 4 do not suffer any of the order relation problems possibly associated with two-point statistics kriging estimates.

The counterpart of this simplicity is the total dependence on the training image(s). The multiple-point statistics and, hence, the geometrical structures are borrowed directly from the training image without any modeling or filtering. One could argue, however, that a training image has the advantage of making fully explicit the structural information implicit to any stochastic imaging algorithm. That prior information can be visually evaluated and then accepted or rejected, whereas it is much more difficult to evaluate the appropriateness of, say, a multi-Gaussian random function model or the random function model associated with a Markov chain simulation algorithm. Moreover, one can account for the uncertainty about the structures expected in the subsurface using several training images reflecting alternative geological scenarios.

## Multiple-point Statistics Inference Problem

Borrowing directly all required cpdfs from a training image is, however, not a trivial proposition. Indeed, given a template $\tau_n$ of $n$ data values, the inference of a probability distribution function conditioned to a data event $d_n$ requires that at least one replicate of $d_n$ can be found in the training image. If each of the $n$ nodes of $\tau_n$ can take $K$ possible states, the total number of possible data events associated with $\tau_n$ is $K^n$; e.g., $K = 4$ and $n = 15$ lead to $K^n > 10^9$, whereas the total number of training data events associated with $\tau_n$ is capped by the size of the training image, which is typically made of $10^5$–$10^7$ grid nodes. This means that only a relatively small number of cpdfs can actually be inferred from the training image. Two avenues of solutions have been considered to alleviate this inference problem. First, Caers and Journel (1998) propose to use an iterative simulation algorithm based on neural networks that proceeds in two steps. First, a neural network is trained to fit a multiple-point parametric cpdf to the few experimental cpdfs that can actually be inferred from the training image for a fixed data template $\tau_n$. Because the conditioning configuration ($\tau_n$) is fixed, the simulation is iterative, based on a Markov chain Monte Carlo approach; the simulation grid is first initialized by freezing the original data values at their locations and filling in the remaining nodes with values drawn from the global target histogram. The nondata nodes are visited repeatedly along a random path and perturbed according to a probabilistic criterion based on the local cpdf retrieved from the neural network. Although this algorithm provides good reproduction of curvilinear features, practical issues remain regarding the internal structure of the neural net and the choice of the conditioning geometric template. More importantly, as any iterative technique, this algorithm suffers from convergence considerations and is computer time demanding.

The second avenue, pioneered by Guardiano and Srivastava (1993) and further developed in this chapter, consists of using only the few experimental cpdfs that can actually be inferred from the training image. If the goal is to reproduce the structures displayed by the training image, one should indeed concentrate on the multiple-point statistics specific to that training image. The approach is then remarkably simple: at each unsampled node, infer the local cpdf by scanning the training image, reducing progressively the size $n$ of the conditioning data event $d_n$ until at least one replicate of $d_n$ is found. Because the simulation is sequential (noniterative), it is potentially fast. Each unsampled grid node is visited only once along a random path, each simulated value becoming a hard datum value and conditioning the simulation of neighboring node values visited later in the sequence (Goovaerts, 1997). Although the simulation is noniterative, Srivastava's original code was extremely computer time demanding because the full training image had to be scanned anew at each unsampled node.

To avoid the repetitive scanning of the training image, we propose to store all training cpdfs ahead of time in a dynamic data structure called the "search tree," which allows their fast retrieval as the simulation proceeds.

## THE SNESIM ALGORITHM

The sequential indicator simulation algorithm described below is called snesim to indicate that the simulation of each unsampled node **u** requires solving only one single normal equation identifying the very definition of equation 3 of a conditional probability.

First, a data search neighborhood needs to be defined. Only the original sample data and previously simulated nodes falling within that neighborhood centered at **u** will be considered for conditioning the simulation at **u**. This search neighborhood is typically an ellipsoid whose dimensions and anisotropy directions account for the overall geometrical characteristics of the training facies patterns. Given the data template $\tau_n$ that covers every node located in the search neighborhood, the first step of snesim consists of building the search tree corresponding to $\tau_n$. This search tree is a dynamic data structure (Strebelle, 2002) that stores the numbers $c(d_n)$ of occurrences of any data event $d_n$ associated with $\tau_n$ that can be actually found in the training image, as well as the corresponding numbers $c_k(d_n)$ of occurrences of $d_n$ associated to a central value equal to $s_k$. A detailed description of the search tree is provided in Strebelle (2000).

During the simulation, the local cpdf at any unsampled node **u** is retrieved directly from the previous search tree, as follows. Let $n' < n$ be the number of locations in $\tau_n$ that are informed by the original sampled data and previously simulated nodes, and let $d_{n'}$ be the corresponding conditioning data event. The number $c(d_{n'})$ of training replicates of $d_{n'}$ is obtained by summing the numbers of replicates of all data events $d_n$ that match the $n'$ conditioning data. These data events $d_n$ may possibly differ from one another at the $(n - n')$ nondata template locations, as given by

$$c(d_{n'}) = \sum_{\substack{d_n \text{ associated with } \tau_n \\ \text{such that } d_{n'} \subset d_n}} c(d_n) \tag{5}$$

Similarly, for the number $c_k(d_{n'})$ of $d_{n'}$ replicates with a central value equal to $s_k$, as given by

$$c_k(d_{n'}) = \sum_{\substack{d_n \text{ associated with } \tau_n \\ \text{such that } d_{n'} \subset d_n}} c_k(d_n) \tag{6}$$

Knowledge of $c_k(d_{n'})$ and $c(d_{n'})$ allows retrieving the local conditional probabilities using relation 4.

Because the number of data events $d_n$ scanned from the training image is necessarily smaller than the size of that image, the search tree, which stores those data events, requires a reasonable amount of memory. In addition, the construction of that search tree requires scanning the training image one single time prior to the image simulation, which makes the whole simulation algorithm a fast process (Strebelle, 2000).

In greater detail, the snesim algorithm proceeds as follows:

1) Define a data search neighborhood, and construct the search tree corresponding to the data template $\tau_n$ covering that neighborhood.
2) Assign the sample hard data to the closest grid nodes. Define a random path visiting only once each unsampled node.
3) At each unsampled location **u**, consider the conditioning data actually present in the template $\tau_n$ centered on **u**. Let $n' < n$ be the number of those conditioning data, and $d_{n'}$ the corresponding data event. Retrieve from the search tree the probability distribution conditional to $d_{n'}$. If no replicate of $d_{n'}$ has been stored in the search tree, i.e., no replicate of $d_{n'}$ can be found in the training image, then the farthest conditioning datum is dropped, reducing the number of conditioning data to $(n' - 1)$; the proportions conditioned to this lesser data event $d_{n'-1}$ are retrieved again from the search tree, and so on. If the number of data drops to $n' = 1$, and no replicate of $d_{n'}$ can be found, the conditional probability $p(\mathbf{u}; s_k|d_{n'})$ is replaced by the marginal probability $p_k$.

4) Draw a simulated $s$ value for node $\mathbf{u}$ from the previous cpdf. That simulated value is then added to the $s$ data to be used for conditioning the simulation at all subsequent nodes.
5) Move to the next node along the random path and repeat steps 3 and 4.
6) Loop until all grid nodes are simulated. One stochastic image has been generated. Reiterate the entire process from step 2 with a different random path to generate another realization.

## Multiple Grid Implementation

The data search neighborhood should not be made too small; otherwise, large-scale structures of the training image could not be reproduced. However, if the search neighborhood is too large, the associated data template $\tau_n$ would include a large number of grid nodes, which would increase the memory taken by the search tree and the computation time required to retrieve the training proportions of equation 4 through summations of types 5 and 6. One solution to capture large-scale structures while considering a data template with a reasonably small number of grid nodes is provided by the multiple grid concept (Tran, 1994). The multiple grid concept consists of simulating a number $G$ of increasingly finer grids, where the $g$th $(1 \leq g \leq G)$ grid comprises each $2^{g-1}$th node of the final simulation grid ($g$ = 1). The data template $\tau_n = \{\mathbf{h}_\alpha, \alpha = 1,\ldots, n\}$ associated with the search neighborhood of the fine grid is scaled proportionally to the spacing of the nodes within the grid to be simulated; thus, $\tau_n^g = \{\mathbf{h}_\alpha^g = 2^{g-1}\mathbf{h}_\alpha, \alpha = 1,\ldots, n\}$ is the data template used for the $g$th grid. The large data templates considered for the initial large grids allow capturing the large-scale structures of the training image.

One search tree needs to be constructed per simulation grid, possibly using a different training image reflecting the geological heterogeneities specific to that scale (Strebelle, 2000). When the $g$th grid simulation is completed, its simulated values are frozen as hard data values to be used for conditioning on the next finer simulation grid.

## Reproduction of Local Marginal Distributions

No explicit constraint ensures reproduction of the sample histogram or any other target marginal distribution $\{p_k, k = 1,\ldots, K\}$. Simulated realizations may display global proportions significantly different from this target distribution, particularly if the probability values of type 4 are borrowed from a training image having a histogram significantly different from the target distribution.

The first recommendation is to use training images with global proportions reasonably similar to the final target proportions. Next, the following servosystem can be implemented so that the simulation always remains close to the target global proportions $\{p_k, k = 1,\ldots, K\}$ as it progresses from one node to another.

Let $\{p_k^c, k = 1, \ldots, K\}$ be the current proportions before simulating node $\mathbf{u}$, that is, the histogram of the sample data plus all values simulated before visiting node $\mathbf{u}$. The observed difference $p_k - p_k^c$ is added to the conditional probability $p(\mathbf{u};s_k|d_n)$ inferred from the training image, with a constant weighting parameter $\mu \geq 0$. This yields the modified local conditional probability $p^*(\mathbf{u}; s_k|d_n) = p(\mathbf{u}; s_k|d_n) + \mu(p_k - p_k^c)$.

For any given category $s_k$, the overestimation of the target proportion $p_k$ by the current proportion $p_k^c$ entails $p_k - p_k^c < 0$, then $p^*(\mathbf{u}; s_k|d_n) < p(\mathbf{u}; s_k|d_n)$. Hence, the probability of drawing the category $s_k$ is reduced to correct for the current overestimation.

The parameter $\mu$ controls the correction applied to the conditional probability at each location $\mathbf{u}$. One should start with $\mu = 0$, i.e., no servosystem, check the global proportions of a few realizations, and if needed, increase the value $\mu$ gradually. Order relations created by that servosystem are easily corrected.

## Integration of Soft Information

Soft data provide indirect information on facies spatial distributions. To pass that information to the facies model, soft data need to be calibrated, i.e., converted into local facies probabilities (Caers et al., 2001; Strebelle et al., 2002). Snesim allows integrating such soft data-derived facies probabilities.

To simplify the notations, we propose to rewrite the probability of a facies inferred from the training image as $P(A|B)$, where $A$ is the facies indicator value at the node $\mathbf{u}$ to be simulated $A = \{I(\mathbf{u}) = 1\}$, and $B$ is the conditioning data event $d_n$. Similarly, the soft data-derived sand probability can be written as $P(A|C)$, where $C$ represents the soft information, which can be the single soft datum located at $\mathbf{u}$, or more generally, a window of soft data centered on **u**. Integrating the soft data-derived information along with the geological prior information (training image) calls for inferring the probability $P(A|B, C)$. Journel (2002) proposes to estimate $P(A|B, C)$ by recombining $P(A|B)$ and $P(A|C)$ as follows:

$$\frac{x}{b} = \left(\frac{c}{a}\right)^{\tau} \tag{7}$$

where $\tau$ is a parameter weighting the information $C$, and

$$x = \frac{1 - P(A|B,C)}{P(A|B,C)}, \quad b = \frac{1 - P(A|B)}{P(A|B)},$$
$$c = \frac{1 - P(A|C)}{P(A|C)}, \quad \text{and } a = \frac{1 - P(A)}{P(A)}$$

$P(A)$ is the global facies proportion. Therefore, the value $a$ can be interpreted as a prior distance to the event $A$, i.e., prior to integrating the information carried by the geological event $B$ and the soft data event $C$. Indeed if $P(A) = 1$, the facies is certain to occur at location **u**; hence, the distance is $a = 0$. Likewise, the values $b$ and $c$ update that distance given the geological information $B$ and the soft information $C$, respectively. Finally, $x$ represents the distance once updated by both $B$ and $C$. The probability $P(A|B, C)$ calculated from equation 1 with $\tau = 1$ is

$$\mathrm{P}(A|B,C) = \frac{a}{a + bc} \tag{8}$$

Journel models equation 8 with $\tau = 1$ and assumes a form of conditional independence in that the relative contribution of the soft information $C$ for predicting the unknown facies $A$ is independent from the information carried by $B$. In that respect (conditional independence), Journel has recently established the parallel between the permanence of ratios in equation 7 and Bayes updating, the latter consisting of the following four steps:

1) Bayes inversion: $P(A|B) = \frac{P(B|A)}{P(B)} \cdot P(A)$
2) Add conditioning to $C$:
$$P(A|B,C) = \frac{P(B|A,C)}{P(B|C)} \cdot P(A|C)$$
3) Under conditional independence:
$$P(B|A,C) = P(B|A) = P(A|B)\frac{P(B)}{P(A)}; \text{ thus,}$$
$$P(A|B,C) = \frac{P(A|B)P(A|C)}{P(A)} \cdot \frac{P(B)P(C)}{P(B,C)}$$
4) The term $\frac{P(B)P(C)}{P(B,C)}$ is eliminated by standardization: $P(A|B,C) = \frac{\varphi(A)}{\varphi(A)+\varphi(\tilde{A})} \in [0,1]$, an expression shown identical to equation 8, with
$$\varphi(A) = \frac{P(A|B)P(A|C)}{P(A)}, \tilde{A} = \text{no } A, \text{ and}$$
$$\varphi(\tilde{A}) = \frac{[1 - P(A|B)][1 - P(A|C)]}{1 - P(A)}$$

## SIMULATION OF A FLUVIAL RESERVOIR

The snesim algorithm is tested on the simulation of a horizontal 2-D section of a fluvial reservoir. Fluvial reservoirs are characterized by sinuous, sand-filled channels in a silty mudstone matrix. A realistic modeling of the curvilinear (high-permeability) channel geometry is critical for reliable connectivity assessment and simulation of the flow performance of such reservoirs.

The reference true image of size $100 \times 100 =$ 10,000 pixels is shown in Figure 1a. Thirty-two sample data were collected at random over that true image, with a 28.1% sand proportion very close to the 28.0% sand proportion of the true image (Figure 1b). The training image, which could have been hand drawn by a geologist, then digitized, depicts the inferred geometry of the sand channels (Figure 1c). To alleviate ergodicity problems related to the long-range continuity of the channels, the training image was taken significantly larger than the simulation grid: $250 \times 250 = 62{,}500$ pixels.

The multiple grid option is used to capture the large-scale channel continuity: four increasingly finer grids are considered. The search neighborhood corresponding to the finest grid is a rectangle window of size $7 \times 5$ pixels. Thus, the corresponding data template $\tau_n$ covers 35 nodes. Generating the conditional simulated realization displayed in Figure 2a using snesim takes 9 sec with a silicon graphics octane. To provide a yardstick for comparison, Figure 2b shows a realization conditioned to the same 32 sample data, generated by the Geostatistical Software Library two-point statistics-based sequential indicator simulation program sisim (Deutsch and Journel, 1998), using an anisotropic spherical semivariogram model, with a range of 25 pixels in the east–west direction and 8 pixels in the north–south direction. This semivariogram model was inferred from the training image.

In contrast to sisim, snesim allows reproducing reasonably well the sinuous channel patterns displayed by the training image. The previously described

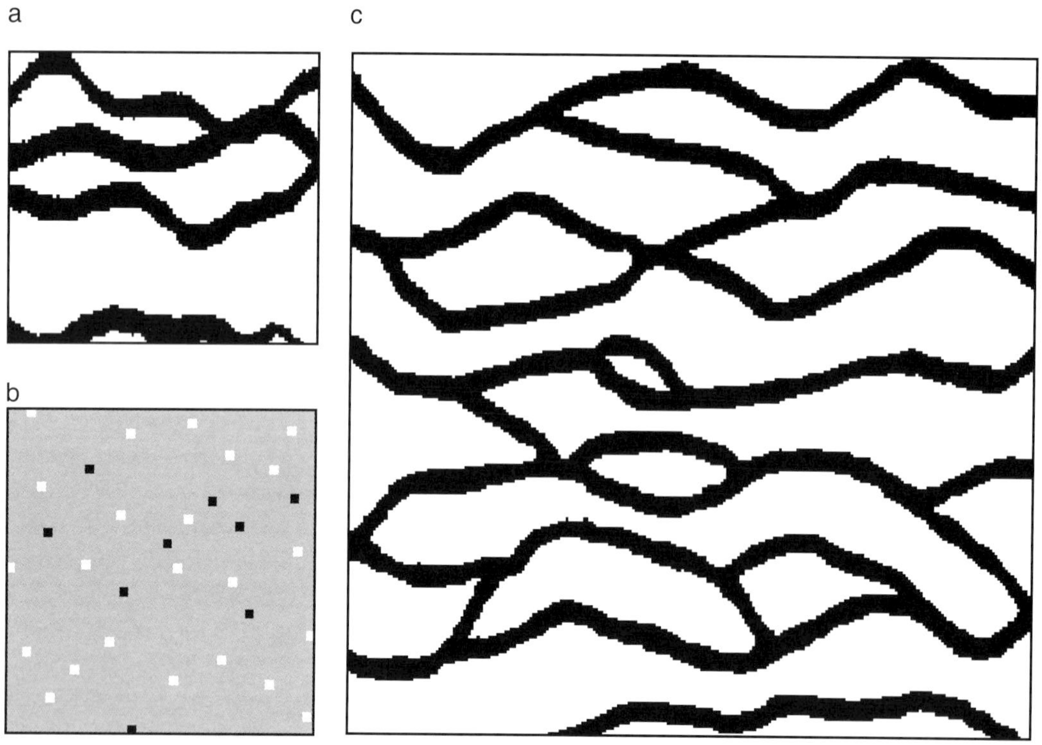


**Figure 1.** **Simulation of a horizontal 2-D section of a fluvial reservoir: (a) a true image (28.0% sand); (b) 32 sample data values (28.1% sand); and (c) a training image (27.7% sand).**

servosystem with $\mu = 1$ allows the proportion of the snesim realization (27.8%) to be reasonably close to the target 28.1% sample sand proportion.

In many applications where seismic data are available, sand geobodies can be interpreted directly from the seismic data. These geobodies, if considered as well known, should be integrated in the model as hard data. In contrast to current object-based techniques, snesim allows such integration, as shown by the simulated realization of Figure 3b, conditioned to the channel pieces of Figure 3a.

## Integration of Soft Information

To illustrate the integration of soft information, a moving window averaging process was applied to the true image to generate a sand probability map (Figure 4a). This map gives a blurred image of where the channels are in the reservoir but without any crisp channel geometry. The integration expression 8 was used in snesim to generate a model conditional to both that sand probability map and the previous 32 sample hard data of Figure 1b. The

**Figure 2.** **Two snesim realizations: (a) 27.8% sand; (b) 27.4% sand.**

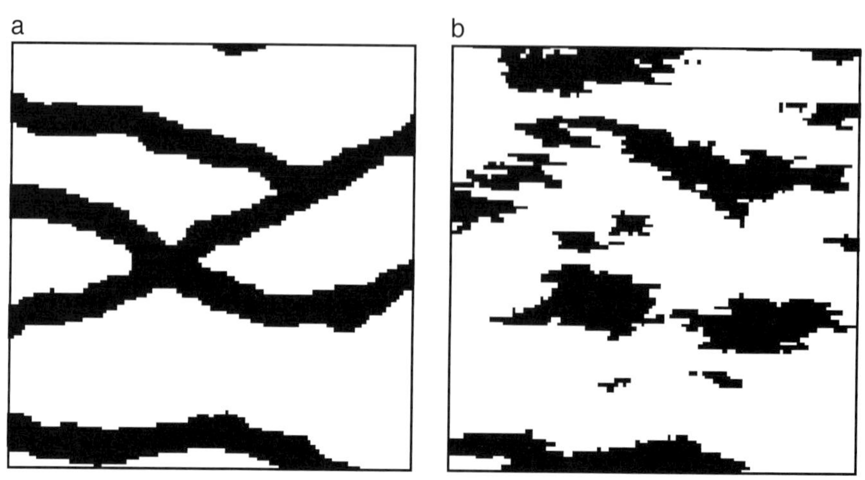

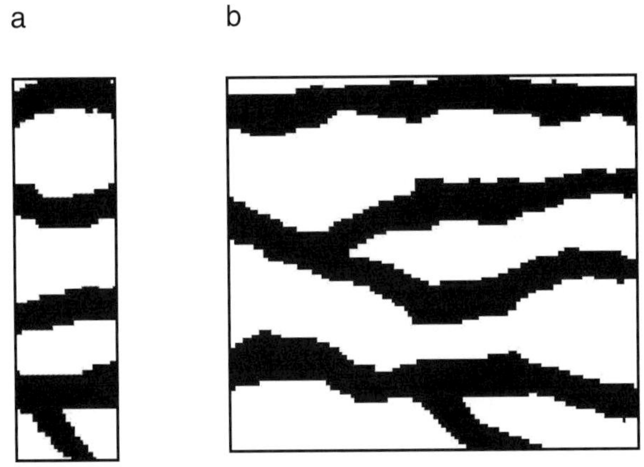


**Figure 3.** **(a) Channel pieces possibly interpreted from seismic data and (b) a corresponding snesim realization conditioned to the interpreted channel pieces (33.4% sand).**

simulated channels are similar to the training channels and preferentially located in the high-sand-probability areas (Figure 4b). The resulting simulated sand proportion is 29.4%.

## Nonstationary Models

All geostatistical structural models such as semivariogram models rely on a stationarity decision (Deutsch and Journel, 1998). For example, in traditional semivariogram-based geostatistics, statistics such as histogram and semivariogram are assumed stationary over the study area. The stationarity decision taken for one- and two-point statistics in semivariogram-based algorithms carries over to higher order statistics in multiple-point geostatistics, implying that patterns extracted from the training image can be exported into any region of the study area for which the training image is deemed relevant. However, actual reservoirs contain many nonstationary local features, such as vertical and areal facies proportion variations or changes in facies pattern directions or shapes.

The servosystem previously described can be extended to reproduce target areal proportion maps or vertical proportion curves (Strebelle and Journel, 2001). Rotation angle or affinity ratio (rescaling factor) data describing local variations in facies pattern shapes can be determined from seismic data (Strebelle et al, 2002) or borrowed from outcrop data (Xu, 1996). A simple practical approach consists of discretizing the interval of variability of the rotation angle and affinity ratio data into $L$ classes. Prior to simulation, $L$ new training images are generated by rotating or stretching the original training image using the rotation angle or affinity ratio of each class. Then, at each unsampled node $\mathbf{u}$, the cpdf is estimated from the transformed training image corresponding to the class that the local rotation angle and affinity ratio value at $\mathbf{u}$ belongs to.

The map of local rotation angles displayed in Figure 5a, of size 250 × 250, was used to illustrate the

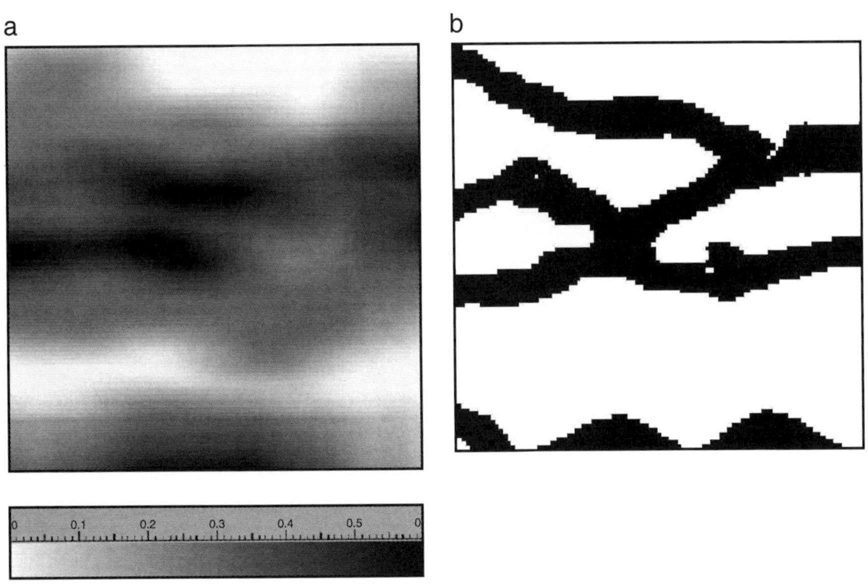


**Figure 4.** **(a) Sand probabilities derived from soft data and (b) a corresponding snesim realization conditioned to 32 sample data and soft data derived from sand probabilities (29.4% sand).**

generation of nonstationary channel-fan models. The angle values, ranging from 0 to 90°, were discretized into five classes. Reproduction of channels is seen to be good (Figure 5b). One realization (62,500 grid cells) takes 71 sec, which is still reasonable.

## CONCLUSIONS

A new geostatistical-simulation algorithm is proposed that imparts into reservoir models complex geological and structural information as provided by the geologist under the form of prior training images. The training image should display the heterogeneity patterns expected to prevail in the subsurface but need not be conditioned to any local or areal data. Under a prior decision of stationarity, the patterns extracted from the training image(s) through multiple-point statistics moments are anchored to the actual subsurface data, both hard and soft. The simulation mode is sequential (noniterative) and, hence, extremely fast.

One major advantage of the proposed approach is its simplicity. The snesim algorithm does not involve any of the classical steps of a semivariogram-based simulation technique such as semivariogram modeling followed by kriging; at each unsampled location, a simulated value is drawn from the local cpdf read directly from the training image. Furthermore, in contrast to object-based algorithms, the geometries to be reproduced in the reservoir model need not be parameterized. Hence, the same algorithm can be used to simulate any type of geological heterogeneity, of any shape, at any scale, as long as a corresponding training image can be provided. An application using geometries other than curvilinear channels is presented in Strebelle (2002). Multiple-point geostatistics aims at complementing, not replacing, two-point geostatistics. The latter is still the best approach for simulating continuous distributions within relatively homogeneous facies.

The straightforwardness of the snesim algorithm has, however, a major drawback. The multiple-point statistics actually exported from the training image to the reservoir model are not well understood. We clearly do not want to reproduce all irrelevant details (idiosyncrasies) of the training image. We would like to control or filter reproduction of each specific training multiple-point statistics. One possible avenue would be to identify then impose the reproduction of those specific multiple-point statistics known to be critical for correct prediction of flow responses (Srinivasan, 2000). Another avenue would consist of regrouping similar training patterns into clusters or prototypes. The prototypes containing too few patterns would be ignored. For the remaining prototypes, one cpdf per prototype could be computed from the central values of the patterns belonging to the prototype. During the simulation, the probability distribution conditional to any data event would then be identified by the cpdf of the prototype most similar to that conditioning data event.

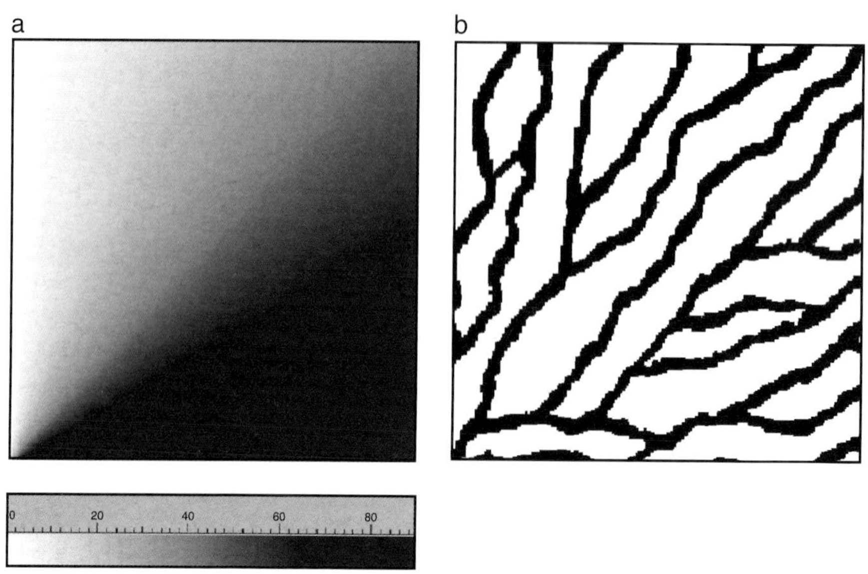


**Figure 5. (a) A map of local channel azimuths (0° = north, 90° = east) and (b) a corresponding snesim realization constrained by the map of channel azimuths.**

## REFERENCES CITED

Caers, J., and A. Journel, 1998, Stochastic reservoir simulation using neural networks trained on outcrop data: Society of Petroleum Engineers Annual Technical Conference and Exhibition, New Orleans (September 27–30), SPE Paper 49026, 16 p.

Caers, J., S. Srinivasan, and A. Journel, 1999, Geostatistical quantification of geological information for a fluvial-type North Sea reservoir: Society of Petroleum Engineers Annual Technical Conference and Exhibition, Houston (October 3–6), SPE Paper 56655, 16 p.

Caers, J., P. Avseth, and T. Mukerji, 2001, Geostatistical integration of rock physics, seismic amplitudes and geological models in North Sea turbidite systems: Society of Petroleum Engineers Annual Technical Conference and Exhibition, New Orleans (September 30–October 3), SPE Paper 71321, 16 p.

Deutsch, C., and A. Journel, 1998, GSLIB: Geostatistical software library and user's guide, 2d ed.: New York, Oxford University Press, 368 p.

Deutsch, C., and L. Wang, 1996, Hierarchical object-based stochastic modeling of fluvial reservoirs: Mathematical Geology, v. 28, p. 857–880.

Dubrule, O., 1993, Introducing more geology in stochastic reservoir modeling, *in* A. Soares, ed., Geostatistics troia: Dordrecht, Kluwer Academic Publications, p. 351–370.

Goovaerts, P., 1997, Geostatistics for natural resources evaluation: New York, Oxford University Press, 376 p.

Guardiano, F., and R. M. Srivastava, 1993, Multivariate geostatistics: Beyond bivariate moments, *in* A. Soares, ed., Geostatistics troia: Dordrecht, Kluwer Academic Publications, p. 133–144.

Haldorsen, H., and E. Damsleth, 1990, Stochastic modeling: Journal of Petroleum Technology, v. 42, p. 404–412.

Journel, A., 1993, Geostatistics: Roadblocks and challenges, *in* A. Soares, ed., Geostatistics troia: Dordrecht, Kluwer Academic Publications, p. 213–224.

Journel, A., 2002, Combining knowledge from multiple information sources: An alternative to traditional data independence hypotheses: Mathematical Geology, v. 34, p. 573–596.

Journel, A., and F. Alabert, 1989, Non-Gaussian data expansion in the earth sciences: Terra Nova, v. 1, p. 123–134.

Srinivasan, S., 2000, Integration of production data into reservoir models: A forward modeling perspective: Ph.D. dissertation, Stanford University, Stanford, California, 224 p.

Strebelle, S., 2000, Sequential simulation drawing structures from training images: Ph.D. dissertation, Stanford University, Stanford, California, 187 p.

Strebelle, S., 2002, Conditional simulation of complex geological structures using multiple-point statistics: Mathematical Geology, v. 34, p. 1–22.

Strebelle, S., and A. Journel, 2001, Reservoir modeling using multiple-point statistics: Society of Petroleum Engineers Annual Technical Conference and Exhibition, New Orleans (September 30–October 3), SPE Paper 71324, 11 p.

Strebelle, S., K. Payrazyan, and J. Caers, 2002, Modeling of a deepwater turbidite reservoir conditional to seismic data using multiple-point geostatistics: Society of Petroleum Engineers Annual Technical Conference and Exhibition, San Antonio (September 29–October 2), SPE Paper 77425, 10 p.

Tran, T., 1994, Improving variogram reproduction on dense simulation grids: Computers and Geosciences, v. 20, p. 1161–1168.

Viseur, S., 1999, Stochastic Boolean simulation of fluvial deposits: A new approach combining accuracy and efficiency: Society of Petroleum Engineers Annual Technical Conference and Exhibition, Houston (October 3–6), SPE Paper 56688, 8 p.

Xu, W., 1996, Conditional curvilinear stochastic simulation using pixel-based algorithms: Mathematical Geology, v. 28, p. 937–949.

Kelley, V. A., and S. Mishra, 2006, An efficient approach for quantifying the uncertainty in groundwater-model predictions, *in* T. C. Coburn, J. M. Yarus, and R. L. Chambers, eds., Stochastic modeling and geostatistics: Principles, methods, and case studies, volume II: AAPG Computer Applications in Geology 5, p. 151–158.

Chapter 12

# An Efficient Approach for Quantifying the Uncertainty in Groundwater-model Predictions

**V. A. Kelley**
**S. Mishra**
*INTERA Inc.*
*Austin, Texas, U.S.A.*

## ABSTRACT

Distributed-parameter models are increasingly being used to incorporate hydrogeologic uncertainty in predictive analyses of groundwater flow and contaminant transport. Geostatistical techniques have proven to be useful for generating multiple equiprobable realizations of subsurface parameters that are consistent with the available data. However, computational constraints commonly preclude detailed transport calculations with all realizations in a Monte Carlo simulation framework for quantifying the uncertainty in model predictions. As a result, several studies have been performed investigating techniques to rank realizations of a stochastic reservoir or groundwater aquifer model. These techniques share the common goal of trying to develop a surrogate measure that will preserve the quantiles of the desired performance measure. This chapter proposes a technique to both rank and weight realizations through the use of a surrogate traveltime measure and to then accurately reproduce the performance-measure statistics with only a few forward simulations. The methodology is demonstrated using a field example from the Waste Isolation Pilot Plant site in New Mexico that is well documented in the groundwater literature.

## INTRODUCTION

Geostatistical techniques, such as conditional simulations, have proven to be useful for generating multiple equiprobable realizations of subsurface parameters that are consistent with the observed field data. In most cases, the computational burden necessary to generate multiple reservoir realizations is very small relative to the computational burden required to calculate the desired performance measure for all simulated reservoir realizations. As a result, detailed forward simulations using all reservoir realizations are


DOI:10.1306/1063813CA53232

not often performed to quantify performance measure uncertainty. For example, in reservoir forecasting, reservoir engineers typically rank realizations using a surrogate-performance measure and perform reservoir calculations using a subset of realizations selected based on the surrogate-performance measure. This chapter presents an efficient approach for ranking and weighting realizations using a surrogate-performance measure to estimate uncertainty in the aquifer or reservoir performance from a limited number of model simulations. In the petroleum industry, the geosystem being modeled is typically called a reservoir. In the groundwater industry, the geosystem being modeled is typically called an aquifer. The methodology presented in this chapter applies to both industries and the terms aquifer and reservoir are used interchangeably.

The method of ranking realizations to assess reservoir-performance uncertainty has received a great deal of attention in recent petroleum literature. Several studies have been performed investigating various techniques to rank realizations of a stochastic reservoir model. In each case, the objective is to define a quantile-preserving surrogate-performance measure to rank the geostatistical realizations from which uncertainty in reservoir performance can be estimated. Ballin et al. (1992) used a single-phase tracer response as the surrogate-performance measure. An estimate of reservoir permeability connectivity was used as a surrogate-performance measure by Alabert and Modot (1992). Saad et al. (1996) used tracer production history as the surrogate-performance measure. Most recently, swept volume estimates through streamline simulation were used as a surrogate-performance measure (Idrobo et al., 2000).

All of these ranking methods seek to identify a few realizations that will exhibit the worst case, median or average, and best case reservoir performance. Inherent in this approach is the assumption that the surrogate-performance measure selected to rank the realizations (i.e., best and worst case) also applies to the performance measure of interest. However, these approaches do not provide information regarding the likelihood of the performance measure of interest for realizations, which may fall between the few simulated. Likewise, they do not provide a means to calculate summary statistics of the performance measure of interest such as its mean and standard deviation. In response to these limitations, this chapter proposes a technique to both rank and weight realizations through the use of a surrogate-performance measure and to then accurately reproduce the performance-measure statistics with only a few forward simulations.

Summary statistics of the performance measure of interest provides important information for a petroleum engineer when performing an economic (e.g., oil recovery) risk analysis. Likewise, in the environmental industry, summary statistics of the performance measure (e.g., contaminant concentration) would be beneficial for performing a probabilistic health-risk analysis. The proposed methodology seeks to provide this information without the computational burden associated with the simulation of all model realizations.

## APPROACH

The proposed approach entails a four-step process that is presented in Figure 1. The key concept here is to provide a strategy for selecting only a few realizations as input to detailed simulations and then extracting information regarding the mean and variance of reservoir performance by appropriately weighting the results of these simulations. In the following paragraphs, the proposed ranking and weighting method is discussed in a generic format so that it could be applied to any surrogate-performance measure. The methodology is demonstrated using a well-documented field example from the Waste Isolation Pilot Plant (WIPP) site in New Mexico.

### Quantify Reservoir Description Uncertainty

The first step, which is not the focus of this chapter, is the development of a stochastic description of the reservoir. Typically, uncertainty in reservoir parameters and geometry is handled through conditional simulations, although several new approaches have also emerged in recent years. Yarus and Chambers (1994) provide a good overview of stochastic methods for reservoir characterization. Once multiple equiprobable images of the subsurface have been generated, a surrogate-performance measure must be defined. As has been discussed earlier in this chapter, there have been several surrogate-performance measures proposed in the petroleum field. Mishra

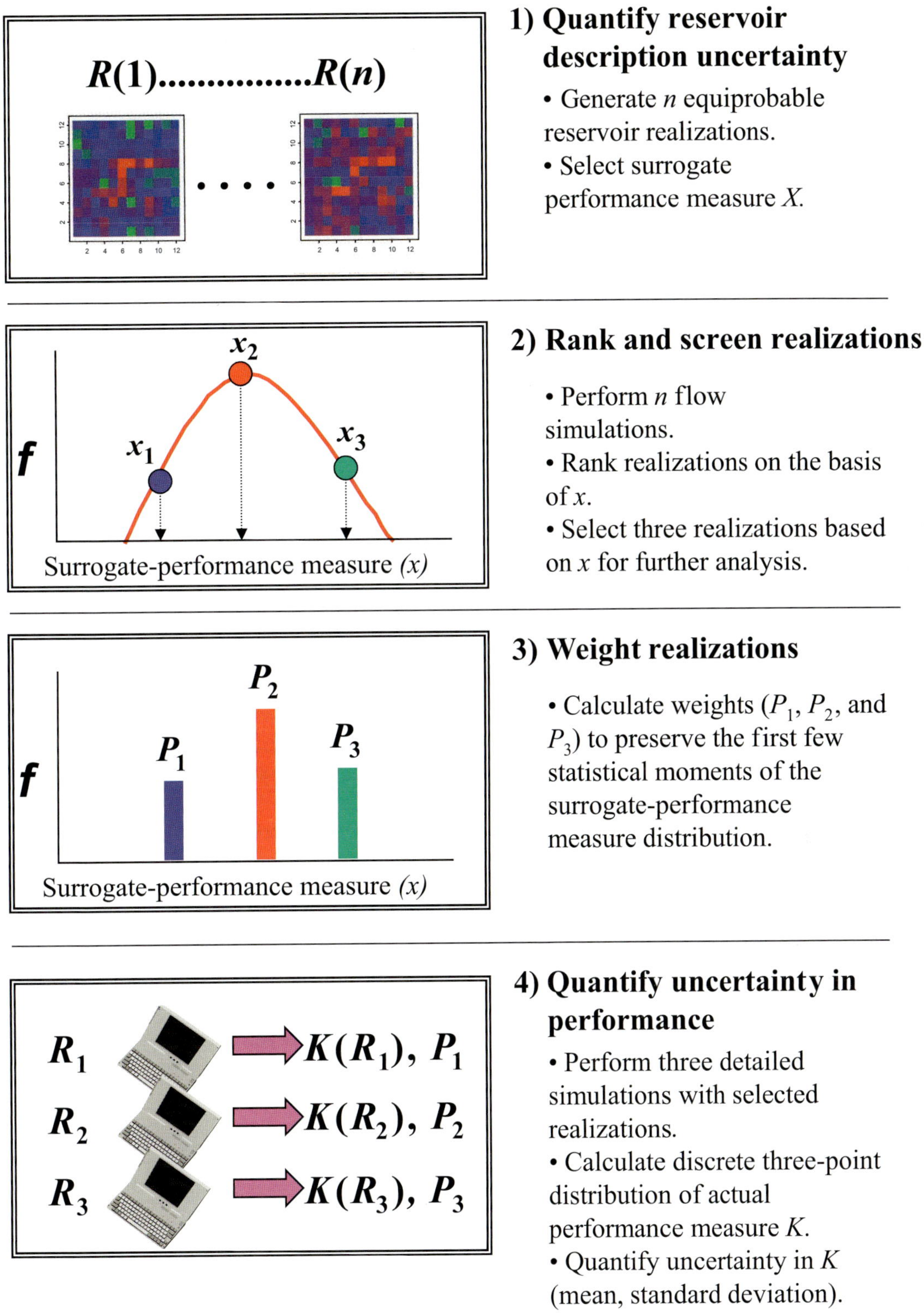


**Figure 1.** **Proposed ranking and weighting methodology.**

et al. (2000) provide an update summary of surrogate-performance measures proposed to rank geostatistical realizations. The surrogate-performance measure should be chosen such that (1) its rank-ordered behavior is equivalent to that of the actual-performance measure of interest and (2) it is much faster to compute than the actual-performance measure.

## Rank and Screen Realizations

Next, forward simulations are carried out for all ($n$) realizations, and the chosen surrogate-performance measure (denoted as $x$) is calculated for each simulation. At this point, the analyst has a distribution of the surrogate-performance measure that will be used to rank all realizations. With realizations ranked, the analyst must select or screen which discrete values of the surrogate measure, and their corresponding realizations, should be chosen for further analyses. This process is subjective. However, one would like to select realizations such that the surrogate-performance measure captures the full range of uncertainty. Consistent with industry practice, we suggest using the median (50th percentile) realization, along with a realization falling between the 5th and 10th percentiles at the low end and the 90th and 95th percentiles at the high end. The geostatistical models (realizations) corresponding to these discrete values then become candidates for carrying out detailed simulations.

## Weight Realizations

Up to step 2, our approach is similar to many other investigators described in the introduction of this chapter. However, this approach differs from others previously proposed in the process of weighting realizations. The idea behind our weighting method goes back to the work of Kaplan (1981) in the field of probabilistic seismic hazard and risk assessment, where decision trees are commonly used for propagating uncertainty. To avoid the problem of combinatorial explosion, which puts a practical limit on the number of uncertain variables, Kaplan suggested that a discrete distribution with multiple (e.g., 10 or greater) values should be replaced by a simpler one with three to five values to make the uncertainty analysis problem tractable. The two distributions are made consistent by requiring that the weights (probabilities) for the new values be chosen so as to preserve the first few statistical moments. The approach described herein extends the method of Kaplan to the limit of a continuous distribution (as represented by an empirical cumulative distribution function, or CDF, generated with ~100 samples).

The weighting methodology proposed in this work is based on the idea that an empirical CDF can be approximated by a discrete distribution such that the statistical moments of the original distribution are preserved. As depicted in Figure 1, this implies that if we chose the values $x_1$, $x_2$, and $x_3$ as discrete representations of the surrogate-performance measure, $x$, then their respective weights, $P_1$, $P_2$, and $P_3$ must satisfy the following moment-matching constraints:

$$P_1 \times x_1 + P_2 \times x_2 + P_3 \times x_3 = \mathrm{E}[x] \quad (1)$$

$$P_1 \times x_1^2 + P_2 \times x_2^2 + P_3 \times x_3^2 = \mathrm{E}[x^2] = \mathrm{E}^2[x] + \mathrm{V}[x] \quad (2)$$

where E[•] denotes the statistical expectation or average, and V[•] denotes the variance. Note that $x$ is some readily computed surrogate measure (indicator) of reservoir performance, and the values $x_1$, $x_2$, and $x_3$ correspond to the screened realizations $R_1$, $R_2$, and $R_3$, respectively.

From the empirical CDF of $x$, we know E[$x$] and V[$x$]. Thus, once the discrete quantities $x_1$, $x_2$, and $x_3$ are chosen, the weights $P_1$, $P_2$, and $P_3$ are determined using equations 1 and 2 and an additional constraint requiring that the discrete weights (probabilities) must sum to unity:

$$P_1 + P_2 + P_3 = 1 \quad (3)$$

## Quantify Uncertainty in Performance

The realizations selected in this process ($R_1$, $R_2$, and $R_3$) are then used as inputs to detailed model simulations for computing the actual-performance measure of interest (e.g., contaminant concentration, oil recovery), denoted hereafter as $K$. The uncertainty in performance measure $K$ can be characterized as follows:

$$\mathrm{M}[K] = P_1 \times K(R_1) + P_2 \times K(R_2) + P_3 \times K(R_3) \quad (4)$$

$$\mathrm{SD}[K] = \{P_1 \times \{K(R_1) - \mathrm{M}[K]\}^2 + P_2 \times \{K(R_2) - \mathrm{M}[K]\}^2 + P_3 \times \{K(R_3) - \mathrm{M}[K]\}^2\}^{1/2} \quad (5)$$

where M[•] denotes the mean, and SD[•] denotes the standard deviation.

# FIELD APPLICATION

The proposed methodology is illustrated using a field example from the Department of Energy's WIPP site located in southeastern New Mexico. The Culebra

Dolomite field case was selected to test our methodology because (1) there already exists 70 calibrated conditional simulations of the Culebra Dolomite transmissivity field (Lavenue et al., 1995), (2) the data set is well documented, and (3) the data set was available to the authors.

The WIPP site is a licensed facility for the disposal of defense-related transuranic wastes. The repository is located within the bedded salt of the Salado Formation at a depth of approximately 650 m (2132 ft). The 7-m (23-ft)-thick Culebra Dolomite is the most transmissive and laterally continuous hydrogeologic unit located above the repository horizon. As a result, this unit is considered the most likely pathway for transport of nuclides to the accessible environment.

## Quantifying Reservoir Description Uncertainty

As part of the repository licensing requirements, the WIPP repository system performance assessment was evaluated throughout a 10,000-yr compliance period. The U.S. Environmental Protection Agency compliance regulations require a probabilistic performance assessment that adequately reflects the variability and uncertainty within all factors that contribute to repository performance. Therefore, for performance assessment, one must account for regional flow system uncertainty in the Culebra Dolomite. In response to this need, a new methodology for the solution of the inverse problem was developed. The method, referred to as the pilot-point methodology, generates several equiplausible conditional simulated transmissivity fields and then automatically calibrates each transmissivity field to steady-state and transient pressure measurements through an iterative process (RamaRao et al., 1995). The resulting method ends with a set of multiple calibrated transmissivity fields that honor the transmissivity measurements at their locations, preserves the semivariogram, and matches the available pressures. Further documentation on this methodology can be found in Lavenue et al. (1995) and Zimmerman et al. (1998).

In the WIPP performance assessment, the Culebra flow field was modeled using the finite-difference code Sandia Waste Isolation Flow and Transport-II with a two-dimensional grid that comprises 2850 grid blocks covering a 23 × 27-km (14 × 16-mi) domain. Seventy conditional simulations of the transmissivity field were conditioned to 41 measured transmissivity values and calibrated against steady-state hydraulic head measurements and transient hydraulic head measurements observed at 30 different locations throughout an 8-yr calibration period. A detailed description of the field study can be found in Lavenue et al. (1995).

The primary performance-assessment scenario for transport within the Culebra is that of an artificial connection between the repository and the Culebra that comprises an exploration borehole. Figure 2 shows a schematic drawing of this possible release scenario. In this scenario, waste is transported to the Culebra through the hypothetical borehole and is then carried to a compliance boundary under regional flow controls. From a performance-assessment perspective, the performance measure of interest is the concentration of waste measured within the Culebra at the compliance boundary. As a surrogate-performance measure for concentration, we have selected advective groundwater traveltime as calculated based on particle tracking. This surrogate-performance measure was also used by Zimmerman et al. (1998) to compare seven geostatistical inverse methods, including the pilot-point method of RamaRao et al. (1995).

## Ranking and Screening Realizations

The 70 calibrated conditional simulations of transmissivity of Lavenue et al. (1995) were simulated to develop steady-state velocity fields. For each velocity field, the advective groundwater traveltime was calculated for a particle dropped within the Culebra Dolomite at the approximate centroid of the WIPP underground waste repository. The compliance boundary was arbitrarily established at a distance of 2 km (1.2 mi) from the centroid and measured perpendicular to the model $y$-axis (see Figure 2 for a schematic representation). The realizations were then ranked based on the surrogate-performance measure advective groundwater traveltime. Figure 3 shows the advective groundwater-traveltime CDF. Three realizations, corresponding to the 7th, 50th, and 93rd percentile values of advective-groundwater traveltime ($x$), were selected for further analyses and detailed simulations.

## Weighting Realizations

The next step is to weight these three realizations such that the three-point discrete distribution matches the 70-point CDF through the preservation

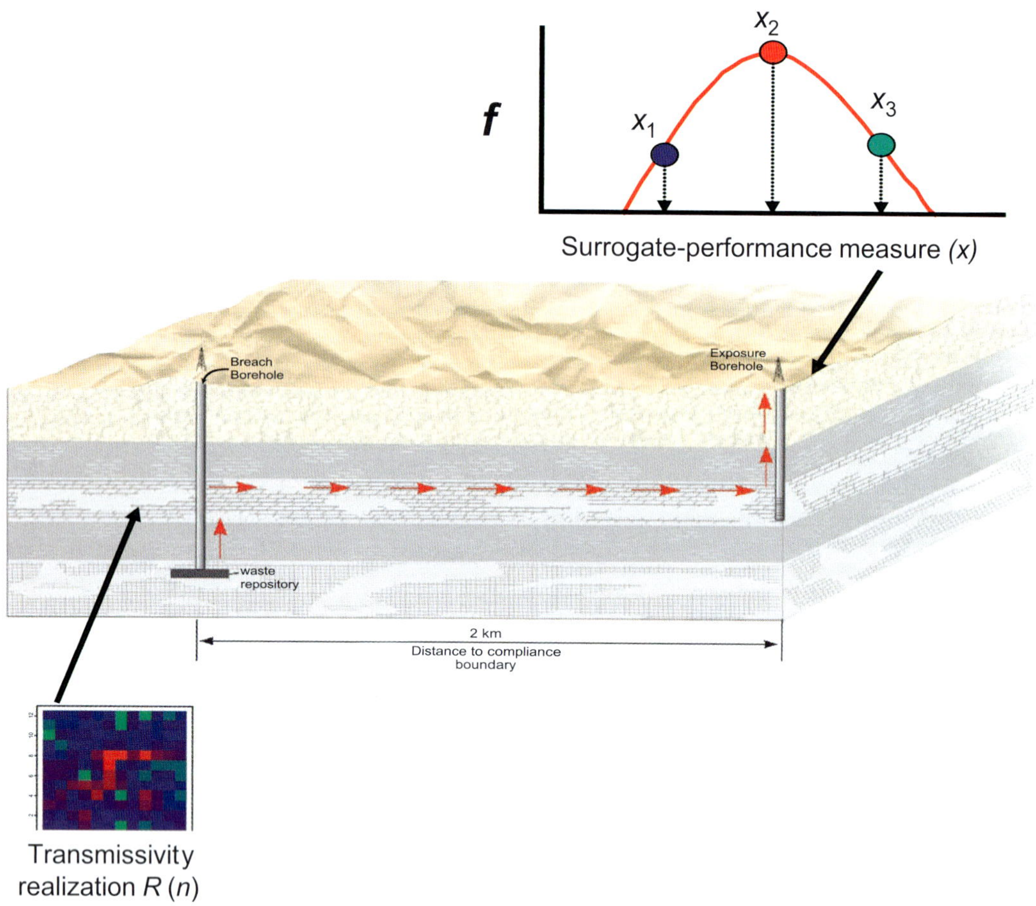


**Figure 2. Performance-assessment release scenario.**

of the first few statistical moments. The three unknown weights corresponding to these three realizations were calculated as per equations 1–3 and are summarized in Table 1.

## Quantifying Uncertainty in Performance

The final step in the proposed methodology is to perform detailed transport simulations for the three selected realizations. The performance measure of interest ($K$) was selected to be the relative concentration ($C/C_o$) measured at the compliance boundary. Full transport simulations were performed for the three selected realizations. The results from these three transport simulations were combined using the weights provided shown in Table 1 and equations 4 and 5 to obtain estimates of the mean and standard deviation of the relative concentration at the compliance boundary as a function of time. To evaluate the accuracy of the proposed weighting methodology, detailed transport simulations were also carried out for all 70 realizations.

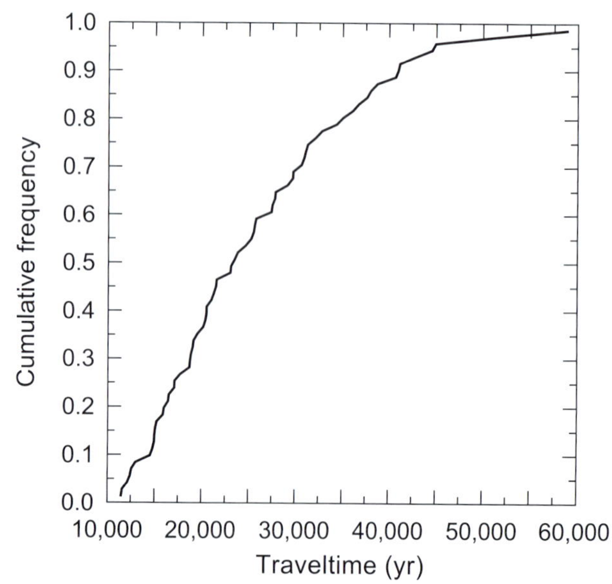


**Figure 3. Advective groundwater traveltime to the compliance boundary.**

**Table 1. Selected realizations and their calculated weights.**

| *Percentile* | *Traveltime (yr)* | *Realization* | *Weight* |
|---|---|---|---|
| 7 | 12,557 | 19z | 0.2190 |
| 50 | 23,496 | 22z | 0.5481 |
| 93 | 42,719 | 34z | 0.2329 |

Figure 4 shows the average relative concentration simulated as a function of time from all 70 simulations to the average calculated using the three transport simulations and the weighting method. Similarly, Figure 5 shows the standard deviation of relative concentration as a function of time for all 70 realizations as compared to the three-simulation estimate. As shown in these figures, the weighted averages from the three screened realizations were computed at only a limited number of equal-time intervals. A comparison between the results from our approximate method and those from detailed simulations shows very good agreement, considering the fact that only three transport simulations are used to reproduce the average behavior of all 70 realizations and their associated uncertainty. In particular, note the robustness of the proposed approach in view of the number-of-orders-of-magnitude variability of relative concentration throughout the 20,000-yr simulation period.

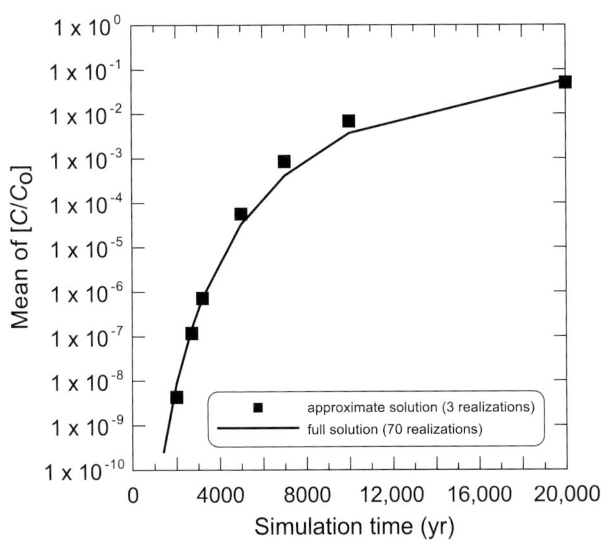


**Figure 4. Comparison between the approximate mean relative concentration and the true mean.**

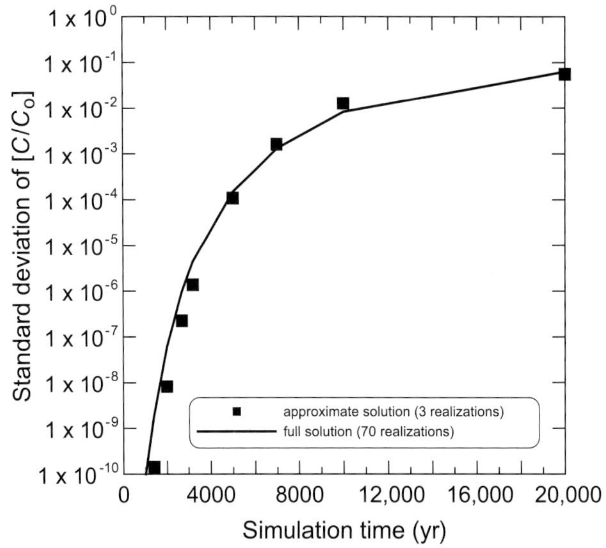


**Figure 5. Comparison between the approximate standard deviation of relative concentration and true standard deviation.**

## SUMMARY AND CONCLUSIONS

In this chapter, a methodology has been proposed to rank and weight multiple reservoir realizations through the use of a surrogate-performance measure. Current realization-ranking methods proposed in the literature do not provide information regarding the likelihood of the performance measure of interest for realizations that may fall between the few simulated. In contrast, the proposed methodology provides a computationally efficient method of calculating the performance measure mean and variance over all realizations through the use of only a few selected forward simulations. This approach can be used with any surrogate-performance measure as long as the surrogate-performance measure statistics adequately reproduce the statistics of the actual-performance measure of interest.

The proposed methodology was applied to a field example taken from the WIPP site performance-assessment studies. In this case, groundwater traveltime was used as a surrogate-performance measure for the contaminant concentration, the actual-performance measure of interest. Of 70 aquifer realizations, 3 were selected for further analysis. Using these three realizations, the performance of all 70 realizations was reproduced as well as defined by the mean and standard deviation of the performance measure of interest (i.e., contaminant concentration). The application of this methodology to an oil reservoir-forecasting problem is documented in Mishra et al. (2000). In that study, the proposed methodology was applied to estimating the uncertainty in oil-reservoir forecasts using streamline time-of-flight volumetric sweep as the surrogate-performance measure for water cut at multiple production wells. Consistent with the results from this study, the ranking and weighting approach using three simulations showed good agreement with the reservoir forecasts from the full suite of reservoir realizations.

## ACKNOWLEDGMENTS

The authors thank Marsh Lavenue and Banda RamaRao for the 70 calibrated transmissivity realizations for the Culebra Dolomite.

## REFERENCES CITED

Alabert, F. G., and V. Modot, 1992, Stochastic models of reservoir heterogeneity: Impact on connectivity and average permeabilities: Society of Petroleum Engineers Annual Technical Conference and Exhibition, Washington, D.C., October 4–7, 1992, SPE Paper 24893, 16 p.

Ballin, P. R., A. G. Journel, and K. A. Aziz, 1992, Prediction of uncertainty in reservoir performance forecasting: Journal of Canadian Petroleum Technology, v. 31, p. 52–58.

Idrobo, E. A., M. K. Choudhary, and A. Datta-Gupta, 2000, Swept volume calculations and ranking of geostatistical reservoir models using streamline simulation: Society of Petroleum Engineers/AAPG Western Regional Meeting, Long Beach, California (June 19–23), SPE Paper 62557, 13 p.

Kaplan, S., 1981, On the method of discrete probability distributions: Risk Analysis, v. 1, p. 189–196.

Lavenue, A. M., B. S. RamaRao, G. de Marsily, and M. G. Marietta, 1995, Pilot-point methodology for automated calibration of an ensemble of conditionally simulated transmissivity fields: 2— Application: Water Resources Research, v. 31, p. 495–516.

Mishra, S., M. K. Choudhary, and A. Datta-Gupta, 2000, A novel approach for reservoir forecasting under uncertainty: Society of Petroleum Engineers Annual Technical Conference, Dallas, Texas, October 1–4, 2000, SPE Paper 62926, 11 p.

RamaRao, B. S., A. M. Lavenue, G. de Marsily, and M. G. Marietta, 1995, Pilot-point methodology for automated calibration of an ensemble of conditionally simulated transmissivity fields: 1— Theory and computational experiments: Water Resources Research, v. 31, p. 475–493.

Saad, N., V. Maroongroge, and C. T. Kalkomey, 1996, Ranking geostatistical models using tracer production data: European 3-D Reservoir Modeling Conference, Stavanger, Norway (April 16–17), SPE Paper 35494, 14 p.

Yarus, J., and R. L. Chambers, eds., 1994, Stochastic modeling and geostatistics: Principles, methods, and case studies: AAPG Computer Applications in Geology 3, 379 p.

Zimmerman, D. A., et al., 1998, Comparison of seven geostatistically-based inverse approaches to estimate transmissivities for modeling advective transport by groundwater flow: Water Resources Research, v. 34, p. 1373–1413.

Lecour, M. A., P. Thore, and R. Cognot, 2006, Consistent structural model simulations for volumetric uncertainty study, *in* T. C. Coburn, J. M. Yarus, and R. L. Chambers, eds., Stochastic modeling and geostatistics: Principles, methods, and case studies: volume II: AAPG Computer Applications in Geology 5, p. 159–172.

*Chapter 13*

# Consistent Structural Model Simulations for Volumetric Uncertainty Study

**M. A. Lecour**
*Earth Decision*
*Nancy, France*

**P. Thore**
*Total*
*Pau, France*

**R. Cognot**
*Earth Decision*
*Nancy, France*

## ABSTRACT

Uncertainty studies become increasingly important as reservoir complexity increases. Although such studies are commonly a key factor in oil and gas exploration, projects are commonly restricted to simulation of facies and petrophysical properties. The structure of the reservoir itself is commonly considered to be a deterministic parameter, and yet it may contain uncertainty that has a major impact on the reserve estimations.

A new methodology based on the P-field technique and focusing on reservoir geometry uncertainty is proposed to tackle this problem. With this method, not only horizon but also fault geometries are simulated around a given reference model.

Because simulating fault geometries is more difficult than simulating horizon geometries, a new data structure has been designed to produce efficient computation of geometries. The simulation method tries to preserve the initial geometry at best. It may affect the location, dip, and shape in map view of all faults or any combination of these three basic modifications. Then, for each fault network realization, several horizon geometry simulations may be performed. The modeling and simulation system keep the geological model consistent after each simulation loop, allowing volumetric studies to proceed in a consistent fashion.


DOI:10.1306/1063814CA53233

## INTRODUCTION

Uncertainties related to oil-field reserve estimates are everywhere. They may be caused by a combination of factors, such as reservoir geometry (horizon and fault geometry), facies distribution, and rock properties. Indeed, the available data for modeling deep reservoirs are commonly uncertain, scattered, and inaccurate. Most of the time, such uncertainties have a strong impact on the field exploitation. If the initial model does not correspond to the real structure, subsequent reserve estimates could be biased, thus having important negative effects on field exploitation costs. To minimize risks, the earth model must include uncertainties to generate not only one model or three models (optimistic, median, and pessimistic), but several models scanning the range of possibilities for the zone of interest (Srivastava, 1994; Yarus, 1994; Vincent et al., 1999; Thore et al., 2002).

Currently, most uncertainty studies concentrate on facies distribution (S. F. Carle, 2000, personal communication) and rock property simulations while considering the structure of the reservoir as a deterministic parameter (O. H. Fjelltun and A. Kihlberg, 2000, personal communication; K. Yang, 2000, personal communication; Williams and Aqrawi, 2006). Conventionally, only one structural model is built to represent the reservoir structure from a geological interpretation of seismic data (Thore et al., 2002). Nevertheless, in many cases, full uncertainty studies showed that the structure of the reservoir has the strongest impact on the reserve estimates uncertainty (Tinker, 2000) (see Figure 1). Some factors affect the accuracy with which the geometry of the reservoir (and particularly, the geometry of faults) may be constrained (Lecour, 2000).

The work presented in this chapter concerns the study of reservoir geometry uncertainties; not only the horizons but also the fault geometries around a given reference model are simulated. A new algorithm is proposed for modeling faults, fault networks, and the associated uncertainty. The simulation method is based on the P-field technique mostly used for reservoir properties simulation. This method has been chosen because, in comparison to other methods, it is fast, it allows one to spatially correlate the selected random numbers, and the computation time is substantially reduced (Srivastava, 1992, 1994). The objective is not to estimate the uncertainty itself. The purpose is to incorporate an estimate of uncertainty in the fault modeling process and to simulate many equiprobable fault network geometries quickly, while preserving the initial geometry.

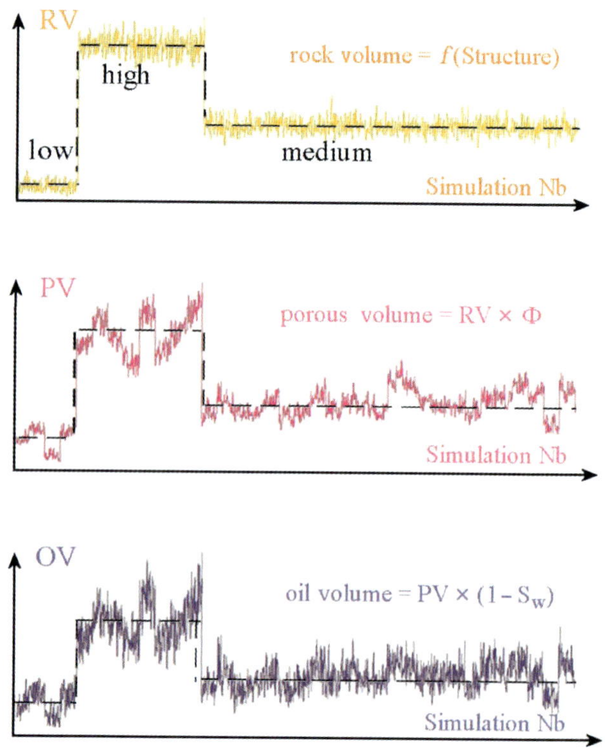


**Figure 1. In many cases, structural uncertainty has the strongest impact on reserve estimates. In this study, the signature of the different volume computations (rock volume, porous volume, and oil volume) is controlled by structural uncertainty (rock volume).**

## FAULT UNCERTAINTIES: PROBLEM TO SOLVE

Structural modeling is accomplished after completing a series of steps to translate raw data into information that can be used in the modeling process (seismic acquisition, seismic processing, seismic interpretation, and time-to-depth conversion). Although techniques have improved significantly during the last few years, uncertainty may occur at each step in this chain. This section describes some of the uncertainties affecting fault modeling.

### Why Study Fault Uncertainties?

In some cases, the knowledge of fault geometry may be critical in representing the real geology in the structural model and also in accurately estimating

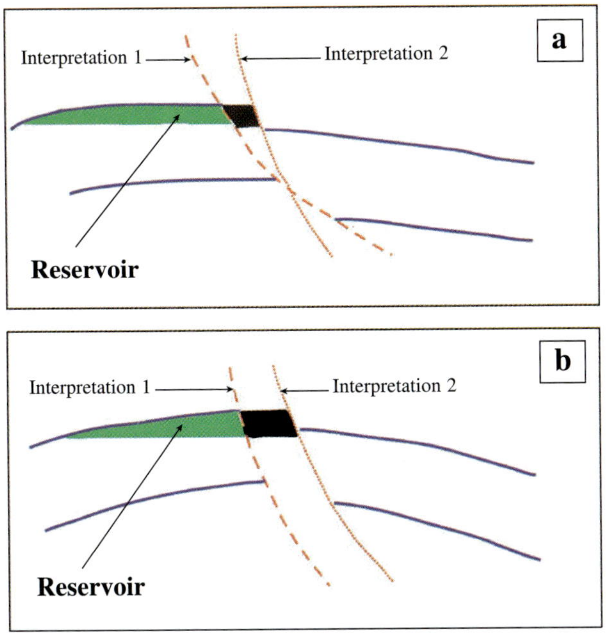


**Figure 2. Two cases that illustrate why studying the uncertainty of fault geometry may be critical: (a) bad dip interpretation and (b) bad interpretation of the real location of the fault. The well may be drilled on the wrong side of the fault if the interpretation is too optimistic, or the production installation may not be well designed if the reserves have been underestimated.**

the reserves. Faults are commonly involved in trapping systems, and a good definition of their geometry may be a key to success (P. Abrahamsen, 1999, personal communication).

Although internal faults have few impacts on reserve estimates (O. H. Fjelltun and A. Kihlberg, 2000, personal communication), a small modification of the dip angle of border faults may have a significant impact on reserve estimates and on production planning (M. Lecour, R. Cognot, P. Thore, and J.-C. Dulac, 1998, personal communication). In the example in Figure 2a, either reserves are overestimated if the real fault corresponds to interpretation 2 and the interpreted fault corresponds to interpretation 1, or reserves are underestimated in the opposite case. A shift between the real fault location and the interpreted one may also lead to bad volume estimates (see Figure 2b).

The well may be drilled on the wrong side of the fault if the interpretation is too optimistic, or the production installation may not be well designed if the reserves have been underestimated. This is why it may sometimes be necessary to have a better estimation of structural, and particularly fault, uncertainties before any facies or property simulation is performed.

## Uncertainty Sources Around Faults

As previously discussed, structural modeling is accomplished late in the exploration process. Several sources of uncertainty from previous steps may be combined at the beginning of the structural modeling step (see Figure 3). Such uncertainties may come from the acquisition, the seismic processing, the migration, the interpretation in time domains, or the model construction in depth domain.

## Data Acquisition and Preprocessing

During these steps, uncertainty affects the seismic quality but have almost no effect on structural uncertainty (Thore, 1999; Thore et al., 2002). To reduce the noise in the raw seismic data, traces are stacked. This step is very important because it corresponds to the creation of the section to be interpreted. Uncertainty at this step is related to the sharpness of the seismic picture, but it has a minor impact on the structure definition itself.

## Migration

Migration is one of the most important steps from a structural point of view. Migration removes, or at least attenuates, diffraction hyperboles caused

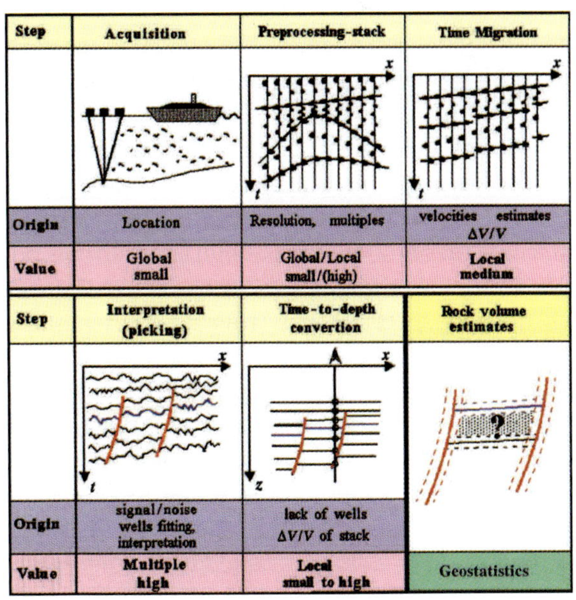


**Figure 3. Summary of uncertainty sources for fault modeling in the depth domain (modified after Thore, 1999). The most important source of uncertainty corresponds to the time migration and the time-to-depth conversion that require well-estimated velocities and to the interpretation steps that are accomplished manually.**

by diffracting points (Henry, 1994). These diffracting points are very common along horizon-fault contacts. If the migration is not done properly (e.g., because the velocity field is not well defined), diffraction zones at horizon-fault contacts will not be fully removed, and the uncertainty zone may increase significantly.

### Interpretation in Time Domain

This step is critical from a structural point of view because it is the last step that uses seismic data as input. Because most of the interpretation is done manually and because seismic data are commonly blurred along fault zones, a lot of uncertainty is linked to this step, especially in complex structures with dipping faults. The uncertainty caused by the interpretation phase may also increase when there is some gas that produces high-amplitude zones or when the lithology corresponds to salt or shale, which are known to be very bad reflectors.

Another problem occurs when the interpretation is done in a highly channelized zone, where braided structures make the seismic reflection too blurred. More than this, the variation of rock properties inside channels sometimes creates fake faults.

### Reconstruction in Depth Domain

Once the structural model has been interpreted and validated in the time domain, it is passed to the depth domain. This step commonly represents the major part of the uncertainty (Thore et al., 2002). Indeed, the transformation from time to depth domain requires a correct velocity model, which is commonly not available in most complex structures. More than this, a simple stretching (depth = velocity × time) is not always possible with subvertical structures like faults. Usually, only horizons are transformed from time to depth, with the fault network being fully reinterpreted and validated in three dimensions (3-D) in the depth domain.

Because seismic data are commonly blurred near faults, horizon-fault contacts are not well defined. The horizon interpretation either is going too far across the fault (when an automatic picking method is used) (see Figure 4a) or is stopped as soon as the interpreter does not see enough on the seismic section (when the interpretation is performed manually). In the second case, the gap left in the horizon is bigger than the real width of the fault (see Figure 4b).

Sometimes, the gap is reduced by drawing the most probable shape of the fault in map view. However, because fault polygons are manually drawn in the depth domain, where raw seismic data are not available anymore, uncertainty about real fault location remains. This uncertainty occurs especially where a secondary fault is linked to a main fault and where uncertainties caused by each fault are combined (see Figure 4b).

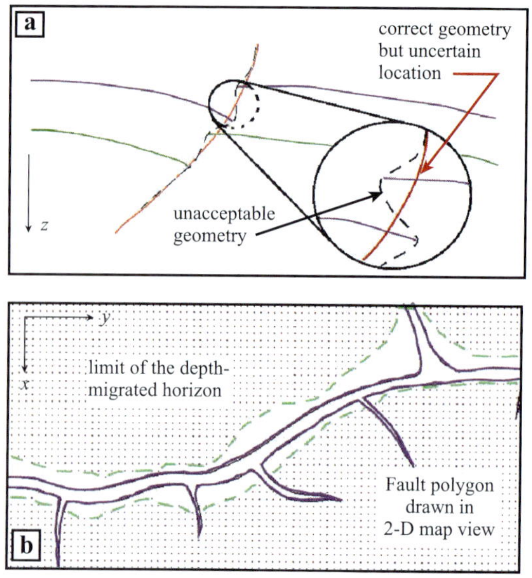


**Figure 4. Most of the time, faults are reinterpreted in the depth domain from the migrated horizon interpretations. When horizon interpretations have been extended too far by autopicking methods, strictly respecting their limits results in an unacceptable fault geometry; but interpreting an acceptable fault geometry leads to larger uncertainty on the exact fault location (a). However, when the interpretation has been accomplished manually, it commonly stops far from the fault location; and the real location of the fault is uncertain as well (b).**

### Limits of the Present Study

Of all the sources of uncertainty previously discussed, the remainder of this chapter will focus on uncertainties caused by the interpretation step. Others have considered velocity model uncertainty (Failly-Berthet and Thore, 1996; Juliard and Thore, 1999; Thore et al., 2002). In the following sections, uncertainty analysis is performed in the context of a constant velocity model. This model is taken to be the reference, with only small variations in fault geometry being introduced.

## UNCERTAINTY ESTIMATION

A simple way to estimate the uncertainty caused by the interpretation step is to compare the limit of the depth-migrated set of points corresponding to

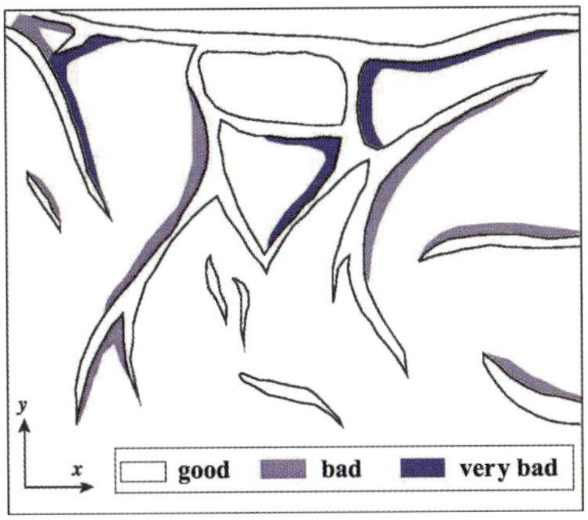


**Figure 5. An example of the seismic quality map that may be used to estimate uncertainty around faults. Even if it is still qualitative, such a map has the advantage of being linked directly to the input data.**

the interpreted horizon and the fault polygon drawn by a structural geologist in the depth domain. When the two curves coincide, the uncertainty may be considered low because the interpretations achieved by two different methods agree. When the difference between the two is large, the shape of the fault is more uncertain (see Figure 4b).

To get a good estimate of the uncertainty, it would be better to analyze the original seismic data as illustrated in Figure 5. Some geophysicists already provide seismic quality maps that can be used to estimate the quality of the interpretation (Thore et al., 2002).

Whatever the method used to determine fault uncertainty, it is very important to check the consistency of the estimates. If the uncertainty concerning the location of a fault is small at one point, this uncertainty cannot be large 20 ft (6 m) away from this point. The consistency of uncertainty estimation needs to be checked before going further in the process.

## GENERAL RULES AND METHODS FOR FAULT SIMULATION

The following discussion presents some general rules and methods for fault simulation.

### Methods Used for Horizon Geometry Simulation

Methods already exist for simulating several equiprobable geometries for horizon surfaces (Samson et al., 1996) (M. Shuguang and A. Journel, 1997, personal communication). The principle of such methods is to assign an interval of uncertainty estimates at each point in the reference horizon. Points are then randomly selected, and new locations are simulated for them. These points are chosen to cover the full extension of the surface in such a way that the average distance between them is similar. Finally, a kriging-based method is used to compute new locations for the points that were not selected in the first step (see Figure 6).

### Fault Specificity

Although horizon geometry can be simulated randomly using the preceding approach (see Figure 7a), it must be further adapted to facilitate fault geometry simulation. Indeed, fault geometries must respect more geological rules than horizons because of rock mechanics (see Figure 7c). Otherwise, a pure random simulation may be unacceptable, yielding unrealistic geometries for the faults (see Figure 7b).

Additional discussion of correct fault geometry simulation focuses on two rules:

1) R1: An inversion of the curvature of the fault along the vertical direction is not allowed.
2) R2: The throw characteristics of the fault are to be retained during the simulation process (i.e., a normal fault should remain a normal fault).

As shown below, the proposed method is general enough to include more geological rules. This will

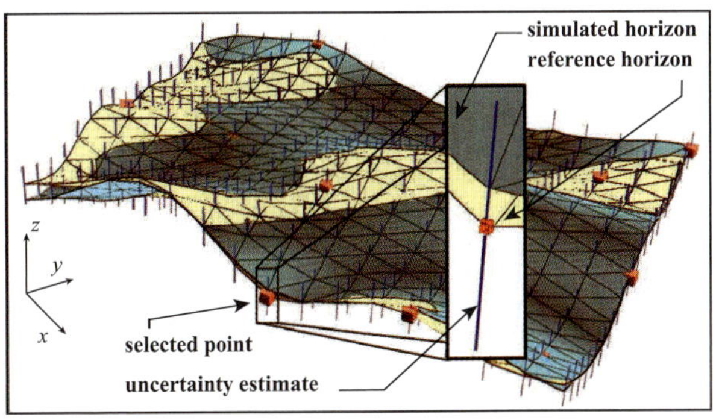


**Figure 6. An illustration of one of the methods commonly used to simulate several realizations of horizon geometry. An interval of estimated uncertainty is assigned at each point in the reference horizon (light color). Points are then randomly selected on this reference horizon, and new locations are simulated for them. Finally, a kriging-based method is used to compute new locations for the points that were not selected in the first step. This defines the geometry of the new realization (dark color).**

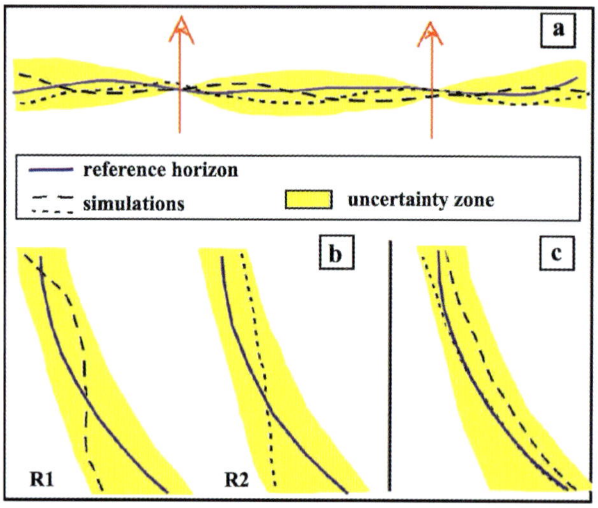


**Figure 7. The previously presented method for simulating horizon geometry (a) should be adapted for faults to get an acceptable fault geometry (c). Otherwise, the two rules R1 (no curvature inversion along the vertical direction) and R2 (throw characteristics of the fault are to be retained during the simulation process) on which the methodology currently focuses may not be respected (b).**

be possible as soon as these rules are defined in a mathematical way instead of relying on a quantitative description.

## FAULT NETWORK SIMULATION

Because of the specificity of faults described in the previous section, a new process has been designed especially for fault geometry simulation. This process, based on the p-field technique described by Srivastava (1992), allows uncertainty estimates to be considered while keeping the geometry as close as possible to the reference fault geometry. For fast and accurate simulations, a new fault data structure specifically dedicated for this task has also been designed.

### Fault Data Structure

The requirements for the fault data structure include

1) integrating all available input data describing fault geometries (in the case of uncertainty, it is indeed critical to consider all available geological information with its own degree of confidence)
2) storing uncertainty estimates that may be asymmetrical across the fault, which allows a full uncertainty vector field to be interpolated along the fault based on few estimates given by the geophysicist
3) flexibility for representing complex fault geometries without inhibiting simulation speed
4) lightness of the structure to insure a fast simulation process

### *Available Input Data Available*

Input data for fault modeling in the depth domain consist of any combination of the following.

#### *Depth-migrated Horizon Interpretations*

These data are important because it is the only link to the original seismic data. It corresponds to the interpretation performed in the time domain and passed to the depth domain, where the final structural model is constructed. Definitions of faults from these data have a medium degree of confidence. They come from the time-to-depth conversion, where the velocity model is not always well defined, and represent a horizon-fault contact seldom seen in the seismic data. Therefore, they may be good indicators of the general shape of the fault in map view but bad indicators of the exact location and geometry for the fault.

#### *Fault Interpretation Curves*

These curves are interpreted in the time domain. They are used to define the limits of the interpreted horizons but are seldom migrated to the depth domain. Indeed, it is much more difficult to migrate subvertical structures like faults than subhorizontal structures like horizons. No such fault interpretation is available for the study presented here.

#### *Fault Polygon Curves*

Fault polygons are commonly drawn by the interpreter on two-dimensional (2-D) map views representing the horizons in the depth domain. These 2-D horizontal curves are then projected at their exact Z-location in 3-D, thereby minimizing the uncertainty of the exact fault location inside the gap left in the point set representing the horizon. Although uncertainty on the fault geometry is limited, these are still merely subjective interpretations of reality and may contain additional uncertainty.

### *Coherency-checking Curves*

These data have a higher degree of confidence than other kinds because they are associated with a final interpretation that encounters all available information. Such data are not available for the study presented here.

### *Well Markers*

Such data are seldom available because wells are drilled as far as possible from uncertain structures like faults. When available, well markers give precise information about the location and slope of the fault. In this study, no well data are available.

### *Structural Geologist's Knowledge*

This type of input is difficult to define and to transform in equations (Yarus, 1994). However, it is one of the most important pieces of information for fault modeling and should be included when possible. The knowledge of the tectonic style for the region may help the geologist to substantially reduce the number of scenarios or geometries for the faults. Consequently, the methods presented here seek to respect the characteristics of the fault designed by the geologist as closely as possible during the simulation process.

The foregoing discussion about input data may be summarized as follows (see Figure 8).

1) Curves represent most of the input data. These curves may be either subhorizontal (limits of the gaps in the depth-migrated point sets representing the horizons and fault polygons), subvertical (fault interpretations), or in any direction along the fault (curves drawn during the coherency-checking phase).
2) These data have varying degrees of reliability. Subhorizontal curves are defined along the full extension of the fault but contain considerable uncertainty; subvertical curves are less numerous and more sparse, although more precise. The data structure representing faults must include this asymmetric aspect of the data in order for these characteristics to be adequately considered during the simulation process.

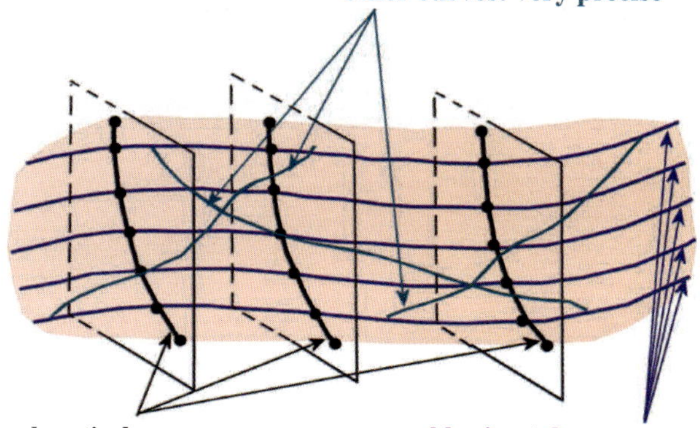


**Figure 8. Most of input data used to model faults in the depth domain are represented by curves. These curves have different orientations and degrees of confidence that need to be considered during the modeling process and the uncertainty assessment.**

## Curve Representation: Flexibility and Lightness

To limit the computing time, fault geometry is discretized along two main directions of the fault (strike and dip). To fulfill the lightness and flexibility constraints of the fault structure, curves are the only geometrical elements used to represent fault geometry in the data structure.

A main curve, called the backbone, defines the fault's geometry in one direction (e.g., along the vertical direction). Secondary curves are attached to each node of this main curve. These secondary curves, called generators, define the geometry in the orthogonal direction (see Figure 9).

To model complex geometries, no limitation is placed on the number of nodes along the main curve or in the number of points along each generator. The only limitation is having the same number of points on each generator for a fast access during the simulation process.

With such a representation, two indices $(i, j)$ are sufficient to represent any point along the fault (a fault point in Figure 9). The first index represents the main curve node to which the generator is attached, and the second index represents the index of the fault point at which the location should be simulated. This 2-D grid representation also allows for a fast navigation and correlation of simulated geometries between neighboring fault points (M. Lecour, P. Thore, J.-C. Dulac, and R. Cognot, 1999, personal communication).

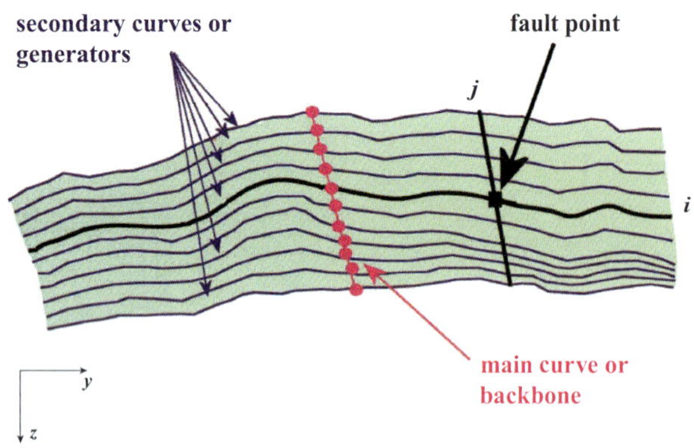


**Figure 9.** The proposed fault data structure is defined by curves only: a main curve (backbone) onto which secondary curves (generators) are attached. Two indices (*i*, *j*) are sufficient to represent any point along the fault and allow for fast navigation among the points.

## Modeling and Editing the Reference Fault

Commonly, the gap representing the fault in the depth-migrated horizon points is greater than the real width of the fault. This problem occurs especially if the interpretation has been done manually and stopped where the seismic data are too blurred along the horizon-fault contact. Even if fault polygons reduce the gap representing the fault, it may still be too big compared to the real width of the fault. Instead of using the borders of the horizon that commonly add a staircase effect to the fault geometry (Dubrule et al., 1997), faults are commonly constructed from the middle lines computed between the borders of the point sets (see Figure 10a, b).

Data are commonly incomplete and contain some inconsistency. Indeed, the interpretation commonly focuses on the reservoir of interest so that the most precise data are available only for a part of the fault. Consequently, editing is commonly necessary to produce a correct geometry and to model the full extension of the fault instead of the reservoir part alone. This is why functions such as lateral and vertical extension are helpful to get a full description of the fault surface (see Figure 10d). By increasing the number of generators along the main curve or by increasing the number of points along the generators, more details about the geometry of the fault can be easily considered. The lightness of the structure presented in the previous section allows for these fast-editing operations.

## Storing Estimates of Uncertainty

Because uncertainty may occur asymmetrically along faults, it may be modeled by a couple of vectors attached to each fault point. Only a few uncertainty vectors need to be estimated by the geophysicist and/or structural geologist. From these few estimates, a

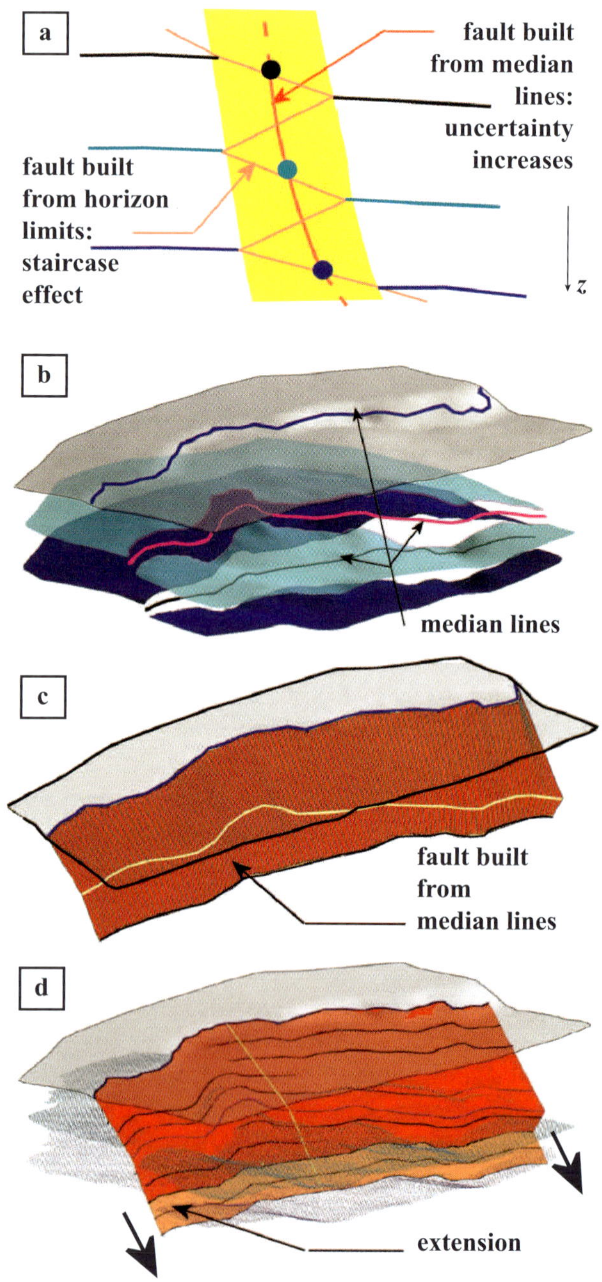


**Figure 10.** To avoid the staircase effect caused by uncertainty on horizon interpretations (a), the fault is commonly constructed from the lines corresponding to the middle of the fault gap (b, c). Then, an extension along the backbone may be applied to get the full extension of the fault across the different horizon surfaces (d).

couple of uncertainty vectors are computed for all the fault points by a bilinear interpolation method. The storage of these uncertainty vectors defines the volume within which the fault geometry must be simulated.

Even if uncertainty is currently seldom estimated on seismic data and, when provided, only one value of uncertainty is given for each fault, the previous definition of uncertainty estimates allows the consideration of more complex uncertainty estimate later: very small uncertainties at well location, asymmetric values for the uncertainties on both sides of the fault, and varying uncertainties along the two main axis (strike and dip) of the fault. As presented before, when such a precise estimate will be available, a post-processing may be needed to ensure that neighboring uncertainty vectors are consistent.

## PROPOSED SIMULATION METHOD

The proposed method is based on the p-field technique (Srivastava, 1992), where the simulation is done in two steps. First, a local probability distribution function (pdf) is generated. Then, a global random function is used to compute numbers correlated in space (Froidevaux, 1993).

A pdf is attached to each point of the discretized fault (see Figure 11a). This local pdf defines the probability for a fault point to be located anywhere on the segment defined by the reference location and the two corresponding uncertainty vectors. For example, an asymmetric triangular pdf may be developed from the uncertainty vectors attached to the point (see Figure 11a).

Next, a global random function $P(u)$ is used to draw random numbers for all fault points defined by their $u$ index (see Figure 11b). These numbers are randomly computed in the [0,1] space and then mapped into a chosen interval along the uncertainty vector thanks to a function defined globally for the fault. The definition of a global random function allows the correlation of the random numbers to be controlled along the main directions of the fault (strike and dip), thereby avoiding any unacceptable geometry for the simulated fault (see Figure 12). The general method can have many variations, either in the definition of the local pdfs, or in establishing the global random function.

Finally, each number selected via $P(u)$ is used to assign a new location for the fault points based on

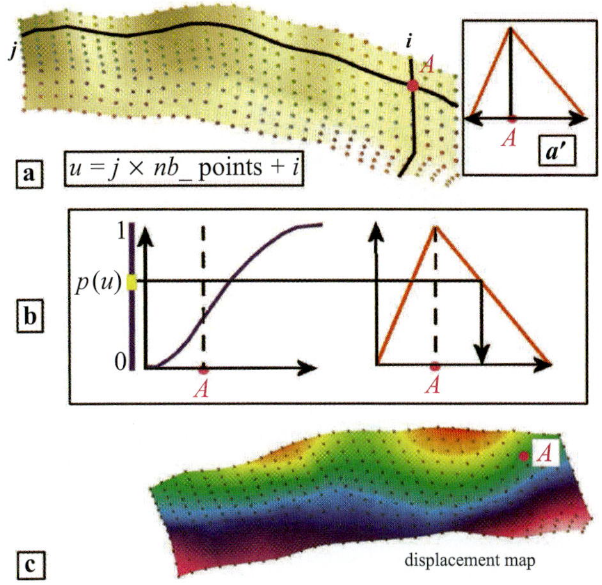


**Figure 11. The proposed simulation method is based on the p-field technique. (a) For each point, a local pdf is defined from the uncertainty vectors ($a'$). A random number $p(u)$ is drawn in the [0,1] space (b), and the new simulated location is derived from the random number along the uncertainty vector for this fault point (c).**

their respective local pdf. This step establishes a translation map, giving the simulated locations for the fault points (see Figure 11c).

### Local Probability Function Definition

In general, the geologist's interpretation should be considered as the reference case. To honor the geologist's choice, a triangular distribution giving a higher probability to the reference position can be used to define the local pdfs. That is, the maximum probability is assigned to the location corresponding to the geologist's interpretation. Minimum probability is assigned to the extreme values of the uncertainty vectors. If the geometry of the reference fault is not considered to be more probable, a uniform distribution can alternatively be used to define the local pdfs. Other distributions can also be considered as long as they are not unbounded (e.g., the Gaussian or normal distribution).

### Global Random Function Definition

The global random function must adhere to the following rules:

1) The random function $P(u)$ must be uniformly distributed on the interval [0,1].

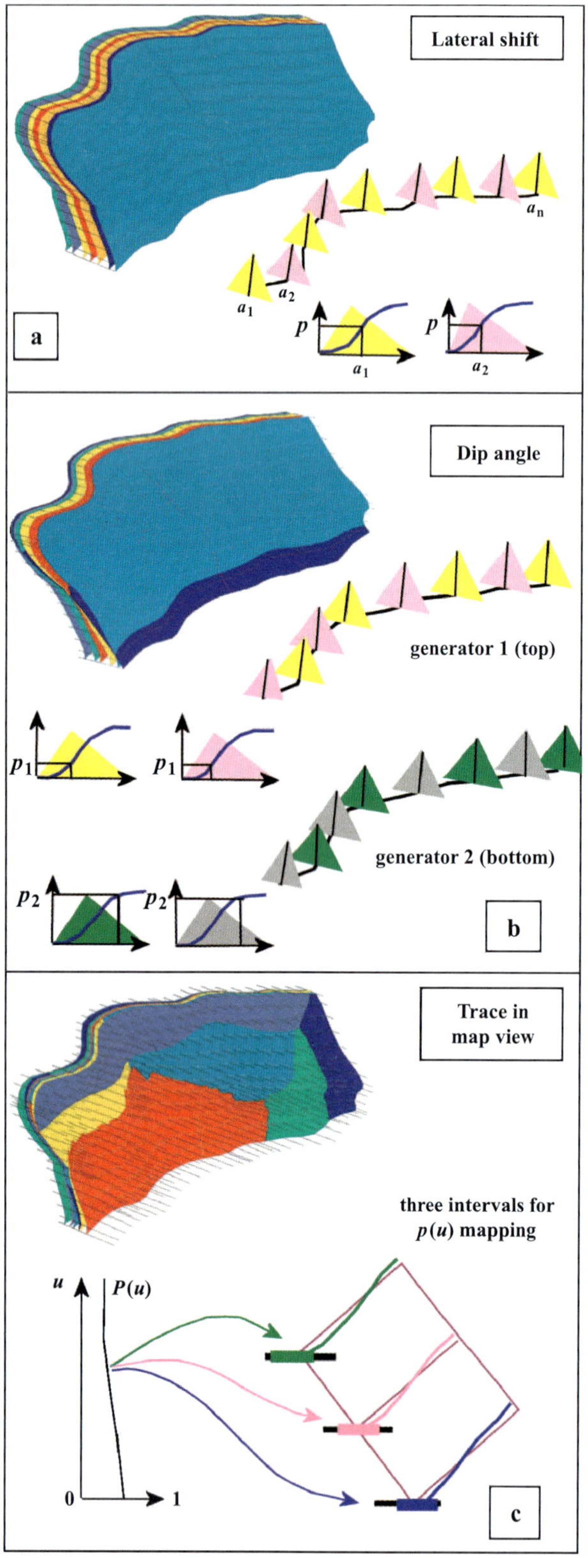


**Figure 12.** Some examples of random functions $P(u)$. A constant random function gives a shifting of the fault geometry (a). A linear function along the dip axis and constant along the strike axis modifies the dip angle of the fault (b). Using a nonlinear function along the strike direction allows the modification of the shape in map view but is more difficult to control (c).

2) The mapping of $P(u)$ in a chosen interval must respect the general rules (R1: noninversion of the curvature; R2: keep the fault throw characteristic normal and inverse fault).

Below is a list of different functions, starting with the simplest and moving to the most complex. If the uncertainty vector field is not homogenous enough, a regularization has to be performed to avoid big deformations of the reference fault geometry (Lecour, 2000; Lecour et al., 2001).

1) a constant function: This function corresponds to a lateral shift of the fault location (see Figure 12a).
2) a function that is constant in the strike direction and linear in the dip direction: This function is a combination of a lateral shift of the reference fault plus a modification of its dip angle of the fault (see Figure 12b).
3) a function that is linear in the strike direction and constant in the dip direction
4) a function that is linear in the two main directions (strike and dip)
5) a function that is constant in the dip direction and nonlinear in the strike direction: Such a function is used to simulate a modification of the shape of the fault in map view (see Figure 12c). It is adapted from a fast method designed to generate equiprobable images of reservoir heterogeneities (J.-L. Mallet and A. Shtuka, 1997, personal communication).
6) a function that is nonlinear in the dip direction and constant in the strike direction
7) a function that is linear in the dip direction and nonlinear in the strike direction: This function simulates not only a modification of the dip angle but also a modification of the shape of the fault in map view.
8) a function that is nonlinear in both the strike and dip directions

To date, only functions of types 1, 2, 5, and 7 have been implemented because they are easily linked to one or several parameters, horizontal trace, dip angle, or location, that describe the fault geometry. Functions of type 6 or 8 are more difficult to control because of the nonlinearity, and satisfying R1 may not be guaranteed.

The method is general enough to incorporate more geological rules like the fact that a fault must be a sliding surface (Thibaut, 1996), some rules for fault throw analysis (Walsh and Watterson, 1987; Maerten et al., 1999), or rules for balancing sections (Gibbs, 1983; White et al., 1986; Vendeville, 1991; White and Yielding, 1991).

## AN APPLICATION OF THE METHOD TO VOLUMETRIC STUDIES

One possible application of the methodology outlined here involves the impact of fault uncertainty on volumetric studies. The goal of this application is to simulate the geometry of all the faults of a reference fault network. Then, all fault-fault and horizon-fault connectivities are updated. Finally, a 3-D model is constructed from the new geometry, and precise volume computations are performed in the different blocks.

A panel showing one histogram per region of the model is updated after each simulation and provides an overview of the variation in volume caused by fault uncertainty (see Figure 13). From this panel, the reservoir engineer may select some of the models. These models may correspond to specific quantiles or may be randomly sampled among all simulated models.

To generate a large number of models in a reasonable amount of time, the method needs to be fast and trustable: after each simulation, the construction of the 3-D model must give exactly the same number of blocks as the reference model.

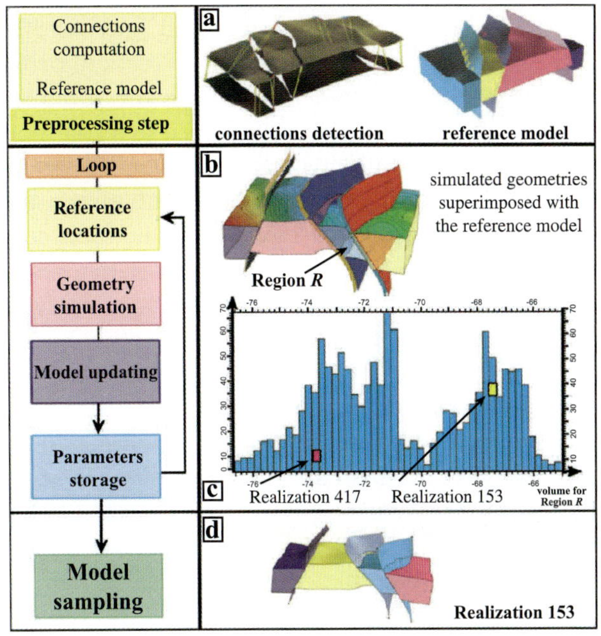


**Figure 13. The goal of the volumetrics application is to compute one histogram per region of the reservoir representing the impact of fault uncertainties on rock volumes. The proposed workflow is the following. (a) All connectivity information is computed, and geometrical constraints are installed once in a preprocessing phase. Then, the process enters in the main loop. First, all geometries are initialized with reference locations to obtain independent simulations. Then, all fault geometries are simulated. One simulation superimposed with the reference model is shown in (b). From the connectivity information, the model is updated, and the volumes are recomputed (c). Finally, only simulation parameters are stored to retrieve some of the simulations (d).**

### Problem

A classical approach for building 3-D models involves extending all horizons across the faults, computing the intersection lines between the different surfaces, and constructing the model from the cut surfaces. This method can be time consuming and can produce numeric instabilities, leaving holes along the contact line (N. Euler, C. H. Sword, and J.-C. Dulac, 1998, personal communication).

To ensure perfect contacts between the different surfaces, two main solutions may be applied. First, the intersection line between the main and the branching faults should always be included in the main fault discretization. This ensures that the contact is perfect. However, the flexibility is reduced because the geometry of main faults is almost fixed. Second, keep the flexibility for the geometry by recomputing the contacts after each simulation. As a result, the mesh of the main fault, including the contact line, is recomputed, so that the contact line is shared by all surfaces in contact (faults or horizon faults).

### Proposed Methodology

Using the second method described above, the proposed methodology is designed as modular as possible. For maximum speed, the process requires the detection of the connectivity in a preprocessing step and the use of this information to update the contacts after each simulation (both fault-fault and horizon-fault contacts).

### Connectivity Detection

In a preprocessing step, all the connectivity information is automatically detected and stored as topological graphs (I. Duvinage, J.-L. Mallet, and J.-C. Dulac, 1999, personal communication).

Subsequently, different geological (gap, width, and orientation) and geometrical (direction, proximity) criteria are used to compare the different graphs and set all borders connected to a common fault together. Using this information, geometrical constraints are automatically imposed between the different surfaces of the model and the fault surface on which they must be connected (Mallet, 1997). Then, an interpolation of the surface geometry is performed, respecting all the constraints. Finally, new fault surfaces are constructed that include all related intersection lines in their mesh. All the nodes corresponding to the intersection between two or more surfaces are shared by all surfaces involved in the contact, thus ensuring that no hole is created. This methodology suggests that reservoir volume can be automatically estimated after each simulation of the fault network geometry (see the preprocessing step in Figure 13).

## Loop: From Simulation to Volumes Computation

Once connectivities have been detected and stored, the simulation enters a loop process. Each loop encompasses the following steps (see Figure 13, loop):

1) Reset reference locations: To produce independent simulations, the geometry of all surfaces is reset to the reference location so that the starting configuration is the same.
2) Fault network simulation: All fault geometries are simulated independently using one of the random functions previously described.
3) Model updating and volume computation: All simulated surfaces are interpolated using the constraints imposed during the preprocessing step. Then, new surfaces, including the new intersection lines in their mesh, are regenerated. A 3-D model is built using these surfaces, and all volumes are computed.
4) Parameter storage: For the sake of efficiency, only simulation parameters are stored after each loop as opposed to the full models. All histograms are subsequently updated.

## Results

The results of applying this procedure to two different situations are described below.

### Simple Model

An initial test has been performed on a simplified real model containing two horizons and three subparallel faults defining only four regions (see Figure 13).

This example was simple enough in terms of number of surfaces and number of contacts to test the robustness of the 3-D model construction and the speed of the process. Each simulation loop takes about 1 min on an SGI O2 with R5000 processor, which is a great improvement compared to building all these models one by one in a nonautomatic process.

It also validates the chosen random function because all simulated models are correct regarding the fault geometry and respect the characteristics of the reference model.

### Synthetic Model with More Contacts

A second model has been tested to determine whether the method remains effective and robust when the number of contacts increases in the model. Only the number of contacts is important in this testing phase. All synthetic surfaces are flat to focus on the impact of contact number on the process (see Figure 14).

This model contains 3 horizons and 5 faults, 36 horizon-fault connectivities, 4 fault-fault connectivities, 6 boundary-fault connectivities, and 36 horizon-boundary connectivities. All 82 contact lines need to be recreated after each simulation. At the end of each simulation, the 12 histograms representing the impact of fault uncertainty on volumes need to be updated (one histogram for each block in the model).

The method has been successfully tested on this model. On Figure 14, the resulting histograms after 50 runs in a fully automatic process are presented. The method produces the expected results and is an improvement for fault geometry uncertainty studies on simple cases. Nevertheless, additional optimization is required to speed up the process and use it on more complex models.

# CONCLUSION

The first tests of the proposed method are promising and suggest that it can be helpful in the study of fault geometry uncertainty, which is seldom studied, because of the difficulty of the task and the slowness of the modeling processes. The new fault

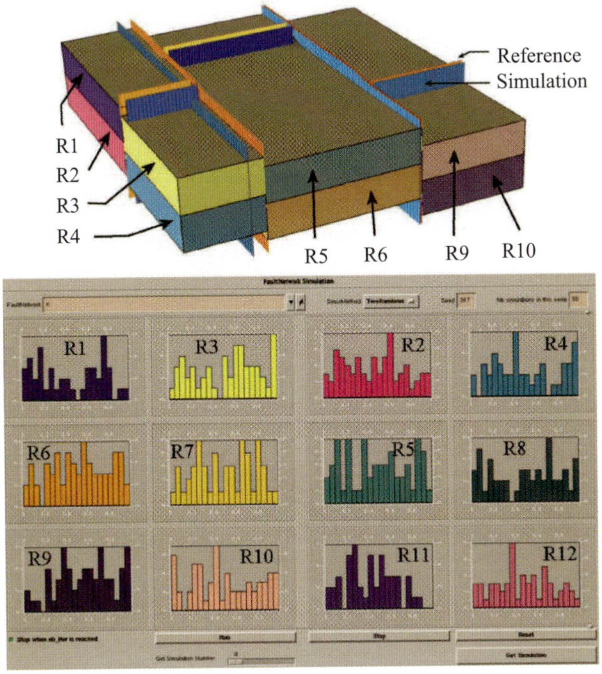


**Figure 14. The process has been tested on a synthetic model containing a larger number of connectivities (82) to test the sensitivity of the process on the number of contacts. The figure shows one set of 50 realizations of this model. Increasing the number of contacts has no impact on the process.**

object presented here considers all available input data, is flexible, and light enough to provide fast access to all fault points during the simulation process. Its design allows easy storage of uncertainty estimates from geophysicists. The proposed simulation method is fast and respects initial geometries with simple random functions. The design of this method is modular so that improvement may be added in the future while keeping the main workflow described in this chapter.

Finally, one application of the method is presented here that includes the monitoring of rock volume variations during the simulation process. This first application allows determining the impact of fault geometry uncertainty on gross rock volumes, which are known to be important in most of the fields. Such methodology could help evaluate the impact of faults on reserve estimates by not oversimplifying fault geometry in the model when this factor is important.

## Current Limits of the Method

Although the proposed automatic methodology is a great improvement over other approaches, additional enhancements are required to speed up the process and validate the method on more complex data sets.

The hierarchy inside the fault network must also be considered. Faults are simulated independently in the current process. It would be better to correlate the simulations of all secondary faults linked to the same main fault to better respect the initial fault network scheme.

Finally, some postprocessing using geological criteria like fault throw analysis (Maerten et al., 1999) or balanced cross sections (Gibbs, 1983; Vendeville, 1991) may be added to validate simulated geometries when a complex random function is used. This part requires translation of the qualitative information given by geologists into equations that may be incorporated in the simulation process, while being validated by the geologists (Coburn, 1994).

## Perspectives

The study of fault uncertainty is not a goal in itself but should be part of global uncertainty studies. Reservoir models are commonly expensive to produce, and the reservoir engineer may try to simplify them to reduce the computation time during flow simulation. If limiting the complexity of the model reduces a lot of the computation time, oversimplifying the structural model in complex structures may be misleading (Yarus, 1994).

Therefore, uncertainty analysis on all parameters is critical to determine which ones may be simplified. The structure of the reservoir is one of the parameters that should not be simplified in most cases because they drive volume computations in a lot of cases.

Some approaches incorporate nested simulations of horizon geometry, facies distribution, petrophysical properties (porosity, permeability), and fluid contacts. Such approaches facilitate identification of uncertainties that have a major impact on the volume computations and allow all sources of uncertainty to be combined. The goal of such an approach is clearly not to provide the best model but a spectrum of possible models that aids in important decisions for field exploitation (Journel, 1994).

The methodology presented in this chapter might be used to add a new level in the uncertainty analysis process: simulating not only horizon geometry, but also fault geometry to study the variations of rock volumes caused by structural uncertainties.

## ACKNOWLEDGMENTS

This work is a result of a Ph.D. dissertation sponsored by TotalFinaElf and realized within the gOcad Consortium. The authors thank the Laboratoire Infographie et Analyse de Donnees, Nancy, France; all members of the gOcad Consortium; and especially TotalFinaElf for their support in this work.

## REFERENCES CITED

Coburn, T. C., 1994, Reflections on the proliferation of geostatistics in petroleum exploration and production, *in* J. M. Yarus and R. L. Chambers, eds., Stochastic modeling and geostatistics: Principle, methods, and case studies: AAPG Computer Applications in Geology 3, p. 21–23.

Dubrule, O., C. Basire, S. Bombarde, P. Samson, D. Segonds, and J. Wonham, 1997, Reservoir geology using 3-D modeling tools, SPE Paper 38659.

Failly-Berthet, M., and P. Thore, 1996, Quantitative estimation of seismic resolution: Society of Exploration Geophysicists Annual Meeting Abstracts, p. 1434–1437.

Froidevaux, R., 1993, Probability field simulation, *in* A. Soares, ed., Geostatistics troia: Dordrecht, Kluwer Academic Publishers, p. 73–84.

Gibbs, A. D., 1983, Balanced cross-section construction from seismic sections in areas of extensional tectonics: Journal of Structural Geology, v. 5, p. 153–160.

Henry, G., 1994, Géophysique des bassins sédimentaires: Paris, Editions Technip, 445 p.

Journel, A., 1994, Geostatistics and reservoir geology, *in* J. M. Yarus and R. L. Chambers, eds., Stochastic modeling and geostatistics: Principle, methods, and case studies: AAPG Computer Applications in Geology 3, p. 19–20.

Juliard, C., and P. Thore, 1999, Velocity anomaly estimation, a cookbook: Geophysical Prospecting, v. 47, p. 299–312.

Lecour, M., 2000, Modélisation des incertitudes concernant la géométrie des failles. Impact sur les études volumétriques des réservoirs pétroliers: Ph.D. dissertation, University of Nancy, France.

Lecour, M., P. Thore, and J.-C. Dulac, 2001, Stochastic fault and fault networks modeling for structural uncertainties study: Petroleum Geoscience, v. 7, May, p. 17–23.

Maerten, L., E. J.-M. Willemse, E. D. Pollard, and K. Rawnsley, 1999, Slip distributions on intersecting normal faults: Journal of Structural Geology, v. 21, p. 259–271.

Mallet, J.-L., 1997, Discrete modeling for natural objects: Mathematical Geology, v. 29, no. 2, p. 199–220.

Samson, P., O. Dubrule, and N. Euler, 1996, Quantifying the impact of structural uncertainties on gross-rock volume estimates, SPE Paper 35535.

Srivastava, M. R., 1992, Reservoir characterization with probability field simulation, SPE Paper 24753.

Srivastava, M. R., 1994, An overview of stochastic methods for reservoir characterization, *in* J. M. Yarus and R. L. Chambers, eds., Stochastic modeling and geostatistics: Principle, methods, and case studies: AAPG Computer Applications in Geology 3, p. 3–16.

Thibaut, M., 1996, Géométrie des surfaces de faille et dépliage 3D. Méthodes et applications: Ph.D. dissertation, Grenoble, University of Grenoble.

Thore, P., 1999, Champs de vitesse et incertitudes géophysiques: Mémoire d'Habilitation à Diriger les Recherches, Pau.

Thore, P., A. Shtuka, M. Lecour, T. Ait-Ettajer, and R. Cognot, 2002, Structural uncertainties: Determination, management and applications: Geophysics, v. 67, p. 840–852.

Tinker, S., 2000, No ground truth: Keynote presentation at the AAPG Hedberg Research Conference, Applied Reservoir Characterization Using Geostatistics (December 3–6), Woodlands, Texas.

Vendeville, B., 1991, Mechanisms generating normal fault curvature: A review illustrated by physical models, *in* A. M. Robert, G. Yielding, and B. Freeman, eds., The geometry of normal faults: Geological Society Special Publication 56, p. 241–249.

Vincent, G., C. Corre, and P. Thore, 1999, Managing structural uncertainty in a mature field for optimal well placement: SPE Reservoir Evaluation and Engineering, v. 2.

Walsh, J. J., and J. Watterson, 1987, Distributions of cumulative displacement and seismic slip on a single normal fault surface: Journal of Structural Geology, v. 9, p. 1039–1046.

White, N. J., and G. Yielding, 1991, Calculating normal fault geometries at depth: Theory and examples, *in* A. M. Robert, G. Yielding, and B. Freeman, eds., The geometry of normal faults: Geological Society (London) Special Publication 56, p. 1–13.

White, N. J., J. A. Jackson, and D. P. MacKenzie, 1986, The relationship between the geometry of normal faults and that of the sedimentary layers in their hanging wall: Journal of Structural Geology, v. 8, p. 897–909.

Williams, D. B., and A. A. M. Aqrawi, 2006, Utility of using acoustic impedance data in the stochastic modeling of a carbonate reservoir, *in* T. C. Coburn, J. M. Yarus, and R. L. Chambers, eds., Stochastic modeling and geostatistics: Principles, methods, and case studies, vol. II: AAPG Computer Applications in Geology 5, p. 253–268.

Yarus, J., 1994, Preface, *in* J. M. Yarus and R. L. Chambers, eds., Stochastic modeling and geostatistics: Principle, methods, and case studies: AAPG Computer Applications in Geology 3, p. ix–x.

McKenna, S. A., and P. C. Reeves, 2006, Fractured continuum approach to stochastic permeability modeling, *in* T. C. Coburn, J. M. Yarus, and R. L. Chambers, eds., Stochastic modeling and geostatistics: Principles, methods, and case studies, volume II: AAPG Computer Applications in Geology 5, p. 173–186.

*Chapter 14*

# Fractured Continuum Approach to Stochastic Permeability Modeling

**S. A. McKenna**
**P. C. Reeves**
*Sandia National Laboratories*
*Albuquerque, New Mexico, U.S.A.*

## ABSTRACT

A new kind of fracture-permeability model called the fractured continuum model (FCM) is presented that incorporates the advantages of discrete fracture network models and the processing speed of effective continuum representations of fracture permeability. Observations of fracture orientation, length, frequency, and transmissivity from boreholes and outcrops are used as input to the FCM approach. Geostatistical realizations of fracture connectivity, represented by the coordination number of a local percolation network, and fracture frequency are combined with object-based simulations of high- and low-permeability classes in the model domain. At each location, these three spatially variable properties are combined into an effective grid-block permeability using an approximation based on the effective medium theory. The resulting realizations of fracture permeability, containing greater than $10^6$ elements, are used as input to a single-phase flow model. A parallel computer platform coupled with a unique groundwater flow code is used to efficiently solve steady-state pressure fields on multiple realizations. The ability to solve many realizations in a short amount of time allows for the evaluation of the effects of two different conceptual models of fracture permeability on particle traveltimes.

## INTRODUCTION

The function of fractures in subsurface fluid-flow problems has long been recognized and has become an especially important issue throughout the past two decades in the areas of petroleum-reservoir engineering and long-term storage of nuclear waste. Several different techniques for modeling the permeability of fracture networks for fluid-flow problems have been developed. One of these techniques is the stochastic continuum model (SCM). Stochastic continuum models have been proposed as a means of representing the spatially heterogeneous nature of fracture permeability at the scale of continuum grid blocks (see Hsieh, 1998). In general, the discrete behavior of individual fractures is represented by an average property over a


DOI:10.1306/1063815CA53234

representative elementary volume (REV). The volume necessary to define a REV within a fracture system can be quite large. The SCM approach has been used previously to model fracture systems at experimental facilities for nuclear waste programs in the United States (see Altman et al., 1996; Finsterle, 2000) and in Sweden (Follin and Thunvik, 1994; Tsang et al., 1996). In these studies, geostatistical-simulation algorithms were used to model continuum properties of discrete fracture systems at the scale of numerical model grid blocks (ones to tens of meters). The geostatistical realizations of these properties were then used as input to flow and transport models to predict the advective traveltime of particles through the heterogeneous systems.

Another technique for modeling fluid flow in groundwater and petroleum systems is the discrete fracture network (DFN) model, in which each individual fracture is modeled as a discrete one- or two-dimensional (1-D or 2-D) feature within a 2-D or three-dimensional (3-D) domain, respectively. Initial work with DFNs focused on techniques for deriving the equivalent permeability of areas or volumes of fractured rock (e.g., Long et al., 1982; Long and Witherspoon, 1985). Recent developments in DFN models allow for the simulation of large, complex discrete fracture systems in 2-D and 3-D domains (e.g., Dershowitz et al., 1998).

Both the DFN and SCM approaches to modeling fracture permeability have strengths and weaknesses. The relative merits of the different modeling approaches are discussed by Dershowitz et al. (1998). The DFN models can explicitly model the fractures as 2-D objects in a 3-D domain. These models can be conditioned to fracture observations and generally reproduce the distributions of observed fracture attributes. However, stochastic realizations of a DFN model require computationally expensive regridding of the finite-element mesh for each realization, and the fracture network must commonly be abstracted to a series of 1-D pipes for solute transport modeling in groundwater problems. The computationally efficient regular grid employed in SCM models allows for multiple realizations and complex solute transport processes to be modeled. Because of the necessary upscaling to the REV-size grid blocks, explicit conditioning of SCMs to observed fracture data can be problematic. In addition, abstractions of discrete fracture patterns to a mean, variance, and spatial covariance of permeability, as commonly done in the SCM approach, overly homogenize the representation of the fracture network. The homogenization employed in SCM approaches may not be able to reproduce potential extremes of groundwater transport behavior. An approach to fracture-permeability modeling that can capitalize on the strengths of both DFN and SCM approaches and can directly use information measured on discrete fractures is presented herein.

This chapter describes a new approach to stochastic modeling of fracture permeability— the fractured continuum model (FCM). The approach is based on a two-step process: critical properties of the DFN are considered to be spatially random variables and are modeled with a combination of geostatistical and object-based simulation. These simulations define the number of fractures, the connectivity of the fractures, and the relative permeability class of the fractures in each grid block of the model domain. The effective permeability for each grid block is derived from the fracture properties using a local effective medium approximation. The simulations also define the larger scale connectivity of the permeability values throughout the model domain through the correlation of properties across adjacent grid blocks. Such a modeling approach is made feasible by the use of a fine-scale grid discretization that necessitates a parallel-flow solver. An example problem demonstrating this modeling approach as applied to the prediction of groundwater flow away from a proposed nuclear waste repository is presented. In this example, single-phase fluid flow and streamline particle tracking are calculated directly on the fine-scale permeability models using a recently developed parallel groundwater flow simulator.

## EFFECTIVE MEDIUM APPROXIMATION

The first step in the FCM approach is to determine the numerical model grid-block-scale fracture permeability given the number of fractures, a measure of fracture connectivity and the permeability of each individual fracture contained within any grid block.

A constitutive approximation based on effective medium theory (EMT) is employed to derive the effective permeability of each grid block in the model domain. This approximation is derived from work done in percolation theory (Kirkpatrick, 1973) and has been previously used to model effective permeability in fractured rock (David et al., 1990;

Zimmerman and Bodvarsson, 1996). A brief overview of this approximation, following the approach of Zimmerman and Bodvarsson (1996), is presented below.

Consider a 3-D, cubic, numerical model grid block with all three dimensions equal to a length, $L$, which contains a single set of parallel, evenly spaced fractures spanning the block in one direction. Along each fracture, a conductance, or transmissivity, $C_i[L^2/T]$, is defined here by invoking the assumption of flow through parallel plates (Snow, 1969):

$$C_i = \frac{gb^3\rho}{12\mu} \tag{1}$$

where $g[L/T^2]$ is the gravitational acceleration, $\mu[M/LT]$ is the viscosity, $\rho[M/L^3]$ is the density of water, and $b[L]$ is the fracture aperture. The length and width of the fractures are equal and span the grid block in both directions.

By placing a pressure difference, $\Delta H[L]$, across the grid block parallel to the fracture set, the volumetric flux, $Q[L^3/T]$, through the block can be defined as

$$Q = N_f C^* \Delta H \tag{2}$$

where $N_f$ is the number of parallel vertical fractures and $C^*$ is an effective conductance value calculated as an average of the $N_f$ values of $C_i$.

Volumetric flux through the cubic block can also be calculated by applying Darcy's law. The volumetric flux through the face of the grid block orthogonal to the fracture orientation is

$$Q = K_b L \Delta H \tag{3}$$

where $K_b[L/T]$ is the continuum hydraulic conductivity for the grid block in the direction parallel to the fracture set. By setting equations 2 and 3 equal to one another, it is possible to solve for $K_b$ of the cubic block using only geometric information on the fracture network, the dimensions of the numerical grid block, and the effective conductance of the fracture segments:

$$K_b = \frac{N_f C^*}{L} \tag{4}$$

For natural fracture systems, multiple fracture sets may be intersecting and variably connected. Therefore, derivation of $C^*$ requires a sophisticated averaging technique. Consider a 2-D fracture map as shown in Figure 1, where the fractures are the result of multiple tectonic events producing several orientations of fractures.

The effective medium approach begins with maintaining the irregular geometry of the fracture network but replacing the individual fracture conductances with a single effective conductance, $C^*$ (upper right image, Figure 1). The geometry of the fracture system is held constant, with the dead-end fractures removed, and the macroscopic conductivity of the fracture system remains the same for both the original conductances and the single effective conductance. $C^*$ can be found by solving the implicit equation (Kirkpatrick, 1973)

$$\sum_{i=1}^{N} \frac{C^* - C_i}{(z_i/2 - 1)C^* + C_i} = 0 \tag{5}$$

where the coordination number, $z_i$, is the number of conductive fracture branches emanating from each fracture intersection.

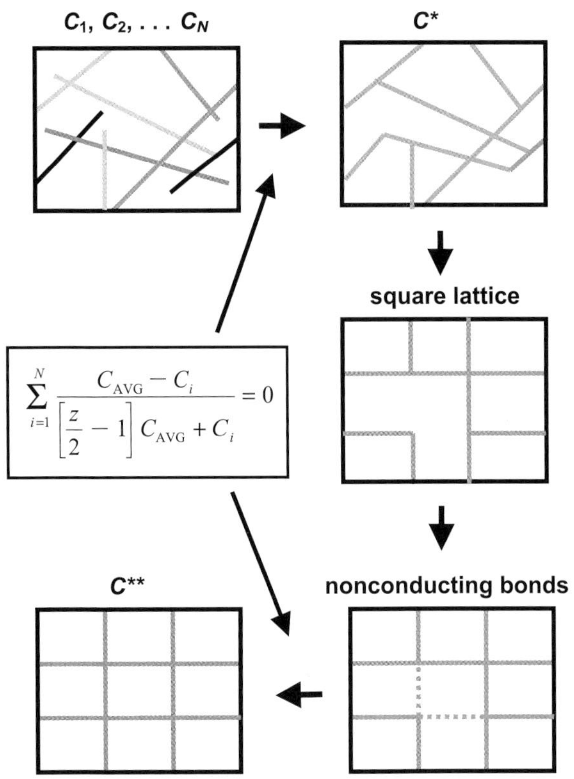


$$\sum_{i=1}^{N} \frac{C_{AVG} - C_i}{\left[\frac{z}{2} - 1\right] C_{AVG} + C_i} = 0$$

**Figure 1. Conceptual example of determining the average fracture conductance for an irregular discrete fracture network in a numerical model grid block. Fracture junctions are the intersection of any two fractures, and fracture segments are the straight-line fracture traces connecting the junctions. See text for details. The gray scale of the fractures indicates the relative permeability.**

As pointed out by Zimmerman and Bodvarsson (1996), equation 5 was originally developed for networks where the coordination number does not vary between intersections. However, equation 5 can also be used with an average coordination number for topologically irregular networks. For the example shown in Figure 1, eight intersections are present, and the average coordination number, $z^*$, is 3.0.

After the effective conductance value, $C^*$, is determined through equation 5 and defined for each fracture segment, the irregular discrete fracture map is transformed to a square network where the mean fracture spacing and mean coordination number are maintained (Figure 1, middle image). The square network requires that $z^* = 4.0$. The transition from an irregular lattice with $z^* < 4.0$ to a square lattice with $z^* = 4.0$ is accomplished by filling in the missing segments with zero conductivity segments (Figure 1, dashed lines in the bottom right image).

A final effective hydraulic conductivity must be calculated for the grid block containing the square fracture network with the zero-conductance segments. Note that in the example shown in Figure 1, a total of eight fracture intersections with the grid-block boundaries are present, or an average of two intersections per side. Maintaining this average results in two vertical and two horizontal fractures through the grid block. $C^*$ is treated in this work as a scalar quantity that applies to flow through the block in all three principal directions. More complex approaches to determining anisotropic measures of conductance and the number of horizontal and vertical fractures have been evaluated by Hestir and Long (1990); however, the relatively simple approach taken here follows the work of Zimmerman and Bodvarsson (1996).

The effective value of fracture transmissivity in the grid block can now be calculated by recognizing that the value of $C^*$ derived from equation 5 must be modified to consider that the actual coordination number is less than 4.0. The fraction of zero-conductance segments in the square lattice is $1 - (z^*/4)$, and the fraction of segments with $C^*$ is $z^*/4$. A second application of equation 5 to the square lattice leads to the relation

$$C^{**} = \left(\frac{z^*}{2} - 1\right)C^* \tag{6}$$

This relationship between the effective conductance calculated for the irregular network and the effective conductance of the resulting square network demonstrates that at average coordination numbers less than or equal to 2.0, the resulting effective conductivity of the grid block is zero. At this point, the scalar value of the effective hydraulic conductivity of the numerical grid block can be calculated using $C^{**}$ as the effective conductance in equation 4.

A strong advantage of using the effective media theory is that three parameters controlling the hydraulic conductivity in any grid block are derived from readily obtainable fracture observations. These parameters are summarized in Table 1. Another advantage of the EMT approach is that fracture connectivity, parameterized as the coordination number, is explicitly considered in the derivation of the effective permeability. This is in contrast to other approaches commonly used to derive effective permeability from discrete fracture information (e.g., Snow, 1969; Oda, 1985). It is noted that fracture length does not explicitly enter the effective medium formulation, but indirectly, fracture length will influence the observed fracture spacing and connectivity (see Renshaw, 1999).

## STOCHASTIC MODELING

The effective grid-block conductivity (equation 4) is a function of three fracture parameters defined locally at the scale of the numerical grid block: (1) the

**Table 1. Listing of the parameters used to define the EMT approximation to permeability and the corresponding parameters observable in discrete fracture systems along with general data sources.**

| *Effective Media Parameter* | *Effective Media Symbol* | *Discrete Fracture Parameter* | *Data Sources* |
|---|---|---|---|
| Fracture spacing | $N_f$ | Fracture spacing | Borehole and outcrop observations |
| Coordination number | $z$ | Fracture connectivity | Outcrop observations |
| Conductivity | $C_i$ | Fracture transmissivity | Downhole pressure testing |

average conductance of the individual fractures; (2) the number of fractures in the grid block; and (3) the average coordination number. A second step in the modeling approach is necessary to populate a larger 3-D domain and address spatial variability in fracture permeability. These three fracture parameters are considered spatial random functions and are modeled using Boolean and geostatistical simulation. The development of this second step of the modeling approach is demonstrated with an example problem. Results of the models are described in terms of permeability $[L^2]$.

## Available Data

For the example problem, information regarding the shape, radius, frequency, orientation, transmissivity, and aperture of two sets of discrete fractures have been summarized by K. Oyamada and H. Takase (1998, 1999, personal communication). The fractures are disk shaped, and the fracture radius is defined by a power-law distribution. For this example problem, the power-law distribution was truncated at 7 and 300 m (23 and 984 ft) (minimum and maximum) (K. Oyamada and H. Takase, 1999, personal communication). The power-law distribution results in a large number of small fractures, 97.8% of which are less than 7-m (23-ft) truncation length. These smaller fractures are not modeled explicitly and are considered to contribute to the background permeability.

The frequency of the fractures over the entire radius distribution is described by a Poisson distribution with an expected value of 0.52 fractures per 2 m (6.6 ft). The frequency is given in these units to be consistent with the grid-block size used in the following models. The Poisson distribution dictates that only 40% of the grid blocks in the model will contain a fracture.

Two fracture sets with mean orientations of north–south and east–west (0.0 and 90.0°) are present. Variation in the orientation about these mean directions is governed by a Fisher distribution. The $\log_{10}$ fracture transmissivities ($m^2/s$) are defined by a normal distribution with a mean of −8.99 and a standard deviation of 1.07. Fracture aperture, $b$, is necessary to calculate fracture porosity in each grid block and is defined by an empirical relationship with the fracture transmissivity, $T$: $b = 2 \times \sqrt{T}$ (K. Oyamada and H. Takase, 1999, personal communication).

## Permeability Classes

The Poisson distribution of fracture frequency dictates that only 40% of the grid blocks will contain fractures. The fracture length distribution shows that 97.8% of the fractures are below the 7-m (23-ft) measurement truncation limit. With such a high proportion of small fractures, it is reasonable to conclude that all grid blocks will have some fracture permeability. To accommodate the truncated small-radius fractures, two permeability classes (high and low) are defined. Boolean simulation is used to define the spatial distribution of the two permeability classes.

The Boolean simulation algorithm is used to generate zones of high-permeability grid blocks that occupy 60% of the model domain and are centered at random locations throughout the model domain. These zones represent high permeability resulting from a single fracture or a heavily fractured zone extending across multiple grid blocks. The Boolean simulation code is an extension of ellipsim (Deutsch and Journel, 1998) that allows the radii and orientations of the elliptical objects to be defined by Monte Carlo draws from user-defined distributions. The radii of the elliptical objects is taken from the truncated power-law distribution of fracture radii with 97.8% of the radii set equal to the lower truncation limit of 7 m (23 ft), and the orientation of the zones is drawn from a triangular distribution approximation of the Fisher distribution of deviations about the mean orientations of north–south and east–west.

An example Boolean simulation is shown in Figure 2 (upper left image). Note that most of the conducting grid blocks (red in Figure 2, upper left image) are arranged in small clusters, but a few large high-permeability zones span the model domain. This pattern results from the truncated power-law distribution of the fracture radii. Most of the clusters are formed using radii at the truncation limit (7 m; 23 ft); however, a small fraction of the features have much longer lengths up to the upper truncation limit of the distribution (300 m; 984 ft). The Boolean model employs a constant anisotropy ratio that results in the width of the features being proportional to the length of the features. By creating the Boolean objects using the distributions of radii and orientations measured on the discrete fractures, the spatial correlation of intergrid-block effective permeability caused by fracture lengths greater than the grid-block size is ensured.

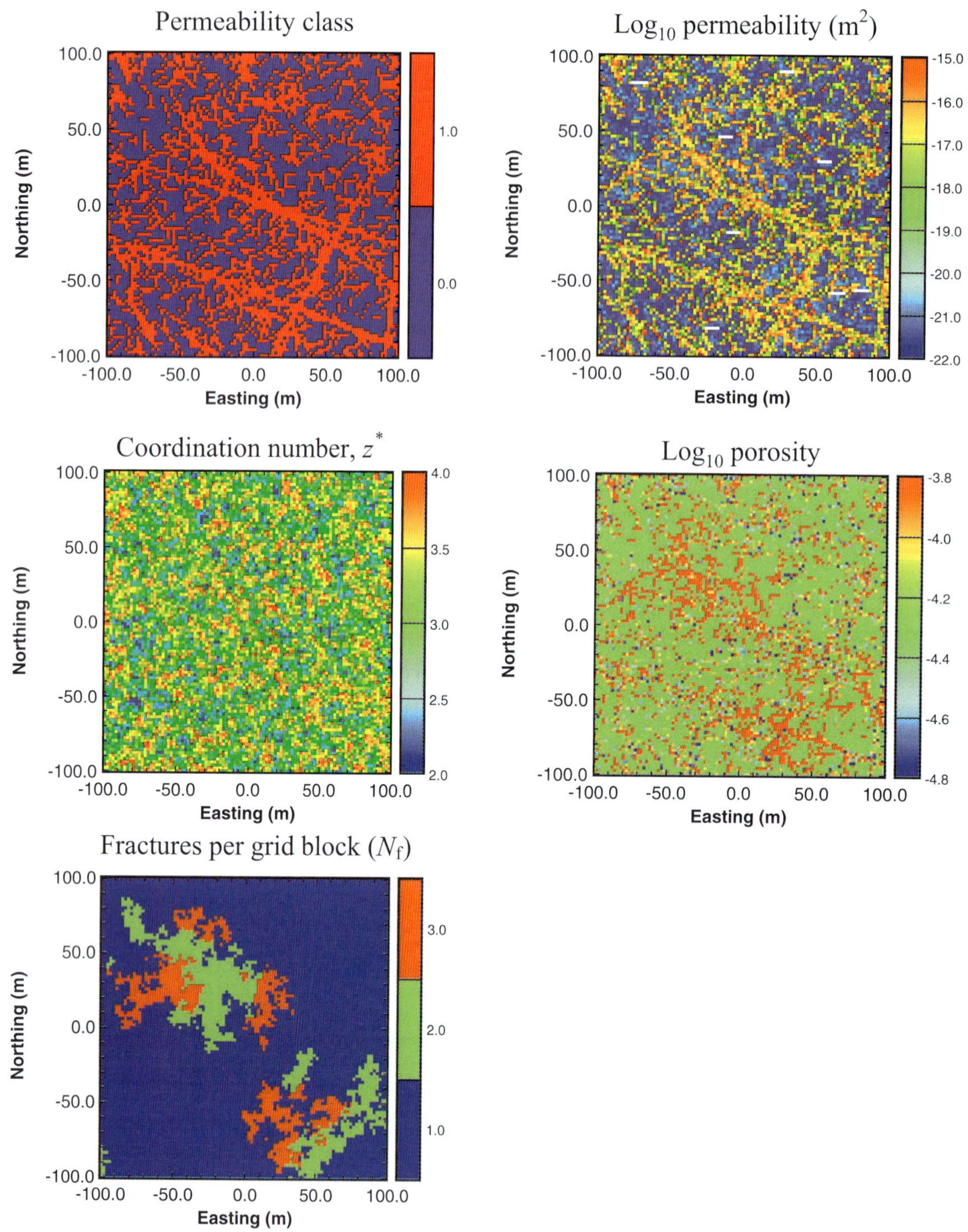


**Figure 2. Two-dimensional example of the stochastic simulations of permeability class, coordination number, and fracture frequency. These three realizations are combined to produce the fracture permeability and porosity shown in the two images on the right.**

## Fracture Frequency

The number of fractures in a grid block, $N_f$, is defined as an integer value for use in equation 4. Because of the isotropy assumption, $N_x = N_y = N_f$, and therefore, only $N_f$ must be simulated. The simulations of $N_f$ are done using the indicator geostatistical-simulation algorithm sisim (Deutsch and Journel, 1998). In these simulations, the minimum value of $N_f$ for a 2-m (6.6-ft) grid block in the high-permeability class is 1, and the maximum is constrained to 5. The probability of occurrence for the different values of

$N_f$ (1, 2, 3, 4, or 5) is taken from the Poisson fracture frequency distribution defined by K. Oyamada and H. Takase (1998, personal communication). These probabilities of occurrence define the proportion of each value of $N_f$ in the indicator geostatistical simulations.

The truncated power-law distribution of fracture radii shows approximately 98% of the fractures to be equal to or less than the truncation limit of 7 m (23 ft). Based on this observation, the integral scale of the model of spatial correlation for $N_f$ is taken to be 14 m (46 ft) (length of fracture is twice the radius). An isotropic, exponential semivariogram model is used with an effective range of 52 m (170 ft) (three times the integral scale). The resulting indicator geostatistical simulations show a mosaic pattern of homogeneous zones of fractures per grid block ($N_f$) (Figure 2, lower left image).

## Average Coordination Number

The average coordination number for a fracture system in a grid block is a complex function of the fracture lengths, orientations, and frequency in that grid block. For the basically orthogonal system considered here, a maximum upper bound on the average coordination number is 4. This situation corresponds to every fracture being continuous across the grid block and thus intersecting every fracture from the opposite set (those in the orthogonal direction). Given the specified fracture length distribution (K. Oyamada and H. Takase, 1999, personal communication), the proportion of fractures in a 2-m (6.6-ft) grid block that are longer than 2 m (6.6 ft), thus leading to a coordination number of 4, is high but less than 1. The absolute lower bound of the average coordination number is 2 corresponding to the case of straight lines cutting across the grid block in a single direction or to elbow joints in a set of roughly orthogonal fractures. The probability of an average coordination number being 2 for this data set is negligible. Given the theoretical constraints on the coordination number of a square lattice as 2 and 4 for the minimum and maximum, respectively, the cumulative distribution function (CDF) is constructed with a triangular distribution to define the univariate distribution of coordination number. The minimum, mode, and maximum of the distribution are set at 2.0, 3.5, and 4.0, respectively. The value of the mode, being closer to the theoretical maximum instead of the minimum, is caused by the relatively high proportion of fractures that are longer than the 2-m (6.6 ft) grid-block dimension.

Simple graphical constructs were examined to determine that coordination number is a much stronger function of fracture length than of fracture frequency. For this example, the average coordination number, $z^*$, and the fracture frequency, $N_f$, are considered to be independent variables and are created through independent geostatistical simulations. The multi-Gaussian geostatistical-simulation algorithm, sgsim (Deutsch and Journel, 1998), is used to create realizations of the coordination number. No information regarding the spatial correlation of the coordination number is present, and an isotropic, spherical semivariogram with a range of 5 m (16 ft) (2.5 grid blocks) is used to create the coordination number simulations (Figure 2, center left image). The values of the coordination number range from above 2 to 4, and the univariate distribution across the 2-D domain matches the triangular distribution defined above. The short range (5 m; 16 ft) employed in the coordination number simulations creates the somewhat spatially random appearance of the simulation (Figure 2, center left image).

## Permeability and Porosity

The final information necessary to create the final image of effective grid-block permeability is the transmissivity of the individual fractures in each grid block. For this study, these transmissivities are drawn independently from two different log-normal distributions. For grid blocks in the high-permeability class, as defined by the Boolean simulation, the fracture transmissivities are drawn randomly from the log-normal distribution (mean = −8.99, standard deviation = 1.09) defined in K. Oyamada and H. Takase (1998, personal communication). $\text{Log}_{10}$ ($m^2/s$) fracture transmissivities for the low-permeability class grid blocks are drawn from a second normal distribution of lower fracture transmissivities with a mean of −14.0 and a standard deviation of 0.5. The combination of the Boolean model of permeability classes, along with the geostatistical simulations of coordination number and $N_f$, as well as the transmissivity of the individual fractures using equation 4, produces the final grid-block permeability map (Figure 2, upper right image).

Fracture porosity in the high-permeability class is calculated as the total volume of the fractures per grid-block volume. Because this modeling approach does not explicitly define the orientation and connectivity of the individual fractures in each grid block, it is necessary to assume that each fracture is fully connected across the grid blocks ($z^* = 4.0$) for the porosity calculations. This assumption will slightly increase the fracture porosity of the grid blocks relative to an explicit porosity calculation using the same fracture parameters. Porosity in the low-permeability class is set to a constant value of $5 \times 10^{-5}$. An example porosity image is shown in the lower right image of Figure 2. From Figure 2, it is seen that porosity is a stronger function of fracture frequency than permeability.

In summary, the 2-D effective media approximation for permeability of fracture networks in a single grid block can be employed with Boolean and geostatistical simulation to develop 3-D models of fracture permeability. This development is made by assuming that the 2-D system in any numerical grid block can be extended uniformly in the third dimension up to the size of that grid block. Three factors allow this extension to three dimensions: (1) the two fracture sets specified for this work are in the $X$ and $Y$ planes, and no third fracture set is present in the $Z$ plane; (2) the fracture sets have a vertical orientation (zero dip angle); and (3) the size of the numerical model grid blocks can be kept small enough (e.g., 2 m [6.6 ft]) that fracture properties can be assumed constant over the size of the grid block. An example permeability realization on a $202 \times 200 \times 202$-m ($662 \times 656 \times 662$-ft) domain is shown in Figure 3.

## Modeling Stages

In the development of the FCM, the intersection of multiple, high-permeability features in a single numerical model grid block does not have an explicit effect on the fracture permeability assigned to that grid block. To examine the effect of increased high-conductivity regions in a grid block on permeability, a second stage of modeling (stage 2) is created with correlation between the high- and low-conductivity regions of the model and the fracture frequency and coordination number. Instead of just assuming that the high-conductivity regions are caused solely by the local presence of relatively higher transmissivity fractures, a stage 2 model considers that the high-conductivity grid blocks are also the result of higher fracture frequency and greater fracture connectivity.

In stage 2 models, the values of both the number of fractures and the average coordination number

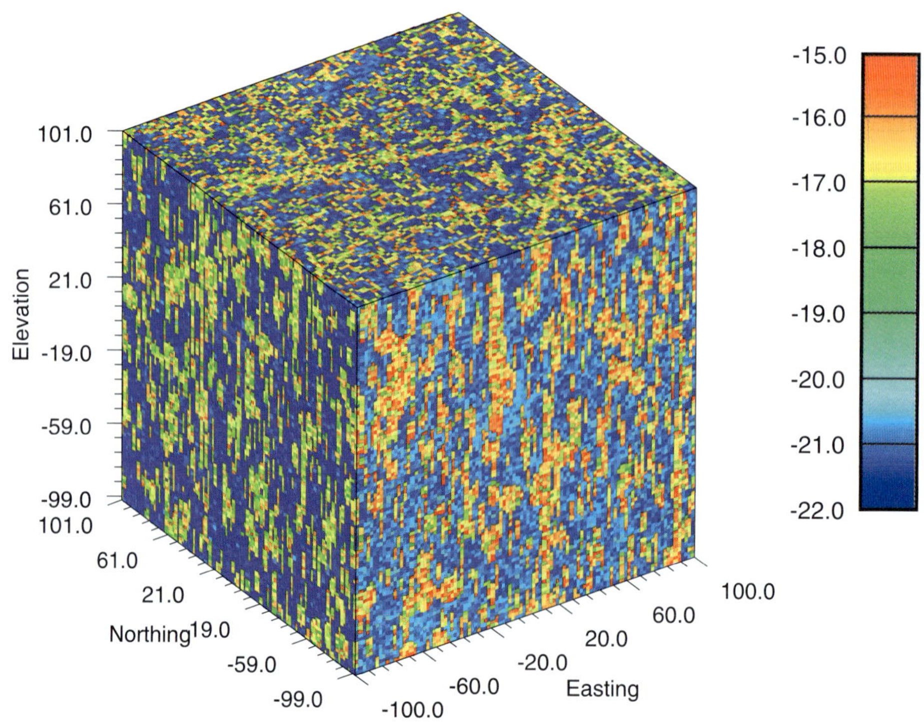


**Figure 3. Example realization (stage 1 approach) of the fracture permeability in $\log_{10}$ ($m^2$) in the model domain.**

**Table 2. Relationship between proportion of high-conductivity grid blocks in the template and fracture frequency.**

| *Proportion of Template in High-conductivity Features* | *Assigned Fractures per Grid Block* |
|---|---|
| <0.222 | 1 |
| 0.222–0.444 | 2 |
| 0.444–0.666 | 3 |
| 0.666–0.889 | 4 |
| >0.889 | 5 |

drawn from the respective distributions are connected to the locations of the high-conductivity features through the use of a 3 × 3 × 3 grid-block template. To assign properties to the central block of the template, the fraction of high-permeability grid blocks defined by the Boolean model in the template is calculated. This fraction is used to determine the cumulative probability of the fracture frequency and the average coordination number when drawing from those distributions. Therefore, in stage 2, the probability value applied to the CDFs of fracture frequency and average coordination number is not randomly determined as in stage 1 but is defined as the fraction of the blocks in the nine-point template that contains high-conductivity features. The template size of 3 × 3 × 3 grid blocks limits the influence of the surrounding model grid blocks to only those that are adjacent to the current grid block being calculated.

For stage 2 permeability models, a nearly uniform distribution of the number of fractures in a grid block is used. Table 2 lists the number of fractures assigned to each grid block as a function of the proportion of the 3 × 3 × 3 template in the high-conductivity zone. The remainder of stage 2 modeling is identical to that used for stage 1, with the exception that the assignment of the average coordination number is based on the proportion of the template in the high-permeability class. Stage 2 models require less computational effort than do the stage 1 models because it is not necessary to create the geostatistical simulations of average coordination number and fracture frequency. The Boolean simulations used to define the regions of high and low conductivity are identical to those used in stage 1. Maintaining a constant Boolean template for each pair of realizations allows for direct comparison of the flow modeling transport results between the two stages. A comparison of stage 1 and stage 2 permeability models is given in Figures 4 and 5.

In general, the stage 2 approach creates a more distinct permeability difference between the areas in the Boolean objects and the background material. Note the greater contrast in permeability values between the two classes in stage 2 (Figure 4, lower image) relative to stage 1 (Figure 4, upper image). The extreme high and low permeability values are of greater magnitude in the stage 2 models (Figure 5). Two ensembles of 50 realizations each (stages 1 and 2) are constructed for the 202 × 200 × 202-m (662 × 656 × 662-ft) 3-D volume.

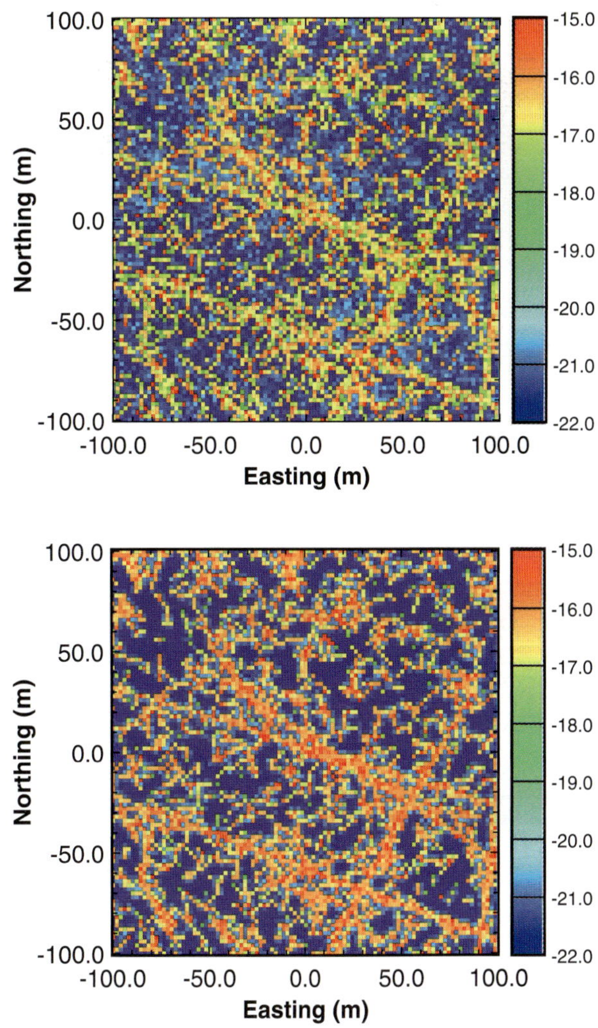


**Figure 4. Permeability models created with the stage 1 approach (upper image) and the stage 2 approach (lower image). Both of these images show the 30th layer of realization number 1. The color scale indicates permeability in $\log_{10}$ ($m^2$).**

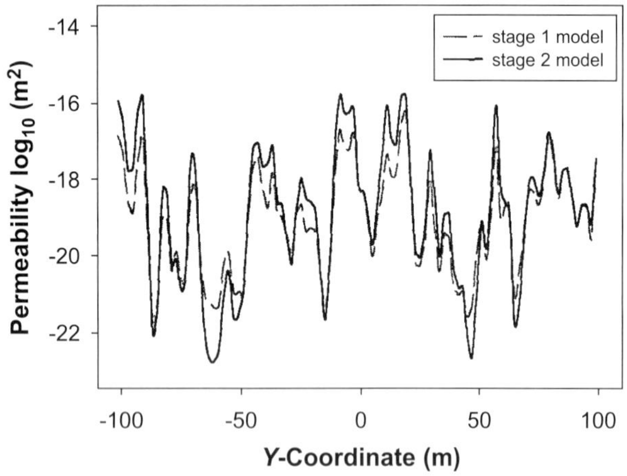


**Figure 5. Stage 1 and stage 2 permeability values along a transect from (0.0,−100) to (0.0,+100.0) along the middle layer (layer 51) of realization 1.**

## MODEL APPLICATION

Two ensembles of 50 realizations each are created using stage 1 and stage 2 modeling approaches. The results of the permeability modeling are evaluated for their ability to reproduce the observed fracture data used as input. These models then serve as input to a set of groundwater flow and streamline transport calculations.

### Model Evaluation

The FCM realizations are checked by comparing the frequency, orientations, and lengths of the Boolean objects against the distributions of the same parameters specified in the input data for the fractures. These three parameters provide the only data against which the models can be directly compared; there are no values of permeability or porosity measured at the 2-m (6.6-ft) grid-block scale.

The proportion of the model domain in the Boolean objects is compared to the target proportion of grid blocks containing conductive fractures. Orientations of the two sets of Boolean objects are compared to the orientations of the two fracture sets, and the proportion of the Boolean objects with lengths less than or equal to the truncation limit of 7 m (23 ft) is compared to the same proportion as specified in the truncated power-law length distribution (Table 3). Comparative statistics (minimum, mean, and maximum) are computed using the results from all 50 realizations. Recall that the same 50 Boolean model realizations are used for both stage 1 and stage 2. Model evaluation results shown in Table 3 indicate that these three fracture attributes are accurately reproduced by the Boolean model across the 50 realizations.

### Flow Model and Particle Tracking

The 100 realizations of fracture permeability and porosity created in stages 1 and 2 are used as input to a parallel flow code, POR-SALSA (Martinez et al., 1997), that is used to obtain a single-phase, steady-state solution to the pressure field with prescribed pressures at two ends and no flow boundaries on the sides of the model (Figure 6).

A total of 600 particles are introduced into the flow field uniformly along the central 100 m (328 ft) of a three-element-wide disturbed zone surrounding a backfilled tunnel that extends through the domain (Figure 6). The tunnel is horizontal and centered along the lateral face of the domain midway between the upstream and downstream boundaries. Particles are tracked along their respective streamlines to the downstream boundary of the model domain. The average hydraulic gradient is $8.0 \times 10^{-3}$,

**Table 3. Comparison of the Boolean model results to the target values of fracture frequency, orientation, and length.**

| *Parameter* | *Target Value* | *Minimum*[a] | *Mean*[a] | *Maximum*[a] |
|---|---|---|---|---|
| Proportion of high-permeability zones | 0.405 | 0.406 | 0.406 | 0.409 |
| Orientation: Set 1 (°) | 0.00 | −0.51 | 0.02 | 0.49 |
| Orientation: Set 2 (°) | 90.00 | 89.57 | 89.96 | 90.41 |
| Radii less than 7 m (23 ft): Set 1 (%) | 97.8 | 97.7 | 98.0 | 98.2 |
| Radii less than 7 m (23 ft): Set 2 (%) | 97.8 | 97.7 | 98.0 | 98.1 |

[a]The minimum, mean, and maximum values are calculated across all 50 realizations. The two different fracture sets are defined by the two different target orientation values of 0.0 and 90.0°.

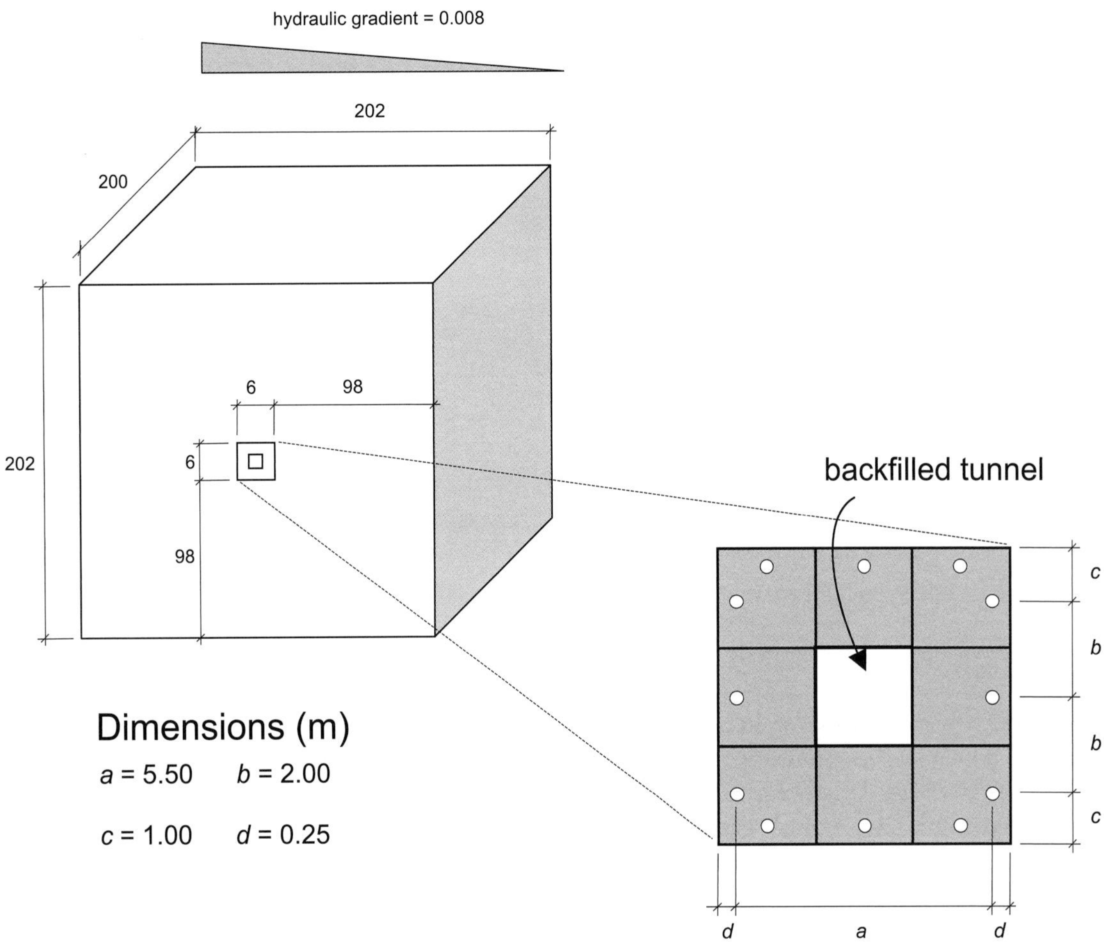


**Figure 6. Schematic diagram of the groundwater flow model setup and the particle starting locations. The particle starting locations are shown as white dots in the expanded view and are not drawn to scale.**

and the total number of 2 × 2 × 2-m (6.6 × 6.6 × 6.6-ft) elements in the flow model is 1,020,100.

The flow solutions for this study were accomplished using 20 Digital Equipment Corporation Alpha processors (612 MHz clock speed) with 4 GB of shared random-access memory. The flow code required 15–25 min to achieve a stable pressure field for each realization. Using a single processor, less than 10 sec were required to track the 600 particles through each realization. The speed at which the pressure solutions are obtained allowed flow and particle tracking across 50 realizations to be accomplished overnight. An obvious advantage of the parallel approach is that multiple realizations of several different conceptual models of the fracture system can be explored in a reasonably short amount of time.

The results of the particle traveltime calculations for stage 1 and stage 2 are shown in Figure 7. For both stages, the cumulative distributions of the minimum, median, and maximum arrival times are shown. Stage 2 permeability models produce longer minimum and median arrival times than stage 1 models. Relative to stage 1 models, stage 2 models have higher permeability in the Boolean objects and lower permeability in the background material. The increase in traveltimes in stage 2 models is caused by the increase in fracture porosity in the high-permeability parts of the domain defined by the Boolean simulation.

The increase in fracture porosity in the high-permeability materials is caused by the functional dependence of the fracture apertures on the measured fracture transmissivity and the more uniform fracture frequency distribution employed in stage 2 models. Both of these factors increase the cross sectional area available for flow, thus decreasing the flow velocities in the model grid blocks relative to the stage 1 approach. The result is an increase of roughly a factor of 2 in the minimum and median arrival times.

**Figure 7. Cumulative distributions of minimum, median, and maximum particle arrival times at the downstream face of the model domain for both stage 1 and stage 2 models.**

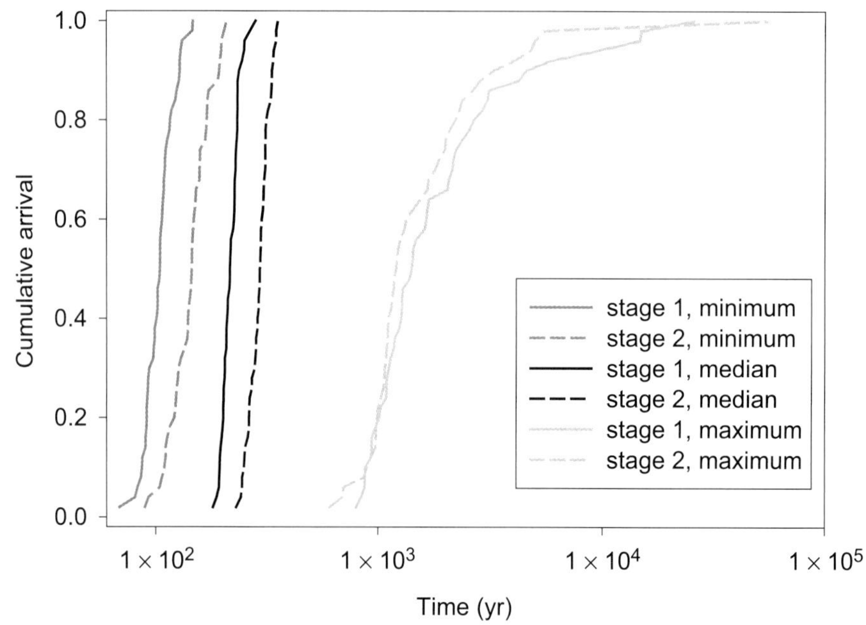


The porosity of the low-permeability class material is set to the same constant value as in the stage 1 models ($5 \times 10^{-5}$). Given the similarity of stage 1 and stage 2 maximum traveltime distributions in Figure 7, it is inferred that the pathways that produce the maximum traveltimes are predominantly within the low-permeability class material. The spatial arrangement of this material did not change from stage 1 to stage 2.

The standard deviation of the 600 traveltimes in each realization was also calculated. One-to-one comparison of the standard deviations for each individual realization shows a strong correlation between stage 1 and stage 2 results. This result indicates that the changes in the approach to modeling the permeability and porosity from stage 1 to stage 2 do not affect the variability of traveltimes in a given realization. The spatial distribution of the Boolean objects remains the same in both modeling stages, and we conclude that the distribution of these objects is the primary control on traveltime variability within a realization.

## CONCLUSIONS

This chapter documents the FCM approach to creating stochastic realizations of fracture permeability. The FCM approach provides for the quantitative inclusion of multiple data types as collected on discrete fracture systems into a constitutive model of fracture permeability. The FCM serves as a practical bridge between DFN and SCM approaches. The combination of approximating the local permeability with an effective medium approximation, based on readily observable fracture attributes and spatial simulation techniques, provides a traceable path from discrete fracture observations to final permeability models. In addition, the FCM approach avoids the large-scale averages commonly employed in the SCM approach in deference to local averages that better honor the behavior of the underlying DFN. The FCM approach is flexible enough to simulate different conceptual models of the fracture permeability.

Flow modeling over multiple realizations of the FCM is simplified by the use of a uniform orthogonal mesh. For the example problem of groundwater flow from a hypothetical nuclear waste repository, these results show that the range, maximum to minimum, of groundwater traveltimes is approximately one order of magnitude for both modeling stages. The second-stage models increase the porosity of the high-permeability regions relative to stage 1 models and thus increase the minimum and median traveltimes. The distribution of the longest groundwater traveltimes and the variation of times in a single realization do not vary between modeling stages and indicate that these two results are controlled by the spatial distribution of the Boolean objects, which is held constant between both modeling stages.

## Extensions of FCM

The example application presented here was based on a relatively simple fracture system, and several extensions of the FCM would be possible for more complex models. If nonvertical fracture sets were included in the modeling, it would be necessary to use an alternative approach (e.g., Chen et al., 1999) to determining the grid-block-scale effective-permeability tensor. The current EMT approach to calculating grid-block permeability can be used to calculate anisotropic fracture permeability. Anisotropic calculations would be based on a different number of fractures and average coordination number in each direction. If specific fracture information is available, the centers of the Boolean objects can be conditioned to specific locations in the model domain, and the FCM approach can be extended to condition the final permeability models to local measurements of permeability made on specific fractures.

## Model Comparison

The discrete fracture data provided for the example problem described in this chapter were also provided to six other modeling groups. Each of these modeling groups employed different approaches to modeling fracture permeability, including several variations of the DFN and SCM approaches. The results of the seven different modeling approaches with respect to several groundwater flow and transport performance measures have been compared (Sawada et al., 2000). This comparison shows that the DFN models produce higher bulk permeability values and faster groundwater flow velocities than do the SCM and FCM models. The bulk permeability calculated for the SCM and FCM models is close to that of the harmonic mean of the fracture permeability distribution, whereas the bulk permeability of the DFN models is better approximated by the arithmetic average permeability. These results appear to be caused by the DFN modeling groups making the assumption of a constant spatially homogeneous permeability value in each individual fracture.

## ACKNOWLEDGMENTS

Funding for this work was provided by the Japanese Nuclear Cycle Development Institute. This chapter was improved through reviews by Steve Carle, Eileen Poeter, Norm Warpinski, and two anonymous reviewers. The original draft of this chapter was completed and sent to the editors in November 1999. Sandia is a multiprogram laboratory operated by Sandia Corporation, a Lockheed Martin Company, for the U.S. Department of Energy's National Nuclear Security Administration under contract DE-AC04-94AL85000.

## REFERENCES CITED

Altman, S. J., B. W. Arnold, R. W. Barnard, G. E. Barr, C. K. Ho, S. A. McKenna, and R. R. Eaton, 1996, Flow calculations for Yucca Mountain groundwater travel time (GWTT-95), SAND96-0819: Albuquerque, New Mexico, Sandia National Laboratories, 170 p., http://infoserve.sandia.gov/sand_doc/1996/960819.pdf (accessed November 15, 1999).

Chen, M., M. Bai, and J.-C. Roegiers, 1999, Permeability tensors of anisotropic fracture networks: Mathematical Geology, v. 31, p. 355–374.

David, C., Y. Gueguen, and G. Pampoukis, 1990, Effective medium theory applied to the transport properties of rock: Journal of Geophysical Research, v. 95, p. 6993–7007.

Dershowitz, B., P. LaPointe, T. Eiben, and L. Wei, 1998, Integration of discrete fracture network methods with conventional simulator approaches: Society of Petroleum Engineers Annual Technical Conference and Exhibition, New Orleans, Louisiana (September 27–30), SPE Paper 49069, 9 p.

Deutsch, C. V., and A. G. Journel, 1998, GSLIB: Geostatistical software library and user's guide, 2d ed.: New York, Oxford University Press, 369 p.

Finsterle, S., 2000, Using the continuum approach to model unsaturated flow in fractured rock: Water Resources Research, v. 36, p. 2055–2066.

Follin, S., and R. Thunvik, 1994, On the use of continuum approximations for regional modeling of groundwater flow through crystalline rocks: Advances in Water Resources, v. 17, p. 133–145.

Hestir, K., and J. C. S. Long, 1990, Analytical expressions for the permeability of random two-dimensional Poisson fracture networks based on regular lattice percolation and equivalent media theories: Journal of Geophysical Research, v. 95, no. B13, p. 21,565–21,581.

Hsieh, P. A., 1998, Scale effects in fluid flow through fractured geologic media, *in* G. Sposito, ed., Scale dependence and scale invariance in hydrology: New York, Cambridge University Press, p. 335–353.

Kirkpatrick, S., 1973, Percolation and conduction: Reviews of Modern Physics, v. 45, p. 574–588.

Long, J. C. S., and P. A. Witherspoon, 1985, The relationship of the degree of interconnection to permeability in fracture networks: Journal of Geophysical Research, v. 90, no. B4, p. 3087–3098.

Long, J. C. S., J. S. Remer, C. R. Wilson, and P. A. Witherspoon, 1982, Porous media equivalents for networks of discontinuous fractures: Water Resources Research, v. 18, p. 645–658.

Martinez, M. J., P. L. Hopkins, and J. N. Shadid, 1997, LDRD final report: Physical simulation of nonisothermal multiphase multicomponent flow in porous media, SAND97-1766: Albuquerque, New Mexico, Sandia National Laboratories, 65 p., http://infoserve.sandia.gov/sand_doc/1997/971766.pdf (accessed November 15, 1999).

Oda, M., 1985, Permeability tensor for discontinuous rock mass: Geotechnique, v. 35, p. 483–495.

Renshaw, C. E., 1999, Connectivity of joint networks with power law length distributions: Water Resources Research, v. 35, p. 2661–2670.

Sawada, A., M. Shiotsuki, K. Oyamada, and H. Takase, 2000, Study of flow model comparison in fractured rock, *in* Proceedings of the 30th Fossil Energy Symposium (in Japanese): Japanese Society for Civil Engineering Monograph 17, p. 123–127.

Snow, D., 1969, Anisotropic permeability of fractured media: Water Resources Research, v. 5, p. 1273–1289.

Tsang, Y. W., C. F. Tsang, F. V. Hale, and B. Dverstop, 1996, Tracer transport in a stochastic continuum model of fractured media: Water Resources Research, v. 32, p. 3077–3092.

Zimmerman, R. W., and G. S. Bodvarsson, 1996, Effective transmissivity of two-dimensional fracture networks: International Journal of Rock Mechanics, Mineral Science and Geomechanical Abstracts, v. 33, p. 433–438.

Torres-Verdín, C., A. Grijalba-Cuenca, and H. W. J. Debeye, 2006, A comparison between geostatistical inversion and conventional geostatistical-simulation practices for reservoir delineation, *in* T. C. Coburn, J. M. Yarus, and R. L. Chambers, eds., Stochastic modeling and geostatistics: Principles, methods, and case studies, volume II: AAPG Computer Applications in Geology 5, p. 187–205.

*Chapter 15*

# A Comparison between Geostatistical Inversion and Conventional Geostatistical-simulation Practices for Reservoir Delineation

**C. Torres-Verdín**
**A. Grijalba-Cuenca**[1]
*The University of Texas at Austin*
*Austin, Texas, U.S.A.*

**H. W. J. Debeye**
*Fugro-Jason*
*Rotterdam, The Netherlands*

## ABSTRACT

Geostatistical inversion provides a quantitative way to integrate the high vertical resolution of well logs with the dense aerial coverage of poststack three-dimensional seismic amplitude data. A systematic field study is presented in this chapter to understand the relative merits of geostatistical inversion over standard geostatistical-simulation procedures that do not make explicit use of three-dimensional seismic amplitude data. It is shown that, by making quantitative use of the poststack seismic amplitude data, geostatistical inversion considerably reduces the space of stochastic realizations that honor both the well-log data and the spatial semivariograms. Sensitivity analysis also shows that geostatistical inversion remains less affected by a perturbation of semivariogram parameters than standard geostatistical simulation. Tests of cross-validation against well-log data show that geostatistical inversion yields additional information over the average trends otherwise obtained with stochastic simulation. In the vicinity of existing wells, geostatistical inversion can potentially infer vertical variations of resolution similar to that of well logs and, at worst, of vertical resolution equal to that of the seismic amplitude data at locations distant from wells. A drawback of geostatistical inversion is the need to convert well-log data from depth to seismic traveltime. In addition, geostatistical inversion may be rendered computationally prohibitive when applied to large seismic and well-log data sets.

[1]*Present address*: Occidental Petroleum Corporation, Houston, Texas, U.S.A.


DOI:10.1306/1063816CA53235

## INTRODUCTION

Geostatistical inference is a popular method used by reservoir specialists to interpolate and/or extrapolate petrophysical variables between, and/or away from, existing wells. One of the most attractive properties of geostatistical inference is its ability to integrate measurements acquired with different degrees of spatial resolution and lateral continuity. Geostatistical techniques also provide a natural means to incorporate structural and stratigraphic information in their estimation algorithms and can be readily used to quantify uncertainty. Lastly, geostatistical estimation techniques can enforce a nondeterministic (statistical) relationship between actual measurements and the unknown underlying petrophysical parameters.

A survey of current hydrocarbon-reservoir applications shows that geostatistical techniques are almost exclusively based on the use of well-log and core data acquired along existing wells. Seldom do these techniques make use of the three-dimensional (3-D) surface seismic amplitude data that are now commonly obtained in the interwell region. In the best cases, seismic amplitude data are used by geostatistical estimation techniques as secondary or ancillary information (e.g., as a way to impose an external drift) in cokriging or collocated cokriging techniques.

Recently, a novel geostatistical technique has emerged with the objective of jointly honoring the wire-line data acquired in control wells and the seismic amplitude data measured at every node of the 3-D seismic cube (see, for instance, Haas and Dubrule, 1994; Debeye et al., 1996; Grijalba-Cuenca et al., 2000). This new technique, referred to as geostatistical inversion, is best described as a general stochastic procedure to estimate property distribution functions in the interwell region.

At the outset of the geostatistical inversion process, the entire seismic cube is populated with realizations of acoustic impedance obtained using conventional sequential Gaussian collocated simulation. The realizations are subsequently tested against the measured seismic amplitude data before being accepted as final realizations. To perform this test, the realizations of acoustic impedances are converted into a reflectivity series that is convolved with the seismic wavelet. This convolution yields a numerical simulation of the poststack seismic amplitude data. A seismic residual (normally a least-squares residual) is then calculated from the difference between the measured and numerically simulated seismic traces. Whenever the seismic residual is in agreement with a reduction of the global seismic residual (i.e., the collective residual that considers all of the traces available in the 3-D seismic cube), the current local realization of acoustic impedance is deemed acceptable. Otherwise, the realization is rejected and replaced with the existing one. A new node in the 3-D seismic cube is then chosen by way of a random walk. An iteration of the stochastic inversion is complete when all of the nodes in the seismic cube (this includes lateral spatial location and time) are visited by this local random draw and check procedure. Subsequent iterations are aimed at reducing the global seismic residual in a monotonic fashion. Several iterations are generally needed before the global seismic residual is brought down to a level consistent with the energy of the noise present in the seismic amplitude data (or else with the energy of the seismic-well tie error). Simulated annealing is used to reach a local minimum of the global seismic misfit by such a sequence of iterations.

Compared to standard geostatistical estimation techniques, geostatistical inversion commonly requires large quantities of computer time to provide satisfactory results for extensive volumes of seismic data. However, in similarity with conventional geostatistical practices, geostatistical inversion provides a great deal of flexibility with which to integrate existing measurements having different lengths of spatial support. Torres-Verdín et al. (1999) and Grijalba-Cuenca et al. (2000) have shown that geostatistical inversion can also be used to increase the vertical resolution of petrophysical variables estimated from 3-D seismic amplitude data, especially in the vicinity of existing wells.

Figure 1 is a flow diagram of the geostatistical inversion algorithm described above. A practical and common example of geostatistical inversion is the estimation of total porosity in the interwell region when porosity and acoustic impedance are statistically related by way of a joint probability density function (pdf) (Pendrel and Van Riel, 1997).

The central objective of this chapter is to provide a quantitative comparison of the performance of both conventional geostatistical estimation and geostatistical inversion. This objective is accomplished by making use of a complete well-log and 3-D seismic amplitude data set acquired in an active hydrocarbon reservoir in San Jorge Basin, Argentina.

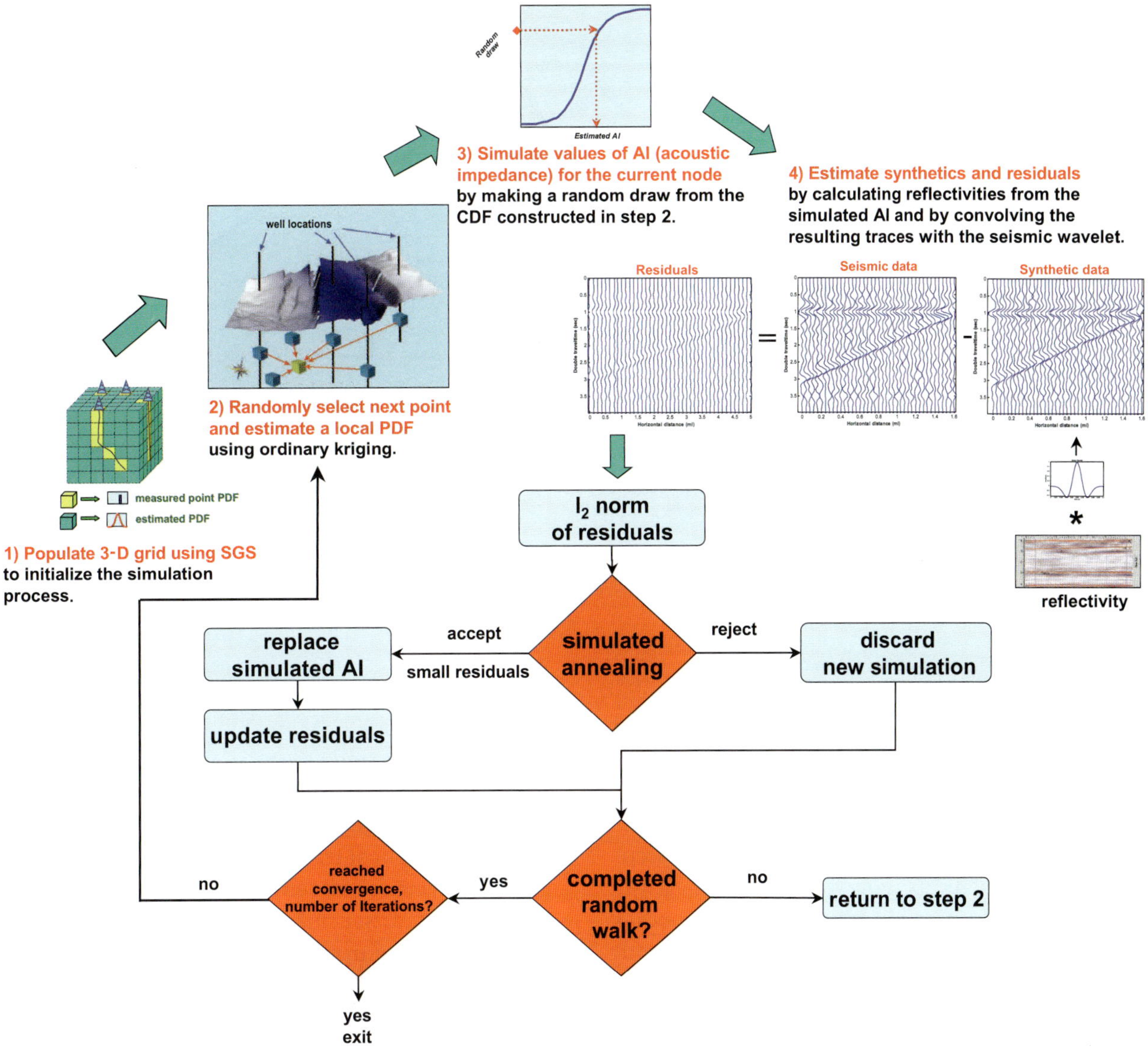


Figure 1. Flow diagram describing how geostatistical inversion yields estimates of acoustic impedance in the interwell region. A stochastic-simulation algorithm is used to generate random realizations of acoustic impedance at every node in the seismic cube. These realizations are consistent with a previously specified degree of spatial correlation and are only accepted if they honor both existing well logs and the 3-D poststack seismic amplitude data. In the diagram, SGS stands for sequential Gaussian simulation, PDF stands for probability density function, and CDF stands for cumulative density function.

## THE DATA SET AND THE LOCAL GEOLOGY

Figure 2 is a graphical rendering of the seismic amplitude and well-log data set used as a test example in this chapter. Three vertical wells are spaced approximately 900 and 250 m (2952 and 820 ft) apart. The three wells are located within a segment of a 3-D seismic cube comprising an area of 12 × 16 km (7 × 10 mi). Only a small part of 1.3 × 0.6 km (0.80 × 0.37 mi) of the complete seismic cube is shown in Figure 2. Seismic traces are separated at a uniform interval of 25 m (82 ft), and a total of 1225 traces are present in the part of the seismic cube shown in Figure 2. One color-coded seismic horizon is shown in Figure 2 to give an indication of the structural framework. The coding in colors indicates two-way vertical traveltimes in seconds as described in the color legend to the left of Figure 2. Seismic amplitude data were acquired with a time sampling interval of 2 msec and

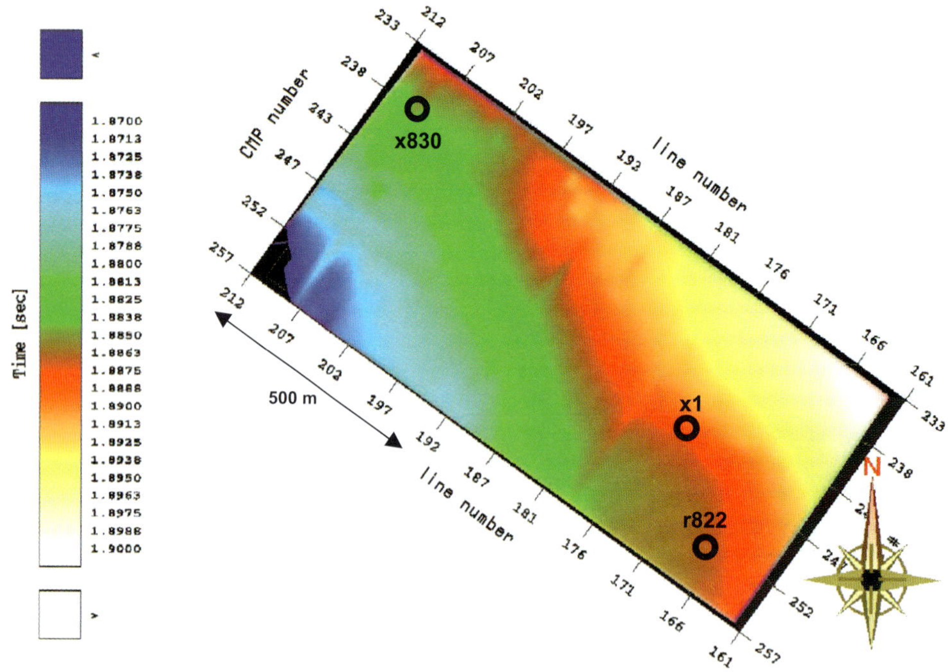


Figure 2. Plan view of the area of study comprising a rectangle of 1.3 × 0.6 km (0.80 × 0.37 mi) and three well locations (identified with circles). A seismic horizon, color-coded in double vertical traveltime (seconds), is here used to give an indication of the underlying structural embedding. Seismic traces are evenly spaced at 25-m (82-ft) intervals along the two horizontal, orthogonal directions of the seismic cube. Well trajectories are vertical. CMP = common mid-point.

with a central frequency of approximately 35 Hz, a lowest frequency of 6 Hz, and a highest frequency of 85 Hz. This frequency bandwidth places an important constraint on the thickness that can be resolved at a given depth. For the case at hand, calculations and observations show that the poststack 3-D seismic amplitude data cannot resolve sand bodies thinner than 12 m (39 ft) when buried at a depth between 2000 and 3000 m (6600 and 9800 ft).

Structurally, the local geological formations are part of a large extensional basin. Hydrocarbon reservoirs in this area consist of 1–15-m (3.3–49-ft)-thick sand bodies of highly anastomosed fluvial architectures interspersed within thick shale deposits of lacustrine origin. The sedimentary column also exhibits a significant presence of volcanic debris and tuffaceous deposits in the zone of interest.

## ACOUSTIC IMPEDANCES

For simplicity, but without sacrifice of generality, the discussion henceforth concentrates on the relatively simple problem of estimating acoustic impedances between the three existing wells. Being the product of bulk density and P-wave velocity, acoustic impedance is the basic building block of 3-D poststack seismic amplitude data. Not only are poststack seismic amplitude data unable to provide separate values of density and velocity (only of their product), but they are also equally unable to provide values of acoustic impedance in the depth domain (only in the double vertical traveltime domain).

Double vertical traveltime is the natural vertical axis of poststack seismic amplitude data, and hence, for the present purposes, the choice has been made to work exclusively in the same vertical domain instead of in the more common depth domain. Such a strategy requires that all of the well-log data be transformed from depth into the seismic double traveltime domain, and this has been done with the aid of existing check-shot data and with the implementation of synthetic seismograms. In all cases, the depth-to-time transformations are deemed to have excellent quality and have been further checked against numerical integration of the sonic logs (in turn, the sonic and density logs have been subjected to a processing step aimed at correcting and reducing the effect of bad caliper readings).

Figure 3 is a time-domain cross section of the well-log acoustic impedance data taken along a polygon that joins the three existing wells in the area of study. Acoustic impedances are described with color-coded vertical stripes posted at the exact location of the three wells. The colors in the scale are encoded as described by the legend shown to the left of the same figure. Two seismic horizons are

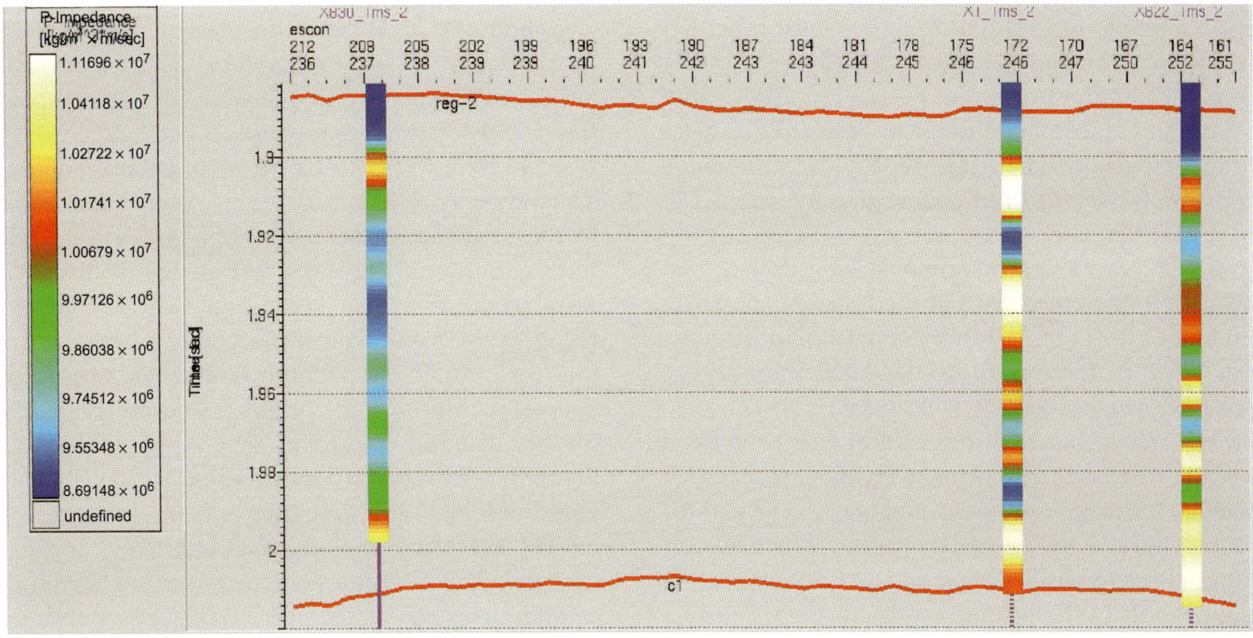


**Figure 3. Time domain cross section of the acoustic impedance logs posted at the location of the three existing wells. Acoustic impedance values are color coded according to the legend at the left. The logs shown have been resampled to a vertical sampling interval of 1 msec. Two seismic horizons, shown here with red lines, provide upper and lower bounds for the time window of study considered here.**

indicated with red lines in Figure 3. One of these horizons is depicted in plan view in Figure 2, and they provide upper and lower bounds in the time domain for all of the estimation results developed here.

A seismic wavelet has also been estimated from the seismic amplitude data via synthetic seismograms. Independent analyses in the three existent wells yield a consistent seismic wavelet that can be used across the complete zone of interest. This wavelet is shown in Figure 4, both in the time and frequency domains.

## TRACE-BASED INVERSION RESULTS

A first step in the analysis consists of transforming the seismic traces into acoustic impedances. This transformation is commonly accomplished with a linear inversion procedure in which the seismic amplitude data are devoid of their wavelet signature. Commonly, the inversion of seismic amplitude data into acoustic impedances is performed with a deconvolution procedure that yields seismic reflectivities. The reflectivities are then transformed into acoustic impedances

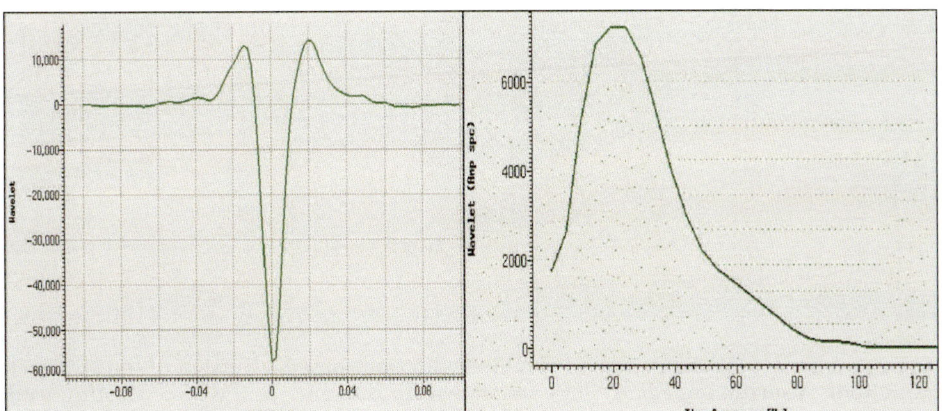


**Figure 4. Wavelet estimated from the well-log data acquired in the three existing wells and from the seismic amplitude data acquired in their immediate vicinity. The time- and frequency-domain counterparts of the seismic wavelet are shown in the left- and right-hand panels, respectively.**

with a simple integration formula. Albeit inherently nonlinear, inverting acoustic impedances directly from seismic amplitude data has proven to be a more robust and efficient method than standard deconvolution, especially when time-dependent value-range constraints are enforced during the inversion (Debeye and Van Riel, 1990).

An incentive for inverting seismic traces into acoustic impedances is that the latter purportedly no longer bears the arbitrary signature of the seismic wavelet. Because of this, acoustic impedances will have a natural affinity with actual values of acoustic impedance gathered in a vertical well, whereas a seismic trace will not. Moreover, the inversion of seismic traces into acoustic impedances can significantly reduce the so-called tuning phenomenon otherwise responsible for spurious laterally continuous events in the seismic amplitude data. Figure 5 is a cross section of acoustic impedance inverted from the seismic data along a polygon that connects the existing three wells in the area of study (the same polygon used to construct the cross section shown in Figure 3).

The inversion of seismic traces into acoustic impedances produces a result devoid of low-frequency components. By construction, such low-frequency components are not present in the seismic wavelet (see Figure 4) and, therefore, cannot be recovered by the inversion. Low-frequency, acoustic impedance components are responsible for the compaction trend (or drift) that causes sonic velocity logs to monotonically increase as a function of depth. The only way to incorporate this compaction trend in the inversion of acoustic impedances is to numerically interpolate it (or extrapolate it) from the compaction trends of sonic logs acquired in existing wells. No universal method can perform the interpolation of compaction trends. One way is, of course, to use ordinary kriging; another is to use inverse distance weighting. The results shown in Figure 5 are based on a compaction trend obtained with a simple local weighting scheme that considers the curvature of the upper and lower seismic horizons. This interpolation procedure has been successfully checked against the compaction trends of other wells available in the area of study.

In the cross section shown in Figure 5, the inverted acoustic impedances exhibit a vertical resolution of 2 msec (i.e., the same vertical resolution of the seismic amplitude data). High values of acoustic impedance correlate well with the presence of packets or groups of shaly sand layers, whereas low values of acoustic impedance correlate with shale beds. Note that the inverted acoustic impedances shown

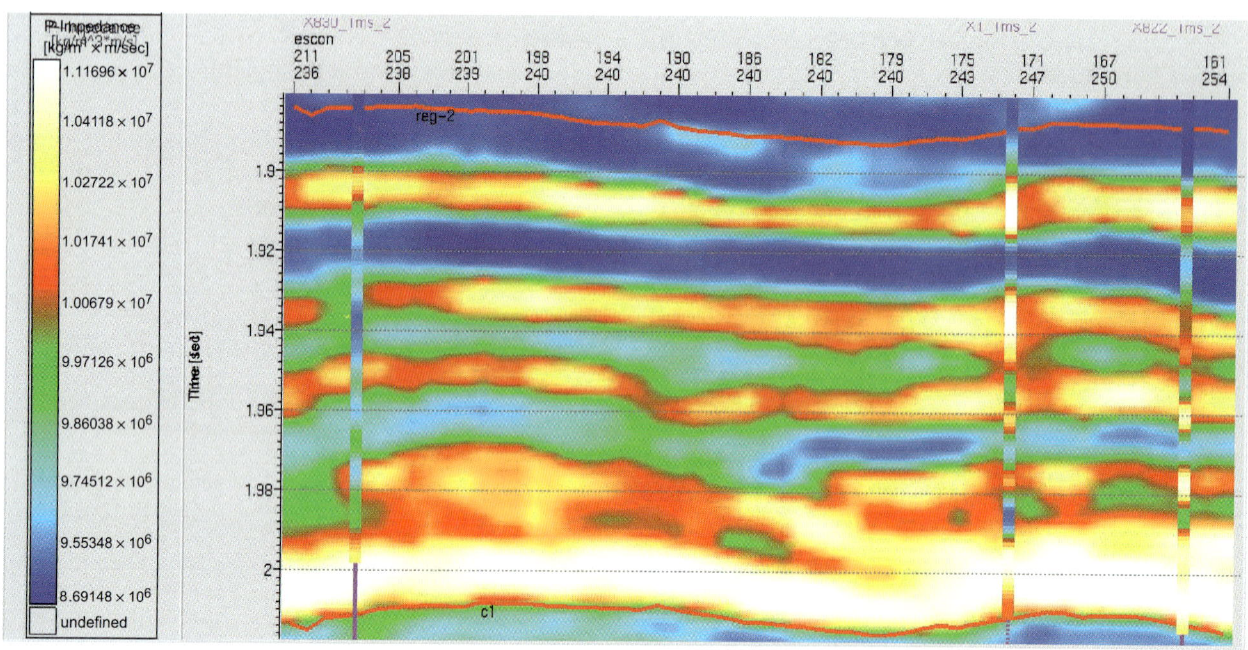


**Figure 5.** Time-domain cross section of the acoustic impedances obtained with trace-based inversion of the poststack seismic data (2-msec vertical resolution). Acoustic impedance logs are shown for reference at the three existing well locations.

in Figure 5 have not been obtained by enforcing a 100% match with the well logs at existing well locations, except in their low-frequency components. In fact, the acoustic impedance logs posted in Figure 5 are shown only to give an indication of the quality of the inversion results. The inversion algorithm simply incorporates both the estimated seismic wavelet and the seismic amplitude data to yield traces of acoustic impedance. In this case, well logs are used and plotted only to verify that the inversion results adequately reproduce the data at the existing wells. In what follows, the deterministic (nonstatistical) results of Figure 5 are used as a qualitative benchmark to compare similar results obtained with geostatistical-simulation and geostatistical-inversion algorithms.

## HISTOGRAMS, SEMIVARIOGRAMS, AND GEOMETRICAL SUPPORT

The acoustic impedance cross section considered in Figure 5 comprises a time window of approximately 150 msec. In theory, there is no maximum limit to the width of the time window used for the analysis presented in this work. However, in an effort to minimize the effects of mechanical compaction on acoustic impedances (quite ubiquitous in the survey area), the choice has been made to concentrate on this relatively narrow time window given that (1) it does not exhibit significant compaction effects, (2) it is framed by well-recognizable upper and lower seismic horizons, and (3) it contains a good number of well-defined hydrocarbon-producing sand reservoirs. Similar analyses have been performed on different upper and lower reservoir zones with approximately equivalent results. Suffice it to say that there is extensive evidence that both reservoir properties and geophysical measurements in the zone of interest in Figure 6 exhibit the desirable statistical property of second-order stationarity that is commonly assumed by geostatistical-simulation techniques.

Figure 6 is a plot of a histogram of acoustic impedance constructed with samples of well-log measurements acquired in the three existing wells. Values of well-log acoustic impedance are obtained by multiplying the velocity and density logs. The best Gaussian pdf fit to the sample histogram is shown on the same figure for comparison. In this particular example, well-log data measured in wells have been resampled with a 1-msec sampling rate (i.e., half the sampling rate of the seismic amplitude data). The well-log data exhibit a sampling rate of 0.5 ft (0.15 m). Hence, to facilitate a larger measure of geometrical support, it is necessary to low-pass filter (upper frequency cut of 500 Hz) and resample the data at the desired 1-msec sampling rate. The 1-msec sampling rate represents a good compromise between the vertical resolutions of the seismic and well-log data. In practice, it is necessary to construct a sample histogram for every sampling rate (i.e., for every measure of geometrical support) at which the simulation is to be performed; but in general, a larger variability is to be expected for a smaller measure

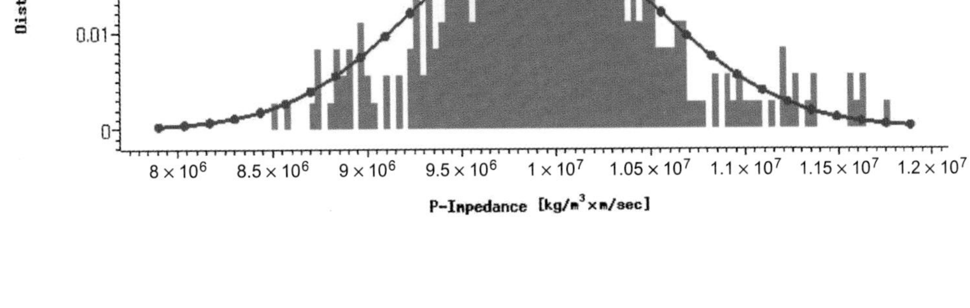


**Figure 6. Sample histogram of the acoustic impedances measured in the three existing well locations. A sampling interval of 1 msec is used to sample the well-log acoustic impedances. The continuous blue line represents the best-fit Gaussian pdf.**

of geometrical support (i.e., for a shorter sampling interval).

An important issue arises as to whether the histogram of Figure 6 is consistent with unavailable lateral measurements of acoustic impedance, especially for distances shorter than the average separation between existing wells. One tempting way to bridge such a spatial sampling gap is to construct the histogram making use of acoustic impedances yielded by trace-based inversion. This would bring the shortest lateral distance between acoustic impedance samples to the uniform separation of traces in the seismic cube (25 m [82 ft] in the present case). Strictly speaking, the difference in length of spatial support between the well-log impedances and the trace-based inverted acoustic impedances invalidates such a procedure. In light of the lack of adequate spatial sampling in the horizontal plane, the initial guess for a sample histogram represents a compromise between the histograms sampled from 1-msec well logs and 2-msec trace-based inverted acoustic impedances. In like fashion, a-priori semivariograms are normalized from the sampled histograms and are estimated from samples of 1-msec well-log and 2-msec trace-based inverted acoustic impedances. Such histograms and semivariograms are then input to the simulation process, and a new, consistent set of 1-msec histograms and semivariograms is produced a posteriori. This process is repeated until a consistent set of lateral and vertical semivariograms and histograms is produced for the simulations. In the present case, the final set of semivariograms is in agreement with the outcrop data and with well-to-well correlation studies previously conducted in the same area. Sensitivity analysis has also been conducted to determine the effect on the simulations of changing the semivariogram parameters.

Figure 7 shows the final a-posteriori histogram of a 1-msec acoustic impedance overlain by the corresponding best fit Gaussian pdf. Final a-posteriori 1-msec acoustic impedance semivariograms are plotted in Figure 8 for the vertical and horizontal directions. The same figure describes the number of acoustic impedance samples used to estimate the semivariogram at a given lag distance.

## SEQUENTIAL GAUSSIAN SIMULATION OF ACOUSTIC IMPEDANCES

Figure 9 shows the cross section of acoustic impedance obtained with a popular geostatistical-simulation procedure referred to as sequential Gaussian simulation (see, for instance, Chilès and Delfiner, 1999) and driven by the acoustic impedance logs sampled in the three well sites. The result shown in the cross section of Figure 9 is the average of 30 independent stochastic simulations. Interpolation of the local pdfs was performed via simple kriging and by making use of the modeled semivariograms described in the previous section. Note that the cross section of Figure 9 bears a poor resemblance to the corresponding cross section of trace-based inverted acoustic impedances shown in Figure 5.

To quantify the agreement of the simulations shown in Figure 9 with the measured poststack

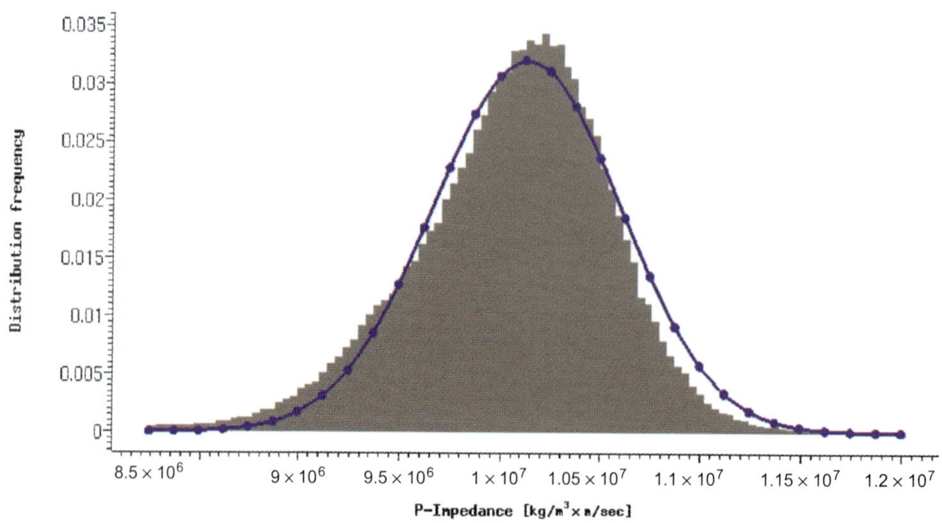


**Figure 7. Histogram of acoustic impedances sampled a posteriori from 1-msec geostatistical inversions of acoustic impedance at trace locations in the seismic cube, and well logs resampled at a resolution of 1 msec. The continuous blue line represents the best-fit Gaussian pdf. These results can be compared to those shown in Figure 6.**

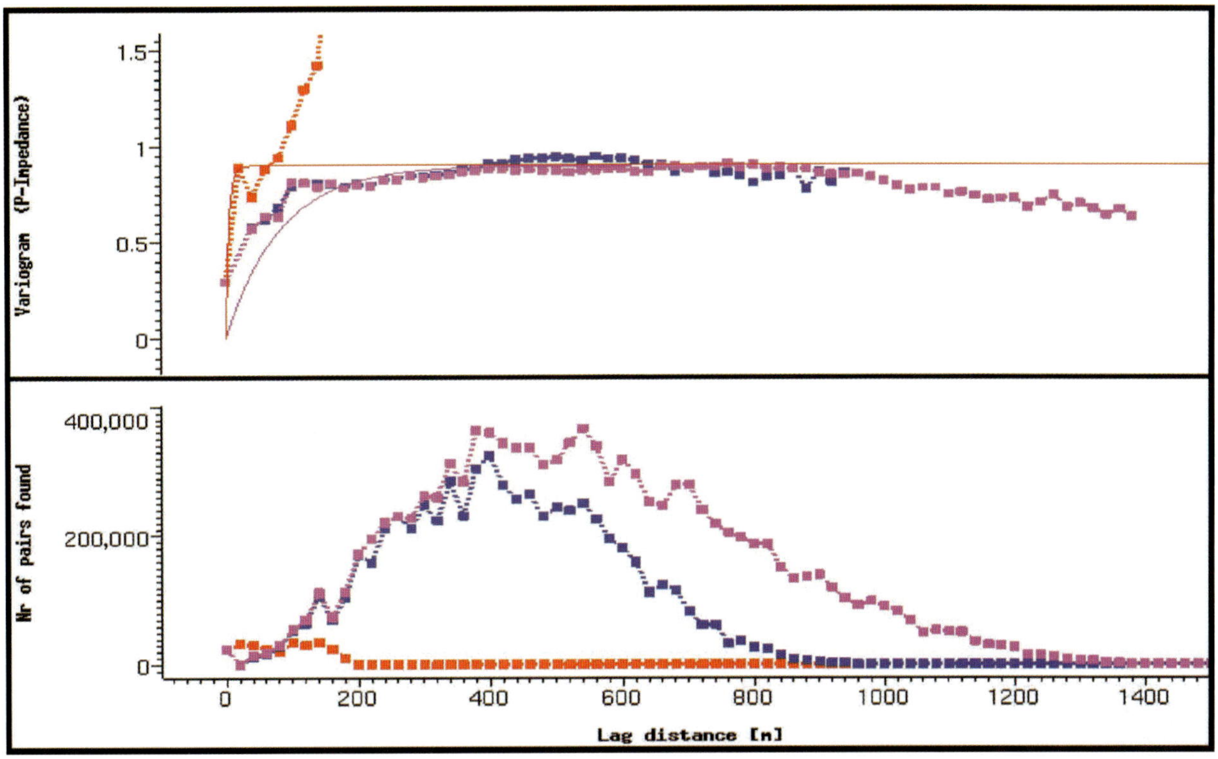


Figure 8. Semivariograms of acoustic impedance. The upper panel shows the vertical (red) and lateral (blue and magenta) semivariograms estimated from samples of acoustic impedance. Lateral semivariograms have been estimated and adjusted a posteriori from 1-msec geostatistical inversions of acoustic impedance at trace locations in the seismic cube. The vertical semivariogram is estimated from resampled 1-msec well-log data. Thin red lines describe the corresponding model semivariograms derived with an exponential parametric model of range equal to 250 m (820 ft) in the horizontal direction and 15 m (49 ft) in the vertical direction, and a sill equal to 0.9. The lower panel describes the number of independent samples used to estimate the three semivariograms at a particular lag distance.

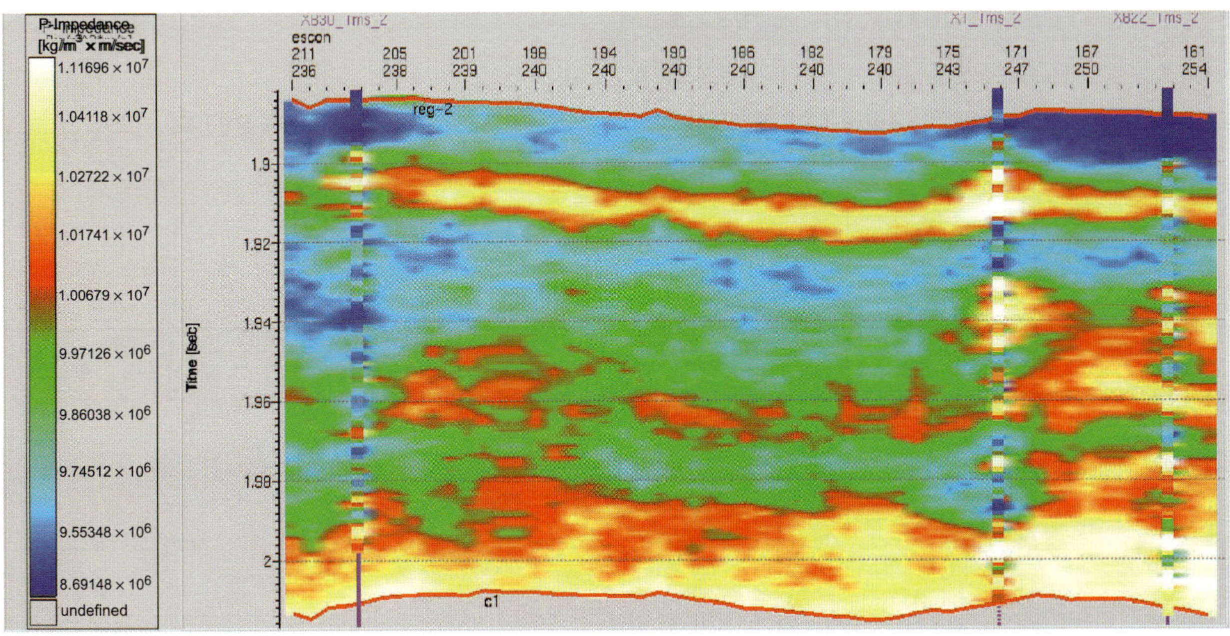


Figure 9. Time-domain cross section of the acoustic impedances obtained via sequential Gaussian simulation with a vertical resolution of 1 msec. The cross section shown is the average of 30 independent realizations.

seismic amplitude data, a simple convolution operation is performed using the wavelet shown in Figure 4. This is accomplished by first calculating seismic reflectivities from the simulated acoustic impedances and then by convolving the latter with the seismic wavelet. Such an operation provides one with a set of synthetic seismic traces. The cross-correlation between the synthetic seismic traces and the measured seismic traces is then computed to establish a measure of agreement between the two sets of traces. Results of this correlation are shown in plan view in Figure 10 for 4 of the 30 independent stochastic simulations of acoustic impedance. A color scale for the correlation values is shown to the left of that figure. Quite clearly, the correlation with the seismic amplitude data is poor and laterally inconsistent except in the vicinity of the existing wells.

## GEOSTATISTICAL INVERSION OF ACOUSTIC IMPEDANCES

Figure 11 shows the acoustic impedance cross section obtained with geostatistical inversion by enforcing a global 1% misfit error between the measured and simulated poststack seismic amplitude data. A total of 10 iterations are used by the simulated annealing algorithm of geostatistical inversion to produce the desired 1% global seismic misfit. The result shown in Figure 11 is the average of 30 independent realizations (the 30 independent realizations required for their computation of approximately 36 hr of CPU time on an Octane SGI (500-MHz) computer). The agreement with the corresponding cross section of trace-based inverted acoustic impedances is not only remarkable, but it also provides slightly more

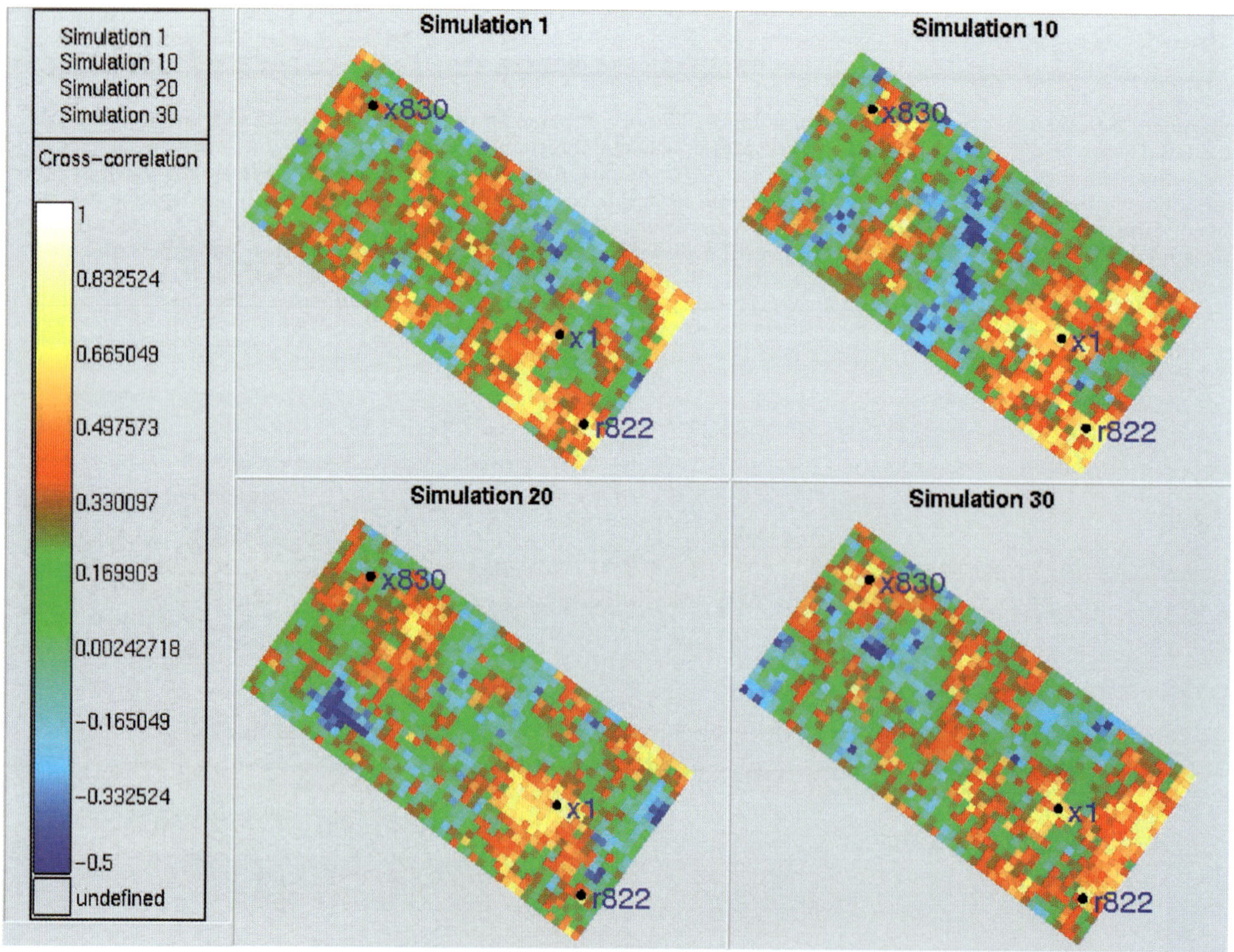


**Figure 10.** Plan views of the cross-correlation between the measured and numerically simulated seismic amplitude data obtained from acoustic impedances via sequential Gaussian simulation (see Figure 9). Each plan view depicts the seismic cross-correlation associated with a given independent realization of acoustic impedance. Cross-correlation values are color coded according to the legend shown in the left.

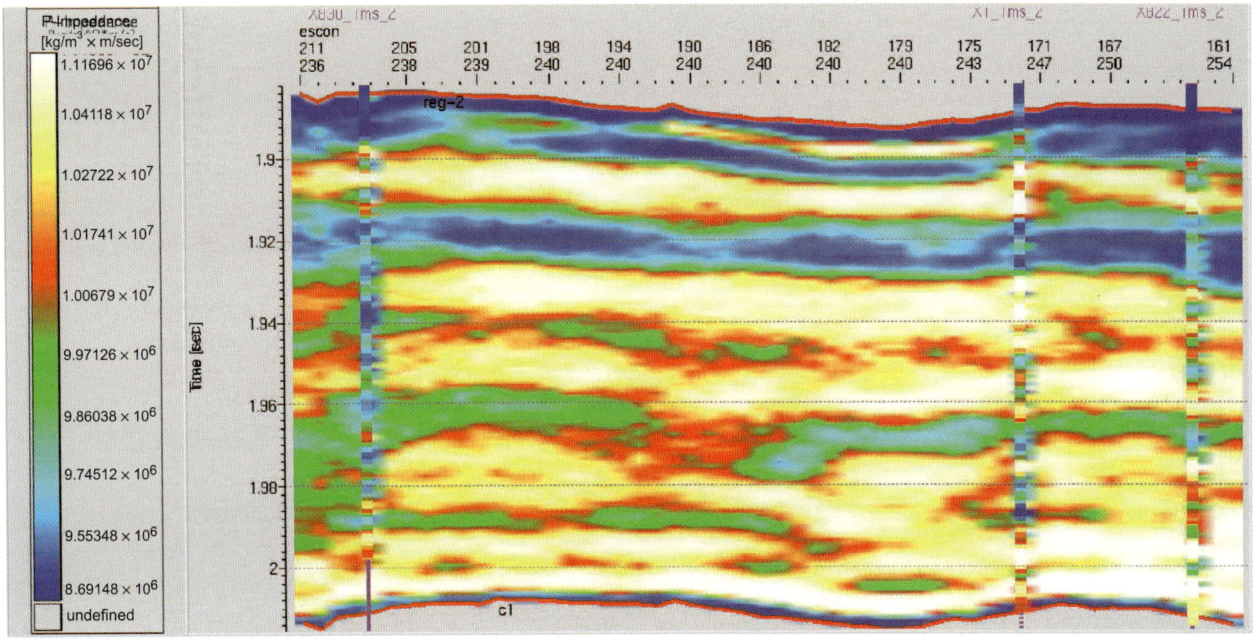


**Figure 11. Time-domain cross section of the acoustic impedances obtained via geostatistical inversion with a vertical resolution of 1 msec. The cross section shown is the average of 30 independent realizations.**

resolution in the vertical direction. The maps shown in Figure 12 are the result of a correlation analysis of the input seismic amplitude measurements with 4 of the 30 independent realizations. Correlation values are consistently high (higher than 0.9) except for a few locations (the lack of correlation at such locations remains consistent for all of the 30 simulations and is perhaps an indication of either amplitude-vs.-offset effects or other biasing effects originating during the acquisition and/or processing of the seismic amplitude data). The agreement with the measured seismic amplitude data is extremely good, especially when compared to the poor agreement associated with the sequential Gaussian simulation algorithm.

Figure 13 illustrates the performance of the simulated annealing algorithm as a function of iteration. The four panels in that figure are maps of the cross-correlation between the simulated and the measured seismic amplitude data at a given iteration. Cross-correlation maps are shown corresponding to iteration numbers 1, 3, 5, and 7, respectively, of the simulated annealing algorithm used to reach a minimum of the global seismic data misfit. After five iterations, the simulated annealing algorithm has practically converged to its final result. Prior to that iteration, the cross-correlation maps evidence significant discrepancies between the simulated and measured seismic amplitude data (the cross-correlation map for iteration number 0 corresponds to one of the panels shown in Figure 10).

## ASSESSMENT OF NONUNIQUENESS

Figures 14 and 15 are cross sections depicting the local standard deviation of acoustic impedance calculated from the 30 independent realizations provided by sequential Gaussian simulation and geostatistical inversion, respectively. For the same semivariograms, it is evident that the local variability associated with sequential Gaussian simulation is much more pronounced than that associated with geostatistical inversion. This important result comes as an immediate consequence of restricting the geostatistical simulations to honor the poststack 3-D seismic amplitude data. In so doing, a strong selection criterion is imposed, so that only a relatively small subset of acoustic impedances is accepted from the large pool of available random realizations.

## SENSITIVITY TO A CHANGE IN SEMIVARIOGRAM PARAMETERS

An extensive analysis was performed to assess the sensitivity of the simulations of acoustic impedances to a small change in semivariogram parameters. Results from this analysis are illustrated in Figures 16 and 17. The cross sections contained in

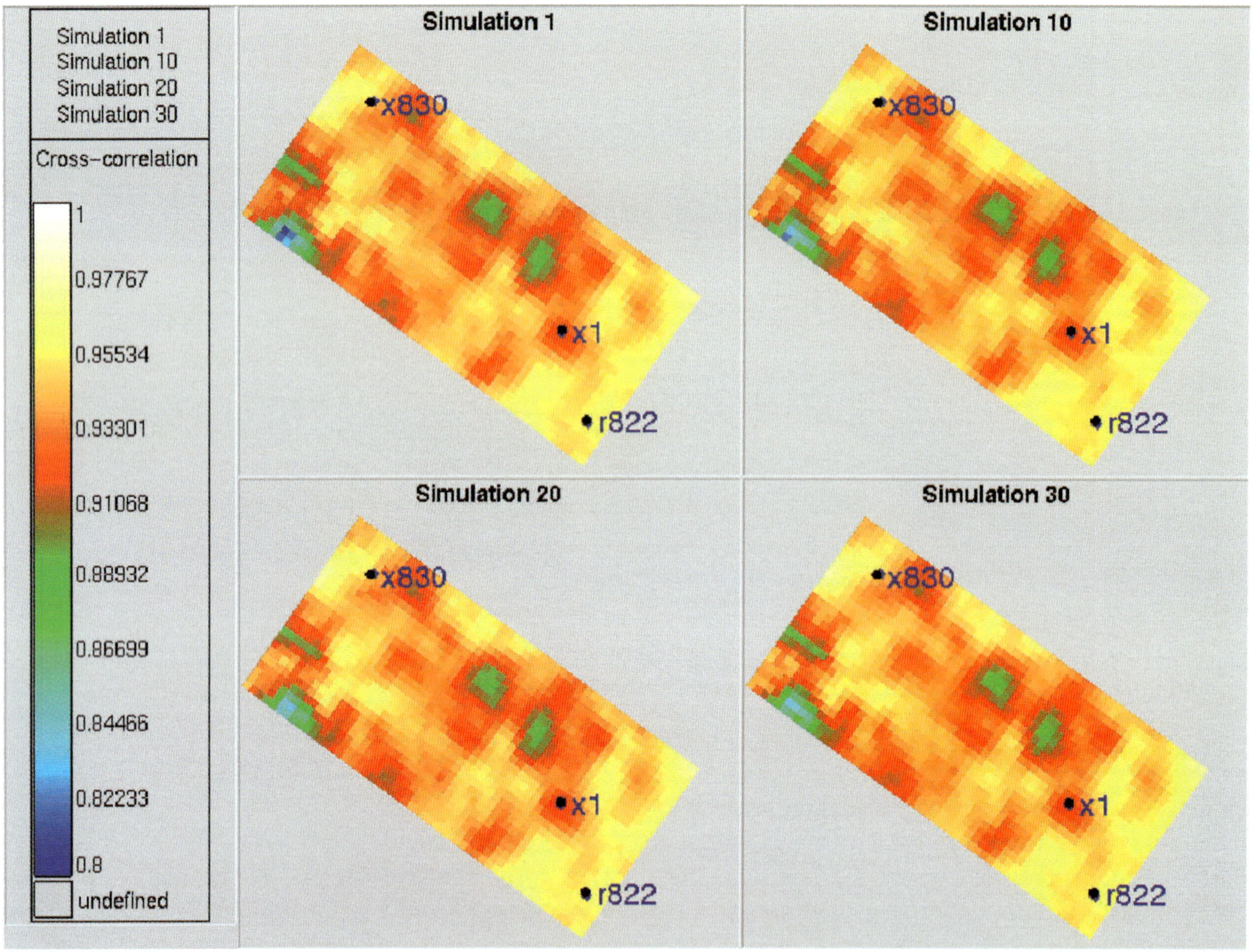


**Figure 12. Plan views of the cross-correlation between the measured and numerically simulated seismic amplitude data obtained from acoustic impedances via geostatistical inversion (see Figure 11). Each plan view depicts the seismic amplitude cross-correlation associated with a given independent realization of acoustic impedance. Cross-correlation values are color coded according to the legend shown in the left. Notice that the cross-correlations shown in Figure 10 are color coded in such a way as to reflect a wider range of values.**

these figures are obtained by decreasing the range of the semivariograms by 10%; all of the remaining semivariogram parameters remain constant during the simulations. Figure 16 shows the results obtained with sequential Gaussian simulation, whereas Figure 17 shows the results obtained with geostatistical inversion. Both cross sections represent the mean of 30 independent realizations. Appreciable variation is present in the realizations obtained with sequential Gaussian simulation but only a relatively small variation in the realizations obtained with geostatistical inversion compared to those described earlier (see also Figures 9 and 11, respectively). Although there seems to be a clear dependence of semivariogram properties on the realizations obtained with geostatistical inversion (as expected), a large number of numerical experiments indicate that constraining the realizations to honor the seismic amplitude data causes relatively low sensitivity to changes in semivariogram parameters. Conversely, the realizations obtained with sequential Gaussian simulation are highly sensitive to changes in semivariogram parameters.

## CROSS-VALIDATION

Several tests of cross-validation have been pursued in an effort to further explore the dependency of the realizations of acoustic impedance on changes in input data. Figure 18 shows a plot of an

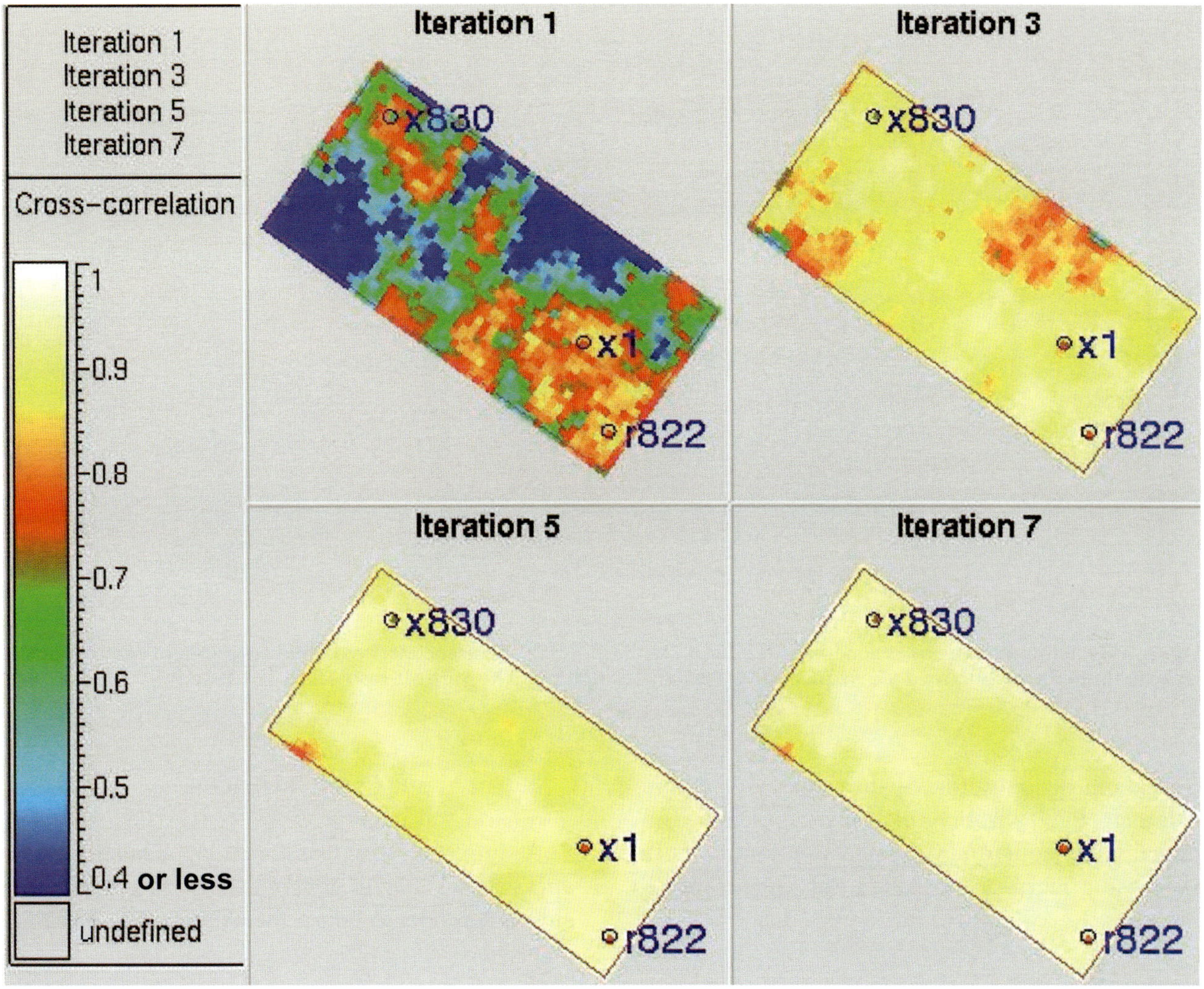


**Figure 13. Plan views of the cross-correlation between the measured and numerically simulated seismic amplitude data obtained from acoustic impedances via geostatistical inversion (see Figure 12). Each plan view depicts the seismic cross-correlation associated with a given iteration of the simulated annealing algorithm (panels show cross-correlation maps for one, three, five, and seven iterations). Cross-correlation values are color coded according to the legend shown in the left. The map for iteration 1 (upper left panel) describes cross-correlation values below 0.4 as dark-blue zones. Some of these blue zones exhibit negative cross-correlations. Notice the convergence of the simulated annealing algorithm practically after iteration 5.**

acoustic impedance log obtained for the central well in the study using sequential Gaussian simulation. The result is peculiar because actual data from the central well were not employed in the simulation process. Figure 18 also shows the actual 1-msec acoustic impedance log acquired along the central well. The log of acoustic impedance obtained with sequential Gaussian simulation agrees well with the average tendency of the actual acoustic impedance log, but it fails to replicate the remaining details. By contrast, Figure 19 shows the acoustic impedance log obtained with geostatistical inversion. This plot more closely replicates the details exhibited by the actual log and therefore gives credence to the realizations of acoustic impedance obtained elsewhere in the seismic cube with geostatistical inversion.

Figures 20 and 21 show the complete cross sections of acoustic impedance obtained with sequential Gaussian simulation and geostatistical inversion, respectively, without the use of actual data from the central well. A strong visual agreement exists between the realizations obtained with geostatistical inversion (Figure 21) and those illustrated in Figure 11 (the central well data is used to construct Figure 11). Conversely, the lack of lateral

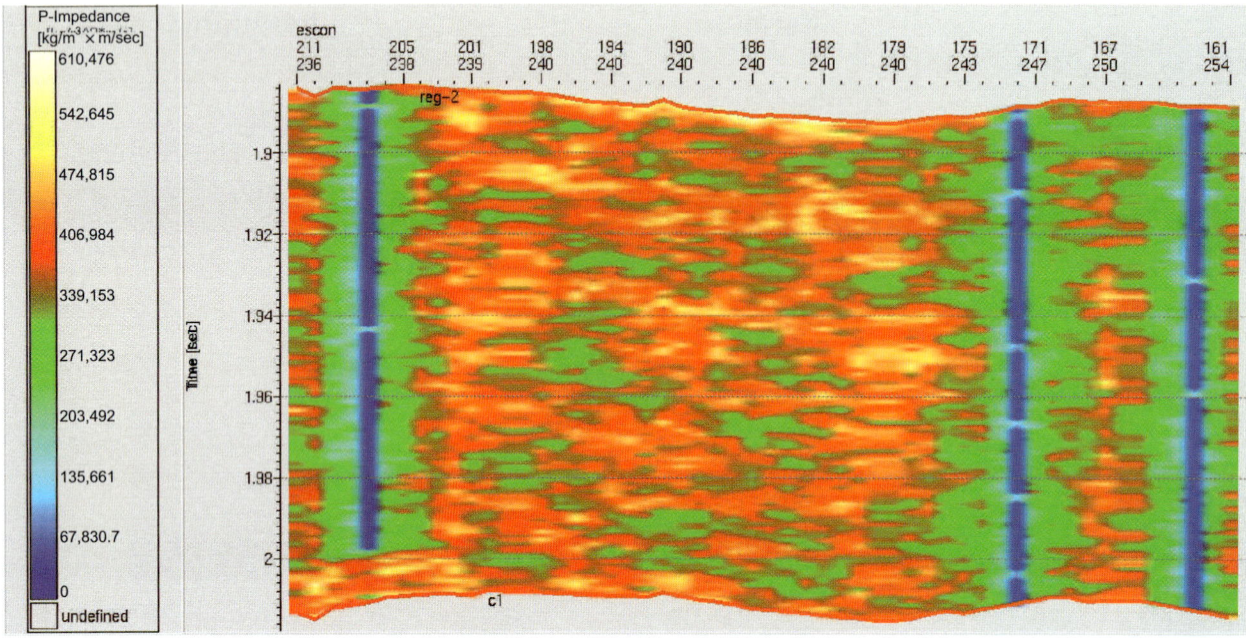


**Figure 14.** **Time-domain cross section of the local standard deviation associated with 30 independent realizations of acoustic impedance obtained via sequential Gaussian simulation. The vertical resolution is 1 msec.**

control otherwise exerted by the central well severely impairs the realizations of acoustic impedance obtained with sequential Gaussian simulation, thereby causing Figure 21 to have little resemblance to Figure 9.

## CONCLUSION

A simple exercise using well-log acoustic impedances shows that, in general, standard geostatistical simulation fails to honor the existing 3-D poststack

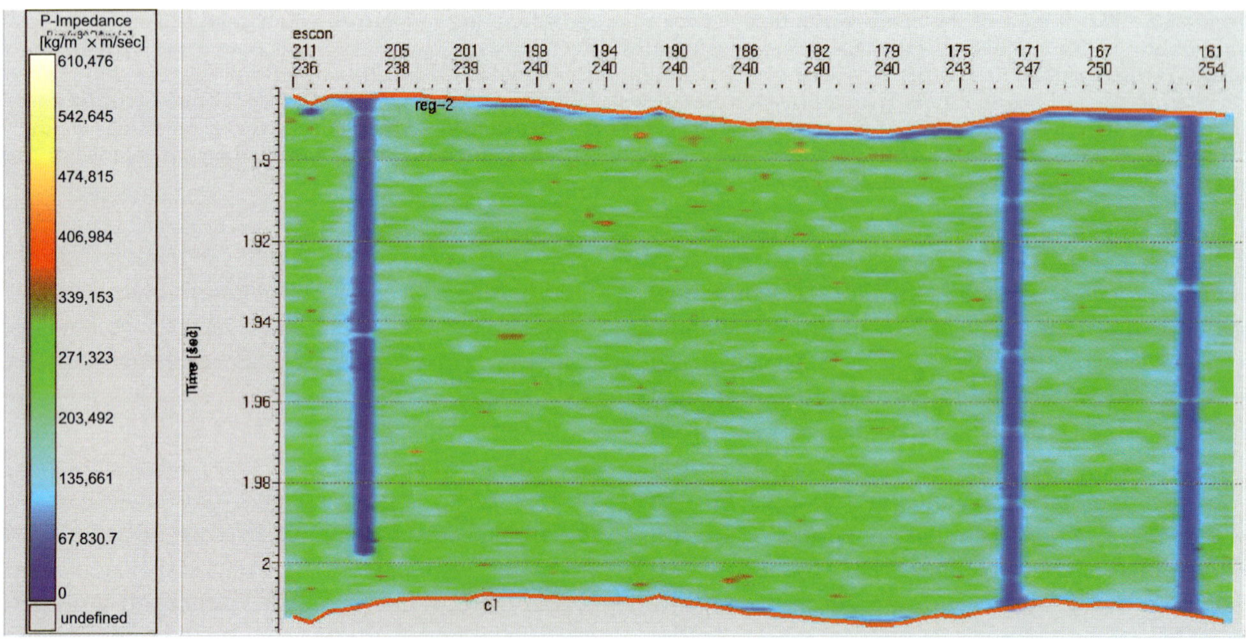


**Figure 15.** **Time-domain cross section of the local standard deviation associated with 30 independent realizations of acoustic impedance obtained via geostatistical inversion. The vertical resolution is 1 msec.**

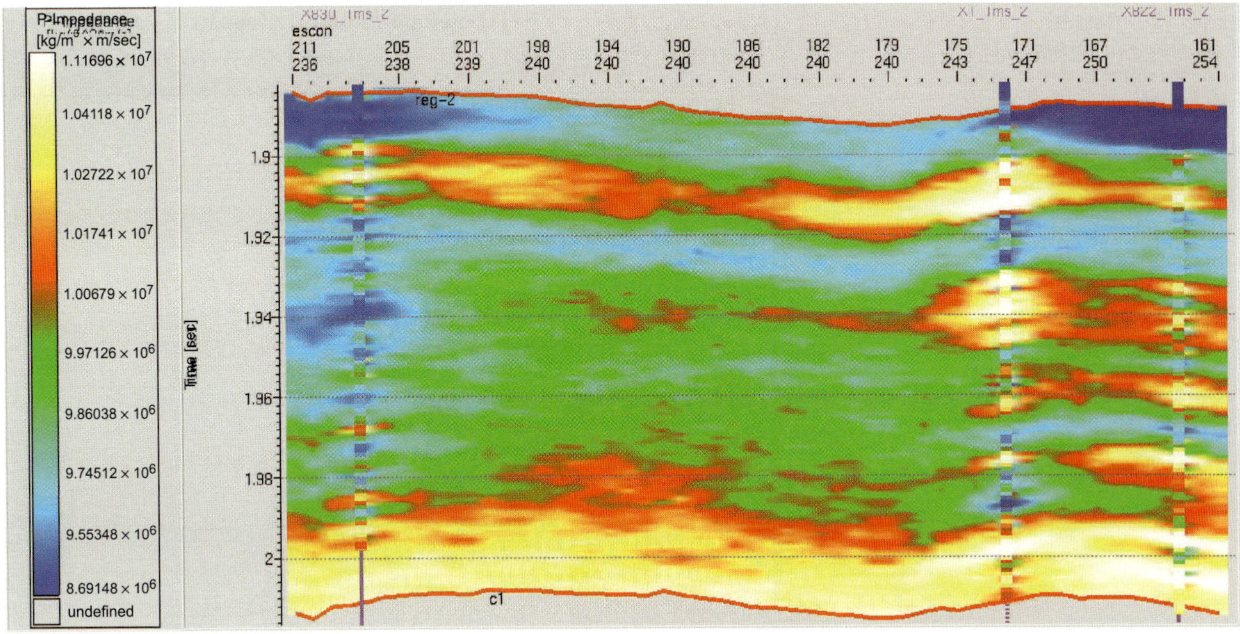


**Figure 16.** Time-domain cross section of acoustic impedance obtained with sequential Gaussian simulation using a vertical resolution of 1 msec. The cross section shown is the average of 30 independent realizations based on a horizontal semivariogram range of 225 m (738 ft). Comparisons can be made to Figure 9.

seismic amplitude data. As a result, conventional geostatistical simulation driven exclusively by well logs and semivariograms may be unable to capture fundamental properties of the underlying spatial distribution of acoustic impedance. Moreover, conventional geostatistical simulation remains extremely sensitive to a perturbation in the parameters of the modeled semivariogram; independent realizations

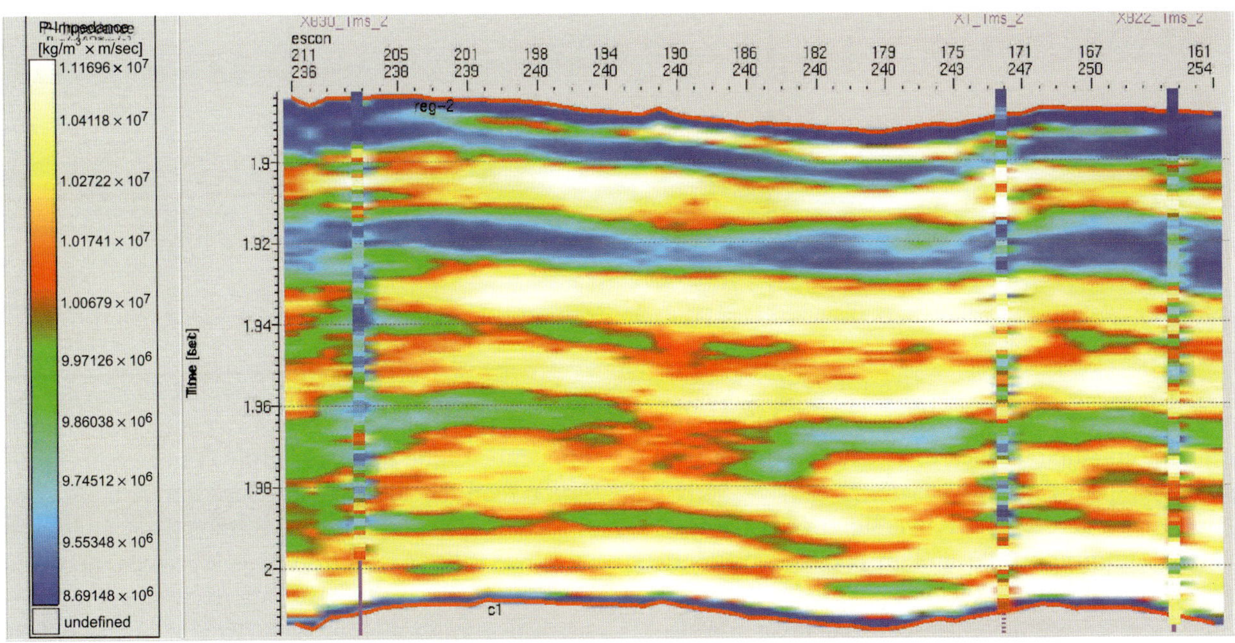


**Figure 17.** Time-domain cross section of acoustic impedances obtained with geostatistical inversion using a vertical resolution of 1 msec. The cross section is the average of 30 independent realizations based on a horizontal semivariogram range of 225 m (738 ft). Comparisons can be made to Figure 11.

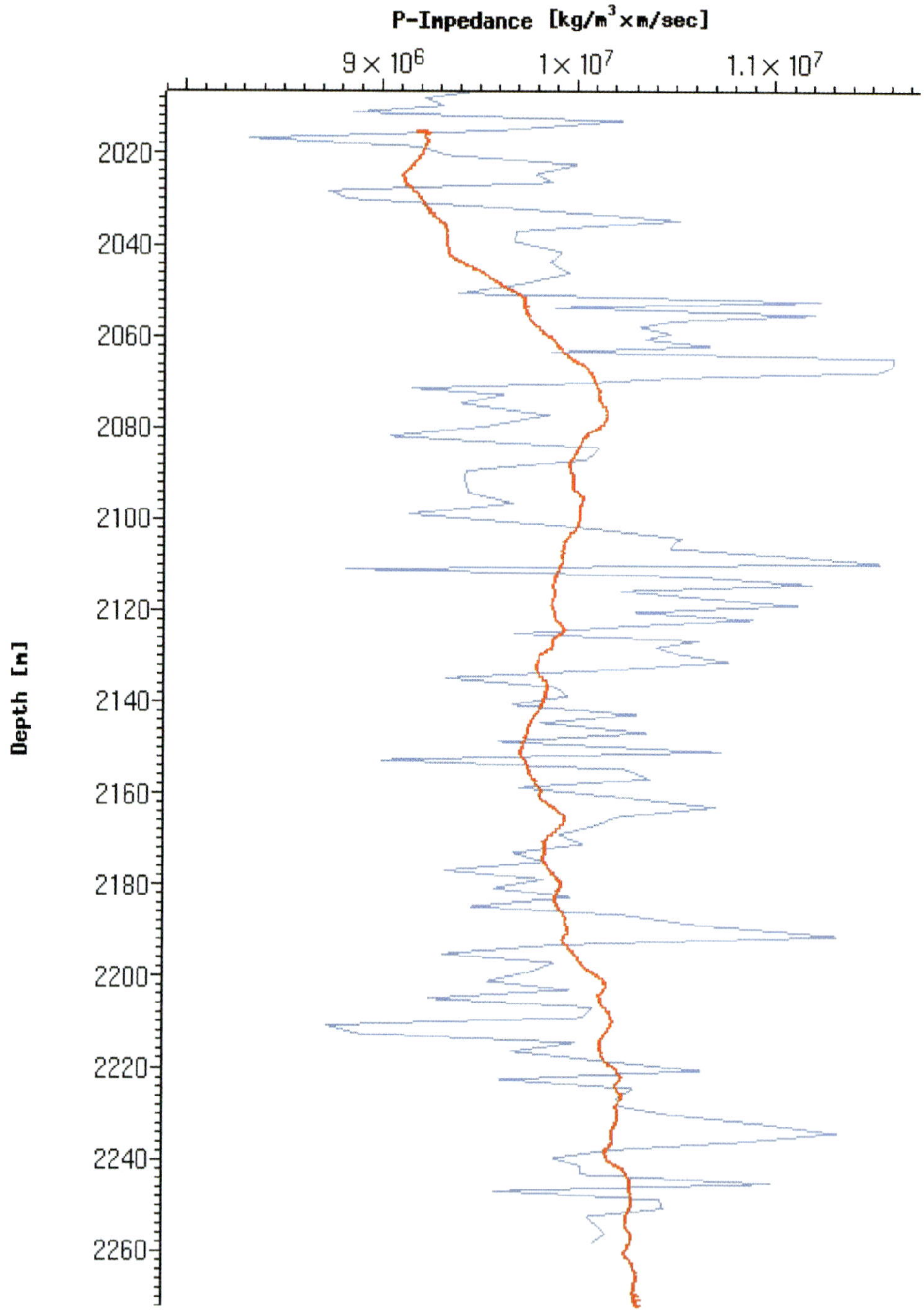


Figure 18. Results of a cross-validation exercise, showing a realization of acoustic impedance at the central well obtained with sequential Gaussian simulation (red line), along with the measured acoustic impedance log (blue line). The realization of acoustic impedance is produced without using data from the central well as input.

may also exhibit a great deal of variability, and this hinders their power to predict petrophysical parameters away from wells.

However, geostatistical inversion does honor the seismic amplitude data and provides realizations consistent with both existing well logs and the assumed semivariograms. Geostatistical inversion also embodies a quantitative means to integrate the high vertical resolution of well logs with the dense spatial sampling of 3-D poststack seismic amplitude data. The end result of geostatistical inversion is an acoustic impedance cube having slightly better vertical resolution than the cube of trace-based inverted acoustic impedance, especially in the immediate vicinity of existing wells. Away from existing wells, geostatistical inversion yields realizations with an effective vertical resolution equivalent to that of the seismic amplitude data. In addition, the standard deviation of the realizations obtained with geostatistical inversion is consistently smaller than the standard deviation of

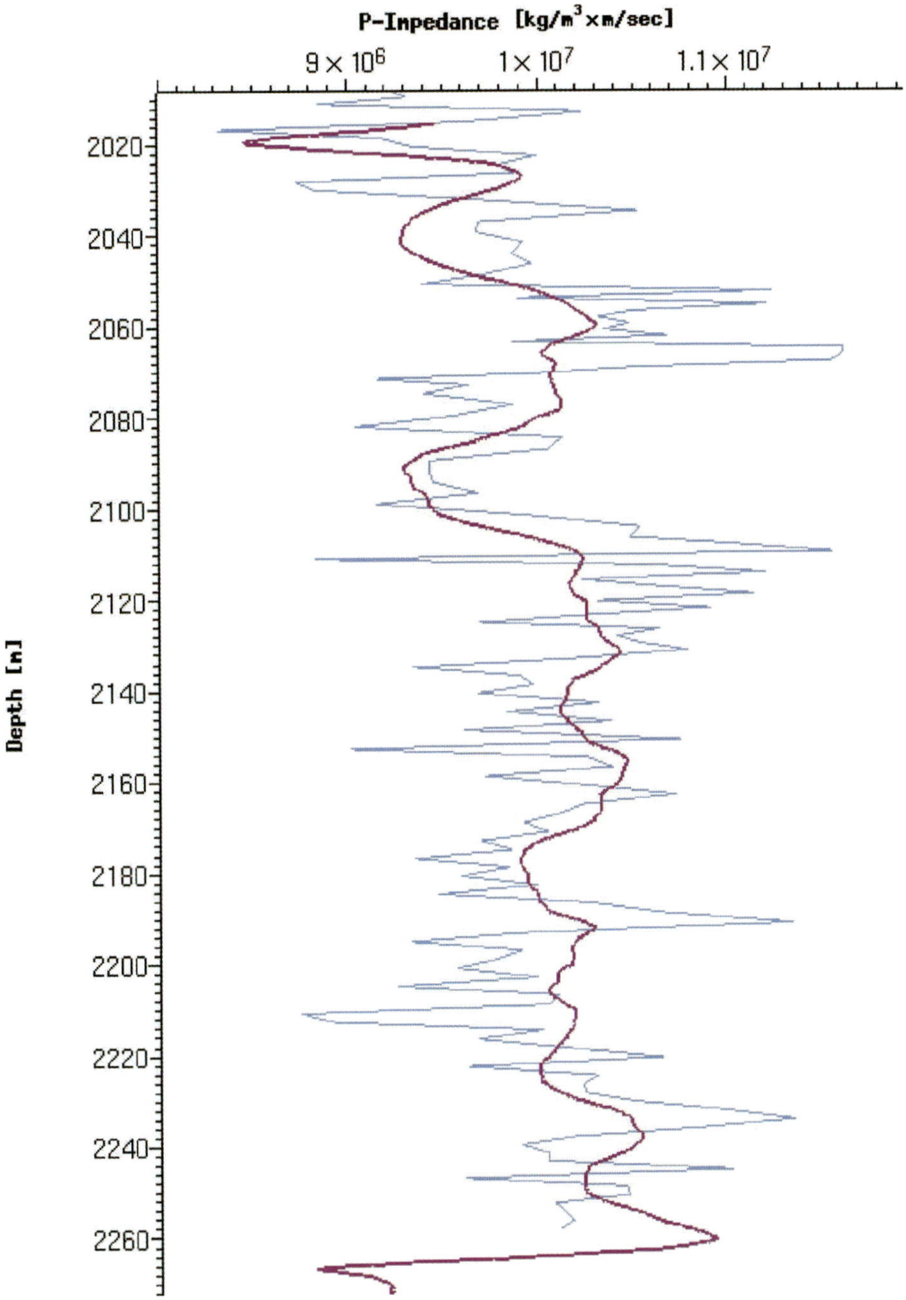


**Figure 19.** Results of a cross-validation exercise, showing a realization of acoustic impedance at the central well obtained with geostatistical inversion (purple line), along with the measured acoustic impedance log (blue line). The realization of acoustic impedance is produced without using data from the central well as input. Compare this result with that shown in Figure 18.

realizations obtained with conventional geostatistical simulation. This favorable result comes as a consequence of constraining the space of possible realizations of acoustic impedance to satisfy both the well logs and the seismic amplitude data. Because of the same reason, realizations obtained with geostatistical inversion are less sensitive to the removal of a well than conventional geostatistical simulation. One main drawback of geostatistical inversion is the commonly exorbitant CPU time required to process a cube of 3-D seismic data, especially when realizations are required at time-sampling intervals finer than that of the seismic amplitude data. Noisy depth-to-time conversions of well logs required by geostatistical inversion may also cause deleterious effects on the ensuing realizations of acoustic impedance. Finally, biases on the estimated seismic wavelet could be responsible for biases in the realizations of acoustic impedance obtained with geostatistical inversions. Undoubtedly, the ability of geostatistical inversion to estimate petrophysical parameters (e.g., effective porosity) becomes practical only when acoustic

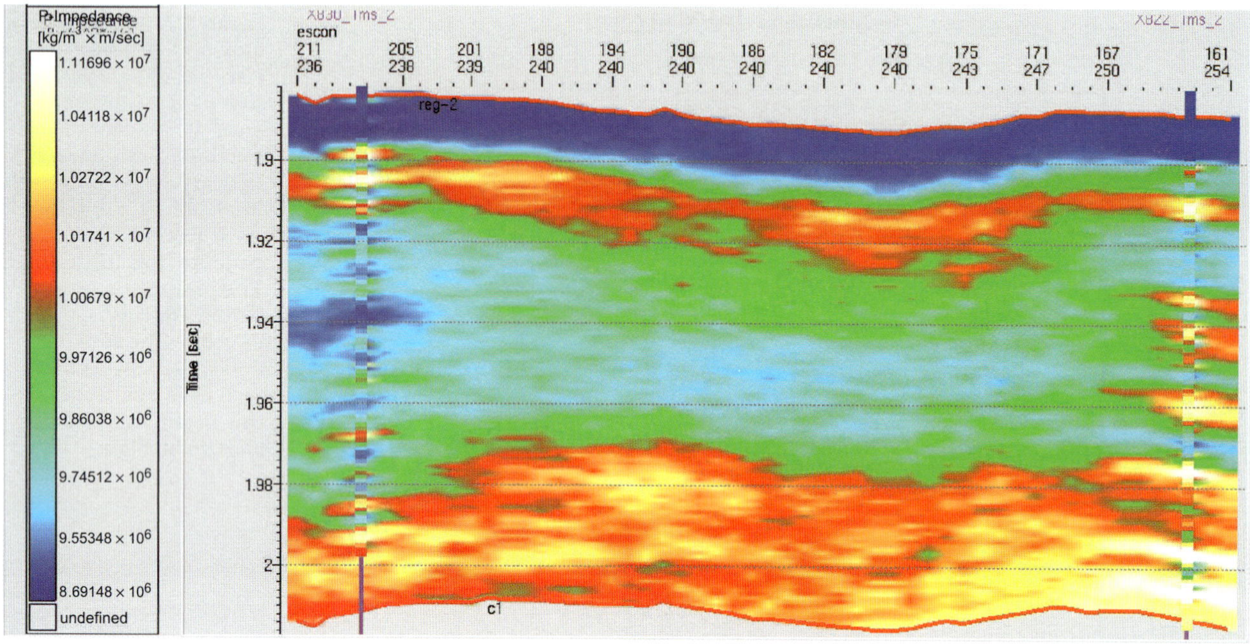


**Figure 20.** Results of a cross-validation exercise shown as a time-domain cross section of the acoustic impedance obtained with sequential Gaussian simulation based on a vertical resolution of 1 msec. Acoustic impedances measured along the central well are not used as input data. The cross section is the average of 30 independent realizations. Comparisons can be made to Figure 9.

impedance bears a clear statistical correlation with that parameter. Such a correlation may also exist through a lithology indicator. However, accurate estimations of petrophysical parameters are not possible with geostatistical inversion if acoustic impedance bears no statistical correlation with petrophysical parameters.

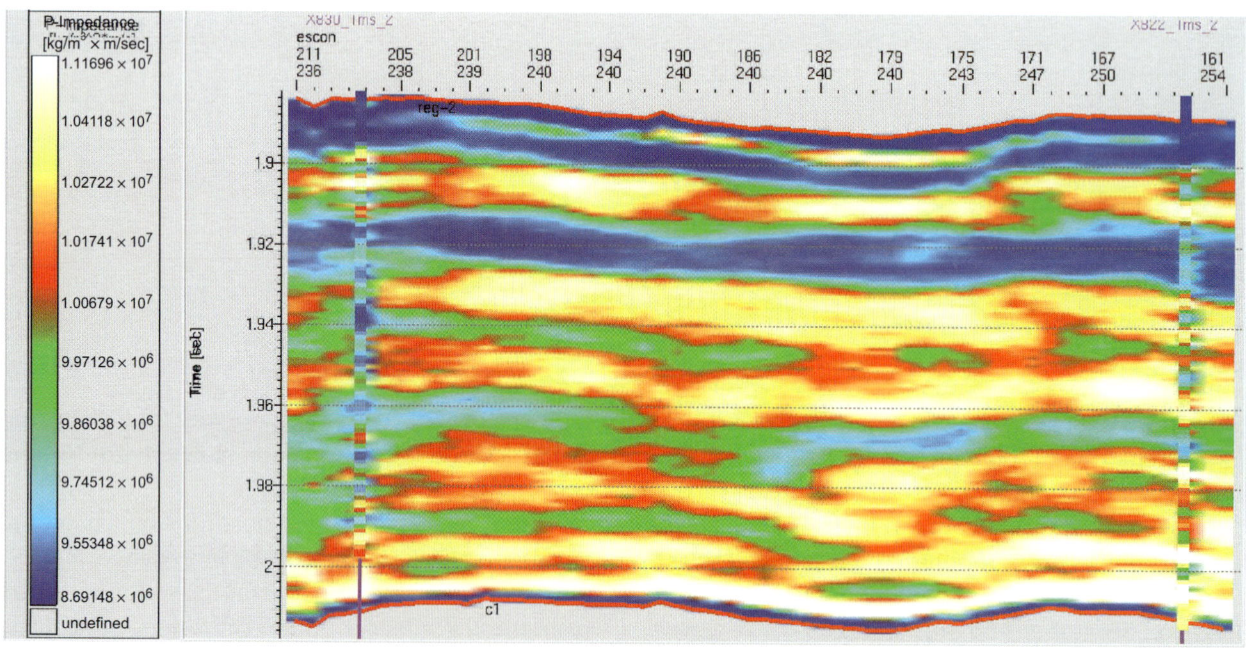


**Figure 21.** Results of a cross-validation exercise shown as a time-domain cross section of acoustic impedance obtained with geostatistical inversion based on a vertical resolution of 1 msec. Acoustic impedances measured along the central well are not used as input data. The cross section is the average of 30 independent realizations. Comparisons can be made to Figure 11.

## ACKNOWLEDGMENTS

The authors express their deepest appreciation to Repsol-YPF for providing the comprehensive data set used to construct the field examples described in this chapter. A note of special gratitude goes to Baker Atlas, Halliburton, and Schlumberger, sponsors of the Center of Excellence in Formation Evaluation at the University of Texas at Austin, for their partial funding of this work. Thanks are extended to Fugro-Jason for a generous donation of its complete line of software to the University of Texas at Austin and used to perform the work described in this chapter. Sagar Ronghe and two anonymous reviewers provided valuable constructive comments that improved the first version of the chapter.

## REFERENCES CITED

Chilès, J.-P., and P. Delfiner, 1999, Geostatistics, modeling spatial uncertainty: New York, John Wiley and Sons, 695 p.

Debeye, H. W. J., and P. Van Riel, 1990, Lp-norm deconvolution: Geophysical Prospecting, v. 38, p. 381–403.

Debeye, H. W. J., E. Sabbah, and P. M. van der Made, 1996, Stochastic inversion: Expanded Abstracts, 66th Annual International Meeting of the Society of Exploration Geophysicists, Denver, Colorado, November 10–15, p. 1212–1215.

Grijalba-Cuenca, A., C. Torres-Verdín, and H. W. J. Debeye, 2000, Geostatistical inversion of 3-D seismic data to extrapolate petrophysical variables laterally away from the well (abs.): Society of Petroleum Engineers Annual International Meeting, Dallas, Texas (October 1–4), SPE Paper 63283, 17 p.

Haas, A., and O. Dubrule, 1994, Geostatistical inversion—A sequential method for stochastic reservoir modeling constrained by seismic data: First Break, v. 12, p. 561–569.

Pendrel, J. V., and P. Van Riel, 1997, Estimating porosity from 3D seismic inversion and 3D geostatistics: Expanded Abstracts, 67th Annual International Meeting of the Society of Exploration Geophysicists, Dallas, Texas, November 9–13, p. 834–837.

Torres-Verdín, C., M. Victoria, G. Merletti, and J. Pendrel, 1999, Trace-based and geostatistical inversion of 3-D seismic data for thin-sand delineation, an application to San Jorge Basin, Argentina: The Leading Edge, v. 18, p. 1070–1076.

van Riel P., P. Mesdag, H. Debeye, and M. Sams, 2006, Full integration of seismic data into geostatistical reservoir modeling, *in* T. C. Coburn, J. M. Yarus, and R. L. Chambers, eds., Stochastic modeling and geostatistics: Principles, methods, and case studies, volume II: AAPG Computer Applications in Geology 5, p. 207–218.

*Chapter 16*

# Full Integration of Seismic Data into Geostatistical Reservoir Modeling

**P. van Riel**
**P. Mesdag**
**H. Debeye**
*Fugro-Jason Netherlands BV*
*Leid Schendam, Netherlands*

**M. Sams**
*Fugro-Jason Malaysia Sdn. Bhd.*
*Kuala Lumpur, Malaysia*

## ABSTRACT

Seismic reflection amplitude data are increasingly used in reservoir modeling to provide information on changes in earth properties away from well locations. In geostatistical reservoir modeling, the most common application is to use seismic data as background data in some form of comodeling. Seismic data image reflectors and not earth layer properties. Therefore, prior to use in comodeling, seismic data must first be transformed into an earth layer property. Typically, the transform is to acoustic impedance using an appropriate seismic inversion method.

Seismic inversion methods generate results that are generally band limited in nature, resulting in limits to vertical resolution. The vertical resolution achieved can be an order of magnitude below the vertical model resolution required from geostatistical reservoir modeling, which is in the order of well-log resolution. Hence, in using seismic data, geostatistical modelers encounter a problem of downscaling, not the more commonly encountered upscaling problem. This difference in scale introduces scatter between the primary data with well-log order resolution and the secondary seismically derived rock property data used in the comodeling. As a result, to preserve vertical heterogeneity, only limited use of the secondary data can be made in comodeling procedures. This results in models that only partially fit the seismic data, i.e., only limited use is made of the seismic information. If the secondary data are more strongly imposed, the fit to the seismic data improves, but the required vertical heterogeneity is not


DOI:10.1306/1063817CA53236

preserved. The inability to overcome this difference in scale issue, therefore, limits the value of the application of comodeling methods to integrate seismic data into reservoir models.

One class of geostatistical methods that overcomes this limitation relies on iterative geostatistical modeling. In these methods, referred to as geostatistical seismic inversion, the iterative modeling process is conditioned such that the final models generated closely match the seismic data while maintaining the required vertical heterogeneity. The application of these methods is computationally expensive relative to comodeling methods but is now practical for large models on today's desktop hardware. Relative to comodeling, geostatistical seismic inversion methods make full use of the information carried in the seismic data, resulting in a significant reduction in model uncertainty away from well control.

## INTRODUCTION

The use of geostatistical methods in reservoir characterization has rapidly evolved throughout the last decade. A wide range of methods is now available to handle different applications in hydrocarbon exploration, development, and production (Yarus and Chambers, 1994; Dubrule et al., 2001). These references illustrate that there are at least four advantages to using a geostatistical approach to reservoir modeling (where a model is a three-dimensional [3-D] description of the reservoir in terms of quantitative parameters). First, the modeling process is commonly carried out in the context of a 3-D geologic framework. Such modeling may incorporate a variety of geologic information, such as depositional setting and trends in reservoir properties. Using geostatistical modeling, much of the available geologic information and data can be integrated and honored. Second, the 3-D models honor well control. Third, the models can be constructed at the level of detail most appropriate for the reservoir. Fourth, structural and stratigraphic heterogeneity is preserved when using appropriate geostatistical-simulation algorithms.

With geostatistical-simulation multiple, equiprobable realizations may be generated. This provides input for sensitivity and uncertainty analysis of the reservoir model itself and of reservoir performance using these model realizations in reservoir-simulation studies.

In practice, these advantages offer ample justification to apply geostatistical modeling methods in most reservoir-characterization studies.

A technical advance that is occurring concurrently with the evolution of geostatistical modeling is the integration of seismic data into the reservoir-characterization process. Three-dimensional seismic data are now widely available, and seismic data quality is such that seismic amplitudes are now routinely used to help determine reservoir lithology and fluid distribution. Modern seismic data uniquely support reservoir characterization by providing reliable spatial control to complement spatially sparse well data.

Because of the parallel, synergistic development of these two technologies, one might expect that seismic data are regularly employed in 3-D geostatistical reservoir modeling. Unfortunately, in the authors' experience, this is generally not the case. Three key reasons for this situation exist. Seismic data are registered in seismic time, and most practical seismic wave propagation modeling methods operate in time also. A sufficiently accurate velocity model is not always available to properly transform between the time and depth domain. In such a case, geostatistical modeling in depth cannot be linked to the seismic data specified in seismic time. In addition, the description of the geologic framework must be sufficiently accurate; that is, horizons used in constructing the geostatistical model must be geologically meaningful and fully consistent with the seismic data. Such consistency is sometimes lacking. Finally, the seismic well ties must be accurate. Well tying involves careful calibration of the time-depth relationships at the wells and estimation of the seismic wavelet. The latter further requires the availability of reliable log suites that are fully calibrated through petrophysical and rock physics modeling studies. Such high-quality log data are not always available.

The preparation of the pieces of the puzzle before the start of the integrated geostatistical modeling phase itself already demands a truly multidisciplinary effort, with contributions from at least petrophysics, rock physics, geology, and geophysics staff. In most cases, contributions from the reservoir engineers are also necessary and desirable. Meeting the demands of the full project requires proper planning and significant time and funding resources. When such a project is considered in isolation, the required budget may be large, and it will only be approved if management is keenly aware of the added value offered by an integrated approach. However, most of the work components in an integrated geostatistical modeling study will be undertaken anyway at some point in the life of a reservoir. Hence, if the work packages are properly coordinated in a work plan, then the incremental cost of integrating seismic data into geostatistical modeling is relatively minor. This cost is readily justified by the increased precision of the estimate of the spatial distribution of reservoir properties away from the wells and the ability to quantify uncertainty.

Assuming that an integrated study is commissioned, the next step is to select an appropriate geostatistical modeling methodology. The most commonly applied geostatistical modeling methods that use seismic information are cokriging and cosimulation. Initially, mainly because of computational limitations, geostatistical modeling was carried out at the map level. In such applications, maps of seismic amplitude attributes that correlate to the properties being modeled are used as a background trend (secondary data) (Chambers et al., 1994). Increased computational performance allows realistic 3-D geostatistical modeling. In 3-D modeling, seismic data cannot be used directly as secondary data because seismic data image reflectors and not earth parameters. Therefore, a transform to the layer rock property domain is required. In the simplest case, where only poststack seismic data are available, the seismic data are inverted (they transform from seismic data back to rock properties) to acoustic impedance with some appropriate seismic inversion algorithm (e.g., Pendrel and van Riel, 1997; Buxton et al., 2000). Acoustic impedance is the product of density and acoustic velocity and, as such, is a rock property that is sensitive to various reservoir parameters, such as lithology, porosity, and fluid saturation. Many examples have been described where a seismically derived impedance volume is used as secondary parameter in 3-D geostatistical modeling (e.g., Doyen and Guidish, 1992; Bashore et al., 1994).

Three-dimensional geostatistical comodeling with seismic-derived impedance (or other seismic-derived rock properties) represents positive progress toward integrating seismic information into geostatistical reservoir modeling. It is, however, only a partial step. This is readily verified by generating synthetic seismic data from the geostatistical comodeling results and comparing these to the measured seismic data. If full use of the seismic data is made, the synthetic data should match the measured seismic data to within the noise level on the seismic data (where noise is taken to include seismic modeling inaccuracies). As will be illustrated below, comodeling methods do not pass this test.

Using comodeling to incorporate seismically derived information has a further, poorly recognized drawback caused by the band-limited nature of seismic data. In the vertical direction, this typically translates into vertical resolution in the order of 20–60 m (66–197 ft), whereas it is commonly desirable to generate reservoir models at a scale of meters. This introduces a difference in scale or support problem where a model is desired at a higher resolution than offered by the secondary data. Interestingly, this difference in scale problem is the reverse of that found in most other applications as, for example, in mining and permeability modeling (Isaaks and Srivastava, 1989; Anguy et al., 1994; Olea, 1994), where the goal is to produce models with lower resolution than that of the measurements. In other words, in geostatistical modeling using seismic data, the requirement relative to the seismic data resolution is to downscale instead of to upscale. This also implies that methods used to address the upscaling issue do not necessarily apply when seismic data are to be used in geostatistical modeling.

One way in which the seismic difference in scale problem could be overcome is to increase the resolution of the acoustic impedance volume used in comodeling. Indeed, today's more powerful seismic inversion methods can successfully reconstruct broadband rock parameter models by leveraging geologic models and/or seismic velocity information. However, recovery of high-frequency information is generally limited (Buxton et al., 2000). As a result, imaging of thin layers, although strongly improved over seismic resolution, remains limited, typically to the 10–30-m (33–99-ft) range. This is still insufficient for most geostatistical reservoir modeling applications. To the authors' knowledge, no geostatistical comodeling methods address the seismic difference in scale issue.

Geostatistical methods that do fully integrate seismic information use iterative geostatistical simulation to match the synthesized seismic response matches to the measured seismic data (Bortoli et al., 1993; Haas and Dubrule, 1994; Torres-Verdin et al., 1999). These methods thus provide an alternative solution to the seismic inverse problem (i.e., from measured data to rock properties) and are hence forth referred to as geostatistical seismic inversion methods. As with conventional geostatistical-simulation multiple, equiprobable models should be generated. Full use of the seismic data results in a reduction of model uncertainty relative to methods that only partially use the seismic data.

Geostatistical seismic inversion algorithms are computationally much more demanding and more difficult to implement than their nongeostatistical equivalents. Until recently, practical application was limited to single-stack seismic data (where such data may have been optimized to exploit seismic amplitude-vs.-offset behavior through range-limited and/or weighted stacking). Conventional seismic inversion has progressed to use prestack seismic data through the application of increasingly sophisticated seismic modeling algorithms (Xia et al., 1998; Mallick et al., 2000; Dubucq et al., 2001). The first experimental algorithms involving geostatistical prestack seismic inversion were reported in 2002 (Torres-Verdin et al., 2002), and more recently by Contreras et al., 2005.

An important additional advantage unique to geostatistical seismic inversion is that it properly solves for the difference in scale between the high-resolution model and the band-limited lower resolution seismic data. In the following, the outline of an algorithm for geostatistical seismic inversion is first provided. This is followed by examples illustrating the improved integration of seismic information achieved with geostatistical seismic inversion relative to the integration of seismic information through cosimulation as measured by the criteria of data fit, reduction in uncertainty, and ability to maintain vertical heterogeneity.

## GEOSTATISTICAL SEISMIC INVERSION METHODOLOGY

The objective of integrated geostatistical reservoir modeling is to produce models that can have a vertical scale in the range of meters or down to well-log vertical resolution, honor well control, honor stratigraphy, have vertical and lateral heterogeneity consistent with well and/or other geologic control, and honor the seismic data. The currently available methods that meet these objectives are iterative in nature and generally consist of at least the following steps:

1) constructing a stratigraphic framework from interpreted seismic horizons. This stratigraphic framework defines the microlayering in each layer defined within the framework, and the layers jointly define a 3-D geologic model with faults, unconformities, and other geologic features.
2) performing a geostatistical analysis of the available data to determine the impedance distribution and semivariograms in each geologic layer in the stratigraphic framework. Here, well log information, geologic knowledge, and collairy data will be used to determine the input variables to the geostatistical inversion.
3) performing iterative geostatistical modeling in the stratigraphic framework conditioned to the seismic data, as illustrated in Figure 1. In this step, starting from an initial simulation result, all points in the volume are revisited several times following a random path through the 3-D model. At each point, the impedance is perturbed by drawing a new impedance value from a distribution calculated by solving a locally determined cokriging system using a neighborhood of points around the point of interest. Each perturbed impedance value is then tested in terms of an improvement of seismic data fit and may be accepted or rejected based on a simulated annealing criterion, or it may simply be accepted only if the seismic data fit improves. The procedure is terminated when the seismic data fit is at a level corresponding to the noise in the data (where noise may include seismic modeling errors).

Repeating the procedure several times leads to multiple, equiprobable realizations.

Application of this scheme is particularly attractive in the case of the delineation and characterization of reservoirs comprising seismically thin layers (Torres-Verdin et al., 1999). Several variations of this scheme have been employed. For example, indicator simulation may be directly incorporated into the scheme to enhance the resolving power for separate lithologies (Sams et al., 1999; Shanor et al., 2002).

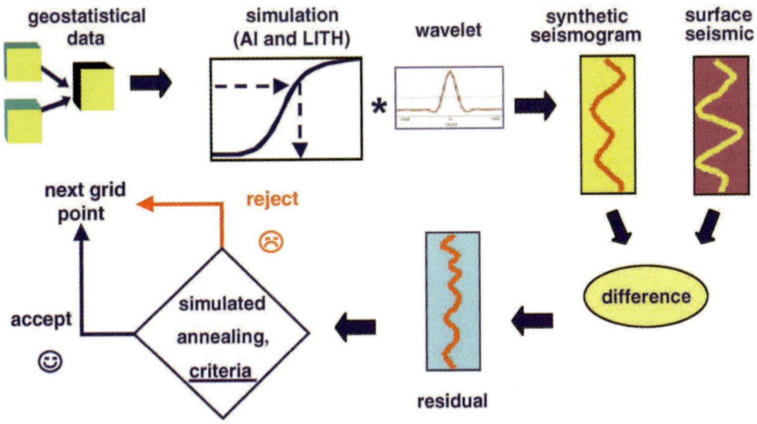


Figure 1. Geostatistical seismic inversion process summary.

Another variation is one whereby complete traces are simulated instead of following a point-by-point procedure (Bortoli et al., 1993; Haas and Dubrule, 1994).

In the above references, the seismic modeling component relies on the simple convolution model of seismic wave propagation to link impedance to seismic reflection data. To improve the accuracy of the results, more sophisticated seismic modeling methods can be applied involving prestack and time-lapse seismic data (Barens et al., 2002; Torres-Verdin et al., 2002).

## THE IMPACT OF DIFFERENCE IN SCALE

To understand the different issues when integrating seismic data into geostatistical modeling, it suffices to study the simplest problem, that of acoustic impedance modeling, where the link between the seismic data and acoustic impedance is provided by the convolution model. In this setup, geostatistical cosimulation of acoustic impedance can be performed using seismic-derived, band-limited acoustic impedance as background. This gives a one-to-one relationship of the primary to secondary properties. The only difference between the two is that the secondary data are band limited relative to the primary data. This permits a straightforward comparison of cosimulation vs. geostatistical seismic inversion measured against the criteria of data match, uncertainty, and handling of difference in scale.

In geostatistical comodeling, well-log data are used as the primary, hard data. Given the typical, desired vertical modeling resolution in the meter range, limited upscaling may be performed on the well-log data. For seismic modeling, this can be done using the Backus averaging (Castagna and Backus, 1993). Irrespective of this well-log upscaling, the desired vertical resolution of the model generally exceeds that of the seismic data by an order of magnitude. Figure 2 illustrates a typical example of this difference in scale. In Figure 2, an acoustic impedance well log is shown (black) and is overlain by the same well log high-cut filtered at 60 Hz. This filtered log represents the optimal result that could be obtained from conventional acoustic impedance inversion at the well location for a typical high-cut frequency. This example nicely shows the importance of modeling at or close to the level of well-log resolution. In the well log, the well reservoir zones are characterized by low acoustic impedance. The yellow line in Figure 2b represents a possible cutoff for pay. Clearly, if this cutoff was applied to the acoustic impedance results derived through conventional seismic inversion, no pay would be detected. Even if the cutoff value is higher, pay would still be underestimated (systematic overestimation is also possible). This example illustrates the need to overcome the issue of difference in scale in using seismic data for reservoir characterization.

As a next step, consider using colocated cosimulation (Chiles and Delfiner, 1999) for the simple problem, where both the primary and secondary variables are acoustic impedances. To perform colocated cosimulation requires only the correlation coefficient between the primary and the secondary data sets. To calculate the correlation coefficient between the original and filtered impedance, a crossplot is made, as illustrated in Figure 3. Ideally, because impedance is being compared to impedance, a perfect correlation might be anticipated. This means

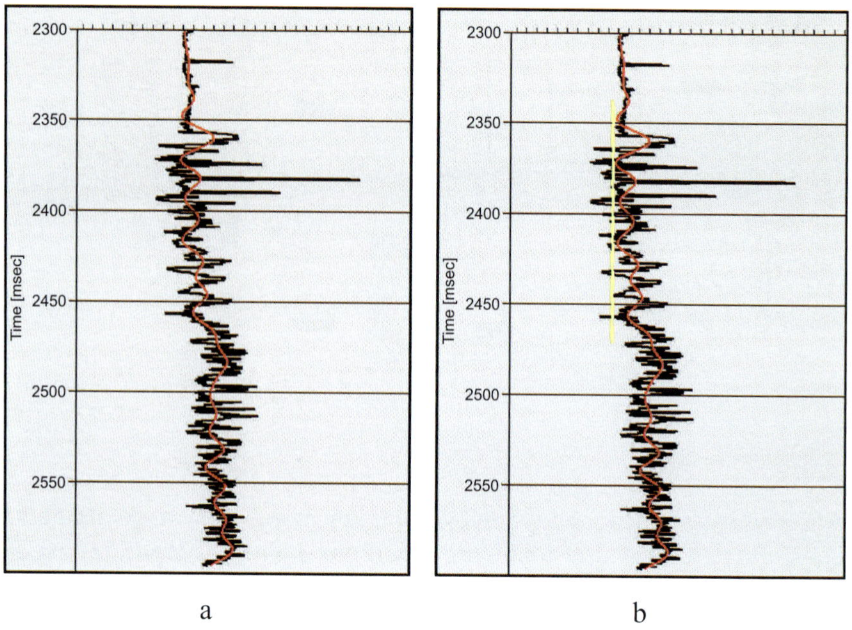


**Figure 2.** (a) Acoustic impedance logs in seismic two-way traveltime (black is at well-log sampling, whereas red is filtered to the seismic frequency content); the figure in (b) is identical to (a), with a possible pay cutoff superimposed (yellow).

that all the data points would lie on a diagonal line in Figure 3. Clearly, they do not, and the correlation appears to be quite poor, a situation caused by the difference in scale between the two data sets.

Another way to explain the situation is to recognize that the (downscaling) transform from data at the seismic scale to data at the well-log scale is nonunique, i.e., data at the well-log scale cannot be uniquely reconstructed from data at the seismic scale. As discussed above, the difference in scale issue impacts the usefulness of comodeling methods. This is illustrated in a series of experiments using a 3-D synthetic data set. Figure 4 illustrates a line through this data set, showing the acoustic impedance model data, generated within a stratigraphic framework using geostatistical simulation, and the corresponding synthetic seismic data, generated with a realistic seismic wavelet. Noise is purposely not added to the seismic data to allow focus on the differences in the geostatistical modeling approaches. Note the rapid vertical variation in the model relative to the slower change in seismic waveform cycles. This provides a first, visual cue of a difference in scale between the model and seismic data. In the following, these data sets are respectively referred to as the reference model and the reference seismic data.

To set a baseline, in the first experiment, a new model is simulated over a subzone of interest and without using any of the seismic data. Synthetic seismic data are generated and compared to the reference seismic data. The results are shown in Figure 5.

As expected, the match to the seismic data is poor. This is further illustrated in Figure 6, showing the cross-correlation between the two seismic data sets over the subzone of interest. Because no guidance from the reference seismic data is present in the modeling, the cross-correlation coefficients are low and can even become negative. Visual comparison of the models in Figures 4 and 5 shows that the vertical

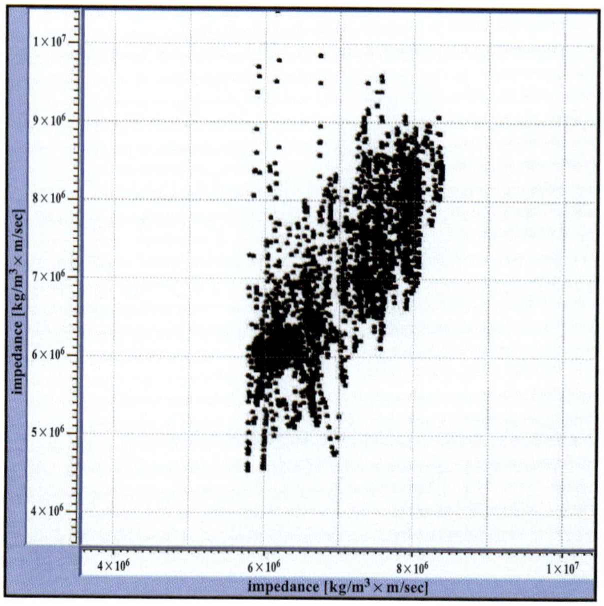


**Figure 3.** A scatter plot of the two logs shown in Figure 2.

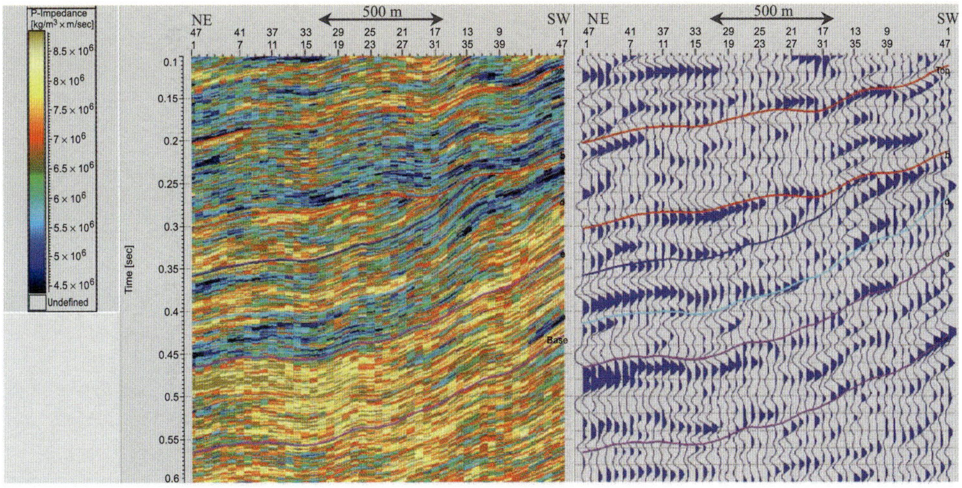


Figure 4. A cross section through the reference model (left) and the corresponding reference seismic data (right).

heterogeneity is similar. Experimental semivariograms are estimated to quantify this. Figure 7 shows the experimental vertical semivariogram of the simulated model in comparison to the semivariogram of the reference model. As the same geostatistical parameters are used in the simulation (except for a different random seed and path), the two semivariograms are effectively identical.

The next experiment evaluates the use of colocated cosimulation to bring in information from the seismic data. The high-cut filtered reference acoustic impedance is used as background data, where the high-cut filter is set at the high end of the reference seismic data bandwidth. Except for the band limitation, this data set equals the reference model exactly and, hence, represents the best possible result that can be obtained with seismic inversion. In the first experiment, colocated cosimulation is run with a realistic correlation coefficient of 0.7, representative of the scatter between the reference model and high-cut filtered reference model, analogous to the result illustrated in Figures 2 and 3. The result is illustrated in Figure 8.

The seismic data match is improved relative to the pure simulation experiment, as is also illustrated in Figure 6, which shows a marked improvement in correlation. This illustrates that seismic data are being used. However, mismatches between the synthetic and reference seismic data clearly show up, and the seismic to synthetic correlation level achieved, although improved, remains low. However, the model, at least visually, continues to exhibit a reasonable level of vertical heterogeneity. The experimental semivariogram plotted in Figure 7 shows that some vertical heterogeneity and overall variability are lost. The slope for the small lags is less steep, and the sill does not reach the same level as that of the reference model. This indicates that

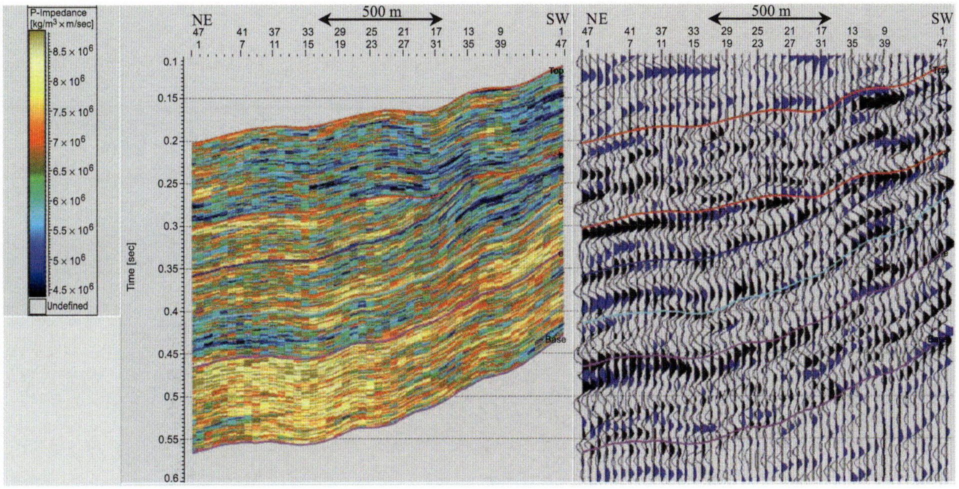


Figure 5. A cross section through the new simulated model (left) and the corresponding synthesized seismic data (black) in overlay of the reference seismic data (right).

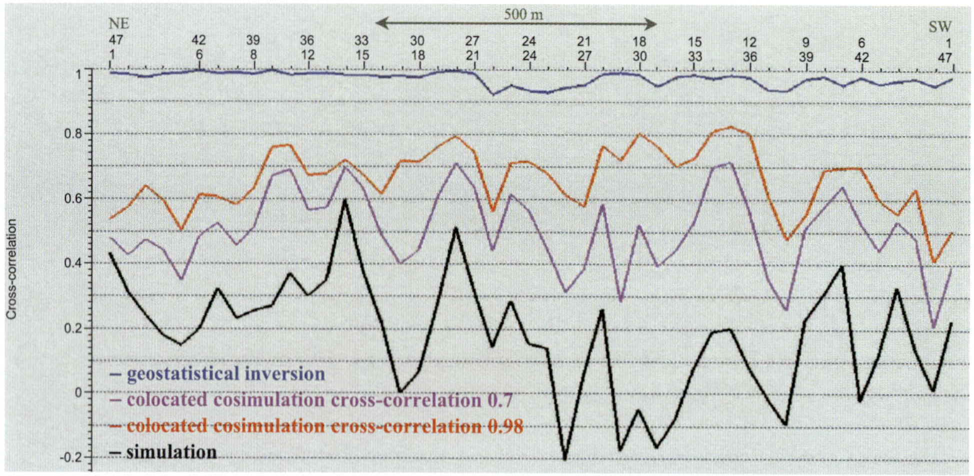


Figure 6. A cross section with the reference seismic data to experiment synthetic data trace cross-correlation coefficients.

comodeling introduces a trade-off between honoring the seismic data and maintaining vertical heterogeneity. This trade-off is caused by the difference in scale between the primary and secondary data.

In the next experiment, colocated cosimulation is again applied. To improve the match to the seismic data, the correlation to the background data is unrealistically increased to 0.98. The results, shown in Figures 6 and 9, demonstrate the seismic data fit improves as expected. However, clear mismatches remain. In addition, as expected, the estimated model is now much too smooth in the vertical direction. This is further demonstrated in Figure 7. The slope of the semivariogram is low at the small lags, indicative of strong smoothing. In addition, the sill is too low, indicating an overall reduction of variability.

These experiments clearly demonstrate that comodeling can only partially incorporate the information carried in the seismic data. In addition, care is required to not bring in seismic information too strongly because this may lead to undesirable smoothing of vertical heterogeneity and overall variability.

Figures 6, 7, and 10 show the result of geostatistical seismic inversion. Clearly, a good seismic data fit is achieved. In addition, the semivariogram from the model is, for practical purposes, consistent with that of the reference model and the simulation model, demonstrating that a result is obtained with the desired level of heterogeneity. This demonstrates that geostatistical seismic inversion can successfully overcome the problem of difference in scale.

Figure 6 shows the cross-correlation coefficients between the reference seismic and synthetic data for each of the experiments.

A closer look at Figure 6 shows that the cross-correlations of the simulation and cosimulation experiments behave much the same, with the main

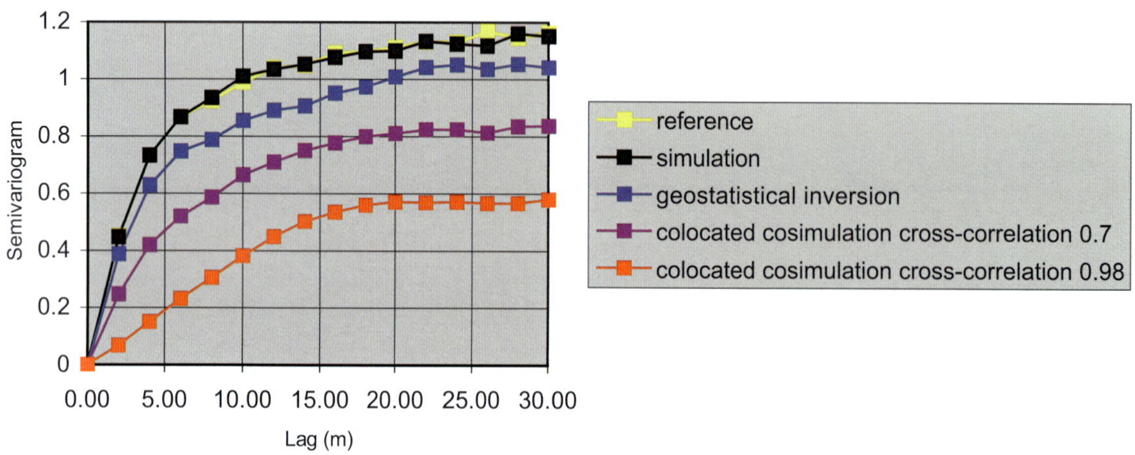


Figure 7. Experimental semivariograms for the different experiments.

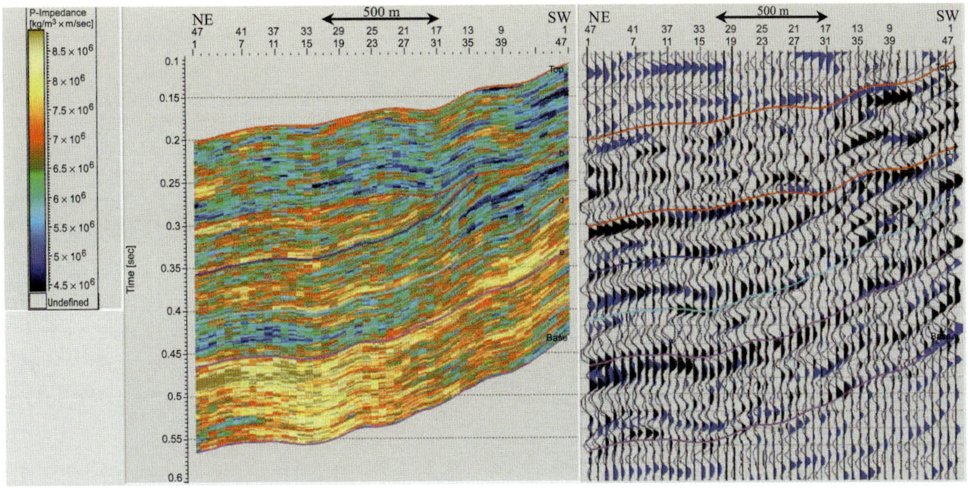


Figure 8. A cross section through the model generated with colocated cosimulation with cross-correlation coefficient 0.7 (left) and the corresponding synthesized seismic data in overlay of the of the reference seismic data (right).

difference being an average vertical offset. This is caused by purposely letting these experiments start from the same random seed and run through the same random path. The differences thus are solely attributable to the level to which seismic data are brought in through the cosimulation process. In the iterative geostatistical seismic inversion process, each iteration is randomized, so that this similarity is lost.

Figure 7 shows the experimental vertical semivariograms for the reference data and the various experiments. All semivariograms have been normalized with the standard deviation of the reference acoustic impedance distribution. Comparison of the semivariograms shows that the simulation experiment is fully consistent with the reference data. The geostatistical inversion experiment is also largely consistent. A minor reduction in vertical heterogeneity and overall variability is caused by the conditioning to the seismic data that occurs in the iterative modeling. The semivariogram of the colocated cosimulation experiment with cross-correlation coefficient 0.7 shows that vertical heterogeneity is reasonably maintained. Interestingly, the sill does not reach that of the reference data. This is also caused by the smoothing induced by the band-limited acoustic impedance background data, which reduce overall variability. Finally, the semivariogram for the colocated cosimulation experiment with cross-correlation coefficient 0.98 clearly illustrates the strong smoothing induced by the band-limited acoustic impedance background data. The slope of the semivariogram for the small ranges is much lower relative to that of the other experiments, showing strong loss of vertical heterogeneity. In addition, the total variability is strongly reduced, as shown by the drop of the normalized sill.

The final experiment concerns the evaluation of the reduction of uncertainty. This experiment is

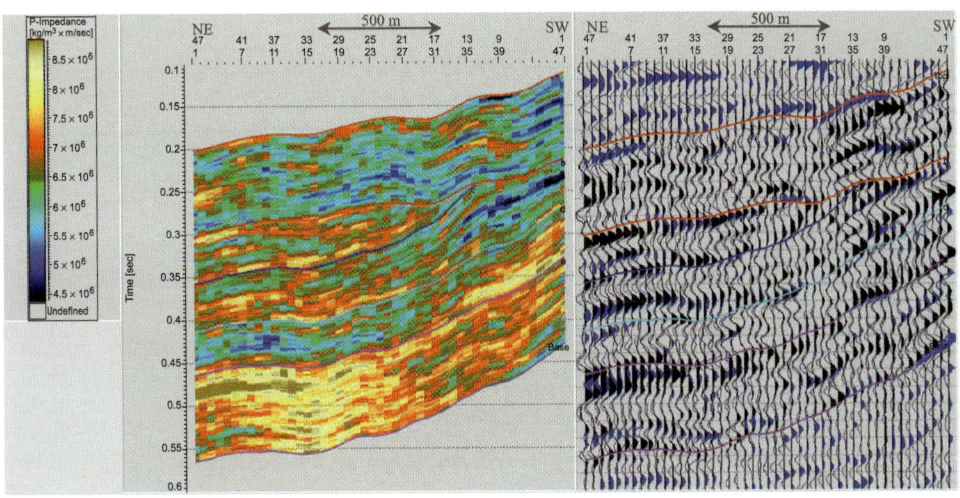


Figure 9. A cross section through the model generated with colocated cosimulation with cross-correlation coefficient 0.98 (left) and the corresponding synthesized seismic data in overlay of the reference seismic data (right).

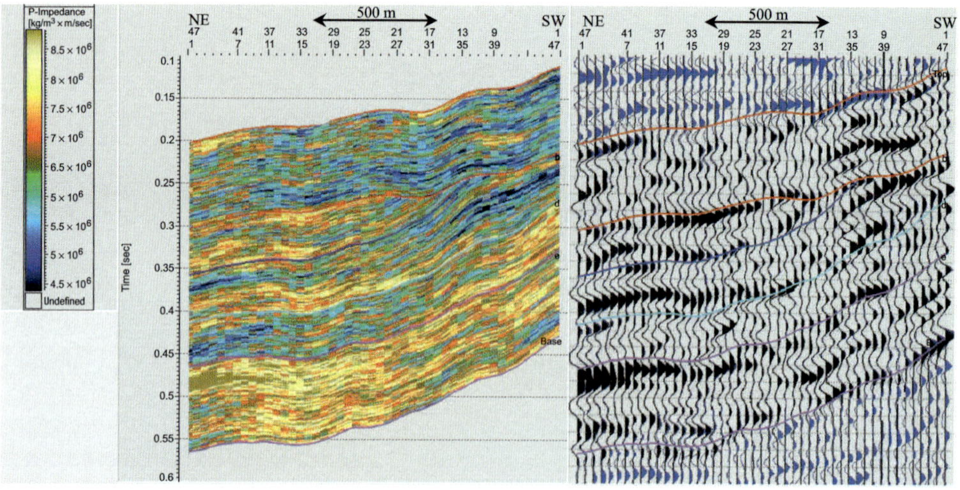


**Figure 10.** A cross section through the model generated with geostatistical seismic inversion (left) and the corresponding synthesized seismic data in overlay of the reference seismic data (right).

performed by running a series of simulations, after which the model standard deviation is estimated. Figure 11 summarizes these experiments for colocated cosimulation using a realistic correlation coefficient to the background model and geostatistical seismic inversion.

As expected, the result clearly shows that uncertainty away from the wells is much reduced by geostatistical seismic inversion relative to cosimulation. This improvement is caused by the fact that geostatistical seismic inversion makes significantly better use of the information carried in the seismic data than cosimulation.

## CONCLUSION

Seismic data are unique in the information it carries about the spatial distribution of reservoir properties. It is therefore a logical step to use seismic data in geostatistical reservoir modeling. The success of a modeling method in bringing in seismic data can be judged by the criteria of preservation of heterogeneity, data fit of seismic synthetic data generated from the model, and reduction of model uncertainty. The latter two criteria are connected in the sense that data fit is a measure of how well seismic data is integrated, and better use of seismic data

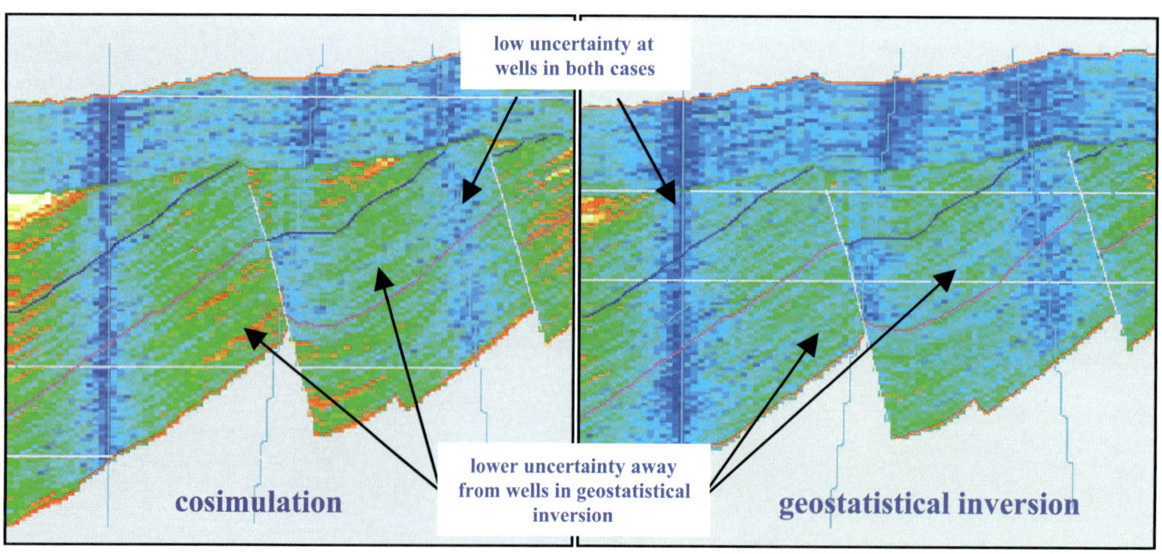


**Figure 11.** Comparison of cosimulation uncertainty vs. geostatistical seismic inversion uncertainty. Low uncertainty is shown in blue, intermediate-level uncertainty is shown in red, and high uncertainty is shown in yellow. Geostatistical seismic inversion significantly reduces uncertainty.

results in the reduction of model uncertainty away from well control.

Most efforts so far have concentrated in bringing in seismic information through cosimulation procedures, where the seismic data serve as a background variable. To do so effectively requires that the seismic data are first transformed to a rock property, for example, to acoustic impedance from poststack seismic data. This transform is performed through seismic inversion. Seismic inversion generally does not recover frequency information much beyond the seismic bandwidth. This results in a limitation to vertical resolution, where the vertical resolution of the seismic layer property data can be an order of magnitude less than required for the reservoir model. This means that reservoir modeling must overcome this difference in scale issue. It is not well recognized that this is a nonunique downscaling problem, instead of the much more commonly encountered upscaling problem.

The consequence of the difference in scale issue in comodeling is that a balance must be sought between the fit to the seismic data and the preservation of vertical heterogeneity and overall variability. If the seismic data are brought in weakly, vertical heterogeneity and overall variability are preserved, but the fit of model synthetics to the seismic data is poor. Vice versa, if the seismic data are brought in strongly, seismic data fit improves, but at the cost of vertical oversmoothing. In conclusion, geostatistical comodeling methods make use of seismic information but are limited in their capability to do so.

Geostatistical seismic inversion is the term used for a new class of methods that fully integrate seismic data with geostatistical modeling. These methods rely on an iterative geostatistical-simulation approach, where the iterations are conditioned such that the seismic data fit is progressively improved to where the model synthetic data adequately match the measured seismic data. The gain is that, relative to comodeling methods, geostatistical seismic inversion produces models that preserve vertical heterogeneity and fit the seismic data. Relative to comodeling, this results in improved control on the spatial distribution of reservoir properties through a significant reduction of model uncertainty away from well control. The gain comes at the cost of computational performance. However, with today's desktop hardware, multiple realizations of large models are readily generated, thus allowing practical application of geostatistical seismic inversion to improve integrated, quantitative reservoir characterization.

## REFERENCES CITED


Anguy, Y., R. Ehrlich, C. M. Prince, V. L. Riggert, and D. Bernard, 1994, The sample support problem for permeability assessment in sandstone reservoirs, *in* J. M. Yarus and R. L. Chambers, eds., Stochastic modeling and geostatistics: Principles, methods, and case studies: AAPG Computer Applications in Geology 3, p. 37–54.

Barens, L. M., P. Williamson, P. Rowbotham, and P. Swaby, 2002, Joint geostatistical inversion of time-lapse data: European Association of Geoscientists and Engineers 64th Annual Conference and Exhibition, v. 1, p. A-22.

Bashore, W. M., U. G. Araktingi, M. Levy, and W. J. Schweller, 1994, Importance of a geological framework and seismic data integration for reservoir modeling and subsequent fluid-flow predictions, *in* J. M. Yarus and R. L. Chambers, eds., Stochastic modeling and geostatistics: Principles, methods, and case studies: AAPG Computer Applications in Geology 3, p. 159–175.

Bortoli, L. J., F. A. Alabert, A. Haas, and A. G. Journel, 1993, Constraining stochastic images to seismic data, *in* A. Soares, ed., Geostatistics troia: Dordrecht, The Netherlands, Kluwer Academic Publishers, p. 325–338.

Buxton, L., R. R. Davison, and P. van Riel, 2000, An interpreter's guide to understanding and working with seismic-derived acoustic impedance data: The Leading Edge, v. 19, no. 3, p. 242–256.

Castagna, J. P., and M. M. Backus, eds., 1993, Offset-dependent reflectivity: Theory and practice of AVO analysis: Society of Exploration Geophysicists, 348 p.

Chambers, R. L., M. A. Zinger, M. C. Kelly, 1994, Constraining geostatistical reservoir descriptions with 3-D seismic data to reduce uncertainty, *in* J. M. Yarus and R. L. Chambers, eds., Stochastic modeling and geostatistics: Principles, methods, and case studies: AAPG Computer Applications in Geology 3, p. 143–157.

Chiles, J. P., and P. Delfiner, 1999, Geostatistics: Modeling spatial uncertainty: New York, John Wiley & Sons, 695 p.

Contreras, A., C. Torres-Verdin, K. Kvien, T. Fastnacht, and W. Chesters, 2005, AVA stochastic inversion of pre-stack seismic data and well logs for 3D reservoir modeling: European Association of Geoscientists and Engineers, Extended Abstracts, p. T-15.

Doyen, P. M., and T. M. Guidish, 1992, Seismic discrimination of lithology and porosity: A Monte Carlo approach, *in* R. E. Sheriff, ed., Investigations in geophysics no. 7: Reservoir geophysics: Society of Exploration Geophysicists, p. 243–250.

Dubrule, O., E. Damsleth, and A. Hurst, eds., 2001, Thematic set on geostatistics: Petroleum Geoscience, v. 7, p. S1–107.

Dubucq, D., S. Busman, and P. van Riel, 2001, Turbidite reservoir characterization, multi-offset stack inversion for reservoir delineation and porosity estimation: A Gulf of Guinea example: Society of Exploration Geophysicists

70th International Exposition and Annual Meeting, Expanded Abstracts 20, p. 609.

Haas, A., and O. Dubrule, 1994, Geostatistical inversion—A sequential method of stochastic reservoir modeling constrained by seismic data: First Break, v. 12, p. 561–569.

Isaaks, E. H., and R. M. Srivastava, 1989, An introduction to applied geostatistics: New York, Oxford University Press, 561 p.

Mallick, S., X. Huang, J. Lauve, and R. Ahmad, 2000, Hybrid seismic inversion: A reconnaissance tool for deepwater exploration: The Leading Edge, v. 19, p. 1230–1237.

Olea, R. A., 1994, Fundamentals of semivariogram estimation, modeling and usage, 1994, *in* J. M. Yarus and R. L. Chambers, eds., Stochastic modeling and geostatistics: Principles, methods, and case studies: AAPG Computer Applications in Geology 3, p. 27–35.

Pendrel, J., and P. van Riel, 1997, Methodology for seismic inversion and modeling: A Western Canadian reef example: Canadian Society of Exploration Geophysicists Recorder, v. 12, p. 5–15.

Sams, M. S., D. Atkins, N. Said, E. Parwito, and P. van Riel, 1999, Stochastic inversion for high resolution reservoir characterization in the Central Sumatra Basin: Society of Petroleum Engineers Asia Pacific Improved Oil Recovery Conference, Kuala Lumpur, Malaysia, SPE Paper 57260.

Shanor, G., M. Rawanchaikul, M. Sams, R. Muggli, G. Tiley, and J. Ghulam, 2002, From seismic to simulator through geostatistical modelling and inversion: Makarem gas accumulation, Sultanate of Oman: First Break, v. 20, p. 110–115.

Torres-Verdin, C., M. Victoria, G. Merletti, and J. Pendrel, 1999, Trace based and geostatistical inversion of 3D seismic data for thin-sand delineation: An application in San Jorge Basin, Argentina: The Leading Edge, v. 18, no. 9, p. 1070–1077.

Torres-Verdin, C., Z. Wu, O. J. Varela, M. K. Sen, and I. G. Roy, 2002, Joint inversion of reservoir production measurements and 3D pre-stack seismic data: Proof concept: Society of Exploration Geophysicists 72nd International Exposition and Annual Meeting, Extended Abstracts, RTC p. 1.2.

Xia, G, M. K. Sen, and P. L. Stoffa, 1998, 1-D elastic waveform inversion: A divide-and-conquer approach: Geophysics, v. 63, no. 5, p. 1670–1684.

Yarus, J. M., and R. L. Chambers, eds., 1994, Stochastic modeling and geostatistics: Principles, methods, and applications: AAPG Computer Applications in Geology 3, 379 p.

## *Section IV*

# Case Studies

The fourth section of the book is devoted to more extensive case studies. Case studies are perhaps the most interesting articles for readers because they often provide many of the details needed to replicate modeling and analytic approaches in similar areas or regimes. Case studies also tend to include specific information about what worked and what didn't work. The six case studies presented here encompass diverse geological settings and depositional environments around the world. Several of them involve the integration of seismic data in the modeling process, while others focus on different approaches.

The first chapter describes a reservoir characterization study conducted at ExxonMobil by Penny Patterson and her colleagues on a braided-stream depositional system. Object-based modeling techniques are used to represent the internal architecture of the reservoir intervals. Several types of data are incorporated. In the second chapter, Raj Shrestha and his collaborators describe a modeling study of Rotliegend Gas Field in Germany that integrates three-dimensional seismic inversion and geostatistical modeling. The next two chapters also present the results of reservoir modeling studies involving the integration of seismic data. Dan Williams and Adnan Aqrawi describe work they performed on a carbonate reservoir in the Middle East, and Wenlong Xu and his colleagues at Unocal present their work on a sandstone reservoir offshore Texas. In the fifth chapter, Peter Frykman considers the spatial variability of porosity in chalk reservoirs, investigating the comparison between an outcrop and a North Sea reservoir. The study specifically focuses on the spatial distribution of petrophysical properties supplemented with geological information. Finally, Hélène Beucher-Darricau, Brigitte Doligez, and Jeffrey Yarus present comparative modeling results for several reservoirs from the Ravenscar Group located along the east coast of Yorkshire in the United Kingdom. This particular chapter demonstrates the use of object-based, pixel-based, and stochastic simulation techniques, and suggests that a combination of techniques may be needed in any particular study to obtain optimum results.

Patterson, P. E., T. A. Jones, A. D. Donovan, J. D. Ottmann, and C. J. Donofrio, 2006, Hierarchical description and geologic modeling of a braided-stream depositional system, *in* T. C. Coburn, J. M. Yarus, and R. L. Chambers, eds., Stochastic modeling and geostatistics: Principles, methods, and case studies, volume II: AAPG Computer Applications in Geology 5, p. 221–236.

*Chapter 17*

# Hierarchical Description and Geologic Modeling of a Braided-stream Depositional System

**P. E. Patterson**
**T. A. Jones**
**A. D. Donovan**[1]
*ExxonMobil Upstream Research Company*
*Houston, Texas, U.S.A.*

**J. D. Ottmann**
**C. J. Donofrio**
*ExxonMobil Exploration Company*
*Houston, Texas, U.S.A.*

## ABSTRACT

The facies architecture, sequence-stratigraphic framework, and structural configuration of a petroleum reservoir were delineated by an integrated study. These interpretations formed the basis of three-dimensional (3-D) geologic models that were constructed for estimation of reserves, flow-simulation studies, and field-development planning. The study incorporated 3-D seismic interpretations, well-log correlations, facies and petrophysical analyses of cored intervals, and interpretations derived from outcrop exposures of the reservoir.

The reservoir intervals are interpreted to represent a fluvial depositional system that varies systematically along an updip to downdip transect. Proximal (updip) fluvial-facies deposits are inferred to represent amalgamated channel complexes that form widespread sheets. Medial fluvial-facies deposits are interpreted as amalgamated to semiamalgamated braid-bar deposits that are thinner and less laterally persistent. Distal (downdip) fluvial facies are inferred to represent thin yet laterally extensive braid-bar deposits.

Object-based modeling techniques were used to model the internal architecture of the reservoir intervals. Proximal channel facies were generated using standard software to populate the zones with channel elements that are clustered to form channel complexes. Medial and distal bar facies, however, required an innovative method that populates the zones with discrete, user-defined, braid-bar elements

[1]*Present address*: BP Upstream Technology, Houston, Texas, U.S.A.


DOI:10.1306/1063818CA53237

that are distributed along thalwegs. Clusters of thalwegs form amalgamated to semiamalgamated bar complexes. This capability, referred to as bar-train modeling, results in a better computer-model representation of the fluvial-sandstone bar geometry and spatial distribution.

The resulting geologic models provide an improved reservoir characterization of the large-scale and small-scale fluvial architecture for the subsurface reservoir. In particular, the geologic models more accurately describe the complex architecture of the lowstand channel and braid-bar fluvial sandstones as well as the internal architecture of the intervening mudstone deposits.

## INTRODUCTION

A detailed study of a petroleum reservoir characterized the facies architecture, sequence-stratigraphic framework, and structural framework for construction of three-dimensional (3-D) geologic models. The focus of this study was to provide fundamental data for the development of 3-D geologic models used for improved estimates of reserves, flow-simulation studies, and field-development planning.

The reservoir consists of approximately 1400 ft (427 m) of intercalated fluvial sandstone, flood-plain sandstones and mudstones, and lacustrine shales. As evident in the well-log signature in Figure 1, intervals of sandstone and shale describe three scales of fining-upward cycles, including small-scale cycles 30–60 ft (9–18 m) in thickness, intermediate-scale cycles 200–300 ft (61–91 m) in thickness, and large-scale cycles of roughly 1800 ft (550 m). In general, the thickness and lateral extent of the small- and intermediate-scale cycles decrease stratigraphically upward and reflect a systematic change in the style of fluvial deposition in the reservoir.

To assist with the description of the reservoir, an integrated study was conducted that incorporated all available data, including 3-D seismic interpretations, well-log correlations, facies and petrophysical analyses of cored intervals, and interpretations derived from outcrop exposures of the reservoir. The results of this study revealed the development of three fluvial-facies belts that systematically vary along the depositional profile of the alluvial system.

The complex fluvial architecture interpreted for the reservoir interval could not be adequately represented with standard, available computer programs. Hence, a new approach was taken, and software capabilities were developed to better characterize the internal reservoir architecture. This software focused on two main areas: (1) the capability to condition the fluvial-channel thalwegs (i.e., the paths followed by the channels) by trend maps, vector fields (which may be defined by depositional flow paths), and deterministic thalwegs; and (2) the ability to model geometries of fluvial braid bars or other depositional elements. The modeler may vary the shape, distribution, and size of the depositional elements along the length of a thalweg.

This chapter describes the lithofacies associations, interpretation of the style of fluvial deposition, and hierarchical, sequence-stratigraphic description of the reservoir. Methods for the construction of 3-D geologic models, specifically the code that was developed for braid-bar deposits, are summarized following the stratigraphic description. It concludes with a description of the application of modeling techniques to the reservoir.

## FLUVIAL-FACIES ARCHITECTURE

Interpretation of the fluvial-facies architecture of the reservoir is based on the hierarchical framework proposed by Patterson et al. (1995) and Sprague et al. (2002). The hierarchical approach, founded on the physical stratigraphy of the strata and their associated bounding surfaces, encompasses genetically related stratal successions at a variety of scales.

In the reservoir interval, small-scale architectural elements are defined from the description of approximately 2000 ft (610 m) of core, characterization of well-log signatures, examination of outcrop exposures of the reservoir interval, and a reconnaissance study of an interpreted modern analog. Delineation of the large-scale, hierarchical elements is based on well-log correlations, 3-D seismic interpretations, and well-test data.

A detailed description of the core reveals the presence of 10 lithofacies associations. Of these, three sandstone lithofacies associations characterize the subsurface reservoir strata. The nonreservoir

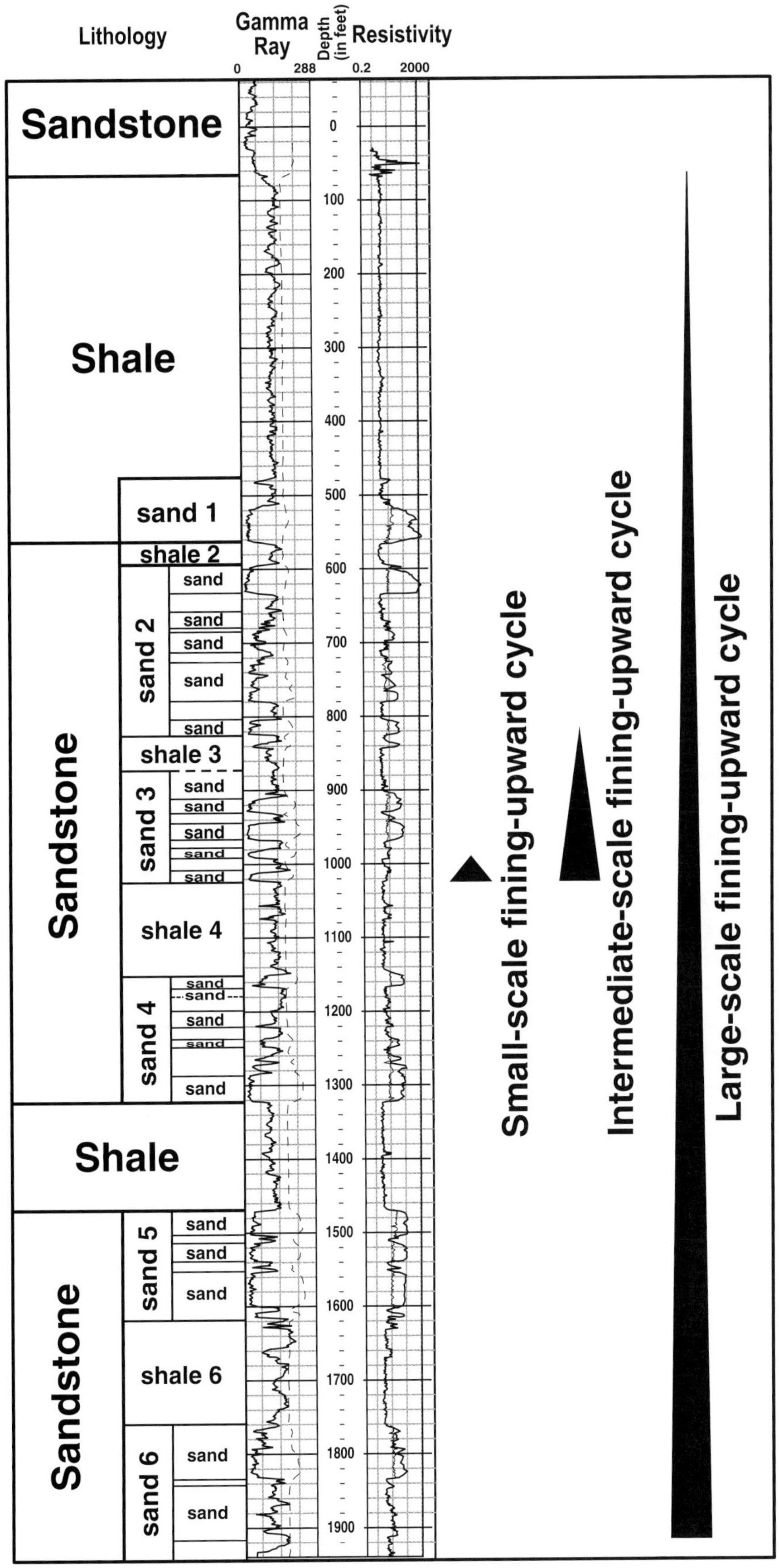


Figure 1. Representative gamma-ray and resistivity well-log signatures of the subsurface reservoir, which consists of approximately 1800 ft (550 m) of intercalated sandstone and shale. Note that three scales of fining-upward cycles describe the hierarchical framework of the reservoir.

intervals consist of interbedded sandstone and mudstone lithofacies associations. Because the geologic-modeling effort primarily focused on description of the reservoir sandstones, these lithofacies associations are summarized here. A synopsis of key modeling issues follows the description of each sandstone lithofacies association.

## Sandstone Lithofacies Association I: Amalgamated Channel Deposits

Sandstone lithofacies association I consists of coarse- to very coarse-grained, trough cross-bedded sandstones (Figure 2). Individual cross-stratified sandstone beds fine upward and, in addition, amalgamate vertically to form fining-upward sandstone packages that have a maximum vertical thickness of about 20 ft (6 m). The fining-upward packages are bounded at their bases by sharp scour surfaces and sparse rip-up clast deposits. Because these sandstone packages are commonly eroded by the overlying sandstone successions, complete stratal successions are not often preserved.

Fining-upward stratal packages in sandstone lithofacies association I are interpreted to represent fill deposits of single channels that formed in low-sinuosity braided streams. The predominance of multiple scour surfaces, the deposition of coarse grain-size fractions, and the lack of interlaminated clay indicate that these channels formed under relatively high-energy conditions. Hence, they are inferred to have formed on an active, laterally extensive braid plain proximal to the source area.

In the hierarchical classification scheme proposed by Patterson et al. (1995) and Sprague et al. (2002), channel-fill deposits amalgamate vertically and laterally to form multistory sandstones, which are referred to as channel complexes. In the reservoir interval, individual channel complexes typically have a maximum thickness of 60 ft (18 m). Because the channel complexes generally lack clay-rich flood-plain strata, they exhibit a blocky, low-response, gamma-ray well-log signature (Figure 2).

The fundamental depositional element for modeling the proximal fluvial facies is a single, low-sinuosity channel. These channels cluster vertically and laterally to form channel complexes that incorporate very minor amounts of flood-plain deposits. As a result, these channel complexes exhibit high vertical and lateral continuity.

## Sandstone Lithofacies Association II: Amalgamated Braid-bar Deposits

Sandstone lithofacies association II consists of medium- to coarse-grained, trough cross-bedded and low-angle cross-stratified sandstones (Figure 3). Individual beds decrease in grain size upward and are stacked vertically to form fining-upward sandstone packages that have a maximum thickness of about 20 ft (6 m). The basal surfaces of these sandstone packages are erosional and are commonly overlain by abundant clay rip-up clasts. In some cases, thin (1–3 in. [2.5–6.3 cm] in thickness), laminated mudstones overlie the upper surfaces of the fining-upward stratal packages.

Sandstone lithofacies association II is inferred to represent downstream to obliquely accreting, midchannel bars or bank-attached bars in low-sinuosity braided streams. The laminated mudstones are interpreted to represent slack-water sedimentation associated with the cessation of bar migration. In many cases, the laminated mudstones possess an erosional upper surface, suggesting variable erosion of these deposits subsequent to deposition of the overlying bar. Braid-bar and slack-water deposits form a slight heterolithic character to the fill of the braid channels. The subtle increase in preservation of braid bars and slack-water deposits suggests a decrease in stream competence and possibly a slightly more episodic discharge in comparison to the fluvial channels of sandstone lithofacies association I. Amalgamated bar deposits are interpreted to have formed in medial regions of the alluvial plain.

Within a hierarchical context, braid-bar channels stack vertically and laterally to form channel complexes (after Patterson et al., 1995; Sprague et al., 2002). In the subsurface reservoir, channel complexes have a maximum thickness of roughly 50 ft (15 m) and contain minor clay laminations, which result in a blocky, low to moderate gamma-ray log response (Figure 3).

An example of a modern stream, which depicts the interpreted style of braided-stream deposition, is the Chari River, Africa (Figure 4). In this section of the river, midchannel bars accrete primarily in a downstream direction. Subsequent to the development of the braid bars, prolonged periods of emergence result in stabilization of the accretionary bars and, in some cases, vegetation of the bar tops. Geomorphologic and hydrodynamic descriptions of modern braided rivers reveal the formation of downstream

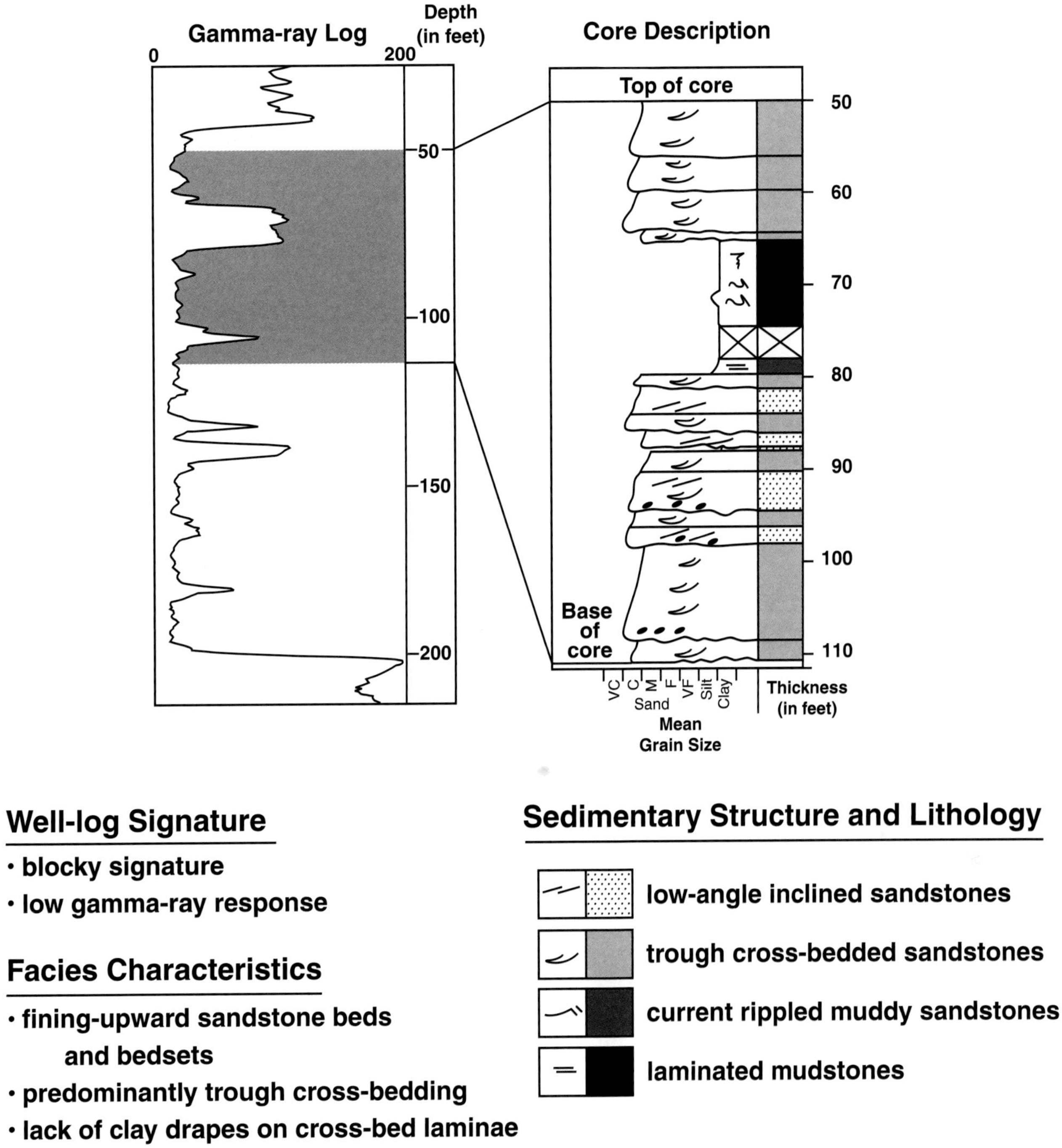


**Figure 2.** Representative gamma-ray, well-log response, and facies characteristics of sandstone lithofacies association I. This lithofacies association primarily consists of fining-upward trough cross-bedded sandstones and is interpreted to represent fills of single, low-sinuosity channels.

and oblique accretion of braid bars, interpreted to result from a decrease in stream competency to transport coarse-grained bed-load material (Leopold and Wolman, 1957; Smith, 1971; Bridge, 1985; Bristow and Best, 1993; Shukla et al., 1999; Bristow et al., 2002). Analogous downstream accretionary bar forms have been interpreted in outcrop exposures (Banks, 1973; Bromley, 1991; Van Wagoner, 1995; Jones et al., 2002).

A single braid bar is the key depositional element for modeling braided fluvial systems. An example of modern bars (Figure 4) suggests that the braid bars

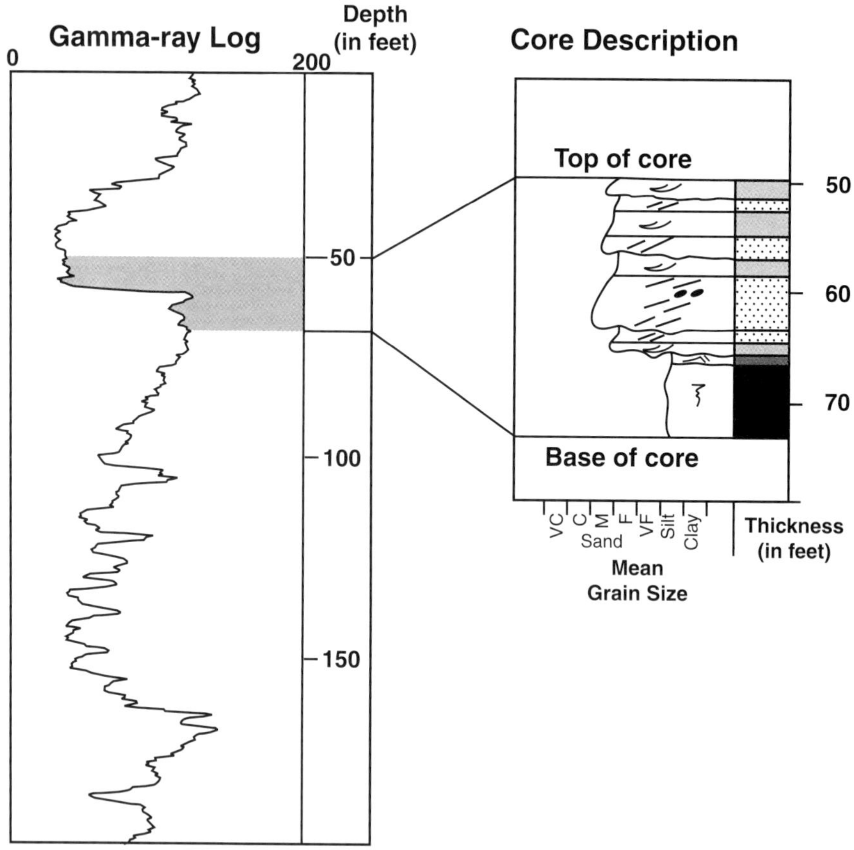


## Well-log Signature

- blocky signature
- low to moderate gamma-ray response

## Facies Characteristics

- fining-upward sandstone beds and bedsets
- interbedded trough cross-bedded and low-angle inclined sandstones
- minor clay drapes on cross-bed laminae

Figure 3. Representative gamma-ray, well-log response, and facies characteristics of sandstone lithofacies association II. This lithofacies association is composed of fining-upward beds and bedsets of trough cross-bed and low-angle inclined cross-stratified sandstones and is interpreted to represent downstream to obliquely accreting midchannel or bank-attached braid bars. (Refer to Figure 2 for the legend of sedimentary structure and lithology.

are elongate, possessing a width-to-length ratio of 1:3. In single-thread braid channels, the braid bars are distributed along the axis of the channel. Additionally, they are separated by sparse, thin clay laminae that reflect episodic deposition and preservation of clay-rich, slack-water deposits. The presence of clay in the reservoir is observed both in core and well-log response. Within the geologic model, the clay acts as minor baffles to fluid flow.

Single-thread channels containing braid bars are clustered vertically and laterally to form channel complexes. As a result of a slight decrease in amalgamation of the channel complexes, these hierarchical elements commonly contain intercalated floodplain deposits and, therefore, exhibit moderate vertical and lateral continuity.

### Sandstone Lithofacies Association III: Isolated Braid-bar Deposits

Sandstone lithofacies association III consists of fine- to medium-grained, low-angle cross-stratified, and to a lesser extent, trough cross-stratified sandstones (Figure 5). Each cross-stratified bed decreases

**Figure 4. Low-flight aerial photograph of the Chari River, Africa. The river is flowing from the lower right to the upper left area of the photograph, and the width of the river is approximately 3500 feet (1.0 km). Note the general downstream accretion of midchannel bars in the braid channel. A prolonged period of emergence has resulted in the stabilization of the bars and, in some regions, subsequent vegetation of the bar surfaces.**

in grain size upward and possesses moderate to abundant clay drapes on lamina surfaces. These cross-stratified beds amalgamate vertically, forming fining-upward sandstone packages that are commonly overlain by thick (as much as 6 in. [15 cm] in thickness), laminated mudstones. The maximum thickness of single fining-upward packages is approximately 15 ft (4 m).

Sandstone lithofacies association III is interpreted to represent isolated bank-attached bars in low-sinuosity braided streams. The prevalence of clay drapes on the cross-bed laminae suggests highly variable flow conditions. Mudstone laminae that cap the fining-upward sandstones are inferred to represent relatively long-lived slack-water sedimentation that blankets the bars subsequent to avulsion of the active braid channel. Hence, these channel deposits contain pronounced heterolithic fills, which reflect repeated development and subsequent abandonment of braid bars. The prevalence of lower flow-regime sedimentary structure suggests that these channels formed on the most distal regions of the braid plains.

In the hierarchical framework, distal braid-bar channels amalgamate to only a minor extent and form relatively thin channel complexes, typically less than 40 ft (12 m) in thickness. In well logs, they exhibit a pronounced serrated signature indicative of ubiquitous clay throughout the channel complex in the form of clay laminae, clay drapes on abandonment surfaces, and clay-rich flood-plain deposits (Figure 5).

Similar to sandstone lithofacies association II, the fluvial bar is the fundamental depositional element for modeling the distal braided streams. Although the braid bars are distributed along the axis of the channel, each bar is typically separated by variable clay laminae that are inferred to act as significant baffles to fluid flow. These distal braid channels are poorly clustered both vertically and laterally, and as a result, they form channel complexes that incorporate variable amounts of flood-plain deposits. The low degree of amalgamation of the channels indicates that these distal fluvial deposits possess poor vertical and lateral continuity.

## Sequence-stratigraphic Architecture

Interpretation of the sequence-stratigraphic architecture for the reservoir interval is based on a hierarchical framework that encompasses genetically related successions and their stratal bounding surfaces (Patterson et al., 1995; Sprague et al., 2002). For

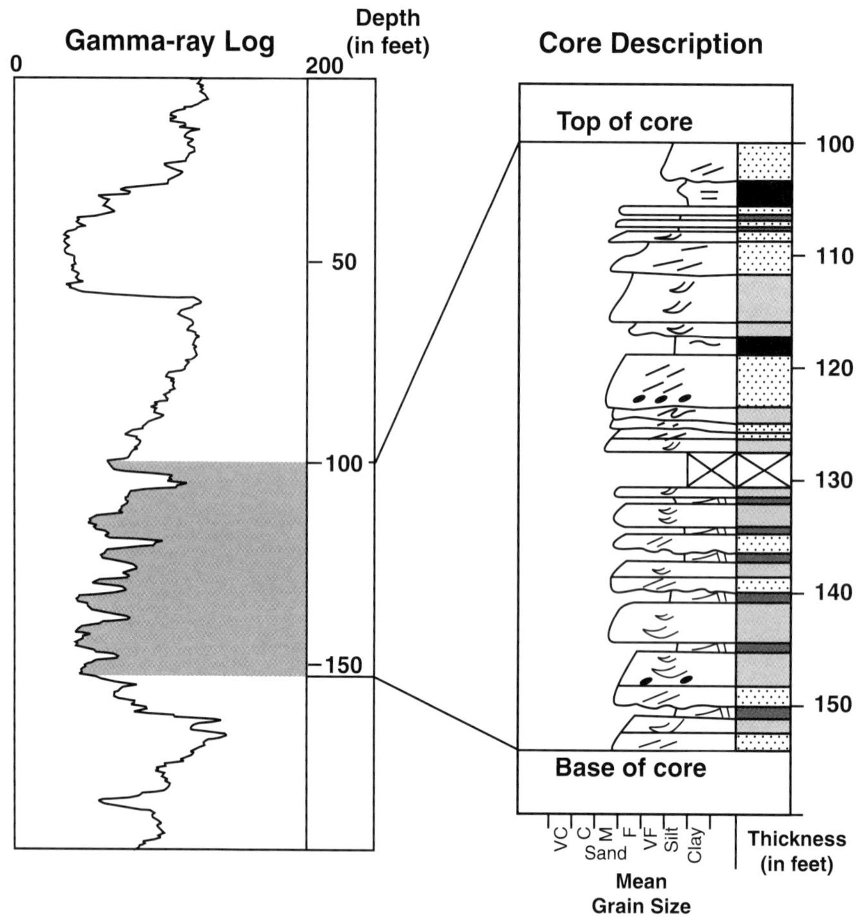


## Well-log Signature

- serrated signature
- moderate gamma-ray response

## Facies Characteristics

- fining-upward sandstones overlain by sandy mudstones
- interbedded low-angle inclined and trough cross-bedded sandstones
- abundant clay drapes on cross-bed laminae

Figure 5. Representative gamma-ray, well-log response, and facies characteristics of sandstone lithofacies association III. This lithofacies association mainly consists of fining-upward beds and bedsets of low-angle inclined and trough cross-stratified sandstones that are overlain by laminated mudstones. These strata are interpreted to represent isolated bank-attached bars in low-sinuosity braided streams. (Refer to Figure 2 for the legend of sedimentary structure and lithology.)

geologic-modeling purposes, the smallest architectural element is the braid bar, which is aligned along the length of the braid channels. Braid bars, in turn, stack to form channel fills that are bounded by basal erosional surfaces and overlying abandonment surfaces.

Channel elements amalgamate vertically and laterally to form a channel complex that constitutes the sand-prone part of the reservoir. Channel-complex elements are overlain by mudstone-prone intervals comprised of flood-plain strata. Combined channel-complex and flood-plain-complex elements constitute a depositional sequence. The sandstone-prone parts of the depositional sequences possess erosional bases (interpreted as sequence boundaries) and horizontal, conformable tops (inferred to represent abandonment surfaces). The mudstone-prone intervals conformably overlie the sandstones and are bounded at their tops by erosional surfaces. Within this context, the sandstone-prone intervals are interpreted as lowstand systems tracts, and the mudstone-prone intervals are interpreted as transgressive and highstand systems tracts.

In the reservoir interval, 24 depositional sequences were interpreted. Each sequence was modeled separately based on its dominant style of fluvial deposition.

## OBJECT-BASED MODELING OF FLUVIAL RESERVOIR FACIES

Most geologic-modeling projects of clastic reservoirs do not simply generate a model of petrophysical properties. Instead, model builders create a model of the distribution of facies or lithofacies associations. This facies model is then used as a template to control the distribution away from well control of petrophysical properties (e.g., porosity, permeability, percent sand). Two approaches are generally used for modeling facies, referred to as cell-based and object-based methods. Cell-based (pixel-based) methods are also known as geostatistical methods because they are based on statistically defined spatial continuity of lithofacies or lithofacies associations. Sequential-indicator simulation (Goovaerts, 1997; Deutsch and Journel, 1998) is the most commonly used cell-based tool. In this method, each cell of the model is assigned a facies property, one at a time, based on statistical and spatial relationships to its neighboring cells and data. The most important controls on indicator simulation are volume fraction and semivariograms that are derived from outcrop, core, seismic, and well-log data.

Object-based methods create geologic objects or bodies (with appropriate geologic characteristics, i.e., size, shape, and orientation) that are thought to represent depositional elements such as sand-filled channels or bars. The objects typically are much larger than a model cell, so numerous cells comprise a given object. All cells that define the object are then placed into the model together, with the body treated as an entire entity. The fundamental tenet in this method is to honor depositional concepts and stratigraphic relations interpreted by the modeler (Haldorsen and Damsleth, 1990). As with cell-based methods, the parameters used to constrain the object-modeling algorithms are interpreted from outcrops, cores, seismic surveys, and well logs.

Two types of object-based modeling capabilities were evaluated for modeling the fluvial depositional system. One commonly used technique involves generation of fluvial channels that occur as continuous ribbons along the length of the 3-D geologic model. The channels may be modeled as either sand-filled or shale-filled bodies (Hatløy, 1994; Bratvold et al., 1995; Roxar, 2000a) and typically are based on geological concepts and rules but are not necessarily based on mathematical theory. The channel elements typically are clustered to form channel complexes or lowstand systems tracts (Hatløy, 1994; Patterson et al., 1995).

The second object-based technique involves the delineation of discrete shapes or objects that are not continuous through the model (Haldorsen and Damsleth, 1990; Holden et al., 1998; Roxar, 2000b). This modeling capability has been applied, for example, to the generation of 3-D models of fluvial point bars. A limitation, however, is the inability to align the objects along a thalweg. This limitation may result in 3-D geologic models that do not accurately portray the fluvial-facies architecture.

The summary of facies architecture and sequence stratigraphy in the previous sections indicates that most of the reservoir interval of interest is characterized by amalgamated to poorly amalgamated bar deposits that formed within braided-stream channels. Therefore, neither of these two object-based modeling capabilities is adequate for this reservoir. To build a more accurate reservoir description of braid bars, enhancements were added to a commercially available computer program, RMS (Roxar, 2000c). Detailed functional specifications and some algorithms were provided to Roxar. These enhancements were specifically designed to allow more realistic modeling of bar stacking and geometries in braided-stream channels. The modifications are informally referred to as the bar-train module, described in the section below.

## OBJECT-BASED MODELING OF BRAID-BAR DEPOSITS

A major goal of the bar-train modeling process is to place bars along fluvial channels, instead of scattering them randomly. Figure 6 diagrammatically shows the elements of the process. The bar-train module uses thalwegs (that is, centerlines) to define the paths of the fluvial channels that span the model. After a thalweg is defined, objects are then distributed along it to emulate downstream-accreting or

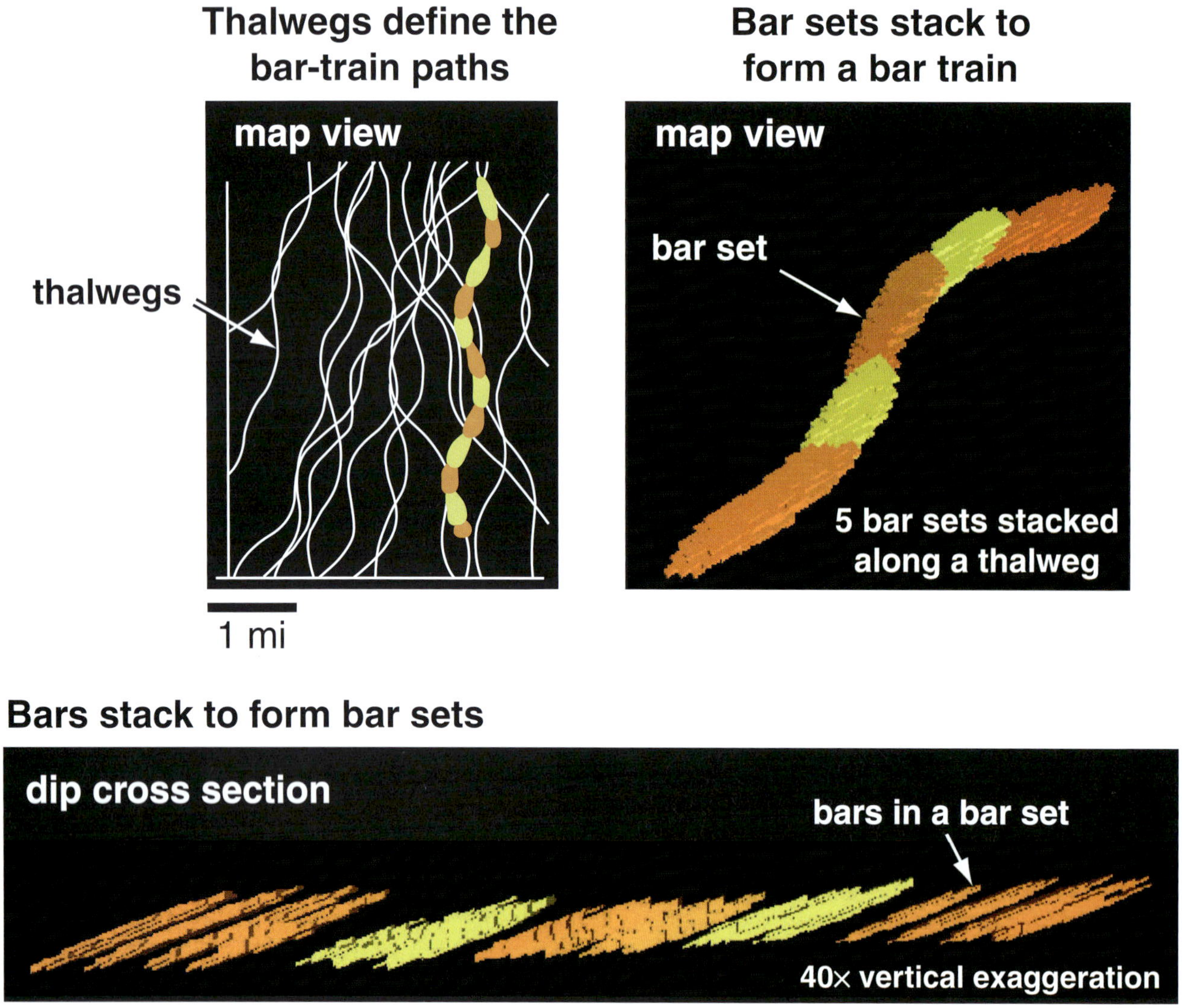


**Figure 6. Object-based modeling of braid-bar deposits. The bar-train module uses a hierarchy of bars, bar sets, and bar trains to model depositional elements of braided fluvial systems.**

longitudinal braid bars. Individual bars cluster to form bar sets, which are aligned with the thalweg. Bar sets are organized into groups to form a bar train that spans the length of a thalweg. Bar trains typically extend from the upstream edge of the model to the downstream edge. Figure 7 shows a simplified flowchart of the process.

Variations to this process are possible. Oblique and laterally accreting fluvial bars may be modeled by adjusting the input parameters. Vertically amalgamated bar complexes can be generated by constraining several thalwegs to be located within a small region of the model. The facies models generated by the bar-train module are architecturally consistent with fluvial depositional processes.

## Thalweg Elements

Thalwegs locate braid channels in the model to control where bars (that is, the facies of interest) are placed. Without special instructions, the bar-train module will generate typical thalwegs that extend from one edge of the model to the other, following a trend that is more or less straight. However, geologic events and features existed that affected river flow and deposition. For example, the development of incised valleys may cause numerous channels to cluster in a restricted region of the reservoir. Similarly, syndepositional growth of structural features and topographically high regions may influence the paleogeographic configuration of fluvial systems.

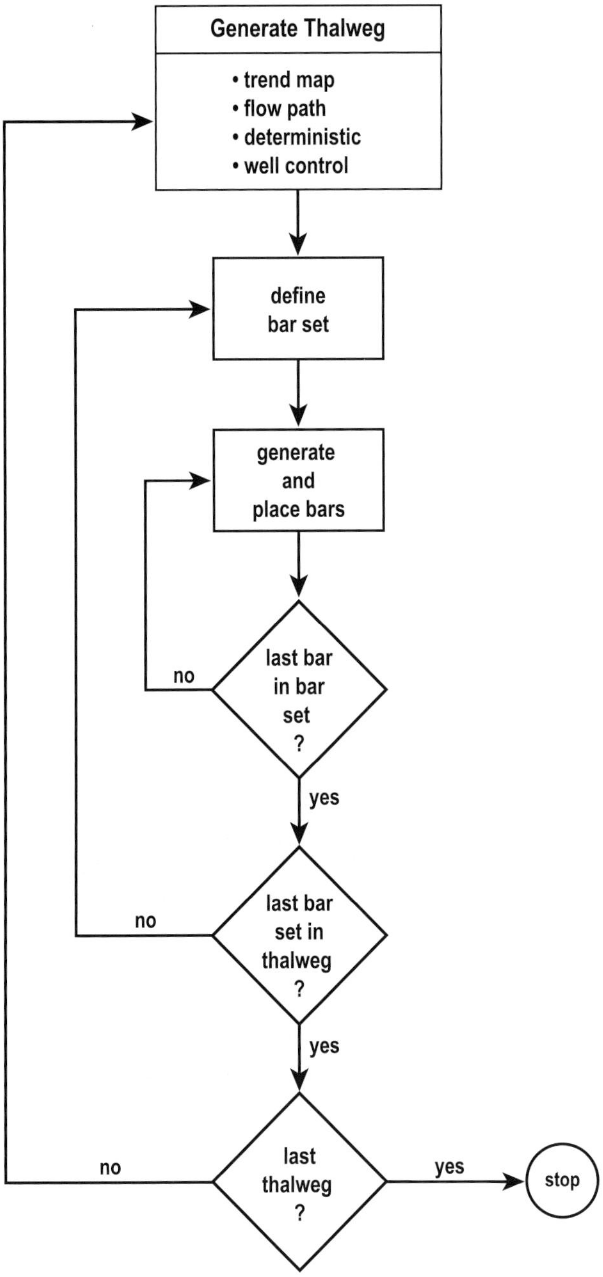


**Figure 7. Flowchart of the process to generate thalwegs, bars, and bar sets.**

Two sources of information discussed below (trend maps and flow paths [vector fields]) may be used to define the locations and paths of the thalwegs. In addition, user-mapped (deterministic) thalwegs may be used to incorporate interpretation.

The bar-train module allows the model builder to define and use deterministically created thalwegs. A series of $(X,Y)$ points that form a curved line defines the thalweg without random elements. Deterministic thalwegs may be based on geologic inference, seismic interpretation, or subsurface data. A typical model contains only a few deterministic thalwegs. More commonly, thalwegs are generated by a random process (see below).

The bar-train module has several specification parameters for delineation of the thalweg elements. The location of a thalweg is defined partly by its intersection with a point. This point may be chosen randomly within the entire model volume or within a selected region specified by the user, or it may be located at a well. The thalweg then is drawn through that point to follow a general trend, defined by a simple azimuth or by more complex information. Smaller scale sinuosity may be superimposed on this general trend by means of mathematical (e.g., sine) or stochastic (e.g., Gaussian) functions. All of these input variables may be defined in terms of geologically realistic probability distributions, from which values are drawn to locate and shape each thalweg.

Two sources of information may be used to stochastically create the thalwegs: trend maps and flow paths.

## Trend Maps

A trend map consists of a two-dimensional (2-D) grid that represents the spatial probability or relative abundance of some property, in this case the distribution of the thalwegs in a geologic model. Thalwegs are preferentially inserted in areas of high probability and, conversely, diverted away from areas of low probability. Because bars occur only in conjunction with thalwegs, the trend-map probabilities may also be considered as representing bar-facies probabilities. The algorithm used is similar to the modeling process described by Holden et al. (1998).

The process stochastically models a thalweg in the 2-D plane, conditioned by a point along the thalweg and influenced by the trend map. The algorithm operates by defining a probability density for the realizations of possible thalwegs and then splitting it into two factors. One defines the 2-D trend represented by the trend map, and the other is a one-dimensional Gaussian field that controls the sinuosity of the thalweg. A Metropolis-Hastings algorithm (Hastings, 1970) is used as follows: (1) form a tentative thalweg with stochastically generated wavelength, amplitude, and curvature; (2) determine how closely it fits the trend map; (3) calculate the existence probability of this tentative thalweg; (4) accept or reject the tentative thalweg based on this probability;

and (5) repeat the previous steps until enough thalwegs are created. Roxar (2001) provides operational guidelines.

### Flow Paths and Vector Fields

Modeling with flow paths and vector fields has been described by Jones and Larue (1997) and Jones (2001). In some depositional settings, a curving trend of a channel complex or lowstand systems tract may be interpreted from geological and geophysical data. The ability to model reservoir intervals interpreted to contain incised valleys requires the use of flow path conditioning in many projects. However, ordinary modeling of thalwegs assumes a generally straight trend, with superimposed sinuosity.

To specify curvature in the thalwegs, one or more flow paths must delineate the approximate route that thalwegs should follow across the area being modeled. Flow paths are defined by connected line segments (polylines). Although flow paths are intuitive, algorithms that use them to define thalwegs are complex and slow to execute. Instead, we use flow paths to calculate a vector field, which is then used to stochastically generate individual thalwegs. Figure 8 shows thalwegs that were influenced by a vector field. Note that the thalwegs follow the general trend of the flow paths but do not parallel them.

A vector field is a 2-D map array of azimuths. An azimuth (vector) is located at each $(X,Y)$ location of the geologic model, and the array of vectors specifies the local direction of flow in the zone being modeled and, hence, the desired orientation of thalwegs. Thalwegs are developed by moving through the vector field in a series of short steps, with the direction of each step provided in part by the local azimuth of the vector field. Randomly modifying the azimuth provides variation in the path of the channel. Jones (2001) describes the generation of the vector field from flow paths and use of the vector field to generate thalwegs. Precise definition of thalweg parameters and use of vector fields can generate thalwegs that mimic those of modern braid channels.

### Braid-bar Elements

Several types of fluvial elements were defined above in a hierarchical framework to generate the 3-D geologic model. The smallest object or depositional element is the braid bar. These are clustered to form bar sets, which are bars of similar geometries that characterize selected reaches of the river. Bar sets, in turn, cluster along the thalweg to form the bar train, which is inferred to extend the entire length of the braided system. Input parameters, however, enable the modeler to terminate bars or bar sets before reaching the downstream edge of the model.

Fluvial elements, including bars, bar sets, and bar trains, are controlled by parameters in the bar-train module. Varying several controls in the bar-train module enables the generation of diverse fluvial-facies architectures. The general shape or geometry of the bar may be selected from a list in the program (e.g., ellipsoid, prism); alternatively, the modeler may define a shape by creating and combining the top and basal surfaces of the bar. Dimensions of the bar element (width, length, and thickness) and the bar orientation relative to the thalweg are specified by the modeler. These may be set as constants, allowed to vary spatially, or specified by probability distributions. Erosional properties between bars may be set to depict either downstream erosion or accretion.

The number and spacing of bars in a bar set are specified by the modeler. These may be set as fixed values or distributions. For example, Van Wagoner (1995) has documented in the Castlegate Sandstone that upstream bars in a bar set are larger than the bars in the downstream end, and the spacing between bars is less at the upstream end of the bar set than downstream. In addition, the bar sets in a bar train typically are separated from each other by gaps that are larger than intrabar-set spacings. The bar-train parameters, however, provide sufficient flexibility to accommodate a range of possible spatial variations.

## APPLICATION OF FLUVIAL MODELING CAPABILITIES

As a result of the complex sequence-stratigraphic architecture of the reservoir, each depositional sequence was modeled separately. The external architecture for each of the 24 sequences was derived from surfaces interpreted from the 3-D seismic data and well-log correlations. In general, the abandonment surfaces above and below the lowstand systems tracts defined the top and base of each modeled reservoir. The sequence boundaries, which bound the base of the lowstand sandstones, were generated stochastically and were largely influenced by the well-log data and paleoflow interpretations. In a

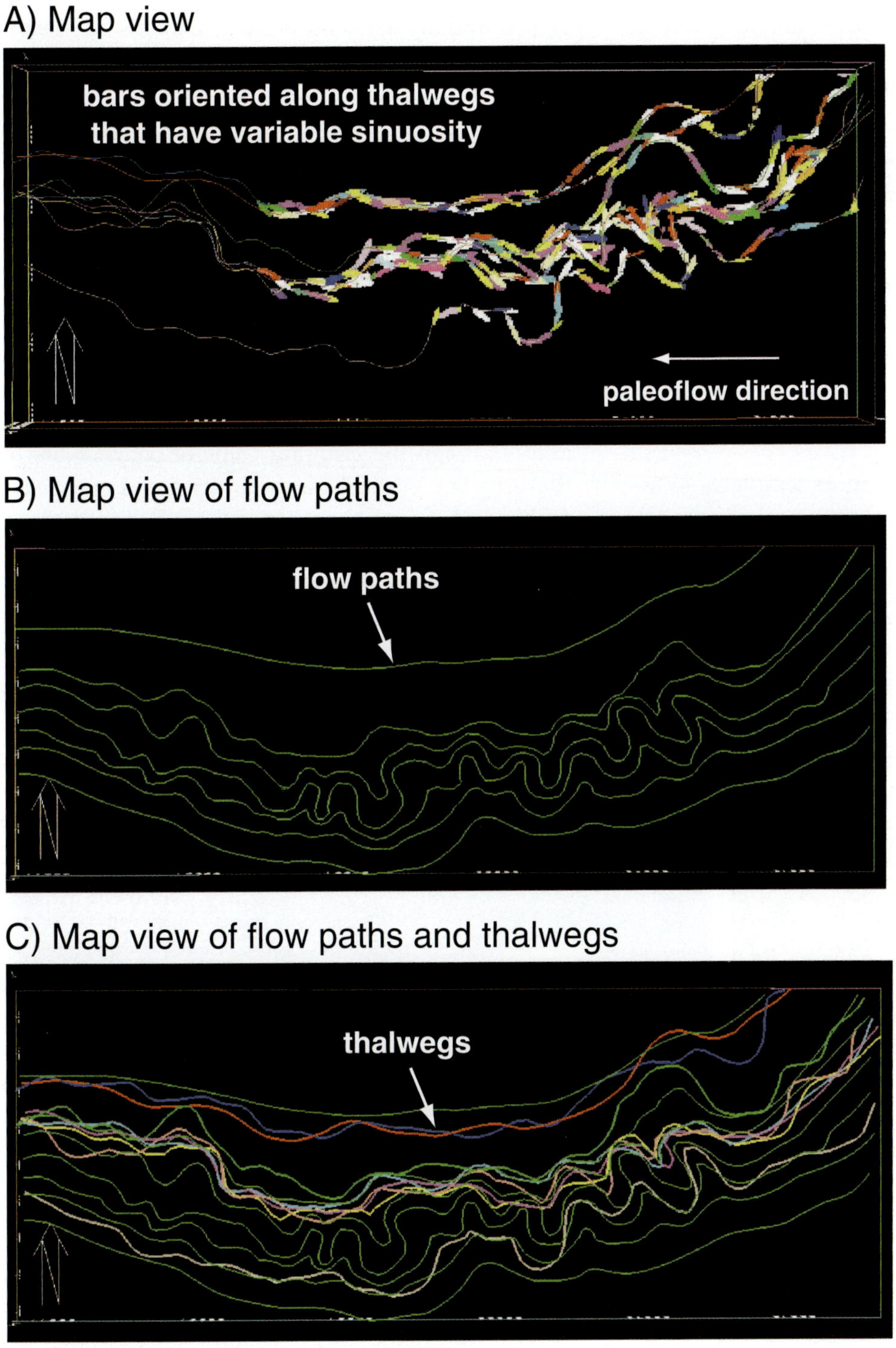


**Figure 8.** (A) An example of fluvial bars conditioned to user-defined flow paths. (B) Highly variable flow paths that were defined by the modeler and were used to generate a vector field. (C) Note the close match of the model-generated thalwegs to those of the user-defined flow paths (shown in green).

few reservoir intervals, the incised valleys containing lowstand sandstones were interpreted from 3-D seismic data to occur in relatively confined regions of the field. In these intervals, probability maps of sandstone occurrence were used to populate the model with the inferred fluvial facies.

The internal architecture of the reservoir intervals was modeled by two techniques. Proximal, amalgamated channels were modeled using standard, object-based modeling software to populate zones containing channel elements that are clustered to form channel complexes. The channel dimensions were obtained from subsurface-reservoir and outcrop-analog data.

Medial, amalgamated braid bars and distal, isolated braid bars were modeled using the bar-train module. Medial bar deposits were modeled using closely spaced bar and bar-set elements. Clustered thalwegs form amalgamated to semiamalgamated bar complexes. Distal braid-bar deposits were modeled using greater distances between individual bars and bar sets than those of the medial bars. Moreover, a divergent distribution of the thalwegs was used to simulate the more isolated character of the distal fluvial bars. Figure 9 shows selected vertical sections through the reservoir model of the distal braid bars. In all the reservoir intervals, the channel and bar-thickness distributions generated within the 3-D geologic model honored distributions determined from core description. Inferred widths and lengths honored those determined from 3-D seismic and outcrop-analog data.

The facies models were used as templates to assign petrophysical properties to the geologic model. Analysis of core and well logs showed distinct vertical trends of permeability in individual fluvial channels and bars, whereas porosity exhibits more poorly defined vertical trends. Hence, permeability was assigned initially to the individual elements (i.e., channels or bars) in the facies models, and porosity followed the permeability trends.

Although the vertical trends are very strong, the lateral continuity of permeability in a bar is of secondary importance. Hence, a general upward-increasing permeability trend was defined, referenced from the base of each fluvial channel or bar. For each cell in a bar, the following steps were performed: (1) establish the height of the cell above the bar's base; (2) interpolate a permeability value from the trend; (3) add to the trend a random, uncorrelated variate value (with standard deviation small relative to the trend); and (4) assign the resulting value of permeability to the cell. The vertical permeability trends modeled within individual channels or bars honored those trends observed from detailed core analyses. To simulate the presence of rip-up lag beds observed in the core, low-permeability values were assigned to the basal layers of some randomly selected fluvial

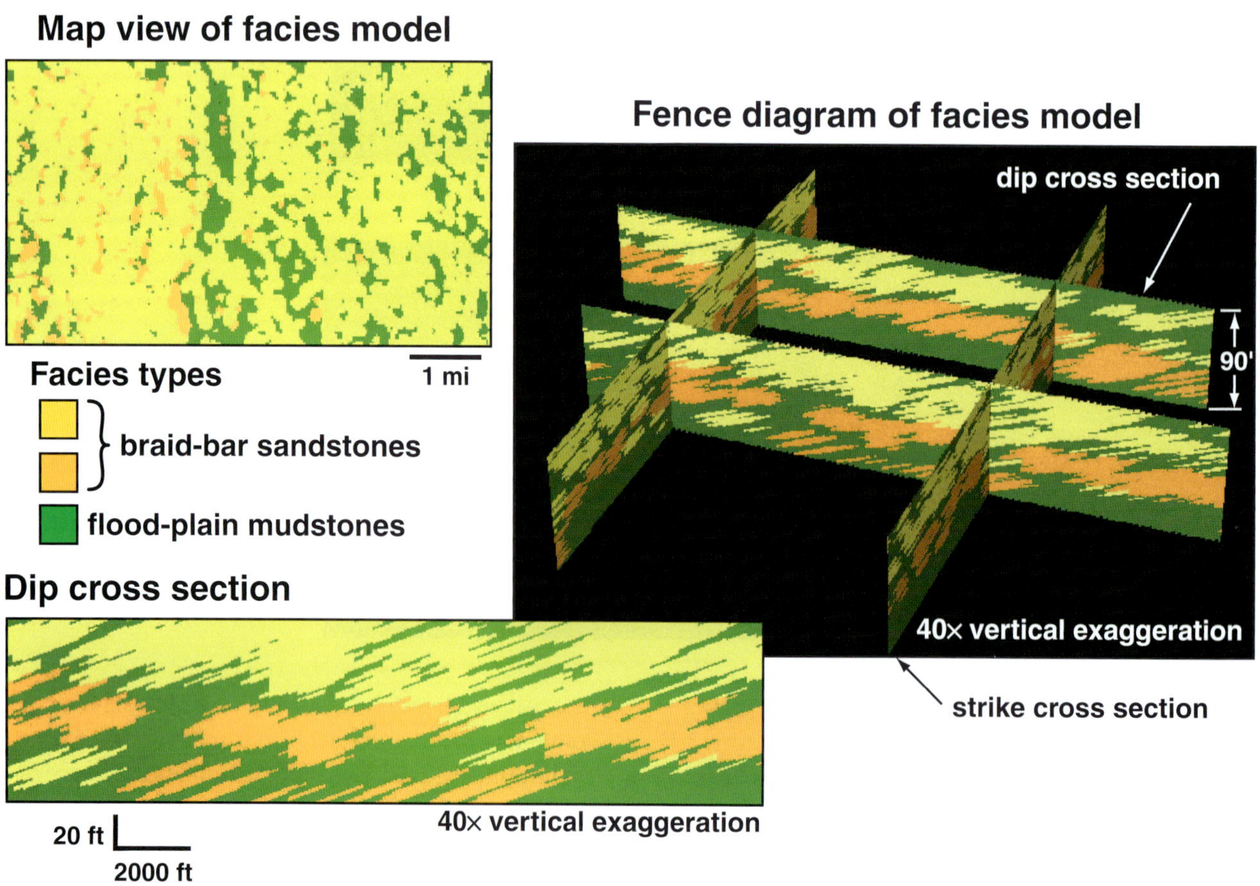


Figure 9. Facies model of isolated braid-bar deposits. Note that the nonreservoir mudstone intervals occur as horizontal layers between vertically stacked channel complexes, as well as inclined layers that parallel the depositional fabric of the braid bars.

elements. Although rip-up clast beds comprise less than 5% of the depositional facies, they act as barriers to fluid flow.

Porosity values were based on the covariance between core permeability and core porosity and the variances of well-log porosity and core permeability. Sampling from conditional distributions of permeability and porosity provided the porosity value assigned to the cells. Cumulative distributions of permeability and porosity, and a crossplot of permeability vs. porosity values from the geologic models, closely match those from core and well-log analyses.

Three-dimensional geologic models of amalgamated channel deposits possess a simplistic internal architecture characterized by relatively high vertical and lateral continuity. The nonreservoir mudstone intervals occur as isolated horizontal layers that bound the relatively high-permeability fluvial-channel sandstones. In contrast, however, the 3-D geologic models of the amalgamated and isolated braid-bar deposits possess a complex internal reservoir architecture that accurately represents the highly variable architecture of these deposits. Nonreservoir mudstone intervals occur as horizontal layers between vertically stacked channel complexes, as well as inclined layers that parallel the depositional fabric of the braid bars (Figure 9). Moreover, because petrophysical properties were assigned individually to bar elements and because a broad range in these values exists, the petrophysical properties in adjacent bars along the thalweg may vary markedly.

Finally, reservoir-simulation models derived from the geologic models exhibit a favorable comparison to values of field-measured permeability from pressure-transient tests. Previous geologic models constructed by means of standard modeling techniques required a fourfold adjustment in permeability to match field measurement. In contrast, the current geologic models required no permeability adjustments in the reservoir interval modeled as amalgamated channel complexes and adjustments ranging from none to twofold when calibrating permeability values in the braided-bar deposits (A. Stackel, 2002, personal communication).

## CONCLUSIONS

A recently developed computer program accurately characterized three interpreted styles of fluvial deposition: (1) amalgamated channels, (2) amalgamated to semiamalgamated braid bars, and (3) isolated braid bars. The bar-train method enables the modeler to generate geologic models that better represent the depositional geometry of braid bars and their spatial arrangement. Furthermore, the resulting realizations better depict the shale beds that drape abandoned braid bars as well as the shale layers bounding the larger order channel complexes. Accordingly, the computer program has the capability to capture details of the heterolithic character of these reservoir intervals.

The resulting object-based geologic models provide an improved description of the external and internal hierarchical framework for the subsurface reservoir intervals under study. Moreover, the models provide a more accurate description of the complex architecture of the lowstand fluvial sandstones and the internal spatial distribution of the nonreservoir mudstones. As a result of the detailed description of the reservoir, the models more closely match pressure-transient well tests than did previous 3-D geologic models generated by other methods.

## ACKNOWLEDGMENTS

We acknowledge the efforts and dedication of Trond Skjerven and Frode Viste of Roxar asa. in development of the bar-train code. We also thank Gordon Moir and AAPG reviewers James Jennings, Jeffrey May, and Alister MacDonald for suggestions that improved the manuscript.

## REFERENCES CITED

Banks, N. L., 1973, The origin and significance of some downcurrent-dipping cross-stratified sets: Journal of Sedimentary Petrology, v. 43, p. 423–427.

Bratvold, R. B., L. Holden, T. Svanes, and K. Tyler, 1995, STORM: Integrated 3-D stochastic reservoir modeling tool for geologists and reservoir engineers: Society of Petroleum Engineers Computer Applications, p. 58–67.

Bridge, J. S., 1985, Paleochannel patterns inferred from alluvial deposits: A critical evolution: Journal of Sedimentary Petrology, v. 55, p. 579–589.

Bristow, C. S., and J. L. Best, 1993, Braided rivers: Perspectives and problems, *in* C. S. Bristow and J. L. Best, eds., Braided rivers: Geological Society of America Special Publication 75, p. 1–11.

Bristow, C. S., J. L. Best, and P. J. Ashworth, 2002, GPR 2002: The use of GPR in developing a facies model for large sandy braided river, Brahmaputra River, Bangladesh, *in* D. Noon, G. Stickley, and D. Longstaff, eds., Proceedings of the Eighth International Conference on Ground Penetrating Radar: Society of Photo-optical

Instrumentation Engineers Proceedings Series, v. 4084, p. 95–100.

Bromley, M. H., 1991, Variations in fluvial style as revealed by architectural elements, Kayenta Formation, Mesa Creek, Colorado, U.S.A.: Evidence for both ephemeral and perennial fluvial processes, *in* A. D. Miall and N. Tyler eds., The three-dimensional facies architecture of terrigenous clastic sediments and its implications for hydrocarbon discovery and recovery: SEPM Concepts in Sedimentology and Paleontology 3, p. 94–102.

Deutsch, C. V., and A. G. Journel, 1998, GSLIB: Geostatistical software library and user's guide, 2d ed.: New York, Oxford University Press, 369 p.

Goovaerts, P., 1997, Geostatistics for natural resources evaluation: New York, Oxford University Press, 483 p.

Haldorsen, H. H., and E. Damsleth, 1990, Stochastic modeling: Journal of Petroleum Technology, v. 42, p. 404–412.

Hastings, W. K., 1970, Monte Carlo sampling methods using Markov chains and their applications: Biometrika, v. 57, p. 97–109.

Hatløy, A. S., 1994, Numerical facies modeling combining deterministic and stochastic methods, *in* J. M. Yarus and R. L. Chambers, eds., Stochastic modeling and geostatistics: Principles, methods, and case studies: AAPG Computer Applications in Geology 3, p. 109–120.

Holden, L., R. Hauge, O. Skare, and A. Skorstad, 1998, Modeling of fluvial reservoirs with object models: Mathematical Geology, v. 30, p. 473–496.

Jones, C. R., J. C. Van Wagoner, T. M. Demko, R. W. Wellner, G. G. McCrimmon, S. T. Hasiotis, R. T. Beaubouef, and H. R. Feldman, 2002, Large, prograding fluvial megafan complexes: Influence of climate cyclicity on reservoir architecture (abs.): AAPG Annual Meeting Program, v. 11, p. A88.

Jones, T. A., 2001, Using flowpaths and vector fields in object-based modeling: Computers and Geosciences, v. 27, p. 133–138.

Jones, T. A., and D. K. Larue, 1997, Object-based modeling and deepwater depositional systems, *in* V. Pawlowsky-Glahn, ed., Proceedings of International Association for Mathematical Geology '97, the Third Annual Conference of the IAMG: Barcelona, Spain, International Center for Numerical Methods in Engineering, p. 438–443.

Leopold, L. A., and G. M. Wolman, 1957, River channel patterns: Braided, meandering and straight: U. S. Geological Survey Professional Paper 282-B, 85 p.

Patterson, P. E., A. R. Sprague, R. E. Hill, and K. M. McDonald, 1995, Sequence stratigraphy and fluvial architecture, Farrer and Tuscher formations (Campanian), Tusher Canyon, Utah (abs.): AAPG Annual Meeting Program, v. 4, p. P74A.

Roxar, 2000a, RMSgeomod user guide, release 6.0: Stavanger, Norway, Roxar asa., 119 p.

Roxar, 2000b, RMSgeoplex user guide, release 6.0: Stavanger, Norway, Roxar asa., 121 p.

Roxar, 2000c, RMSbase user guide, release 6.0: Stavanger, Norway, Roxar asa., 175 p.

Roxar, 2001, Facies: Elementary bar-train modeling user guide, release 6.0: Stavanger, Norway, Roxar asa., 32 p.

Shukla, U. K., I. B. Singh, P. Srivastava, and D. S. Singh, 1999, Paleocurrent patterns in braid-bar and point-bar deposits: Examples from the Ganga River, India: Journal Sedimentary Research, v. 69, p. 992–1002.

Smith, N. D., 1971, Transverse bars and braiding in the Lower Platte River, Nebraska: Geological Society of America Bulletin, v. 82, p. 3407–3420.

Sprague, A. R., et al., 2002, The physical stratigraphy of fluvial strata: A hierarchical approach to the analysis of genetically related stratigraphic elements for improved reservoir prediction (abs.): AAPG Annual Meeting Program, v. 11, p. A167–A168.

Van Wagoner, J. C., 1995, Sequence stratigraphy and marine to nonmarine facies architecture of foreland basin strata, Book Cliffs, Utah, U.S.A., *in* J. C. Van Wagoner and G. T. Bertram, eds., Stratigraphy of foreland basin deposits: AAPG Memoir 64, p. 137–224.

Shrestha, R. K., S. Faruqi, M. Boeckmann, and M.-K. Wang, 2006, Three-dimensional seismic inversion and geostatistical reservoir modeling of the Rotliegend gas field, *in* T. C. Coburn, J. M. Yarus, and R. L. Chambers, eds., Stochastic modeling and geostatistics: Principles, methods, and case studies, volume II: AAPG Computer Applications in Geology 5, p. 237–251.

*Chapter 18*

# Three-dimensional Seismic Inversion and Geostatistical Reservoir Modeling of the Rotliegend Gas Field

**R. K. Shrestha**[1]
*Halliburton Digital and Consulting Solutions*
*Houston, U.S.A. (formerly Veritas Digicon, Inc.)*

**S. Faruqi**
*Occidental Oil and Gas Corporation*
*Houston, U.S.A. (formerly Veritas Digicon, Inc.)*

**M. Boeckmann**
*RWE-Dea AG Exploration and Production*
*Hamburg, Germany*

**M.-K. Wang**
*Eni Petroleum*
*Houston, U.S.A. (formerly Veritas Digicon, Inc.)*

## ABSTRACT

This chapter describes a study of the Rotliegend gas field in Germany in which impedance-constrained facies and property models were constructed to reevaluate the field and guide future development.

Three-dimensional (3-D) stochastic seismic inversion produces a high-resolution acoustic impedance volume by integrating well data with fine vertical sampling and seismic data with dense areal sampling. This acoustic impedance volume can be used as soft data in geostatistical reservoir property modeling. The dense areal sampling in the 3-D seismic grid provides valuable information in interwell areas. Impedance data are particularly useful where well data are sparse.

[1]*Present address*: ConocoPhillips Company, Houston, Texas, U.S.A.


DOI:10.1306/1063819CA53238

## INTRODUCTION

This reservoir characterization study created detailed three-dimensional (3-D) reservoir models to aid the reevaluation of the Upper Permian Rotliegend gas field located to the north of Hannover in northwestern Germany. The project was carried out in two phases: (1) the seismic inversion phase, to construct acoustic impedance models using deterministic and stochastic inversion; and (2) the reservoir modeling phase, implemented to generate impedance-constrained facies models and facies-dependent porosity and permeability models. The seismic inversion and geostatistical reservoir modeling were performed using the Reservoir Characterization Research and Consulting $(RC)^2$ software.

## RESERVOIR GEOLOGY

The two main producing zones of the Rotliegend gas field are the Wustrow and Ebstorf members of the Hannover Formation, composed mostly of interbedded sand-shale sequence, situated at the depths of 4500–5000 m (14,700–16,400 ft). The Wustrow Member, containing a thick shoreline sandstone unit, is the primary gas reservoir in this field. Reservoir quality in the Rotliegend gas field is governed primarily by the diagenetic alteration of the sandstones (Hock et al., 1995), including cementation of the primary pore spaces by illite, kaolinite, quartz, and calcite. The presence of chlorite helps preserve the primary porosity in the Wustrow sand. The Rotliegend reservoirs, sourced from the carboniferous coal, are immediately overlain by the Zechstein salt, which serves as the top seal.

## INPUT DATA

The migrated 3-D seismic data volume covers about 510 $km^2$ (196 $mi^2$) and consists of 1273 in-lines and 848 cross-lines from 2.0 to 3.7 sec, both spaced 25 m (82 ft) apart (Figure 1). Six seismic time markers (top A2, base Zechstein, top Wustrow, top Ebstorf, top Dethlingen, and base Dethlingen) define the reservoir framework. In addition, two pseudo-time markers, one at 100 msec above the top A2, and the other at 100 msec below the base Dethlingen horizon, were generated to serve as the upper and lower limits of seismic inversion (Figure 2). Twenty-six wells possessed sonic and density logs and were used for deterministic and stochastic inversions.

The facies, impedance, porosity, and permeability logs, along with the time and depth markers from 35 wells, the seismic inversion model, and the time and depth surfaces were used in the stochastic modeling of the Rotliegend gas field.

## DATA PREPARATION

Considerable time was spent in quality checking and preparing the seismic and well data, bearing in mind that the quality and integrity of data determine the accuracy of the product. In most wells, the density logs were recorded only in the reservoir zone; therefore, pseudodensity curves at these wells were generated from the available sonic logs using empirical relations. These pseudologs were subsequently spliced into the measured density curves for use in seismic inversion. Check-shot corrections were applied to the sonic curves in 15 wells having check-shot data.

## SEISMIC INVERSION

Deterministic inversion produces an absolute impedance model with resolution defined by the inherent seismic bandwidth. The deterministic inversion process includes wavelet estimation and the generation of 3-D low-frequency reflectivity balancing and impedance models by integrating associated well and seismic information (Shrestha and Boeckmann, 2002).

Stochastic inversion builds upon the tasks accomplished in the deterministic inversion to create high-resolution acoustic impedance models. The process generates multiple realizations of the impedance pseudologs at each seismic trace location using sequential simulation algorithms. The realization having the synthetic seismic response that best correlates with the actual seismic trace at that location is retained (Bortoli et al., 1992).

## WAVELET ESTIMATION

Wavelet extraction facilitates synthetic seismic generation, well-seismic tie analysis, time-depth relation derivation, time log generation, and wavelet estimation at well locations.

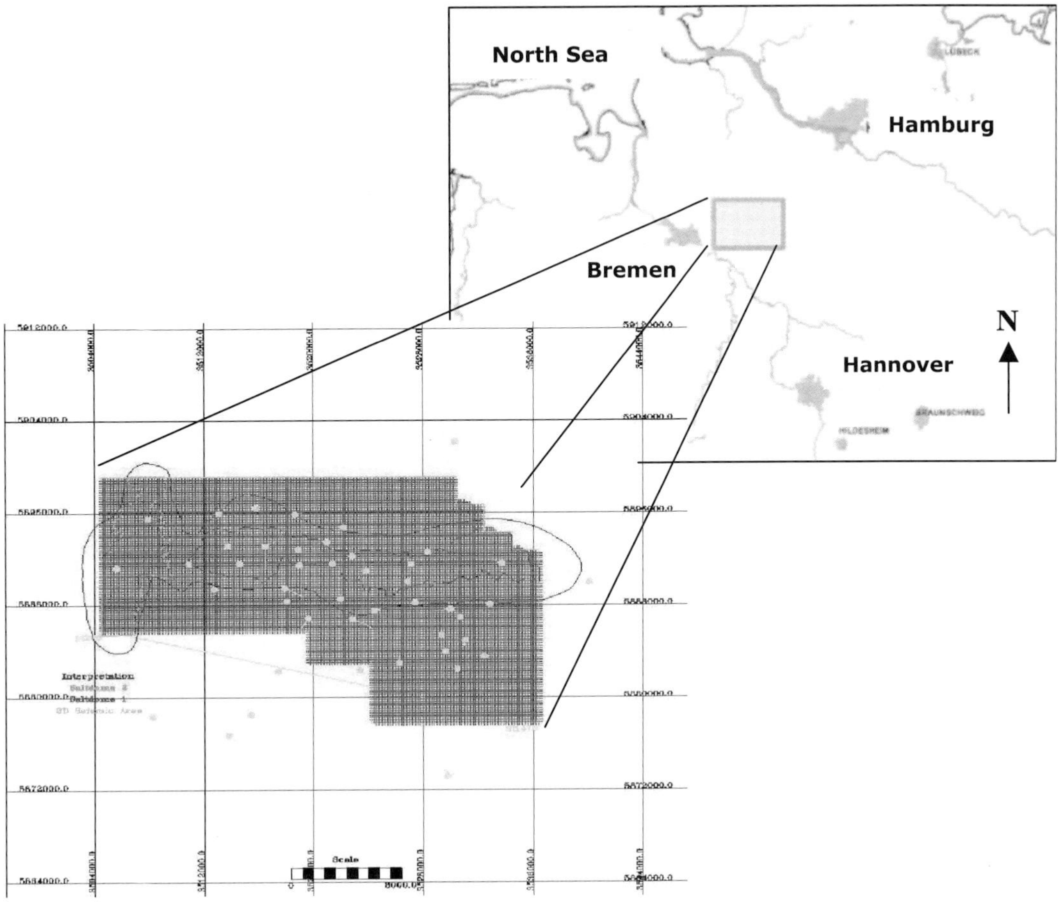


**Figure 1. Base map of the area of interest showing wells, 3-D seismic coverage, and polygon in purple color representing salt-influenced area.**

Synthetic seismograms were generated using sonic and density logs at 26 wells. The seismic data, according to a-priori information, were reversed in polarity. Therefore, a phase correction of 180° was applied to the entire 3-D seismic volume before wavelet extraction.

The seismic-to-well-tie and the estimated wavelet at well B are shown in Figure 3. The wavelets estimated at all 26 well locations are consistent in their amplitude and phase characteristics. A composite wavelet constructed using all estimated wavelets displays a near-zero-phase wavelet with a peak frequency at 30 Hz.

## DETERMINISTIC INVERSION

Seismic spectral analysis at every 20 lines revealed an 8–67-Hz bandwidth in most of the seismic data. Spectral whitening was accomplished by applying a zero-phase deconvolution algorithm to the entire seismic data set.

The spectral extrapolation or sparse-spike deconvolution algorithm used a prediction error filter to predict the high-frequency amplitudes based on the pass-band frequency of the seismic data (Oldenburg et al., 1983). The process assumed the existence of a reflectivity series having the fewest number of reflection coefficients or sparse spikes, such that, when band limited, it would approximate the seismic data. The spectral extrapolation also prevented the occurrence of inverted traces having large side lobes as a result of seismic bandwidth reflectivities and thus helped preserve a blocky appearance of the inversion.

The next step was to construct a 3-D amplitude balancing model by interpolating reflectivity envelopes at each seismic trace location using envelopes of the synthetic seismograms derived during wavelet extraction at well locations. It was assumed that the well information (i.e., synthetic seismogram)

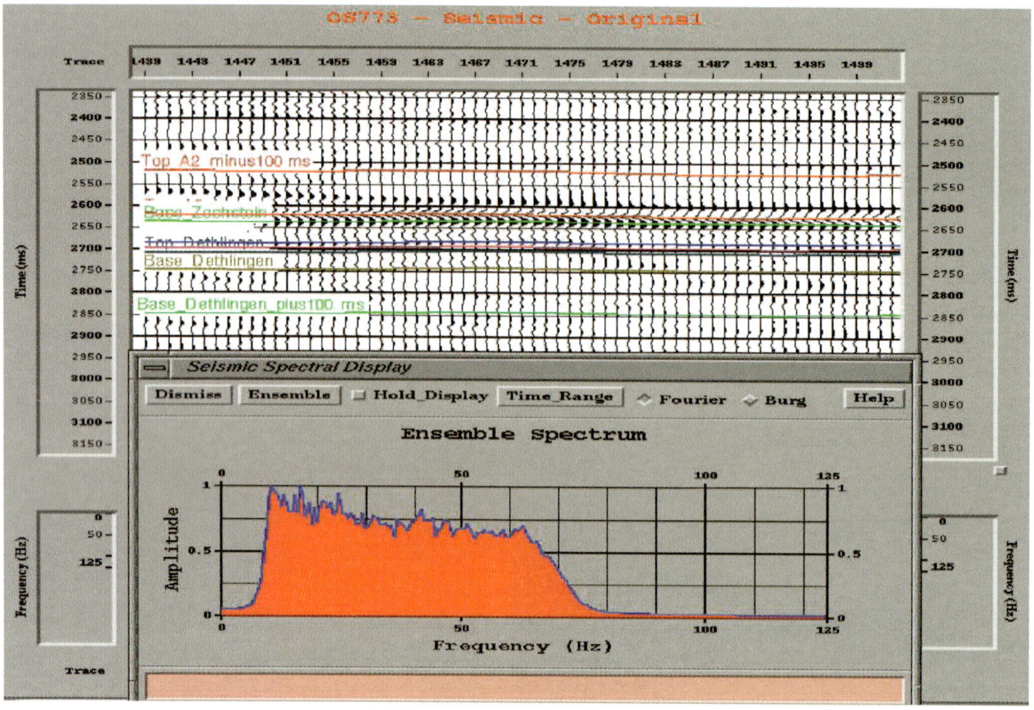


Figure 2. Line 773 of the original seismic data displaying the interpreted horizons (top) and the frequency spectrum (bottom) using $(RC)^2$ software.

was a better predictor of amplitude trends than the seismic data. Again, the objective was to produce impedance traces with absolute values.

Marker horizons from both the wells and seismic interpretation were used to guide the estimation process as well as to ensure that the extracted balancing

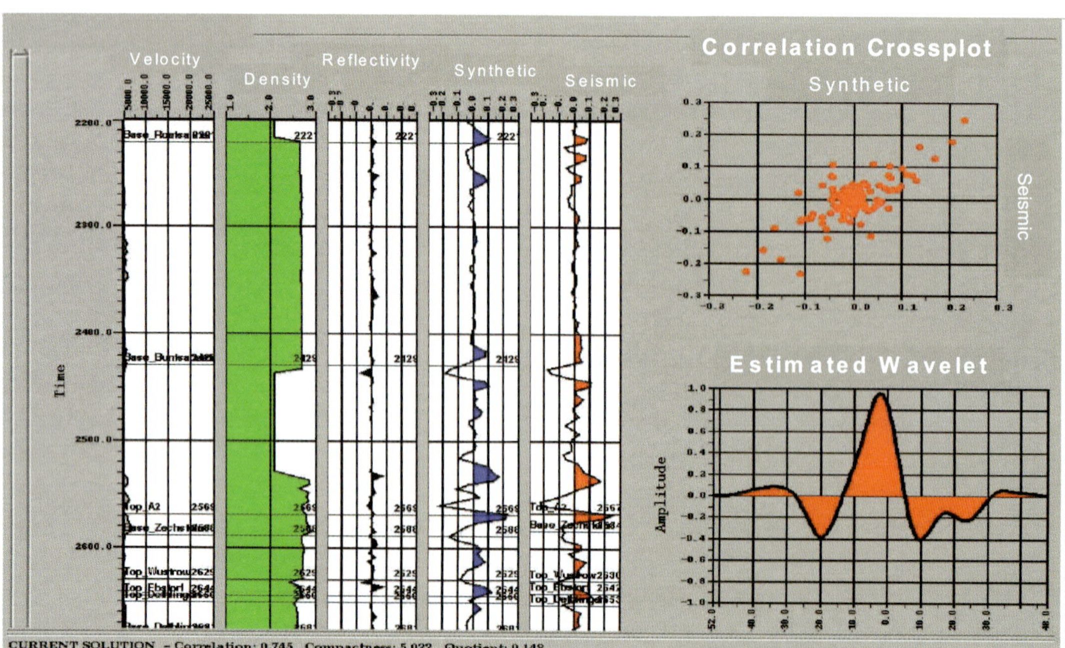


Figure 3. Correlation between well and seismic data at well B and line 1020. Also displayed are the synthetic-to-seismic crossplot (upper right) and the estimated wavelet (bottom right).

traces tied the seismic reflection. The balancing model was then applied to the sparse-spike data to obtain the balanced traces scaled to the modeled reflectivity envelopes.

The low-pass filtered impedance logs in the time domain saved during wavelet extraction at 26 well locations were used to build the 3-D background impedance model through interpolation.

The deterministic inversion produced pseudologs of acoustic impedance values at each seismic trace location by spectrally combining the inverted balanced reflectivity trace and the low-frequency background model about the transition frequency of 8 Hz. The deterministic inversion results of North Hannover data consist of lines and traces with a 25-m (82-ft) spacing bounded by markers at top A2 minus 100 msec and base Dethlingen plus 100 msec.

## STOCHASTIC INVERSION

Stochastic inversion integrates the fine vertical sampling of the log data with the dense areal sampling of the seismic data to produce a high-resolution earth model of acoustic impedance, density, or velocity using geostatistical techniques (Bortoli et al., 1992; Haas and Dubrule, 1994). A large number of equiprobable impedance traces are simulated at a seismic trace location selected through a random sequencing within a 3-D framework. These impedance traces are converted to a reflectivity series and then convolved with an assigned wavelet to generate the synthetic seismograms. The synthetic seismogram that matches best with the actual seismic trace is retained as the inversion solution at that location. This procedure is repeated at the next location and is continued until the entire seismic volume is filled with pseudologs of acoustic impedances at all trace locations.

The stochastic inversion employs sequential Gaussian simulation with colocated cokriging (Xu et al., 1992) to produce an acoustic impedance model incorporating the high vertical resolution of the acoustic impedance well logs. This is accomplished using the deterministic inversion model as soft data to constrain the simulation of hard acoustic impedance well-log data.

For data preparation, the phase-corrected seismic data were first balanced using the same amplitude-balancing model constructed for use in deterministic inversion. Next, both the balanced seismic and deterministically inverted data were resampled at 0.5 msec to match the impedance logs generated at the 26 well locations.

Both directional areal ($X$, $Y$) and vertical ($Z$) semivariograms for the impedance data were then modeled for each of the seven zones. The northwest direction of the major axis of anisotropy roughly coincides with the trend of the Rotliegend Graben system (Gast, 1988).

One hundred realizations of impedance data were simulated at each seismic trace location in the grid using the resampled impedance well data as the hard data and the resampled deterministically inverted impedance trace as the soft data. Synthetic seismograms for all impedance traces were generated after converting to reflectivity traces and convolving them with a zero-phase 30-Hz Ricker wavelet having amplitude and frequency characteristics similar to the composite wavelet. The synthetic seismogram giving the best match (based on correlation coefficient of synthetic-seismic amplitudes) with the actual seismic data was retained as the solution at that particular trace location. A random walk through the 3-D framework selected the next location and repeated the same procedure to produce the seismic inversion solution at that location. The procedure continued until the entire 3-D volume was filled with pseudologs of acoustic impedance.

A correlation coefficient map representing a measure of correlation between the synthetic or model trace and the actual seismic trace (Figure 4) is output at the completion of stochastic inversion. Although not a direct measure of uncertainty, the correlation coefficient map is an indicator of uncertainty in stochastic inversion (Bortoli et al., 1992). A high degree of correlation between synthetic and seismic traces enhances confidence and indicates a good calibration between seismic and synthetic data during wavelet estimation and vice versa. Figure 4 displays correlation coefficients as high as 95% outside the salt influence area that support the results of stochastic inversion. Because the seismic data quality in the area influenced by the salt body is poor, the stochastic inversion algorithm is only permitted to generate one impedance curve at each trace location. In other words, the poor-quality seismic traces were not allowed to contribute to the simulation.

Figures 5 and 6 show the stochastic and deterministic inversion results for line 1202. Stochastic inversion leads to enhanced vertical resolution, albeit nonunique, in the interwell space. The Wustrow

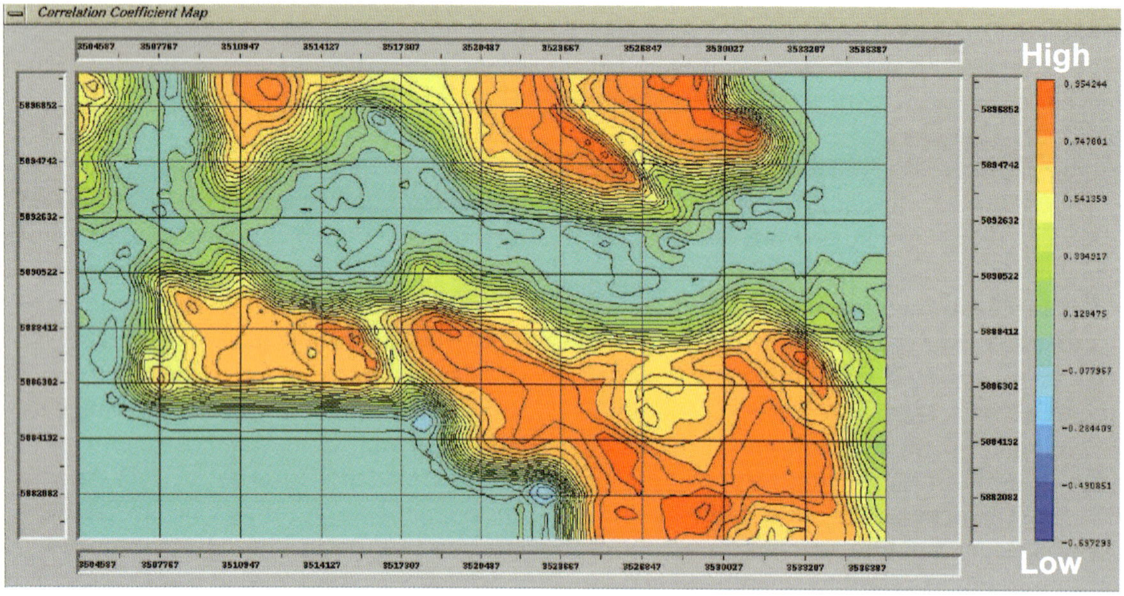


Figure 4. Contour map of the correlation between synthetic and seismic data after stochastic inversion of the 3-D seismic data.

reservoir interval, characterized by lower impedance values as compared to the encasing rocks, is better defined in the stochastic inversion. Variation in acoustic impedance values indicates possible change in porosity and/or fluid type in the reservoir interval.

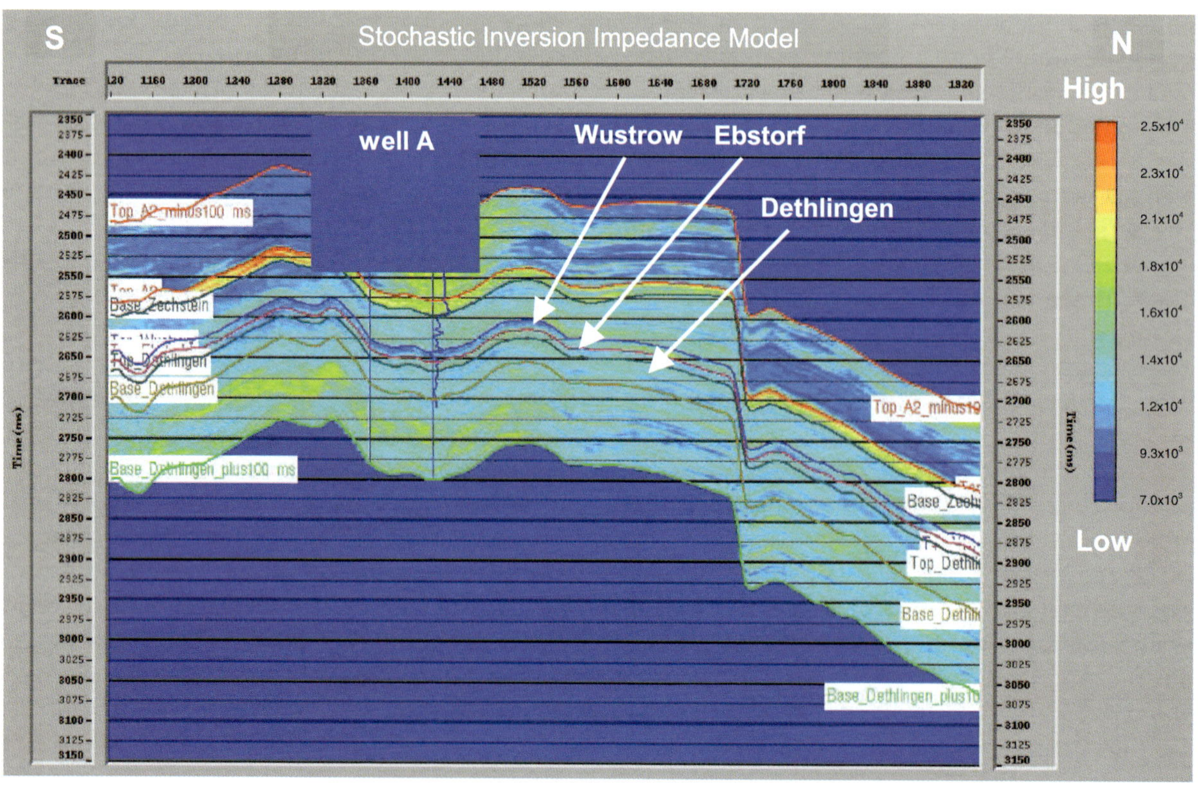


Figure 5. Results of stochastic inversion at line 1202 with the well A impedance log overlay along a south–north cross section.

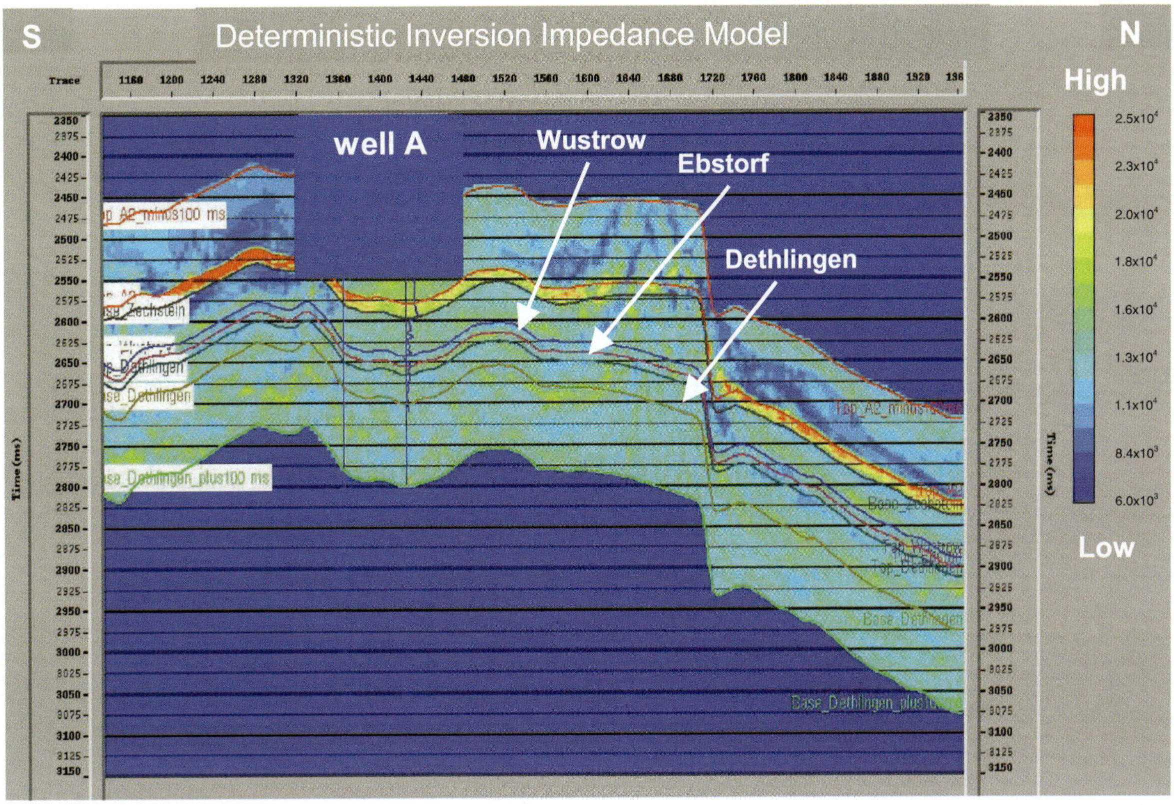


**Figure 6.** Results of deterministic inversion at line 1202 with the well A impedance log overlay along a south–north cross section.

## IMPEDANCE-DEPENDENT FACIES MODELING

Many petrophysical properties are dependent on lithofacies. Lithofacies are characterized by similar petrophysical and fluid-saturation properties.

Using a facies-based approach to modeling continuous petrophysical properties, such as porosity and permeability, has many advantages over treating the data as a single population. First, it can account for varying spatial continuity of reservoir properties in different facies. Second, facies-based property models preserve the facies-specific univariate distribution of the property (e.g., porosity) and bivariate relationships between two properties (e.g., porosity and permeability). Facies classification also facilitates distinction between the reservoir and nonreservoir rocks.

Thirty different lithofacies were grouped into six categories of lithotypes representing shoreline sand (A), illite-coated sand (B), aeolian sand (C), mudstones (D), fanglomerates (E), and anhydrite, carbonates, and volcanics (F). Facies A and B are reservoir sands; facies C is nonreservoir sand; and lithotypes D, E, and F are nonreservoir rocks. These lithotypes will hereafter be referred to as facies A to F.

### Facies-modeling Workflow

The generalized workflow for facies modeling consists of defining facies and creating facies logs, extracting facies fractions for reservoir zones, modeling areal and vertical semivariograms, generating facies-based impedance histograms, building facies proportion curves, snapping stochastic inversion data to the depth grid, and constructing a facies model employing sequential indicator simulation (SIS) with a local proportions algorithm.

### Facies Fractions

The proportions of each facies in each of the three zones were first estimated (Table 1).

Table 1. Facies proportions for the three zones.

| *Zone* | *Facies* | *Fractions* |
|---|---|---|
| Wustrow | A | 0.28 |
| | B | 0.29 |
| | C | 0.09 |
| | D | 0.34 |
| Ebstorf | A | 0.043 |
| | B | 0.26 |
| | C | 0.12 |
| | D | 0.57 |
| | E | 0.007 |
| Dethlingen | A | 0.061 |
| | B | 0.8068 |
| | C | 0.017 |
| | D | 0.04 |
| | E | 0.072 |
| | F | 0.0032 |

## Facies Semivariogram Modeling

Areal (directional) and vertical facies semivariograms were modeled from the facies indicator logs. As an example, the zone-directional semivariograms and vertical facies semivariogram for facies A in the Wustrow interval are shown in Figures 7 and 8, respectively. The northwest–southeast orientation of the major axis of anisotropy roughly coincides with the trend of the Rotliegend Graben system (Gast, 1988).

## Facies-dependent Impedance Histograms

Impedance histograms for various facies were constructed from the impedance and facies indicator log data. Figure 9 contains example histograms for facies A to D of the Wustrow zone. Considerable overlap is observed in the univariate statistics representing facies C and D, whereas the histograms representing facies A to C exhibit good separation.

## Facies Calibration Proportion Curves

The data represented in the histograms (Figure 9) were normalized and combined to produce a facies-vs.-impedance calibration diagram (Lo and Bashore, 1999) for the Wustrow interval (Figure 10). The vertical and horizontal axes in this figure represent the facies probability and impedance, respectively. Facies proportion curves for the Ebstorf and Dethlingen intervals were also constructed in the same manner.

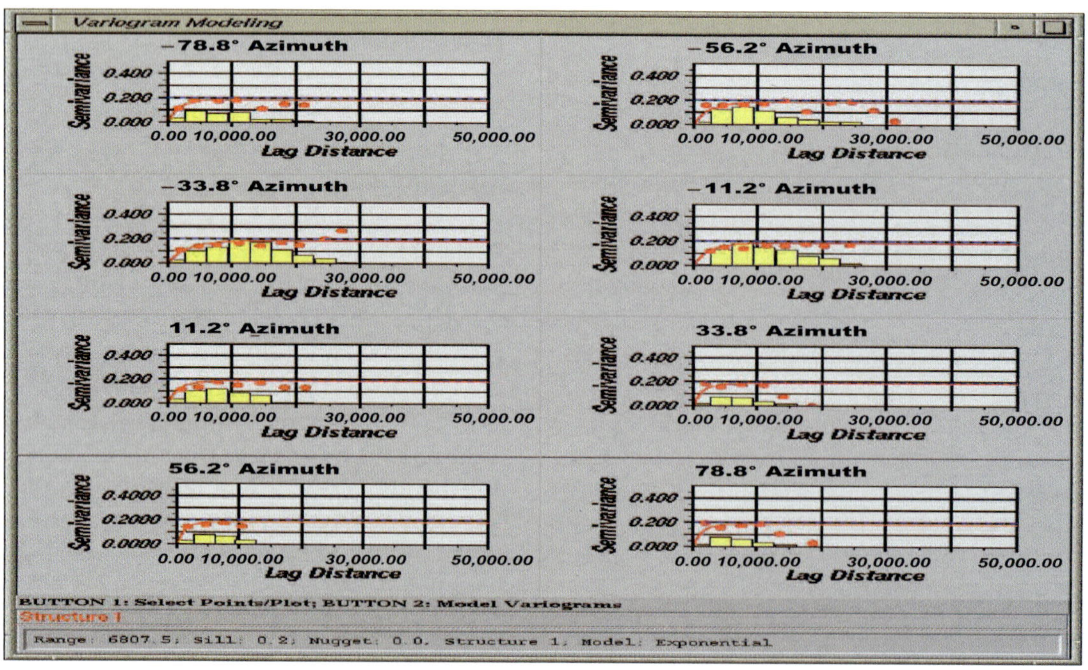


Figure 7. Areal (directional) semivariograms for facies A in the Wustrow interval using $(RC)^2$ software. Lag distance is in meters.

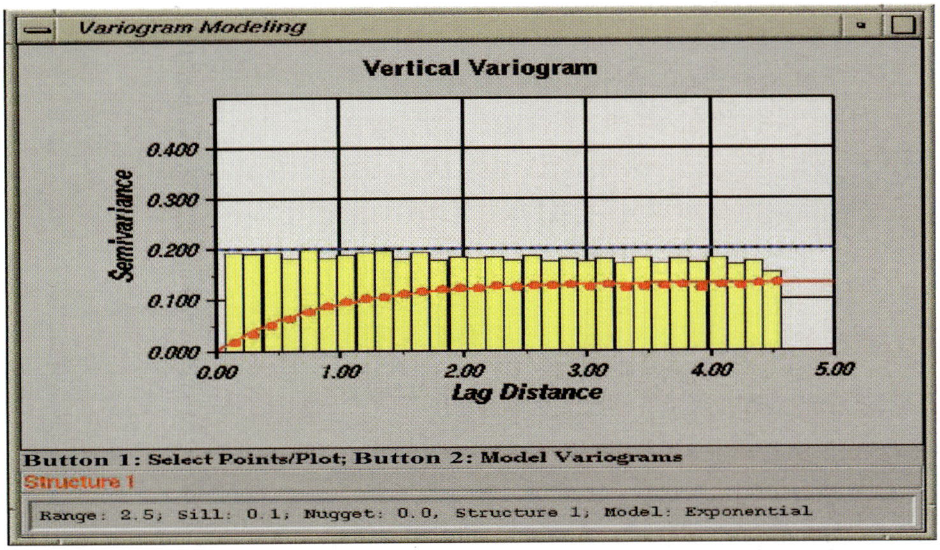


**Figure 8.** Vertical facies semivariogram for facies A in the Wustrow interval. The yellow vertical bars represent the number of data points at each lag and indicate the reliability of the variogram. Lag distance is in meters.

## Stochastic Inversion Data for Facies Modeling

The stochastic inversion results were snapped to the depth grid for each of the three zones of interest for calibration purposes during facies modeling. The predicted impedance values in the Wustrow interval were crossplotted against the impedance log at well location and showed excellent correlation (Figure 11).

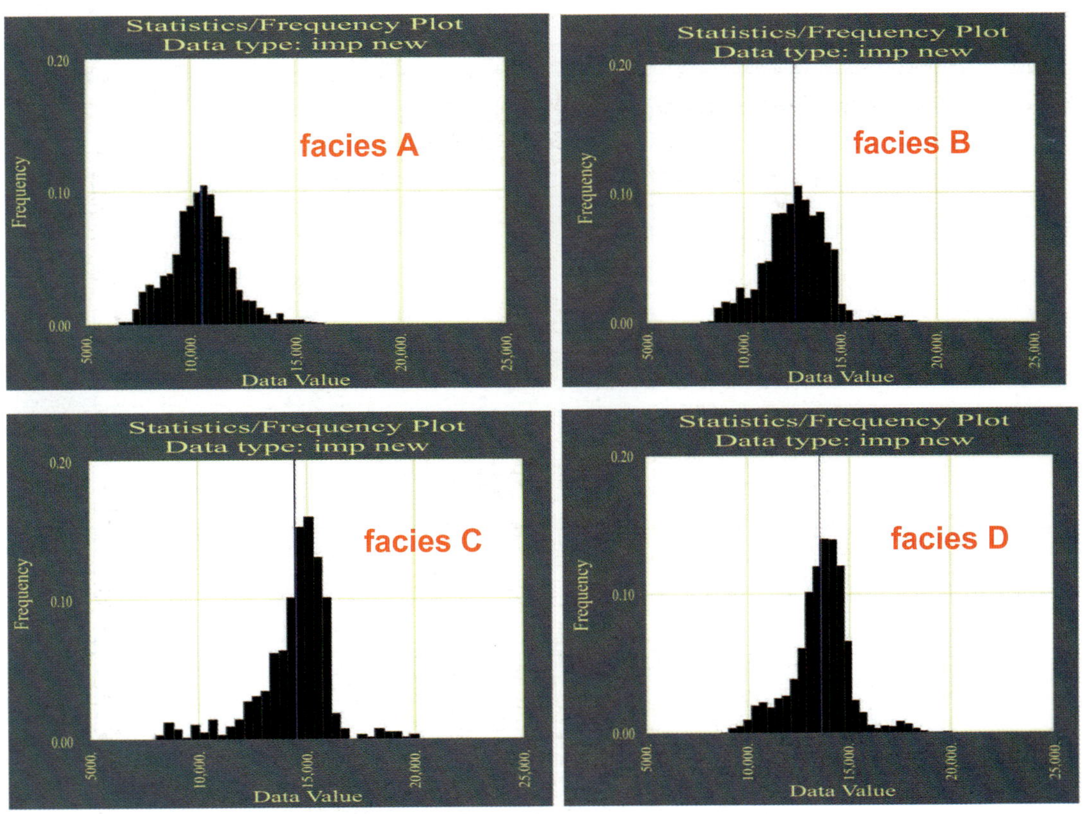


**Figure 9.** Impedance histograms for facies A (mean = 10,438; standard deviation = 1500), facies B (mean = 12,618; standard deviation = 1692), facies C (mean = 14,392; standard deviation = 1826), and facies D (mean = 13,553; standard deviation = 1525) in the Wustrow interval.

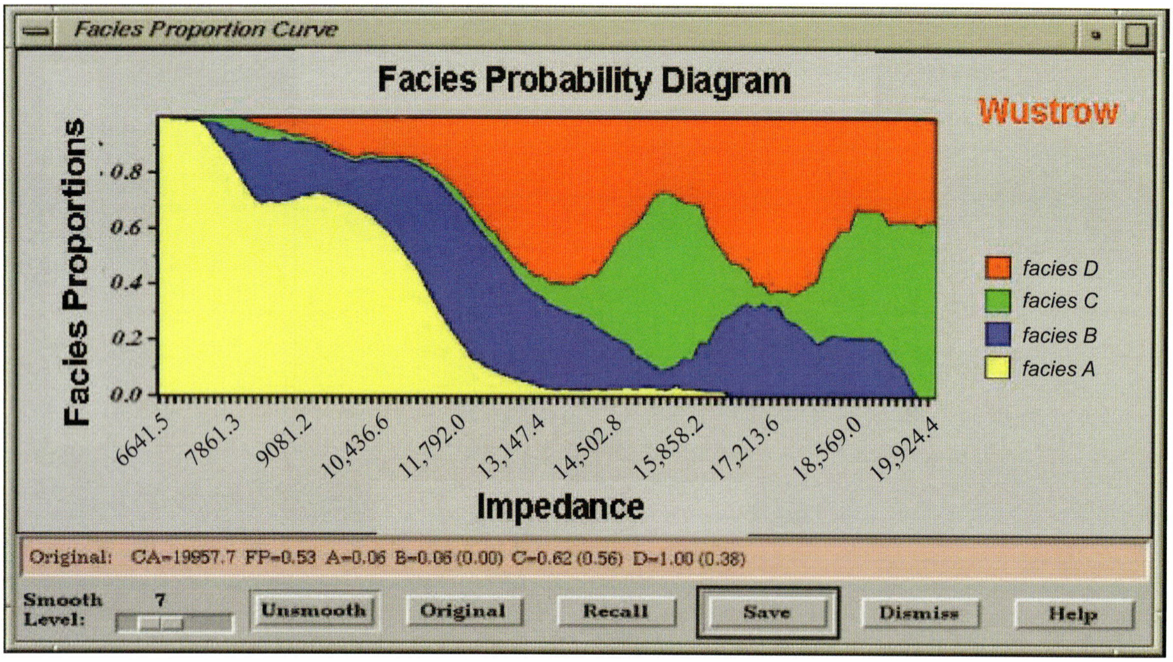


Figure 10. Facies vs. impedance calibration diagram for the Wustrow interval.

## Facies Modeling Using SIS with Local Proportions

Sequential indicator simulation (Journel and Alabert, 1988; Xu et al., 1992) was used for facies modeling. Facies indicator logs were used as hard data, and the facies proportion curves were used as soft data (Lo and Bashore, 1999). The facies proportion curves were tied to the 3-D impedance volume through the facies calibration. A separate model was constructed for each of the Wustrow, Ebstorf, and Dethlingen zones. A 3-D framework was constructed on an orthogonal grid of 306 × 216 cells ($\Delta x = \Delta y = 100$ m [330 ft]) using the available depth

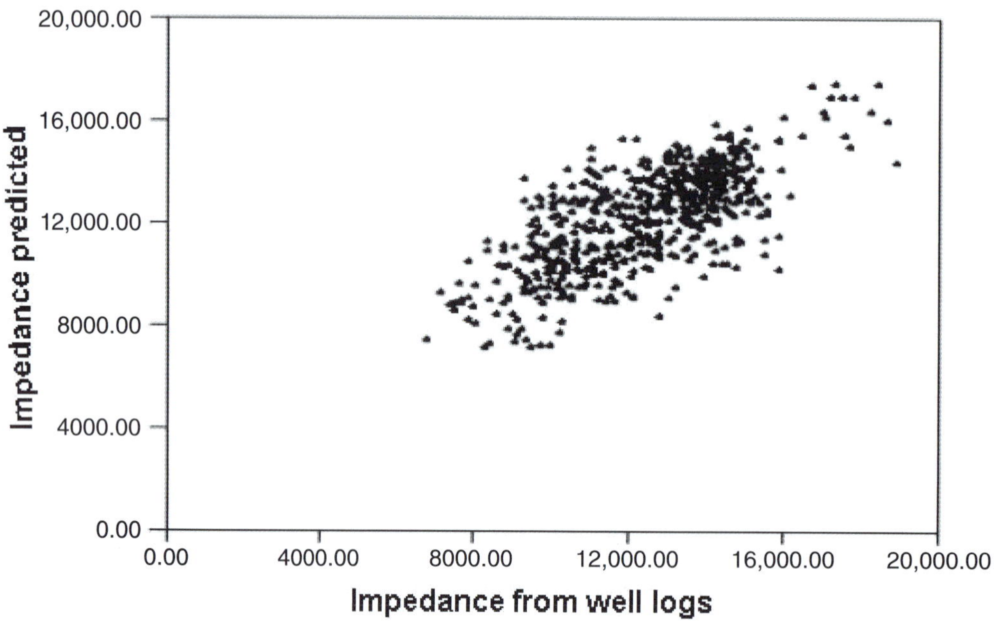


Figure 11. Crossplot of impedance data from well logs and predicted impedance represented in the 3-D cube for the Wustrow interval.

horizons over the seven zones. The Wustrow, Ebstorf, and Dethlingen intervals in this framework were vertically subdivided into 24, 21, and 96 layers, respectively. The Wustrow, Ebstorf, and Dethlingen facies models were subsequently combined using the 3-D framework. This full facies 3-D cube is presented in Figure 12.

## POROSITY MODELING

The next step in the workflow was the geostatistical simulation of porosity in each facies (lithotype) in such a way as to honor both large- and small-scale heterogeneity. The porosity distributions of the three sands (facies A, B, and C) showed a good separation, whereas a significant overlap in porosity distributions were observed between mudstones (facies D) and nonreservoir sands (facies C).

### Porosity Semivariogram Modeling

Areal and vertical semivariograms of porosity were developed for each facies in the Wustrow, Ebstorf, and Dethlingen intervals.

### Porosity Modeling Using Sequential Gaussian Simulation by Facies

Sequential Gaussian simulation (Ripley, 1987; Xu et al., 1992) was used to geostatistically simulate the porosity in each facies in each of the three reservoir intervals. Porosity values from well log were used as hard data, along with the corresponding facies cubes that were previously generated for each interval. Three simulations were produced using the area and vertical semivariograms constructed in the earlier step. Figure 13 exhibits the first realization of estimated porosity in layer 15 of the Wustrow interval. The 3-D porosity grids for the three zones were combined to construct a unified 3-D porosity volume for each realization using the 3-D framework created during facies modeling (Figure 14).

## PERMEABILITY MODELING

The facies-based conditional cloud transform algorithm takes a continuous property model, e.g., porosity, which is facies dependent, and applies a cloud transform to it. The resulting continuous property model, e.g., permeability, honors the bivariate

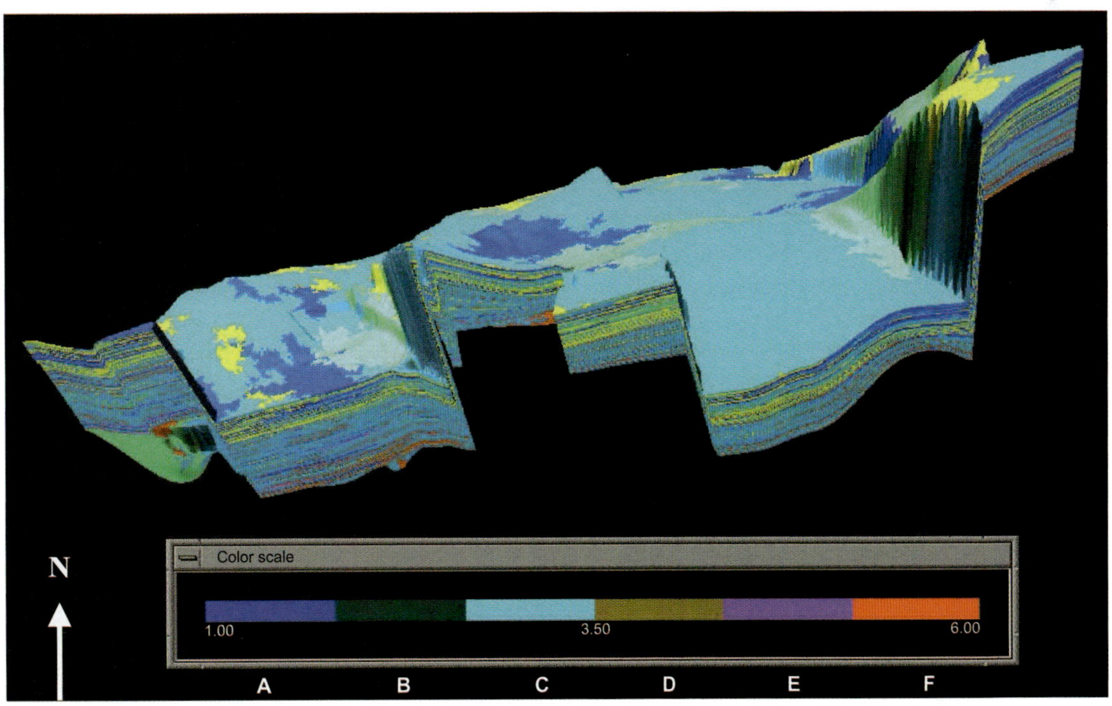


**Figure 12.** Three-dimensional representation of simulated facies for the combined Wustrow, Ebstorf, and Dethlingen intervals (vertical exaggeration 20×).

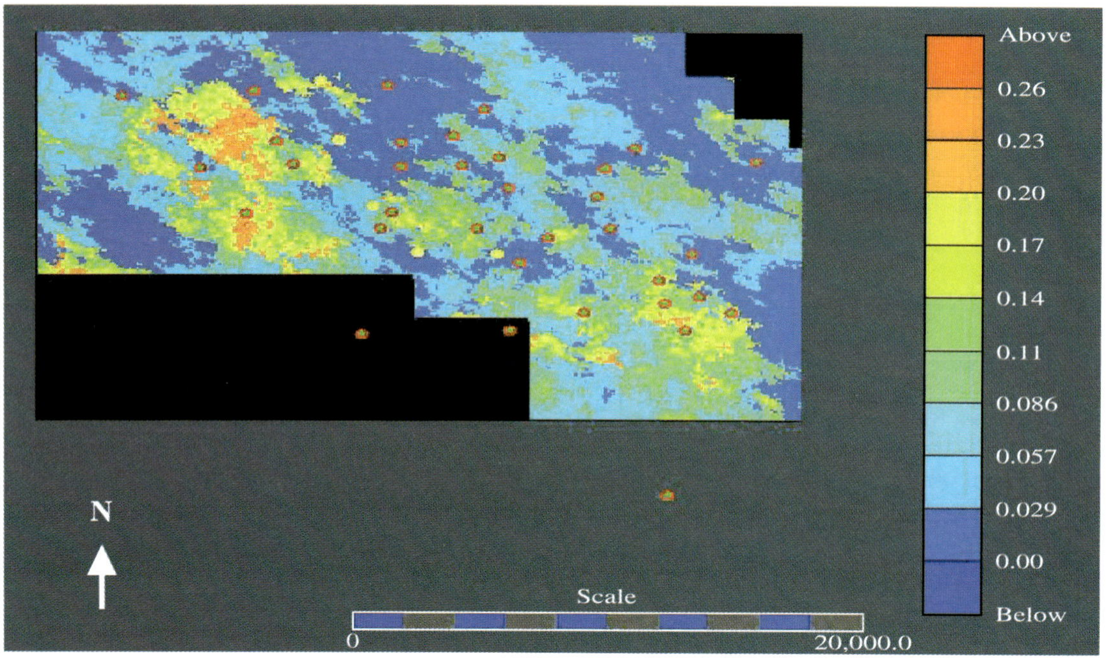


**Figure 13. First realization of simulated porosity (in fraction) in layer 15 of the Wustrow interval.**

scatter or cloud (Bashore et al., 1994). The cloud transform consists of a crossplot of porosity and permeability obtained from core or log data, instead of a simplistic regression function. The cloud transform helps preserve the extreme permeability values and facilitates reproduction of the flow behavior in the reservoir. This algorithm requires a cloud transform for each facies present in the facies model. If the property data of the continuous property model that is being transformed are

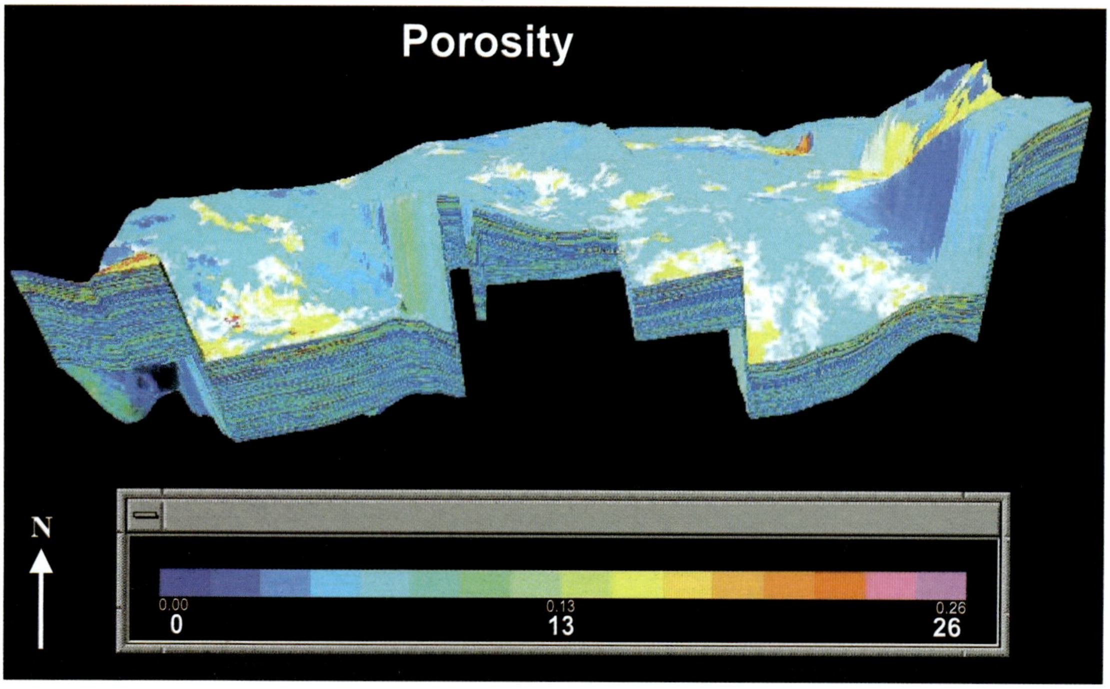


**Figure 14. Three-dimensional representation of simulated porosity (in percentage) of the Wustrow interval based on the first realization (vertical exaggeration 20×).**

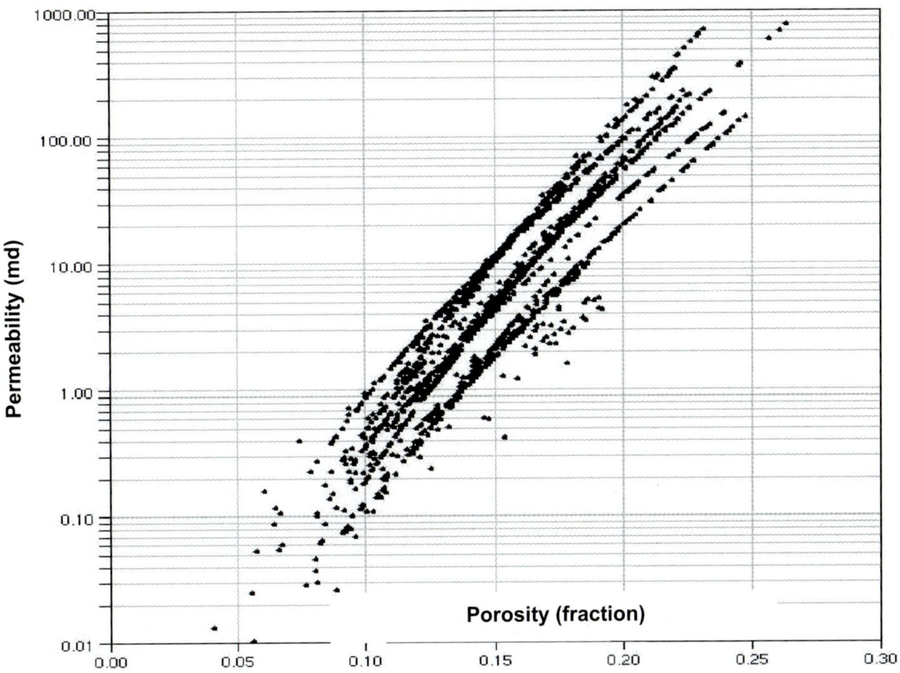


**Figure 15. Facies-based cloud transform of permeability and porosity logs of facies A of the Wustrow interval.**

available, then the resulting property model will be conditioned to those data. As an example, if the permeability log data are available, then the resulting permeability model will honor the permeability data.

Statistical transforms (cloud transform) were developed for each facies relating permeability to porosity on the basis of well-log data (Figure 15). Simulation was conducted using the permeability log data, the porosity-permeability cloud transform for each facies, the areal and vertical semivariograms for each facies, and the facies and porosity cubes.

Three realizations were generated for each of the Wustrow, Ebstorf, and Dethlingen zones using the stochastic facies-based conditional cloud transform algorithm. The first realization of simulated permeability for layer 15 of the Wustrow interval is shown in Figure 16. As in the case of porosity, the

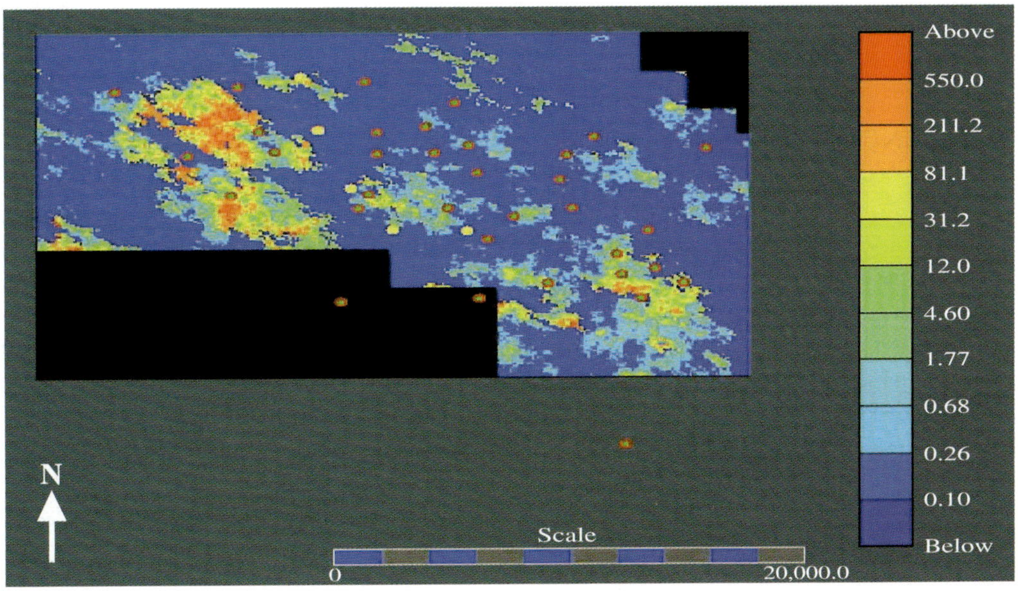


**Figure 16. First realization of simulated permeability (in millidarcys) in layer 15 of the Wustrow interval.**

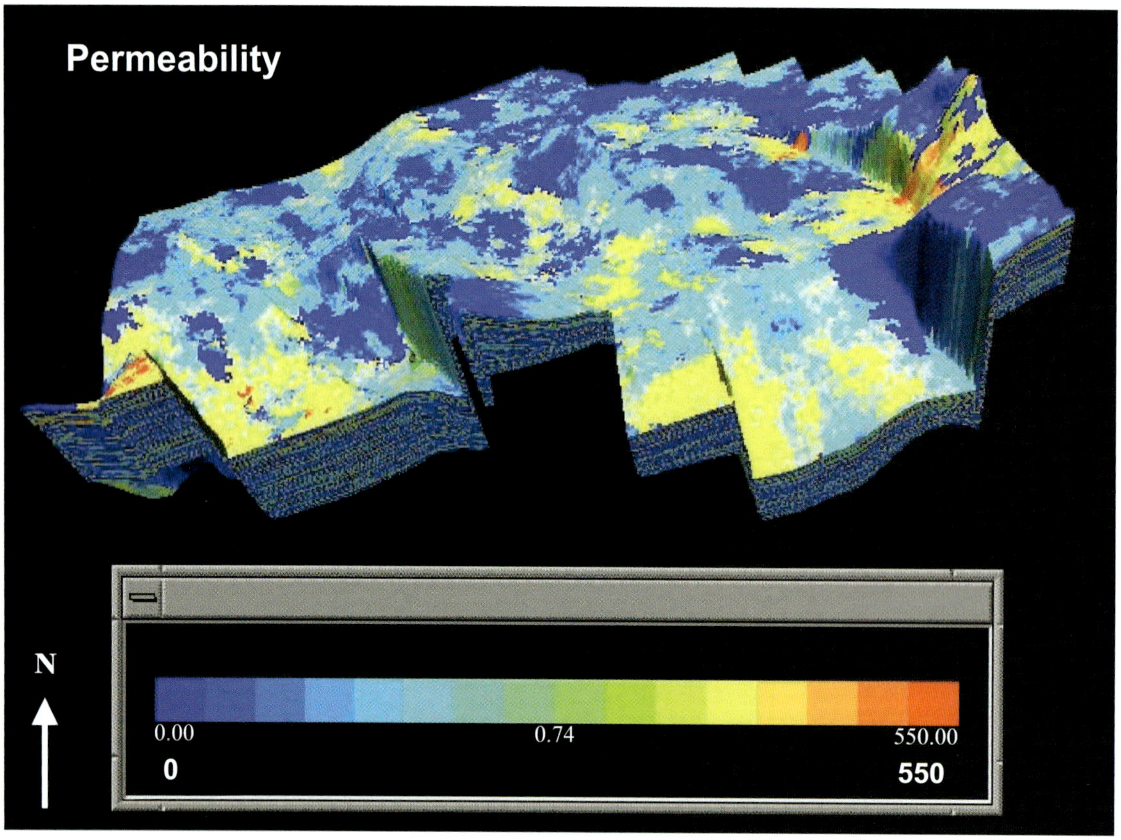


**Figure 17.** Three-dimensional representation of simulated permeability (in millidarcys) for the combined Wustrow, Ebstorf, and Dethlingen intervals based on the first realization (vertical exaggeration 20×).

combined 3-D volume of simulated permeability values, including all three zones for each realization, was subsequently constructed using the 3-D framework based on the permeability logs and the depth surfaces (Figure 17).

## CONCLUSIONS

This study presents a geostatistical approach to reservoir property modeling using seismic data in the Rotliegend gas field. The seismic data with dense areal sampling and well data with fine vertical sampling were integrated to build a high-resolution 3-D impedance model that improved characterization of Wustrow, Ebstorf, and Dethlingen reservoirs. The 3-D stochastic inversion approach encompassed a vertical and lateral resolution necessary for petrophysical modeling.

The use of the stochastic inversion impedance model as a soft data is critical to the construction of facies (lithotype) model because of sparse well control in the study area. The facies (lithotype) is characterized by a group of rocks with similar petrophysical and fluid properties.

A facies-to-property approach was adopted to model continuous petrophysical properties of porosity and permeability. These models can account for varying spatial continuity of reservoir properties in different facies. Facies-based property models preserve the facies-specific statistical distribution of the property. Facies classification also facilitates distinction between the reservoir and nonreservoir rocks. Stochastic simulation by facies produces more realistic porosity and permeability realizations. For example, two facies might have similar porosity but different permeability values, which is resolved by building facies-dependent property models.

The facies-based 3-D porosity models provide an improved prediction of porosity distribution and reservoir heterogeneity. The permeability simulations based on facies and the cloud transform between porosity and permeability allow better control of spatial connectivity patterns across the reservoir that could be used for improved reservoir performance prediction.

## ACKNOWLEDGMENTS

The authors gratefully appreciate BEB Erdgas und Erdoel GmbH, Mobil Erdoel-Erdgas GmbH, RWE-DEA AG, and Wintershall AG Erdoelwerke, Hamburg, Germany, for their permission to publish this study. Thanks are due to William M. Bashore, Udo G. Araktingi, Tienwhen Lo, Brad Bankhead, and Sean Carey for reviewing this chapter and providing valuable comments.

## REFERENCES CITED

Bashore, W. M., U. G. Araktingi, M. Levy, and W. J. Scweller, 1994, Importance of a geological framework and seismic data integration for reservoir modeling and subsequent fluid-flow predictions, *in* J. M. Yarus and R. L. Chambers, eds., Stochastic modeling and geostatistics: Principles, methods, and case studies: AAPG Computer Applications in Geology 3, p. 159–175.

Bortoli, L. J., F. Alabert, A. Haas, and A. G. Journel, 1992, Constraining stochastic images to seismic data, *in* A. O. Soares, ed., Geostatistics tróia: The Netherlands, Kluwer Academic Publishers, p. 325–337.

Gast, R. E., 1988, Rifting im Rotliegenden Niedersachsens: Die Geowissenschaften, v. 6, p. 115–122.

Haas, A., and O. Dubrule, 1994, Geostatistical inversion—A sequential method of stochastic reservoir modeling constrained by seismic data: First Break, v. 12, p. 561–569.

Hock, M., T. Kraft, F. Kloas, and I. Stowe, 1995, Lithology and sandstone diagenesis types from petrophysical well logs—Tool for improved reservoir characterization in the Rotliegend Formation, Permian basin, northwestern Germany: First Break, v. 13, p. 441–450.

Journel, A. G., and F. G. Alabert, 1988, Focusing on spatial connectivity of extreme-valued attributes: Stochastic indicator models of reservoir heterogeneities: Society of Petroleum Engineers 63rd Annual Technical Conference, SPE Paper 18324, p. 821–832.

Lo, T., and W. M. Bashore, 1999, Seismic constrained facies modeling using stochastic seismic inversion and indicator simulation, a North Sea example: Society of Exploration Geologists 69th Annual Meeting Expanded Abstracts, p. 923–925.

Oldenburg, D. W., T. Schuere, and S. Levy, 1983, Recovery of the acoustic impedance from reflection seismograms: Geophysics, v. 48, p. 1318–1337.

Ripley, B. D., 1987, Stochastic simulation: New Jersey, John Wiley and Sons, 237 p.

Shrestha, R. K., and M. Boeckmann, 2002, Stochastic seismic inversion for reservoir modeling: Society of Exploration Geologists 72nd Annual Meeting Expanded Abstracts, p. 902–905.

Xu, W., T. T. Tran, R. M. Srivastava, and A. G. Journel, 1992, Integrating seismic data in reservoir modeling: The co-located co-kriging alternative: Society of Petroleum Engineers 67th Annual Technical Conference, SPE Paper 24742, p. 833–842.

Williams, D. B., and A. A. M. Aqrawi, 2006, Utility of using acoustic impedance data in the stochastic modeling of a carbonate reservoir, *in* T. C. Coburn, J. M. Yarus, and R. L. Chambers, eds., Stochastic modeling and geostatistics: Principles, methods, and case studies, volume II: AAPG Computer Applications in Geology 5, p. 253–268.

*Chapter 19*

# Utility of Using Acoustic Impedance Data in the Stochastic Modeling of a Carbonate Reservoir

**D. B. Williams**
*Envision AS*
*Stavanger, Norway*

**A. A. M. Aqrawi**
*Statoil ASA*
*Stavanger, Norway*

## ABSTRACT

A Middle Eastern Cretaceous rudist-bearing carbonate reservoir is chosen for this study. This limestone sequence (about 55 m [180 ft] thick) consists of tight argillaceous mudstones and wackestones at the base, grading upward into a more porous section dominated by bioclastic grainstones and packstones.

Applying sequence-stratigraphic concepts, the reservoir is classified into two main sequences. In addition, the upper zone is further divided into two sequences of smaller accommodation cycles. Each of the latter includes a distinctive reservoir type (i.e., dominated by either barrier or shoal facies). The lower main sequence is considered as one unified nonreservoir cycle in the study area of the field. The three-dimensional seismic and sequence-stratigraphic analyses are used in conjunction with designing a deterministic structural model. A three-layer model is used, each layer representing a stratigraphic sequence previously recognized.

Describing the facies and flow units of carbonate reservoirs for reservoir-simulation purposes is a critical task that needs careful study of both the depositional textures and the diagenetic overprints in a sequence-stratigraphic framework. Eight main depositional facies are recognized for facies-modeling purposes. However, only two of them are high-quality reservoirs, dominated by grain-supported textures.

The reservoir architecture is generated using a combination of grid- and object-based simulation techniques to accurately reproduce the facies- and reservoir-type distributions. Petrophysical parameters porosity and permeability are


DOI:10.1306/1063820CA53239

stochastically simulated within facies bodies using a Gaussian cosimulation algorithm. Well data are used as hard information that must be honored, whereas acoustic impedance information is used to define porosity maps, which are used as conditioning trends in the simulation. These trends are believed to represent the effects of diagenesis.

Statistical analysis of the relationship between acoustic impedance and porosity allowed us to use impedance-derived porosity trends as conditioning data in petrophysical simulation. These porosity trends, adjusted to the range of data values from the wells, are believed to represent the effects of diagenetic activity. Using these trends alters the resultant data distribution compared with the distribution as inferred from well data. This is a consequence of the assumption that these porosity trends do reflect diagenesis and, in addition, does so not only at the wells but in the interwell areas as well.

## INTRODUCTION

Reservoir characterization using stochastic techniques is moving into mainstream use among oil companies around the world. The application of geostatistical techniques is recognized as a powerful approach to the problem of adequately characterizing reservoir heterogeneities (e.g., Yarus and Chambers, 1994), particularly in carbonate reservoirs, which are characterized by higher heterogeneities compared to sandstones. Rudist-bearing carbonate reservoirs are a dominant type in most of the Middle East fields, especially those producing from the Cretaceous (e.g., Stonley, 1990; Aqrawi et al., 1998; Sadooni and Aqrawi, 2000). As a result, such a reservoir has been chosen for this study.

Two unique benefits of stochastic simulation exist. First is the ability to integrate multisource data to produce detailed heterogeneity models of depositional facies and the corresponding distributions of petrophysical variables. Distributions of petrophysical variables can subsequently be used as direct input to flow simulators, which, in combination with facies models, can be used in well-planning scenarios. Second is the ability to generate several realizations of structure, facies, or petrophysical variables that allow quantification of the span of uncertainty.

The use of information from seismic inversion (acoustic impedance) is an important method in the stochastic modeling arsenal. Incorporating such information allows subsurface characteristics to be estimated away from the immediate vicinity of the wellbore and over large areas. Seismic inversion provides the geoscientist a more geological, layer-oriented view of a reservoir, instead of the interface-oriented image of a standard seismic section, making it easier to interact with other disciplines.

A normal workflow for the simulation part of this type of project typically involves an ordered series of tasks, the next task depending on or directly using the results of the previous one. Stochastic modeling of structural surfaces may be necessary to establish a structural framework for the reservoir (if a deterministic model is not deemed acceptable). Stochastic modeling of facies is normally the next step, followed by simulation of petrophysical attributes, such as porosity, within sedimentary facies and bodies.

Carbonates contain about 50% of hydrocarbons around the world. These rocks, when compared to clastics, are very fragile and subject to postdepositional diagenesis and fracturing to a greater degree. Carbonate reservoirs generally have a higher degree of heterogeneity than sandstones, with commonly erratic distributions of porosity and permeability.

This chapter describes the characterization of a carbonate sequence in a less-productive part of a field representing the nose of a giant anticline in the Middle East using seismic and acoustic impedance information and sedimentological and well-log data. The purpose of the study is to generate several realizations of porosity and permeability, primarily as input to a reservoir flow simulator, using all available information. In a traditional stochastic modeling workflow, facies simulation and then petrophysical simulation using the generated realizations of facies distributions are applied. Multiple realizations of facies and petrophysical attributes are generated.

In addition, the chapter emphasizes the importance of sedimentology and sequence stratigraphy in various stages of modeling a carbonate reservoir.

## AVAILABLE DATA

A three-dimensional (3-D) seismic data set is available and has been interpreted. Inversion processing is conducted to produce an impedance volume. This is a forward modeling process that uses an initial macromodel defined by the time horizons, an initial impedance volume, and iterative updating to produce a final model. Among the first steps in the inversion process is the generation of synthetic seismograms, which also produce logs of impedance in the wells. Validation of the inversion process is provided by comparison to well logs, which are not explicit inputs to the process.

As noted above, inversion results produce a more geologic-looking output that is tied to depth. Because of the cross-validation, the exercise described above the well data, including porosity and impedance, is tied to the inversion volume. The reservoir interval is represented by two impedance layers (dictated by seismic resolution). An artifact of the model-based approach is that within each impedance layer, values vary laterally but not vertically. We were able, therefore, to sample within each impedance layer from which two-dimensional (2-D) impedance maps are produced.

Measurements of porosity and permeability are derived from interpreted well-log data, or they are calculated from other information. Nine wells constitute the key input to the study. The available sedimentological data are taken from two major (proprietary) reservoir petrological and biostratigraphic studies.

The material used in these studies include about 150 ft (45 m) of core data from a key well (within the study area), together with other petrographic results from some 300 samples in another four wells (across the field). The original sedimentological results include a detailed core description, thin-section analysis (of almost every 1-ft [0.3-m] interval sample), and biostratigraphic analysis (about every third interval sample), in addition to the x-ray diffraction mineralogical study and scanning electron microscopy analysis of more than 100 selected samples.

All the sedimentological and biostratigraphic results have been reviewed critically, and a representative facies scheme in a sequence-stratigraphic framework has been generated over the study area. The available wire-line logs in eight wells are also used in this construction. Two basic objectives are borne in mind during the review:

- textural classification of the reservoir and nonreservoir facies (using the Dunham, 1962, classification, which is modified by Jordan and Wilson, 1994), considering the diagenetic overprints to classify the flow units (e.g., Bebout et al., 1979; Tucker and Bathurst, 1990)
- integrating the resulting depositional and diagenetic model with the biostratigraphic results and wire-line logs to build up a sequence-stratigraphic framework using the principles and procedures of Vail et al. (1977), Haq et al. (1987), Van Wagoner et al. (1990), and Loucks and Sarg (1993).

## SEQUENCE STRATIGRAPHY

Two main shallowing-upward depositional cycles have been recognized in the studied successions. Although each of them could be divided into two sequences of smaller scale accommodation cycles, only the upper one is subdivided for modeling purposes because the lower main sequence only consists of nonreservoir facies. However, in other parts of the field, this sequence could contain some productive shoal and barrier facies. The three identified sequences represent shallowing-upward sedimentary cycles in the main reservoir limestone. Each of these sequences ends with an unconformity representing an erosional (or nondepositional) surface formed during the end of each shallowing upward or local sea level fall.

Because of shallowing upward (e.g., Loucks and Sarg, 1993), each sequence has been dominated by deposition during highstand sea level, starting with a finer facies as a transgressive systems tract and ending with deposition of a coarser facies as a highstand systems tract. Every sequence has been identified and correlated across the study area of the field. Because the study area represents the structural nose of the field, only variations in thickness of each sequence can be detected in various cross sections. Time lines and seismic markers, which can lead to the establishment of a better sequence-stratigraphic framework, are not available across the entire structure.

Finally, each of the differentiated sequences is considered to be a structural zone for subsequent 3-D modeling. This allows the distribution of the reservoir facies (i.e., flow units) to be stochastically predicted between the boreholes. Impedance layers correspond to the two main depositional cycles. Two upper zones (comprising the upper depositional cycle) use information from the same impedance layer for both upper structural zones. The lower zone, corresponding to the lower depositional cycle and structural zone, is not subdivided.

## FACIES DESCRIPTION AND DEPOSITIONAL HISTORY

Based on information contained in the proprietary petrographic studies, the limestone sequence can be subdivided into eight main depositional facies according to their texture using Dunham's (1962) classification of carbonate rocks, modified by Jordan and Wilson (1994). However, only two of them are high-quality reservoirs characterized by grain-supported texture (e.g., Tucker and Wright, 1990; Zimmerle, 1995; Kupecz et al., 1997).

The depositional environment of this limestone begins with outer-shelf planktonic mudstone-wackestones (facies 8) and deep-slope bioclastic wackestone-mudstones (facies 7). The porosigins with outer-shelf planktonic mudstone-wackestones (facies 8) and deep-slope bioclastic wackestone-mudstones (facies 7). The porosigins with outer-shelf planktonic mudstone-wackestones (facies 8) and deep-slope bioclastic wackestone-mudstones (facies 7). The porosity in these two facies (commonly less than 10%) is the lowest among the identified facies, and the permeability is less than 0.5 md. They are considered nonreservoir rock based on their petrophysical properties. Coarse-grained bioclastic rudstone-grainstones (facies 5) are deposited from a barrier (or larger shoals) at the shelf margins, overlying the forebarrier facies 6 of bioclastic packstone-grainstone. These two coarser facies of grain-supported textures mark a new depositional cycle, which commonly ends with the deposition of fossiliferous packstone-wackestones (facies 4) deposited in a back-barrier setting (not detected in the studied wells). In addition, another specific facies representing an intrabarrier channel (facies 9) fining upward into the lagoonal-back-barrier (facies 4) is only detected in one borehole. After a short nondepositional and erosional period, the last cycle starts with the deposition of lagoonal benthic wackestone-mudstones (facies 3). Then, smaller shoals prograde over this facies, depositing some grain-supported high-quality grainstone-packstones (facies 2) from more agitated environments. Although such a cycle should end with intertidal-subtidal deposits (facies 1), this facies is commonly absent here, most probably because of later erosion. Figure 1 shows a conceptualized diagram of the facies and depositional environment in the study area.

A reservoir-typing scheme has been devised for the well from which detailed sedimentological data are available. This typing scheme has also been expanded to include the other eight wells in the study area for which only wire-line–log information exists. The following reservoir types are defined: shoal (facies 2), lagoonal (facies 3), back-barrier (facies 4), barrier (facies 5), forebarrier (facies 6), downslope (facies 7), basinal (facies 8), and channel (facies 9). A facies-reservoir type log is defined for each well and is used as the primary facies-conditioning information. The well data are hard information because each stochastic realization must honor the well information.

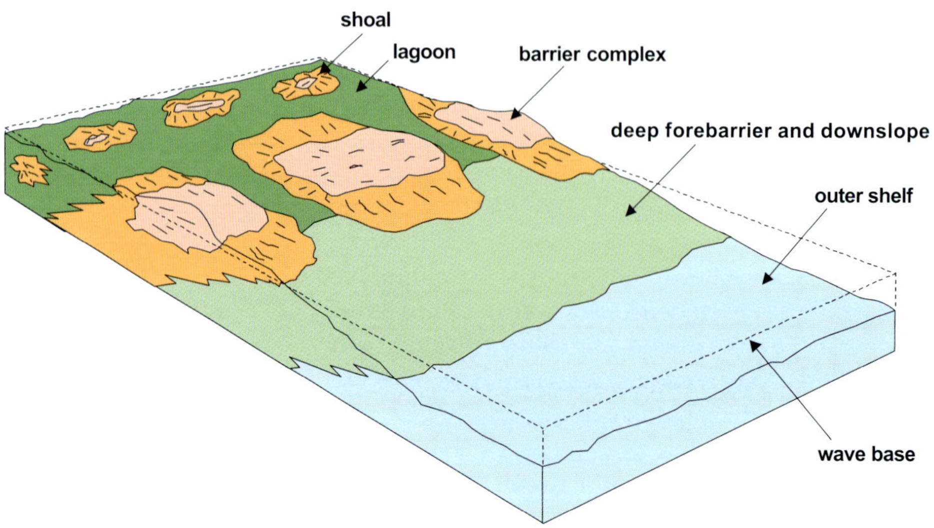


**Figure 1. Conceptualized diagram of facies and depositional environment.**

## FLOW-UNIT CHARACTERIZATION

The description and characterization of the depositional facies in terms of reservoir quality, particularly permeability, lead naturally to the identification of flow units. The high-quality units in the reservoir are mainly restricted to grain-supported rudist-bearing facies, which are limited to the two upper sequences in the area of the field. However, one of these facies has been partly affected by cementation, reducing its quality based on the available well data.

Most of the highly porous and permeable depositional facies have been further enhanced in their reservoir quality through the diagenetic history, particularly by near-surface meteoric dissolution. The latter process has resulted in the creation of molds and vugs and the improvement of communication between the pores. Such an improvement results from the enlargement of the existing pore throats and connection of the isolated pores in the facies framework, particularly in the barrier facies.

An original and primary pore network is required for subsequent diagenetic dissolution. This mostly exists within packstones and grainstones. However, the latter has been partly affected by early isopachous rim cementation. Although this cementation prevented these reservoir facies from later compaction by overburden thickening with time, together with compaction, it reduced the pore throats and damaged the effective porosity (i.e., permeability) to some extent (for example, in facies 2). As a consequence, these flow units may perform poorly during production and may lead to the creation of diagenetic compartmentalization in the reservoir, particularly in specific areas of high cementation in the field.

Finally, although high porosity exists in most fine-grained and mud-supported facies such as wackestones, these facies are classified as nonreservoir with very low-flow or nonflow characteristics because of poor communication among the pores.

## 3-D SEISMIC STRUCTURAL MODEL

Because of seismic resolution and reservoir thickness, only three reflectors are interpreted across the interval of interest: top of the reservoir, base of the reservoir, and an intrareservoir horizon thought to represent a distinct change in porosity from tight to porous (based on well information). These horizons have been exported to a mapping program to facilitate gridding and depth conversion.

Based on the seismic interpretation, a three-layer zonation scheme has been developed. Each zone represents one of the stratigraphic sequences previously identified, although the middle zone is not seismically interpreted. Instead, isochore maps have been generated from well data for this zone and added to the base of the uppermost zone. All surfaces are matched to tops at the well locations. The depth surfaces are then exported to the stochastic modeling software and integrated into a deterministic structural model using a base-conformable framework for the top zone and a proportional framework for the other two zones.

## RESERVOIR FACIES MODELING

Two different stochastic simulation methods are used to model facies. First, an object-based, or Boolean, technique is employed that incorporates a marked-point process model. A marked-point process describes the position and shape of objects of all facies except the facies designated as the background. Length, width, and thickness values are drawn from multi-Gaussian distributions, with the parameters describing the distributions being given as constants or as trends. Intensity functions can be specified to introduce trends in the number of bodies. Simulation is accomplished using a Metropolis-Hastings algorithm and simulated annealing to fulfill conditioning of well observations and volume restrictions, respectively. Second, a grid-based approach is employed that uses truncated Gaussian simulation. This method is based on simulating a continuous Gaussian variable and discretizing this variable into facies associations. A trend is established on the basis of the conceptual geologic model. Residuals are also simulated to correctly condition the well observations, with residuals being determined by a semivariogram model. The simulated continuous variable (combination of the trend and residual) is then discretized using specified rules to create transitions between the modeled facies. This approach is commonly used to characterize broad, laterally extensive facies or depositional framework architectures.

Each reservoir zone is modeled independently. Zone 1 (uppermost zone), containing shoal and lagoonal facies, is modeled using the object-based

(Boolean) method. Shoal facies are modeled as ellipsoidal shapes in a background facies defined as the lagoonal facies. Object dimensions (length, width, and thickness) are specified as a constant mean and standard deviation. This gives a distribution from which dimensions are taken as each object is placed within the background facies during the simulation process. Dimension estimates are determined through statistical analysis of each facies, using primarily the facies- and reservoir-type log as input data. Additional conditioning information for placement of the objects and relative volume intensity derives from one-dimensional (1-D) trends in *X* and *Z* (depth) directions, respectively. The trend in *X* is a slight seaward-increasing thickness trend, as indicated by statistical analysis, and the trend of decreasing intensity (fewer objects) in depth is an expectation based on our geologic knowledge of the depositional environment. Figure 2 shows a facies realization for zone 1.

As described below, zone 2, which contains forebarrier, barrier, and channel facies, is modeled using a combination of the grid- and object-based methods, with individual realizations being merged to create a final integrated realization. Truncated Gaussian (grid-based) simulation is first used to create a two-facies realization consisting of forebarrier (below) and a temporary barrier complex facies (above). As part of the simulation, a 3-D trend is output. Boolean simulation is then used to populate a background with ellipsoidal-shaped objects representing barrier facies. The 3-D trend serves as conditioning information to ensure that barrier objects appear only in the upper barrier complex part of the model. These two realizations are then merged, with barrier objects replacing the temporary barrier complex facies. The result consists of barrier superimposed over the forebarrier facies. Next, object-based modeling is used again to simulate the channel facies at a very low volume percentage. A 1-D trend in depth is specified

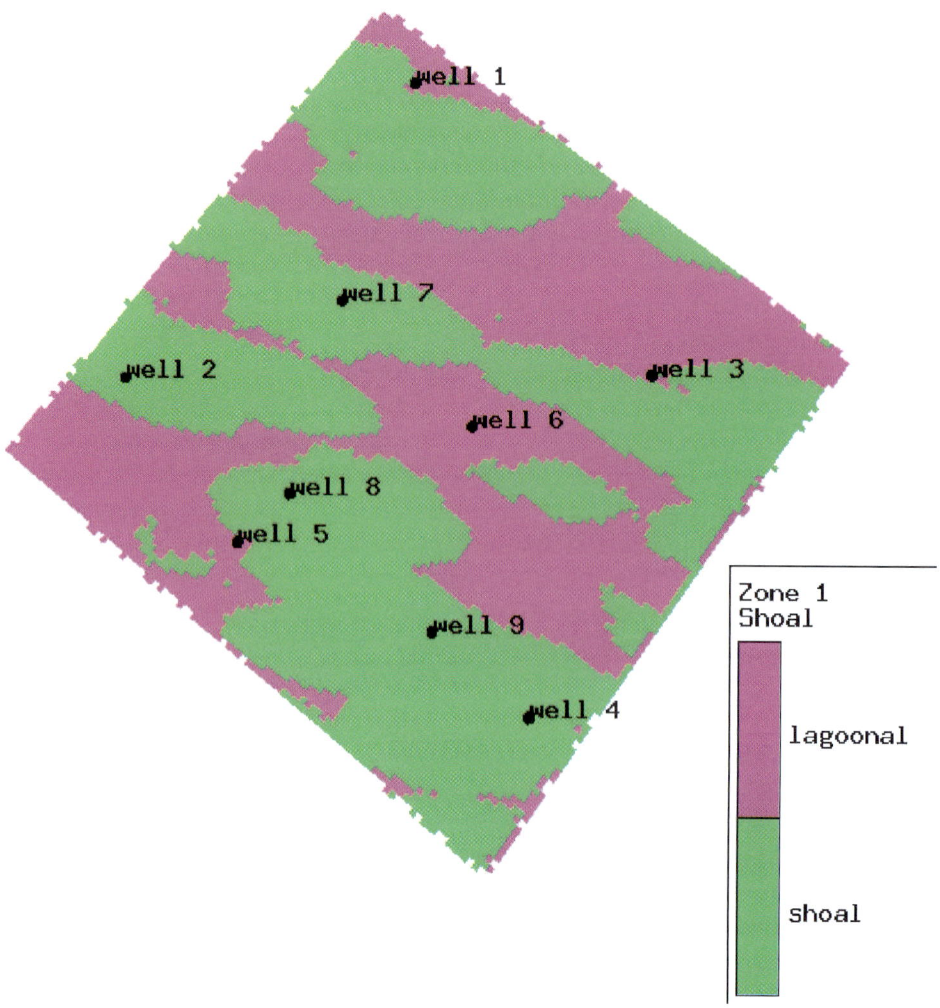


**Figure 2. Zone 1 facies realization.**

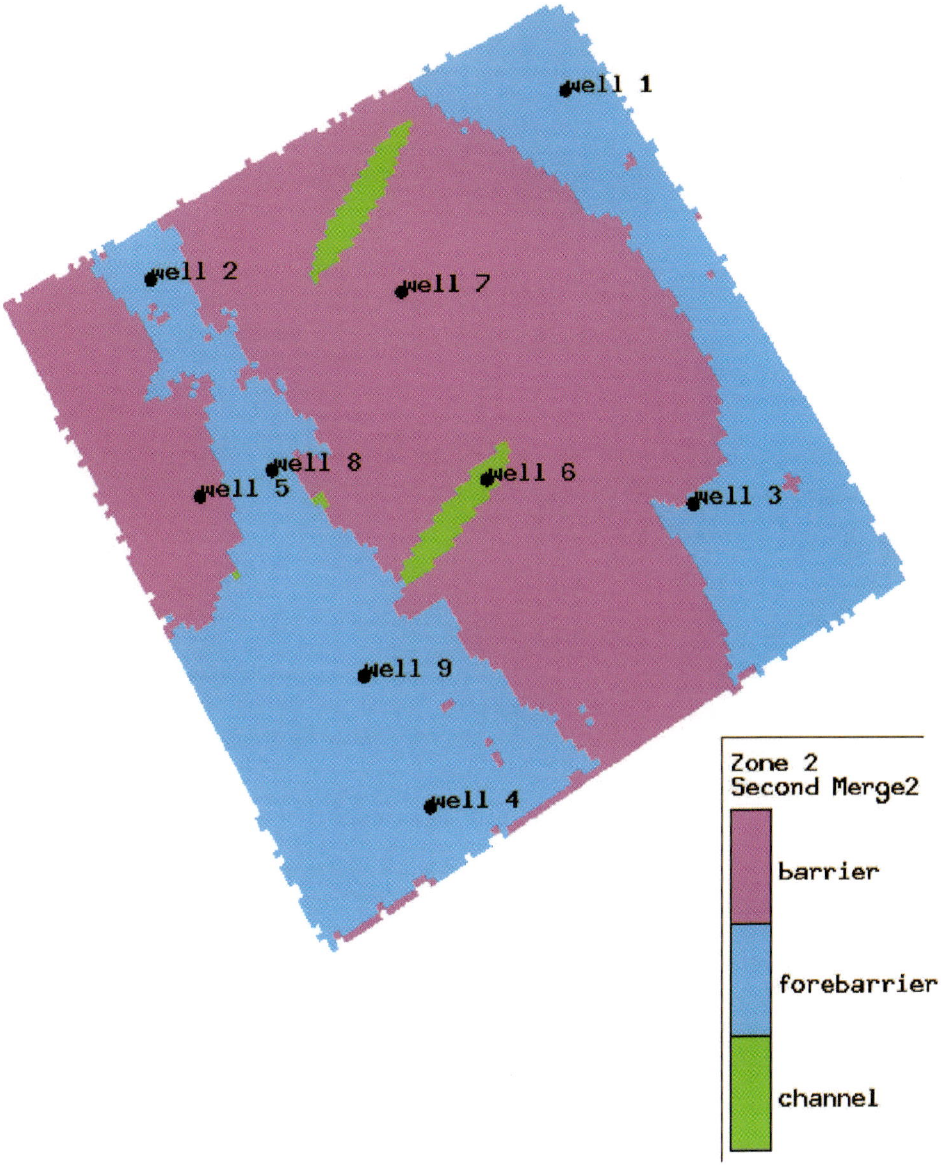


Figure 3. Zone 2 facies realization.

to ensure appropriate vertical placement of the objects. A user-defined shape for the object is created to mimic a more natural channel shape that is somewhat flatter than an ellipsoid on top. Finally, this channel realization is merged with the result of the previous merge, with channel facies taking precedence. The final result reflects the most appropriate technique for each depositional environment and accurately portrays the distribution of forebarrier, barrier, and channel facies. Figure 3 shows a facies realization for zone 2.

Zone 3, containing the nonreservoir downslope and basinal facies, is modeled using a straightforward application of a truncated Gaussian simulation to create a two-layer sequence. As in zones 1 and 2, model parameters are estimated from data analysis and knowledge of the depositional environment and sequence stratigraphy. These estimates are chosen to represent a gentle, prograding sequence. Visually rather simple with minor interfingering, the downslope facies is simulated over the basinal facies, expected in a regressive setting.

## PETROPHYSICAL MODELING

Simulation of petrophysical parameters is based on the assumption that they are approximately normally (Gaussian) distributed after the application of appropriate transformations. As a result, a Gaussian field model is defined for each combination of facies type and petrophysical variable. The Gaussian field

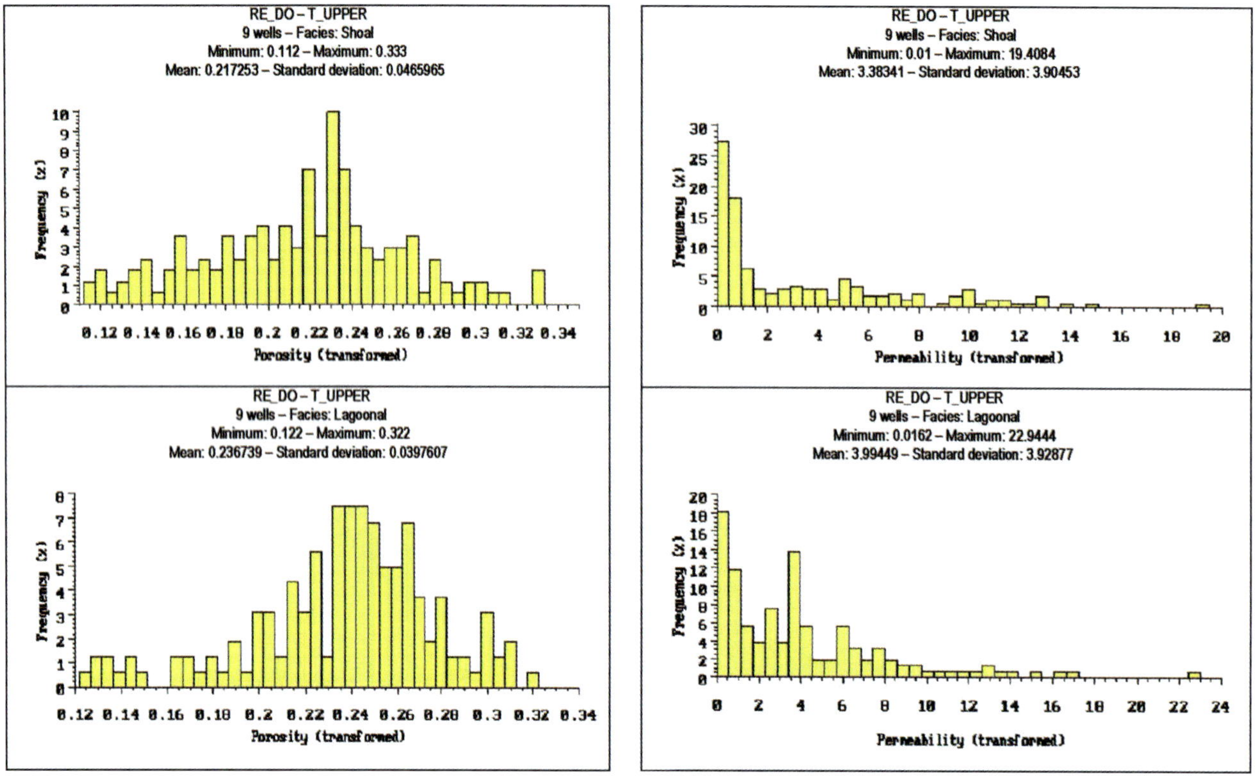


**Figure 4.** Histograms of raw values of porosity and permeability in various facies of zone 1. Figures combine data from nine wells.

model is characterized by statistical estimates that reflect, for instance, the spatial variability of the variables. Once stochastic modeling is completed, the transformations are applied in reverse order to regenerate the original data distribution.

Stochastic simulation of porosity and permeability is conditioned to the generated facies realizations in each zone. Well data (hard data) are available on all three variables in nine wells. As in the case of facies simulation, petrophysical simulation occurs zone by zone. To properly derive estimates of model parameters and ensure that spatial trends are reproduced in the output realizations, statistical analysis of the input data is a critical step. Figures 4–6 show histograms of raw values of porosity and permeability for various facies in zones 1–3, respectively.

Spatial and/or vertical trends are determined through inspection of scatter plots of each variable against *X*, *Y*, or *Z*. As noted, the task is to identify trends so that they can be removed in the transformation process. The objective is to ensure that these trends are reproduced in the output simulations. The task of identifying trends is somewhat subjective, and scatter plots need to be examined carefully for data irregularities. Figure 7 shows a scatter plot of porosity vs. *X* for one facies in one zone, suggesting a potential trend.

Applying the transformations resulted in the histograms depicted in Figures 8–10 for zones 1–3, respectively. The red curve on each plot represents an idealized normal distribution. The degree of match between the histograms of the transformed data and a normal distribution varies, but perfect matches rarely, if ever, occur. The gaps that occur between the red line and the actual data represent areas where simulated attributes may be slightly over- or underestimated. Areas where discrepancies occur are minimal and are within ranges expected for real data.

As previously noted, acoustic impedance data are available in four wells. These values are plotted against the corresponding porosity values for each facies. Porosity values for the shoal and lagoon facies in zone 1 are poorly correlated with impedance possibly because of partial cementation. The porosity-impedance correlation for barrier and forebarrier facies in zone 2 is somewhat better at values of about 0.41 and 0.43, respectively. The porosity-impedance correlation is best for the downslope and basinal facies in zone 3 with values of about 0.67. Figure 11 shows crossplots of porosity vs. impedance for facies

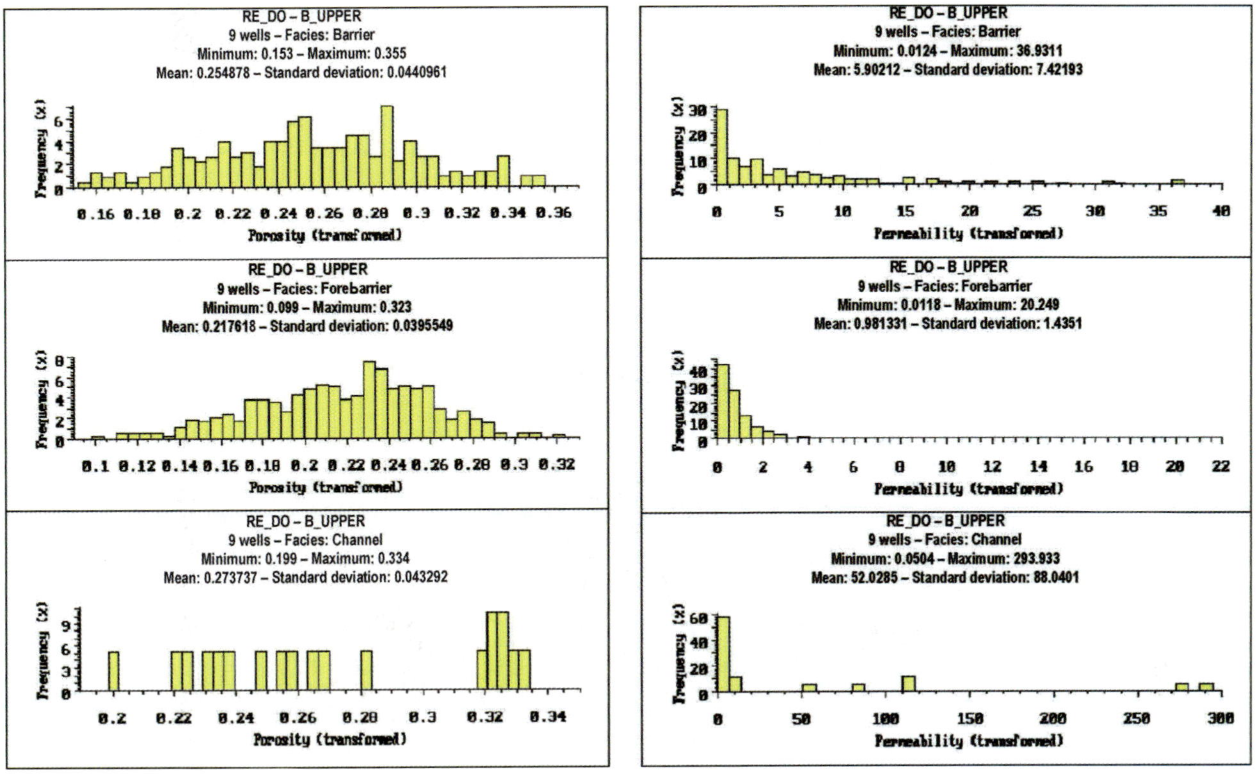


Figure 5. Histograms of raw values of porosity and permeability in various facies of zone 2. Figures combine data from nine wells.

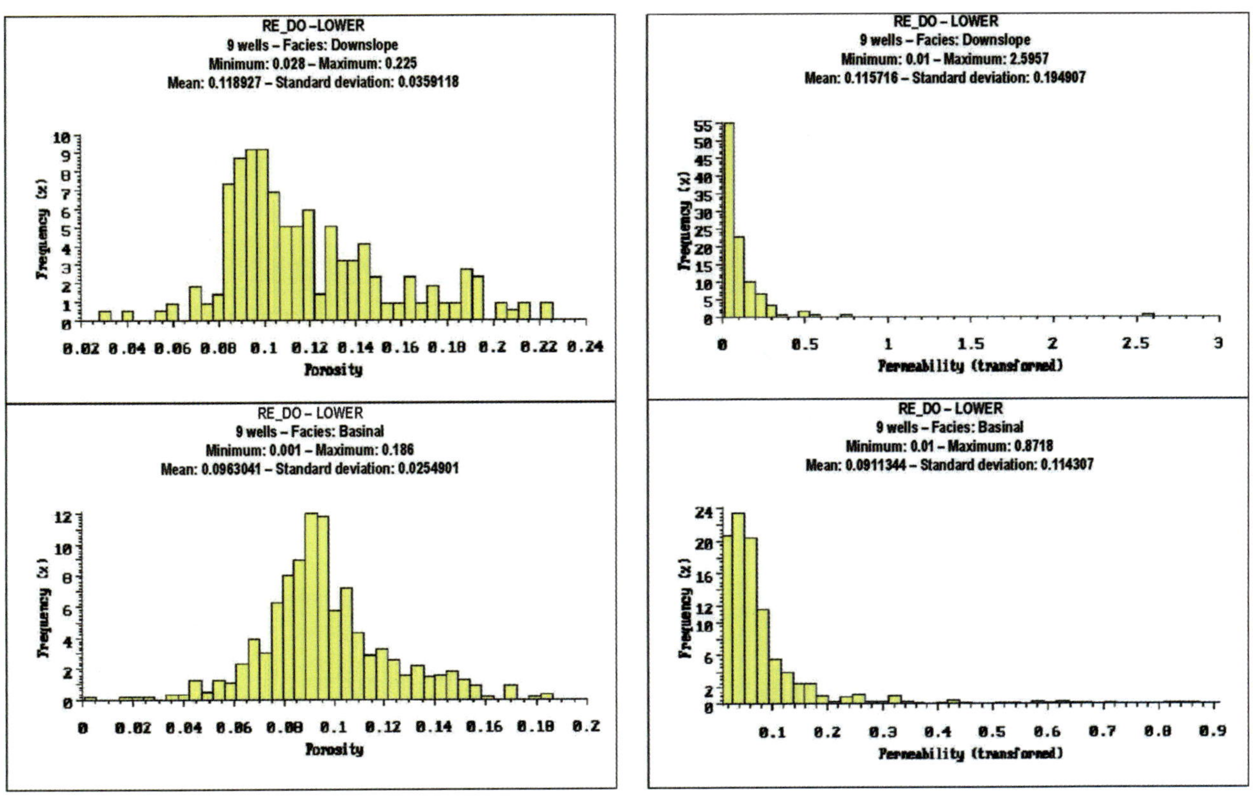


Figure 6. Histograms of raw values of porosity and permeability in various facies of zone 3. Figures combine data from nine wells.

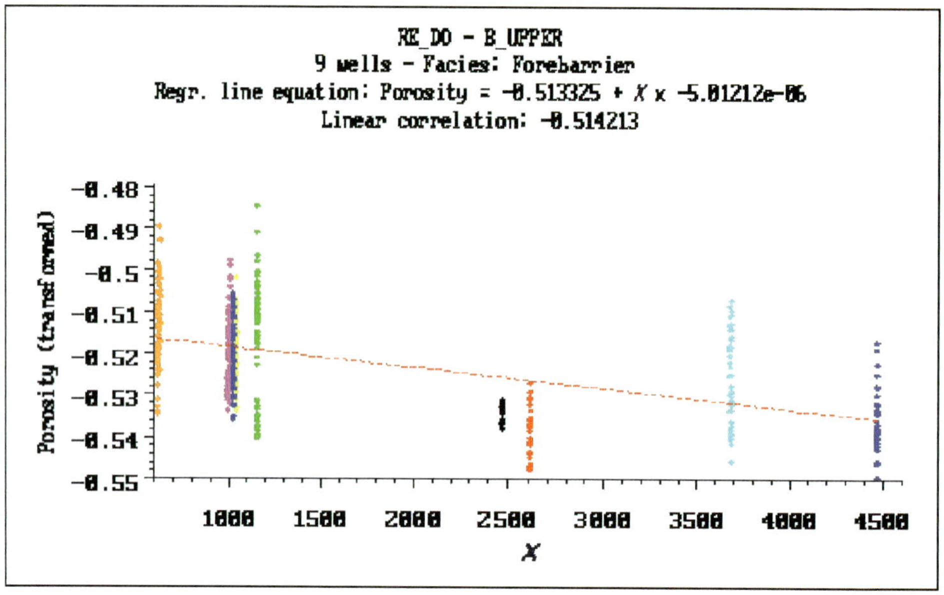


Figure 7. Plot of porosity vs. distance in the $X$ direction in the forebarrier facies of zone 2. The figures combine data from nine wells. The clustered look of the plot is because each well has many data points, but each data point has the same distance. Interpretation takes experience and is somewhat subjective.

in zones 2 and 3. On the basis of these findings, impedance information is used to condition porosity in the barrier and forebarrier facies of zone 2 and in the downslope and basinal facies of zone 3. The result is incorporated as a 2-D grid transformation. The 2-D map of impedance is first smoothed slightly to reflect broad changes, and then a linear relationship between impedance and porosity values is applied. Ultimately, a trend map of porosity is produced for zones 2 and 3 (Figure 12).

Simulation of permeability is linked to porosity in all zones. Log permeability is first plotted against porosity for all facies in each zone, and the resulting relationship is used to generate a 3-D volume of

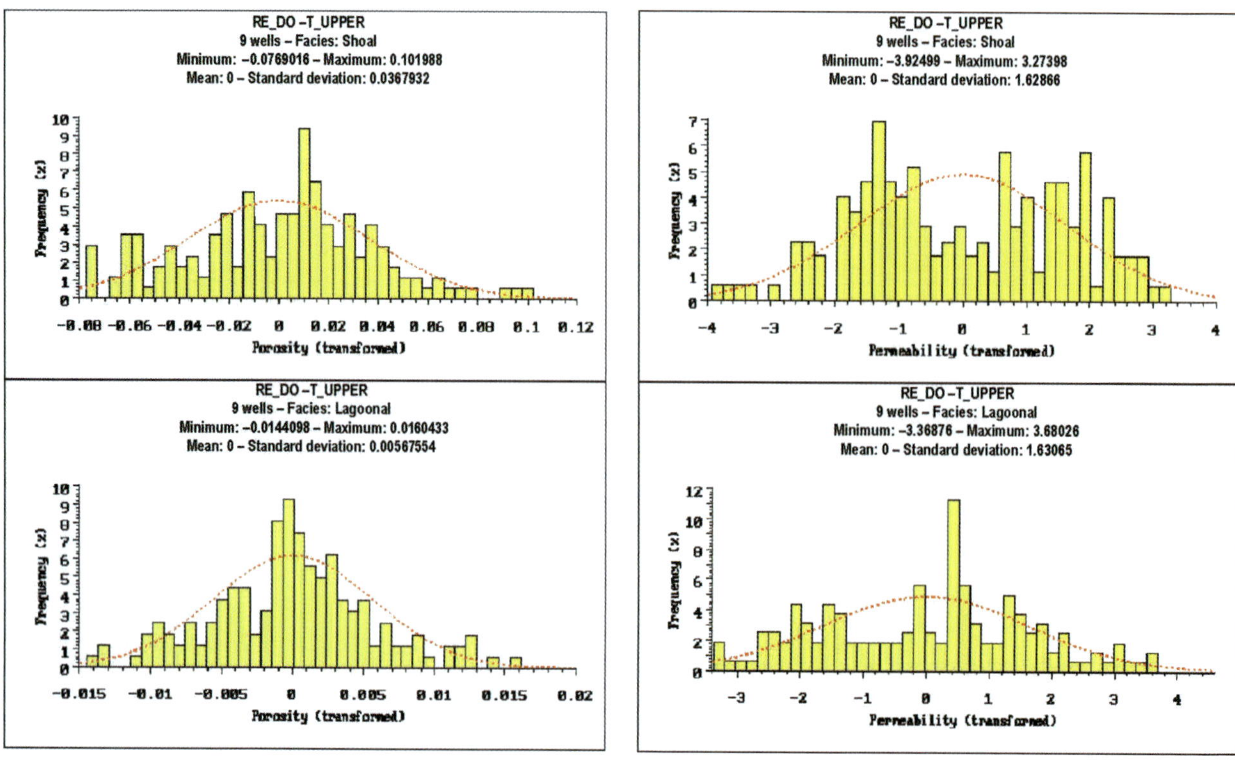


**Figure 8. Histograms of transformed values of porosity and permeability for various facies of zone 1. Figures combine data from nine wells and show an idealized normal distribution superimposed.**

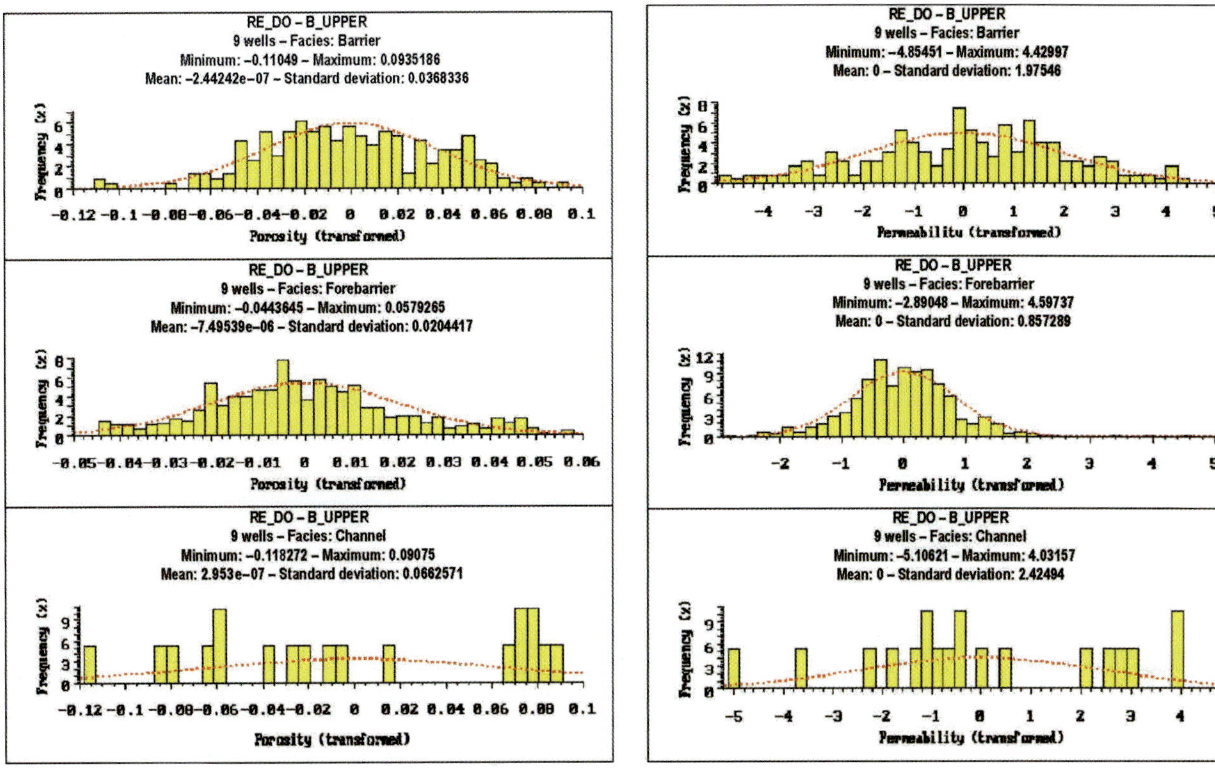


Figure 9. Histograms of transformed values of porosity and permeability for various facies of zone 2. Figures combine data from nine wells and show an idealized normal distribution superimposed.

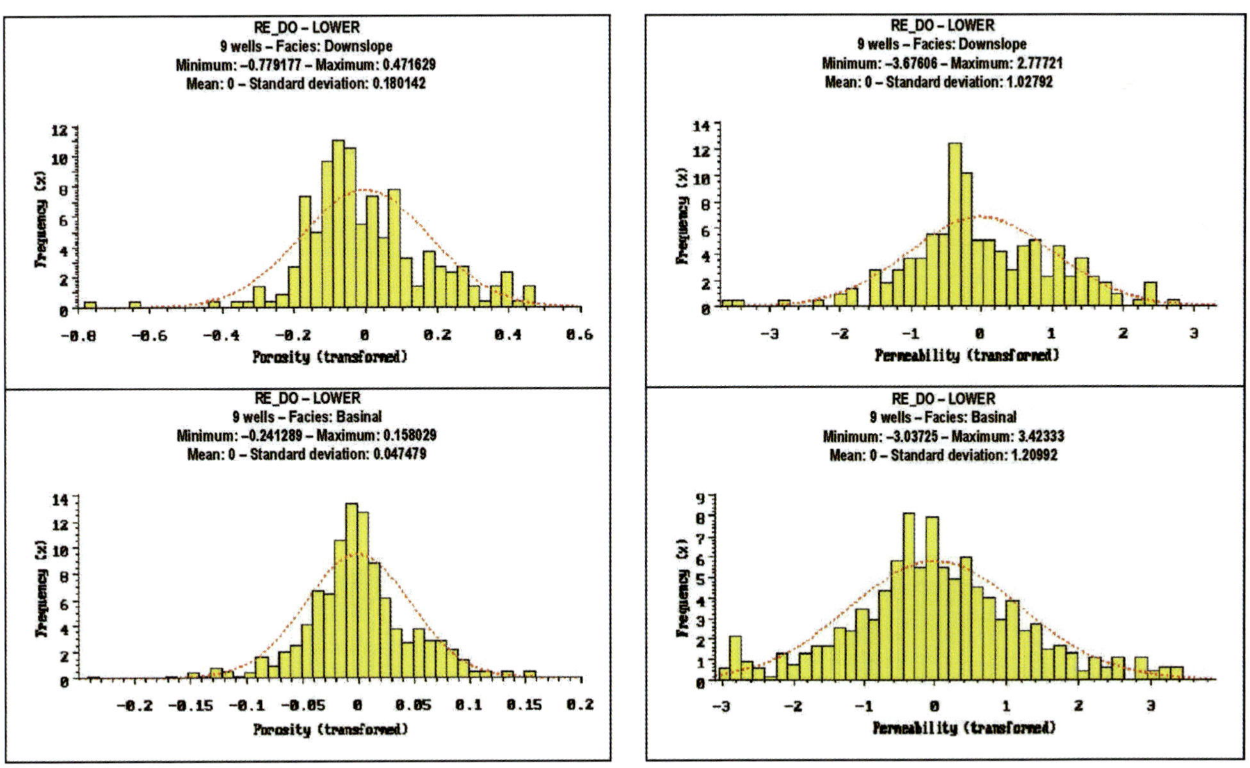


Figure 10. Histograms of transformed values of porosity and permeability for various facies of zone 3. Figures combine data from nine wells and show an idealized normal distribution superimposed.

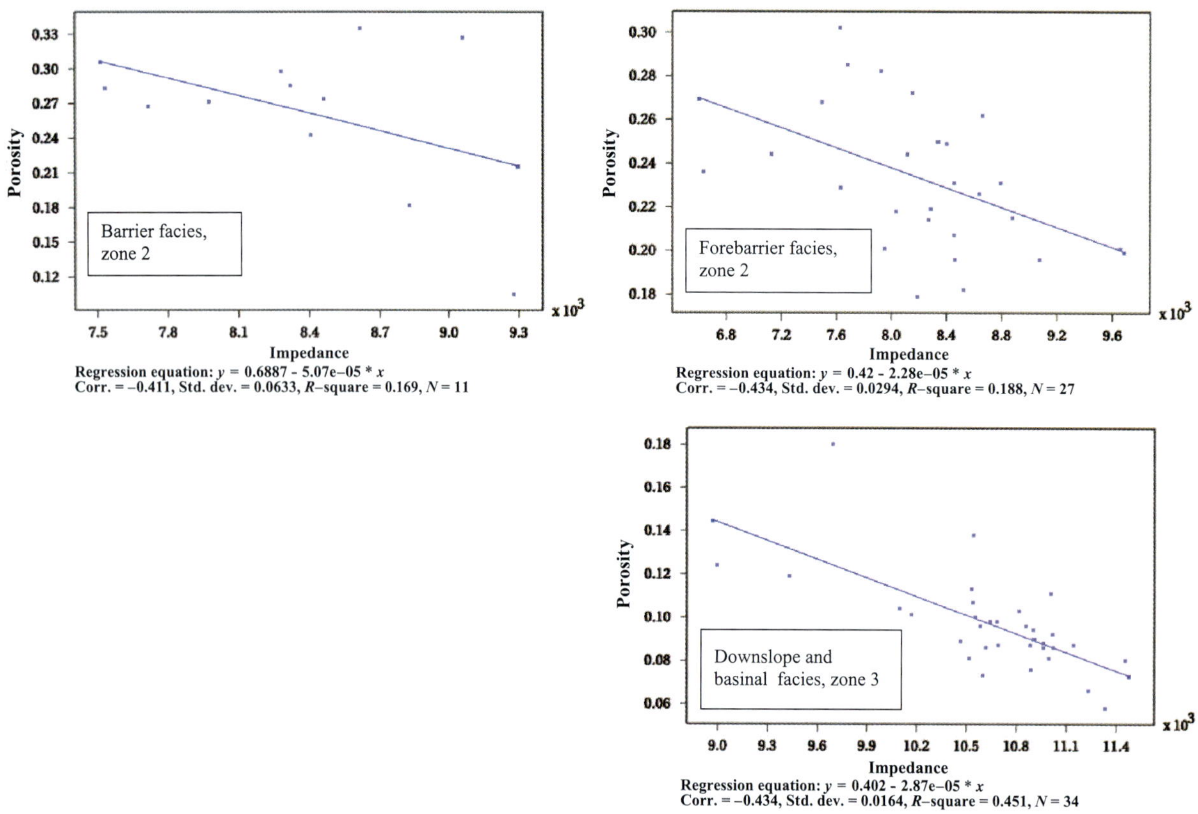


**Figure 11.** Crossplots of porosity vs. impedance for barrier and forebarrier facies in zone 2 and combined downslope and basinal facies in zone 3. Corr. = correlation coefficient; Std. dev. = standard deviation.

permeability, which is used as trend-conditioning (soft) information in the simulation process. Figure 13 shows crossplots of log permeability vs. porosity for each zone.

The semivariogram model is used to simulate values of variables at particular geographic locations. Based on knowledge of the depositional environment, a semivariogram orientation of N15°W

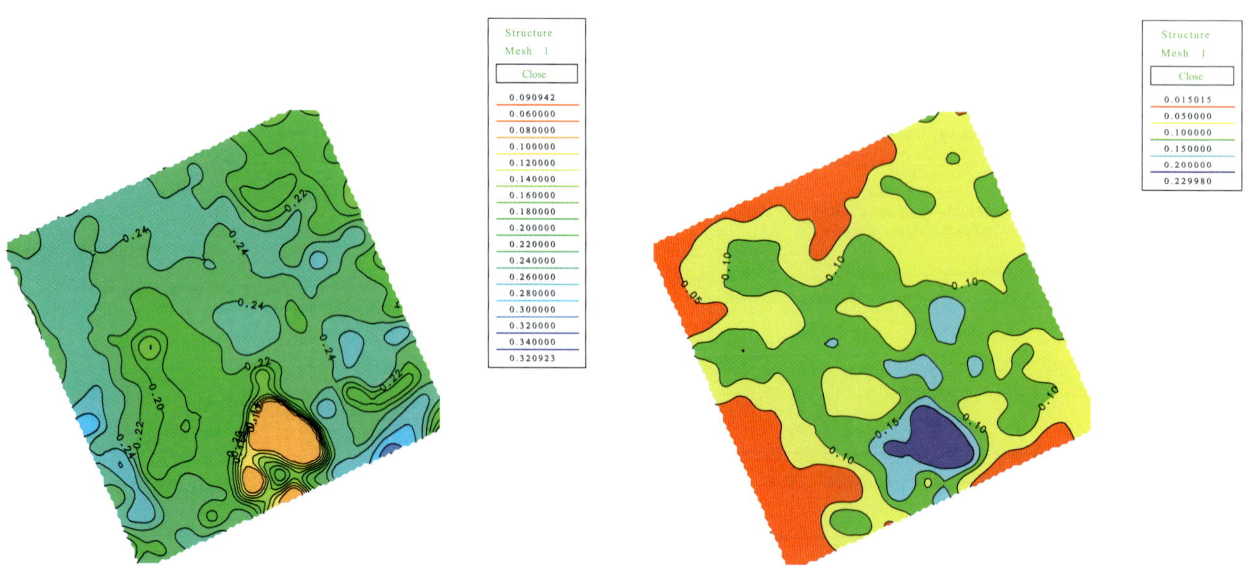


**Figure 12.** Porosity trend maps derived from impedance for zone 2 (left) and zone 3 (right).

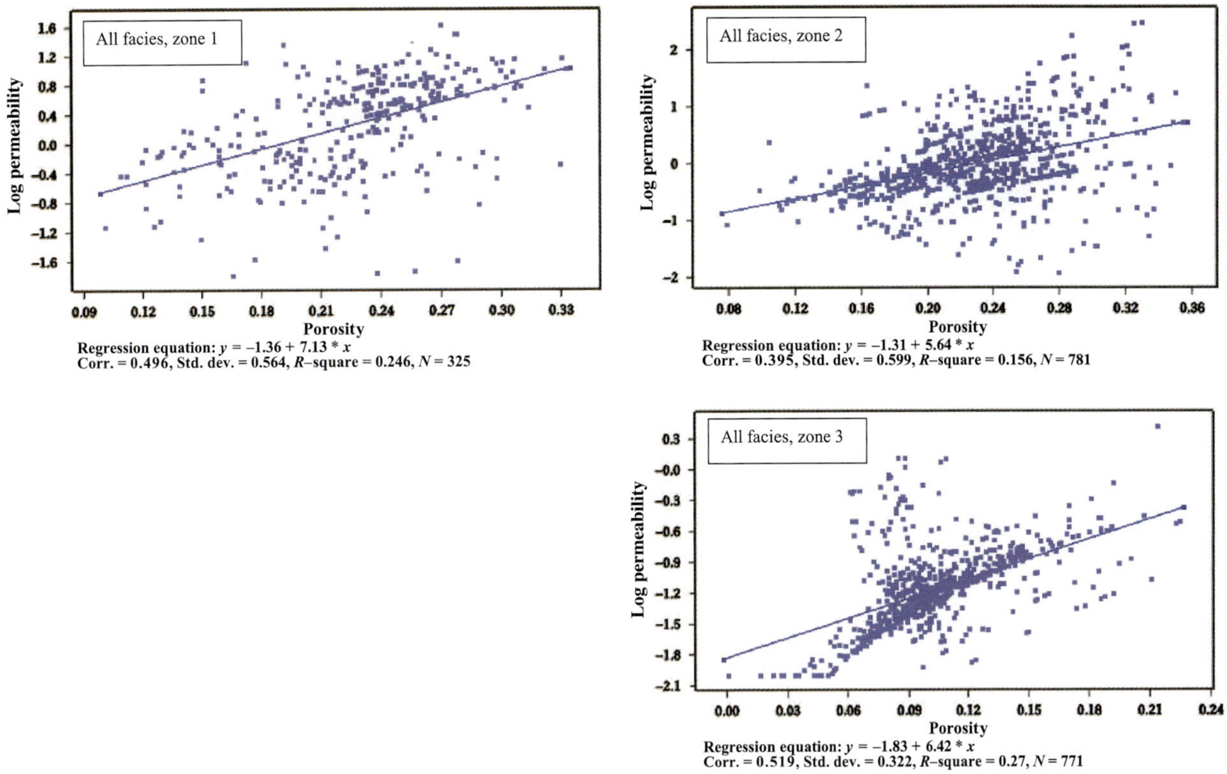


**Figure 13. Crossplots of porosity vs. log permeability per zone, all facies combined. Corr. = correlation coefficient; Std. dev. = standard deviation.**

analyzed is considered to be most appropriate. Anisotropy is indicated in the analysis and is reflected in a slightly higher range in the $X$ direction. This anisotropy is expected based on the interpretation of the depositional environment and the expected continuity of facies. In the $Z$ direction, the analysis indicates that a range equal to 1/2 the zone thickness is suited to the data. A spherical semivariogram type is used for all zones.

## DISCUSSION

An important facet of stochastic simulation is the discretization of petrophysical data prior to statistical analysis and facies simulation. Discretization can be achieved through reservoir typing. Such discretization constitutes the separation of well data into specific reservoir types based on a combination of lithofacies, knowledge of the depositional environment, and flow unit concepts. In the special case of carbonate reservoirs, the effects of diagenesis must also be considered.

A key supposition in this study is that the analysis of the relationship between impedance and porosity at well locations can be used to condition porosity in the areas away from wells. In this instance, it is believed that the porosity trends (derived from impedance) reflect the effects of diagenesis, which cuts across facies, enhancing porosity in some areas and degrading it in others. It is assumed that enhanced porosity also leads to enhanced permeability. Therefore, a 3-D trend was developed by crossplotting porosity and log permeability and deriving a linear relationship. This was then used to create the 3-D trend of permeability and, along with well data, used to simulate permeability.

For comparison, porosity and permeability realizations are generated with and without conditioning to a 2-D and 3-D trend, respectively. Figure 14 shows two such porosity realizations (with and without trend), each averaged over all layers (30) for zone 2. Comparing the porosity realizations shows the influence that the trend has on the distribution and range of values. Note the area of low porosity on the trend-based realization. Comparing this to the 2-D trend maps in Figure 12, it is apparent that this area of low porosity is not reflected in the well data. Figure 15 shows histograms, by facies, of the porosity realizations contained in Figure 13. The main difference

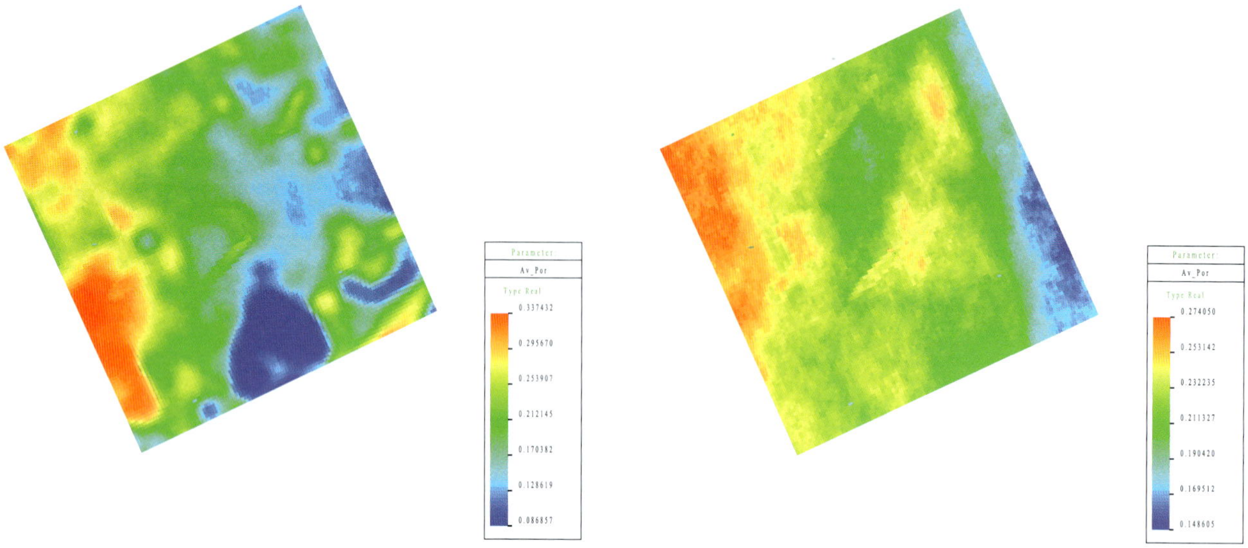


**Figure 14. Porosity realizations for zone 2, with a porosity trend derived from impedance (left) and without the trend (right).**

between the histograms is the change in the distribution of values in the trend-based realizations, most notably at the tails of the distributions. However, the minimum, maximum, and mean values are very close to the original data distributions. This shows that using the trend information derived from impedance, we have introduced values that are not indicated from the well data but, nevertheless, are believed to be within the range of data values outside well control.

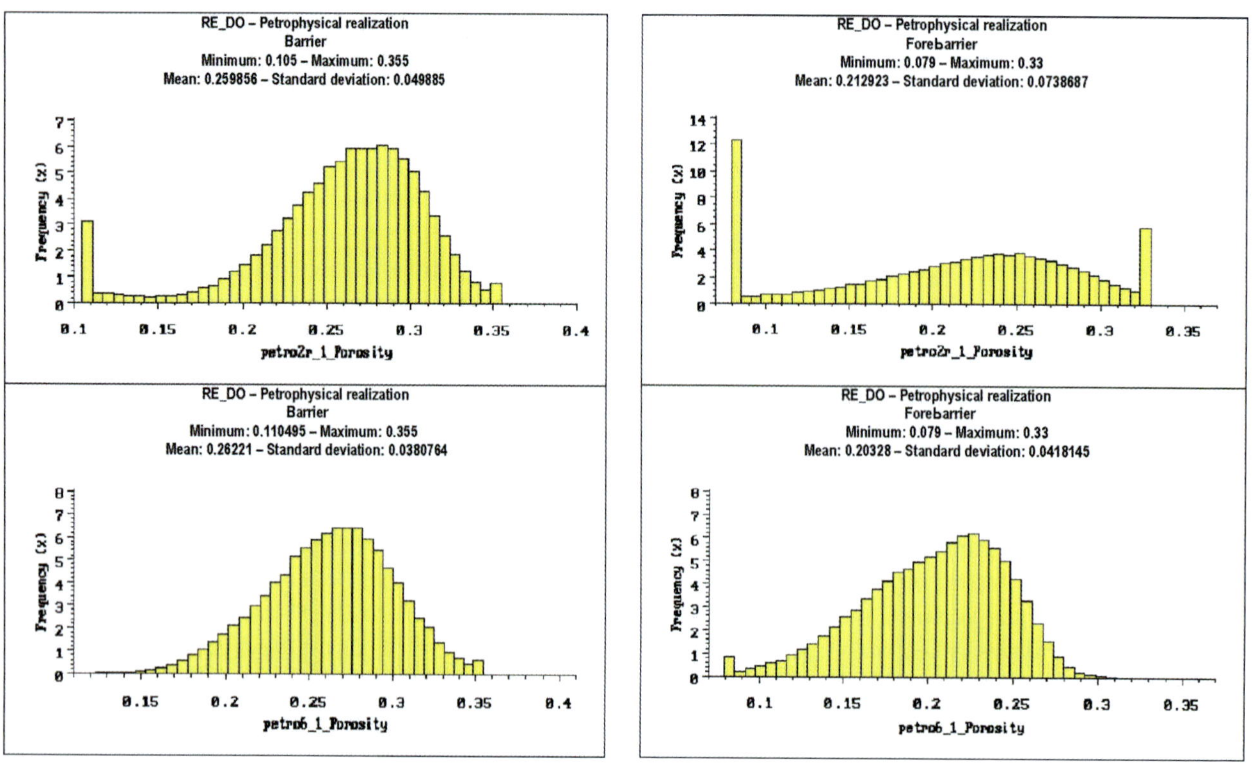


**Figure 15. Histograms of simulated porosity values for two facies. The top histogram in each pair summarizes porosity values based on trend information, whereas the bottom histogram depicts porosity values without trend information.**

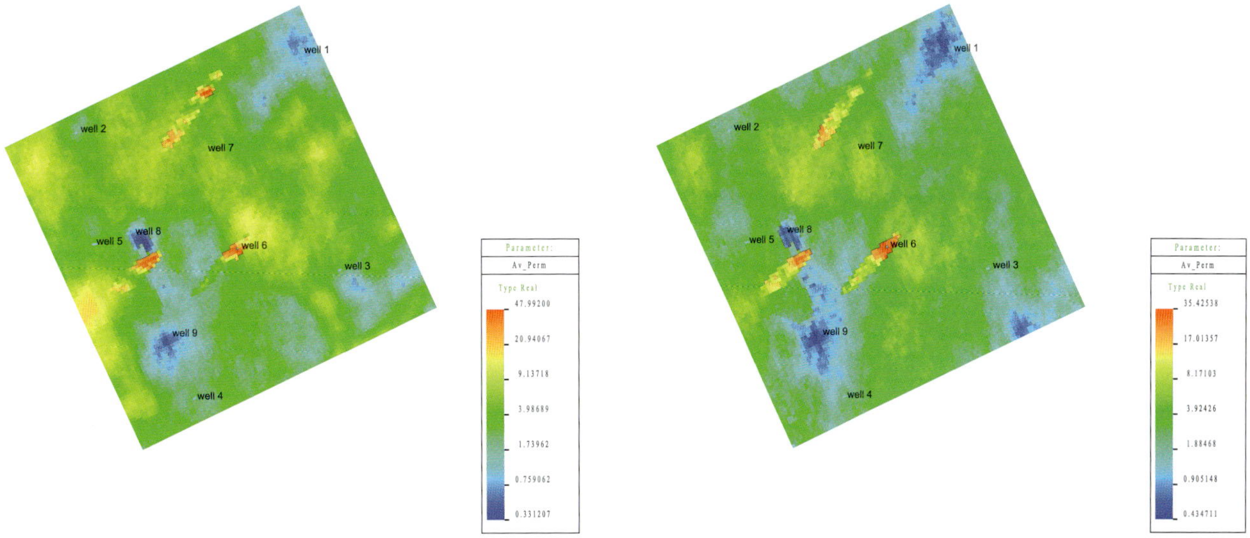


**Figure 16. Simulated permeability in zone 2 using 3-D trend information (left) and without 3-D trend information (right).**

Figure 16 shows permeability averaged over all layers (30) for zone 2 with and without trend information being used. The differences are not as distinct as those exhibited by porosity. Clearly, higher values of permeability are concentrated in areas characterized by higher porosities.

## CONCLUSIONS

The use of detailed sedimentological information results in the definition of a meaningful reservoir-typing scheme, including depositional facies and their diagenetic overprints. Sequence-stratigraphic analysis leads to effective subdivision of the reservoir into three depositional units that are used as a framework for subsequent structural modeling. Within each sequence, the facies ordering has a logical and predictable pattern. This facilitates the development of 3-D stochastic models of facies and petrophysical attributes in a logical and consistent manner, accurately reproducing patterns seen in well data and expected from sequence stratigraphy.

Using object- and grid-based simulation techniques results in a more accurate model of the facies architecture in the reservoir. A key assumption is that impedance data reflect changes in porosity caused by diagenesis. This assumption allows impedance-derived porosity trends to be used as conditioning information in the stochastic simulation of porosity.

## ACKNOWLEDGMENTS

The authors thank Roxar, Inc., for allowing us to use valuable resources in preparing this chapter. We thank Jeffrey Yarus for steering us in the right direction and colleagues at Roxar for numerous fruitful discussions.

## REFERENCES CITED

Aqrawi, A. A. M., G. A. Theni, G. H. Sherwani, and B. M. A. Kareem, 1998, Mid-Cretaceous rudist-bearing carbonates of the Mishrif Formation: An important reservoir sequence in the Mesopotamian basin, Iraq: Journal of Petroleum Geology, v. 21, p. 57–82.

Bebout, D., G. Davies, C. H. Moore, and N. C. Wardlaw, 1979, Geology of carbonate porosity: AAPG Education Course Notes 11, 247 p.

Dunham, R. J., 1962, Classification of carbonate rocks according to depositional texture, *in* W. E. Ham, ed., Classification of carbonate rocks: AAPG Memoir 1, p. 108–121.

Haq, B. U., J. Hardenbol, and P. R. Vail, 1987, Chronology of fluctuating sea levels since the Triassic (250 million years to present): Science, v. 235, p. 1156–1167.

Jordan, C. F., and J. L. Wilson, 1994, Carbonate reservoir rocks, *in* L. B. Magoon and W. G. Dow, eds., The petroleum system—from source to trap: AAPG Memoir 60, p. 141–158.

Kupecz, J. A., J. Gluyas, and S. Bloch, eds., 1997, Reservoir quality prediction in sandstones and carbonates: AAPG Memoir 69, 311 p.

Loucks, R. G., and J. F. Sarg, eds., 1993, Carbonate sequence stratigraphy, recent developments and applications: AAPG Memoir 57, 545 p.

Sadooni, F. N., and A. A. M. Aqrawi, 2000, Cretaceous sequence stratigraphy and hydrocarbon potential of the

Mesopotamian basin, Iraq, *in* A. S. Alsharhan and R. W. Scott, eds., Middle East models of Jurassic/Cretaceous carbonate systems: SEPM Special Publication 69, p. 315–334.

Stonley, R., 1990, The Middle East basin: A summary overview, *in* J. Brooks, ed., Classic petroleum provinces: Geological Society (London) Special Publication 50, p. 293–298.

Tucker, M. E., and R. G. C. Bathurst, 1990, Carbonate diagenesis: Blackwell Scientific Publications, International Association of Sedimentologists Reprint Series 1, 312 p.

Tucker M. E., and V. P. Wright, 1990, Carbonate sedimentology: Oxford, Blackwell Scientific Publications, 482 p.

Vail, P. R., R. M. Mitchum Jr., R. G. Todd, J. M. Wildmier, S. Thompson III, J. B. Sangree, J. N. Bubb, and W. G. Hartfield, 1977, Seismic stratigraphy and global changes of sea-level, *in* Seismic stratigraphy—Applications to hydrocarbon exploration: AAPG Memoir 26, p. 49–212.

Van Wagoner, J. C., R. M. Mitchum, K. M. Campion, and V. D. Rahmanian, 1990, Siliciclastic sequence stratigraphy in well logs, cores and outcrops: Concepts for high resolution correlation of time and facies: AAPG Methods in Exploration 7, 55 p.

Yarus, J. M., and R. L. Chambers, 1994, Stochastic modeling and geostatistics: Principles, methods, and case studies: AAPG Computer Applications in Geology 3, 379 p.

Zimmerle, W., 1995, Petroleum sedimentology: London, Kluwer Academic Publishers, 413 p.

Xu, W, K. Wrolstad, D. Kramer, P. Dooley, K. Domingue, and D. T. Vo, 2006, Brazos A-105 D-sand reservoir modeling by integration of seismic elastic inversion results with geostatistical techniques, *in* T. C. Coburn, J. M. Yarus, and R. L. Chambers, eds., Stochastic modeling and geostatistics: Principles, methods, and case studies, volume II: AAPG Computer Applications in Geology 5, p. 269–278.

*Chapter 20*

# Brazos A-105 D-Sand Reservoir Modeling by Integration of Seismic Elastic Inversion Results with Geostatistical Techniques

**W. Xu**
*Unocal Thailand, Ltd.*
*Bangkok, Thailand*

**K. Wrolstad**
*Unocal Exploration and Production Technology*
*Sugar Land, Texas, U.S.A.*

**D. Kramer**
*International Reservoir Technologies, Inc.*
*Lakewood, Colorado, U.S.A.*

**P. Dooley**
*Unocal Thailand, Ltd.*
*Bangkok, Thailand*

**K. Domingue**
*Nexen Petroleum U.S.A. Inc.*
*Dallas, Texas, U.S.A.*

**D. T. Vo**
*Unocal Thailand, Ltd.*
*Bangkok, Thailand*

## ABSTRACT

An integrated reservoir modeling study of the Bigenerina humblei (Big Hum) Miocene D-sand at Brazos A-105 field, offshore Texas, was conducted to predict the lateral extent of the reservoir, to build a porosity model for use in flow simulation and reserve evaluation, and to assess the uncertainty of the reserve estimation. Several geostatistical techniques for integrating well-log porosity with quantitative average porosity derived from a forward elastic model-based inversion method for three-dimensional seismic data were applied in this reservoir modeling study. Elastic modeling was necessary to predict the correct porosity-amplitude relationship for this reservoir because it is a class 2 type amplitude-vs.-offset reflection. The results of the study showed that if a reservoir is seismically resolved and properly imaged, elastic model-based inversion of the type employed can be used in conjunction with geostatistical methods to obtain a more complete reservoir description. These techniques were determined to have direct application to reservoir-flow modeling and hydrocarbon reserve volume estimation.


DOI:10.1306/1063821CA53240

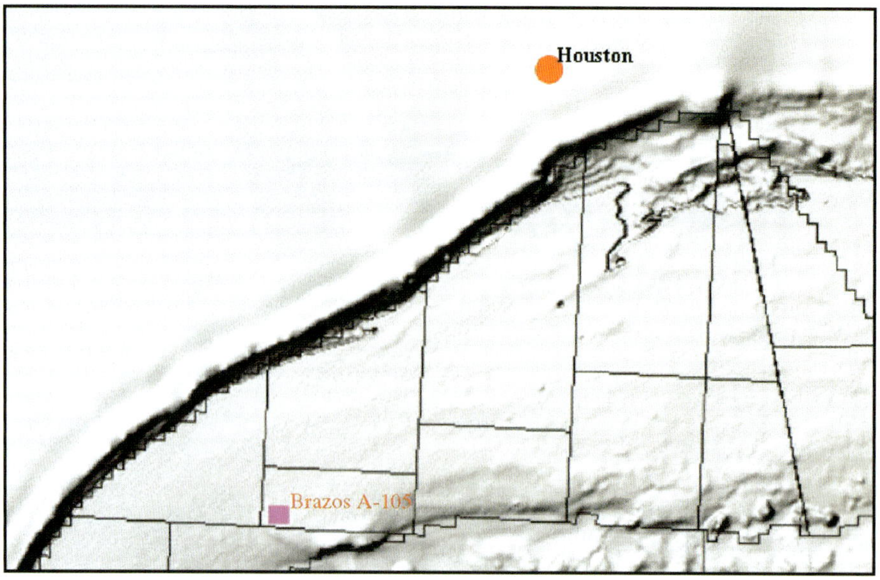


Figure 1. Location map of Brazos A-105 gas field.

## INTRODUCTION

The Brazos A-105 field is a faulted, four-way rollover structure located downthrown along the Miocene Corsair fault trend in the Gulf of Mexico (Figure 1). Brazos A-105 is one of several major gas fields along this trend that produce from Textularia (W) sands and deeper Bigenerina humblei (Big Hum) sands. The Big Hum sand series is dominated by lowstand systems tracts with minor intervals of mud-prone intervening transgressive and highstand systems tracts. A simplified cross section of the Brazos A-105 field is shown in Figure 2.

At present, eight wells have penetrated the Big Hum D sand unit (D-sand) reservoir imaged by the three-dimensional (3-D) seismic survey in productive and nonproductive areas, and it is appropriate to integrate reservoir-property estimates derived from seismic data with log data using geostatistical methods. However, in the interest of time and expense, only the poststack field seismic data were made available to us. The correlation method described below was then applied to the poststack data with the goal of mapping reservoir porosity as a measure of reservoir continuity. Thickness is constrained by isopachs already determined by Landmark interpretation.

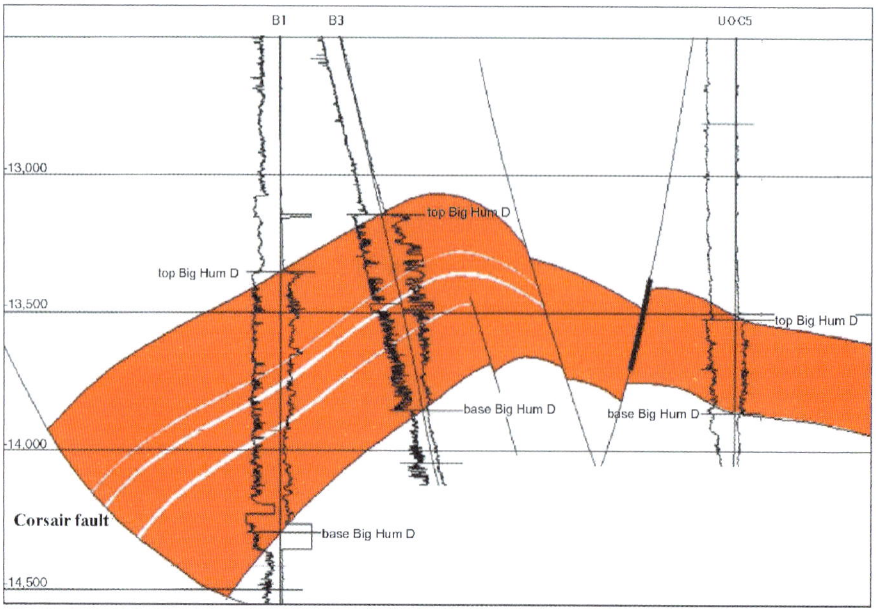


Figure 2. Schematic cross section of Big Hum D sand (D-sand) reservoir. Red represents sandstone units, whereas minor mudstone units within D-sand are in white.

The gas-filled D-sand is a Rutherford and Williams (1989) class 2 reflector with higher P-wave velocity and lower density than the overlying shale. This gives a phase reversal with offset at the top D interval. Elastic modeling is required to accurately predict the stacked amplitudes at the top D-sand. The relationship of reservoir porosity to seismic amplitude cannot be obtained from well-log impedance-porosity crossplots.

## MODEL-BASED SEISMIC INVERSION

To use seismic attributes for 3-D reservoir modeling, it is necessary to transform the information contained within the seismic trace into a recognizable and quantitative reservoir description parameter. The model-based inversion procedure consists of comparing a theoretical geophysical response, computed from a plausible geological model, with the observed response until the difference between the two responses is minimized.

### Seismic Characteristics and Elastic Modeling

Most seismic information that is integrated with well data using geostatistical techniques such as collocated cokriging (Xu et al., 1992) is related to stacked trace amplitudes or inverted impedance. Inverted impedance is related to porosity through well-seismic crossplots. In the present case, seismic amplitudes at the top D-sand are always dim or variable. Elastic modeling shows that the D-sand has a compressional (P-wave) velocity increase but bulk density decrease, relative to the overlying shale. This causes a mid-offset polarity reversal and a near-zero stack trace amplitude (a class 2 reflector). The velocity picking tries to optimize coherency and commonly misses the actual polarity reversal, yielding variable stack trace responses. Thus, acoustic modeling is not appropriate for this target sand. In fact, as finer grained sediments occur downdip, the porosity decreases, and the polarity reversal generally moves to larger offsets, giving a brighter stack trace response. Thus, porosity will be inversely related to seismic amplitudes for the class 1 sand in the reservoir area. The seismic amplitudes of class 3 bright spot sands are directly related to porosity, giving traditional bright spot response to high porosity. Elastic modeling is necessary to determine the seismic amplitude-porosity relationship because the impedance of the overlying shale is a crucial factor, as is the velocity-density relationship in the sand.

The advantages of this technique are that the effect of elastic reflection coefficients, interbed reflections, converted waves, and bed tuning are included in the seismic signatures. Differential changes to the signatures can then be related to specific petrophysical or fluid effects as modeled. Nonuniqueness is minimized by constraining the model array and the lag range in the cross-correlation.

Unocal's Litho-Strat Workstation (a proprietary software package developed to perform lithological analysis such as amplitude-vs.-offset [AVO] modeling, crossploting, etc.) (Aminzadeh et al., 1992) was used to load log volumetric data along with sonic transit time, density, and true vertical depth. Additional curves were created in the log volumetric data file including a lithology code for calcite, the density curve of wet clay, the sonic curve of wet clay, and the fraction of calcite relative to the total solid volume of the rock. Then, the sonic curve was upscaled (or blocked), giving about 150 layers in the interval from 11,000 or 12,000 to 15,000 or 16,000 ft (3352 or 3657 to 4572 or 4876 m). Shear-wave velocities were computed from log volumetric data as described by Wrolstad (1986). These in-situ elastic synthetics were computed at each well.

An example of a well-log volumetrics and an in-situ log response model with the normal move-out (NMO)–corrected prestack elastic gather is shown in Figure 3. The top and base D-sand locations are labeled in the figure. A red line tracks the top D-sand reflection in each synthetic gather. The AVO behavior along the red line is shown inside the box to the lower left. Notice that the synthetic gather predicts the phase reversal with offset. Perturbing the porosity in this well from its in-situ average value of 13.4 to 11% moves the polarity reversal out to 13,300 ft (4053 m). The offset, at which the polarity reversal occurs, increases with decreasing D-sand porosity. This will cause the top D-sand event to appear more coherent and brighter downdip in the distal, finer grained, low-porosity region.

### Model-based Inversion Procedure

Because stratigraphic interpretation suggests that the stratigraphic correlation style for this reservoir is comformable, thickness changes can be accomplished by simply stretching or contracting the top to base D-sand interval. Saturation changes, if any,

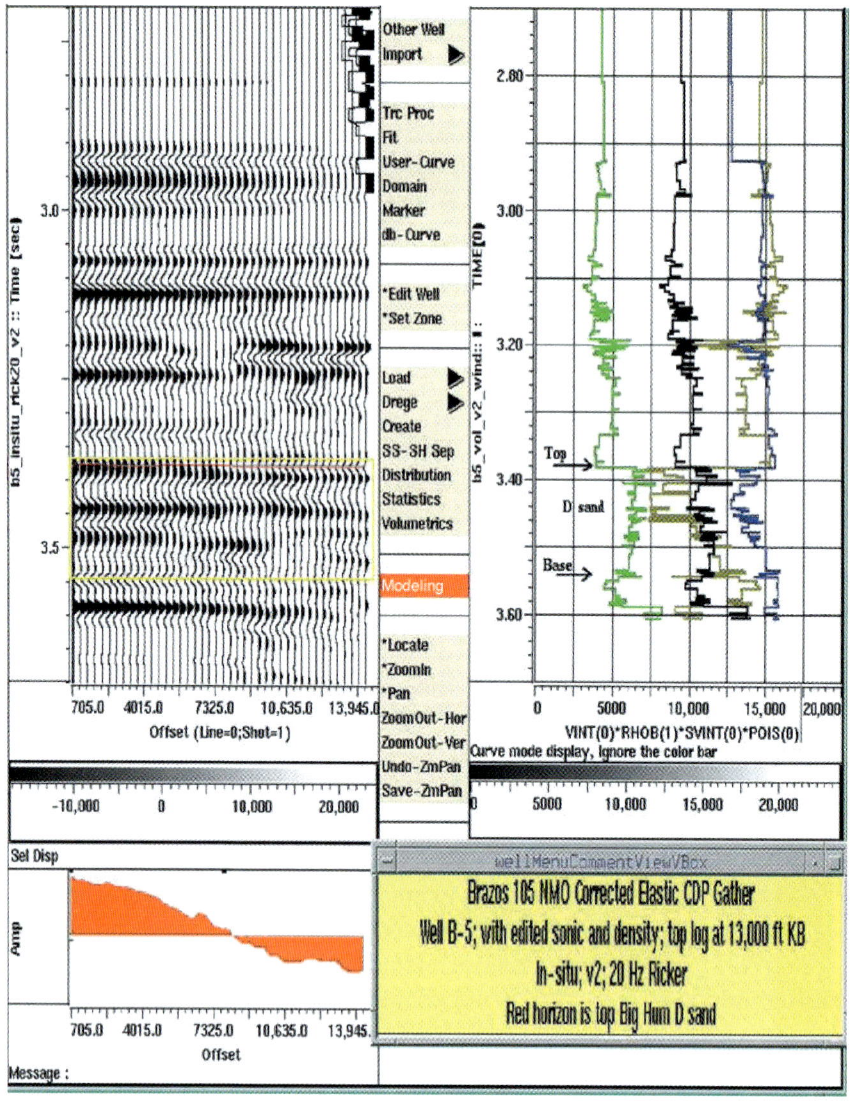


Figure 3. An in-situ model with the normal move-out–corrected prestack elastic gather.

require the nonlinear zero-frequency limit of the Biot (1962) equation or, equivalently, the Gassmann's (1951) equation. However, gas is encountered to some degree of saturation in all eight wells. Therefore, all models assume gas saturation. Porosity changes are made using the Raymer equation, an empirical equation that is considered the best relationship for most clastic rocks (Raymer et al., 1980). These changes are accomplished on a sample-by-sample basis in the original volumetric data, and velocities and densities are properly averaged according to the Backus (1962) equations in blocked layers. Perturbations in top to base D-sand thickness and porosity are made, incrementally yielding 10 thicknesses and 20 porosities, or 200 models per well. A total of 1600 perturbation models are generated for the eight wells. Figure 4 shows the diagram of our model-based inversion procedure.

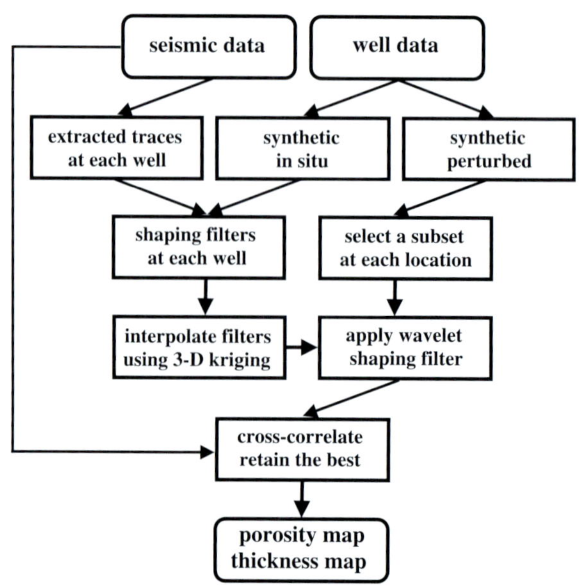


Figure 4. Flowchart of model-based inversion process.

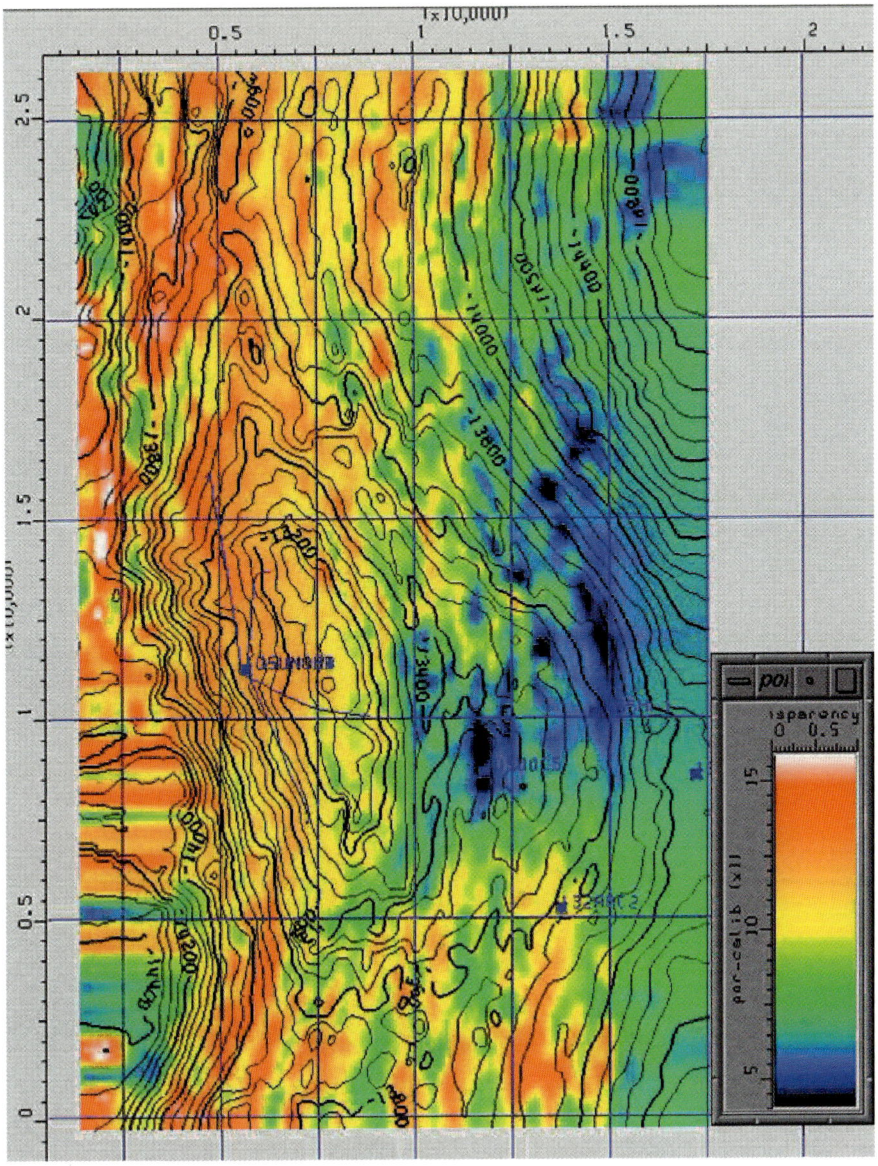

**Figure 5.** D-sand average porosity distribution derived from model-based inversion. Porosity ranges from 5% (dark blue) to 15% (red). The values in the upthrown side of Corsair fault (in lower left of map) are ignored.

At each field seismic trace location, the two nearest wells are chosen according to their distance to the trace. The interval-perturbed synthetic gathers from these wells are selected and further constrained by the known D-sand thickness at that location plus and minus one interval of thickness. The stacked, resampled, and polarity-reversed synthetic traces are each convolved with the interpolated filter for that location, and they are cross-correlated with the field trace in a small window encompassing the D-sand. The lag range is constrained to prevent cycle skipping. This is possible because of the close tie that has already been established. An interval average porosity and interval thickness from the model with the best correlation are saved along with the correlation magnitude. The final results are maps of interval-averaged porosity, modeled D-sand thickness, and correlation magnitude (Figures 5–7). The correlation magnitude map is a direct measure of uncertainty in the estimation at every trace location. The thickness map provides useful additional information because the thickness directly interpreted from seismic data may not represent actual thickness because of the difficulty of identifying the top of D-sand on seismic sections. In this case, the top of D-sand may be responded by very small amplitude or just an inflection in phase. The map of the porosity is smoothed with an eight-point running average filter in both inline and cross-line directions. The synthetic trace with the highest correlation is also retained at each trace

Figure 6. D-sand gross thickness map derived from model-based inversion. Thickness ranges from 400 (in dark blue) to 1500 ft (in red) (121 to 457 m). The values in the upthrown side of Corsair fault (in lower left of map) are ignored.

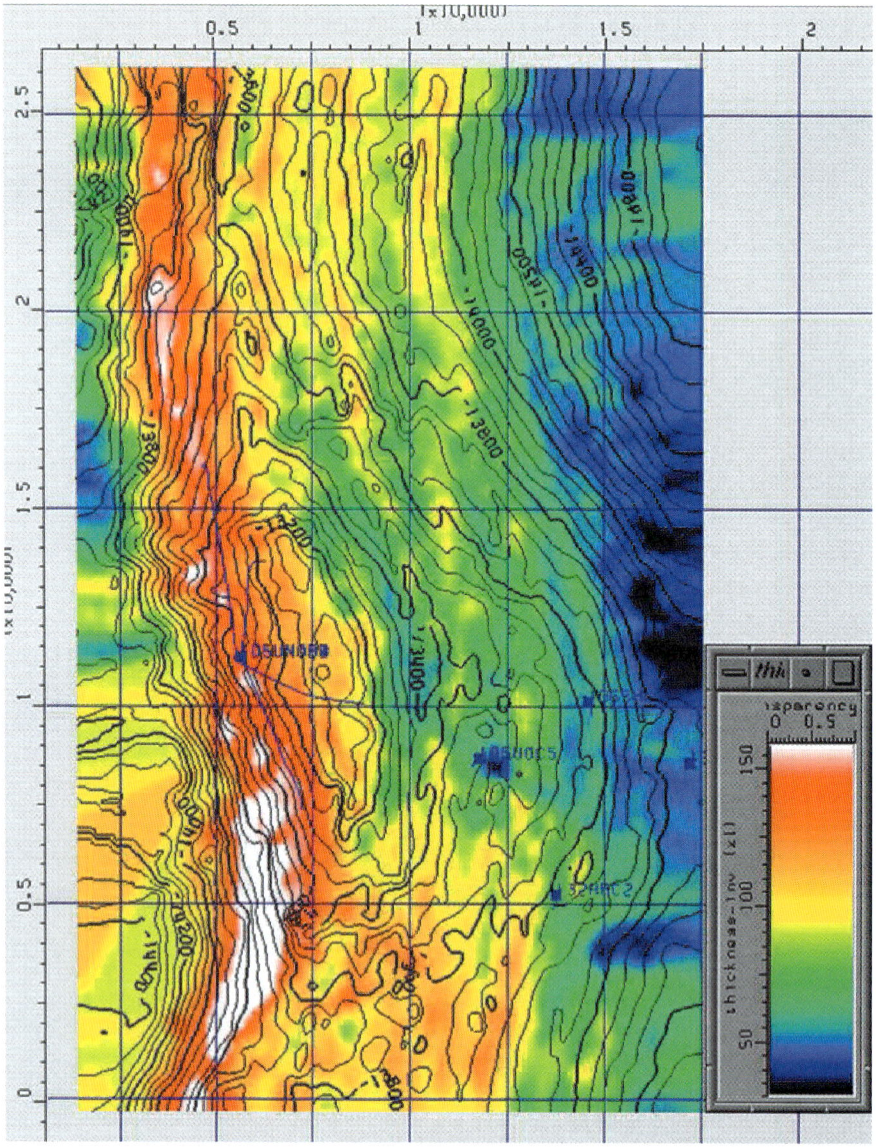


location, giving a 3-D volume of best match synthetic traces. Examples of section and time slices of synthetic and corresponding field data are shown in Figure 8. Notice that many details of the signature are explained very well by the 1600 thickness and porosity perturbations, providing further confidence in the resulting porosity map.

## RESERVOIR MODELING

Reservoir modeling consists of three phases: analysis of the reservoir facies and porosity data, stochastic simulation using a probability field-simulation approach, and evaluation of the modeling results.

### Data Analysis of Reservoir Facies and Porosity

Based on log analysis, the reservoir is mainly separated into two lithological groups (or facies): sandstone and shale. If 25% or more shale content is present in a sample location, then this location is classified as shale; otherwise, it is classified as sandstone. Facies data are obtained in all eight wells using the above separation. Facies proportions are obtained from facies data at all well locations. Sand (facies) proportions at wells on the crest of the structure are higher than those at four flank wells. Well data show that sand proportion closely correlates to the vertically averaged porosity.

For the present case, detailed log correlation analysis and production data analysis suggest that

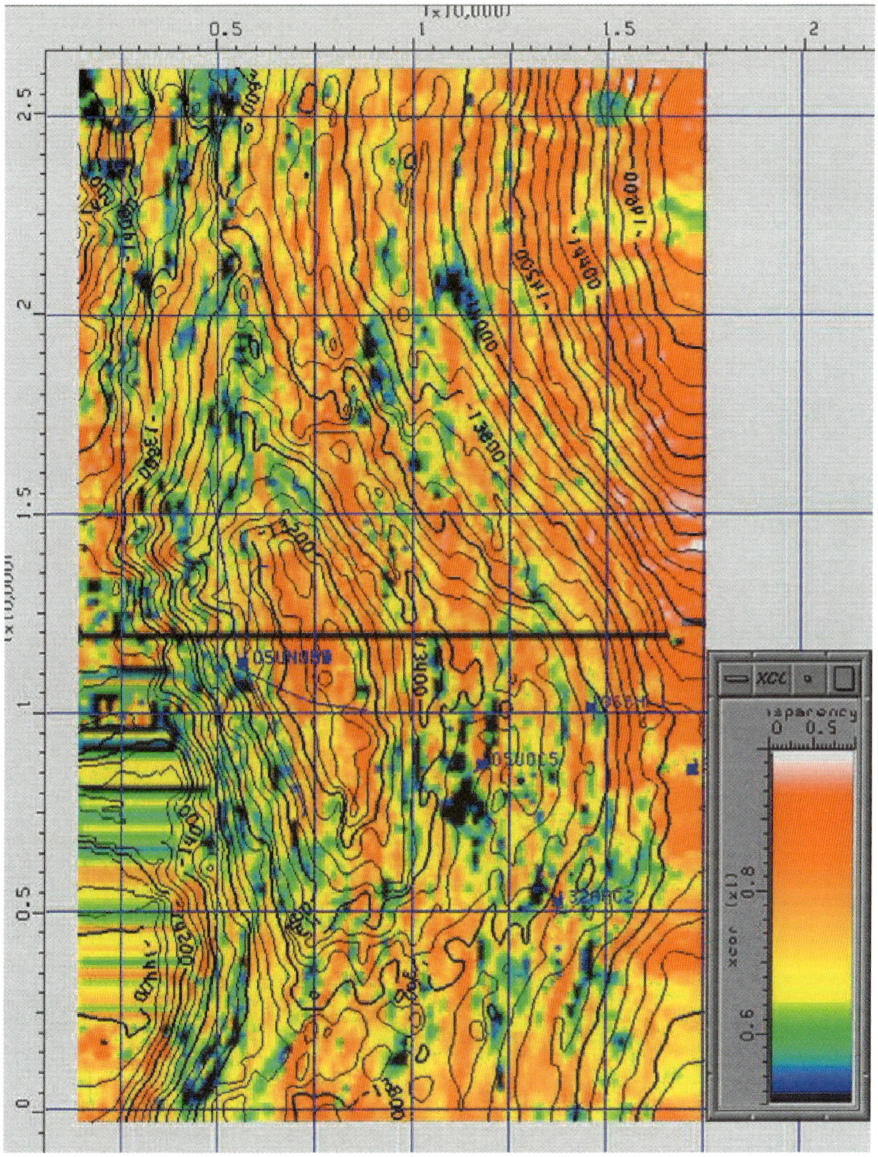


**Figure 7.** Cross-correlation magnitude map from model-based inversion. Correlation ranges from 0.5 (in dark blue) to 1.0 (in red). The dark line in the middle is caused by a missing seismic line.

stratigraphic conformity (proportional) to the existing top and base of the reservoir interval can be adopted. The experimental directional (areal and vertical) semivariograms of sand facies are calculated and modeled (Figure 9). In a two-facies system, one facies always complements another in spatial variability. Therefore, the semivariogram of the shale facies is identical to that of sand. The directional semivariograms of porosity are shown in Figure 10.

## Stochastic Simulations of Facies and Porosity

The seismic-derived average porosity map (Figure 5) and reservoir sand facies proportions at the eight wells were integrated using collocated cokriging to obtain a facies proportion map (Figure 11). A two-stage stochastic modeling approach has been used to construct a reservoir-porosity model. A 3-D facies model, which controls the large-scale reservoir heterogeneities, is generated by an improved version of the probability field-simulation technique (Xu, 1995), conditioned to both the well facies and the facies proportion map derived from seismic inversion. Figure 12 shows four northwest–southeast cross sections extracted from one of 3-D facies-simulated realizations. The 3-D porosity simulations are performed within sand facies using the porosity data from the eight wells. A very low effective porosity (0.1%) is assigned to the shale facies cells. By combining the porosity distributions of reservoir and nonreservoir facies, the whole 3-D porosity distribution was obtained (Figure 13).

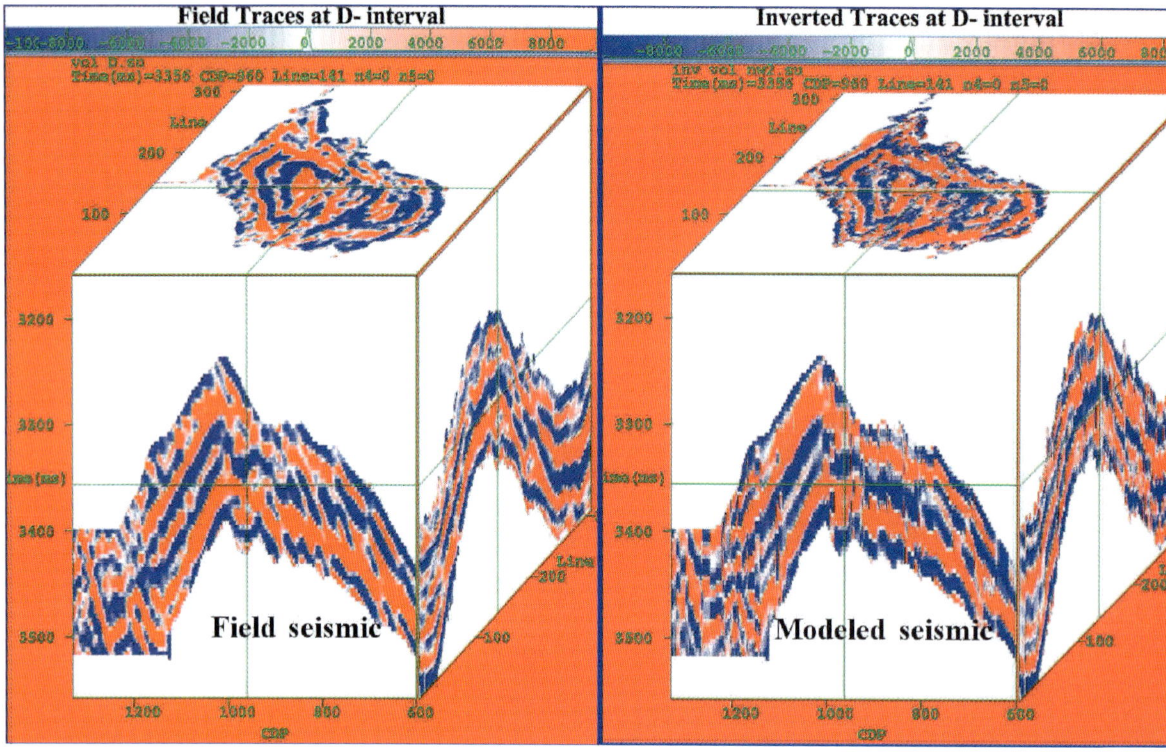


Figure 8. Comparison of 3-D seismic volume (left) and 3-D synthetic volume retained from inversion in the D-sand reservoir interval.

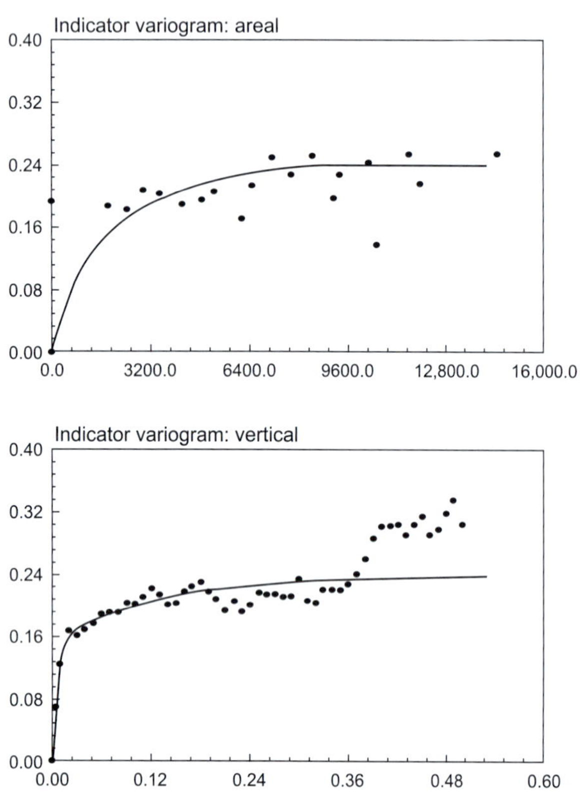


Figure 9. Areal (upper) and vertical (lower) semivariograms of sand facies. Dots: experimental semivariograms; line: fitted exponential models.

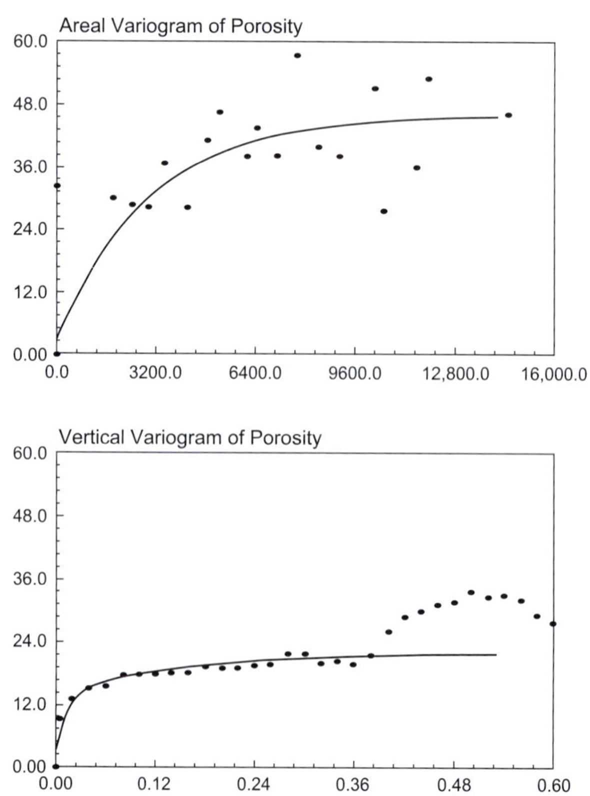


Figure 10. Areal (upper) and vertical (lower) semivariograms of porosity. Dots: experimental semivariograms; line: fitted exponential models.

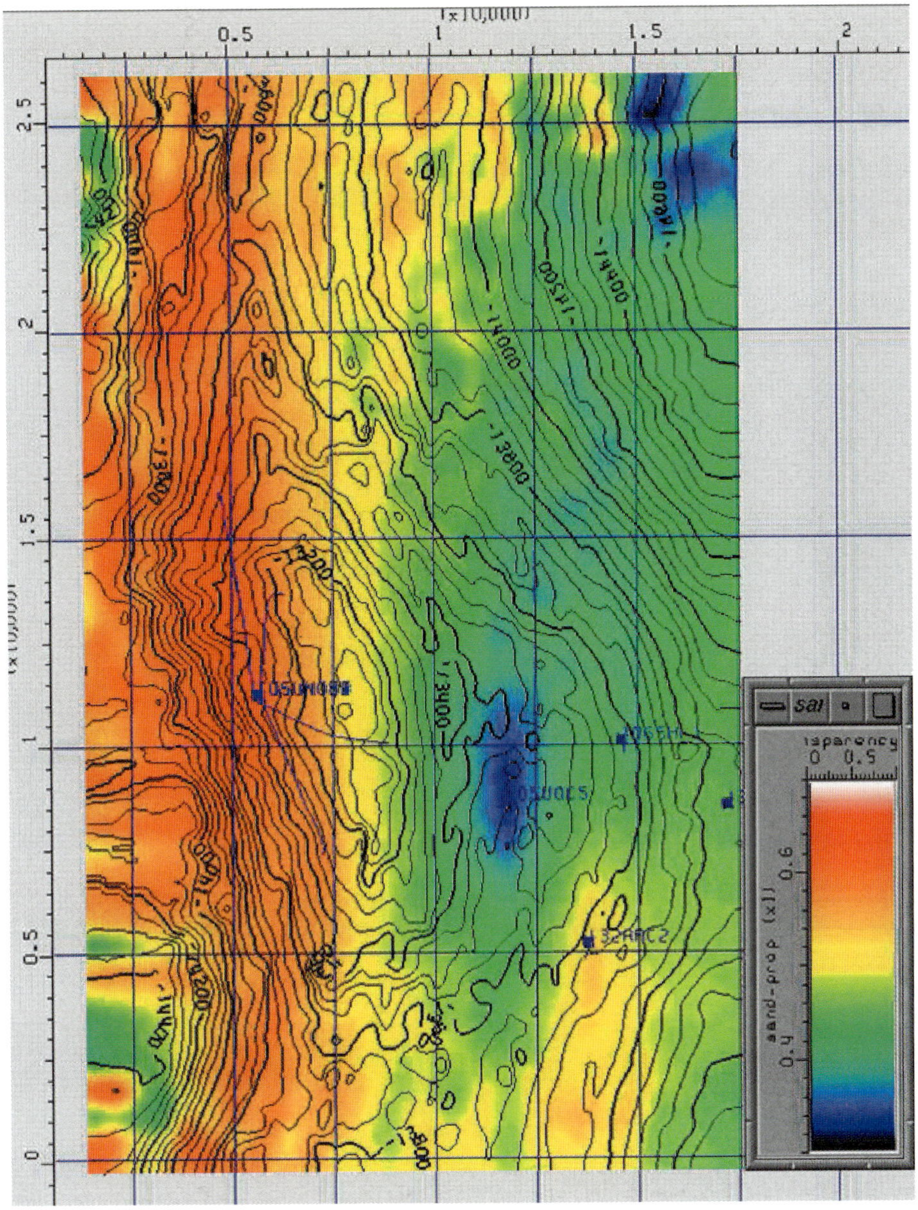


**Figure 11.** Estimated sand proportion in D-sand interval. Sand proportion ranges from 0.3 (dark blue) to 0.7 (red). The values in the up-thrown side of Corsair fault (in lower left of map) are ignored.

## Modeling Results

Net pore volume and gas reserves in place are estimated based on 100 realizations of the porosity model. Pore volume and reserve uncertainties are also assessed. Reservoir-flow simulations are performed based on the upscaled 3-D porosity model. The original gas in place obtained from stochastic volumetric analysis matches nicely with the results from full-field-flow reservoir simulation. The flow-simulation model is being used successfully as a reservoir-management tool to guide production and exploitation activities.

## CONCLUSIONS

Elastic instead of conventional acoustic inversion can extract more geological content from seismic data, particularly in a dim spot reservoir. The elastic model-based inversion technique provides a direct estimate of average porosity over the reservoir area from post-stack 3-D seismic data. The integrated average porosity map, when further calibrated with well porosity data using collocated cokriging, provides a reasonable determination of the lateral limits of the reservoir.

The use of seismic data as an additional source of information for conditioning data reduces the

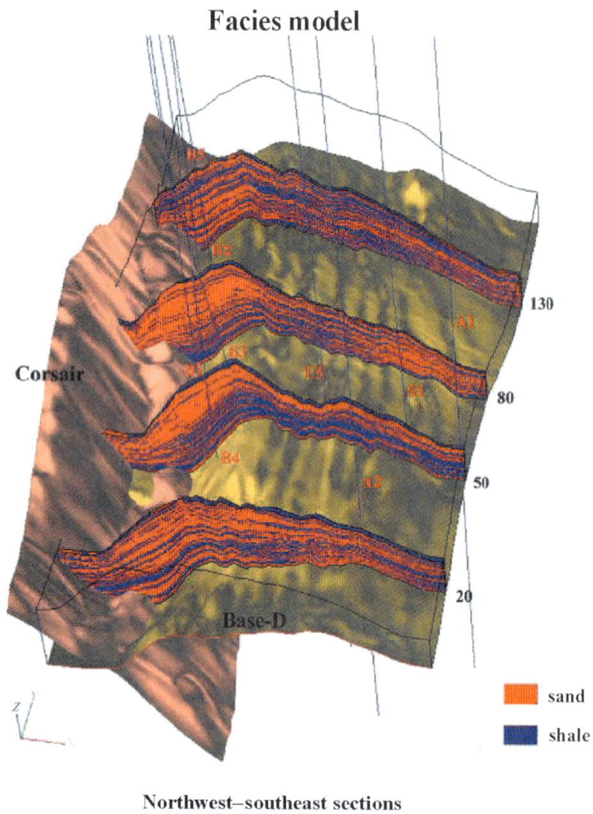


**Figure 12. Cross section views of the 3-D facies model from stochastic simulation.**

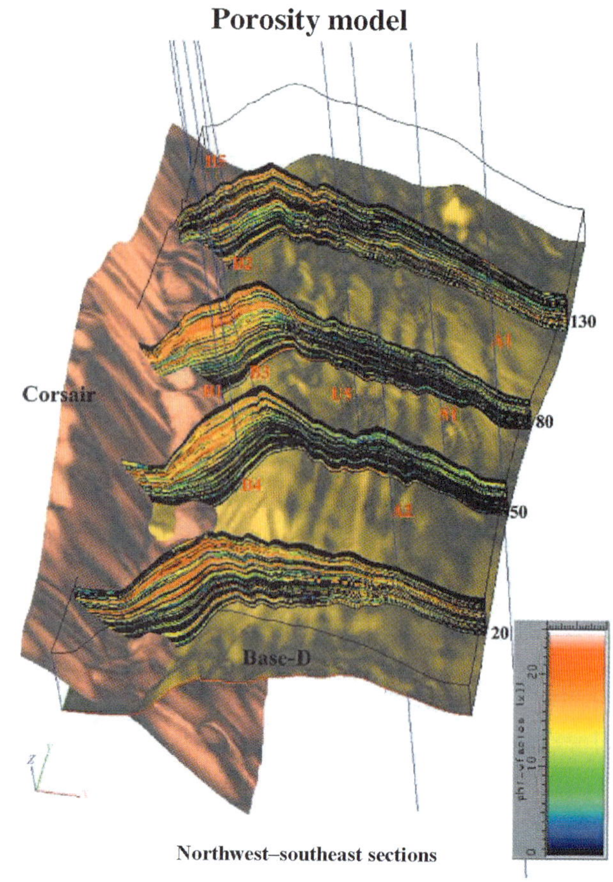


**Figure 13. Cross section views of the 3-D porosity model from stochastic simulation. High porosity values are shown in red; low porosity values are in blue. No porosity (in shale units) is in black.**

uncertainty in the stochastic reservoir model because the uncertainty of the reservoir description in the interwell areas is reduced.

## ACKNOWLEDGMENTS

The authors are grateful for management and partner support for this integrated project. Special thanks go to Dave Bacchus (Unocal E&PT) for the petrophysical analyses performed in conjunction with this work. The Brazos A-105 Partnership comprises Spirit Energy 76 (Unocal), The Northwestern Mutual Life Insurance Company, and Mariner Energy. We also thank Petter Abrahamsen and another reviewer for their comments, which improved the submitted manuscript.

## REFERENCES CITED

Aminzadeh, F., C. K. Von Kahrs, and K. H. Wrolstad, 1992, Seismic ray tracing method and apparatus: U.S. Patent 5,079,749.

Backus, G. E., 1962, Long-wave elastic anisotropy produced by horizontal layering: Journal of Geophysical Research, v. 67, p. 4427–4440.

Biot, M. A., 1962, Mechanics of deformation and acoustic propagation in porous media: Journal of Applied Physics, v. 33, p. 1482–1498.

Gassmann, F., 1951, Elastic waves through a packing of spheres: Geophysics, v. 16, p. 673–685.

Raymer, L. L., E. R. Hunt, and J. S. Gardner, 1980, An improved sonic transit time-to-porosity transform, *in* Transactions of the Society of Professional Well Log Analysts 21st Annual Logging Symposium: Society of Professional Well Log Analysts, Paper P.

Rutherford, S. R., and R. H. Williams, 1989, Amplitude-versus-offset variations in gas sands: Geophysics, v. 54, p. 680–688.

Wrolstad, K. H., 1986, Offset-dependent amplitude analysis by viscoelastic lithological modeling: Society of Exploration Geophysicists Expanded Abstracts, Paper S2.5, p. 341.

Xu, W., 1995, Stochastic modeling of reservoir lithofacies and petrophysical properties: Ph.D. dissertation, Stanford University, Stanford, California, 214 p.

Xu, W., T. T. Tran, R. M. Srivastava, and A. G. Journel, 1992, Integrating seismic data in reservoir modeling: The collocated cokriging alternative: 67th SPE Annual Technical Conference and Exhibition, 4–7 October, Washington, D. C., SPE Paper 24742, 10 p.

Frykman, P., 2006, Spatial variability of porosity in chalk: Comparison of an outcrop and a North Sea reservoir, *in* T. C. Coburn, J. M. Yarus, and R. L. Chambers, eds., Stochastic modeling and geostatistics: Principles, methods, and case studies, volume II: AAPG Computer Applications in Geology 5, p. 279–288.

*Chapter 21*

# Spatial Variability of Porosity in Chalk: Comparison of an Outcrop and a North Sea Reservoir

**P. Frykman**
*Geological Survey of Denmark and Greenland (GEUS)*
*Copenhagen, Denmark*

## ABSTRACT

The study of the variability of porosity and permeability, and of the spatial distribution of petrophysical parameters, has received increasing interest for chalk deposits, which form important reservoirs for both hydrocarbons and groundwater. A common activity is to construct numerical reservoir models for different types of flow simulation. These models must be based on a diversity of data, including outcrop information when subsurface information is sparse. This chapter presents a comparison of the spatial variability of outcrop and reservoir petrophysical properties in chalk and a discussion on the use of information from horizontal wells.

## INTRODUCTION

The objective of this study is to supply quantitative data from outcrop characterization that might form an analog for parts of chalk sequences studied elsewhere. Analysis of data from analog outcrops is commonly the only way to obtain knowledge about the correlation structure in the horizontal direction, which has large consequences for reservoir models and their flow performance (Kupfersberger and Deutsch, 1999). Uncertainty about the horizontal correlation ranges is already known from prior analysis and use of vertical and horizontal well information (Almeida and Frykman, 1994; Vejbæk and Kristensen, 2000).

The study specifically focuses on the spatial distribution of petrophysical properties supplemented with geological information from the investigated sequence. The usual approach for clastic sequences of first outlining rock types as a foundation for the characterization of the petrophysical properties is not routine in chalk investigations. A major obstacle is that distinct facies types are difficult to outline in most chalk successions because of the relative homogeneity of the lithology; and the coupling between facies and petrophysical properties is not well quantified or understood. The present study, in fact, involves analysis of the spatial distribution of porosity and permeability in a very homogeneous chalk sequence and is therefore meant to


DOI:10.1306/1063822CA53241

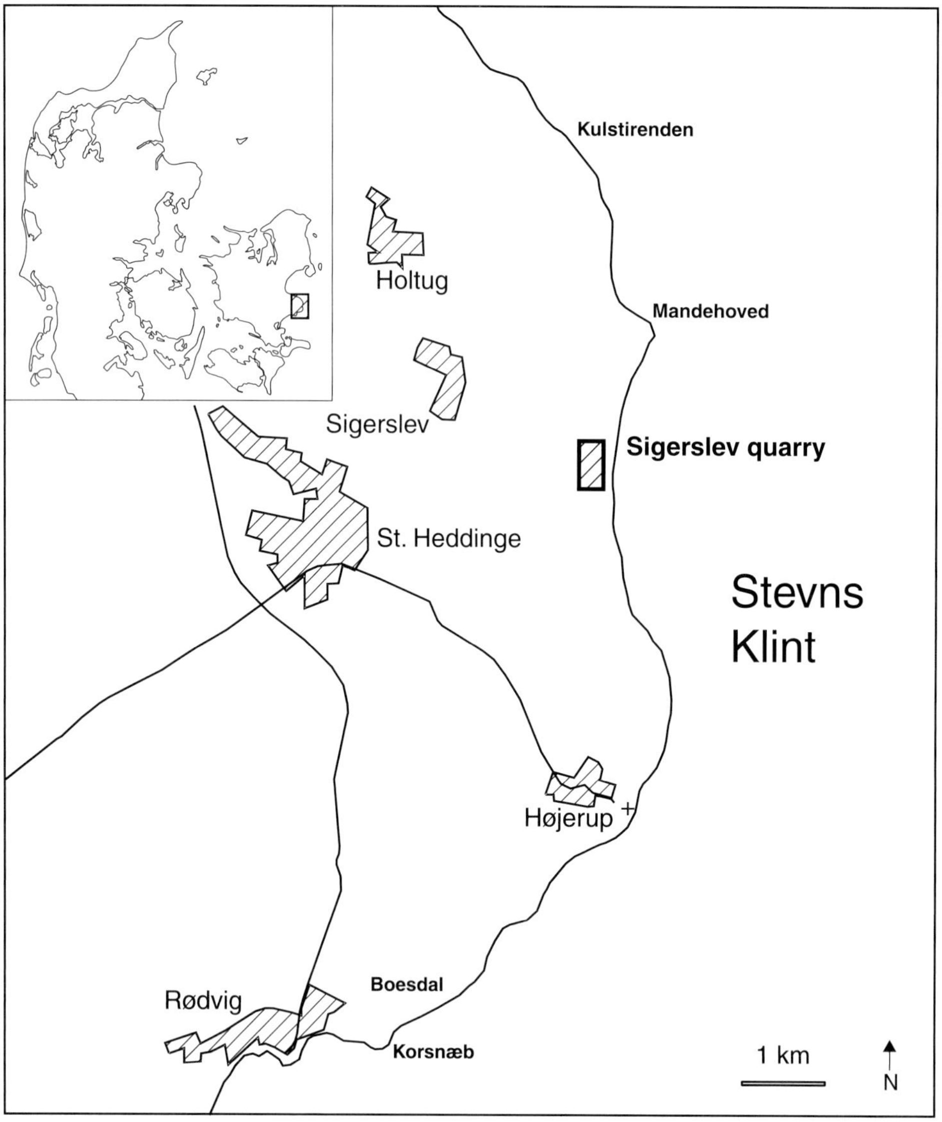


**Figure 1.** Map of the Stevns Klint area showing the position of the Sigerslev quarry. The classical Stevns outcrop cliff sections extend from Rødvig in the south to Kulstirenden in the north.

provide information about sequences of this specific nature.

The area of interest is a section in the uppermost Maastrichtian chalk exposed in the Sigerslev quarry at Stevns Klint, Denmark. The details of the locality and the interpretations of the variability and spatial correlations have already been published (Frykman, 2001). The interval of investigation is of comparable stratigraphic age to part of the uppermost reservoir zones in the Tor Formation in the North Sea chalk reservoirs. The porosity and permeability values found in the outcrop are generally comparable to those of the most porous reservoir chalks. However, reservoir chalk commonly exhibits lower values and larger variability, and the cyclic layering described from some reservoir sections is not observed at the study location. This absence is ascribed partly to primary facies differences between this locale and the chalk in the North Sea and partly to a diagenetic enhancement of the contrasts in the reservoirs. Geostatistical analysis indicates a subtle layering in the outcrop chalk and provides estimates of both the vertical and horizontal variability in porosity and permeability.

## MATERIAL

The outcrop is located in an active chalk quarry close to the classical Stevns area chalk outcrops (Figure 1). The geological context has been described elsewhere, and the succession is comparable to the

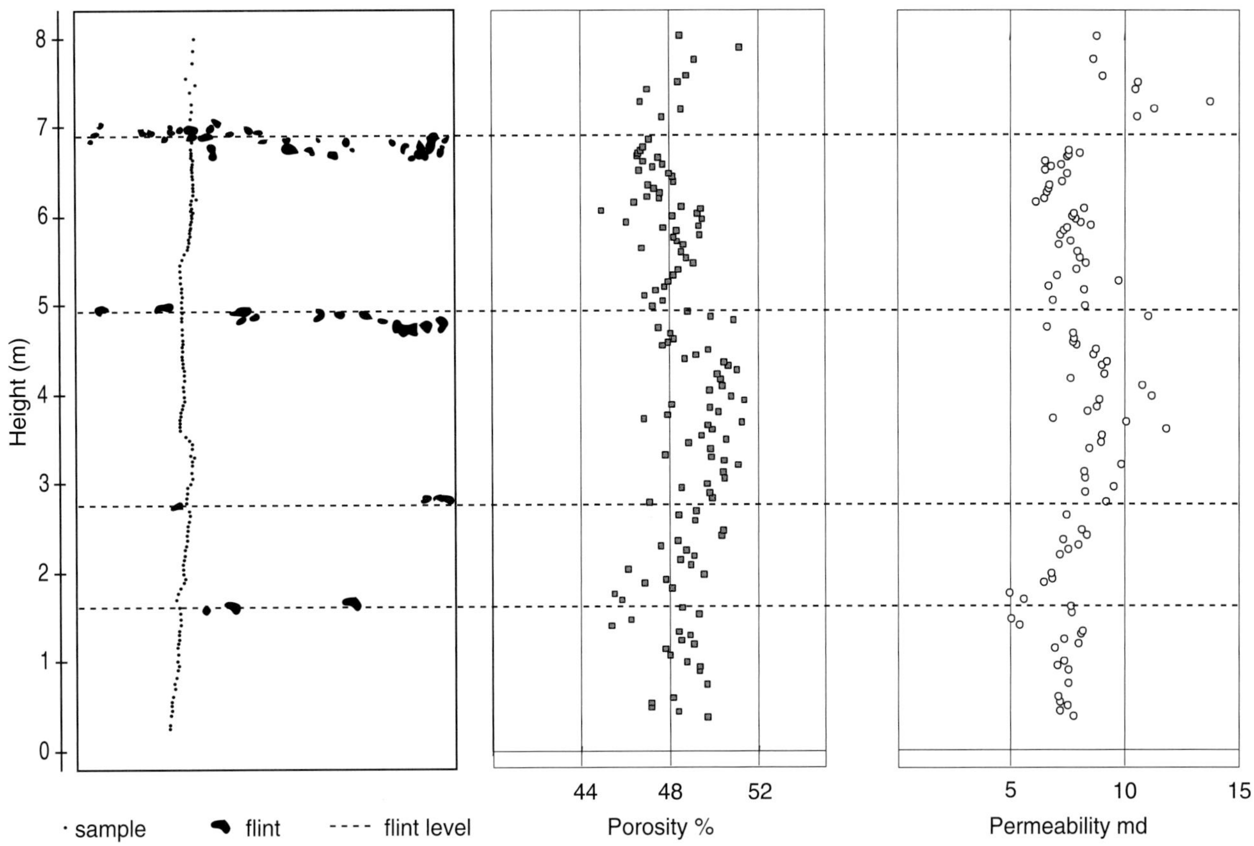


**Figure 2. The vertical outcrop section with porosity and permeability sampling at the Sigerslev-1 locality. Some of the larger fluctuations in porosity and permeability are possibly linked to levels with occurrence of flint nodules.**

general sequence in the Stevns area (Håkansson and Surlyk, 1997; Surlyk and Håkansson, 1999). The uppermost Maastrichtian chalk in the Sigerslev quarry has been sampled by drilling tightly spaced, 1-in. (2.5-cm) plugs from the exposed vertical wall. A total of 134 samples have been taken, representing a continuous vertical section of 8 m (26 ft) (Figure 2). Along a horizontal level, a total of 231 samples represent a lateral section of 28 m (91 ft). Measurements of porosity and permeability have been obtained in the laboratory by conventional core analysis methods (He porosimetry and gas permeameter). The results of porosity and air-permeability measurements for the vertical sample suite in the outcrop section are shown in Figure 2.

Semivariogram analysis of outcrop data has been conducted to analyze the spatial variability and possible correlation structure in the data (Frykman, 2001). The semivariograms are calculated according to normal standards (Isaaks and Srivastava, 1989; Deutsch and Journel, 1998). Basic familiarity with the calculation and significance of semivariograms is assumed in the discussion below.

The semivariograms for the porosity data are shown in Figure 3. Following the conventional approaches, the structures are modeled starting with the short-scale structures and proceeding to larger scale, with the sample variance from the data set being the limiting sill (Gringarten and Deutsch, 2001). Analysis of the permeability data suggests a similar structure (Frykman, 2001).

## SEMIVARIOGRAM ANALYSIS AND THE LINK TO GEOLOGY

### Nugget Effect

As indicated in the semivariogram modeling in Figure 3, the nugget effect is fairly high and accounts for approximately 35% of the total variance in the vertical data set. The nugget effect is the expression of very fine-scale random variations in porosity. These variations might be related to textural differences associated with different types of trace fossils or bioturbation and/or the influence by early

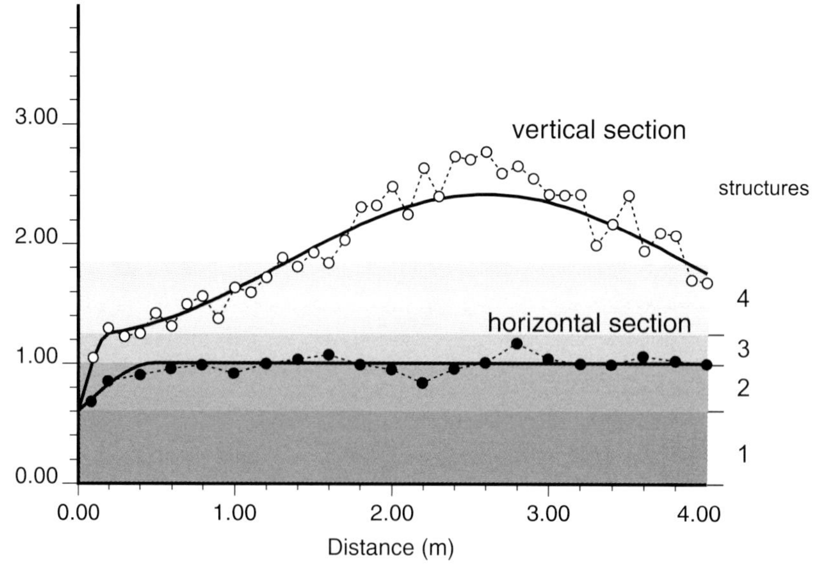


**Figure 3. Semivariograms for the vertical (upper) and horizontal (lower) outcrop porosity data. Both the experimental (with points) and the modeled (line) semivariograms are shown for each data suite. The interpreted structures for the different parts of the variability are shown.**

diagenesis. The size and spacing of the sampled plug volumes are large compared to these features, which, therefore, cannot be resolved in detail.

## Short- to Intermediate-scale Variance

The model for the vertical semivariogram includes a spherical structure with range 0.20 m (0.66 ft), which accounts for 35% of the variability. Being at the limit of the resolution given by the closely spaced plug samples, some uncertainty exists in the modeling of this range, but a structure with a vertical correlation range within 15–25 cm (6–10 in.) seems to be evident in the data. The short-range correlation of 15–25 cm (6–10 in.) detected in the vertical semivariogram cannot be supplemented with any data on grain size, texture, trace fossils, clay content, or other detailed sedimentological features. The only supporting information available arises from the fairly equidistant horizontal fractures or cleavage planes in the outcrop, which might indicate the presence of heterogeneities, however subtle, in the chalk.

The horizontal data set exhibits a vague, but similar, correlation structure; yet it only accounts for 22% of the total variance. If the choice is made to model this horizontal structure explicitly instead of incorporating it into the nugget effect, it has a range of approximately 0.40 m (1.3 ft). This is a longer range than the vertical range, a difference that obeys the principle of geometric anisotropy for layered geological systems. This principle is connected to the well-known geological Walther's law (Middleton, 1973), which implies that the horizontal and the vertical semivariograms are dependent, and that longer correlation lengths are expected in the horizontal direction and shorter in the vertical direction. Furthermore, the horizontal structure with 0.40-m (1.3-ft) range does not have the same variance contribution as the vertical structure, and this difference is the effect of what is called zonal anisotropy. The zonal anisotropy arises because the horizontal semivariogram samples are compared only within the horizontal layer that has more continuity. Therefore, the full variability is not represented in the horizontal direction but is only expressed in the vertical direction going across layers.

## Larger Scale Variance

The larger scale fluctuations in porosity, which are clearly observed in the vertical data set (Figure 3), can be modeled using a hole effect with a wavelength of 5.2 m (17.0 ft), which fits the undulations of the semivariogram. A hole effect is normally incorporated when the phenomena are periodic in nature, but it is not clear whether data are periodic because there is only a limited amount (8 m; 26 ft) of data. As a rule of thumb, the hole effect should be expressed across three to four wavelengths of the semivariogram to fully support the interpretation of periodicity. Therefore, this part of the semivariogram probably reflects a deterministic feature that should not be modeled explicitly.

The cyclic development of the chalk at meter scale, as seen in many reservoir sections, is not a pronounced feature in the outcrop chalks. The meter-scale cyclicity observed in the North Sea reservoirs (Scholle et al., 1998; Stage, 1999) has not been documented so far in Danish outcrops. Figure 4 compares porosity measurements from the vertical outcrop section to porosity data from a well in the Dan field in the North Sea. The well data exhibit a generally lower level of porosity but a clearer fluctuation pattern in both the core and log measurements. Any indication of cyclicity in the outcrop succession awaits detailed investigations of other parameters, such as trace-fossil assemblages, or isotopes, which might aid in resolving any possible bedding or periodicity in these deposits. It should also be noted that the two different data sets have different resolution. The core and outcrop measurements have the resolution given by the plug volume and the sample spacing, whereas the porosity log data represent averages more than approximately 0.60 m (1.9 ft) because of the tool and logging procedure. This difference will influence the comparison of the histograms and semivariograms for these data sets (Frykman and Deutsch, 2002).

The variation in the vertical outcrop data over several meters may be caused by the fluctuation in porosity and permeability, which seems to be related to the occurrence of flint nodules in certain layers (Figure 2). The specific processes governing flint occurrence and the connection to the petrophysical properties have only been sparsely studied, and further investigations are needed to establish this link.

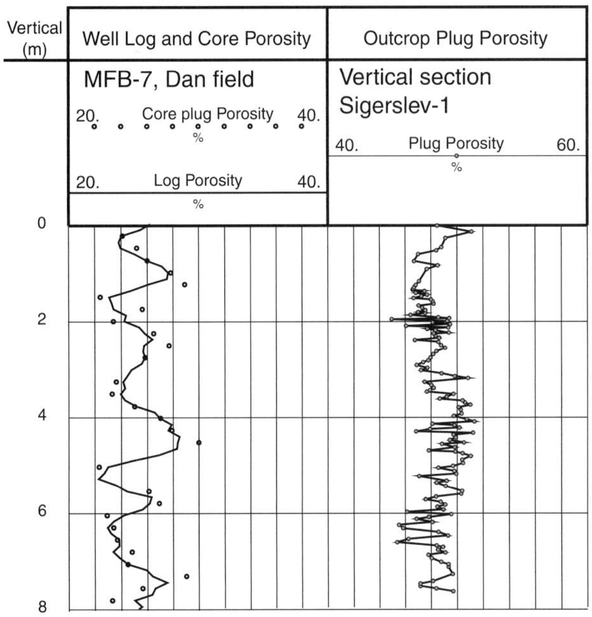


**Figure 4. Comparison of the outcrop porosity data in the vertical section with data from a section in a well from a North Sea chalk reservoir showing cyclic variations in the porosity. Note the difference in the porosity range and the variability for the two data sets.**

As previously noted, the horizontal variability seen in the available section is smaller than the vertical variability caused by the layering in the succession. Any estimate of a long-range horizontal correlation structure is impossible and is restricted by the limited extent of the section. Because the particular onshore succession analyzed here has abundant mound-shaped features and associated irregular bedding, longer lateral sections showing continuity are not expected from this setting. These mound structures are dominant features in the east Denmark Maastrichtian chalk, and other localities must be explored for more continuous bedding to investigate the lateral correlation patterns.

## ANALYSIS OF RESERVOIR DATA

The chalk successions observed in the North Sea reservoirs, and especially in core and log data in horizontal wells, seem to be very layered systems having only small lateral variations and long correlation ranges in the horizontal direction. This is a result of the hemipelagic or pelagic nature of the main depositional processes in the North Sea, which create homogeneity and continuity along layers. In some chalk reservoirs, the pattern is occasionally interrupted by layers or lenses of redeposited material and/or disturbances by sliding and slumping processes, a phenomenon that is most frequent in the northern North Sea. In the southern North Sea (Danish sector), redeposition of chalk seems to be dominated by more gentle winnowing processes (Scholle et al., 1998; Stage, 1999). These processes are able to create thickness differences in the individual layers at the horizontal scale of 100–1000 m (330–3300 ft), still leaving a great deal of lateral continuity. Although horizontal correlation ranges have been estimated from well data in a chalk reservoir, some degree of uncertainty still exists (Vejbæk and Kristensen, 2000; Vejbæk, 2002). This uncertainty can be limited by carefully selecting data from horizontal wells in the reservoir as supplementary information.

Luckily, horizontal wells have become exceedingly popular, especially in chalk reservoirs, and a wealth of information can be derived from them.

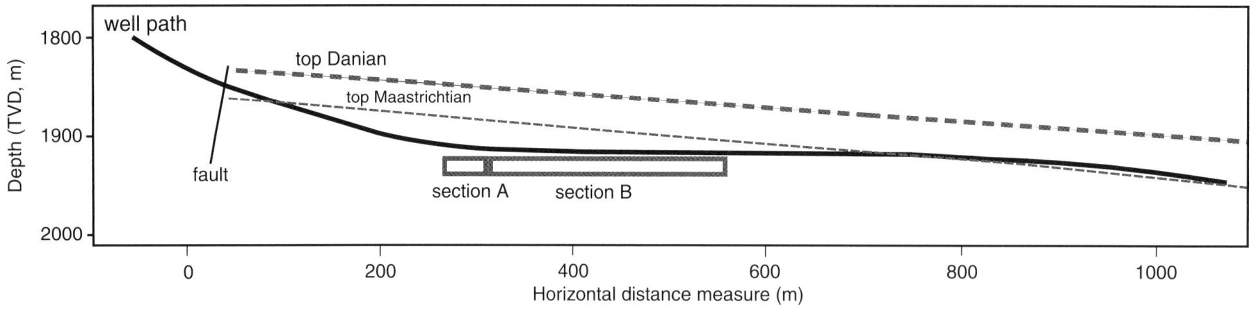


**Figure 5. Profile of the well MFA-17 along the southern flank of the Dan field showing the entry through the fault face and transect in the Danian, upper Maastrichtian, and reentry in the lower Danian section, where the final depth is reached. The two subsections A and B studied in more detail are indicated.**

However, horizontal well data must be given careful consideration to avoid misinterpretation because artifacts can arise from the drilling and logging process. For example, horizontal wells have a higher probability of encountering small faults along the wellbore, which can give rise to sudden shifts in the signal level. In addition, in some wells, a minor artifact can be seen as a seemingly cyclic porosity signal because of the influence of the corkscrew effect on density logging (Bellay et al., 1996). Nonetheless, careful selection of designated intervals along the wellbore can contribute valuable information to the analysis of lateral variability. The horizontal well MFA-17 in the southern part of the Dan field is a case in point as illustrated in Figure 5. The entry point into the chalk is a fault face in the Danian Ekofisk Formation, continuing down into the main reservoir interval in the Tor Formation of Maastrichtian age. The wellbore reenters the Danian succession in the last part of the wellbore. As illustrated, the wellbore is near parallel to the stratigraphic surfaces for a small part of the section, and this subsection of 40 m (131 ft) length has been analyzed specifically for correlation structure. The decision as to whether a specific section is bedding parallel relies very much on the mapped surfaces for the reservoir units and the uncertainties associated with the map. Especially on the flank area, which has limited well control, this uncertainty cannot be neglected.

Figure 6 shows a trace of porosity derived from the density log in well MFA-17. Two sections (A and B) of interest are indicated. Note that the 40-m (131-ft)-long A section, being close to bedding parallel, has limited variability. Interval B is longer (240 m; 787 ft) and has a nearly constant angle relative to the stratigraphy.

To compare different variability measures, porosity data from stratigraphically similar intervals

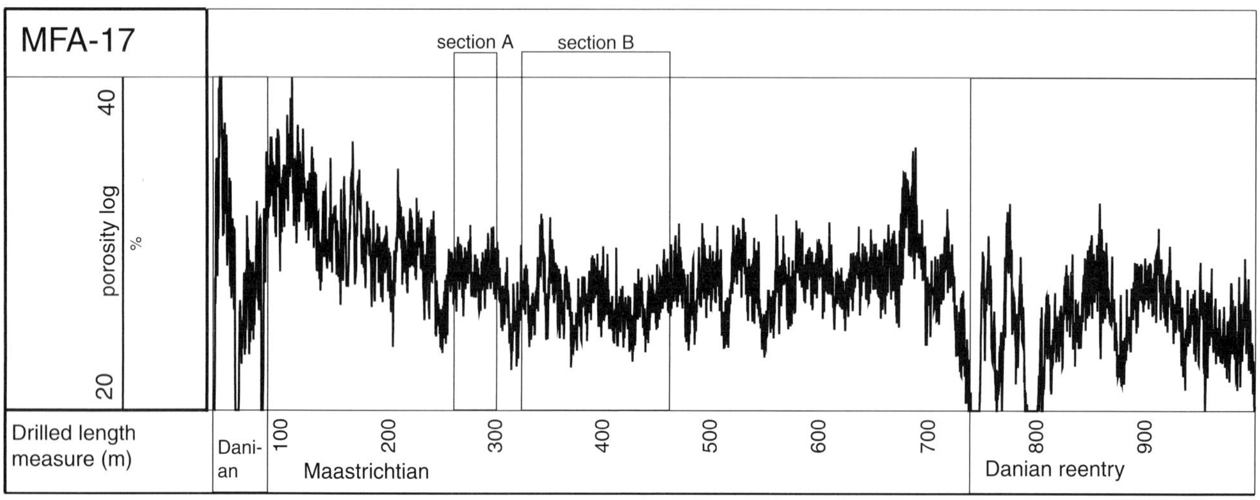


**Figure 6. Porosity log through the horizontal well MFA-17. The two parts of the well data selected for analysis are shown.**

in two vertical wells close to the horizontal well position (1100 m [3608 ft] away) have been included (Figures 7, 8). Semivariograms have been calculated for the different sections. The semivariogram analysis shows that the bed-parallel section A has a low sill value of 1.08, which is reached at a short range of 0.8 m (Figures 9, 10). In addition, the two vertical wells have a short-range structure of about 0.8-m (2.62-ft) range, evidently associated with the bedded nature of the succession. Because the vertical and horizontal correlation structures of 0.8 m (2.6 ft) do not seem to reflect the same feature or to be connected via zonal anisotropy, the 0.8-m (2.6-ft) range in the horizontal well data could be caused by a corkscrew effect or other forms of artifacts that should be considered structured noise. For the 40-m (131-ft)-long A section, there is no indication of variability or any trend caused by lateral changes in porosity over that distance.

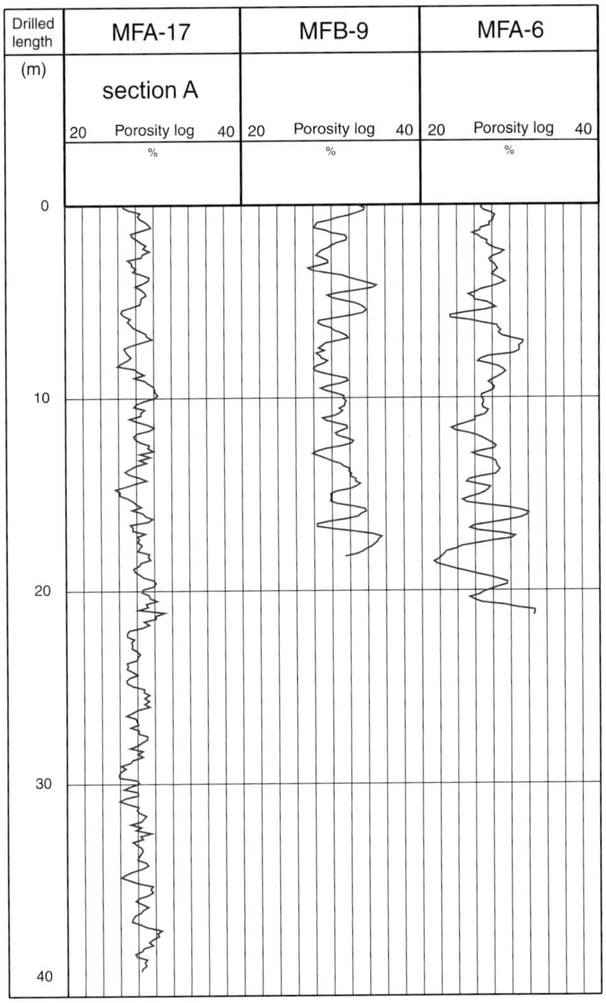


**Figure 7. Comparison of the porosity logs from the low-variability section A in the MFA-17 well and selected parts from the same stratigraphic position in two other wells located 1100 m (3608 ft) away.**

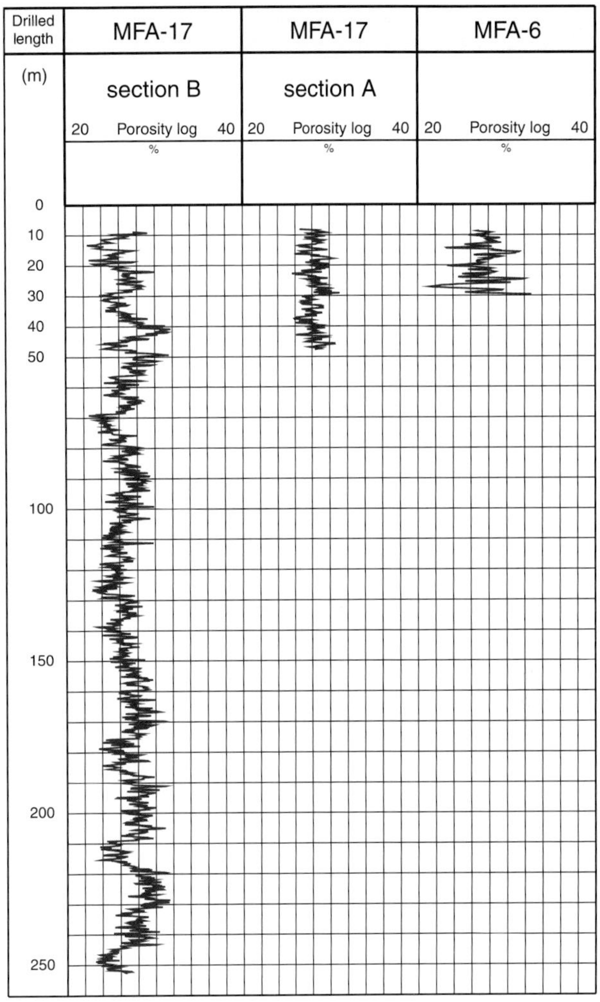


**Figure 8. Comparison of the oblique section B in the MFA-17 well, the low-variability section A, and one of the vertical wells.**

The semivariogram for the oblique section B in the horizontal well exhibits a nested structure reflecting the 0.8-m (2.6-ft) range, as well as a longer 11-m (36-ft) correlation range that appears before the semivariogram reaches the sill value. This range is considered the zonal anisotropy equivalent to the 0.8-m (2.6-ft) range in the vertical wells, considering that the section is drilled at an angle of approximately 4° to the layering. The difference in sill values is caused by the fact that the horizontal well does not exhibit the same amount of contrasts as the two vertical wells. The absence of very long-range correlation and the stable sill value in the oblique well is illustrated by the semivariogram in Figure 11.

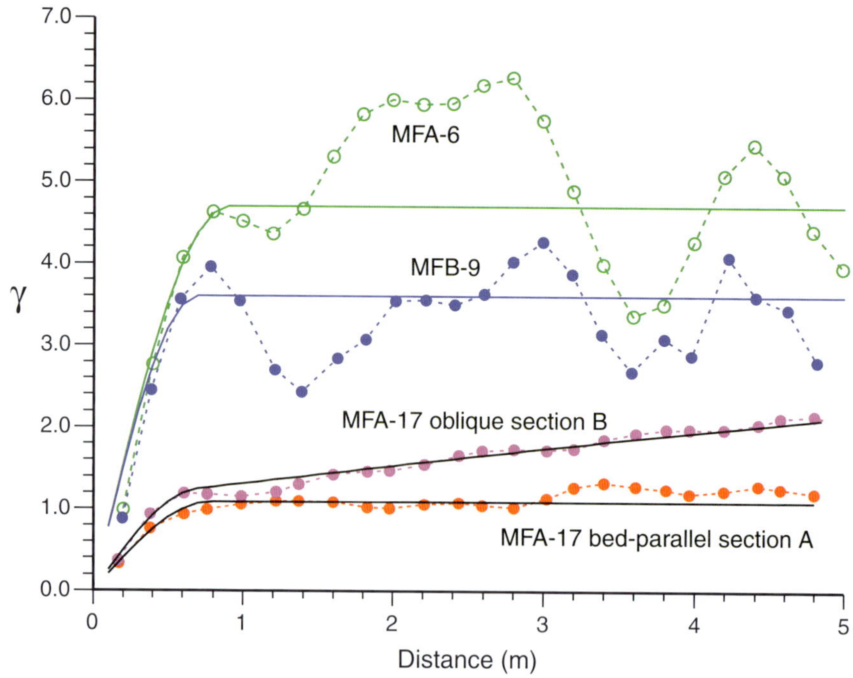


Figure 9. Experimental semivariograms for log porosity showing the short-range structures for the two sections A and B in the horizontal MFA-17 well compared to semivariograms from two selected vertical wells. The MFA-17 A section is modeled with a spherical structure with range 0.8 m (2.6 ft) and sill 1.08, whereas the B section is modeled with two spherical structures having ranges 0.8 and 11 m (2.6 and 36 ft), respectively, and sills of 1.08 and 2.73, respectively. MFB-9 is modeled with a spherical structure having range 0.8 m (2.6 ft) and sill 3.6, whereas MFA-6 is modeled with a spherical structure having range 0.8 m (2.6 ft) and sill 4.7.

When looking for long-range horizontal correlations, the illustration in Figure 12 shows that it is difficult to recognize repeated porosity patterns on the two sides of the assumed bed-parallel section A. Ideally, there should be some amount of symmetry around the section A interval if there were no lateral porosity changes in the layers. Part of the difference is ascribed to the general depth-dependent decrease in porosity level for the chalk, but the contrasts between the two adjoining sections seem so dramatic that other factors must influence the porosity distribution. Whether faulting has dislocated the layers or limited correlation exists over longer distances is not clear.

## VARIABILITY IN POROSITY AND PERMEABILITY

The porosity and permeability values from outcrop have been compared to core measurements taken from wells in the uppermost Maastrichtian chalk

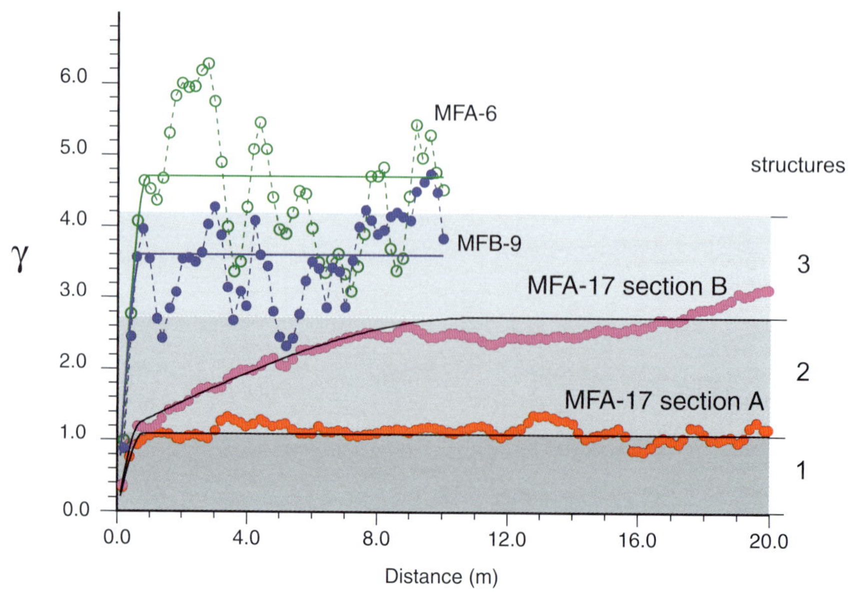


Figure 10. Experimental semivariograms showing the longer range structures. The two sections, A and B, in the MFA-17 well are compared to semivariograms from two selected vertical wells. The interpreted structures are (1) short range; (2) medium-range, showing an apparent geometric anisotropy in the MFA-17 data, which is caused by drilling at a small angle to the layering; and (3) zonal anisotropy, which is caused by a difference in sill value caused by the limited sequence penetrated by the horizontal well compared to the vertical wells. The same semivariogram models as for Figure 9 are used.

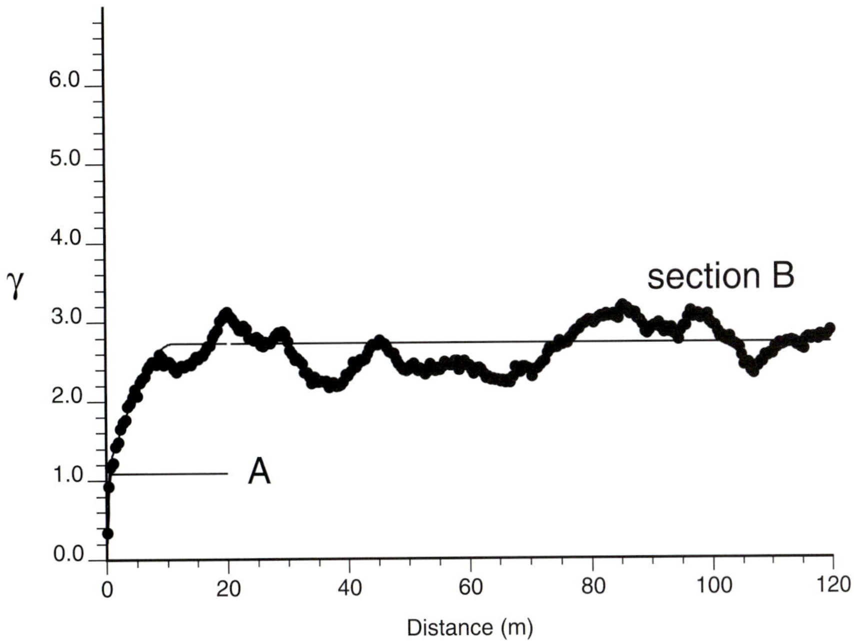


**Figure 11.** Experimental semivariogram showing the long-range behavior for section B in the MFA-17 well and the stable sill value for that long segment. The sill for section A is shown for comparison. The B section is modeled with two spherical structures having ranges 0.8 and 11 m (2.6 and 36 ft), respectively, and sills of 1.08 and 2.73, respectively.

reservoir zone in the Dan and Valhall fields in the North Sea (Figure 13). This comparison shows that the upper Maastrichtian samples from the chalk outcrop overlap the data from the highly porous upper Maastrichtian Tor-1 unit in the Valhall field. Together, these two data sets form a high-end endpoint for the porosity and permeability trend. The data from the Dan field reflect lower porosity and permeability values, the trend of which extends from the very high-porosity Valhall and outcrop chalk. These contrasts are attributed to differences between the primary facies and to changes in burial diagenesis in combination with highly overpressured systems.

## CONCLUSION

Compared to reservoir chalks, outcrop chalks exhibit less variability and only very vague signs of cyclic development in their petrophysical properties. Important differences between outcrop and reservoir chalk exist in both depositional processes and patterns, as well as in the diagenetic history of the two types of chalk. Semivariogram analysis suggests the presence of small-scale heterogeneity on the order of 0.15–0.25 m (0.49–0.82 ft), which is supported by a regular horizontal fracture pattern in the outcrop. Information about variability

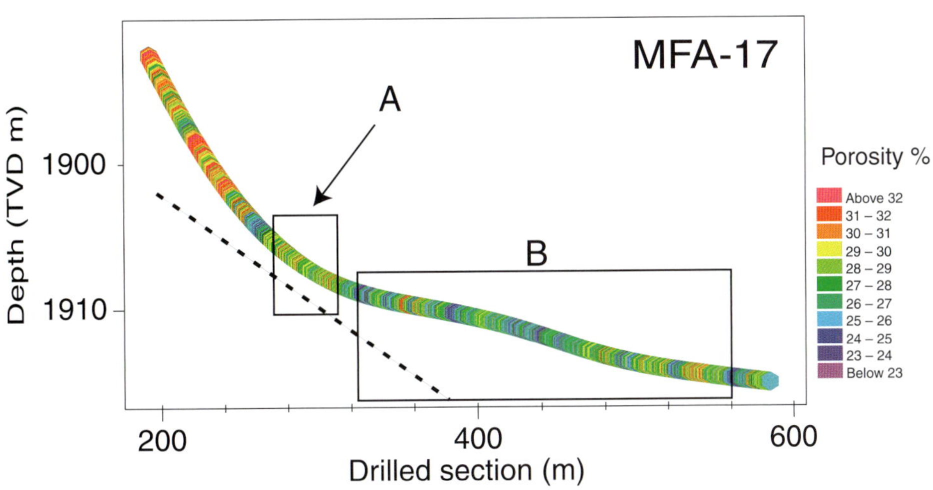


**Figure 12.** Profile for the horizontal well MFA-17 with posted porosity values as color coding along the well transect. The dashed line signifies the interpreted bedding orientation from the structural map that shows a dip of approximately 4°. The relatively stable interval A is outlined and matches the interpreted bedding dip. The section B with higher variability is indicated.

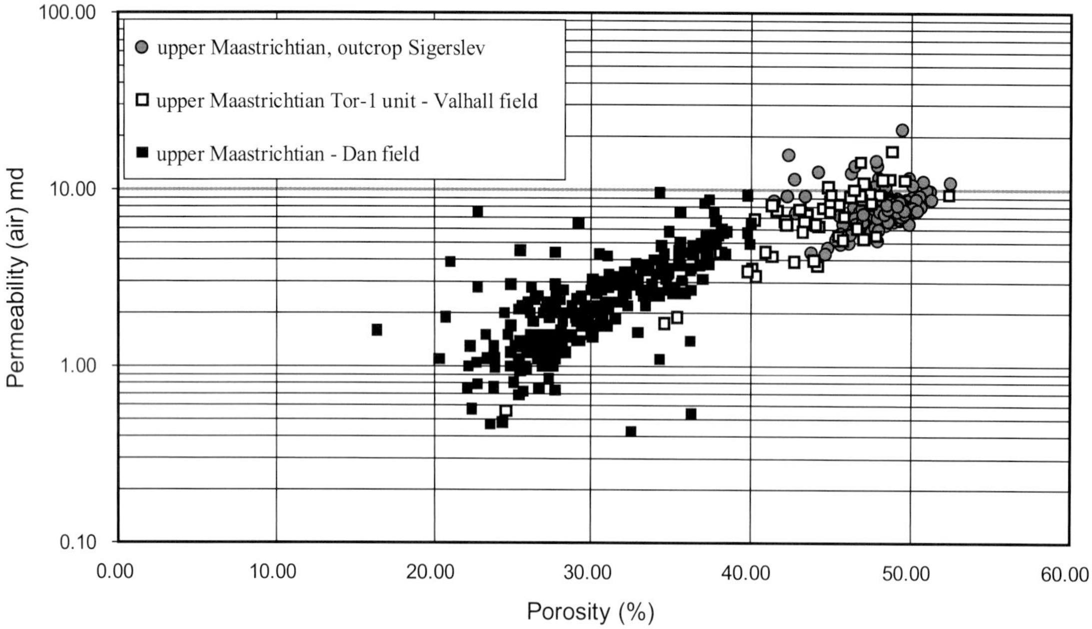


**Figure 13. Crossplot of core analysis porosity and permeability (logarithmic scale), comparing the outcrop and North Sea reservoir data. The crossplot of data from the Valhall field overlaps the corresponding crossplot of the outcrop data.**

derived from wells drilled in the reservoir chalks, particularly horizontal wells, is essential to a complete understanding of chalk variability.

## REFERENCES CITED

Almeida, A. S., and P. Frykman, 1994, Geostatistical modeling of chalk reservoir properties in the Dan field, Danish North Sea, *in* J. M. Yarus and R. L. Chambers, eds., Stochastic modeling and geostatistics: Principles, methods, and case studies: AAPG Computer Applications in Geology 3, p. 273–286.

Bellay, G., H. Al-Waheed, and T. Audah, 1996, Cyclic borehole effects in deviated wells: Presented at the 7th Abu Dhabi International Petroleum Exhibition and Conference, Abu Dhabi, SPE Paper 36266, 10 p.

Deutsch, C. V., and A. G. Journel, 1998, GSLIB—geostatistical software library and user's guide: New York City, Oxford University Press, 369 p.

Frykman, P., 2001, Spatial variability in petrophysical properties in upper Maastrichtian chalk outcrops at Stevns Klint, Denmark: Marine and Petroleum Geology, v. 18, p. 1041–1062.

Frykman, P., and C. V. Deutsch, 2002, Practical application of the geostatistical scaling laws for data integration: Petrophysics, v. 43, p. 153–171.

Gringarten, E., and C. V. Deutsch, 2001, Variogram interpretation and modeling: Mathematical Geology, v. 33, p. 507–534.

Håkansson, E., and F. Surlyk, 1997, D-Denmark. Geology of Denmark, *in* E. M. F. R. W. Moores, ed., Encyclopedia of European and Asian regional geology: New York, Chapman & Hall, p. 183–191.

Isaaks, E. H., and R. M. Srivastava, 1989, An introduction to applied geostatistics: New York, Oxford University Press, 561 p.

Kupfersberger, H., and C. V. Deutsch, 1999, Methodology for integrating analog geologic data in 3-D variogram modeling: AAPG Bulletin, v. 83, p. 1262–1278.

Middleton, G. V., 1973, Johannes Walther's law of the correlation of facies: Geological Society of America Bulletin, v. 84, p. 979–988.

Scholle, P. A., T. Albrechtsen, and H. Tirsgaard, 1998, Formation and diagenesis of bedding cycles in uppermost Cretaceous chalks of the Dan field, Danish North Sea: Sedimentology, v. 45, p. 223–243.

Stage, M., 1999, Signal analysis of cyclicity in Maastrichtian pelagic chalks from the Danish North Sea: Earth and Planetary Science Letters, v. 173, p. 75–90.

Surlyk, F., and E. Håkansson, 1999, Maastrichtian and Danian strata in the southeastern part of the Danish basin: 19th Regional European Meeting of Sedimentology, Copenhagen, Excursion A3: Copenhagen, Geological Institute, p. 29–58.

Vejbæk, O. V., 2002, Reservoir characterization of the Roar gas field, Danish North Sea: Petroleum Geoscience, v. 8, p. 71–87.

Vejbæk, O. V., and L. Kristensen, 2000, Downflank hydrocarbon potential identified using seismic inversion and geostatistics: Upper Maastrichtian reservoir unit, Dan field, Danish Central Graben: Petroleum Geoscience, v. 6, p. 1–13.

Beucher-Darricau, H., B. Doligez, and J. M. Yarus, 2006, Modeling complex reservoirs with multiple conditional techniques: A practical approach to reservoir characterization, *in* T. C. Coburn, J. M. Yarus, and R. L. Chambers, eds., Stochastic modeling and geostatistics: Principles, methods, and case studies, volume II: AAPG Computer Applications in Geology 5, p. 289–299.

## *Chapter 22*

# Modeling Complex Reservoirs with Multiple Conditional Techniques: A Practical Approach to Reservoir Characterization

**H. Beucher-Darricau**
*Centre de Géostatistique, Ecole des Mines de Paris*
*Fontainbleau, France*

**B. Doligez**
*Institut Français du Pétrole*
*Rueil-Malmaison, France*

**J. M. Yarus**
*Quantitative Geosciences*
*Houston, Texas, U.S.A.*

## ABSTRACT

Modeling petroleum reservoirs and the intricacies of their inherent morphological and structural characteristics requires sophisticated tools. These tools may consist of deterministic or stochastic methods that allow the interwell space to be filled with representative reservoir properties. Integrating a variety of modeling tools may be necessary for characterizing complex reservoirs. In such cases, the relationships among the lithofacies, the shapes and sizes of the sedimentary bodies, and the depositional mode of the litostratigraphic units must be defined.

It is also important to incorporate other types of information. Including conceptual geological information or three-dimensional seismic data enhances the model particularly when borehole and core data are limited. This chapter demonstrates the use of object-based, pixel-based, and hybrid stochastic-simulation techniques using a case study involving data from the Ravenscar Group located along the east coast of Yorkshire in the United Kingdom.

## INTRODUCTION

There are many ways to constrain geological images for the development of a static reservoir model. The problem is to choose the best one for a given geological setting. Different depositional and diagenetic environments require the use of different approaches. For instance, in a fluviodeltaic system in the continental domain, the lithofacies are organized as three-dimensional (3-D) objects with specific


DOI:10.1306/1063823CA53242

shapes like channels or crevasse splays. In the littoral domain, the lithofacies vary progressively from continental to marine deposits. It is obvious that different modeling methods must be used to characterize these different environments and to satisfy specific properties such as connectivity (Allard, 1993).

The geostatistical approaches commonly used in reservoir modeling are of two types: pixel based and object based (Galli and Beucher, 1997). The pixel-based approach simulates the lithofacies point by point on the basis of spatial correlation. One example is the truncated Gaussian method, where the parameters estimated from the well data are the lithofacies proportions, which can be either laterally homogeneous (horizontal stationarity) or can vary in 3-D (nonstationarity), and the semivariograms of the lithofacies indicators. The object-based approach, however, simulates the lithofacies as 3-D objects. One example is the Boolean model in which the objects are defined by their geometry and their spatial distribution.

In addition, the two approaches can be combined in various ways: heterogeneities can be simulated using the pixel approach inside one object, or the objects can be simulated within a pixel background.

This chapter illustrates the use of these approaches to simulate a complex reservoir. A case study is presented that involves well and core data from an area along the east coast of Yorkshire in the United Kingdom.

## BRIEF PRESENTATION OF THE METHODS

Among the pixel-based approaches, the truncated Gaussian method is a very flexible one (Matheron et al., 1987). It has the ability to produce more and more complex images of the reservoir when the number of input parameters increases. Beucher et al. (1993) describe its application in the case of a single vertical lithofacies sequence or when the vertical lithofacies sequences vary throughout the field (represented by a 3-D proportion matrix).

Le Loc'h et al. (1994) and Le Loc'h and Galli (1997) suggest enhancements that can be realized when truncating more than one underlying Gaussian random variable to obtain different relationships among lithofacies. Moulière et al. (1997) and Doligez et al. (1999) show how to add constraints, such as seismic-derived information or qualitative hypotheses, during the computation process.

Among the object-based approaches, the Boolean technique is the most popular one. Lantuéjoul (1997a, b, 1998) has developed a conditional version, and the general method has also been extended to nonstationary cases.

### Truncated Gaussian Method

Using the truncated Gaussian method, the lithofacies simulations are produced in two steps as shown in Figure 1. First, a stationary Gaussian random function is simulated using a semivariogram model. Second, this function is truncated using thresholds: the areas where the values of the Gaussian variables are smaller than the threshold belonging to the gray lithofacies.

Figure 2 illustrates the link between thresholds and lithofacies proportions. In practice, the lithofacies proportions are estimated from well data and then projected over the whole domain (Figure 3). Such estimates can be constrained by additional information derived from geological or seismic interpretations. The additional information can be either qualitative, when it outlines areas with homogeneous lithofacies proportions, or quantitative, when seismic attributes are transformed into lithofacies proportions (Moulière et al., 1997; Beucher et al., 1999; Doligez et al., 1999).

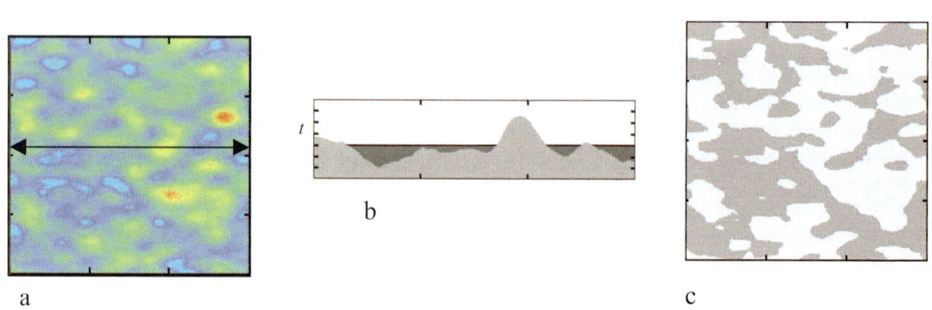

**Figure 1.** Truncated Gaussian method: (a) simulation of a Gaussian random function; (b) constant threshold (*t*) in a cross section; (c) lithofacies simulation after applying threshold *t*.

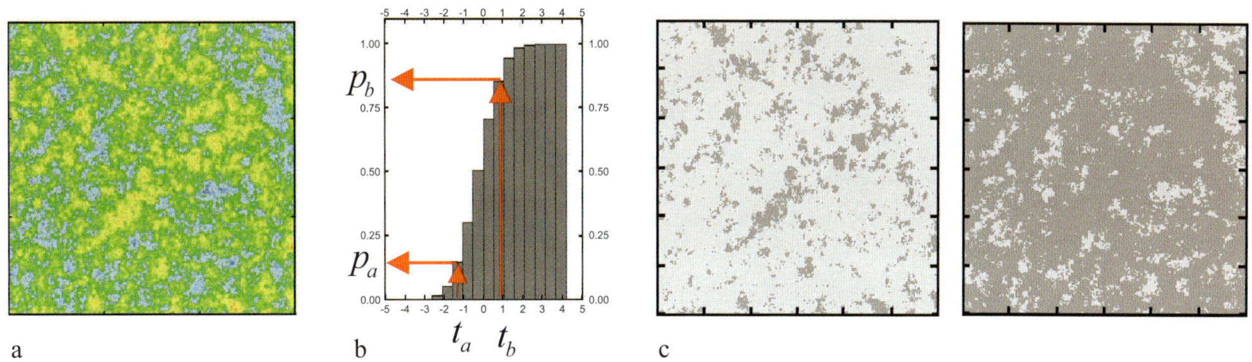


**Figure 2.** **Thresholds and lithofacies proportions: (a) simulated Gaussian random function; (b) cumulative density function; (c) lithofacies simulations using thresholds $t_a$ (left) and $t_b$ (right).**

The estimated proportions represent the mean behavior of the lithofacies in the reservoir. They are commonly very close to the geological interpretation and, thus, can be easily checked by the geologist.

Stochastic simulation of the reservoir architecture proceeds in the stationary framework (when the proportions are constant) or nonstationary framework (when the proportions vary). Depending on the 3-D variations of these proportions, the characteristics of the semivariogram model of the underlying Gaussian random function have an impact on the resulting lithofacies image: the type of the semivariogram model is related to the shape of the lithofacies bodies, and the range of the model is linked to the degree of heterogeneity (Figure 4).

Experimentally, the semivariograms (simple and cross) are calculated from the lithofacies indicators. They must be fitted by the single model of the underlying Gaussian random function through the lithofacies proportions estimated beforehand (Beucher et al., 1993).

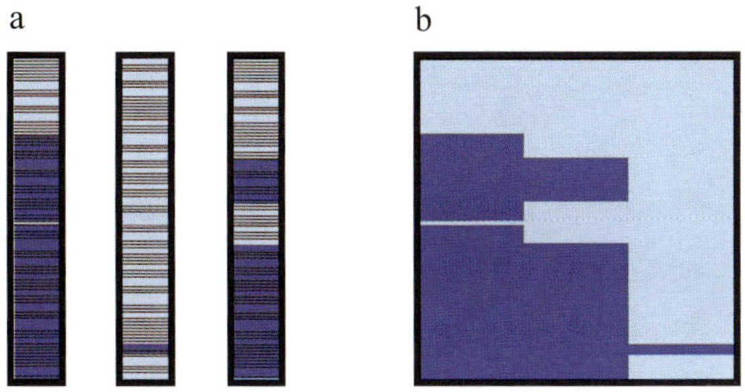


**Figure 3.** **Estimation of a vertical proportion curve from wells: (a) discretized wells in the working system; (b) proportions computed for each horizontal level.**

## Boolean Method

The Boolean method is commonly chosen when the lithofacies are well organized, such as channels in a deltaic plain or shale lenses in a sandy matrix. In this case, the geological bodies (channel or lenses) are considered as objects, and lithofacies simulations are produced by drawing these objects randomly

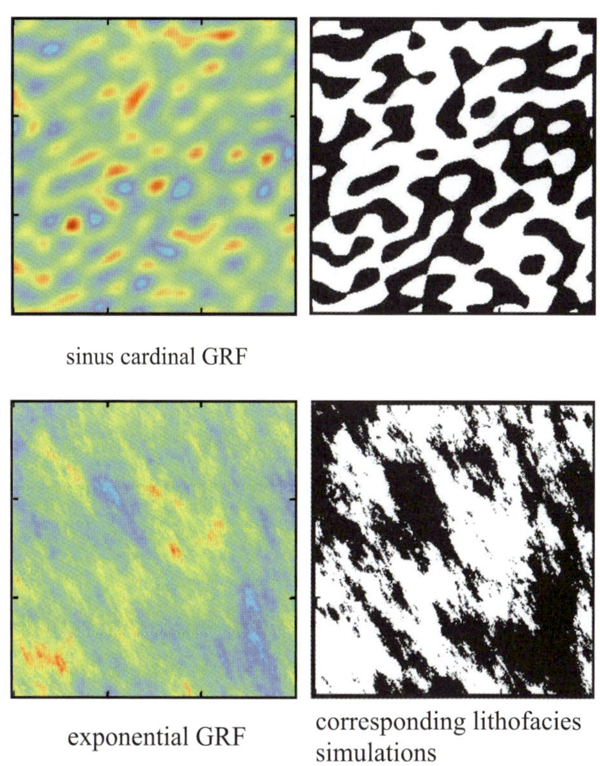


**Figure 4.** **Semivariogram models of the underlying Gaussian random function (GRF) and lithofacies shapes.**

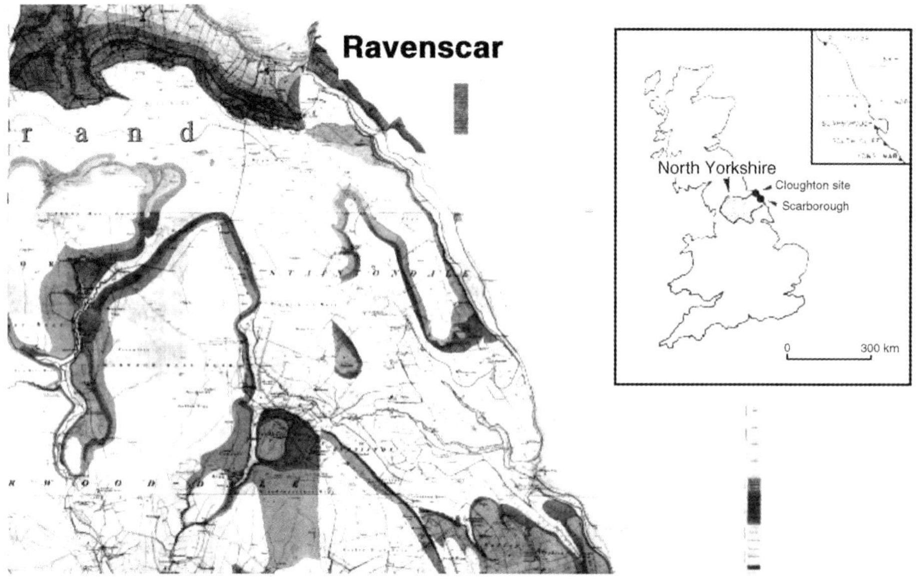


Figure 5. Ravenscar location map.

(according to a 3-D Poisson point process) in the domain. The main parameters associated with the Boolean method are the geometry of the objects (shape, size, and orientation) and the spatial distribution of the object locations (intensity of the Poisson point process). These parameters are fitted for each geological body that will be simulated by the technique.

In the following example, the different families of objects corresponding to different lithofacies are simulated independently, with an order chosen for reproducing a part of their relationships. The shapes are generally chosen to reproduce the main characteristics of the geological bodies (for instance, sinusoidal shapes for channels). All the shapes can be imaged; however, computation time increases rapidly with the complexity of the objects and the number of constraining points.

The dimensions of the objects are estimated from the information available on the corresponding lithofacies or chosen from analogs. Their spatial distributions are estimated from the corresponding lithofacies proportions (Schmitt, 1997; Schmitt and Beucher, 1997).

## CASE STUDY

The Ravenscar Group, located along the east coast of Yorkshire in the United Kingdom (Figure 5), is composed of a 200-m (660-ft)-thick siliciclastic sediment section deposited during the Middle Jurassic in the Cleveland basin. This reservoir unit results from the progradation of a deltaic system southward into a marine system (Figure 6). Seven lithostratigraphic

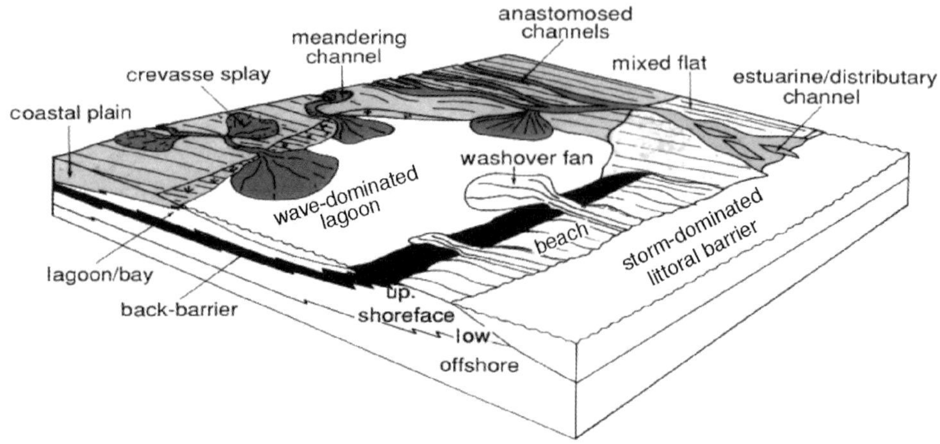


Figure 6. Conceptual geological model.

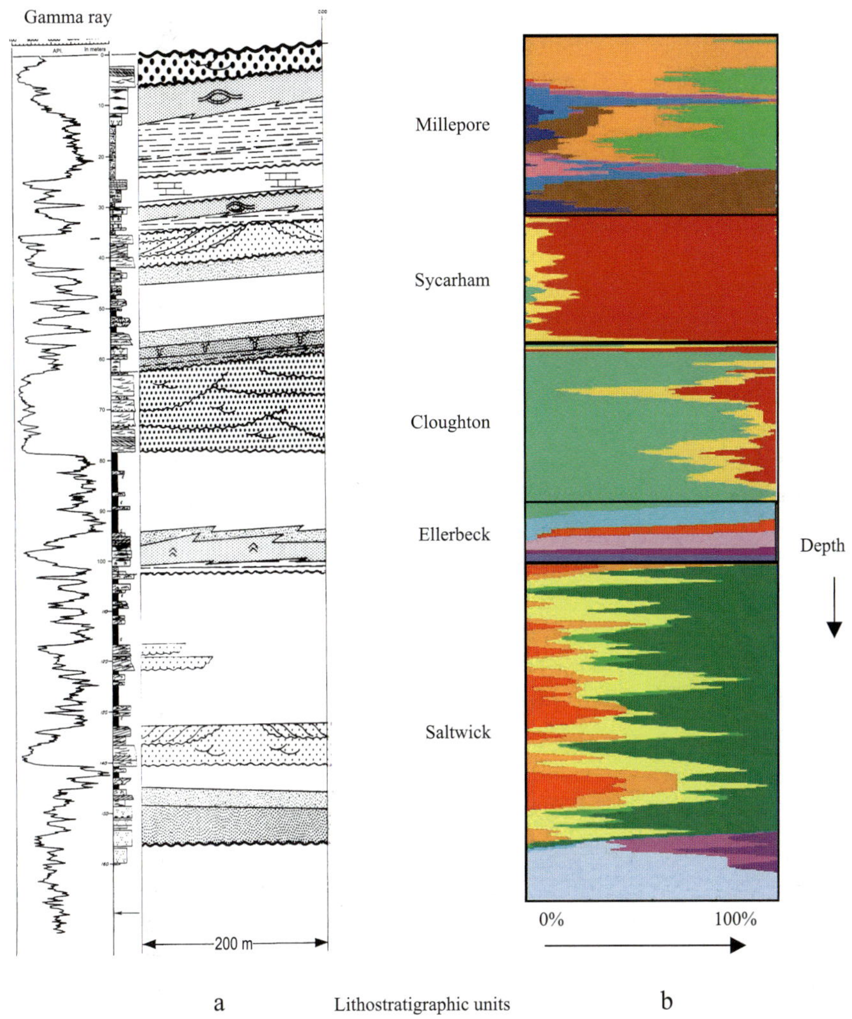


**Figure 7. Stratigraphic column in the Ravenscar Group, with associated vertical proportions curve: (a) sequence stratigraphy; (b) lithofacies proportions. The color scales are presented in the following figures.**

units have been defined after a detailed geological interpretation of the available data (Figure 7).

From bottom to top, the depositional environment varies from marine to continental. The vertical distribution of the lithofacies is determined for all the lithostratigraphic units according to their chronostratigraphic reference (Volpi et al., 1997). Vertical proportion curves summarize the sequences of the different lithofacies. They are different for various depositional environments, which is apparent when comparing the two basal Saltwick and Ellerbeck units. In the Ellerbeck unit, the vertical lithofacies changes are regular from the base with transgressive sand lithofacies (violet) to the top with continental sediments (green). This unit corresponds to a marine environment with regular and continuous changes in sea level. In the Saltwick unit, the vertical changes, with marine lithofacies (mostly blue) on the bottom to continental lithofacies (red to green) at the top, are also visible. Moreover, three main episodes of channel deposits (in red) are well individualized. The different lithofacies must be simulated using the different methods described below.

## Ellerbeck Unit: Truncated Gaussian Method, Stationary Approach

For all the methods, the first problem is the choice of a working system that is similar to the one at the time of the deposition (the main correlation is horizontal), where the different parameters are estimated and the simulations are performed. Two main systems are possible: either parallel to a reference surface (a maximum flooding surface in general) or proportional between the top and bottom of the unit (for differential compaction). When the available data are sufficient, there is no problem in choosing the working system and the reference surface. However, in the appraisal case, these choices may be problematic. Note that this step is very important because different reference surfaces lead to different simulation characteristics.

In Figure 8, the vertical proportion curves are calculated for three possible reference surfaces (top Scarborough, bottom reservoir, and maximum flooding surface), which correspond to different deposit hypotheses. The use of these different vertical proportions leads to different lithofacies relationships in the simulations as illustrated through vertical cross sections. The first choice produces an unrealistic vertical sequence. The second one shows several oscillations that are not compatible with the geological model. The correct choice is the last one for which the lithofacies proportion curve is very regular. Moreover, this vertical curve remains more or less the same throughout the field. This is why the stationary approach has been chosen for the simulation.

In sedimentary deposits, the lithofacies commonly have a strong anisotropy between the horizontal plane and the vertical axis (in the working system). Therefore, the model for the underlying Gaussian random function is commonly factorized at least along the vertical axis. For practical reasons, only the vertical model can accurately be fitted; the horizontal model must be specified in advance. In the studied unit, the indicator semivariograms have been fitted with a factorized exponential model having a vertical range of 5 m (16 ft) (Figure 9) and 400 m (1300 ft) horizontally. The simulated cross section (Figure 10) shows good continuity and is in agreement with the foregoing observations.

## Millepore Gristhorpe Unit: Truncated Gaussian Method, Nonstationary Approach

The depositional environment of the Millepore Gristhorpe unit is more complex than that of the Ellerbeck unit because it contains marine, littoral, lagoon, and continental sediments. In particular, the vertical lithofacies sequence computed with all the wells (Figure 3) presents three main levels containing marine deposits (in dark blue).

Looking at the lithofacies proportions accumulated vertically for each well, a constant distribution of the lithofacies cannot be assumed. These apparent variations can be interpolated using two-dimensional

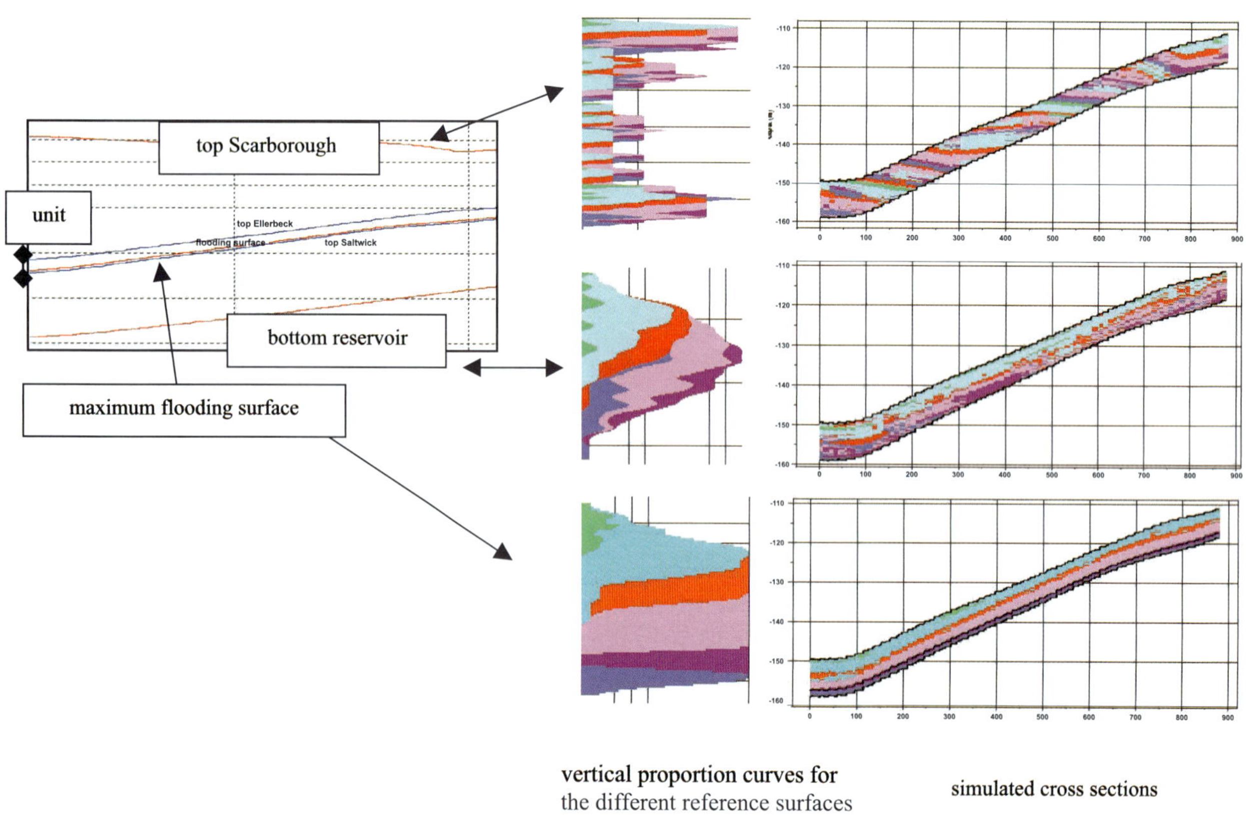


**Figure 8. Choice of the reference surface: Ellerbeck example.**

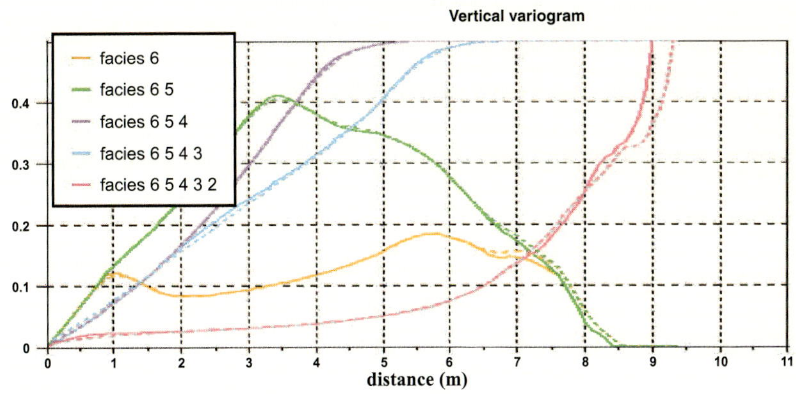


Figure 9. Vertical semivariogram fitting: Ellerbeck unit. Experimental semivariograms are in continuous lines, and fitted semivariograms are in dotted lines.

maps that present the proportion of a given lithofacies accumulated over the total thickness of the unit. Such maps are very useful because they describe the global distribution of the lithofacies and are directly linked to the geological interpretation.

In the Millepore Gristhorpe unit, the lateral lithofacies distribution varies from north to south at the observation scale: the argillaceous sandstone lithofacies is more abundant in the south, the flood plain is more abundant in the central eastern part of the field, and the marine lithofacies is more abundant in the north (Figure 11). In this case, the nonstationary approach must be used.

First, the vertical proportion curves are computed for various groups of wells considered as representative of areas (Figure 12a). Each vertical proportion curve is located at the center of gravity of the corresponding wells. Then, the lithofacies proportions are estimated on the whole 3-D domain. During this estimation phase, it is easy to add an external constraint such as the cumulative proportion of a given lithofacies deduced from seismic attributes (Johann et al., 1996; Beucher et al., 1999) or other quantified geological knowledge.

The result can be displayed as a regular grid on which each cell corresponds to a vertical proportion curve (Figure 12b). Some lithofacies are location dependent. In particular, the channel sandstone (red) and the marine mudstone (dark blue) show three peaks in the north and just one in the middle.

The 3-D simulation results obtained using the previous proportions satisfy all the input constraints. The lateral variation from continental to marine lithofacies is clearly visible on a north–south vertical cross section (Figure 13a). Moreover, the horizontal distribution (Figure 13b) is also in agreement with the conceptual knowledge.

## Cloughton Unit: Boolean Approach

In contrast to the two previous units, the environment of the Cloughton unit is more continental, being composed of a deltaic plain with meandered channels and crevasses. Because the meandered

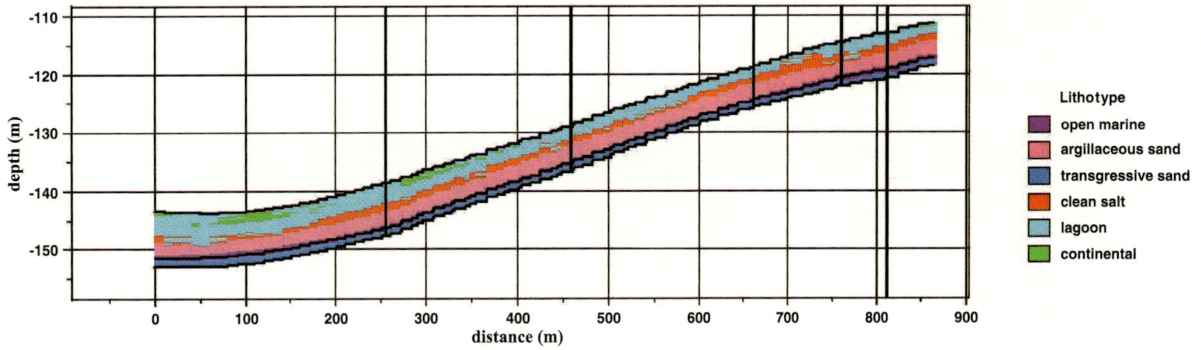


Figure 10. Simulated cross section: Ellerbeck unit.

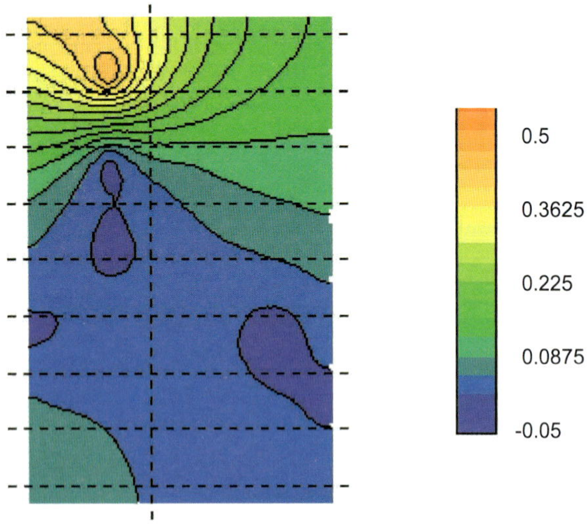


**Figure 11. Two-dimensional map of the marine lithofacies proportions accumulated vertically.**

channel deposits are more individualized here than in the other units, it is preferable to use an object-based approach to reproduce the geology.

For this approach, the parameters of the objects (shapes, size, and orientation) must be specified in advance. In the Cloughton unit, two families of objects have been defined. The first family corresponds to the meandering channels, which can be described as sinusoidal objects with half-ellipsoidal vertical sections (characterized by their width, length, orientation, sinuosity, and thickness). The second family corresponds to crevasse splays represented by elliptical objects (characterized by their length, width, orientation, and thickness). Only the mean thickness can be indirectly deduced from the well data. The mean values of the other parameters are chosen from analogs in conjunction with the conceptual geological model.

The distributions of the objects are given by the Poisson intensity, which is estimated from well data. The peaks of the curve correspond to levels with a large number of objects. For example, the channel lithofacies have two main levels of high intensity (Figure 14b).

The two families of objects are simulated independently and sequentially. The crevasse splays are simulated first, and the channels are simulated second (possibly eroding the objects of the first family). A vertical cross section from a conditional simulation illustrates the distribution of the different objects (Figure 14a).

## Saltwick Unit: Nested Simulations

The Saltwick unit is even more complex because it is composed of deltaic plain sediments with channels and crevasses splay deposits, as well as transgressive sands and lagoon. The complexity of the unit is reflected in the fact that some of the lithofacies represent individualized bodies having specific shapes,

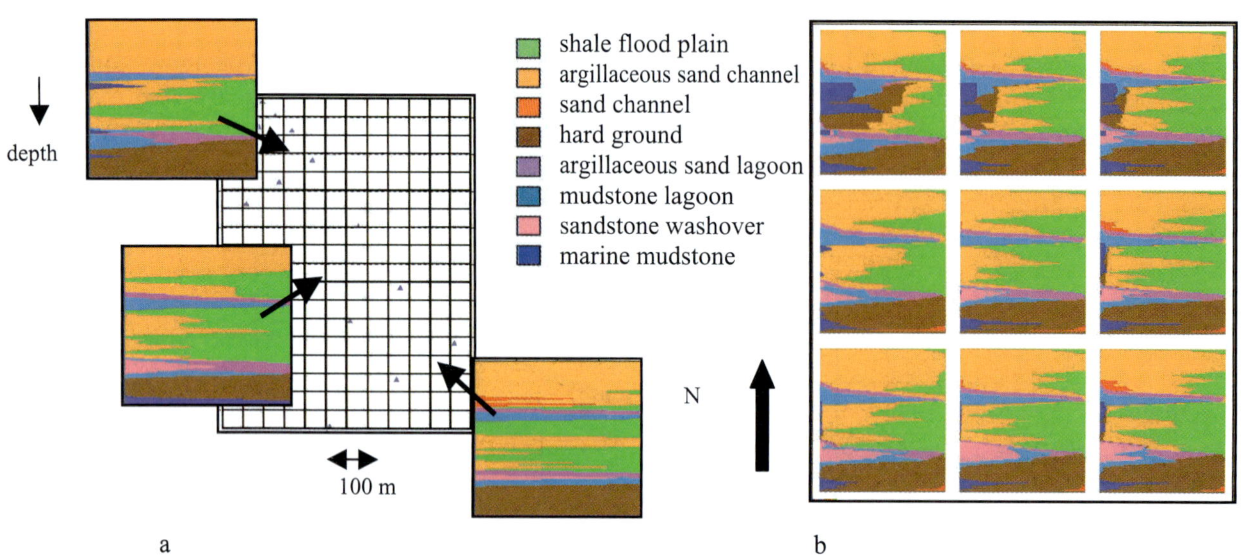


**Figure 12. Variations of lithofacies proportions in the domain of investigation (Millepore unit): (a) local vertical proportions of lithofacies from groups of wells; (b) sampling from the estimated 3-D grid of local vertical proportions of lithofacies (display of one cell every four cells).**

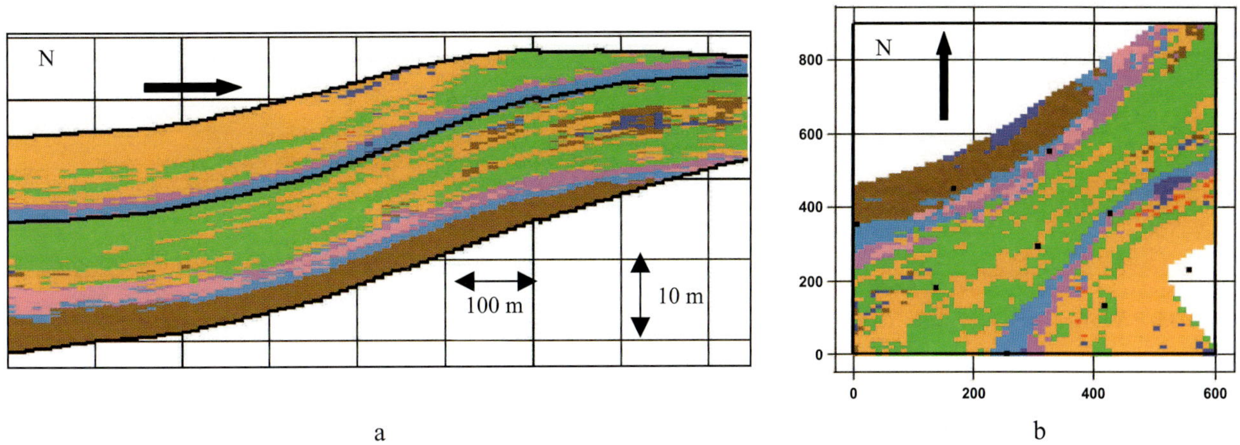


**Figure 13.** Vertical cross section and horizontal plane from a nonstationary simulation (Millepore unit): (a) north–south vertical cross section; (b) horizontal section.

whereas others do not have distinct patterns. Hence, the simulation must be performed in three steps.

First, a pixel-based approach is used to simulate the background represented by a transition from marine to continental. Second, the channels and crevasse splays are simulated inside the continental lithofacies using an object-based approach. Finally, the two different sandstones are simulated within the channel objects using a pixel-based approach. For each realization, the lithofacies are regrouped into two families: the background and the fluvial ones. A first simulation process consists of simulating the fluvial family with the Boolean method, considering the two fluvial lithofacies as one object and the other lithofacies as the background. Within the objects previously obtained, a truncated Gaussian simulation separates the two fluvial lithofacies. Beyond the objects truncated, Gaussian simulation is used to generate the remaining lithofacies. One of the realizations of this simulation process is shown in Figure 15.

## Combined Ravenscar Group

When all the lithostratigraphic units have been simulated, they are stacked together to construct a full image of the reservoir (Figure 16a). The relationships between the lithofacies are clearly visible in two horizontal sections at different depths (Figure 16b). In particular, Figure 16b compares the continental domain, having contacts between sand (red), shaly sands (orange), and shale (green), to the marine domain that exhibits a more regular transition between lithofacies (blue, pink and purple).

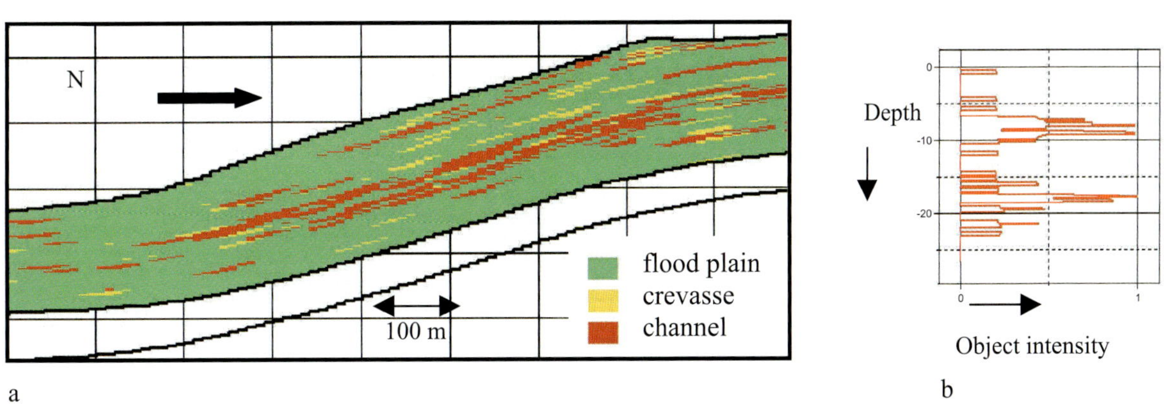


**Figure 14.** Boolean simulation of the Cloughton unit: (a) north–south vertical cross section; (b) channel intensity distribution (the peaks correspond to levels with high density of channels in the simulation).

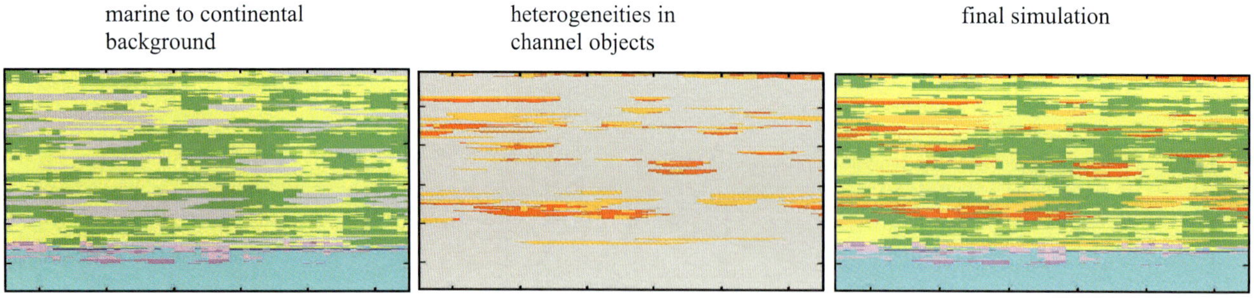


**Figure 15. Vertical cross section corresponding to the three steps of simulation.**

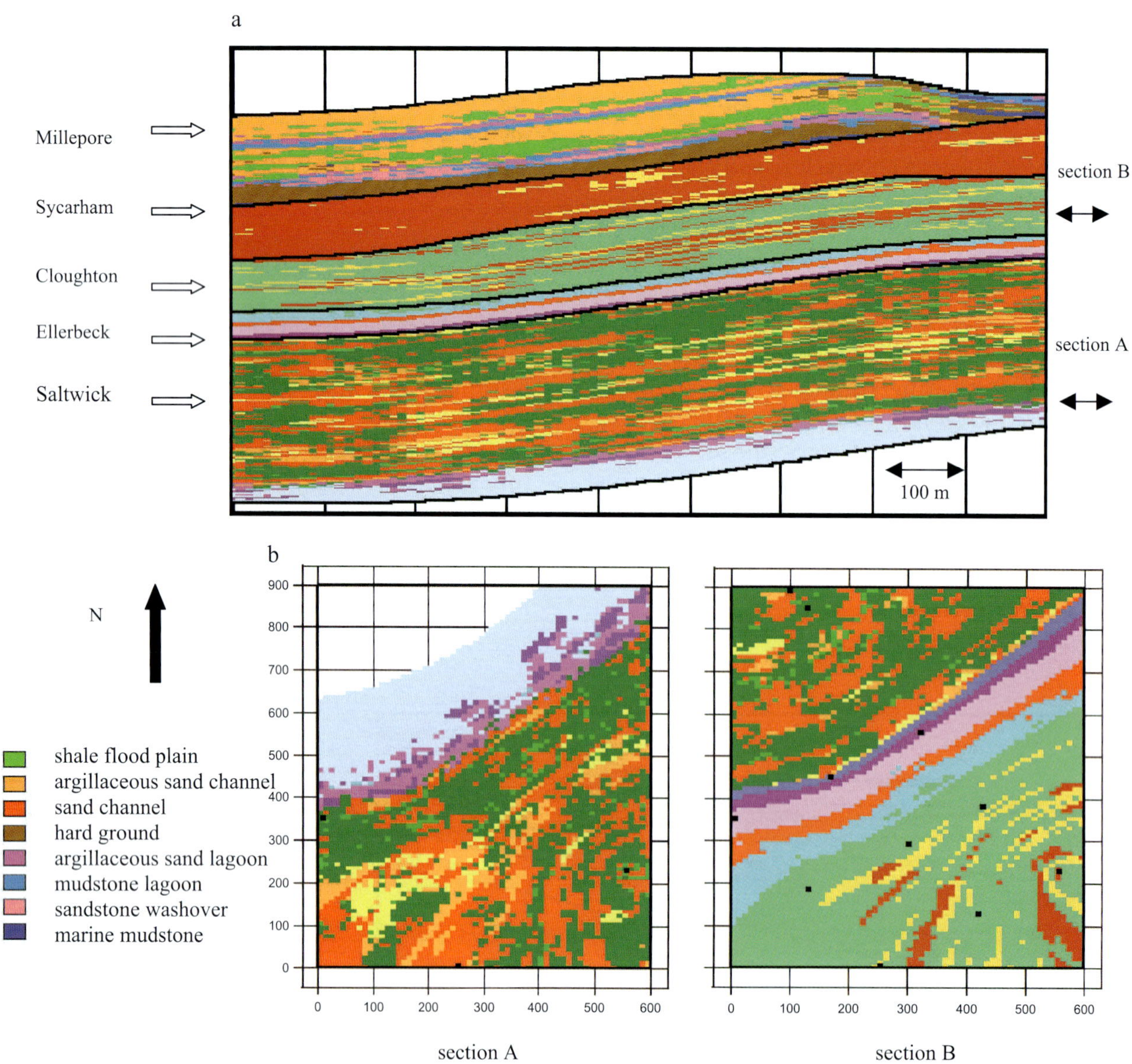


**Figure 16. (a) Vertical cross section from the Ravenscar simulation. (b) Two horizontal planes from the Ravenscar simulation (see depths of sections in Figure 12a).**

## DISCUSSION

This study demonstrates that different simulation methods may be necessary when simulating the depositional environment of a complex reservoir. The two simulation approaches described here, truncated Gaussian and Boolean, are capable of representing many different kinds of relationships among lithofacies. When the quality of the available information is sufficient, these methods can also be combined into more complex setups. Other enhancements can be obtained as well. In particular, two Gaussian random functions may be employed in the truncated method to represent more complex lithological arrangements, or a nonstationary 3-D distribution can be used with the Boolean method.

In all the cases, the main objective is to find estimates of the parameters that will give the best representation of the reservoir under investigation. Because the quantitative information is commonly limited, it is crucial to account for any qualitative information when it is available. Both simulation methods discussed here allow for such a possibility.

## REFERENCES CITED

Allard, D., 1993, On the connectivity of two random set models: The truncated Gaussian and the Boolean, *in* A. Soares, ed., Geostatistics tróia '92, v. 1: Dordrecht, The Netherlands, Kluwer Academic Publishers, p. 467–478.

Beucher, H., A. Galli, G. Le Loc'h, and C. Ravenne, 1993, Including a regional trend in reservoir modelling using the truncated Gaussian method, *in* A. Soares, ed., Geostatistics tróia '92, v. 1: Dordrecht, The Netherlands, Kluwer Academic Publishers, p. 555–566.

Beucher, H., F. Fournier, B. Doligez, and J. Rozanski, 1999, Using 3D seismic-derived information in lithofacies simulations: A case study: SPE Annual Technical Conference and Exhibition, Houston, SPE Paper 56736, p. 581–592.

Doligez, B., H. Beucher, F. Geffroy, and R. Eschard, 1999, Integrated reservoir characterization: Improvement in heterogeneous stochastic modeling by integration of additional external constraints, *in* R. Schatzinger and J. Jordan, eds., Reservoir characterization: Recent advances: AAPG Memoir 71, p. 333–342.

Galli, A., and H. Beucher, 1997, Stochastic models for reservoir characterization: A user-friendly review: Fifth Latin American and Caribbean Petroleum Engineering Conference and Exhibition, Rio de Janeiro, Brazil, August 30–September 3, SPE Paper 38999, 11 p.

Johann, P., F. Fournier, O. Souza, R. Eschard, and H. Beucher, 1996, 3-D stochastic reservoir modeling constrained by well and seismic data on a turbidite field: SPE Annual Technical Conference and Exhibition, Denver, SPE Paper 36501, p. 51–66.

Lantuéjoul, C., 1997a, Iterative algorithms for conditional simulations, *in* E. Y. Baafi and N. A. Schofield, eds., Geostatistics Wollongong '96, v. 1: Dordrecht, The Netherlands, Kluwer Academic Publishers, p. 27–40.

Lantuéjoul, C., 1997b, Conditional simulation of object-based models, *in* D. Jeulin, ed., Proceedings of the International Symposium on Advances in Theory and Applications of Random Sets: Singapore, World Scientific Publishing Company, p. 289–318.

Lantuéjoul, C., 1998, Conditional simulation of random sets, *in* Proceedings of the Section on Statistics and the Environment: Dallas, Texas, American Statistical Association, August 9–13, p. 37–45.

Le Loc'h, G., and A. Galli, 1997, Truncated plurigaussian method: Theoretical and practical points of view, *in* E. Y. Baafi and N. A. Schofield, eds., Geostatistics Wollongong '96, v. 1: Dordrecht, The Netherlands, Kluwer Academic Publishers, p. 27–40.

Le Loc'h, G., H. Beucher, A. Galli, and B. Doligez, 1994, Improvement in the truncated Gaussian method: Combining several Gaussian functions, *in* ECMOR IV, Proceedings of the 4th European Conference on the Mathematics of Oil Recovery, Røros, Norway, 13 p.

Matheron, G., H. Beucher, C. De Fouquet, A. Galli, D. Guérillot, and C. Ravenne, 1987, Conditional simulation of the geometry of fluvio-deltaic reservoirs: 62nd Annual Technical Conference and Exhibition of the Society of Petroleum Engineers, Dallas, SPE Paper 16753, p. 591–599.

Moulière, D., H. Beucher, L. Y. Hu, F. Fournier, P. Terkish, F. Mechiori, and F. Griffi, 1997, Integration of seismic derived information in reservoir stochastic modeling using truncated Gaussian approach, *in* E. Y. Baafi and N. A. Schofield, eds., Geostatistics Wollongong '96, v. 1: Dordrecht, The Netherlands, Kluwer Academic Publishers, p. 374–385.

Schmitt, M., 1997, Estimation of intensity and shape in a non-stationary Boolean model, *in* D. Jeulin ed., Proceedings of the International Symposium on Advances in Theory and Applications of Random Sets: Singapore, World Scientific Publishing Company, p. 251–267.

Schmitt, M., and H. Beucher, 1997, On the inference of the Boolean model, *in* E. Y. Baafi and N. A. Schofield, eds., Geostatistics Wollongong '96, v. 1: Dordrecht, The Netherlands, Kluwer Academic Publishers, p. 200–210.

Volpi, B., A. Galli, and C. Ravenne, 1997, Vertical proportion curves: A qualitative and quantitative tool for reservoir characterization: Memorias del I Congreso Latinamericano de Sedimentologia, p. 351–358.

## *Section V*

# Systems and Resources

It is all well and good to talk, read, and study about geostatistics and stochastic modeling; but when it comes time to actually kick off a project and then bring it to fruition, a lot more is involved than what initially meets the eye. As many of the authors of chapters in this book realize, effective geostatistical and stochastic modeling requires a fairly extensive system of supporting technology and human resources, a strong management commitment, and an internal operating philosophy that promotes success of the project.

The last section in the book contains a collage of three chapters that fall in the general realm of systems, supporting concepts, and resources. The ideas and information presented here can greatly facilitate an understanding of the processes involved in a modeling project and help clarify the business implications of such endeavors within the corporate environment.

In the first chapter, Luca Cosentino and Amaud Pelgrain de Lestang of Beicip-Franlab, along with Jose Edmundo Gonzalez at PDVSA, describe a novel use of geostatistical modeling as a fast-response tool for reservoir modeling. The authors showcase a successful application to the Bachaquero fields on the east coast of Lake Maracaibo in which geostatistical modeling has been fully integrated into the exploitation activities that affect corporate decision-making about field development. In the second chapter, Joost Herweijer, Jeffrey Yarus, and M. J. V. Suana present an overview of current knowledge management practices in upstream oil and gas exploration and production that is specifically aimed at multidisciplinary project teams working with spatial modeling techniques. The chapter describes the development and application of a workflow-based knowledge management system for use in three-dimensional reservoir modeling that can enhance the multidisciplinary collaboration that is so important to the success of modeling projects. In the third chapter Tim Coburn describes an extensive bibliography of references for geostatistics, spatial analysis, and stochastic modeling in petroleum exploration and development that has been compiled for the readers and is included as the last section of this book. This is arguably the most extensive bibliography of its kind, bringing together information from numerous sources that should prove extremely beneficial to practitioners and researchers alike.

Cosentino, L., A. Pelgrain de Lestang, and J. E. Gonzalez, 2006, Geostatistics: A fast-response tool for reservoir management, *in* T. C. Coburn, J. M. Yarus, and R. L. Chambers, eds., Stochastic modeling and geostatistics: Principles, methods, and case studies, volume II: AAPG Computer Applications in Geology 5, p. 301–311.

*Chapter 23*

# Geostatistics: A Fast-response Tool for Reservoir Management

**L. Cosentino**[1]
**A. Pelgrain de Lestang**
*Beicip-Franlab, Rueil-Malmaison*
*France*

**J. E. Gonzalez**
*Petroleo de Venezuela, Sociedad Anonima*
*Exploration, Production and Marketing*
*Maracaibo, Venezuela*

## ABSTRACT

Extensive geostatistical modeling of data from the Bachaquero fields (east coast of Lake Maracaibo, Venezuela) has been conducted within the framework of a large-scale integrated study. These stochastic models have been used for several reservoir-management applications, where a fast and effective response is necessary to satisfy operational requirements. Such requirements include identification of infill drill locations, planning of horizontal well trajectories, and thermal-simulation studies for the optimization of cyclic steam injection projects. Because reservoir management applications commonly concern individual wells (small scale), a specific procedure has been established to extract and statistically analyze the geologic information from geostatistical models (large scale). The procedure has worked very effectively, with most of the operational requirements of corporate field exploitation units being addressed in the required timeframe. In fact, after almost 2 yr of fast-response applications in the Bachaquero fields, most of the operations currently performed by the exploitation unit are supported by geostatistical models. Recommendations based on such fast-response studies have been implemented, and in all cases, very good agreement between predictions and actual results has been observed.

## INTRODUCTION

Geostatistics has entered a new age. Until few years ago, geostatistics could still be considered a leading-edge technology, hardly applicable to real studies, developed and applied by a handful of specialists. Today's situation is different, and geostatistics not only is being routinely applied in integrated reservoir studies (Cosentino, 2001), but it is also entering the domain of reservoir management.

[1]*Present address*: Eni Exploration and Production, Milan, Italy.


DOI:10.1306/1063824CA53243

Reservoir management has been defined as the proper use of the available resources, with the aim of minimizing capital and operating expenditure (CAPEX and OPEX) and maximizing recovery (Thakur, 1996). Its purpose is to control operations to obtain the maximum recovery from a reservoir on the basis of facts, information, and knowledge. Sound reservoir management therefore requires the availability of knowledge-based tools that allow for flexible, time-consistent, and effective solutions to operational needs.

One of the main challenges of reservoir management is the timing of the required response. In the last few years, crude oil price variations, such as those illustrated in Figure 1, have been very important and have heavily affected the development strategies of most oil companies. In countries like Venezuela, additional perturbing factors exist (political instability, Organization of Petroleum Exporting Countries decisions on production cuts, rig availability, social problems, etc.), which commonly determine sudden changes in strategic decisions and ultimately influence the planned operational activity. In such contexts, one of the main challenges of an efficient reservoir-management tool is therefore the capability to provide exploitation units of petroleum companies with fast responses for operational decisions.

Nowadays, geostatistics represents an efficient tool for reservoir management. In recent years, the development of new software applications and the availability of more powerful computing machines have changed the way geostatistics is applied in the context of reservoir studies. In the current practice, when a sound geological well-log database is available, a geostatistical model can be built in a matter of days. Likewise, sensitivity runs for uncertainty evaluation can be quickly performed through batch routines. In addition, most of the available commercial software allows for fast updating of existing geostatistical models with new data, coming, for example, from new wells or even from while-drilling data. This chapter presents the application of geostatistics as a reservoir-management tool for the Bachaquero fields, Venezuela.

## THE BACHAQUERO CONTEXT

The Bachaquero fields are located offshore along the east coast of Lake Maracaibo, Venezuela (Figure 2). They are part of a much larger oil accumulation, mainly located onshore, which has been producing since the late 1930s. To date, more than 3000 wells have been drilled in the Bachaquero fields, including sidetracks and redrills. In recent years, the development has been mainly based on horizontal wells.

Almost 100 reservoirs have been identified in the region, belonging to both the Eocene and Miocene series. The total original oil in place approaches

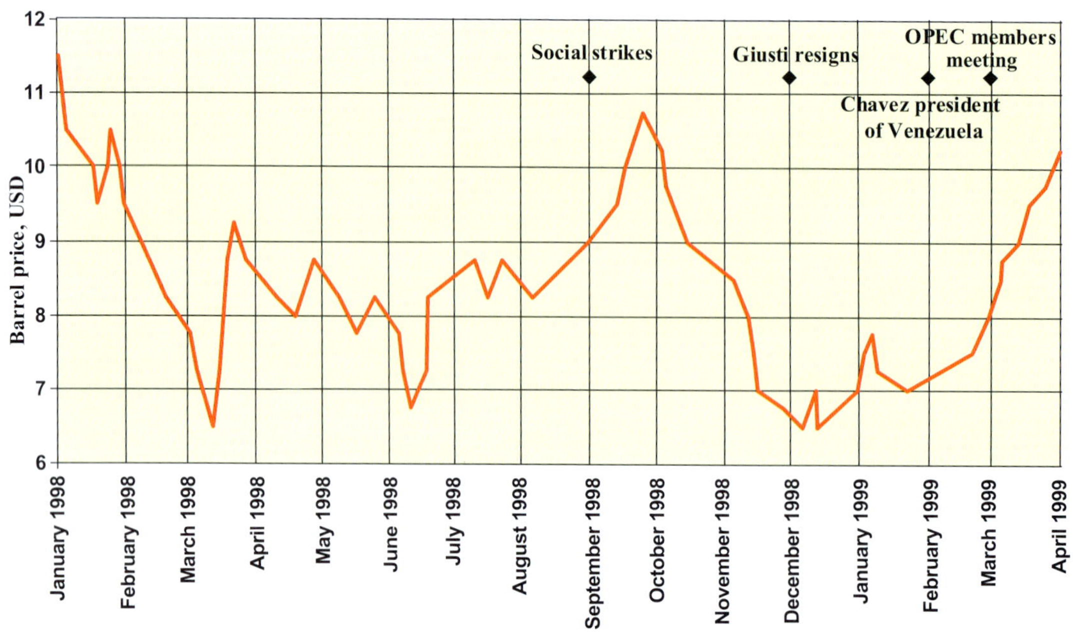


Figure 1. Venezuelan crude oil price fluctuations in recent years.

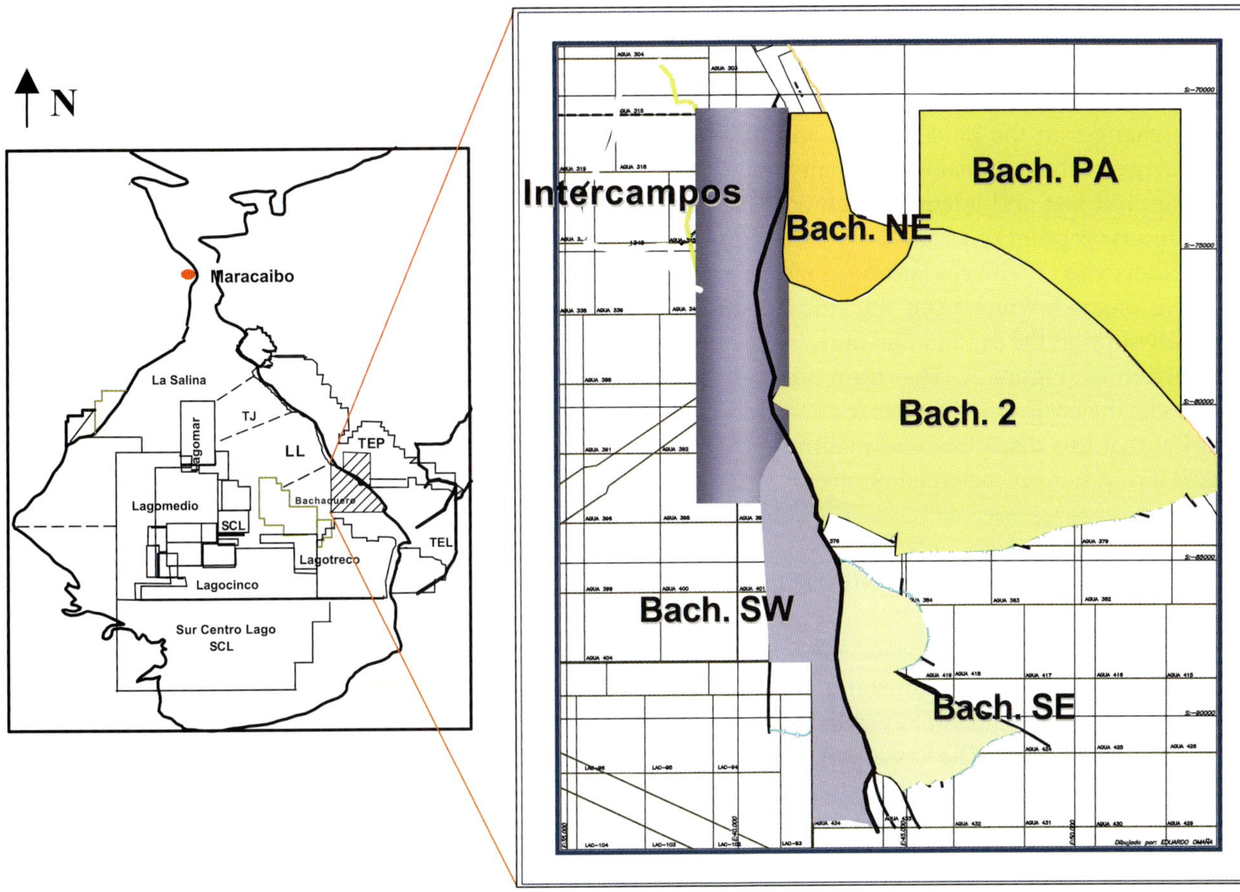


**Figure 2. Bachaquero fields location map. TJ = Tia Juana; LL = Lagunillas; TEP = Teplice; TEL = Tierra Este Liviano; SCL = Sur Centro Lago; NE = Noreste; SW = Suroeste; SE = Sureste; Bach. = Bachaquero.**

19 billion bbl. API gravity of the produced oil ranges from 14 to 40°, depending on the area. The Miocene reservoirs represent the most important part of the hydrocarbon accumulation. The predominant production mechanisms of these reservoirs are water drive, solution gas drive, and compaction, the latter accounting for more than 50% of the current recovery in some cases (Merle et al., 1976). Water- and gas-injection projects have been initiated in most of the fields since the 1960s. Thermal methods, especially cyclic steam injection, are also used in the area on a local basis as a production acceleration process.

The Bachaquero fields have been the object of an ambitious integrated reservoir study that started back in 1995. The general objective of the project was to update the existing static and dynamic models of the reservoirs, and to establish short-, medium-, and long-term exploitation strategy for all the reservoirs. Several stages of the project have been completed to date, and the recommendations of the study have generated a significant increase in the production profiles of these fields. Results of the integrated study have been reported in previous articles (Lopez et al., 1999; Pelgrain de Lestang et al., 1999).

The integrated study involved several phases, including a complete revision of the existing structural, depositional, and petrophysical descriptions of the geological units, as well as the construction of dynamic simulation models for most of the reservoirs. Serveral geostatistical models were also constructed for the main fields, including the giant Bachaquero 2 field.

A further objective of the integrated reservoir study, in addition to establishing the development plans, was assisting the exploitation units in terms of what was referred to as fast response. This assistance was required for reservoir-management purposes and included planning and locating new wells, optimization of conventional and thermal operations (injection and producing rates, soak time, etc.), and general assistance for all such activities that must be supported by a reservoir study.

This task was particularly challenging for the reservoir group because the required response time was very short in many instances. In fact, the variability of Venezuelan crude oil price, in addition to the current political and social instability, generated continuous changes in the development and exploitation strategy of the national oil company, which, in turn, required fast and adequate technical support from a reservoir point of view.

To achieve this objective, a dedicated fast-response team was created, which used the study results to effectively support the exploitation units in the operational activities (Figure 3). This team adopted geostatistics as the basic tool for their routine studies in the belief that this discipline was mature enough to be used as a reservoir-management tool.

## THE GEOLOGICAL MODEL

From a structural point of view, the Bachaquero area can be divided in two main zones, separated by a major north–south regional feature, the Pueblo Viejo reverse fault (Figure 2). The eastern area is a gently south-dipping monocline (6–8°), where the interpreted faults do not have significant vertical throw and do not act as permeability barriers. The western area, in contrast, is a complex reactivated zone, where numerous existing faults determine a significant degree of reservoir compartmentalization.

From a stratigraphic point of view, the reservoirs consist of unconsolidated fine sands and shale, belonging to the Bachaquero Member of the lower Lagunillas Formation (Miocene). The depositional environment is marine to coastal in the lower part (shoreface bar, tidal flat) and fluvial (braided to meandering channels) in the upper section (Figure 4). Overall, the reservoir units belong to a third-order regressive siliciclastic sequence. Reservoir heterogeneity is related to the depositional architecture of the sand bodies and is more pronounced in the fluvial section (channels belt, meandering channels, point bar, crevasse splay). A conceptual model based on core description and genetic sequence stratigraphy concepts has been established for each depositional environment. In the vertical direction, seven stratigraphic units have been identified and named, from bottom to top, units 1 to 7. These units are separated by shaly horizons (flooding surfaces) of variable thickness and continuity. Reservoir sands have very good petrophysical properties, with porosity at about 30% and permeability in the order of hundreds to thousands of millidarcies.

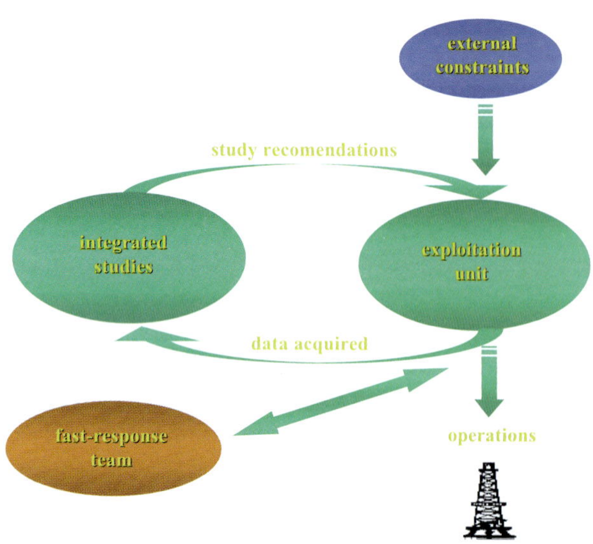


**Figure 3. Organization of technical support to the exploitation unit.**

## STOCHASTIC MODELING

Several geostatistical models were built in the Bachaquero area, the main one being associated to the giant Bachaquero 2 field (Pelgrain de Lestang et al., 1999). The database for the Bachaquero 2 field consists of more than 1300 wells having a regular spacing of 150 or 300 m (490 or 984 ft) (Figure 5). Geological markers (top and thickness of reservoir units), as well as facies descriptions and petrophysical properties, are available for all wells. Seven sedimentological facies, as defined on core description, are grouped into four lithotypes (clean fine sandstone, shaly fine sandstone, siltstone, and shale) on the basis of the petrophysical characteristics.

Vertical proportion curves (Ravenne and Beucher, 1988) and traditional semivariogram functions were used to analyze the spatial variability of the various parameters. The vertical proportion curves of the six genetic units showed very distinct sedimentological patterns (Figure 6). However, the high density of the available information allowed spatial correlation functions (semivariogram) to be unambiguously estimated in both the vertical and the horizontal directions.

The technique of stochastic modeling was used to develop a visual and numerical representation of the reservoir that incorporates the estimated spatial correlation functions. A pixel-based, truncated

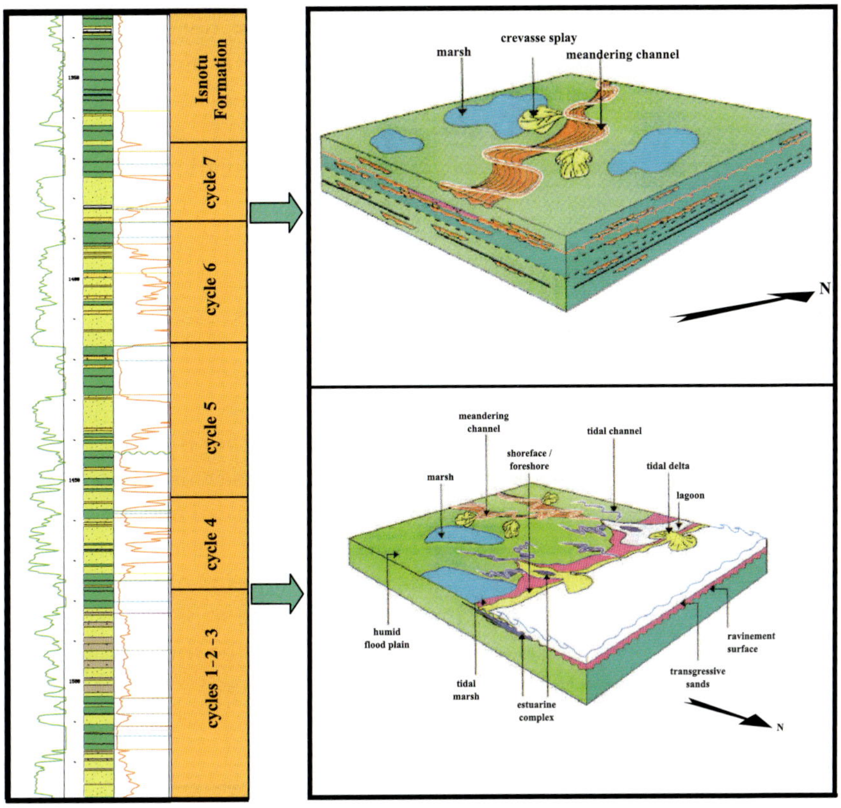


Figure 4. Sedimentological model of the Bachaquero area.

Gaussian algorithm (Matheron et al., 1987) was used to produce individual realizations. This approach was preferred over the Boolean approach because of the high sand/shale ratio of the reservoir units.

Each of the six genetic units was simulated independently. Different grid construction modes were used for the individual units, depending on the depositional conceptual model. In some cases, grids were constructed parallel to a paleohorizontal layer (e.g., a flooding surface), and in others, they were constructed proportional to total thickness. These grids have been subsequently stacked to generate the final geostatistical model. The global simulation grid has about 64 millions cells, whose dimensions are 30 × 30 m (98 × 98 ft) in the *X*- and *Y*-directions and approximately 1 m (3.3 ft) in the vertical direction.

Several full-field simulations were conducted. A comparison of the different realizations showed that the relative differences were minor and on a local scale. This result was expected, because of the high density of conditioning wells and the relatively long correlation ranges. Figure 7 shows a simulated north–south cross section in the reservoir, depicting the various degrees of heterogeneity in different genetic units.

In addition to the Bachaquero 2 full-field model, several smaller geostatistical models (in terms of geographical extent) were generated, both in this reservoir and in other Bachaquero fields, to support the fast-response activity required by the exploitation units. These local models were primarily used to assess various aspects of reservoir uncertainty, e.g., facies presence and distribution and petrophysical properties. In these cases, a large number of realizations (typically 50–100) were generated. As it will be shown later, these realizations were subsequently combined and statistically analyzed to express the uncertainty in terms of cumulative probability (e.g., the existence of a given facies).

Finally, it should be noted that the existing stochastic models are continuously refined and updated as new data are available to increase their accuracy and reliability.

## RESERVOIR-MANAGEMENT APPLICATIONS

As previously discussed, the various geostatistical models were primarily developed to support several reservoir-management activities, which included

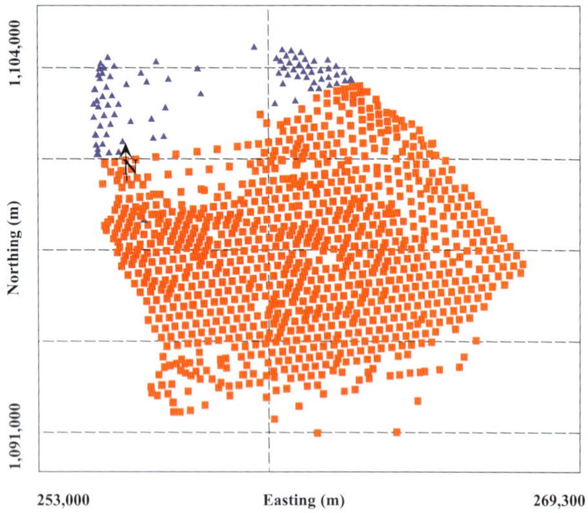


Figure 5. Bachaquero 2 field, wells included in the geostatistical modeling.

selecting infill drilling locations, optimizing the trajectory of horizontal wells, and simulating reservoir performance (phenomenological modeling, two- and three-dimensional [2-D and 3-D] black oil simulation, 3-D thermal simulation, etc.). Each of these activities, and the way in which geostatistical modeling is incorporated, are described more fully below.

## Infill Drilling

In the Bachaquero area, the selection of new well locations is realized by integrating the results of the structural and depositional reservoir models with the dynamic information (pressure and production data). This integrated body of information allows the most interesting areas in terms of undrained oil and sufficient reservoir pressure to be located. In this context, geostatistics is used in the final definition of the well location, with the objective of minimizing

Figure 6. Vertical proportion curves of the six genetic units.

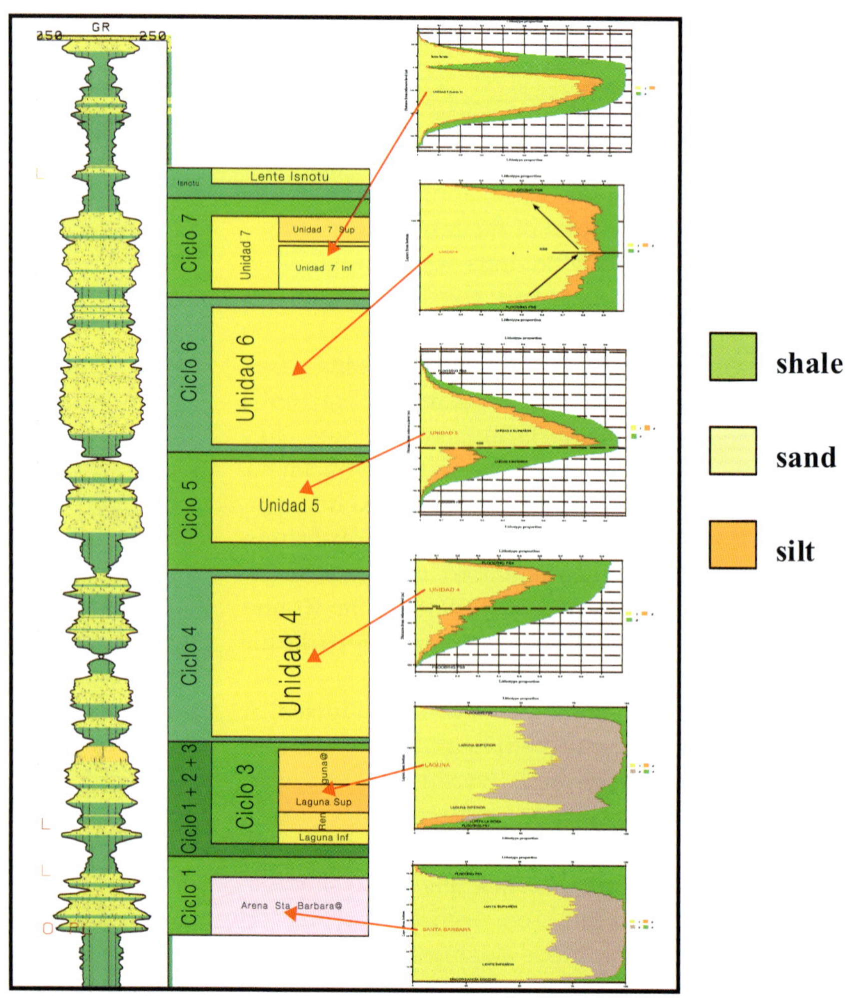

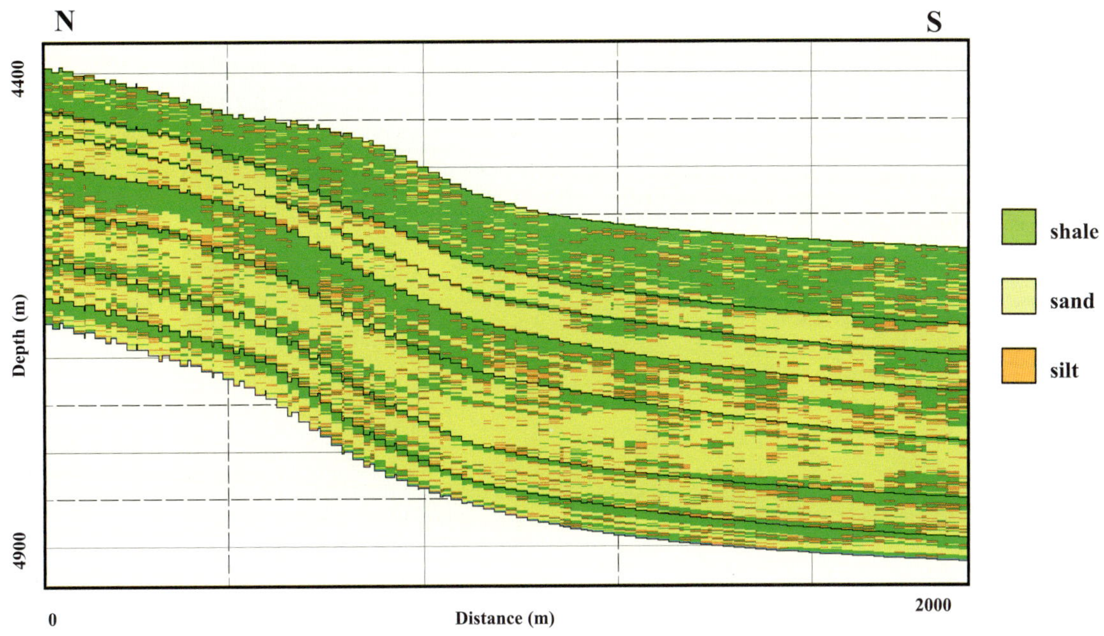


**Figure 7. North–south geostatistical cross section in the Bachaquero 2 field, facies simulation. The different stacked stratigraphic units are recognizable.**

the geological risk and maximize the expected performance of the well. In particular, a statistically significant number of stochastic realizations (normally 50) are combined to obtain average cumulative thickness maps for each lithotype. The target areas for infill drilling are those that show the highest cumulative thickness for the best facies.

Results so far have been satisfactory. During 1999, two vertical producer and four vertical injector locations have been proposed through this approach, and all of them have performed as expected.

## Horizontal Wells Trajectory Optimization

In the Bachaquero area, horizontal well trajectories need to comply with several operational constraints (infill spacing, presence of other wells, drilling and completion practices, etc.). Once the operational constraints have been defined, geostatistical modeling results are used to find an optimum well trajectory, with the objective of reducing the geological risk and maximizing the net/gross sand ratio of the horizontal section. Facies probability maps and cross sections are used in this context, in addition to average thickness maps. A facies probability volume of data is obtained by statistically analyzing all the stochastic realizations and computing the probability of presence of each facies in each grid cell. Therefore, this volume of data can be used to quantitatively assess the risk related to a given well trajectory. Alternative well paths are considered, and the recommendation is issued for the well trajectory that has the highest cumulative probability.

Figure 8 shows an example relative to the optimization of a horizontal well trajectory in stratigraphic unit 6. Note that the initial proposed trajectory (dotted line) tagged a rather poor sandy section in the average sand thickness map (green areas). An alternative well path was therefore proposed (solid line), which is shown to have a high probability of encountering the best sand facies all along the proposed trajectory (yellow and red areas). Figure 9 shows a presence probability cross section for the same revised location, where the high probability of tagging the sand facies can be appreciated.

Results provided by these studies were very satisfactory: out of the eight horizontal well trajectories analyzed in 1999, three have been drilled, and all have encountered the expected lithological section.

## Numerical Reservoir Simulations

Several different types of numerical reservoir simulations have been performed in the framework of the project, including phenomenological studies, 2-D and 3-D black oil, and thermal models. Most

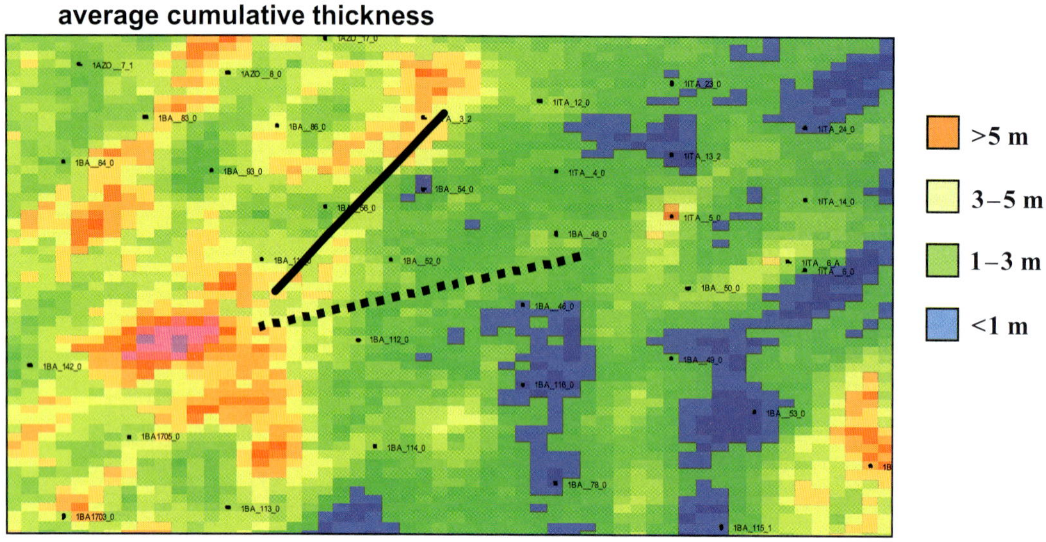


Figure 8. Illustration of an optimized horizontal well trajectory in the Bachaquero 2 field superimposed on a map of average cumulative thickness of the sand facies. The dotted line shows the original well path, and the solid line represents the revised well path.

of these numerical models have been used for conventional purposes (i.e., production profiles and reserve estimation). In addition, some smaller models have been constructed for specific applications like drive mechanism studies, optimization of injection patterns and rates, and simulation of thermal-recovery processes (i.e., cyclic steam injection). These models have been used frequently for reservoir-management purposes, especially for the optimization of operating procedures in thermal processes.

The general approach in this case includes the following two major steps:

1) selecting a representative area in the stochastic models, from which 3-D grids of petrophysical properties are extracted for several realizations (the areal extent of these models is commonly much smaller than the area covered by the stochastic model) and transferring these grids to the numerical simulator.

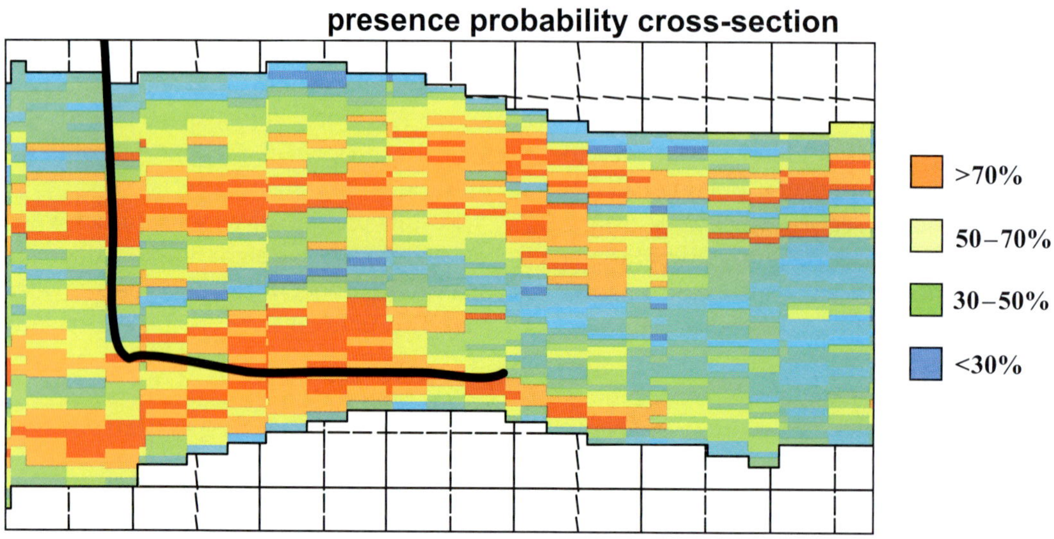


Figure 9. Illustration of an optimized horizontal well trajectory in the Bachaquero 2 field through a cross section of presence probability in the sand facies. The solid line shows the revised well path.

2) running reservoir simulations using different realizations to perform sensitivities on defined parameters (typically porosity and permeability) and to quantify the associated uncertainty in the production profiles.

Some time was dedicated to the generation of quick links among the different software applications involved, but eventually, the procedure allowed for a considerable time reduction in the construction of the model with respect to a conventional 2-D approach.

Figure 10 shows the results of the dynamic simulation of a north–south cross section in the Bachaquero 2 field that was used to investigate the observed water fingering induced by the injector wells. Geostatistical modeling was used in this case to obtain a detailed representation of reservoir heterogeneity that was essential in the proper simulation of the oil-displacement process. Note that no vertical upscaling was applied to the layers under study to maintain the most realistic degree of geological heterogeneity. Results of such a model were used in the optimization of injection rates and vertical allocation.

Other typical applications of geostatistics to dynamic simulation concerned the building of thermal models to optimize cyclic steam operations in the Bachaquero 2 field. Several reservoir-simulation models were built, mainly involving horizontal wells, which allowed for the definition of the optimum operational parameters in terms of production and injection rates, steam quality, and soak time (R. Muñoz, A. Lopez, L. Cosentino, A. Pelgrain de Lestang, and N. Clement, 1997, personal communication).

The most complete of such studies regarded the simulation of cyclic steam injection on parallel horizontal wells perforated in different adjacent geological units (Cosentino et al., 1998). The objective in this case was to verify the expected dynamic and thermal interference between the wells and to define, if possible, an optimum combined exploitation strategy in terms of alternating injection and production cycles in the two wells. The actual configuration of the two horizontal wells is shown in Figure 11.

The dynamic performance of each well and their mutual interference are dependent on the presence and the thickness of the shale separating the two sand bodies (stratigraphic units U6 and U7). Therefore, geostatistical modeling results were used in this case to study the expected thickness of the interbedded shale and its impact on the production performance of the wells. Figure 12 shows a slice of one geostatistical simulation at the stratigraphic depth of the shale that separates the two units U6 and U7. Yellow spots in this slice represent the presence of sand (i.e., discontinuities in the shale layer),

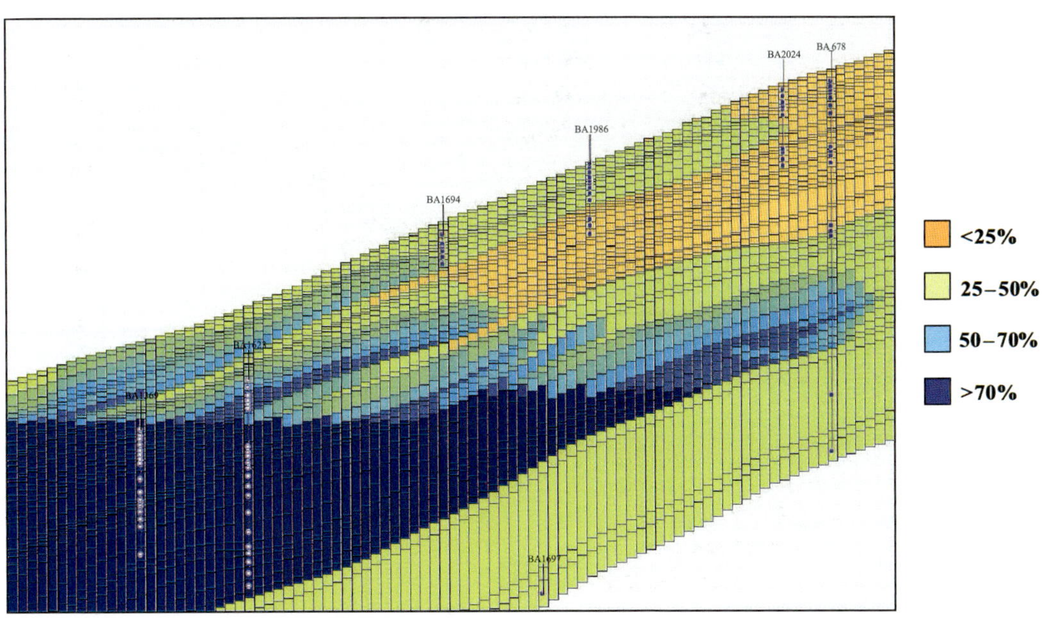


**Figure 10. Results of a numerical simulation study of a north–south cross section in the Bachaquero 2 field. The figure depicts the water-saturation distribution.**

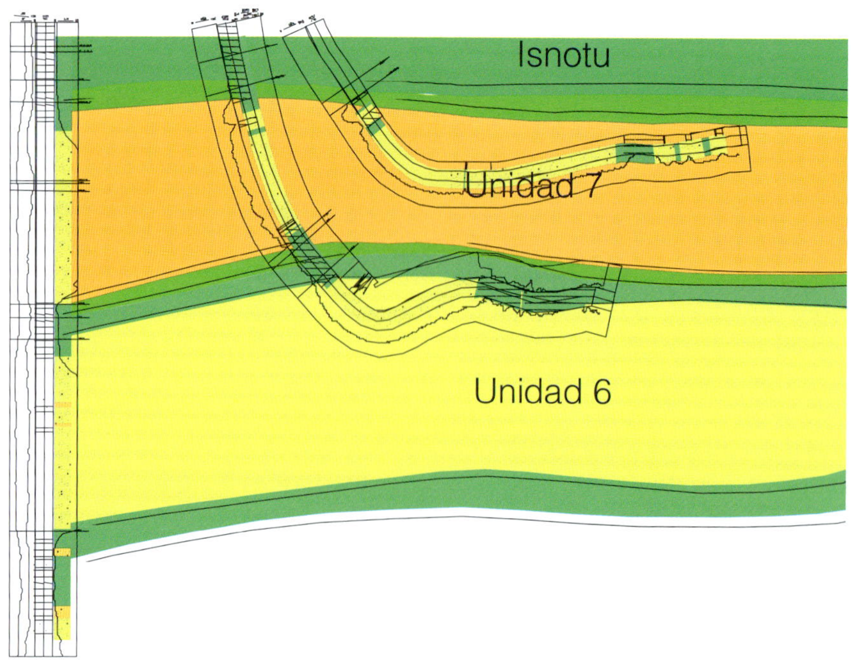


Figure 11. Cyclic steam injection along parallel horizontal wells. The cross section shows the actual results of the drilling phase. The vertical well to the left is the pilot hole. Green layers represent shale horizons.

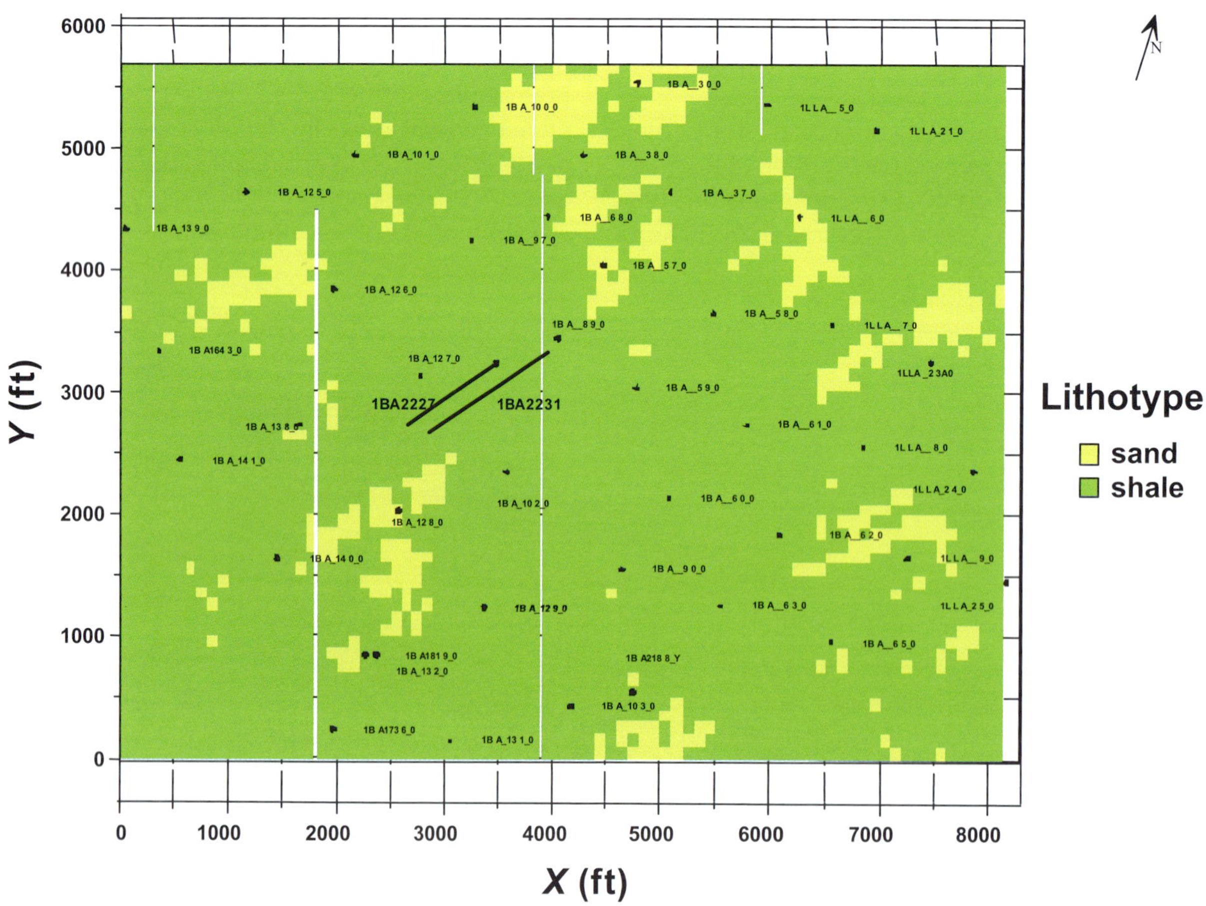


Figure 12. Cyclic steam injection along parallel horizontal wells. The figure shows a horizontal slice of the geostatistical simulation indicating the presence of potential vertical communication between units 6 and 7. The two black lines show the projected location of the two horizontal wells.

which could represent windows of communication between the two units in terms of pressure and temperature interference. The probabilistic study of the thickness of such shale, combined with the dynamic simulation results, eventually provided the required guidelines for the actual drilling and testing of these wells.

## CONCLUSION

Extensive geostatistical studies have been performed in the Bachaquero fields within the framework of a large-scale integrated study. The results of these models have been used for a variety of reservoir-management applications, including the identification of infill drill well locations, optimization of horizontal well trajectories, and thermal dynamic simulation of cyclic steam injection projects, all of which require detailed and time-sensitive responses. The experience gained in this project demonstrates that geostatistics can be an effective tool for reservoir management, resulting in successful field implementation of project recommendations based on modeling results.

## ACKNOWLEDGMENTS

We thank all the geoscientists who participated in these studies over the last few years. We also thank Beicip-Franlab and Petroleo de Venezuela, Sociedad Anonima management for permission to publish these results.

## REFERENCES CITED

Cosentino, L., 2001, Integrated reservoir studies: Paris, Editions Technip, 310 p.

Cosentino, L., G. Spotti, J. E. Gonzalez, Y. Araujo, and J. Herrera, 1998, Cyclic steam injection on parallel horizontal wells: Geostatistical description, thermal simulation and field experience: Society of Petroleum Engineers Annual Technical Conference and Exhibition, New Orleans, SPE Paper 49017, 12 p.

Lopez, D., R. Muñoz, J. E. Gonzalez, Y. Gou, J. C. Pascual, and L. Cosentino, 1999, Large scale integrated reservoir study: The Bachaquero Intercampos experience: Society of Petroleum Engineers Latin American and Caribbean Petroleum Engineering Conference, Caracas, Venezuela, SPE Paper 53996, 8 p.

Matheron, G., H. Beucher, A. Galli, D. Guerillot, and C. Ravenne, 1987, Conditional simulation of the geometry of fluvio-deltaic reservoirs: Society of Petroleum Engineers Annual Technical Conference and Exhibition, Dallas, SPE Paper 16753, 7 p.

Merle, H. A., C. J. P. Kentie, G. H. C. van Opstal, G. M. G. Schneider, 1976, The Bachaquero study: A composite analysis of the behavior of a compaction drive-solution gas drive reservoir: Journal of Petroleum Technology, p. 1107–1115.

Pelgrain de Lestang, A., L. Cosentino, D. Lopez, and J. E. Gonzalez, 1999, A large scale geostatistical study: The Bachaquero 2 field: Society of Petroleum Engineers Annual Technical Conference and Exhibition, Houston, SPE Paper 56657, 8 p.

Ravenne, C., and H. Beucher, 1988, Recent development in description of sedimentary bodies in a fluvio-deltaic reservoir and their 3D conditional simulation: Society of Petroleum Engineers Annual Technical Conference and Exhibition, Houston, SPE Paper 18310, 8 p.

Thakur, G. C., 1996, What is reservoir management?: Journal of Petroleum Technology, v. 48, p. 520–525.

Herweijer, J. C., J. M. Yarus, and M. J. V. Suana, 2006, Process-based knowledge management: Overview and application of Iterative 3-D modeling workflows, *in* T. C. Coburn, J. M. Yarus, and R. L. Chambers, eds., Stochastic modeling and geostatistics: Principles, methods, and case studies, volume II: AAPG Computer Applications in Geology 5, p. 313–324.

*Chapter 24*

# Process-based Knowledge Management: Overview and Application of Iterative 3-D Modeling Workflows

**J. C. Herweijer**
*Reservoir Team Ltd.*
*Rundle Mall, Australia*

**J. M. Yarus**
*Quantitative Geosciences, LLP*
*Houston, Texas, U.S.A.*

**M. J. V. Suana**
*Updip-EP*
*Brugg, Switzerland*

## ABSTRACT

This chapter presents an overview of current knowledge management practices in upstream oil and gas exploration and production, specifically for multidisciplinary project teams working with spatial modeling techniques. A process-based knowledge management methodology is defined that allows for continuous work on complex multiscenario models in a consistent, flexible, and shared manner under conditions where data are less than optimal, modeling techniques are poorly understood, and staffing assignments change before project completion.

## INTRODUCTION

In recent years, three-dimensional (3-D) modeling has taken center stage in petroleum reservoir management. As a consequence of a specific 3-D modeling project, a considerable body of knowledge is generated about what makes the model work or not work for the reservoir in question. Such knowledge, which is essential to systematic uncertainty assessment during a current effort or in subsequent efforts, needs to be effectively managed.

The 3-D modeling approaches currently in vogue already provide for retention of model inputs and outputs, as well as some representation of the processing steps (model-specific workflows). These workflows commonly consist of detailed processing chains as well as the values assigned to software-related parameters. In addition, efforts are sometimes undertaken to retain modeling best practices, which typically entail general know-how (e.g., procedural logic, human-machine-software interaction, where to find


DOI:10.1306/1063825CA51040

data and whom to contact, and metadata) associated with model development.

With more rapid access to increasing amounts of information and with better tools to transform that information into knowledge, even more effective means of knowledge management (KM) are essential to sound decision making. Establishing a formal approach to KM ensures that inclusion of preexisting knowledge will be included in any reservoir model, including controversial, or even absent, knowledge reflected in various measures of uncertainty. It further ensures that the modeling process is comprehensive in nature, encompassing all the issues that may impact reservoir performance. Finally, such an approach guarantees point-forward storage and retention of knowledge generated during the modeling process to facilitate adequate and efficient model updating over time.

This chapter specifically focuses on the notion of KM in the exploration and production (E&P) sector of the petroleum industry, particularly on how process- or workflow-based KM approaches add value to 3-D reservoir modeling projects. Such projects are the basis for significant investment decisions, which can lead to considerable technical, financial, commercial, and environmental risk. This chapter demonstrates that an iterative modeling approach, in combination with process-based KM, is a very powerful tool for uncertainty delineation and risk assessment.

## KM IN OIL AND GAS E&P

Up to the mid-1980s, many E&P entities in the oil and gas industry typically adhered to a discipline-oriented hierarchical organizational structure, including research and training branches and divisions for each discipline. This organizational model allowed for good knowledge rotation and retention in disciplines. However, by the 1990s, the discipline-oriented structure was more and more pushed aside. An important factor that greatly contributed to this organizational shift was the evolution of technology and modern problem-solving techniques, such as 3-D reservoir modeling, that could only be effectively employed in an integrated multidisciplinary work environment.

This fundamental structural change resulted from a transformation in business and industry to more information-driven organizations (Drucker, 1988), the transformation being driven by rapid advances in information technology, increased access to data, and expanded analytical capabilities, as well as the ensuing demand for KM. The upshot has been the establishment of modern information and knowledge business or support units that may exist both internal and external to larger organizations and that comprise specialists who direct and discipline their own performance by turning data into relevant information through knowledge. Powered by ever-increasing competitiveness on both regional and global scales, the effective leverage of organizational knowledge has become as much a key to corporate success as managing capital, inventory, and human resources. A survey of chief executive officers attending the World Economic Forum in 1999 found that 97% of senior executives see KM as a critical issue for their organizations (Abramson, 1999).

Many companies have developed their own KM systems and/or have drawn from a fast-growing number of commercially available solutions. The observed approaches to KM can be broadly grouped into two categories, namely, people-to-document and people-to-people strategies (Hansen et al., 1999).

The first, people-to-document approach, deals primarily with the management of explicit knowledge. The corresponding KM strategy is to provide high-quality, reliable, and fast electronic information systems, which collect, codify, store, and disseminate documented knowledge for reuse anywhere in the organization. This approach, which principally relies on data warehousing, knowledge mapping, and electronic library mechanisms, requires investment in sophisticated information technology systems.

The key advantage of people-to-document approaches is the effective capture of knowledge, with employees being encouraged to turn tacit personal knowledge into explicit organizational knowledge for consumption by the corporate community. A drawback is that such approaches also require a cultural change. Because knowledge gives power to people, a natural tendency to hoard instead of share knowledge exists (Daft, 2001). An important challenge of document-based knowledge capture is how to codify and interlink knowledge documents to avoid a "Google" effect (i.e., the return of too much or too little information).

In contrast, the people-to-people approach focuses on leveraging tacit knowledge by connecting people face to face or through interactive media (this approach is also known as a community of practice). The strategy is to develop networks for linking people such that tacit knowledge can be shared and individual expertise sought to gain advice. Information technology is primarily there to support and facilitate conversation. Mechanisms for knowledge sharing typically include dialog, story telling, framing workshops, peer reviews, and communities of disciplines and practice.

The key advantage of a people-to-people approach is that it leverages tacit knowledge, which is difficult to capture in an explicit form. This includes professional insights, judgment, rules of thumb, intuition and general understanding, and issues that tend not to be formally documented. People-to-people approaches are powerful because they allow knowledge to be efficiently and effectively drawn from a collective brain pool, and they facilitate the creation of new knowledge on the spot. Although much of the tacit knowledge sharing happens in an ad-hoc manner, a lot of it is actually explicitly written down through the use of e-rooms, intranets, e-mail, and other informal means. A drawback is that, for the community as a whole, much of the developed knowledge becomes practically irretrievable once discussions are closed out. The tacit knowledge, although shared in pockets, remains largely tacit and is lost upon departure of its human carriers.

## KM AND MULTIDISCIPLINARY PROJECT TEAMS

As previously noted, 3-D reservoir modeling is a multidisciplinary effort typically carried out by a project team, a situation that provides an extra set of challenges and opportunities for KM. Multidisciplinary project teams are expected to deliver technically sound, economically robust, and commercially viable investment proposals in the shortest timeframe possible. From a project management standpoint, two value-driven knowledge aspects are considered: (1) knowledge relating to the deliverable (e.g., how sound, robust, and competitive is a given field development plan); and (2) knowledge related to the work process-related knowledge (i.e., how efficiently and at what cost has the deliverable been produced).

One business requirement is to ensure that all relevant and up-to-date project-related knowledge residing in (and outside) an organization, in whatever form, flows into the project and is adequately processed. At the core of this effort lies the challenge of identifying, accessing, selecting, and using key knowledge, which may be either technical or nontechnical in nature. Some questions that need to be addressed are the following: What needs to be known to accomplish the objective? What explicit (documented) knowledge exists? Who carries important tacit knowledge? What are the best practices? What are the known pitfalls and key issues?

Another requirement is efficient achievement of each specific task. Some key questions to be addressed here are as follows: How quickly and with what degree of thoroughness is relevant knowledge accessed and processed? How can paralysis by analysis, unnecessary trial and error, or reinventing the wheel be avoided? How quickly can a project be set up? How are workflow and task interdependencies structured? How does the team collaborate and interact with other sources of knowledge in the organization? What is the vulnerability to staff rotation, and how is the resulting risk of knowledge leakage mitigated? How quickly can a project be closed out and new knowledge be fed back into the corporate knowledge system?

From an information management point of view, the development of an appropriate information storage and retrieval system and facility is absolutely necessary given the anticipated inflow of data and information. In concert with this necessity is the desire to have instantaneous access to key data and knowledge when and where it is needed in the work process and across the business enterprise. Unfortunately, this issue has not yet been fully addressed by commercially available KM systems that typically resemble data and document management systems.

This situation suggests that knowledge is better associated with the work process (see Figure 1) instead of with data management systems. In a work-and-results-oriented approach, the owners and administrators of knowledge are the experts themselves, organized in discipline communities, whose responsibility it is to ensure that relevant and up-to-date knowledge flows continuously into established best practice workflows and processes.

## KM IN THE CONTEXT OF SPATIAL MODELING AND 3-D STOCHASTIC RESERVOIR MODELING

Along with general trends outlined in the previous sections, KM is believed to have a specific function in the context of spatial modeling and

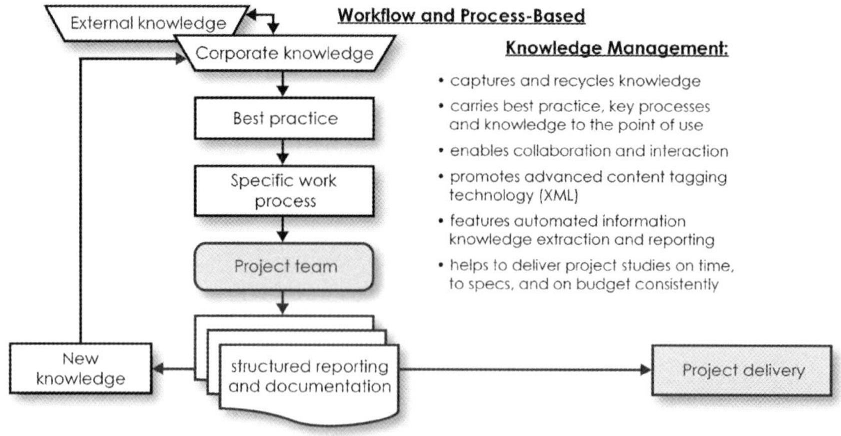


**Figure 1. Conceptual flow in a process-based KM system.**

stochastic reservoir modeling. Over the past 10 yr, 3-D reservoir modeling has moved from the domain of relative experts working in a focused manner on very specific issues to the domain of generalists creating and continuously updating models in a more or less routine matter. Increasingly, these models go beyond the focus of a single data type and integrate numerous pieces of multidisciplinary data each with their own complexity and commonly only fully understood by discipline specialists.

For example, 3-D modeling software will commonly include numerous options for analyzing and applying a model of spatial covariance, commonly referred to as the semivariogram. Many modelers have limited expertise in variography, and questions arise regarding the parameters and options to select for their data. A variety of answers and solutions exist, such as

1) suggest training material that explains the application of the different variogram models
2) provide examples of what works well (and what does not work) under similar conditions or analogs from similar reservoirs along with answers to additional questions, such as the following: Who did it? What pitfalls have been reported for application of the different variogram models? What quirks have been reported for the software that is being used?
3) provide a gateway to a corporate expert (or mentor) and generate a report with a very structured overview of the issue at hand in the holistic context of the entire modeling effort
4) provide a report to another set of modelers who are planning to rerun (update) the model after the next development round and require the historical considerations for choosing the specific variogram model
5) suggest an uncertainty analysis of variogram models from different scenarios to document the final selection and prevent future misunderstandings

The above examples are but a few that point to the value added to a well thought out KM program for modern 3-D reservoir modeling. A 3-D model without KM may solve some immediate problems but will likely fail in terms of providing an efficient segue into the next project step and accurate project reporting; and it will certainly fail in terms of providing future benefit to the company as a source of information to other modelers working on similar projects.

## IMPLEMENTATION OF WORKFLOW-BASED KM

No two reservoir modeling projects are the same such that they can be addressed with an identical workflow process. Nevertheless, a workflow KM system should be able to leverage knowledge from one reservoir modeling project to another in a strictly logical way. This requires such a system to be both flexible and repeatable; that is, workflows must be similar but not rigidly identical, as is typical for some computerized processes. CommonKnowledge™ (a product developed and marketed by Reservoir Team, Ltd.) is one example of a commercially available system that is flexible and whose workflow processes can be easily repeated.

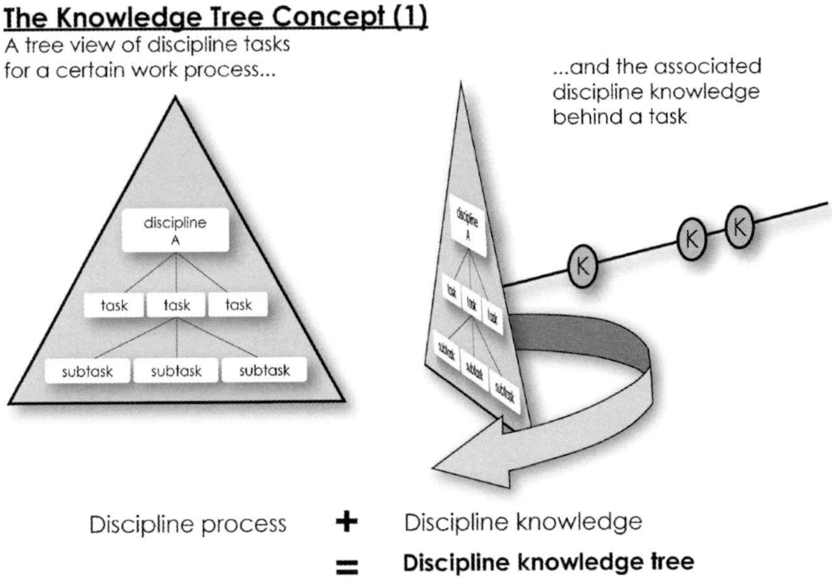


**Figure 2. Discipline knowledge trees contain all possible knowledge for a discipline. Specific tasks (including links to discipline best practice knowledge) can be selected from various discipline trees to compose a given multidisciplinary work process.**

## Basic Architecture

The purpose of the CommonKnowledge system is to construct multidisciplinary workflows from basic knowledge components, and its structural architecture facilitates this goal. These building blocks are obtained by breaking down a given discipline workflow into tasks and subtasks, which are common elements of any workflow, irrespective of the final work objective. Individual components from various disciplines can then be rearranged into customized workflows associated with different work objectives.

The main purpose of these building blocks, also referred to as knowledge nodes, is to serve as the principal resources of knowledge associated with the specific task in question. These resources can contain explicit knowledge, such as best practice instructions, risks, pitfalls, learning from previous experiences, and possible opportunities, as well as links to suppliers and carriers of tacit knowledge associated with the task.

Under this concept, the overall KM goal for each discipline, which embraces the collective brain for a certain field of expertise, is to build and maintain an entire set of knowledge nodes, called a discipline knowledge tree (Figure 2). For multidisciplinary processes, this procedure leads to the establishment of a collection of discipline knowledge trees (i.e., one for each discipline). In addition, nondiscipline-based knowledge, such as geographic areas of expertise and critical project steps, may be established.

## Project-specific Tasks Linked to Corporate Best Practice

A key advantage of a modular knowledge base, such as the one described above, that encompasses all corporate knowledge on a certain topic is the ability to easily compose specific workflow maps for new projects or to modify active project plans with minimal effort. Operationally, this is achieved on the computer screen by clicking and dragging knowledge nodes from shared knowledge trees into specific project trees. This clicking and dragging process will result in the associated knowledge being added to the new project process map (Figure 3), where it is readily available to the user.

Conceptually, a project-specific workflow process consisting of several discipline tasks is created, as illustrated by the knowledge lattice shown in Figure 4. Each discipline has its own lane, but the workflow process meanders across those lanes (as it does in reality). Along the vertical axis for each task is the knowledge for the same task obtained in other projects. Hence, this dimension represents common knowledge and implies that tasks are treated consistently even if they occur in different settings or project workflows. As such, a lattice of both project-specific and general corporate knowledge can be constructed.

The result is this: when project team members are confronted with a specific work task in the new workflow, they are automatically prompted with the appropriate corporate knowledge and are able to accomplish the task satisfactorily by effectively

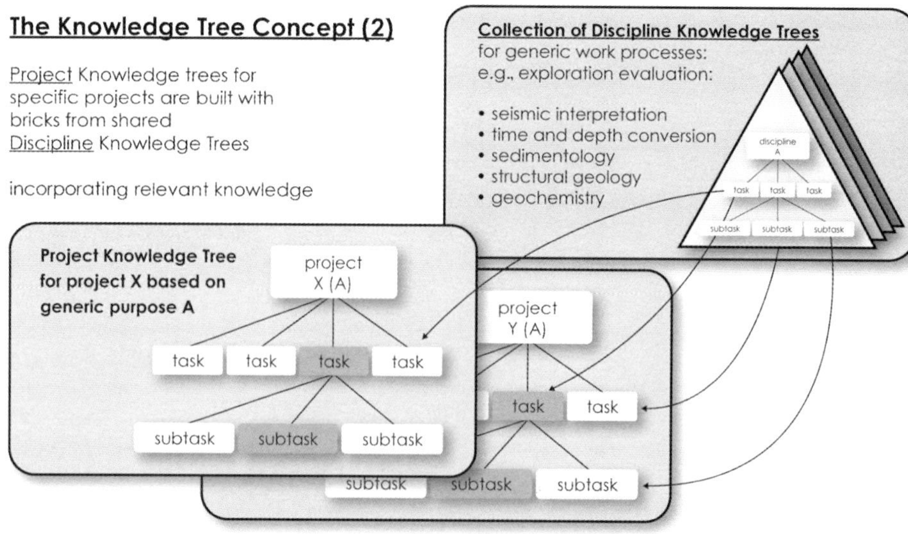


Figure 3. Project-specific knowledge trees are composed from individual shared discipline knowledge tree nodes. By dragging individual nodes to compose a new workflow, all previously associated best practice knowledge is linked into the new workflow.

leveraging the available organizational knowledge. This concept of dynamically delivering knowledge to the user via a node in the workflow process is superior to the traditional and less-efficient approach of retrieving knowledge from one or more static knowledge sources (library, corporate web, or document repository).

## Multiple Types of Relations between Workflow Tasks

In the above examples, the relations between the workflow tasks (workflow nodes) have been hierarchical, or parent-child, in nature (e.g,. milestone, task, subtask). In reality, workflow entails a much more complex web of interconnected tasks and issues. The CommonKnowledge system allows two types of special relations to be established, namely,

1) uncle or aunt-niece or nephew (secondary child) relations, which are secondarily hierarchical
2) dependency relations, which point to the next logical step in the workflow

The function of these two types of relations is to support extendable relations to other nodes that have their own functionality in a knowledge framework. When information is tracked along the workflow

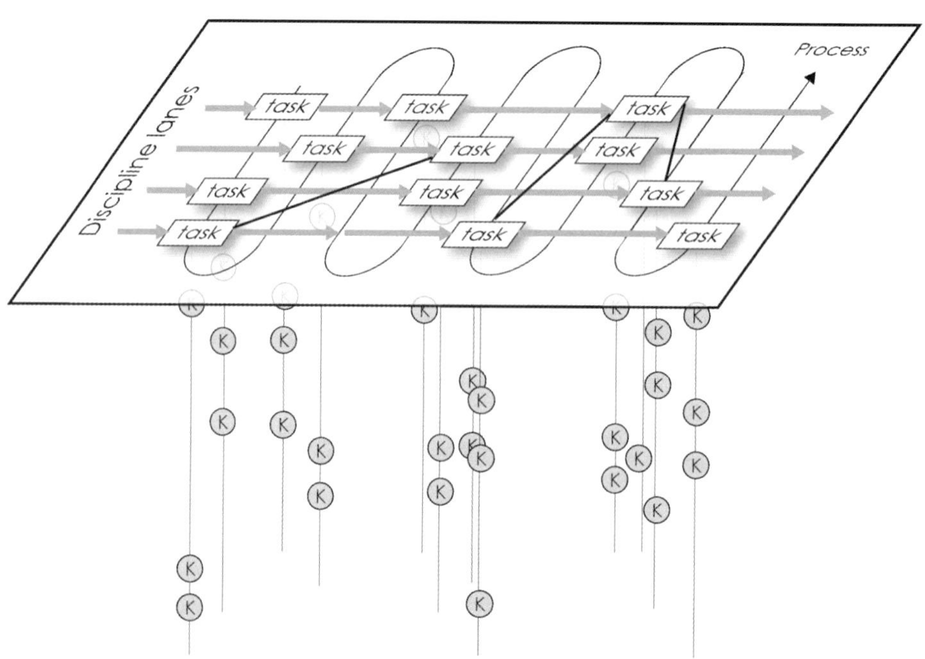


Figure 4. Knowledge lattice consisting of tasks along a project-specific workflow process that meanders across discipline lanes. The vertical dimension represents knowledge sharing with similar tasks as these occurring in other projects perhaps in different positions in the workflow.

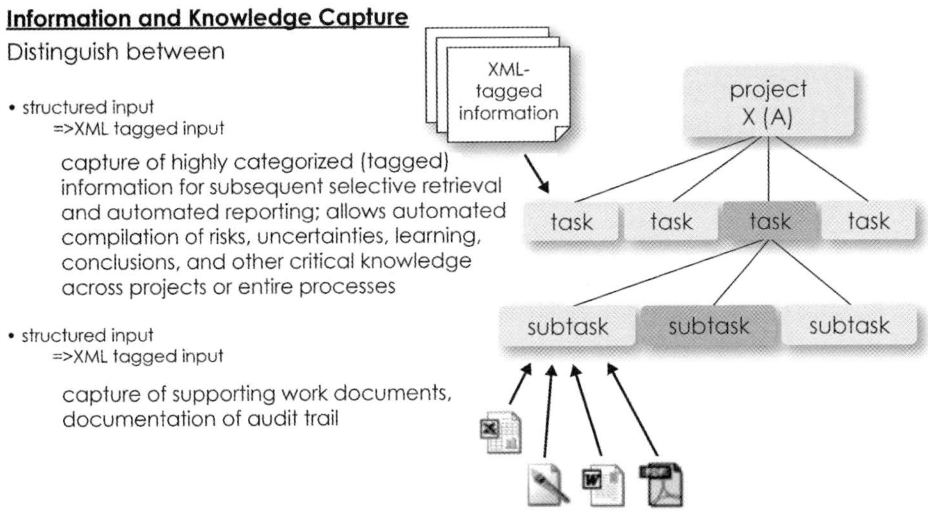


**Figure 5. Work documents, information, results, new insights, and knowledge are fed back into the system through the individual nodes. Note the differentiation between structured and unstructured input.**

process map, the information can be processed differently for each type of relation. For example, the uncertainty of a parent may be compounded by the uncertainty of its children, whereas the uncertainty of a niece or nephew may be processed in another context. Uncertainty may then be passed along to a dependent node.

## APPLICATION OF WORKFLOW-BASED KM

The next sections describe a range of applications of workflow-based KM to 3-D modeling. These applications straddle each stage of a 3-D modeling project ranging from systematically designing workflows at the initial stage of a project to creating a quality assurance (QA) or quality control (QC) audit trail and documentation at the final stage.

### Workflow Visualization and Customization

Ideally, any multidisciplinary effort such as 3-D modeling should carefully consider the workflow at any stage of the project. It should be very clear that the workflow is not linked to specific technology but, instead, to an intricate pattern combining analysis, conceptual design, and modeling. Initially, a basic workflow template, perhaps a copy from a previous project, can be applied. Such a workflow can be visualized step-by-step (see below) and described in detail.

At this stage, some of the basic discipline workflow building blocks can be added or edited, creating a project-specific version based on the previous template. Of course, the focus should be on those discipline tasks that contribute most to resolving or delineating uncertainty. This process facilitates a careful consideration of all discipline tasks and also promotes multidisciplinary awareness and understanding, without the risk of overdoing things by simply engaging discipline specialists to conduct whatever analysis is needed.

As discipline tasks are selected from a standard basket of possibilities, different projects will exhibit significant uniformity, although individual project teams have the option to pursue a customized solution. This flexible-standard approach is juxtaposed against fixed workflows for which it is commonly very difficult to obtain consensus. The application of fixed workflows is contrary to the understanding that no two reservoir studies are exactly the same.

### Documentation

A central obvious function for any KM system is to serve as the primary repository for project documentation of any kind and nature (Figure 5). This may include narratives and supporting working files, which are tagged and codified for appropriate storage and retrieval by other systems. The captured information puts the activities relative to the overall workflow into context and indicates how the outputs can be reproduced. In addition to providing an audit trail, it also significantly reduces vulnerability to staff rotation.

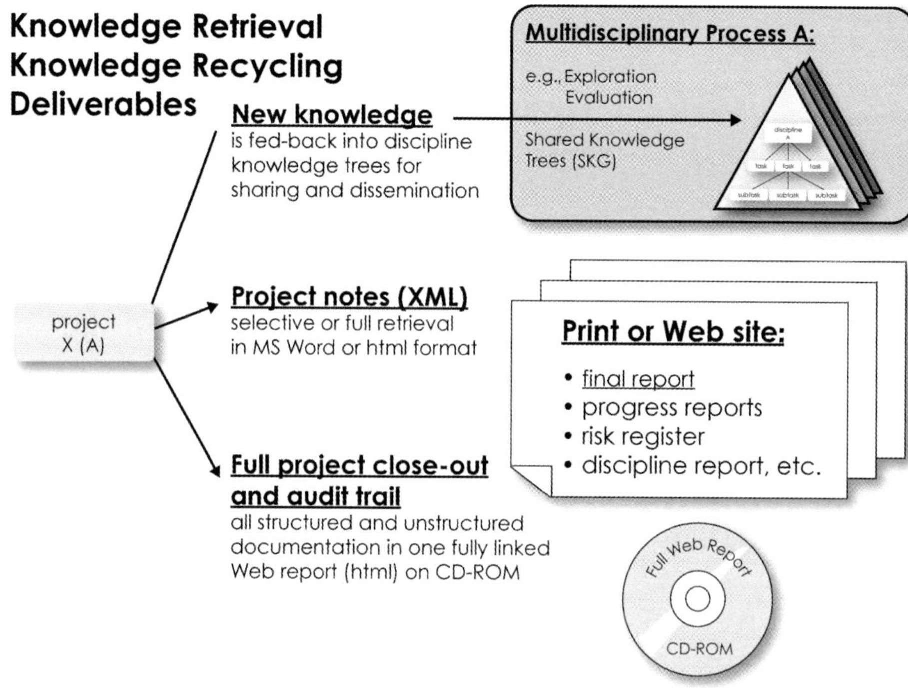


Figure 6. Full or partial reports can be automatically compiled in printable as well as Web format. New knowledge, following validation by the corresponding discipline knowledge community, is fed back into the shared discipline knowledge trees, from where it is incorporated into new projects.

## Collaboration Platform and Technical Supervision

The CommonKnowledge approach (implemented as a Web-based application) also provides an open platform for team collaboration and knowledge sharing without having to use a specific desktop application. Team members can simultaneously oversee the entire work process and project status at any time. This facilitates interdisciplinary work considerably, especially when addressing project tasks, which require collaboration among and contributions from various disciplines. The system can also be used as a vehicle to integrate outside team specialists and mentors into the workflow, to capture their contributions systematically, and to facilitate greater continuity and co-ownership of these outside contributors by virtual means in addition to face-to-face contributions.

## Work Efficiency, QA/QC, and Audit Trail

The capability of constructing virtually any workflow from preexisting shared knowledge building blocks reduces project setup time and prevents repetitive inefficiencies. Furthermore, the nodal structure allows and encourages on-the-fly reporting directly into knowledge nodes following preset formatting. In conjunction with the automated features of content selection and reporting, such an approach leads to reduced time to final reporting and project close-out (Figure 6). Interface capabilities with commonly used office, project management, and scientific software applications further enhance work efficiency.

## Knowledge Recycling

From an active KM perspective, the most important feature is perhaps the concept of feeding back newly created knowledge into the same workflow (knowledge) nodes. New knowledge is created by applying existing information and knowledge to new project work. Team members have the opportunity to alert, formulate, tag, and codify new insight and to feed it back into the system.

Such new knowledge, although entered through an active project-specific knowledge node, is available to the specific discipline community as a whole through incorporation into the shared knowledge trees (i.e., incorporation into the building blocks or knowledge nodes that are used to construct new workflows). This facility ensures that new knowledge acquired during one project is brought to bear for other projects almost instantaneously.

A key advantage of this knowledge-recycling approach is that corporate knowledge is constantly updated based on the results and insights gained from

**Table 1. Example of three typical phases of a 3-D reservoir modeling project, with associated milestones and issues and tasks.**

| *Phase* | *Milestone* | *Issues and Tasks* |
|---|---|---|
| 1 | Project kickoff | Team building and problem definition |
| | Select a 3-D geomodel concept | Review various modeling techniques vs. the reservoir issues at hand and available time and resources for the modeling effort |
| 2 | Data transfer and data QC and QA | Data retrieval and data management |
| | Premodeling and presimulation tasks | Conduct interpretations relevant to 3-D modeling (e.g., sedimentological analog, electrofacies analysis, seismic attribute analysis; see Figure 8) |
| | First-pass (iteration) 3-D geomodel (see Figure 7) | Various simulation tasks, e.g., preparing well-connection data, reconcile production schedule, history matching, and forecasting |
| | First-pass (iteration) simulation model | |
| 3 | New geological and geophysical interpretations | Revise interpretations based on insights resulting from the first-pass geomodel and simulation |
| | Peer review previous results and plan second-pass model | Document and review results and, if necessary or feasible, conduct next iteration of modeling |

ongoing work. The communities of experts themselves, as envisaged by Drucker (1988), can manage this system through their own field of expertise and add to the knowledge base on a routine basis. Such actions include deleting redundant knowledge as well as adding, or pointing to, new development and knowledge advances.

## 3-D RESERVOIR MODELING WORKFLOW CASE STUDY

Presented here is a case study using the CommonKnowledge system as the backbone for a 3-D modeling project. The objective of this project is to create a 3-D geomodel that can be used as input for subsequent simulation of a mature field. The process proceeds in several phases, tracking the gradual buildup of the 3-D modeling effort while simultaneously addressing data issues (e.g., data QA and control, interpretation adequacy). Table 1 shows an example that summarizes three such phases found early in the workflow process.

Figure 7 shows the first level of issues and tasks associated with the construction of the first-pass 3-D geomodel (second shaded row in Table 1). Most of these tasks originate from the 3-D geomodeling discipline tree (acronym 3M in upper left of workflow node boxes). Information entered here can be shared with other projects that have selected similar tasks from the 3-D geomodeling discipline tree.

Figure 8 shows a second level of tasks associated with facies modeling. Note again that all tasks originate from the 3-D geomodeling (3M) discipline tree. Also shown in the lower part of Figure 8 are related workflow nodes (niece and nephew nodes). These workflow nodes (tasks) originate from the geology discipline tree (acronym GL in upper left of workflow node) and are originally parented in the previous milestone, "Premodeling and presimulation tasks," associated with phase 2 in Table 1 (first shaded row). They are added as related nodes associated with facies modeling to create awareness of such tasks (and supporting documentation) and to ensure that the uncertainty assessment of facies modeling includes aspects of uncertainty related to geology.

Figure 9 is a dependency view, which basically consists of a flow diagram illustrating the relationship among all tasks in the context of predecessors and successors. Such a dependency diagram graphically displays the multidisciplinary sequencing of tasks and, as such, can be instrumental in overall project planning.

The complete example workflow for the 3-D model consists of about 80–100 workflow nodes (tasks). Such a complete workflow can be constructed in a

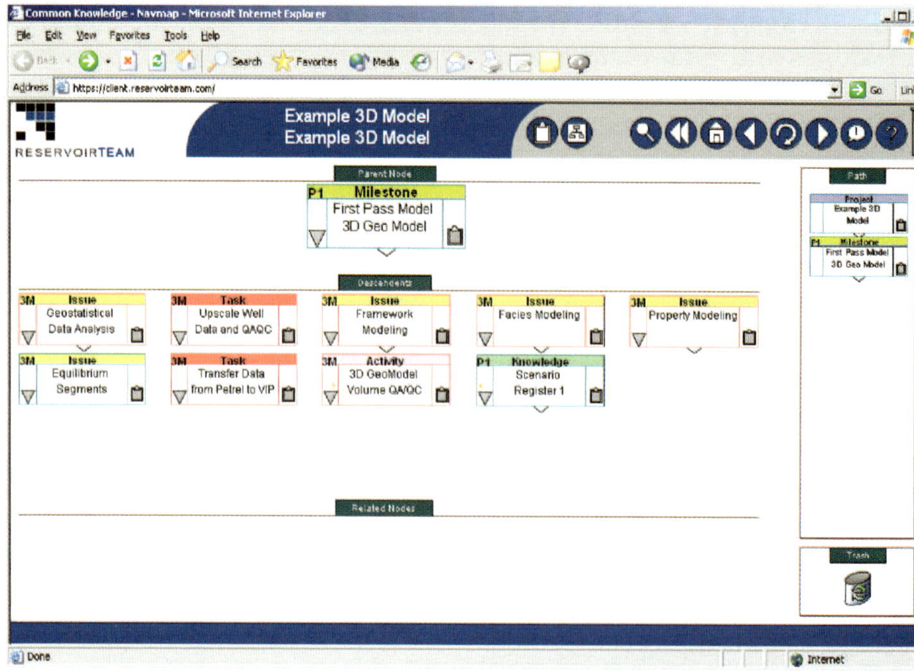


Figure 7. Example of a milestone and related descendants (issues) using Common Knowledge™.

matter of days and customized in a matter of hours. A graphical depiction of workflow (workflow map) conveys extensive information about the specific project approach employed, which can subsequently be used for peer review and mentoring. Moreover, on an individual project basis, the workflow can be assembled in a highly flexible manner because the workflow nodes are derived from well-defined, standard discipline trees. Use of standard discipline trees ensures that the specific project workflow is firmly anchored in a standard set of tasks (workflow nodes), which allows them to be easily located. Of course, templates can be used (and customized) to rapidly execute the workflow steps based on a similar effort.

The workflow maps as outlined above also contain the necessary information to automatically generate

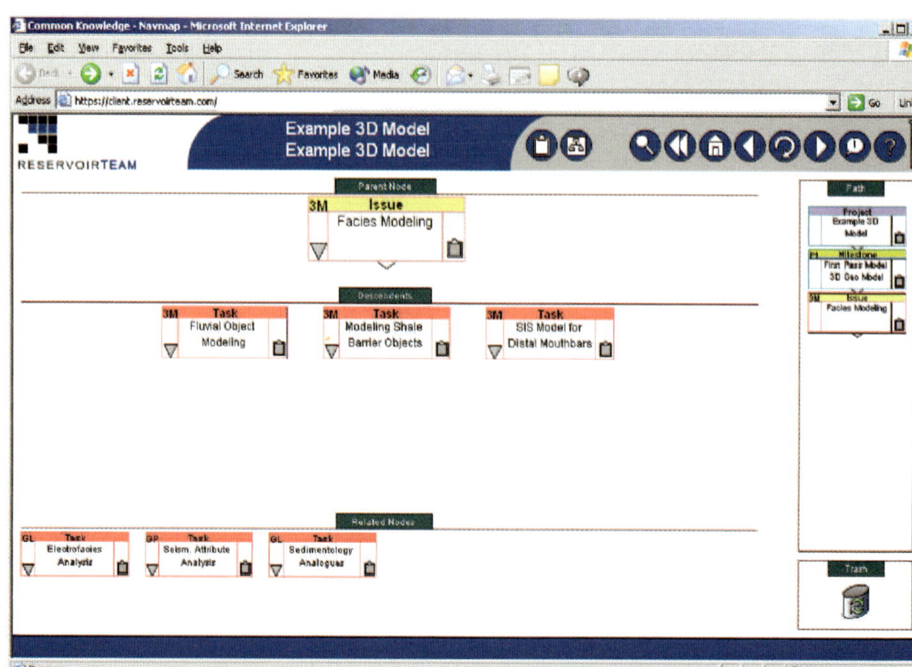


Figure 8. Example of a specific issue and related descendants (issues) using Common Knowledge™.

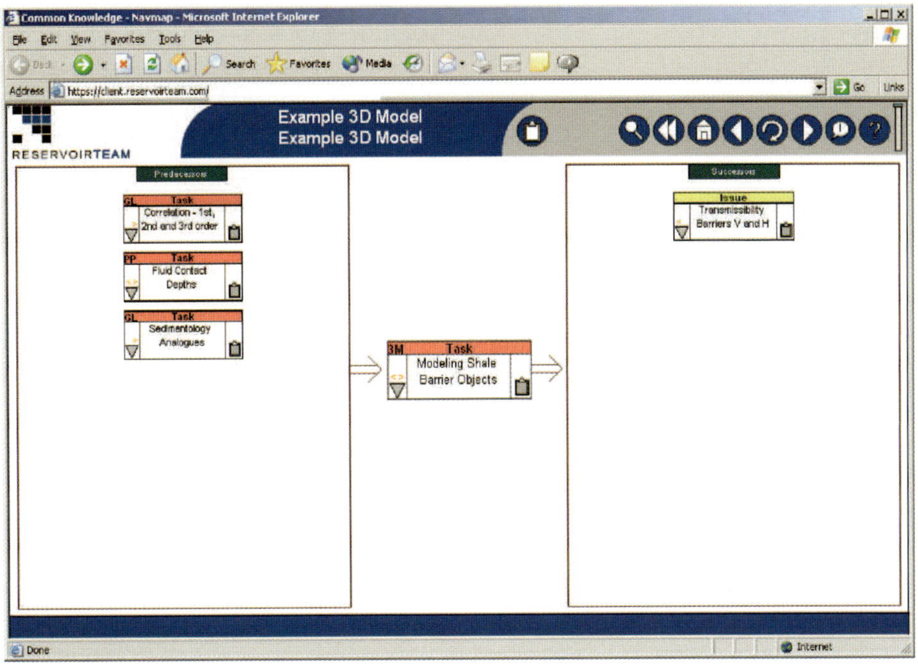


Figure 9. Example showing the dependent workflow nodes (tasks) required to complete prior to resolving a related issue using Common Knowledge™.

project management information for the 3-D modeling effort (e.g., Gant charts). As discussed above, reporting is based on the structure encompassed by the workflow model, and hence, the reported information becomes structurally more transparent and is consistently related to corporate best practices.

## KM AS AN UNCERTAINTY FRAMEWORK FOR 3-D MODELS

Within the framework of workflow-based KM, several concepts and tools are available to systematically record estimated uncertainty and to assess risk. The various relations between workflow nodes (tasks and issues) allow uncertainties and risks for a given task to be combined with uncertainties and risks for related tasks. The goal of this approach is to facilitate the association of uncertainty with a specific task (typically by the specialist who has executed that task and best understands the uncertainty of the results). This approach strives to prevent passing results to subsequent steps of the processing chain without incorporating the related uncertainty.

As an example, consider a simple, two-step workflow process: the collection and investigation of data result in the formation of opinions, followed by conducting analyses that lead to interpretations (Figure 10). An interpretation implies that uncertainty has been narrowed or better understood. For example, suppose task or issue A in Figure 10 consists of determining a facies model. Further, suppose that it has been established that a given sand sequence represents distal interfingered distributary channels and mouth bars, with the possibility that the same sequence could also represent proximal

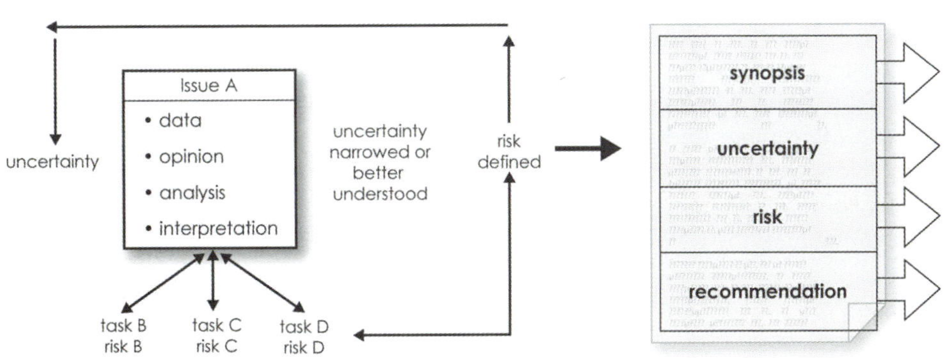


Figure 10. Uncertainty definition and risk assessment for a given workflow node (issue/task) that has relationships with other workflow nodes (issues/tasks).

fluvial channels and large crevasse splays. Hence, an uncertainty associated with the specification of the type of sand (dis)continuity exists. Some tasks or issues typically related to the facies model determination might consist of electrofacies analysis (e.g., task B in Figure 10) and pressure data analysis (e.g., task C in Figure 10). Each of these tasks carries its own uncertainty. For example, inability to clearly discern electrofacies adds to the basic uncertainty in determining the facies model, and ambiguity in the results of pressure data analysis may render the distinction between the proximal fluvial model and the distal interfingered distributary model more indistinguishable. Thus, uncertainty associated with actions related to the main tasks can be compounded (as illustrated by the red lines in Figure 10).

Currently, no operational concept exists for systematically processing uncertainties in the framework of typical 3-D modeling workflows other than documenting them to establish scenarios. The right-hand side of Figure 10 shows how uncertainties can be documented in extensible markup language (XML) format (Yergeau et al., 2004), which at least allows uncertainties to be systematically audited and compiled on an individual project basis or on a multiproject (corporate) basis in repeated workflow situations.

## DISCUSSION AND CONCLUSION

This chapter describes the development and application of a workflow-based KM system for use in 3-D reservoir modeling. The objectives of such a system are to enhance multidisciplinary collaboration and to systematically mine uncertainties along a given workflow path.

The proposed approach has been used in various pilot applications involving 3-D modeling studies, where it has successfully enabled project teams to better communicate results and preserve critical modeling information. The approach has also been applied more broadly within a larger organization, where it has proven to be successful for purposes of workflow training and for rapidly identifying and documenting project progress. However, within larger organizations, formidable cultural changes may be necessary to successfully implement the KM approach.

Application of four-dimensional modeling and the living model approach based on a constant stream of high-density data is expected to become routine in the near future (Chow et al., 2000; Bogan et al., 2003). Much attention is being paid to the impending data-processing requirements and the need for improved throughput of models to handle such large volumes of data. In this situation, the KM approach, as outlined in this chapter, provides an operational mechanism and decision tool by which project teams can avoid information overflow when sifting through even more data and modeling techniques.

Although the implementation costs of a workflow-based KM system are not negligible, such costs are relatively modest in comparison to the typical multimillion-dollar corporate expenditures for information technology. Given the fact that financial performance of assets is related to an appropriate assessment of uncertainty in asset management, uncertainty accountability via KM could certainly be qualified as important enough to warrant the expenditures associated with the methods and suggestions discussed here.

## REFERENCES CITED


Abramson, G., 1999, On the KM midway: Chief Information Officers Enterprise Magazine, Framingham, Massachusetts, May 15, p. 63–70.

Bogan, C., D. Johnson, M. Litvak, and D. Stauber, 2003, Building reservoir models based on 4D-seismic and well data in Gulf of Mexico oil fields: Society of Petroleum Engineers Annual Technical Conference and Exhibition Proceedings, SPE Paper 84370, 11 p.

Chow, C. V., M. C. Arnondin, K. D. Wolcott, and N. D. Ballard, 2000, Managing risks using integrated production models: Journal of Petroleum Technology, v. 52, no. 4, p. 94–99.

Daft, R. L., 2001, Organization theory and design, 7th ed.: Cincinnati, Ohio, South-Western Thomson Learning, p. 239–267.

Drucker, P. F., 1988, The coming of the new organization, Harvard Business Review, January–February, 1988, p. 45–53.

Hansen, M. T., N. Nohria, and T. Tierney, 1999, What's your strategy for knowledge management?: Harvard Business Review, p. 106–116.

Yergeau, F., T. Bray, J. Paoli, C. M. Sperberg-McQueen, E. Maler, 2004, Extensible markup language (XML) 1.0, 3d ed., *in* W3C Recommendation 04 February 2004: http://www.w3.org/TR/2004/REC-xml-20040204 (accessed January 15, 2005).

Coburn, T. C., 2006, Bibliographic resources for geostatistics, spatial analysis, and stochastic modeling in petroleum exploration and development, *in* T. C. Coburn, J. M. Yarus, and R. L. Chambers, eds., Stochastic modeling and geostatistics: Principles, methods, and case studies, volume II: AAPG Computer Applications in Geology 5, p. 325–327.

*Chapter 25*

# Bibliographic Resources for Geostatistics, Spatial Analysis, and Stochastic Modeling in Petroleum Exploration and Development

**T. C. Coburn**
*Abilene Christian University*
*Abilene, Texas, U.S.A.*

## ABSTRACT

The literature on geostatistics has rapidly expanded in the past decade because geostatistical techniques and methodologies have become more mainstream throughout the business, scientific, and engineering communities. The earth sciences continue to be a proving ground for many new geostatistical ideas and developments, building on knowledge and early notions developed throughout a period of 40 yr. The following bibliography contains approximately 1600 citations spanning the history of geostatistics that have a particular bearing on petroleum exploration and development, petroleum engineering and reservoir characterization, and the oil and gas aspects of geological investigation. The bibliography is organized into the seven general categories: case studies; methodological applications; theoretical developments; dissertations and theses; software, software applications, and user manuals; general-interest articles, discussions, and commentaries; and books, reviews of books, and course notes.

## INTRODUCTION

As is the case for other spatial information technologies, interest in geostatistics has experienced phenomenal growth throughout the scientific, engineering, and business communities. That growth continues to be fueled by ongoing research in the earth sciences from which the original notions first emanated. The rate of growth in development, usage, and application of geostatistical methodologies has also been enhanced by the fact that geostatistics has come to be more of a mainstream discipline in its own right, closely linked to the conventional discipline of statistics, yet somewhat distinct. With the expansion of computing capabilities and a pent-up interest in spatial data relationships that are geographic and volumetric in nature, geostatistics has secured an important, influential, and expanding niche. The power and potential of geostatistics and stochastic modeling is nowhere more evident than in the petroleum industry, where it seems that every major service company has implemented major software packages to capture the growing demand for geostatistical analysis of upstream data.


DOI:10.1306/1063826CA51306

The approximately 1600 citations contained in the following bibliography provide extensive evidence of the growing influence of geostatistics in the collection, integration, management, analysis, and interpretation of geoscience data and its particular application in the decision-making aspects of petroleum exploration and development. Selections are taken from a range of sources, but several journals (e.g., *Mathematical Geology, Computers and Geosciences*) and major compilations of articles have received particular emphasis. This collection provides the reader with an extensive array of information that has been disseminated throughout approximately the last 40 yr through active research, scholarly publications, industry reports, project reviews, discussions and presentations, and other professional activities. The bibliography especially focuses on the more recent developments that have not yet been extensively indexed. In addition, the collection includes several entries from the petroleum engineering literature, reflecting the collaborative nature that geostatistical projects commonly assume in actual petroleum exploration, exploitation, and development scenarios.

To assist readers in identifying the most appropriate bibliographic resources for their use, citations are organized into seven broad categories: case studies; methodological applications; theoretical developments; dissertations and theses; software, software applications, and user manuals; general-interest articles, discussions, and commentaries; and books, reviews of books, and course notes. Every attempt has been made to categorize the citations into the best group; but the boundaries, by their very nature, are somewhat fuzzy. In fact, the terms "geostatistics" and "stochastic modeling" are, themselves, sometimes difficult to consistently define and categorize.

Although the geostatistics literature is quite vast, only those citations judged to be most pertinent to petroleum exploration and development have been included. This does not mean, however, that they have no relevance in other business, scientific, or engineering applications (e.g., theoretical and methodological developments are widely applicable to several disparate settings). Likewise, not all the citations are specifically petroleum related, and the relevance of some of them may not be obvious to every reader from the petroleum industry. Many of the articles are theoretical in nature, yet represent important work that has contributed to the evolution of geostatistics as a discipline.

The field of geostatistics, even in the context of earth science, incorporates considerable breadth of application, and so, to enforce some sort of limiting boundary on the extent of the collection, works that are primarily related to specific subdisciplines, such as mining, hydrology, water resources, satellite imagery, environmental science, etc., are omitted. To be sure, the bibliography does include some articles whose origins are in mining, hydrology, and other areas because they also have important applications to petroleum; but as a general rule, this part of the geostatistics literature is not largely represented. Similarly, whereas some aspects of fractal analysis are covered, no attempt has been made to include the entire scope of the fractal literature.

In addition, the terms "spatial analysis" and "stochastic modeling" (as opposed to deterministic modeling) assume different meanings in different contexts, even in the geosciences. For purposes of this bibliography, the terms are taken to be synonymous with geostatistics, and the literature associated with the broader scope of their meanings is not included.

Along with the numerous references to entire articles and reports, several abstracts are cited throughout the bibliography to facilitate the widest possible coverage of information. However, oral and poster presentations from professional or technical meetings for which no published record exists have largely been omitted because of the somewhat incomplete and inconsistent nature of the citations and because of the difficulty of obtaining access to the material in its original form. Every attempt has been made to verify the correctness and completeness of each citation, but some discrepancies do undoubtedly remain.

Finally, not all references cited elsewhere in this text are included here. Similarly, not all references cited in other previously published works are included. The bibliography is meant to be as comprehensive as possible, but it is by no means exhaustive. It is principally targeted at professional geoscientists who want to learn more about the evolution of geostatistics as a discipline and about the practice of geostatistics in petroleum-related ventures. Excellent printed resources and documentation are available elsewhere. Readers are also encouraged to check the Internet, where more information can be found (e.g., www.ai-geostats.org).

For those who require more detailed explanation of terms, excellent glossaries, dictionaries, lists of nomenclature, and explanations of acronyms are already available, and so, no effort has been made in the present context to provide such information. Readers are referred to Alldredge and Alldredge (1978), Journel and Huijbregts (1978), Olea (1991, 1999), Goovaerts (1997), Deutsch and Journel (1998), and Chilès and Delfiner (1999), all of which are cited in the bibliography, as well as the various geostatistical software user guides.

## REFERENCES CITED

Alldredge, J., and N. G. Alldredge, 1978, Geostatistics: A bibliography: International Statistics Review, no. 46, p. 77–88.

Chilès, J.-P., and P. Delfiner, 1999, Geostatistics: Modeling spatial uncertainty: New York, John Wiley & Sons, 695 p.

Deutsch, C. V., and A. G. Journel, 1998, GSLIB, geostatistical software library and user's guide, 2d ed.: New York, Oxford University Press, 369 p.

Goovaerts, P., 1997, Geostatistics for natural resources evaluation: New York, Oxford University Press, 483 p.

Journel, A. G., and C. J. Huijbregts, 1978, Mining Geostatistics: New York, Academic Press, 600 p.

Olea, R. A., 1991, Geostatistical glossary and multilingual dictionary: New York, Oxford University Press, 177 p.

Olea, R. A., 1999, Geostatistics for engineers and earth scientists: Dordrecht, Kluwer Academic Publishers, 303 p.

# Bibliography

## CASE STUDIES

Abbaszadeh, M., O. Takano, H. Yamamoto, T. Shimamoto, N. Yazawa, F. M. Sandria, D. H. Zamora Guerrero, and F. Rodriguez de la Garza, 2003, Integrated geostatistical reservoir characterization of turbidite sandstone deposits in Chicontepec basin, Gulf of Mexico: Society of Petroleum Engineers Annual Technical Conference and Exhibition, SPE Paper 84052.

Abry, C. G., 1975, Geostatistical model for predicting oil: Tatum basin, New Mexico: AAPG Bulletin, v. 59, p. 2111–2122.

Adams, R. D., M. Nikravesh, J. W. Collister, D. D. Ekart, R. A. Levey, M. Sippel, B. J. Wilson, and S. K. Raeuchle, 1999, Integrated geoscience evaluation of carbonate reservoirs in the Permian basin: Ellenberger phase II: Chicago, Gas Research Institute, GRI-99/0065 (+CD-ROM), 324 p.

Alabert, F. G., and B. Corre, 1991, Heterogeneity in a complex turbiditic reservoir: Impact on field development: Society of Petroleum Engineers Annual Technical Conference and Exhibition, SPE Paper 22902.

Alabert, F. G., and G. J. Massonnat, 1990, Heterogeneity in a complex turbiditic reservoir: Stochastic modeling of facies and petrophysical variability: Society of Petroleum Engineers Annual Technical Conference and Exhibition, SPE Paper 20604, 15 p.

Alexander, D. W., and R. W. Jeffrey, 1998, Geostatistical integration of 3-D seismic with gravity and well bore data to identify and estimate additional reserves in the Ghawar field, Saudi Arabia (abs.), *in* Conference abstracts, GEO '98, 3rd Middle East Geoscience Conference and Exhibition: GeoArabia (Manama), v. 3, p. 58.

Almeida, A. S., and P. Frykman, 1994, Geostatistical modeling of chalk reservoir properties in the Dan field, Danish North Sea, *in* J. M. Yarus and R. L. Chambers, eds., Stochastic modeling and geostatistics: Principles, methods, and case studies: AAPG Computer Applications in Geology 3, p. 273–286.

Aly, A., W. J. Lee, A. Datta-Gupta, K. Mowafi, M. Prida, and M. Latif, 1999, Application of geostatistical modeling in an integrated reservoir simulation study of the lower Bahariya reservoir, Egypt: Society of Petroleum Engineers Middle East Oil Show and Conference (Bahrain), SPE Paper 53118.

Arnold, B. W., J. M. Herrin, and P. M. Basinski, 2004, Statistical and geostatistical analyses of coalbed methane production in the Raton basin, New Mexico and Colorado (abs.): AAPG Annual Meeting Program, v. 13, http://www.searchanddiscovery.com/documents/abstracts/annual2004/Dallas/Arnold.htm (accessed June 22, 2006).

Arpat, B., J. Caers, and S. Strebelle, 2002, Feature-based geostatistics: An application to a submarine channel reservoir: Society of Petroleum Engineers Annual Technical Conference and Exhibition, SPE Paper 77426.

Ates, H., A. Bahar, S. El-Abd, M. Charfeddine, M. Kelkar, and A. Datta-Gupta, 2005, Ranking and upscaling of geostatistical reservoir models using streamline simulation: A field case study: SPE Reservoir Evaluation of Engineering, v. 8, p. 22–32.

Azpiritxaga, I., A. Correa, E. Hernandez, A. Galli, C. Ravenne, C. Castillo, E. Gonzalez, G. Gedler, and C. Vasquez, 1998, Seismic constraints for a stochastic reservoir model, C6 Eocene, block I, Lake Maracaibo, Venezuela (abs.): AAPG Bulletin, v. 82, p. 1888–1889.

Balke, S. C., and J. C. Currie, 2000, Modeling of Bayu-Undan field (abs.): AAPG Bulletin, v. 84, p. 1401.

Baruffini, L., F. Fonnesu, E. Tabaldi, and B. Doligez, 1993, Geological and geostatistical analysis of the Cajigar fluvio-lacustrine reservoir system: An outcrop study (abs.), *in* R. Eschard and B. Doligez, eds., Subsurface reservoir characterization from outcrop observations, Proceedings of the 7th Instituto Français du Pétrole Exploration and Production Research Conference: Paris, Editions Technip, p. 167.

Batista, A. C., A. J. Sousa, M. J. Batista, and L. Viegas, 1997, Factorial kriging with external drift: A case study on the Penedono region, Portugal, *in* V. Pawlowsky-Glahn, ed., Proceedings of IAMG '97, 3rd Annual Conference of the International Association for Mathematical Geology: Barcelona, International Center for Numerical Methods in Engineering, p. 687–692.

Bekele, E. B., 1998, Can geostatistical methods be used to infer long-range oil migration through carrier beds? A case study from the Alberta Basin, Canada (abs.): AAPG Annual Meeting Program, v. 1, p. A50.

Benkendorfer, J. P., C. V. Deutsch, P. D. LaCroix, and L. H. Landis, 1995, Integrated reservoir modelling of a major Arabian carbonate reservoir: Society of

Petroleum Engineers Middle East Oil Show, Bahrain, SPE Paper 29869.

Beucher, H., F. Fournier, B. Doligez, and J. Rozanski, 1999, Using 3D seismic-derived information in lithofacies simulations: A case study: Society of Petroleum Engineers Annual Technical Conference and Exhibition, SPE Paper 56736.

Biver, P. Y. A., R. J. Norris III, J. J.-M. Vittori, and F. G. Alabert, 1994, Geostatistical modelling of heterogeneities in a mixed carbonate platform: The characterization of complex progradational and retrogradational mechanisms: Society of Petroleum Engineers Annual Technical Conference and Exhibition, SPE Paper 28418.

Boemer, S. T., S. C. Leininger, A. Sunaryo, W. Alexander, J. Gould, M. Sulitra, B. Sabin, and M. W. Dobin, 1997, Analysis, inversion, geostatistical simulation, and visualization of porosity in a carbonate gas reservoir, *in* Proceedings, 29th Annual Offshore Technology Conference, Richardson, Texas, Offshore Technology Conference, OTC 8288, p. 127–140.

Bridges, R. A., and J. W. Castle, 2004, Application of geostatistical methods to reservoir characterization: An example from the West Coalinga field, San Joaquin basin, California (abs.): AAPG Pacific Section Meeting Abstracts, May 9–12, Bakersfield, California, http://www.searchanddiscovery.com/documents/abstracts/2004pacific/bridges.pdf (accessed June 22, 2006).

Brown, J. M. G., and N. Exakwe, 1998, Improved reservoir management using geostatistical reservoir description, 24Z Stevens reservoir, Elk Hills oil field, California: Society of Petroleum Engineers Western Regional Meeting (Bakersfield, California), SPE Paper 46232.

Brummert, A. C., S. E. Pool, M. E. Portman, J. S. Hancock, and J. R. Ammer, 1991, Determining optimum estimation methods for interpolation and extrapolation of reservoir properties: A case study, *in* L. W. Lake, H. B. Carroll Jr., and T. C. Wesson, eds., Reservoir characterization II: San Diego, Academic Press, p. 644–650.

Caers, J., S. Srinivasan, and A. G. Journel, 2000, Geostatistical quantification of geological information for a fluvial-type North Sea reservoir: SPE Reservoir Evaluation and Engineering, v. 3, p. 457–467.

Caers, J., P. Avseth, and T. Mukerji, 2001, Geostatistical integration of rock physics, seismic amplitudes, and geologic models in North Sea turbidite systems: The Leading Edge, v. 20, p. 308–312.

Caers, J., S. Strebelle, and K. Payrazyan, 2003, Stochastic integration of seismic data and geological scenarios: A west Africa submarine channel saga: The Leading Edge, v. 22, p. 192–197.

Calatayud, P. M., F. M. Petit, P. Tessier, R. J. Norris, and F. G. Alabert, 1994, Combined stochastic and deterministic reservoir characterization leads to faster history matching of a horizontal well in a complex fluvio-lacustrine environment: Society of Petroleum Engineers Annual Technical Conference and Exhibition, SPE Paper 28585.

Carnes, A., J. M. Yarus, P. Cordova, M. Delgado, H. Rodriguez, P. J. Black, R. L. Brown, K. Kramer, K. Yang, and L. D. Green, 1999, The Barua field, Venezuela: Comparison of the results and process of deterministic versus stochastic reservoir characterization: Society of Petroleum Engineers Annual Technical Conference and Exhibition, SPE Paper 56656.

Carr, L. A., R. I. Bentreau, M. P. Corrigan, and G. G. Van Doorne, 1987, The successful characterization of a complex reservoir using 3-D seismic, geostatistical reservoir description and sponge core analysis: Society of Petroleum Engineers Annual Technical Conference and Exhibition, SPE Paper 16780.

Casar-Gonzalez, R., and V. Suro-Perez, 2004, Stochastic simulation of vuggy porous media in the KT boundary carbonate breccia from the Campeche Sound (abs.): AAPG Annual Meeting Program, v. 13, http://www.searchanddiscovery.com/documents/abstracts/2004intl_cancun/short/A89999.pdf (accessed June 22, 2006).

Chavez-Cerna, M., and C. Bianchi-Ramirez, 1998, Geostatistics applied to a reservoir study in northwestern Peru, Talara basin (abs.): AAPG Bulletin, v. 82, p. 1901.

Chidsey Jr., T. C., P. B. Anderson, T. H. Morris, J. A. Dewey Jr., A. Mattson, C. B. Foster, S. H. Snelgrove, and T. A. Ryer, 1998, Geology and petrophysical characterization of the Ferron sandstone for 3-D simulation of a fluvial-deltaic reservoir: Bartlesville, Oklahoma, U.S. Department of Energy, Bartlesville Project Office, Report DOE/BC/14896-22, 77 p.

Chihi, H., A. Galli, D. Renard, and C. Ravenne, 1994, From sequence and seismic stratigraphy to sedimentary quantification and modelling using nonstationary geostatistics (Rhone delta, France), *in* Proceedings of the 4th National Conference on Law, Computers, and Artificial Intelligence: Exeter, University of Exeter, p. 58–59.

Chu, J., W. Xu, H. Zhu, and A. G. Journel, 1991, The

Amoco case study: Stanford Center for Reservoir Forecasting Report 4, 77 p.

Chu, J., W. Xu, and A. G. Journel, 1994, 3-D implementation of geostatistical analyses—The Amoco case study, *in* J. M. Yarus and R. L. Chambers, eds., Stochastic modeling and geostatistics: Principles, methods, and case studies: AAPG Computer Applications in Geology 3, p. 201–216.

Clifton, P. M., and S. P. L. Neuman, 1982, Effects of kriging and inverse modeling on conditional simulation of the Avra Valley acquifer in southern Arizona: Water Resources Research, v. 18, p. 1214–1234.

Coburn, T. C., J. M. Yarus, and J. A. May, 1996, A case study in geostatistical modeling for petroleum reservoir description, *in* H. T. Mowrer, R. L. Czaplewski, and R. H. Hamre, eds., Proceedings, 2nd International Symposium on Spatial Accuracy Assessment in Natural Resources and Environmental Sciences: Ft. Collins, U.S. Department of Agriculture, Forest Service, General Technical Report RM-GTR-277, p. 4–15.

Coburn, T. C., J. M. Yarus, R. D. Winn, and R. M. Srivastava, 1998, An application of geostatistical simulation techniques to petroleum reservoir characterization, *in* Proceedings of the section on statistics and the environment: Alexandria, American Statistical Association, p. 124–129.

Cox, D. L., S. J. Lindquist, C. L. Bargas, K. G. Havholm, and R. M. Srivastava, 1994, Integrated modeling for optimum management of a giant gas condensate reservoir, Jurassic eolian Nugget Sandstone, Anschutz Ranch east field, Utah overthrust (U.S.A.), *in* J. M. Yarus and R. L. Chambers, eds., Stochastic modeling and geostatistics: Principles, methods, and case studies: AAPG Computer Applications in Geology 3, p. 287–321.

Damsleth, E., V. Sangolt, and G. Aamodt, 1998, Subseismic faults can seriously affect fluid flow in the Njord field off western Norway—A stochastic fault modeling case study: Society of Petroleum Engineers Annual Technical Conference and Exhibition, SPE Paper 49024.

Damsleth, E., C. B. Tjolsen, H. Omre, and H. H. Haldorsen, 1992, A two-stage stochastic model applied to a North Sea reservoir: Journal of Petroleum Technology, v. 44, p. 402–408.

de Fouquet, C., H. Beucher, A. Galli, and C. Ravenne, 1998, Conditional simulation of random sets: Application to an argillaceous sandstone reservoir, *in* M. Armstrong, ed., Geostatistics, Proceedings of the 3rd International Geostatistics Congress: Dordrecht, The Netherlands, Kluwer Academic Publishers, p. 517–530.

Dehghani, K., P. M. Harris, K. A. Edwards, and W. T. Dees, 1999, Modeling a vuggy carbonate reservoir, McElroy field, west Texas: AAPG Bulletin, v. 83, p. 19–42.

Desbarats, A. J., 1988, Estimation of effective permeabilities in the lower Stevens Formation of the Paloma field, San Joaquin Valley, California: SPE Reservoir Engineering, v. 3, p. 1301–1307.

de Souza, M. J., and D. Marcotte, 1990, In-situ geostatistical oil reserve estimation, zone 2 of Carmopolis oil field, Brazil: Society of Petroleum Engineers of the American Institute of Mining Metallurgical and Petroleum Engineers, SPE Paper 21616.

Di Julio, S. S., 1993, North Vickers East waterflood performance prediction with fractal geostatistics: SPE Formation Evaluation, v. 8, p. 89–95.

Dixon, M. W., S. C. Meyer, and P. S. Hansel II, 1998, Reservoir modeling and simulation of the Main Pass 299 salt dome, Gulf of Mexico (abs.): AAPG Annual Meeting Program, v. 1, p. A162.

Doligez, B. M., F. Fournier, G. Jolivet, S. Gancarski, and H. Beucher, 2002, Seismic facies map integration geostatistical geological model—A field case (abs.): European Association of Geoscientists and Engineers (EAGE) Annual Conference and Technical Exhibition, p. 215.

Dowd, P. A., and A. G. Royle, 1977, Geostatistical applications in the Athabasca tar sands, *in* APCOM '77, Proceedings of the 15th International Symposium on The Application of Computers and Operations Research in the Mineral Industry: Australasian Institute of Mining and Metallurgy, p. 235–242.

Doyen, P. M., D. E. Psaila, and R. M. Strandenes, 1994, Bayesian sequential indicator simulation of channel sands from 3-D seismic data in the Oseberg field, Norwegian North Sea: Society of Petroleum Engineers Annual Technical Conference and Exhibition, SPE Paper 28382.

Doyen, P. M., L. D. den Boer, and W. R. Pillet, 1996a, Seismic porosity mapping in the Ekofisk field using a new form of collocated kriging: Society of Petroleum Engineers Annual Technical Conference and Exhibition, SPE Paper 36498.

Doyen, P. M., L. D. den Boer, and W. R. Pillet, 1996b, Seismic porosity mapping in the Ekofisk field using Bayesian stochastic simulation (abs.): European Association of Geoscientists and Engineers, 58th Annual Conference and Technical Exhibition, Extended Abstracts Book, p. L030.

Dull, D. W., 1995, Reservoir characterization and the application of geostatistics to three-dimensional modeling of a shallow-ramp carbonate, Mabee San Andres field, Andrews and Martin counties, Texas, in Hydrocarbon reservoir characterization: Geologic framework and flow unit modeling, *in* E. L. Stoudt and P. M. Harris, eds., SEPM Short Course Notes 34, p. 155–198.

Dull, D. W., W. S. Meddaugh, and P. Montgomery, 2003, Wafra first Eocene reservoir model—Application of sequence stratigraphy and geostatistics (abs.): AAPG Annual Meeting Program, v. 12, http://www.searchanddiscovery.com/documents/abstracts/annual2003/short/78086.PDF (accessed June 22, 2006).

Eidsvik, J., H. Omre, T. Mukerji, G. Mavko, and P. Avseth, 2002, Seismic reservoir prediction using Bayesian integration of rock physics and Markov random fields: A North Sea example: The Leading Edge, v. 21, p. 290–294.

Eisenberg, R. A., P. M. Harris, C. W. Grant, D. J. Goggin, and F. J. Conner, 1994, Modeling reservoir heterogeneity within outer ramp carbonate facies using an outcrop analog, San Andres Formation of the Permian basin: AAPG Bulletin, v. 78, p. 1337–1359.

Eschard, R., C. Bacchiana, G. Désaubliaux, J. M. Chautru, P. Lemouzy, and J. Parpant, 1993, Geostatistical simulations of alluvial sandbodies in the Triassic series of the Chaunoy field (abs.): AAPG Bulletin, v. 77, p. 1621.

Eschard, R., P. Lemouzy, C. Bacchiana, G. Désaubliaux, J. Parpant, and B. Smart, 1998, Combining sequence stratigraphy, geostatistical simulations, and production data for modeling a fluvial reservoir in the Chaunoy field (Triassic, France): AAPG Bulletin, v. 82, p. 545–568.

Fanchi, J. R., Hai-Zui Meng, R. P. Stoltz, and M. W. Owen, 1996, Nash reservoir management study with stochastic images—A case study: SPE Formation Evaluation, v. 11, p. 155–161.

Feasel, C. T., H. H. Nielsen, B. Agarwal, and S. C. Key, 1999, Redevelopment of Ekofisk field, North Sea, based on reservoir characterization and innovative well design (abs.): AAPG Bulletin, v. 83, p. 1311.

Fisher, R. S., N. Tyler, M. D. Barton, M. A. Miller, K. Sepehrnoori, J. Holder, and K. E. Gray, 1992, Quantification of flow unit and bounding element properties and geometries, Ferron sandstone, Utah: Implications for heterogeneity in Gulf Coast Tertiary deltaic reservoirs: Chicago, Gas Research Institute, GRI-92/0072, 171 p.

Fisher, W. C., D. J. Bond, G. Skeith, and N. Bonnett, 1994, Targeting the Guinevere horizontal well: A case study involving geostatistics, reservoir description, reservoir simulation, and drilling engineering: Society of Petroleum Engineers European Petroleum Conference (London), p. 109–124.

Flores, D., and R. Gonzalez, 1998, Geostatistical analysis of petrophysical variables influencing reservoir quality in the fractured "O limestone," Maporal field, western Venezuela, *in* Transactions of the 39th Annual Logging Symposium: Society of Professional Well Log Analysts, v. 39, Paper EEE, 12 p.

Fogg, G. E., 1989, Stochastic analysis of acquifer interconnectedness: Wilcox Group, Trawick area, East Texas: Bureau of Economic Geology Report of Investigations 189, 68 p.

Fogg, G. E., and F. J. Lucia Jr., 1990, Reservoir modeling of restricted platform carbonates: Geologic/geostatistics characterization of interwell-scale reservoir heterogeneity, Dune field, Crane County, Texas: Bureau of Economic Geology Report of Investigations 190, 66 p.

Fogg, G. E., F. J. Lucia, and R. K. Senger, 1991, Stochastic simulation of interwell-scale heterogeneity for improved prediction of sweep efficiency in a carbonate reservoir, *in* L. W. Lake, H. B. Carroll Jr., and T. C. Wesson, eds., Reservoir characterization II: San Diego, Academic Press, p. 355–381.

Fournier, F., 1995, Integration of 3-D seismic data in reservoir stochastic simulations: A case study: Society of Petroleum Engineers Annual Technical Conference and Exhibition, SPE Paper 30564, 14 p.

Galikeev, T., and T. Davis, 2003, Reservoir modeling of a thin carbonate reservoir using geostatistics and 4-D seismic attribute, Weyburn field, Saskatchewan, Canada (abs.): Society of Exploration Geophysicists Expanded Abstracts 22, p. 1354.

Galli, A., and G. Meunier, 1987, Study of a gas reservoir using the external drift method, *in* G. Matheron and M. Armstrong, eds., Geostatistical case studies: Dordrecht, The Netherlands, D. Reidel Publishing Company, p. 105–109.

Gastaldi, C., 2000, Reservoir characterization from seismic attributes; an example from Peciko field (Indonesia) (abs.): AAPG Bulletin, v. 84, p. 1429.

Giudicelli, C. B., F. G. Alabert, G. J. Massonnat, and B. Corre, 1992, Anguille Marine, a deep-sea fan reservoir offshore Gabon: From stochastic modelling toward history matching: Society of Petroleum Engineers Annual Technical Conference and Exhibition, SPE Paper 24894.

Goggin, D. J., R. A. Eisenberg, C. W. Grant, and P. M. Harris, 1993, Stratigraphic controls on permeability and fluid flow patterns in ramp carbonates: An outcrop study of the San Andres Formation, New Mexico, U.S.A., *in* R. Eschard and B. Doligez, eds., Subsurface reservoir characterization from outcrop observations: Institut Francais du Petrole, Collection Colloques et Séminaires 51, p. 65–95.

Goldhammer, R. K., P. J. Lehmann, and P. A. Dunn, 1993, The origin of high-frequency platform carbonate cycles and third-order sequences (Lower Ordovician El Paso gap, west Texas): Constraints from outcrop data and stratigraphic modeling: Journal of Sedimentary Petrology, v. 63, p. 318–359.

Gomez-Hernandez, J. J., 1991, Three dimensional multiple indicator conditional simulations of Florida's Jay oilfield: Society of Petroleum Engineers International Petroleum Equipment and Technology Exhibition, Beijing, SPE Paper 22331.

Gomez Nacarino, C., and M. Rodrigo Viñuela, 2003, 3D seismic attributes and geostatistical techniques designed to enhance interpretation: An experience in the Illizi basin (abs.): AAPG Hedberg Research Conference Abstracts, February 18–20, Algiers, http://www.searchanddiscovery.com/documents/abstracts/hedberg2003algeria/allabstracts/Gomez.pdf (accessed June 22, 2006).

Gould, G. E., D. W. Green, G. P. Willhite, R. A. Phares, and A. W. Walton, 1993, Reservoir description and modelling of the Pen field, Graham County, Kansas, *in* B. Linville, T. E. Burchfield, and T. C. Wesson, eds., Reservoir characterization III: Tulsa, PennWell Books, p. 715–744.

Grant, C. W., D. J. Goggin, and P. M. Harris, 1994, Outcrop analog for cyclic-shelf reservoirs, San Andres Formation of Permian basin: Stratigraphic framework, permeability distribution, geostatistics, and fluid-flow modeling: AAPG Bulletin, v. 78, p. 23–54.

Guargena, C. G., G. B. Smith, J. Wardell, B. V. Nystrand, T. H. Nilsen, and T. M. Hegre, 2002, Baffles or barriers—Stretching the recovery envelope by introducing stochastic heterogeneity to the analysis of Palaeocene deep-marine fans in the Jotun field, offshore Norway (abs.): AAPG Annual Meeting Program, v. 11, http://www.searchanddiscovery.com/documents/abstracts/annual2002/DATA/2002/13ANNUAL/EXTENDED/24994.pdf (accessed June 22, 2006).

Guerreiro, L., A. da Costa e Silva, and A. Soares, 2000a, Integrated reservoir characterization of a fractured carbonate reservoir: Society of Petroleum Engineers International Petroleum Conference and Exhibition (Villahermosa, Mexico), SPE Paper 58995.

Guerreiro, L., A. Costa Silva, V. Alcobia, and A. Soares, 2000b, Integrated reservoir characterization of a fractured carbonate reservoir: Society of Petroleum Engineers Annual Technical Conference and Exhibition, SPE Paper 58995, 4 p.

Harper, W. V., and J. M. Furr, 1986, Geostatistical analysis of potentiometric data in the Wolfcamp aquifer of the Palo Duro basin, Texas: Technical Report BMI/ONWI-587, Batelle Memorial Institute, Columbus, Ohio, 106 p.

Hernandez Garcia, A. M., A. Hevia, P. Sanchez, J. Crespo, and R. Viloria, 1999, Facies architecture and reservoir heterogeneity in the B-7-X.14 oil reservoir, Tia Juana/Lagunillas fields, western Venezuela (abs.): AAPG Annual Meeting Program, v. 8, p. A58.

Hirsche, K., L. Mewhort, J. Porter-Hirsche, and R. Davis, 1996, Geostatistical reservoir characterization of a Canadian reef or the use and abuse of geostatistics: Canadian Society of Exploration Geophysicists Recorder, December, p. 26–28.

Hirsche, K., G. Bitoun, C. Gastaldi, and P.-L. Pichon, 1998, Geostatistical characterization of fluvial reservoir system; an example from Southeast Asia, *in* B. Kazokas and B. Luyties, chairmen, Proceedings of the 30th Annual Offshore Technology Conference: Richardson, Texas, Offshore Technology Conference, p. 71–81.

Hohn, M. E., 1985, Geostatistical analysis of Devonian shale stratigraphic, production, and completion data in West Virginia: Annual report (1 July 1984–30 June 1985): Chicago, Gas Research Institute, GRI-85/0228, PB90-178930, 36 p.

Hohn, M. E., 1986, Geostatistical analysis of Devonian shale stratigraphic, production, and completion data in West Virginia: Annual report (1 July 1985–30 June 1986): Chicago, Gas Research Institute, GRI-85/0243, PB87-134086, 35 p.

Hohn, M. E., 1993a, Measuring and predicting reservoir heterogeneity in complex deposystems: The fluvial-deltaic Big Injun sandstone in West Virginia: Morgantown, West Virginia University, Appalachian Oil and Gas Research Consortium, DOE/BC/14657-15, U.S. Department of Energy Contract DE-AC22-90BC14657, 362 p.

Hohn, M. E., 1993b, Petroleum geology and reservoir characterization of the Big Injun sandstone (Price Formation) in the Granny Creek field, Clay and Roane counties, West Virginia: Morgantown,

West Virginia Geological and Economic Survey, B-44, 91 p.

Hohn, M. E., and D. W. Neal, 1986, Geostatistical analysis of gas potential in Devonian shales of West Virginia: Computers and Geosciences, v. 12, p. 611–617.

Hohn, M. E., R. R. McDowell, D. L. Matchen, and A. G. Vargo, 1997, Heterogeneity of fluvial-deltaic reservoirs in the Appalachian basin: A case study from a lower Mississippian oil field in central West Virginia: AAPG Bulletin, v. 81, p. 918.

Høye, T., E. Damsleth, and K. Hollund, 1994, Stochastic structural modeling of Troll West, with special emphasis on the thing oil zone, *in* J. M. Yarus and R. L. Chambers, eds., Stochastic modeling and geostatistics: Principles, methods, and case studies: AAPG Computer Applications in Geology 3, p. 217–239.

Hu, L. Y., R. Basquet, S. Jenni, M. C. Cacas, B. Bourbiaux, and J. M. Daniel, 2004, An integrated workflow for history matching of stochastic models of faulted and fractured reservoirs (abs.): AAPG Hedberg Research Conference Abstracts, March 15–18, El Paso, Texas, http://www.searchanddiscovery.com/documents/abstracts/hedberg2004elpaso/extended/hu/hu.htm (accessed June 22, 2006).

Hwang, L.-F., and D. McCorkindake, 1994, Troll field depth conversion using geostatistically derived average velocities: The Leading Edge, v. 13, p. 262–269.

Idrobo, E. A., A. Malalla, A. Datta-Gupta, and J. O. Parra, 1999, Characterizing fluid saturation distribution using cross-well seismic and well data: A geostatistical study: Society of Petroleum Engineers Annual Technical Conference and Exhibition, SPE Paper 56515.

Jackson, S. R., L. Tomutsa, M. Szpakiewicz, M. M. Chang, M. M. Honarpour, and R. A. Schatzinger, 1991, Construction of a reservoir model by integrating geological and engineering information—Bell Creek field, a barrier/strandplain reservoir, *in* L. W. Lake, H. B. Carroll Jr., and T. C. Wesson, eds., Reservoir characterization II: San Diego, Academic Press, p. 526–556.

Janele, P. T., T. H. Tankersley, G. H. Schmit, B. C. Wibowo, H. A. Rahardja, and W. C. Dawson, 2000, Stochastic facies modeling at Kotabatak field—central Sumatra basin, *in* Proceedings of the 27th Annual Convention: Indonesian Petroleum Association, p. 19–38.

Jean, G., J. M. Yarus, M. K. Harris, G. P. Flach, M. R. Millings, and F. H. Syms, 2003, Geological 3-D modeling of southeastern tertiary coastal plain sediments, Savannah River site, South Carolina: An applied geostatistical approach (abs.): AAPG Annual Meeting Program, v. 12, http://www.searchanddiscovery.com/documents/abstracts/annual2003/short/78512.PDF (accessed June 22, 2006).

Jennings Jr., J. W., 2000, Spatial statistics of permeability data from carbonate outcrops of west Texas and New Mexico: Implications for improved reservoir modeling: Bureau of Economic Geology Report of Investigations 258, 50 p.

Johann, P., F. Fournier, O. Souza, R. Eschard, and H. Beucher, 1996, 3D stochastic reservoir modeling constrained by well and seismic data on a turbidite field: Society of Petroleum Engineers Annual Technical Conference and Exhibition, SPE Paper 36501, 16 p.

Joseph, P., L. Y. Hu, O. Dubrule, D. Claude, P. Crumeyrolle, J. L. Lesuer, and H. J. Soudet, 1993, The Roda deltaic complex (Spain); from sedimentology to reservoir stochastic modeling, *in* R. Eschard and B. Doliguez, eds., Subsurface reservoir characterization from outdrop observations: Institut Francais du Petrole, Collection Colloques et Séminaires 51, p. 97–109.

Joseph, P., O. G. de Souza Jr., R. Eschard, D. Granjeon, O. Lerat, and C. Ravenne, 1999, Characterization of the 3-D architecture of deep-water reservoirs from outcrop analogues in the French Alps and application to the Namorado field (offshore Brazil) (abs.): AAPG Bulletin, v. 83, p. 1321.

Journel, A. G., and J. J. Gomez-Hernandez, 1993, Stochastic imaging of the Wilmington clastic sequence. SPE Formation Evaluation, v. 8, p. 33–40.

Kabbabe, T. M., J. Rendon, J. Betancourt, R. Muñoz, and V. Murgueytio, 2003, Stochastic model of the S5-T sand, Melones field: A new image of an old heavy oil reservoir (abs.): AAPG Annual Meeting Program, v. 12, http://www.searchanddiscovery.com/documents/abstracts/annual2003/short/72700.PDF (accessed June 22, 2006).

Kara, B., E. Kasap, and R. W. Tillman, 1993, Quantitative description of Muddy Sandstone: Geostatistical parameters, automated facies generations, and measurement methodologies: Society of Petroleum Engineers Annual Technical Conference and Exhibition, SPE Paper 26488.

Kelkar, M., and D. Richmond, 1996, Implementation of reservoir management plan—Self unit, Glenn Pool field: Society of Petroleum Engineers SPE/

Department of Energy 10th Symposium on Improved Oil Recovery, SPE/DOE 35407, 7 p.

Kittridge, M. G., L. W. Lake, F. J. Jucia, and G. E. Fogg, 1990, Outcrop/subsurface comparisons of heterogeneity in the San Andres Formation: SPE Formation Evaluation, v. 5, p. 233–240.

Knight, J., B. Slevinsky, R. Benjamin, and J. Costello, 1998a, Reservoir characterization and non-stationary modeling of an incised valley-fill, Crystal Viking field (Lower Cretaceous), Alberta, Canada (abs.): AAPG Annual Meeting Program, v. 2, p. A367.

Knight, J., B. Slevinsky, H. Wishart, M. Skaug, L. Wilcox, and B. Haverslew, 1998b, Reservoir characterization of a braid-plain/braid-delta using a non-stationary geostatistical approach, Terra Nova oil field, Newfoundland, Canada (abs.): AAPG Annual Meeting Program, v. 2, p. A365.

Kongwung, B., and S. Ronghe, 2000, Reservoir identification and characterization through sequential horizon mapping and geostatistical analysis; a case study from the Gulf of Thailand: Petroleum Geoscience, v. 6, p. 47–57.

Kupecz, J. A., T. R. Svor, G. E. Drake, G. P. Haughey, N. J. Andrews, and K. Habib, 2004, Modeling dual porosity carbonate reservoirs: Integration of core, well log, 3D seismic and well bore image data for stochastic modeling of fault and fracture systems, Al Rayyan field, Qatar (abs.): AAPG International Conference Abstracts, October 24–27, Cancun, http://www.searchanddiscovery.com/documents/abstracts/2004intl_cancun/short/A91515.pdf (accessed June 22, 2006).

Langlais, V., and J. Doyle, 1993, Comparison of several methods of lithofacies simulation on the fluvial Gypsy Sandstone of Oklahoma, *in* A. Soares, ed., Geostatistics troia '92: Dordrecht, The Netherlands, Kluwer Academic Publishers, p. 299–310.

Lemouzy, P., 1999, Evaluation of multiple geostatistical models using scaling-up with coarse grids: A practical example, *in* R. Schatzinger and J. Jordan, eds., Reservoir characterization—Recent advances: AAPG Memoir 71, p. 277–292.

Lemouzy, P. M., J. Parpant, R. Eschard, C. Bacchiana, I. Morelon, and B. Smart, 1995, Successful history matching of Chaunoy field reservoir behavior using geostatistical modeling: Society of Petroleum Engineers Annual Technical Conference and Exhibition, SPE Paper 30707, 6 p.

Lerat, O., B. Doligez, R. Eschard, J. Knight, B. Slevinsky, and G. Sullivan, 1998, Outcrop study and quantification of the heterogeneities of a reservoir analog; the Castlegate Sandstone (Book Cliffs, Utah) (abs.): AAPG Annual Meeting Program, v. 2, p. A397.

Levander, A., J. A. Goff, K. Holliger, R. W. England, S. K. Smith, and R. W. Hobbs, 1994, Stochastic characterization and seismic response of upper and middle crustal rocks based on the Lewisian gneiss complex, Scotland: Geophysical Journal International, v. 199, p. 243–259.

Li, H., 2003, Geostatistical shale models for a deltaic reservoir analog: From 3D GPR data to 3D flow modeling (abs.): AAPG Annual Meeting Program, v. 12, http://www.searchanddiscovery.com/documents/abstracts/annual2003/extend/74337.PDF (accessed June 22, 2006).

Li, H., and C. D. White, 2003, Geostatistical models for shales in distributary channel point bars (Ferron sandstone, Utah): From ground-penetrating radar data to three-dimensional flow modeling: AAPG Bulletin, v. 87, p. 1851–1868.

Liu, J. S., R. E. Enwall, R. R. Davis, J. L. Douglas, and R. M. Barber, 1993, Integration of engineering data and geostatistics in simulation of a complex carbonate reservoir, *in* Proceedings, 8th Middle East Oil Show and Conference: Society of Petroleum Engineers, p. II.11–II.28.

Liu, K., P. Boult, S. Painter, and L. Paterson, 1996, Outcrop analog for sandy braided stream reservoirs: Permeability patterns in the Triassic Hawkesbury Sandstone, Sydney basin, Australia: AAPG Bulletin, v. 80, p. 1850–1866.

Lo, T.-W., C. W. Grant, B. Keyser, and T. Chen, 1998, Geostatistical porosity and permeability modeling using 3-D seismic and log data, a carbonate reservoir example (abs.), *in* Conference abstracts, GEO '98, 3rd Middle East Geoscience Conference and Exhibition: GeoArabia (Manama), v. 3, p. 116, 118.

Lomando, A. J., H. A. Legarre, and E. Gischler, 1998, The application of modern carbonate depositional systems in reservoir geostatistics (abs.), *in* Conference abstracts, GEO '98, 3rd Middle East Geoscience Conference and Exhibition: GeoArabia (Manama), v. 3, p. 121–122.

Lucia, F. J., and G. E. Fogg, 1990, Geologic/stochastic mapping of heterogeneity in a carbonate reservoir: Dune (Grayburg) field, Crane County, west Texas: Journal of Petroleum Technology, v. 42, p. 1298–1303.

MacDonald, A. C., and E. K. Halland, 1993, Sedimentology and shale modeling of a sandstone-rich fluvial reservoir: Upper Statfjord Formation, Statfjord field, northern North Sea: AAPG Bulletin, v. 77, p. 1016–1040.

MacDonald, A. C., T. H. Høye, P. Lowry, T. Jacobsen, J. O. Aasen, and A. O. Grindheim, 1992, Stochastic flow unit modelling of a North Sea coastal-deltaic reservoir: First Break, v. 10, p. 124–133.

Maharaja, A., 2005, Stochastic modeling of the Rhine-Meuse delta using multiple-point geostatistics, *in* O. Leuangthong and C. V. Deutsch, eds., Geostatistics Banff 2004, Proceedings of the Seventh International Geostatistics Congress: New York, Springer, p. 653–660.

Mak, N.-K., 2000, Application of geostatistics in reservoir delineation program for the UP-Ford sands in the Long Beach unit, Wilmington field (abs.): AAPG Bulletin, v. 84, p. 883.

Mao, N., 1986, A geostatistical (kriging) study of geological and gas production data from West Virginia, Livermore, CA, Lawrence Livermore National Laboratory: Report prepared for Morgantown Energy Technology Center, Unconventional Gas Program, Eastern Devonian Shales Research, UCID-20614, 35 p.

Massonnat, G. J., F. G. Alabert, and C. B. Guidicelli, 1992, Anguille Marine, a deepsea-fan reservoir offshore Gabon: From geology to stochastic modeling: Society of Petroleum Engineers Annual Technical Conference and Exhibition, SPE Paper 24709.

Mathieu, Y., F. Verdier, P. Houel, J. Delmas, and H. Beucher, 1993, Reservoir heterogeneity in fluviatile Keuper facies: A subsurface and outcrop study, *in* R. Eschard and B. Doligez, eds., Subsurface reservoir characterization from outcrop observations: Institut Francais du Petrole, Collection Colloques et Séminaires 51, p. 145–160.

Matthews, J. L., A. S. Emanuel, and K. A. Edwards, 1988, A modeling study of the Mitsue stage 1 miscible flood using fractal geostatistics: Society of Petroleum Engineers Annual Technical Conference and Exhibition, SPE Paper 188327.

Matthews, J. L., A. S. Emanuel, and K. A. Edwards, 1989, Fractal methods improve Mitsue miscible predictions: Journal of Petroleum Technology, v. 41, p. 1126–1142.

Mavko, G. M., 1998, Integrated seismic/rock physics characterization of naturally fractured tight gas reservoirs: Chicago, Gas Research Institute, GRI-98/0347, 166 p.

McKinley, J. M., C. D. Lloyd, and A. H. Ruffell, 2004, Use of variography in permeability characterization of visually homogeneous sandstone reservoirs with examples from outcrop studies: Mathematical Geology, v. 36, p. 761–779.

Meddaugh, W. S., 2005, Evaluation of stochastic earth model workflows, vertical up-scaling and areal up-scaling using data from the Eunice Monument South unit (New Mexico) and the LL-652 central fault block (Venezuela) reservoirs, *in* O. Leuangthong and C. V. Deutsch, eds., Geostatistics Banff 2004, Proceedings of the Seventh International Geostatistics Congress: New York, Springer, p. 743–750.

Meddaugh, W. S., D. Dull, S. D. Griest, P. Montgomery, and G. McNaboe, 2005, Stochastic reservoir model for the first Eocene reservoir, Wafra field, partitioned neutral zone (PNZ), *in* O. Leuangthong and C. V. Deutsch, eds., Geostatistics Banff 2004, Proceedings of the Seventh International Geostatistics Congress: New York, Springer, p. 757–762.

Meehan, D. N., and S. K. Verma, 1994, Improved reservoir characterization in low-permeability reservoirs with geostatistical methods: Society of Petroleum Engineers Annual Technical Conference and Exhibition, SPE Paper 28608.

Mijnssen, F. C. J., N. Tyler, and K. J. Weber, 1993, Knowledge base development for the estimation of reservoir rock properties in the interwell area: Examples from the Texas Gulf Coast, *in* S. S. Flint and I. D. Bryant, eds., The geological modelling of hydrocarbon reservoirs and outcrop analogues: International Association of Sedimentologists Special Publication 15, p. 169–180.

Minck, R., G. King, R. Dacosta, F. Domingos, B. Carleton, C. Laidlaw, and E. Leggett, 1998, N'sano Pinda; characterization of a finely inter-bedded carbonate siliciclastic reservoir, offshore Cabinda, Angola (abs.): AAPG Bulletin, v. 82, p. 1944.

Moinard, L., 1987, Application of kriging to the mapping of a reef from wireline logs and seismic data: A case history, *in* G. Matheron and M. Armstrong, eds., Geostatistical case studies: Dordrecht, The Netherlands, D. Reidel Publishing Company, p. 93–103.

Montgomery, P., W. S. Meddaugh, W. C. Dawson, M. J. Hegmann, and W. K. Gibbs, 2004, Cyclostratigraphy, lithostratigraphy and geostatistical modeling of the Upper Jurassic Smackover and Norphlet formations, Hatter's Pond field, Mobile County, Alabama: AAPG Annual Meeting Program, v. 13, http://www.searchanddiscovery.com/documents/abstracts/annual2004/Dallas/Montgomery02.htm (accessed June 22, 2006).

Montgomery, S. L., and C. D. Morgan, 1998, Bluebell field, Uinta Basin; reservoir characterization for improved well completion and oil recovery: AAPG Bulletin, v. 82, p. 1113–1132.

Mostaghel, B., A. Bashari, A. Kakzad, and A. A. Sobhani, 2004, Application of stochastic analysis and modeling through integrated reservoir characterization in Godbadli gas field (northeastern Iran) (abs.): AAPG International Conference Abstracts, 24–27 October, Cancun, http://www.searchanddiscovery.com/documents/abstracts/2004intl_cancun/short/A90194.pdf (accessed June 22, 2006).

Mottahedeh, R., 2001, A case study: Successful well trajectory planning of a horizontal injector using 3D geostatistics and 3D visualization on SMART drilling technologies AEC Grande Prairie Halfway V oil reservoir, Alberta, Canada, *in* Proceedings of the 16th World Petroleum Congress: The Congress, p. 310–314.

Norris, B., C. V. Deutsch, and D. Leckie, 2000, Deterministic geological control on a probabilistic, 3-dimensional geostatistical reservoir model of the Lower Cretaceous S1A marine sandstone, upper Qishn Formation, Heijah field, Yemen: GeoCanada 2000, Calgary, June, 4 p.

Norris, R. J., G. J. Massonat, and F. G. Alabert, 1993, Early quantification of uncertainty in the estimation of oil-in-place in a turbidite reservoir: Society of Petroleum Engineers Annual Technical Conference and Exhibition, SPE Paper 26490.

Olea, R. A., and J. C. Davis, 1977, Regionalized variables for evaluation of petroleum accumulation in Magellan Basin, South America: AAPG Bulletin, v. 61, p. 558–572.

Omre, H., H. Tjelmeland, Y. Qi, and L. Hinderaker, 1993, Assessment of uncertainty in the production characteristics of a sandstone reservoir, *in* B. Linville, T. E. Burchfield, and T. C. Wesson, eds., Reservoir characterization III: Tulsa, Pennwell Books, p. 556–604.

Pawar, R., 2000, Comprehensive reservoir modeling of a mature, geologically complex reservoir (abs.): AAPG Bulletin, v. 84, p. 890.

Payne, D. V., K. A. Edwards, and A. S. Emanuel, 1991, Examples of reservoir simulation studies utilizing geostatistical models of reservoir heterogeneity, *in* L. W. Lake, H. B. Carroll Jr., and T. C. Wesson, eds., Reservoir characterization II: San Diego, Academic Press, p. 497–523.

Pearson, R., and B. Hart, 1998, Sequence stratigraphy and seismic attribute analysis of the Bone Spring Formation, Delaware basin, New Mexico (abs.): New Mexico Geology, v. 20, p. 52.

Perez, G., A. K. Chopra, and C. D. Severson, 1997, Integrated geostatistics for modeling fluid contacts and shales in Prudhoe Bay: SPE Formation Evaluation, v. 12, p. 213–221.

Pramanik, A. G., V. Singh, R. Vig, A. K. Srivastava, and D. N. Tiwary, 2004, Estimation of effective porosity using geostatistics and multiattribute transforms: A case study: Geophysics, v. 69, p. 352–372.

Prieditis, J., A. P. Yang, and M. D. Wilkins, 1998, Simulation of a $CO_2$ flood in the Slaughter field with geostatistical reservoir characterization: Society of Petroleum Engineers Annual Technical Conference and Exhibition, SPE Paper 49168.

Pyrcz, M. J., O. Catuneanu, and C. V. Deutsch, 2005, Stochastic surface-based modeling of turbidite lobes: AAPG Bulletin, v. 89, p. 177–191.

Qi, L., 2004, Geostatistical 3D reservoir modeling of Mississippian St. Louis carbonate reservoir systems, Kansas: AAPG Annual Meeting Program, v. 13, http://www.searchanddiscovery.com/documents/abstracts/annual2004/Dallas/Qi.htm (accessed June 22, 2006).

Rached, G., and R. Marschall, 1999, Time-lapse seismic in the Sabiriyah oilfield—Kuwait: Journal of Seismic Exploration, v. 8, p. 383–401.

Ratchkovski, A., D. O. Ogbe, and A. S. Lawal, 1999, Application of geostatistics and conventional methods to derive hydraulic flow units for improved reservoir description: A case study of Endicott field, Alaska: Society of Petroleum Engineers Western Regional Meeting, Anchorage, Alaska, SPE Paper 54587.

Ravenne, C., R. Eschard, A. Galli, Y. Mathieu, L. Montadert, and J. L. Rudkiewicz, 1989, Heterogeneities and geometry of sedimentary bodies in a fluvio-deltaic reservoir: SPE Formation Evaluation, v. 4, p. 239–246.

Ravenne, C., A. Galli, H. Beucher, R. Eschard, and D. Guérillot, 1991, Outcrop studies and geostatistical modeling of a Middle Jurassic Brent analogue, *in* G. Imarisio, M. Frias, and J. M. Bemtgen, Proceedings of the European Oil and Gas Conference—A multidisciplinary approach in exploration and production R&D: London, Graham & Trotman, p. 497–520.

Ravnaas, R. D., R. F. Strickland, A. P. Yang, M. A. Malik, D. R. Prezbindowski, T. Mairs, and L. W. Lake, 1992, Three-dimensional conditional simulation of Schneider (Buda) field, Wood County, Texas: Society of Petroleum Engineers Permian

Basin Oil and Gas Recovery Conference, Midland, Texas, SPE Paper 23970.

Ray, T., J. Holbrook, C. Sodergren, and Y. Cui, 1999, Dakota Sandstone in central Kansas using architectural-element analysis, *in* J. W. Harbaugh, W. L. Watney, E. C. Rankey, R. Slingerland, R. H. Goldstein, and E. K. Franseen, eds., Numerical experiments in stratigraphy: Recent advances in stratigraphic and sedimentologic computer simulations: Society for Sedimentary Geology, p. 279–294.

Ringrose, P., G. Pickup, J. Jensen, and M. Forrester, 1999, The Ardross reservoir gridblock analog: Sedimentology, statistical representivity, and flow upscaling, *in* R. Schatzinger and J. Jordan, eds., Reservoir characterization—Recent advances: AAPG Memoir 71, p. 265–276.

Ritts, B. D., C. L. Johnson, R. Archer, and A. Taylor, 2004, Deterministic and geostatistical characterization and modeling of the geometry, heterogeneity, and connectivity of lacustrine-deltaic sandstone, Green River Formation, southwest Uinta Basin, Utah (abs.): AAPG Hedberg Research Conference Abstracts, May 18–21, Baku, Azerbaijan, http://www.searchanddiscovery.com/documents/abstracts/2004hedberg_baku/extended/ritts/ritts.htm (accessed June 22, 2006).

Ronghe, S., and B. Kongwung, 1998, Reservoir identification and characterization on 3D seismic data using geostatistics and attribute analyses, Gulf of Thailand (abs.): AAPG Annual Meeting Program, v. 2, p. A559.

Ross, J. M., J. Mansoori, and R. Chambers, 1993, Reservoir characterization and performance prediction by modeling and identifying uncertainties; a field case study (abs.): AAPG Bulletin, v. 77, p. 1659–1660.

Rossini, C., F. Brega, L. Piro, M. Rovellini, and G. Spotti, 1994a, Combined geostatistical and dynamical simulations for developing a reservoir management strategy: A case history: Society of Petroleum Engineers Annual Technical Conference and Exhibition, SPE Paper 26625.

Rossini, C., F. Brega, L. Piro, M. Rovellini, and G. Spotti, 1994b, Combined geostatistical and dynamic simulations for developing a reservoir management strategy: A case history: Journal of Petroleum Technology, v. 46, p. 979–985.

Rudkiewicz, J. L., D. Guerillot, and A. Galli, 1990, An integrated software for stochastic modelling of reservoir lithology and property prediction with an example from the Yorkshire Middle Jurassic formation, *in* A. T. Buller, E. Berg, O. Hjelmeland, J. Kleppe, O. Torsæter, and J. O. Aasen, eds., North Sea oil and gas reservoirs II: London, Graham and Trotman, p. 399–406.

Ruppel, S. C., F. J. Lucia, and J. W. Jennings, 2003, Outcrop-based stratigraphic, petrophysical, and geostatistical characterization and modeling of a carbonate platform reservoir succession: Middle Permian, Permian basin (abs.): AAPG Annual Meeting Program, v. 12, http://www.searchanddiscovery.com/documents/abstracts/annual2003/short/80507.PDF (accessed June 22, 2006).

Russell, B., 2002, Combining geostatistics and multiattribute transforms: A channel sand case study, Blackfoot oilfield (Alberta): Journal of Petroleum Geology, v. 25, p. 97–118.

Sahin, A., S. G. Shori, A. Z. Ali, H. F. El-Sahn, H. M. Hassan, and A. Al-Sanounah, 1998, Geological controls of variograms in a complex carbonate reservoir, Eastern Province, Saudi Arabia: Mathematical Geology, v. 30, p. 309–322.

Sams, M. S., and V. C. Vahrenkamp, 2000, Porosity modeling in a complex carbonate reservoir using geostatistical inversion of seismic data (abs.): AAPG Bulletin, v. 84, p. 1485.

Seifert, D., and J. L. Jensen, 1997, Object and pixel-based reservoir modelling of a braided fluvial reservoir, *in* V. Pawlowsky-Glahn, ed., Proceedings of IAMG '97, The 3rd Annual Conference of the International Association for Mathematical Geology: Barcelona, International Center for Numerical Methods in Engineering, p. 719–724.

Seifert, D., and J. L. Jensen, 2000, Object and pixel-based reservoir modeling of a braided fluvial reservoir: Mathematical Geology, v. 32, p. 581–603.

Senger, R. K., F. J. Lucia, C. Kerans, M. Ferris, and G. E. Fogg, 1992, Geostatistical/geological permeability characterization of carbonate ramp deposits in San Andres outcrop, Algerita Escarpment, New Mexico: Society of Petroleum Engineers Permian Basin Oil and Gas Recovery Conference, Midland, Texas, SPE Paper 23967.

Senger, R. K., F. J. Lucia, C. Kerans, M. A. Ferris, and G. E. Fogg, 1993, Dominant control on reservoir-flow behavior in carbonate reservoirs as determined from outcrop studies, *in* B. Linville, T. E. Burchfield, and T. C. Wesson, eds., Reservoir characterization III: Tulsa, PennWell Books, p. 107–150.

Shi, T.-C., A.-P. Yang, C. T. Cheng, Y. Quzn, G. F. Lu, B. Wang, D. L. Martin, and E. M. Hussey, 1998, Geostatistical evaluation of QHD 32-6 oil field:

A DRO in China: Society of Petroleum Engineers International Meeting on Petroleum Engineering, Beijing, SPE Paper 48879.

Simms, S. C., 1993, Integration of 3-D seismic data with reservoir modeling of stratigraphically complex reservoir, central Saudi Arabia (abs.): AAPG Bulletin, v. 77, p. 1664.

Sinding-Larsen, R., R. Wen, and G. Caillet, 1992, Nonparametric geostatistical studies of fault parameters in the Snorre Area, North Sea, *in* J. O. Aasen, ed., North Sea oil and gas reservoirs III: Dordrecht, The Netherlands, Kluwer Academic Publishers, p. 167–172.

Singdahlsen, D. S., 1993, Reservoir characterization and geostatistical modeling—The integration of geology into reservoir simulation: East Painter reservoir field, Wyoming, *in* B. Linville, T. E. Burchfield, and T. C. Wesson, eds., Reservoir characterization III: Tulsa, PennWell Books, p. 973–982.

Soares, A., J. A. Almeida, A. da Costa e Silva, and J. Carvalho, 1991, Geostatistical simulation of the geometry of a shale/limestone sequence in an oil reservoir: Society of Petroleum Engineers, SPE Paper 21633.

Soegaard, J., J. C. Rivenæs, A. Alme, and T. Dreyer, 1999, 3D stochastic modeling of fluvial channels in Ness Formation at Oseberg: Challenges in integration of geologic data with modeling parameters, *in* S. J. Lippard, A. Næss, and R. Sinding-Larsen, eds., IAMG '99, Proceedings of the 5th Annual Conference of the International Association for Mathematical Geology: Trondheim, Tapir, p. 659–664.

Srivastava, R. M., 1990, An application of geostatistical methods for risk analysis in reservoir management: Society of Petroleum Engineers Annual Technical Conference and Exhibition, SPE Paper 20608, 10 p.

Stalkup, F. I., 1986, Permeability variations observed at the faces of crossbedded sandstone outcrops, *in* L. W. Lake and H. B. Carroll Jr., eds., Reservoir characterization: San Diego, Academic Press, Inc., p. 141–175.

Stanley, K. O., K. Jorde, N. Raestad, and C. P. Stockbridge, 1990, Stochastic modelling of reservoir sand bodies for input to reservoir simulation, Snorre field, northern North Sea, Norway, *in* A. T. Buller, E. Berg, O. Hjelmeland, J. Kleppe, O. Torsæter, and J. O. Aasen, eds., North Sea oil and gas reservoirs II: London, Graham and Trotman, p. 91–101.

Stølum, H.-H., 1991, Fractal heterogeneity of clastic reservoirs, *in* L. W. Lake, H. B. Carroll Jr., and T. C. Wesson, eds., Reservoir characterization II: San Diego, Academic Press, p. 579–612.

Strebelle, S., K. Payrazyan, and J. Caers, 2002, Modeling of a deepwater turbidite reservoir conditional to seismic data using multiple-point geostatistics: Society of Petroleum Engineers Journal, v. 8, p. 227–235.

Taheri, M., M. Chin-A-Lien, and E. Rodriguez, 1991, Geostatistical characterization of reservoirs for optimal field development, *in* L. W. Lake, H. B. Carroll Jr., and T. C. Wesson, eds., Reservoir characterization II: San Diego, Academic Press, p. 698–700.

Tetzlaff, D., R. Davies, D. McCormick, C. Signor, P. Mirowski, N. Williams, D. Hodgson, and J. Brady, 2005, Application of multipoint geostatistics to honor multiple attribute constraints applied to a deepwater outcrop analog, Tanqua Karoo basin, South Africa: SEG Expanded Abstracts 24, p. 1370–1373.

Thadani, S. G., and M. Novak, 1996, Non-stationary stochastic modeling techniques for reservoir description: An application from the Alaskan North Slope: Society of Petroleum Engineers Annual Technical Conference and Exhibition, SPE Paper 36502.

Thompson, T. W., and P. E. Harness, 1998, Integrated reservoir characterization using geostatistics and 3D visualization in the Kern River field, Bakersfield, California (abs.): AAPG Bulletin, v. 82, p. 860.

Tillman, R. W., E. Kasap, and B. Kara, 1998, Geostatistical analysis of incised paleovalley reservoir analogs, Wyoming (abs.): AAPG Annual Meeting Program, v. 2, p. A656.

Tomutsa, L., S. Jackson, M. Szpakiewicz, and I. Palmer, 1986, Geostatistical characterization and comparison of outcrop and subsurface facies: Shannon Shelf sand ridges: Society of Petroleum Engineers California Regional SPE Meeting, April 2–4, SPE Paper 15127.

Tonn, R., 1998, Seismic reservoir characterization on Montney Sand in the Peace River Arch area, Canada: The Leading Edge, v. 17, p. 643–646.

Torres-Verdín, C., M. Victoria, G. Merletti, and J. Pendrel, 1999, Trace-based and geostatistical inversion of 3-D seismic data for thin-sand delineation: An application to San Jorge Basin, Argentina: The Leading Edge, v. 18, p. 1070–1077.

Trappe, H., and C. Helmich, 2005a, Seismic characterization of Rotliegend reservoirs: From bright spots to stochastic simulation: Trappe Erdöl Erdgas Consultants Technical Note 2: www.teec.de/tech_note/Tech-Note_2/technical_note_2.html (accessed January 26, 2005).

Trappe, H., and C. Helmich, 2005b, Image processing and geostatistics, Upper Carboniferous reservoir description: Trappe Erdöl Erdgas Consultants Technical Note 3: www.teec.de/tech_note/Tech-Note_2/technical_note_3.html (accessed January 2, 2005).

Tribuanawati, D., 2000, Reservoir characterization in Kaji-Semoga field, south Sumatra, Indonesia, using seismic inversion and geostatistics (abs): AAPG Bulletin, v. 84, p. 1507.

Tyler, K. J., T. Svanes, and A. Henriquez, 1994, Heterogeneity modeling used for production simulation of a fluvial reservoir: SPE Formation Evaluation, v. 9, p. 85–92.

Vejbaek, O. V., 1999, Clastic reservoir characterization using seismic impedances and geostatistics; Middle Jurassic unit of the Harald field, Danish North Sea (abs.): AAPG Annual Meeting Program, v. 9, p. A142.

Vejbaek, O. V., and L. Kristensen, 2000, Downflank hydrocarbon potential identified using seismic inversion and geostatistics, upper Maastrichtian reservoir unit, Dan field: Petroleum Geoscience, v. 6, p. 1–13.

Verly, G., 1986, MultiGaussian kriging—A complete case study, *in* R. V. Ramani, ed., Proceedings of the 19th International Symposium of the Application of Computers and Operations Research in the Mineral Industry: Society of Mining Engineers, p. 283–298.

Voelker, J., and J. Caers, 2005, A combined geostatistical and source model to predict superpermeability from flowmeter data: Application to the Ghawar field, *in* O. Leuangthong and C. V. Deutsch, eds., Geostatistics Banff 2004, Proceedings of the Seventh International Geostatistics Congress: New York, Springer, p. 591–600.

Vogt, J. P., 1993, Comparison of geostatistical characterization methods for the Kern River field, *in* B. Linville, T. E. Burchfield, and T. C. Wesson, eds., Reservoir characterization III: Tulsa, PennWell Books, p. 353–369.

Wajda-Karpoluk, B., and L. Karski, 2002, Recovery of seismically invisible top Ca2 via geostatistics—A case study from the Polish Zechstein Basin (abs.): AAPG Annual Meeting Program, v. 11, http://www.searchanddiscovery.com/documents/abstracts/annual2002/DATA/2002/13ANNUAL/SHORT/40689.pdf (accessed June 22, 2006).

Wang, F. P., J. Dai, and C. Kerans, 1998a, Modeling dolomitized carbonate-ramp reservoirs: A case study of the Seminole San Andres unit: Part II—Seismic modeling, reservoir geostatistics, and reservoir simulation: Geophysics, v. 63, p. 1876–1885.

Wang, F. P., F. J. Lucia, and C. Kerans, 1998b, Integrated reservoir characterization study of a carbonate ramp reservoir: Seminole San Andres unit, Gaines County, Texas: SPE Reservoir Evaluation and Engineering, v. 1, p. 105–113.

Wang, L., P. Wong, and S. A. R. Shibli, 1998, Modelling porosity distribution in the A'nan oilfield: Use of geological quantification, neural networks and geostatistics: Society of Petroleum Engineers Annual Technical Conference and Exhibition, SPE Paper 35854/50438, 8 p.

Wang, S.-W., L. E. Bailey, and J. W. Smetanka, 1998d, Application of geostatistical methods improves results of reservoir engineering study for northwest feature, Hemlock reservoir, McArthur River field, Alaska: Society of Petroleum Engineers, Western Regional Meeting, Bakersfield, California, SPE Paper 46233.

Weissmann, G. S., and G. E. Fogg, 1999, Multi-scale alluvial fan heterogeneity modeled with transition probability geostatistics in a sequence stratigraphic framework: Journal of Hydrology, v. 226, p. 48–66.

White, C. D., D. Novakovic, S. P. Dutton, and B. J. Willis, 2003, A geostatistical model for calcite concretions in sandstone: Mathematical Geology, v. 35, p. 549–576.

Wolcott, D. S., and A. K. Chopra, 1991, Incorporating reservoir heterogeneity using geostatistics to investigate waterflood recoveries for drillsite 1E, A4 sandstone body, Kuparuk River field, Alaska: Society of Petroleum Engineers, International Arctic Technology Conference, Anchorage, Alaska, SPE Paper 22164.

Xu, W., and A. G. Journel, 1993, Gtsim: Gaussian truncated simulations of reservoir units in a west Texas carbonate field: Society of Petroleum Engineers Annual Technical Conference and Exhibition, SPE Paper 27412.

Yang, A. P., 2003, Geostatistical prediction of a sand distribution of gas reservoir in Jilin, China: Society of Petroleum Engineers Annual Technical Conference and Exhibition, SPE Paper 84055.

Yang, C. T., A. K. Chopra, J. Chu, X. Huang, and

M. G. Kelkar, 1995, Integrated geostatistical reservoir description using petrophysical, geological, and seismic data for Yacheng 13-1 gas field: Society of Petroleum Engineers Annual Technical Conference and Exhibition, SPE Paper 30566.

Yang, K., 2000, Stochastic 3-D geological modeling of deep water reservoirs: Methodology and case studies: Gulf Coast Area Geological Society, 50th Annual Convention of the Gulf Coast Area Geological Society, Houston, Texas, October 25–27: AAPG Bulletin, v. 84, p. 1696.

Yang, K., J. M. Yarus, and W. P. Cantanese, 1999, Characterization and 3-D modeling of turbidite reservoir: A case study in Miocene slope deposits, Main Pass area, offshore Gulf of Mexico: Gulf Coast Association of Geological Societies Transactions, v. 49, p. 54–60.

Yao, T., and A. G. Journel, 2000, Integrating seismic attribute maps and well logs for porosity modeling in a west Texas carbonate reservoir: Addressing the scale and precision problem: Journal of Petroleum Science and Engineering, v. 28, p. 65–81.

Yarus, J. M., K. Yang, S. Sriisraporn, N. Chuemthaisong, and K. Sangwongwanich, 2000, Integrating 3D seismic and geostatistics: Building a 3D model of a tertiary deltaic and shallow marine deposit, Malay Basin, offshore Gulf of Thailand, *in* Proceedings of the 32nd Annual Offshore Technology Conference, v. 1 (Geology, Earth Sciences and Environmental Factors): Richardson, Texas, Offshore Technology Conference, p. 61–66.

Yeung, K., 2000, Application of 3-D geostatistical models to characterize and simulate an oil sands reservoir using the steam assisted gravity drainage (SAGD) recovery process in northeastern Alberta, Canada (abs.): AAPG Bulletin, v. 84, p. 904.

Yeung, K. C., 2000, Application of 3-D geostatistical models to characterize and simulate an oil sands reservoir using the steam assisted gravity drainage (SAGD) recovery process in northeastern Alberta, Canada: AAPG/SPE Conference, California, June, CD-ROM.

Yose, L. A., T. L. Davis, S. J. Brown, G. S. Kompanik, T. Eiben, and P. Lapointe, 1999, 3-D permeability model of a fractured carbonate reservoir, Norman Wells, NWT, Canada (abs.): AAPG Annual Meeting Program, v. 8, p. A156–A157.

Zwicky, R. W., 1977, Preliminary geostatistical investigations of tar-bearing sands, Lease-13, Athabasca Tar Sands: Journal of Canadian Petroleum Technology, v. 16, p. 25–30.

## METHODOLOGICAL APPLICATIONS

Aasen, J. O., J. K. Silseth, L. Holden, H. Omre, K. B. Halvorsen, and J. Hoiberg, 1990, A stochastic reservoir model and its use in evaluation of uncertainties in the results of recovery processes, *in* A. T. Buller, E. Berg, O. Hjelmeland, J. Kleppe, O. Torsæter, and J. O. Aasen, eds., North Sea oil and gas reservoirs II: London, Graham and Trotman, p. 425–436.

Aasum, Y., M. G. Kelkar, and S. P. Gupta, 1991, An application of geostatistics and fractal geometry for reservoir characterization: SPE Formation Evaluation, v. 6, p. 11–19.

Abrahamsen, P., 1993, Bayesian kriging for seismic depth conversion of a multi-layer reservoir, *in* A. Soares, ed., Geostatistics tróia '92: Dordrecht, The Netherlands, Kluwer Academic Publishers, p. 385–398.

Abrahamsen, P., 2005, Combining methods for subsurface prediction, *in* O. Leuangthong and C. V. Deutsch, eds., Geostatistics Banff 2004, Proceedings of the Seventh International Geostatistics Congress: New York, Springer, p. 601–610.

Abrahamsen, P., and H. Omre, 1994, Random functions and geological subsurfaces, *in* P. Abrahamsen, H. Omre, and P. Thomassen, eds., ECMOR IV, Proceedings of the 4th European Conference on the Mathematics of Oil Recovery: Røros, Delft University Press, 23 p.

Agterberg, F., 1976, New problems at the interface between geostatistics and geology, *in* M. Guarascio, M. David, and C. Huijbregts, eds., Advanced geostatistics in the mining industry: Dordrecht, The Netherlands, D. Reidel Publishing Company, p. 403–421.

Agterberg, F. P., 1970, Autocorrelation functions in geology, *in* D. F. Merriam, ed., Geostatistics—A colloquium: New York, Plenum Press, p. 113–141.

Alabert, F., 1989, Constraining description of randomly heterogeneous reservoirs to pressure test data: A Monte Carlo study: Society of Petroleum Engineers Annual Technical Conference and Exhibition, SPE Paper 19600.

Alabert, F. G., and V. Modot, 1992, Stochastic models of reservoir heterogeneity: Impact on connectivity and average permeabilities: Society of Petroleum Engineers Annual Technical Conference and Exhibition, SPE Paper 24893.

Allard, D., 1994, Simulating a geological lithofacies with respect to connectivity information using the truncated Gaussian model, *in* M. Armstrong and P. A. Dowd, eds., Geostatical simulations: Dordrecht, The Netherlands, Kluwer Academic Publishers, p. 197–211.

Almeida, A. S., 1993, Joint simulation of multiple petrophysical properties using a Markov-type coregionalization model: Stanford, Stanford University, Stanford Center for Reservoir Forecasting Report 6, 81 p.

Almeida, A. S., T. Tran, and P. Ballin, 1993, An integrated approach to reservoir studies using stochastic simulation techniques, *in* A. Soares, ed., Geostatistics tróia '92: Dordrecht, The Netherlands, Kluwer Academic Publishers, p. 371–384.

Almeida, J. A., A. Soares, M. J. Pereira, and T. S. Daltaban, 1996, Upscaling of permeability: Implementation of a conditional approach to improve the performance in flow simulation: Society of Petroleum Engineers European 3-D Reservoir Modeling Conference, Stavanger, SPE Paper 35490.

Al-Qahtani, M. Y., and I. Ershaghi, 1999, Characterization and estimation of permeability correlation structure from performance data, *in* R. Schatzinger and J. Jordan, eds., Reservoir characterization—Recent advances: AAPG Memoir 71, p. 343–358.

Anantraksakul, N., B. Doligez, P. Joseph, P. Pouclee, and J. Pouzet, 2000, A methodology to build a 3D geological model of reservoir from well data using seismic information (abs.): AAPG Bulletin, v. 84, p. 1397–1398.

Araktingi, U., 1993, A sensitivity analysis of geostatistical parameters on oil recovery predictions, *in* A. Soares, ed., Geostatistics troia '92: Dordrecht, The Netherlands, Kluwer Academic Publishers, p. 493–504.

Araktingi, U. G., W. M. Bashore, T. T. B. Tran, and T. A. Hewett, 1993, Integration of seismic and well log data in reservoir modeling, *in* B. Linville, T. E. Burchfield, and T. C. Wesson, eds., Reservoir characterization III: Tulsa, PennWell Books, p. 513–554.

Baafi, E. Y., 1997, Modeling of facies architecture by marked point models, *in* E. Y. Baafi and N. A. Schofield, eds., Geostatistics Wollongong '96: Dordrecht, The Netherlands, Kluwer Academic Publishers, p. 386–397.

Bahar, A., H. Ates, M. Kelkar, and M. Al-Deeb, 2001, Methodology to incorporate geological knowledge in variogram modeling: Society of Petroleum Engineers Asia Pacific Oil and Gas Conference and Exhibition, Jakarta, SPE Paper 68704.

Bahralolom, I., and J. Heller, 1991, Core sample heterogeneity from laboratory flow experiments, *in* L. W. Lake, H. B. Carroll Jr., and T. C. Wesson, eds., Reservoir characterization II: San Diego, Academic Press, p. 77–101.

Bakr, A. A., L. W. Gelhar, A. L. Gutjahr, and J. R. MacMillan, 1978, Stochastic analysis of spatial variability in subsurface flows: 1. Comparison of one- and three-dimensional flows: Water Resources Research, v. 14, p. 263–271.

Ballin, P. R., A. Journel, and K. Aziz, 1992, Prediction of uncertainty in reservoir performance forecast: Journal of Canadian Petroleum Technology, v. 31, p. 52–62.

Ballin, P. R., K. Aziz, A. G. Journel, and L. Zuccolo, 1993, Quantifying the impact of geological uncertainty on reservoir performing forecasts: Society of Petroleum Engineers 12th Symposium on Reservoir Simulation, SPE Paper 25238, 10 p.

Bardossy, A., I. Bogardi, and L. Duckstein, 1987, A geostatistical model of reservoir deposition: Water Resources Research, v. 23, p. 510–514.

Bardossy, A., G. Bardossy, and I. Bogardi, 1989, Application of geological information to kriging, *in* M. Armstrong, ed., Geostatistics, Proceedings of the 3rd International Geostatistics Congress: Dordrecht, The Netherlands, Kluwer Academic Publishers, p. 591–602.

Barton, C. C., and C. H. Scholz, 1995, The fractal size and spatial distribution of hydrocarbon accumulations: Implications for resource assessment and exploration strategy, *in* C. C. Barton and P. R. LaPointe, eds., Fractals in petroleum geology and earth sciences: New York, Plenum Press, p. 13–34.

Bashore, W. M., U. G. Araktingi, M. Levy, and W. J. Schweller, 1993, Importance of the geological model for reservoir characterization using geostatistical techniques and the impact on subsequent fluid flow: Society of Petroleum Engineers Annual Technical Conference and Exhibition, SPE Paper 26474, 16 p.

Bashore, W. M., U. G. Araktingi, M. Levy, and W. J. Schweller, 1994, Importance of a geological framework and seismic data integration for reservoir modeling and subsequent fluid-flow predictions, *in* J. M. Yarus and R. L. Chambers, eds., Stochastic modeling and geostatistics: Principles, methods, and case studies: AAPG Computer Applications in Geology 3, p. 159–175.

Bean, C. J., and J. McCloskey, 1995, Seismic constraints

on statistical models of upper crustal heterogeneity: First Break, v. 13, p. 105–110.

Belcher, R. C., 1994, Geospatial modeling techniques for understanding internal geometries of complexly faulted reservoirs: Society of Petroleum Engineers Annual Technical Conference and Exhibition, SPE Paper 27543.

Benito, M., G. Morales, H. Beucher, F. Geffroy, and B. Doligez, 2003, Object based simulations in fluvial environment at the reservoir scale and at a local scale (abs.): AAPG Annual Meeting Program, v. 12, http://www.searchanddiscovery.com/documents/abstracts/annual2003/extend/78409.PDF (accessed June 22, 2006).

Bennett, F., 2003, Quantification of model uncertainty by automated calibration of dynamic models to multiple geostatistical realizations (abs.): AAPG Annual Meeting Program, v. 12, http://www.searchanddiscovery.com/documents/abstracts/annual2003/short/80091.PDF (accessed June 22, 2006).

Berteig, V., K. B. Halvorsen, and H. Omre, 1988, Prediction of hydrocarbon pore volume with uncertainties: Society of Petroleum Engineers Annual Technical Conference and Exhibition, SPE Paper 18325.

Beucher, H., A. Galli, G. Le Loc'h, and C. Ravenne, 1993, Including a regional trend in reservoir modeling using the truncated Gaussian method, *in* A. Soares, ed., Geostatistics troia '92: Dordrecht, The Netherlands, Kluwer Academic Publishers, p. 555–566.

Beucher, H., B. Dolilgez, and F. Geffroy, 2003, Complex features in sedimentology and truncated pluri-Gaussian simulations (abs.): AAPG Annual Meeting Program, v. 12, http://www.searchanddiscovery.com/documents/abstracts/annual2003/extend/78655.PDF (accessed June 22, 2006).

Bierkens, M., and P. Burrough, 1993, Modeling of map impurities using sequential indicator simulation, *in* A. Soares, ed., Geostatistics troia '92: Dordrecht, The Netherlands, Kluwer Academic Publishers, p. 637–648.

Biver, P., A. Haas, and C. Bacquet, 2002, Uncertainties in facies proportion estimation: II. Application to geostatistical simulation of facies and assessment of volumetric uncertainties: Mathematical Geology, v. 34, p. 703–714.

Blanc, G., D. Guerillot, D. Rahon, and F. Roggero, 1996, Building geostatistical models constrained by dynamic data—A posteriori constraints: Society of Petroleum Engineers European 3-D Reservoir Modeling Conference, Stavanger, SPE Paper 35478.

Borgman, L. E., M. Taheri, and R. Hagan, 1984, Three-dimensional, frequency-domain simulations of geological variables, *in* G. Verly, M. David, A. G. Journel, and A. Marechal, eds., Geostatistics for natural resources characterization: Dordrecht, The Netherlands, D. Reidel Publishing Company, p. 517–541.

Bortolli, L.-J., F. Alabert, A. Haas, and A. G. Journel, 1993, Constraining stochastic images to seismic data, *in* A. Soares, ed., Geostatistics tróia '92: Dordrecht, The Netherlands, Kluwer Academic Publishers, p. 325–338.

Boutemy, Y., C. Clavier, and R. F. Simond, 1979, Field studies: A progress report on the contribution of logging: Society of Petroleum Engineers Annual Technical Conference and Exhibition, SPE Paper 8178, 13 p.

Brunk, R. G., J. R. Maestas, and P. Parsons, 1995, Multistrata exploration and production study: Washington, U.S. Department of Energy, DOE/MC/26026-95/C0482, CONF-950494-27, 7 p.

Burger, H., M. Eder, A. Mannaa, and W. Skala, 1990, Conditional simulation in oil exploration, *in* G. Gaál and D. F. Merriam, eds., Computer applications in resource estimation: Prediction and assessment for metals and petroleum: Oxford, Pergamon Press, p. 239–252.

Caers, J., 1999, Stochastic reservoir characterization using multiple-point geostatistics, *in* S. J. Lippard, A. Næss, and R. Sinding-Larsen, eds., IAMG '99, Proceedings of the 5th Annual Conference of the International Association for Mathematical Geology: Trondheim, Tapir, p. 467–472.

Caers, J., 2001, Geostatistical reservoir modeling using statistical pattern recognition: Journal of Petroleum Science and Engineering, v. 29, p. 177–188.

Caers, J., 2002, Geostatistics: From pattern recognition to pattern reproduction, *in* M. Nikravesh, F. Aminzadeh, and L. Zadeh, eds., Soft computing and intelligent data analysis in oil exploration: Amsterdam, Elsevier, p. 44–60.

Caers, J., and A. G. Journel, 1998, Stochastic reservoir simulation using neural networks trained on outcrop data: Society of Petroleum Engineers Annual Technical Conference and Exhibition, SPE Paper 49026.

Caers, J., and X. Ma, 2002, Modeling conditional distributions of facies from seismic using neural nets: Mathematical Geology, v. 34, p. 143–168.

Caers, J., and S. Srinivasan, 2002, Statistical pattern recognition and geostatistical data integration, *in* P. Wong, F. Aminzadeh, and M. Nikravesh, eds., Soft computing for reservoir characterization and modelling: Heidelberg, Physica-Verlag, p. 355–386.

Caers, J., and S. Strebelle, 2003, Multiple-point geostatistics: Integrating geological and seismic data in the deepwater environment (abs.): AAPG Pacific Section and SPE Western Regional Joint Meeting, May 19–24, Long Beach, California, Abstracts, CD-ROM.

Caers, J., and T. Zhang, 2004, Multiple point geostatistics: A quantitative vehicle for integrating geologic analogs into reservoir models, *in* G. M. Grammer, P. M. Harris, and G. P. Eberli, eds., Integration of outcrop and modern analogs in reservoir modeling: AAPG Memoir 80, p. 383–394.

Campbell, K., 1986, Kriging for interpolation of sparse and irregularly distributed geologic data: National Technical Information Service, No. DE87013293, 19 p.

Campozana, F. P., L. W. Lake, and K. Sepehrnoori, 1999, How incorporating more data reduces uncertainty in recovery predictions, *in* R. Schatzinger and J. Jordan, eds., Reservoir characterization—Recent advances: AAPG Memoir 71, p. 359–368.

Carle, S. F., 1997, Integration of geologic interpretation into geostatistical simulation, *in* V. Pawlowsky-Glahn, ed., Proceedings of IAMG '97, 3rd Annual Conference of the International Association for Mathematical Geology: Barcelona, International Center for Numerical Methods in Engineering, p. 711–718.

Carr, J. R., 1992, A treatise on the use of geostatistics for the characterization of nonrenewable resources: Nonrenewable Resources, v. 1, p. 61–73.

Carr, J. R., and R. A. Prezbindowski, 1986, An application of co-conditional simulation for petroleum exploration: Computers and Geosciences, v. 12, p. 477–483.

Caumon, G., and A. G. Journel, 2005, Early uncertainty assessment: Application to hydrocarbon reservoir development, *in* O. Leuangthong and C. V. Deutsch, eds., Geostatistics Banff 2004, Proceedings of the Seventh International Geostatistics Congress: New York, Springer, p. 551–558.

Caumon, G., S. Strebelle, J. K. Caers, and A. G. Journel, 2004, Assessment of global uncertainty for early appraisal of hydrocarbon fields: Society of Petroleum Engineers Annual Technical Conference and Exhibition, SPE Paper 89943.

Caumon, G., O. Grosse, and J. L. Mallet, 2005, High resolution geostatistics on coarse unstructured flow grids, *in* O. Leuangthong and C. V. Deutsch, eds., Geostatistics Banff 2004, Proceedings of the Seventh International Geostatistics Congress: New York, Springer, p. 703–712.

Chambers, R. L., 1993, Geostatistical application for exploration and development: Porosity estimation from 3-D seismic data calibrated to well data, *in* Proceedings, 8th Middle East Oil Show and Conference: Society of Petroleum Engineers, p. II.481–II.490.

Chambers, R. L, M. A. Zinger, and M. C. Kelly, 1994, Constraining geostatistical reservoir descriptions with 3-D seismic data to reduce uncertainty, *in* J. M. Yarus and R. L. Chambers, eds., Stochastic modeling and geostatistics: Principles, methods, and case studies: AAPG Computer Applications in Geology 3, p. 143–157.

Chambers, R. L., D. H. Zamora Guerrero, E. O. Maldonado, and J. M. Yarus, 2002, Implementing a geostatistical stochastic modeling methodology to compute volumetrics and original-oil-in-place (abs.): AAPG Annual Meeting Program, v. 11, http://www.searchanddiscovery.com/documents/abstracts/annual2002/DATA/2002/13ANNUAL/SHORT/41248.pdf (accessed June 22, 2006).

Chen, Z., K. G. Osadetz, and H. Gao, 2005, The stochastic simulation of undiscovered petroleum accumulations, *in* O. Leuangthong and C. V. Deutsch, eds., Geostatistics Banff 2004, Proceedings of the Seventh International Geostatistics Congress: New York, Springer, p. 661–670.

Chiasson, P., J. Lafleur, M. Soulie, and K. T. Law, 1995, Characterizing spatial variability of a clay by geostatistics: Canadian Geotechnical Journal, v. 32, p. 1–10.

Chihi, H., A. Galli, C. Ravenne, M. Tesson, and G. de Marsily, 2000, Estimating the depth of stratigraphic units from marine seismic profiles using nonstationary geostatistics: Natural Resources Research, v. 9, p. 77–95.

Chilès, J.-P., 1988, Fractal and geostatistical methods for modeling of a fracture network: Mathematical Geology, v. 20, p. 631–654.

Chilès, J.-P., 1993, Geostatistical modeling of a single fracture, *in* A. Soares, ed., Geostatistics troia '92: Dordrecht, The Netherlands, Kluwer Academic Publishers, p. 95–103.

Chilès, J. P., 2005, Stochastic modelling of natural fractured media: A review, Journel, A. G., 2005,

Beyond covariance: The advent of multiple-point geostatistics, *in* O. Leuangthong and C. V. Deutsch, eds., Geostatistics Banff 2004, Proceedings of the Seventh International Geostatistics Congress: New York, Springer, p. 285–294.

Chilès, M. P., and P. Delfiner, 1977, Conditional simulations: A new Monte Carlo approach to probabilistic evaluation of hydrocarbon in place: Fontainebleau, Ecole de Mines de Paris, Centre de Geostatistique et de Morphologie Mathematique, Notes et Etudes, N526, 30 p.

Chopra, A. K., C. D. Severson, and S. R. Carhart, 1990, Evaluation of geostatistical techniques for reservoir characterization: Society of Petroleum Engineers Annual Technical Conference and Exhibition, SPE Paper 20734.

Cojan, I., A. Galli, S. Lopez, and J. Rivoirard, 2003a, Process-based stochastic modeling of meandering channelized reservoirs (abs.): AAPG Annual Meeting Program, v. 12, http://www.searchanddiscovery.com/documents/abstracts/2005annual_calgary/abstracts/cojan.htm (accessed June 22, 2006).

Cojan, I., S. Lopez, and J. Rivoirard, 2003b, Spatial and temporal distribution of facies in process-based stochastic fluvial reservoir modeling: AAPG Annual Meeting Program, v. 12, http://www.searchanddiscovery.com/documents/abstracts/annual2003/short/79417.PDF (accessed June 22, 2006).

Cojan, I., O. Fouché, S. Lopéz, and J. Rivoirard, 2004, Representation of geological heterogeneities by a forward and stochastic modeling of meandering fluvial system (abs.): AAPG Annual Meeting Program, v. 13, http://www.searchanddiscovery.com/documents/abstracts/annual2004/Dallas/Cojan.htm (accessed June 22, 2006).

Cojan, I., O. Fouché, S. Lopéz, and J. Rivoirard, 2005, Processed-based reservoir modeling in the example of meandering channel, *in* O. Leuangthong and C. V. Deutsch, eds., Geostatistics Banff 2004, Proceedings of the Seventh International Geostatistics Congress: New York, Springer, p. 611–620.

Cranwell, R. M., and F. A. Donath, 1980, An application of geometric probability to the existence of faults in anisotropic media, *in* C. J. M. Northrup, ed., Scientific bases for nuclear waste management: Proceedings of the Materials Research Society Annual Meeting: New York, Plenum Press, p. 787–794.

Czubek, J. A., 1976, Comparison of nuclear well logging data with the results of core analysis: The significance of nuclear well logging in gelogical exploration, *in* Nuclear techniques in geochemistry and geophysics: Vienna, International Atomic Energy Agency, p. 93–106.

da Costa e Silva, A. J., 1985, A new approach to the characterization of reservoir heterogeneity based on the geomathematical model and kriging technique: Society of Petroleum Engineers Annual Technical Conference and Exhibition, SPE Paper 14275.

da Costa e Silva, A. J., L. Ramos, R. Guerra, J. Rodrigues, A. Soares, and E. Pires, 1996a, Use of seismic to improve reservoir characterization: Society of Petroleum Engineers European Reservoir Modeling Conference, Stavanger, SPE Paper 35484.

da Costa e Silva, A. J., F. P. Silva, A. Soares, and K. J. Weber, 1996b, Fluviatille reservoir modelling-integration of geological, seismic and production data with stochastic modelling techniques for reservoir management optimization: Society of Petroleum Engineers Abu Dhabi International Petroleum Exhibition and Conference, SPE Paper 36190.

da Costa e Silva, A. J., A. Soares, J. A. Almeida, L. Ramos, and J. Carvalho, 1997, A multi-step approach for modeling oil reservoir lithologies and petrophysical attributes based on the integration of geostatistics and geology, *in* E. Y. Baafi and N. A. Schofield, eds., Geostatistics Wollongong '96: Dordrecht, The Netherlands, Kluwer Academic Publishers, p. 327–338.

da Cruz, P. S., R. N. Horne, and C. V. Deutsch, 2004, The quality map: A tool for reservoir uncertainty quantification and decision making: SPE Reservoir Evaluation and Engineering, v. 7, p. 6–14.

Dagbert, M., M. David, D. Crozel, and A. Desbarats, 1984, Computing variograms in folded strata-controlled deposits, *in* G. Verly, M. David, A. G. Journel, and A. Marechal, eds., Geostatistics for natural resources characterization: Dordrecht, The Netherlands, D. Reidel Publishing Company, p. 71–89.

Daly, C., and G. W. Verly, 1994, Geostatistics and data integration, *in* R. Dimitrakopoulos, ed., Geostatistics for the next century: Dordrecht, The Netherlands, Kluwer Academic Publishers, p. 94–107.

Damsleth, E., and L. Holden, 1994, Mixed reservoir characterization methods: Society of Petroleum Engineers University of Tulsa Centennial Petroleum Engineering Symposium, August 29–31, SPE Paper 27969.

Damsleth, E., and H. Omre, 1996, Geostatistical approaches in reservoir evaluation: Journal of Petroleum Technology, v. 49, p. 498–501.

Dargahi-Houbary, G. R., 1992, Stochastic modeling based on deterministic formulation: An example: Mathematical Geology, v. 24, p. 161–169.

Datta-Gupta, A., L. W. Lake, and G. A. Pope, 1995, Characterizing heterogeneous permeable media with spatial statistics and tracer data using sequential simulated annealing: Mathematical Geology, v. 27, p. 763–787.

Datta-Gupta, A., Xue, G., and S. H. Lee, 1999, Nonparametric transformations for data correlation and integration: From theory to practice, *in* R. Schatzinger and J. Jordan, eds., Reservoir characterization—Recent advances: AAPG Memoir 71, p. 381–396.

Davis, J. M., F. M. Phillips, and J. L. Wilson, 1993, Using geological information in geostatistical modeling (abs.): Geological Society of America Abstracts with Programs, v. 25, p. 37.

Davis, M. W., and P. G. Culhane, 1983, Contouring very large datasets using kriging, *in* G. Verly, M. David, A. G. Journel, and A. Marechal, eds., Geostatistics for natural resources characterization: Dordrecht, The Netherlands, D. Reidel Publishing Company, p. 599–619.

Delfiner, P., J. P. Delhomme, and J. Pelissier-Combescure, 1983, Application of geostatistical analysis to the evaluation of petroleum reservoirs with well logs, *in* Proceedings of the 1983 SPWLA Annual Logging Symposium: Society of Professional Well Log Analysts, Paper WW, 26 p.

Derksen, R., and W. Nachtmann, 1995, Geostatistics—A tool to map thin reservoirs, *in* Proceedings of the 8th European Symposium on Improved Oil Recovery: European Association of Petroleum Geoscientists, p. 96–104.

Dershowitz, W., P. LaPointe, and T. Cladouhos, 1998, Derivation of fracture spatial pattern parameters from borehole data, *in* J. Orozco and J. Schmitter, eds., NARMS '98, Proceedings of the 3rd North American Rock Mechanics Symposium: Amsterdam, Elsevier, p. 508.

Desbarats, A. J., 1994, Geostatistical analysis of water saturation in heterogeneous unsaturated media under capillary-gravity equilibrium, *in* R. Dimitrakopoulos, ed., Geostatistics for the next century: Dordrecht, The Netherlands, Kluwer Academic Publishers, p. 359–370.

Deutsch, C., 1988, Estimating block effective permeability with geostatistics and power averaging: Society of Petreoleum Engineers, SPE Paper 15991.

Deutsch, C., 1989, Calculating effective absolute permeability in sandstone/shale sequences: SPE Formation Evaluation, v. 4, p. 343–348.

Deutsch, C. V., 1993, Conditioning reservoir models to well test information, *in* A. Soares, ed., Geostatistics tróia '92: Dordrecht, The Netherlands, Kluwer Academic Publishers, p. 505–518.

Deutsch, C. V., 1994, Algorithmically-defined random function models, *in* R. Dimitrakopoulos, ed., Geostatistics for the next century: Dordrecht, The Netherlands, Kluwer Academic Press, p. 393–406.

Deutsch, C. V., 1998, Cleaning categorical variable (lithofacies) realizations with maximum a-posteriori selection: Computers and Geosciences, v. 24, p. 551–563.

Deutsch, C. V., and P. W. Cockerham, 1994, Geostatistical modeling of permeability with annealing cosimulation (ACS): SPE Annual Technical Conference and Exhibition, SPE Paper 28413.

Deutsch, C. V., and T. A. Hewett, 1996, Challenges in reservoir forecasting: Mathematical Geology, v. 28, p. 829–843.

Deutsch, C. V., and A. G. Journel, 1991a, The application of simulated annealing to stochastic reservoir modeling: Society of Petroleum Engineers Annual Technical Conference and Exhibition, SPE Paper 23565.

Deutsch, C. V., and A. G. Journel, 1991b, The application of simulated annealing to stochastic reservoir modeling: SPE Applied Technology Series, v. 2, p. 222–227.

Deutsch, C. V., and A. G. Journel, 1994, Integrating well test-derived effective absolute permeabilities in geostatistical reservoir modeling, *in* J. M. Yarus and R. L. Chambers, eds., Stochastic modeling and geostatistics: Principles, methods, and case studies: AAPG Computer Applications in Geology 3, p. 131–142.

Deutsch, C. V., and R. Lewis, 1992, Advances in the practical implementation of indiator geostatistics, *in* Proceedings of the 23rd International Symposium of the Application of Computers and Operations Research in the Mineral Industry: Society of Mining Engineers, p. 169–179.

Deutsch, C. V., and S. Srinivasan, 1996, Improved reservoir management through ranking stochastic reservoir models, *in* Proceedings, Society of Petroleum Engineers–Department of Energy 10th Symposium on Improved Oil Recovery: Society of Petroleum Engineers, SPE/DOE Paper 35411, p. 105–113.

Deutsch, C. V., and L. Wang, 1996, Hierarchical object-based stochastic modeling of fluvial reservoirs: Mathematical Geology, v. 28, p. 857–880.

Deutsch, C. V., S. Srinivasan, and Y. Mo, 1996, Geostatistical reservoir modeling accounting for precision and scale of seismic data: Society of Petroleum Engineers Annual Technical Conference and Exhibition, SPE Paper 36497.

Deutsch, C. V., M. J. Pyrcz, and T. T. Tran, 2002, Geostatistical assignment of reservoir properties on unstructured grids: Society of Petroleum Engineers Annual Technical Conference and Exhibition, SPE Paper 77427.

Dimitrakopoulos, R., 1989, Characterization of petroleum reservoirs using geostatistics (abs.), *in* M. Armstrong, ed., Geostatistics, Proceedings of the 3rd International Geostatistics Congress: Dordrecht, The Netherlands, Kluwer Academic Publishers, p. 1010.

Dimitrakopoulos, R., 1993, Artificial intelligence developments in geostatistical reservoir characterization, *in* B. Linville, T. E. Burchfield, and T. C. Wesson, eds., Reservoir characterization III: Tulsa, PennWell Books, p. 453–470.

Dimitrakopoulos, R., and M. Dagbert, 1993, Sequential modelling of relative indicator variables: Dealing with multiple lithology types, *in* A. Soares, ed., Geostatistics tróia '92: Dordrecht, The Netherlands, Kluwer Academic Publishers, p. 413–424.

Dimitrakopoulos, R., and A. J. Desbarats, 1993, Geostatistical modeling of grid block permeabilities for three-dimensional reservoir simulators: SPE Reservoir Engineering, v. 8, p. 13–18.

Ding, L. Y., R. K. Mehra, and J. K. Donnelly, 1989, Stochastic modeling in reservoir simulation, *in* Proceedings of the 10th Symposium on Reservoir Simulation: Society of Petroleum Engineers, p. 303–320.

Doligez, B., C. Ravenne, P. Lemouzy, and A. Galli, 1992, Une méthodologie pour une étudé intégrée des réservoirs: Des données de puits aux ximulations d'écoulement en utilisant des outils géostatistiques: Pétroles et Techniques, v. 372, p. 43–47.

Doligez, B., H. Beucher, and F. Geffroy, and R. Exchard, 1999a, Integrated reservoir characterization: Improvement in heterogeneous stochastic modeling by integration of additional external constraints, *in* R. Schatzinger and J. Jordan, eds., Reservoir characterization—Recent advances: AAPG Memoir 71, p. 330–342.

Doligez, B., D. Granjeon, P. Joseph, R. Eschard, and H. Beucher, 1999b, How can stratigraphic modeling help constrain geostatistical reservoir simulations? *in* J. W. Harbaugh, W. L. Watney, E. C. Rankey, R. Slingerland, R. H. Goldstein, and E. K. Franseen, eds., Numerical experiments in stratigraphy: Recent advances in stratiraphic and sedimentologic computer simulations: SEPM Special Publication 62, p. 239–244.

Doligez, B., G. Perrine, A. Benoit, F. Frederique, and H. Beucher, 2003, Use of seismic to constrain geostatistical reservoir models: A quantitative approach using proportions of facies (abs.): AAPG Annual Meeting Program, v. 12, http://www.searchanddiscovery.com/documents/abstracts/annual2003/extend/78165.PDF (accessed June 22, 2006).

Donath, F. A., and R. M. Cranwell, 1981, Probabilistic treatment of faulting in geologic media, *in* J. Handin and N. L. Carter, eds., Mechanical behavior of crustal rocks: American Geophysical Union Geophysical Monograph 24, p. 231–241.

Dowd, P. A., 1983, Conditional simulation of interrelated beds in an oil deposit, *in* G. Verly, M. David, A. G. Journel, and A. Marechal, eds., Geostatistics for natural resources characterization: Dordrecht, The Netherlands, D. Reidel Publishing Company, p. 1031–1043.

Dowd, P. A., 1986, Geometrical and geological controls in geostatistical estimation, *in* R. V. Ramani, ed., Proceedings of the 19th International Symposium of the Application of Computers and Operations Research to the Computer Industry: Society of Mining Engineers, p. 81–84.

Dowd, P. A., 1993, Geological and structural control in kriging, *in* A. Soares, ed., Geostatistics troia '92: Dordrecht, The Netherlands, Kluwer Academic Publishers, p. 923–936.

Doyen, P., A. Journel, and A. Nur, 1984, Porosity mapping in petroleum reservoirs using seismic data: A geostatistical approach (abs.): Society of Exploration Geophysicists Abstracts, v. 3, p. 328.

Doyen, P. M., 1988, Porosity from seismic data: A geostatistical approach: Geophysics, v. 53, p. 1263–1275.

Doyen, P. M., T. M. Guidish, and M. De Buyl, 1991, Lithology prediction from seismic data, a Monte-Carlo approach, *in* L. W. Lake, H. B. Carroll Jr., and T. C. Wesson, eds., Reservoir characterization II: San Diego, Academic Press, p. 557–564.

Doyen, P. M., D. E. Psaila, L. D. den Boer, and K. Hirsche, 1997, Seismic porosity mapping in a 3D curvilinear grid using a new form of cokriging (abs.): European Association of Geoscientists and Engineers (EAGE), 59th Conference and Technical Exhibition, p. C006.

Dubois, G., and M. Saisana, 2002, Optimizing spatial declustering weights—Comparison of methods, *in* U. Bayer, H. Burger, and W. Skala, eds., IAMG 2002, Proceedings of the 8th Annual Conference of the International Association for Mathematical Geology: Berlin, Terra Nostra, p. 473–478.

Dubrule, O., 1993, Introducing more geology in stochastic reservoir modeling, *in* A. Soares, ed., Geostatistics troia '92: Dordrecht, The Netherlands, Kluwer Academic Publishers, p. 351–370.

Dubrule, O., and H. H. Haldorsen, 1986, Geostatistics for permeability estimation, *in* L. W. Lake and H. B. Carroll, eds., Reservoir characterization: San Diego, Academic Press, Inc., p. 223–247.

Dubrule, O., A. Haas, and J. C. Herweijer, 1994, Geostatistics: A way towards data integration and uncertainty quantification, *in* Proceedings of the 14th World Petroleum Congress: New York, John Wiley & Sons, p. 265–276.

Dubrule, O., M. Thibaut, and A. Haas, 1998, Geostatistical reservoir characterization constrained by 3D seismic data: Petroleum Geoscience, v. 4, p. 121–128.

Easley, D. H., L. E. Borgman, and P. N. Shive, 1990, Geostatistical simulation for geophysical applications: Part I—Simulation: Geophysics, v. 55, p. 1435–1440.

Eide, A., H. Omre, and B. Ursin, 1997, Stochastic reservoir characterization conditioned on seismic data, *in* E. Y. Baafi and N. A. Schofield, eds., Geostatistics Wollongong '96: Dordrecht, The Netherlands, Kluwer Academic Publishers, p. 442–453.

Eide, A., H. Omre, and B. Ursin, 2002, Prediction of reservoir variables based on seismic data and well observations: Journal of the American Statistical Association, v. 97, p. 18–28.

Einstein, H. H., and G. B. Baecher, 1983, Probabilistic and statistical methods in engineering geology: Part I. Exploration: Rock Mechanics and Rock Engineering, v. 16, p. 39–72.

Emanuel, A. S., G. K. Alameda, R. A. Behrens, and T. A. Hewett, 1989, Reservoir performance prediction methods based on fractal geostatistics: SPE Reservoir Engineering Journal, v. 4, p. 311–318.

Emanuel, A. S., G. K. Alameda, R. A. Behrens, and T. A. Hewett, 1990, Authors' reply to discussion of "Reservoir performance prediction methods based on fractal geostatistics:" SPE Reservoir Engineering, v. 5, p. 68.

Eschard, R., B. Doligez, D. Rahon, C. Ravenne, and G. Le Loc'h, 1991, A new approach for reservoir description and simulation using geostatistical methods, *in* Proceedings, Advances in Reservoir Technology: Characterization, modeling and management: London, IBC Technical Services Ltd., 3 p.

Eschard, R., B. Doligez, and H. Beucher, 2002, Using quantitative outcrop databases as a guide for geological reservoir modeling, *in* M. Armstrong, C. Bettini, N. Champigny, A. Galli, and A. Remacre, eds., Geostatistics Rio 2000: Dordrecht, The Netherlands, Kluwer Academic Publishers, p. 7–18.

Fang, J. H., H. C. Chen, and P. B. Visscher, 1992, Reservoir characterization via geostatistical and fractal simulations: Society of Petroleum Engineers Annual Technical Conference and Exhibition, SPE Paper 25525.

Farmer, C. L., 1988, The generation of stochastic fields of reservoir parameters with specified geostatistical distributions, *in* S. Edwards and P. King, eds., Mathematics in oil production: Oxford, Clarendon Press, p. 235–252.

Fayers, F. J., and T. A. Hewett, 1992, A review of current trends in petroleum reservoir description and assessing the impacts on oil recovery, *in* T. F. Russell, R. E. Ewing, C. A. Brebbia, W. G. Gray, and G. F. Pinder, eds., Computational methods in water resources IX, v. 2: Mathematical modeling in water resources: Boston, Computational Mechanics Publications, p. 3–33.

Felgueiras, C. A., S. D. Fuks, and A. M. V. Monteiro, 2001, Raster representations of spatial attributes with uncertainty assessment using nonlinear stochastic simulation, *in* D. V. Pullar, ed., Proceedings, 6th International Conference on Geocomputation: Brisbane, University of Queensland, 6 p.

Felletti, F., R. Bersezio, H. Beucher, A. Galli, and R. Potenza, 1999, Vertical and horizontal variograms as a tool to quantify facies trends and sequences in an heterogeneous turbiditic lithosome, *in* S. J. Lippard, A. Naess, and R. Sinding-Larsen, eds., IAMG '99, Proceedings of the 5th Annual Conference of the International Association for Mathematical Geology, Trondheim, Tapir, p. 615–620.

Fichtl, P., and J. J. Royer, 1999, Facies based sequential indicator cosimulation: A collocated cokriging approach to integrating seismic data, *in* S. J. Lippard, A. Naess, and R. Sinding-Larsen, eds., IAMG '99, Proceedings of the 5th Annual Conference of the International Association for Mathematical Geology, Trondheim, Tapir, p. 473–478.

Fichtl, P., F. Fournier, and J. J. Royer, 1997, Cosimulations of lithofacies and associated reservoir properties using well and seismic data: Society of Petroleum Engineers Annual Technical Conference and Exhibition, SPE Paper 38680, 13 p.

Fontanilla, J. P., A. A. Al-Khalawi, and S. G. Johnson, 1995, Performance prediction using geostatistics and window reservoir simulation: Society of Petroleum Engineers 9th Middle East Oil Conference, p. 617–642.

Fournier, F., and J. F. Derain, 1997, A statistical methodology for deriving reservoir properties from seismic data: Geophysics, v. 60, no. 5, p. 1437–1450.

Frykman, P., and C. Deutsch, 1999, Geostatistical scaling laws applied to core and log data: Society of Petroleum Engineers Annual Technical Conference and Exhibition, SPE Paper 56822.

Galli, A., and G. Meunier, 1987, Study of a gas reservoir using the external drift method, *in* G. Matheron and M. Armstrong, eds., Geostatistical case studies: Dordrecht, The Netherlands, D. Reidel Publishing Company, p. 105–109.

Galli, A., D. Guérillot, and C. Ravenne, 1990, Combining geology, geostatistics, and multiphase flow for 3D reservoir studies, *in* D. Guérillot and O. Guillon, eds., Proceedings, 2nd European Conference on the Mathematics of Oil Recovery: Paris, Editions Technip, p. 11–19.

García, M., D. Allard, D. Foulon, and S. O. Delisle, 2005, Fine scale rock properties: Towards the spatial modeling of regionalized probability distribution functions, *in* O. Leuangthong and C. V. Deutsch, eds., Geostatistics Banff 2004, Proceedings of the Seventh International Geostatistics Congress: New York, Springer, p. 579–590.

Garcia, M. H., A. G. Journel, and K. Aziz, 1992, Automatic grid generation for modeling reservoir heterogeneities: SPE Reservoir Engineering, v. 7, p. 278–284.

Gardner, E., P. Duncan, and D. Nester, 1998, 3D reservoir description at multiple scales (abs.): AAPG Annual Meeting Program, v. 1, p. A221.

Garner, D., 1997, Geostatistical depth mapping: Integration of seismic times with well tops: Canadian Society of Exploration Geophysicists Recorder, February, 1997, p. 5–12.

Gastaldi, C., D. Roy, P. Doyen, and L. den Boer, 1998, Using Bayesian simulations to predict reservoir thickness under tuning conditions: The Leading Edge, v. 17, p. 539.

Georgsen, F., T. Egeland, R. Knarud, and H. Omre, 1994, Conditional simulation of facies architecture in fluvial reservoirs, *in* M. Armstrong and P. A. Dowd, eds., Geostatical simulations: Dordrecht, The Netherlands, Kluwer Academic Publishers, p. 235–250.

Geraets, D., P. Ruffo, and A. Galli, 2003, Stochastic modeling of seismic velocity fields (abs.): AAPG Annual Meeting Program, v. 12, http://www.searchanddiscovery.com/documents/abstracts/annual2003/short/79183.PDF (accessed June 22, 2006).

Gerdes, M. L., L. P. Baumgartner, and M. Person, 1995, Stochastic permeability models of fluid flow during contact metamorphism: Geology, v. 23, p. 945–948.

Ghori, S. G., 1992, Effect of four geostatistical methods on reservoir description and flow mechanism: Society of Petroleum Engineers Annual Technical Conference and Exhibition, SPE Paper 24755.

Ghose, S., 1989, Use of geostatistics in planning optimum drilling program (abs.): AAPG Bulletin, v. 73, p. 1031.

Gomez-Hernandez, J. J., and C. V. Deutsch, 1999, Foreword to the special issue: Modeling subsurface flow: Mathematical Geology, v. 31, p. 747.

Gomez-Hernandez, J. J., and C. Guardiola-Albert, 2005, Inverse-conditional simulation of relative permeabilities, Journel, A. G., 2005, Beyond covariance: The advent of multiple-point geostatistics, *in* O. Leuangthong and C. V. Deutsch, eds., Geostatistics Banff 2004, Proceedings of the Seventh International Geostatistics Congress: New York, Springer, p. 325–332.

Gomez-Hernandez, J. J., and A. G. Journel, 1990, Stochastic characterization of grid-block permeabilities: From point values to block tensors, *in* D. Guérillot and O. Guillon, eds., Proceedings of the 2nd European Conference on the Mathematics of Oil Recovery: Paris, Editions Technip, p. 83–90.

Gomez-Hernandez, J. J., and A. G. Journel, 1994, Stochastic characterization of gridblock permeabilities: SPE Formation Evaluation, v. 9, p. 93–99.

Goovaerts, P., 1999, Combining minimum error variance and spatial variability in the modeling of petrophysical properties, *in* S. J. Lippard, A. Naess, and R. Sinding-Larsen, eds., IAMG '99, Proceedings of the 5th Annual Conference of the International Association for Mathematical Geology, Trondheim, Tapir, p. 29–48.

Gorell, S. B., 1995a, Using geostatistics to aid in reservoir characterization: The Leading Edge, v. 14, p. 967–976.

Gorell, S. B., 1995b, Creating 3-D reservoir models using aerial geostatistical techniques combined with vertical well data: Society of Petroleum Engineers Annual Technical Conference and Exhibition, SPE Paper 29670.

Gotway, C. A., and G. W. Hergert, 1997, Incorporating spatial trends and anisotropy in geostatistical mapping of soil properties: Soil Science Society of America Journal, v. 61, p. 298–309.

Gotway, C. A., R. B. Ferguson, G. W. Hergert, and T. A. Peterson, 1996, Comparison of kriging and inverse-distance methods for mapping soil parameters: Soil Science Society of American Journal, v. 60, p. 1237–1248.

Granjeon, D., J. Philippe, and B. Doligez, 1998, Using a 3-D stratigraphic model to optimize reservoir description—A new model called Dionisos not only simulates reservoir deposition, but helps quantify certain reservoir parmeters needed for adequate characterization: Petroleum Engineer International, v. 71, p. 51–58.

Grijalba-Cuenca, A., C. Torres-Verdin, and P. van der Made, 2000, Geostatistical iniversion of 3D seismic data to extrapolate wireline petrophysical variables laterally away from the well: Society of Petroleum Engineers Annual Technical Conference and Exhibition, SPE Paper 68283.

Gringarten, E., 1996, 3-D geometric description of fractured reservoirs: Mathematical Geology, v. 28, p. 881–893.

Gringarten, E., and C. V. Deutsch, 1999, Methodology for variogram interpretation and modeling for improved reservoir characterization: Society of Petroleum Engineers Annual Technical Conference and Exhibition, SPE Paper 56654.

Guerillot, D. R., and I. F. Morelon, 1992, Sorting equiprobable geostatistical images by simplified flow calculations: Society of Petroleum Engineers Annual Technical Conference and Exhibition SPE Paper 24891.

Guerillot, D. R., P. Lemouzy, C. Ravenne, and A. Galli, 1990, 3D fluid flow behavior in heterogeneous porous media characterized by geostatistical method: Society of Petroleum Engineers Latin American Petroleum Engineering Conference, Rio de Janeiro, SPE Paper 21081.

Guerreiro, L., A. J. Almeida, and A. Soares, 1998, Permeability simulation in poor sampling conditions: Society of Petroleum Engineers Abu Dhabi International Petroleum Exhibition and Conference, SPE Paper 49451.

Guo, Q., Y. Kang, Q. Zhu, and G. Zhao, 1998, New methods for supporting decision-making in petroleum exploration and development: Fractal geometry and geostatistics: Acta Petrolei Sinica, v. 19, p. 36–39.

Haas, A., and O. Dubrule, 1994, Geostatistical inversion—A sequential method of stochastic reservoir modeling constrained by seismic data: First Break, v. 12, p. 561–569.

Haas, A., and C. Jousselin, 1976, Geostatistics in petroleum industry, *in* Proceedings of the 1st North Atlantic Treaty Organization Advanced Study Institute (advanced geostatistics in the mining industy): Dordrecht, The Netherlands, D. Reidel Publishing Company, p. 333–350.

Haas, A., and B. Noetinger, 1997, Stochastic reservoir modelling constrained by well test permeabilities, *in* E. Y. Baafi and N. A. Schofield, eds., Geostatistics Wollongong '96: Dordrecht, The Netherlands, Kluwer Academic Publishers, p. 501–511.

Haas, A., and J. R. J. Viallix, 1975, Krigeage applied to geophysics: Geophysical Prospecting, v. 24, p. 49–69.

Haas, A., J. Herweijer, and O. Dubrule, 1994, Geostatistics for integration of reservoir data (abs.): European Association of Petroleum Geoscientists and Engineers Annual Conference and Technical Exhibition Extended Abstracts, p. C012.

Haas, A., P. Biver, and K. Altisen, 1998, Constrained kriging and simulation of a well layering towards a geophysical thickness map, *in* Proceedings of the 6th European Conference on Mathematics and Oil Recovery (ECMOR VI): European Association of Geoscientists and Engineers, Paper C-31, 5 p.

Haldorsen, H. H., 1986, Simulator parameter assignment and the problem of scale in reservoir engineering, *in* L. W. Lake and H. B. Carroll, eds., Reservoir characterization: San Diego, Academic Press, p. 293–340.

Haldorsen, H. H., and E. Damsleth, 1990, Stochastic

modeling: Journal of Petroleum Technology, v. 42, p. 404–412.

Haldorsen, H. H., and C. J. MacDonald, 1987, Stochastic modeling of underground reservoir facies (SMURF): Society of Petroleum Engineers Annual Technical Conference and Exhibition, SPE Paper 16751.

Hall, D. L., and R. L. Chambers, 2002, 3-D stochastic modeling of Norwegian Sea FIS data depicts regional petroleum system (abs.): AAPG Annual Meeting Program, v. 11, http://www.searchanddiscovery.com/documents/abstracts/annual2002/DATA/2002/13ANNUAL/SHORT/41300.pdf (accessed June 22, 2006).

Halvorsen, K., V. Berteig, and H. Omre, 1989, Prediction of hydrocarbon pore volume by use of Bayesian conditional simulation (abs.), *in* M. Armstrong, ed., Geostatistics, Proceedings of the 3rd International Geostatistics Congress: Dordrecht, The Netherlands, Kluwer Academic Publishers, p. 1011.

Handyside, D. D., O. K. Karaoquz, R. H. Deskin, and G. A. Mattson, 1992, A practical appliation of stochastic modeling techniques for turbidite reservoirs: Society of Petroleum Engineers Annual Technical Conference and Exhibition, SPE Paper 24892.

Harding, A., S. Strebelle, M. Levy, J. Thorne, D. Xie, S. Leigh, and R. Preece, 2005, Reservoir facies modeling: New advances in MPS, *in* O. Leuangthong and C. V. Deutsch, eds., Geostatistics Banff 2004, Proceedings of the Seventh International Geostatistics Congress: New York, Springer, p. 559–568.

Harris, P. M., 1998, Methods for determining interwell facies boundaries (and constraining geostatistical correlation) (abs.): AAPG Annual Meeting Program, v. 1, p. A275.

Hatløy, A., 1994, Numerical facies modeling combining deterministic and stochastic methods, *in* J. M. Yarus and R. L. Chambers, eds., Stochastic modeling and geostatistics: Principles, methods, and case studies: AAPG Computer Applications in Geology 3, p. 109–120.

Hayes, S. J., D. Mundy, and K. Safton, 2002, Improving recovery predictions from fluvial and shallow marine reservoirs using combined Boolean and Gaussian facies models (abs.): AAPG Annual Meeting Program, v. 11, http://www.searchanddiscovery.com/documents/abstracts/annual2002/DATA/2002/13ANNUAL/SHORT/41281.pdf (accessed June 22, 2006).

He, N., D. S. Oliver, and A. C. Reynolds, 2000, Conditioning stochastic reservoir models to well-test data: SPE Reservoir Evaluation and Engineering, v. 3, p. 74–79.

Hegstad, B. K., and H. Omre, 1997, Uncertainty assessment in history matching and forecasting, *in* E. Y. Baafi and N. A. Schofield, eds., Geostatistics Wollongong '96: Dordrecht, The Netherlands, Kluwer Academic Publishers, p. 585–596.

Hegstad, B. K., and H. Omre, 2001, Uncertainty in production forecasts based on well observations, seismic data and production history: Society of Petroleum Engineers Journal, v. 6, p. 409–424.

Helwick Jr., S. J., and G. R. Luster, 1983, Fluid-flow modeling using a conditional simulation of porosity and permeability, *in* G. Verly, M. David, A. G. Journel, and A. Marechal, eds., Geostatistics for natural resources characterization: Dordrecht, The Netherlands, D. Reidel Publishing Company, p. 635–650.

Hernandez-Garcia, G., 1989, Automatic history matching in oil reservoir simulation: A geostatistical approach, *in* M. Armstrong, ed., Geostatistics, Proceedings of the 3rd International Geostatistics Congress: Dordrecht, The Netherlands, Kluwer Academic Publishers, p. 1012.

Herweijer, J. C., and O. Dubrule, 1994, Detailed screening of geostatistical models by comparing simulated three-dimensional pressure diffusion with production tests: Society of Petroleum Engineers Annual Technical Conference and Exhibition, SPE Paper 28434.

Herweijer, J. C., and O. Dubrule, 1995, Screening of geostatistical reservoir models with pressure transients: Journal of Petroleum Technology, v. 47, p. 973–987.

Herzfeld, U. C., O. Zahner, and J. Kole, 1997, Geostatistical surface classification using deterministic and connectivistic algorithms, *in* V. Pawlowsky-Glahn, ed., Proceedings of IAMG '97, 3rd Annual Conference of the International Association for Mathematical Geology: Barcelona, International Center for Numerical Methods in Engineering, p. 659–664.

Hewett, R. A., and R. A. Behrens, 1990, Conditional simulation of reservoir heterogeneity with fractals: SPE Formation Evaluation, v. 5, p. 217–225.

Hewett, T. A., 1986, Fractal distributions of reservoir heterogeneity and their influence on fluid transport: Society of Petroleum Engineers Annual Technical Conference and Exhibition, SPE Paper 15386.

Hewett, T. A., 1993, Modelling reservoir heterogeneity with fractals, *in* A. Soares, ed., Geostatistics tróia '92: Dordrecht, The Netherlands, Kluwer Academic Publishers, p. 455–466.

Hewett, T. A., 1994, Fractal methods for fracture characterization, *in* J. M. Yarus and R. L. Chambers, eds., Stochastic modeling and geostatistics: Principles, methods, and case studies: AAPG Computer Applications in Geology 3, p. 249–260.

Hird, K. B., and M. G. Kelkar, 1992, Conditional simulation method for reservoir description using spatial and well performance constraints: Society of Petroleum Engineers 67th Annual Technical Conference and Exhibition, SPE Paper 24750.

Hoffman, K. S., B. F. J. Kelly, D. R. Taylor, and L. Wang, 1997, Ordinary kriging in faulted domains without structural influence (abs.): European Association of Geoscientists and Engineers Annual Conference and Technical Exhibition, p. C013.

Hohn, M. E., and M. V. Fontana, 1986a, Practical application of geostatistics and rule processing to an oil and gas database (abs.), *in* D. C. Peters and S. A. Krajewski, eds., Denver geotech '86, Computer aided methods in geology and engineering: American Institute of Petroleum Geologists, Colorado Section, p. 70–73.

Hohn, M. E., and M. V. Fontana, 1986b, Geostatistics and artificial intelligence applied to stratigraphic correlation (abs.): AAPG Bulletin, v. 70, p. 601.

Hohn, M. E., and M. V. Fontana, 1993, Geostatistical simulation of reservoir heterogeneity in a lower Missisippian sandstone, Appalachian basin (abs.): AAPG Annual Meeting Program, v. 2, p. 120.

Hohn, M. E., and R. R. McDowell, 1994, Geostatistical analysis of oil production and potential using indicator kriging, *in* J. M. Yarus and R. L. Chambers, eds., Stochastic modeling and geostatistics: Principles, methods, and case studies: AAPG Computer Applications in Geology 3, p. 121–130.

Hoiberg, J., H. Omre, and H. Tjelmeland, 1992, A stochastic model for shale distribution in petroleum reservoirs, *in* P. A. Dowd and J. J. Royer, eds., Proceedings of the 2nd CODATA Conference on Geomathematics and Geostatistics: Nancy, Association Scientifique pour la Géologie et ses Applications, p. 135–154.

Houlding, S., 1993, Advances in 3D computer-assisted geological characterization (abs.): Geological Society of America Abstracts with Programs, v. 25, p. 37.

Hove, K., G. Olsen, A. Nilsson, M. Tonnesen, and A. Hatløy, 1992, From stochastic geological description to production forecasting in heterogeneous layered reservoirs: Society of Petroleum Engineers Annual Technical Conference and Exhibition, SPE Paper 24890.

Hu, L. Y., P. Joseph, and O. Dubrule, 1994, Random genetic simulation of the internal geometry of deltaic sand bodies: SPE Formation Evaluation, v. 9, p. 245–250.

Husby, E. G., H. Tjelmeland, and O. Lia, 1999, A stochastic facies model using sequential deposition of facies objects, *in* S. J. Lippard, A. Naess, and R. Sinding-Larsen, eds., IAMG '99, Proceedings of the 5th Annual Conference of the International Association for Mathematical Geology, Trondheim, Tapir, p. 632–647.

Ichikawa, Y., K. Koike, and N. Nakamura, 2000, Spatial distribution modeling of multi-scale fracture system using geostatistics: Geoinformatics, v. 11, p. 78–79.

Ioannidis, M., M. J. Kwiecien, and I. Chatzis, 1995, Computer generation and application of 3-D model porous media: From pore-level geostatistics to the estimation of formation factor, *in* Proceedings of the Petroleum Computer Conference: Society of Petroleum Engineers, p. 185–194.

Jain, A. K., and A. Carnegie, 2000, Value addition through stochastic evaluation of gamma rays; a geostatistical approach to geological modeling and characterization of the reservoir (abs.): AAPG Bulletin, v. 84, p. 1442.

Jaquet, O., 1989, Factorial kriging analysis applied to geological data from petroleum exploration: Mathematical Geology, v. 21, p. 683–692.

Jeffery, R., I. Stewart, and D. Alexander, 1996, Geostatistical estimation of depth conversion velocity using well control and gravity data: First Break, v. 14, p. 313–320.

Jennings Jr., J. W., S. C. Ruppel, and W. B. Ward, 2000, Geostatistical analysis of permeability data and modeling of fluid-flow effects in carbonate outcrops: SPE Reservoir Evaluation and Engineering, v. 3, p. 292–303.

Jensen, J. L., and P. W. M. Corbett, 1999, Methodology for integrating analog geologic data in 3-D variogram modeling; discussion: AAPG Bulletin, v. 85, p. 749–750.

Jensen, J. L., P. W. M. Corbett, and P. S. Ringrose, 1994, Permeability semivariograms, geological structure and flow performance, *in* C.-J. F. Chang, ed., Proceedings of IAMG '94: International Association for Mathematical Geology, p. 181–186.

Jensen, J. L., P. W. M. Corbett, G. E. Pickup, and P. S. Ringrose, 1996, Permeability semivariograms, geological structure, and flow performance: Mathematical Geology, v. 28, p. 419–436.

Jones, T. A., and Y. Z. Ma, 2001, Geologic characteristics of hole-effect variograms calculated from lithology-indicator variables: Mathematical Geology, v. 33, p. 615–630.

Jordan, D. L., and D. J. Goggin, 1995, An application of categorical indicator geostatistics for facies modeling in sand-rich turbidite sytems: Society of Petroleum Engineers Annual Technical Conference and Exhibition, SPE Paper 30603.

Joseph, P., R. Eschard, B. Doligez, and D. Granjeon, 1996, 3D stratigraphic modeling—A new way to constrain geostatistical reservoir simulation (abs.): European Association of Geoscientists and Engineers, 58th Annual Conference and Technical Exhibition, Extended Abstracts, p. L045.

Journel, A., 1975, Geological reconnaissance to exploitation—A decade of applied geostatistics: Canadian Institute of Mining and Metallurgy Bulletin, v. 10, p. 75–84.

Journel, A., and F. Alabert, 1990, New method for reservoir mapping: Journal of Petroleum Technology, v. 42, p. 212–218.

Journel, A. G., 1973, Geostatistics and sequential exploration: Mining Engineering, v. 25, p. 44–48.

Journel, A. G., 1980, Geostatistical simulation methods for exploration and mine planning, *in* A. G. Royle, ed., Geostatistics: New York, McGraw-Hill, p. 93–108.

Journel, A. G., 1986, Geostatistics: Models and tools for the earth sciences: Mathematical Geology, v. 18, p. 110–140.

Journel, A. G., 1987, Reply to comments by J. Serra: Mathematical Geology, v. 19, p. 357–360.

Journel, A. G., 1989, Geostatistical characterization of reservoir heterogeneities (abs.): Society of Exploration Geophysicists Expanded Abstracts, v. 8, p. 593–594.

Journel, A. G., 1990, Geostatistics for reservoir characterization: Society of Petroleum Engineers Annual Technical Conference and Exhibition, SPE Paper 20750.

Journel, A. G., 1994, Geostatistics and reservoir geology, *in* J. M. Yarus and R. L. Chambers, eds., Stochastic modeling and geostatistics: Principles, methods, and case studies: AAPG Computer Applications in Geology 3, p. 19–20.

Journel, A. G., 1996, Conditional simulation of geologically averaged block permeabilities: Journal of Hydrology, v. 183, p. 23–25.

Journel, A. G., 2000, Geostatistics and petroleum geology: Mathematical Geology, v. 32, p. 139–143.

Journel, A. G., and C. V. Deutsch, 1997, Rank order geostatistics: A proposal for a unique coding and common processing of diverse data, *in* E. Y. Baafi and N. A. Schofield, eds., Geostatistics Wollongong '96: Dordrecht, The Netherlands, Kluwer Academic Publishers, p. 174–187.

Journel, A. G., C. Deutsch, and A. J. Desbarats, 1986, Power averaging for block effective permeability: Society of Petroleum Engineers 56th California Regional Meeting, SPE Paper 15128.

Kawahara, H., and S. Miyake, 1993, An application of geostatistical approach to the limestone reservoir with heterogeneity in the reservoir simulation (abs.): Journal of the Japanese Association of Petroleum Technology, v. 58, p. 492–493.

Kelkar, M., and S. Shibli, 1994, Description of reservoir properties using fractals, *in* J. M. Yarus and R. L. Chambers, eds., Stochastic modeling and geostatistics: Principles, methods, and case studies: AAPG Computer Applications in Geology 3, p. 261–272.

Kheloufi, A., and P. Sonnet, 1997, Stochastic simulation of petrophysical properties using sequential Gaussian simulation with and without trend model, *in* V. Pawlowsky-Glahn, ed., Proceedings of IAMG '97, 3rd Annual Conference of the International Association for Mathematical Geology: Barcelona, International Center for Numerical Methods in Engineering, p. 705–710.

Kincare, K. A., and S. M. Aulenbach, 1993, Use of geostatistics to quantify the reliability of mapped data: Geological Society of America Abstracts with Programs, v. 25, p. 211.

Kiraly, L., 1969, Statistical analysis of fractures (orientation density): Geologische Rundschau, v. 59, p. 125–151.

Kolbjørnsen, O., and H. Omre, 1997, Bayesian disjunctive kriging applied to prediction of reservoir volume, *in* E. Y. Baafi and N. A. Schofield, eds., Geostatistics Wollongong '96: Dordrecht, The Netherlands, Kluwer Academic Publishers, p. 609–620.

Kostov, C., and A. G. Journel, 1986, Coding and extrapolating expert information for reservoir characterization, *in* L. W. Lake and H. B. Carroll, eds., Reservoir characterization: San Diego, Academic Press, p. 249–264.

Krivoruchko, K., 2001, Using linear and non-linear kriging interpolators to produce probability maps, *in* IAMG 2001, Proceedings of the 7th Annual Conference: International Association for Mathematical Geology, 18 p.: http://campus.esri.com/campus/library/ConfProc/GeostatisticsTeam/Krivoruchko_2001_UsingLinear.pdf (accessed April 19, 2005).

Krzanowski, R. M., 1989, GIS and geostatistics: Geostatistics (an Interdisciplinary Newsletter), v. 3, no. 3, p. 9–10.

Kupfersberger, H., and C. V. Deutsch, 1999a, Ranking stochastic realizations for improved aquifer response uncertainty assessment: Journal of Hydrology, v. 223, p. 54–65.

Kupfersberger, H., and C. V. Deutsch, 1999b, Methodology for integrating analog geologic data in 3-D variogram modeling: AAPG Bulletin, v. 83, p. 1262–1278.

Lake, L., S. Srinivasan, and A. John, 2005, Statistical scale-up: Concepts and application to reservoir flow simulation practice, *in* O. Leuangthong and C. V. Deutsch, eds., Geostatistics Banff 2004, Proceedings of the Seventh International Geostatistics Congress: New York, Springer, p. 681–690.

Lake, L. W., A. J. Scott, and G. A. Kocurek, 1986, Reservoir characterization for numerical simulation—Final report: Washington, U.S. Department of Energy, Bartlesville Project Office, DOE/BC/10744-8, 226 p.

Lamy, P., P. A. Swaby, P. S. Rowbotham, and O. Dubrule, 1998, From seismic to reservoir properties using geostatistical inversion: Journal of Petroleum Technology, v. 51, p. 38–39.

Lamy, P., P. A. Swaby, P. S. Rowbotham, O. Dubrule, and A. Haas, 1999, From seismic to reservoir properties with geostatistical inversion: SPE Reservoir Evaluation, v. 2, p. 334–340.

Langlais, V., J. D. Doyle, M. L. Sweet, and G. W. Geehan, 1992, An additional geological input to the sequential indicator simulation (SIS): The vertical organization of lithofacies, *in* R. Eschard and B. Doligez, eds., Subsurface reservoir characterization from outcrop observations: Paris, Institut Francais du Petrole, Collection Colloques et Seminaires, p. 111–123.

Lantuéjoul, H. Beucher, J.-P. Chilès, P. Elion, C. Lajaunie, and H. Wackernagel, 2005, Estimating the trace length distribution of fractures from line sampling data, *in* O. Leuangthong and C. V. Deutsch, eds., Geostatistics Banff 2004, Proceedings of the Seventh International Geostatistics Congress: New York, Springer, p. 165–174.

Lanzarini, W. L., C. A. Poleteto, G. Tavares, S. Pesco, and H. Lopes, 1997, Stochastic modeling of geometric objects and reservoir heterogeneities: Society of Petroleum Engineers Fifth Latin American and Caribbean Petroleum Engineering Conference and Exhibition (Rio de Janeiro), SPE Paper 38953.

Lanzarini, W. L., A. B. Barreto, M. B. dos Santos, G. Tavares, H. Lopes, and S. Pesco, 1998, Object-based stochastic modeling of turbidite reservoirs geometry: AAPG International Conference and Exhibition (Rio de Janeiro) Extended Abstracts, p. 750–751.

LaPointe, P. R., 1980, Analysis of the spatial variation in rock mass properties through geostatistics, *in* D. A. Summers, ed., Proceedings of the 21st Symposium on Rock Mechanics: Rolla, University of Missouri, p. 570–580.

LaPointe, P. R., and C. C. Barton, 1995, Creating reservoir simulations with fractal characteristics, *in* C. C. Barton and P. R. LaPointe, eds., Fractals in petroleum geology and earth processes: New York, Plenum Press, p. 263–276.

Larrondo, P., and C. V. Deutsch, 2005, Accounting for geological boundary for simulation of multiple rock types, *in* O. Leuangthong and C. V. Deutsch, eds., Geostatistics Banff 2004, Proceedings of the Seventh International Geostatistics Congress: New York, Springer, p. 3–12.

Lecour, M., R. Cognot, J.-C. Dulac, and P. Thore, 1999, From a-priori to a stochastic models: How to incorporate uncertainties in fault modeling, *in* S. J. Lippard, A. Naess, and R. Sinding-Larsen, eds., IAMG '99, Proceedings of the 5th Annual Conference of the International Association for Mathematical Geology, Trondheim, Tapir, p. 639–644.

le Ravalec, M., and L. Y. Hu, 1999, Stochastic reservoir modeling constrained to dynamic data: Local calibration and inference of the structural parameters: Society of Petroleum Engineers Annual Technical Conference and Exhibition, SPE Paper 56556.

Leung, J., and S. Srinivasan, 2005, Direct assessment of uncertainty using stochastic flow simulation, *in* O. Leuangthong and C. V. Deutsch, eds., Geostatistics Banff 2004, Proceedings of the Seventh International Geostatistics Congress: New York, Springer, p. 631–642.

Lia, O., H. Omre, H. Tjelmeland, L. Holden, and T. Egeland, 1997, Uncertainties in reservoir production forecasts: AAPG Bulletin, v. 81, p. 775–802.

Liu, X., and S. Srinivasan, 2005, Field scale stochastic modeling of fracture networks: Combining pattern statistics with geomechanical criteria for fracture growth, *in* O. Leuangthong and C. V. Deutsch, eds., Geostatistics Banff 2004, Proceedings of the Seventh International Geostatistics Congress: New York, Springer, p. 75–84.

Liu, X., S. Srinivasan, and D. W. Wong, 2002, Geological characterization of naturally fractured reservoirs using multiple point geostatistics: Society of Petroleum Engineers, Society of Petroleum Engineers–Department of Energy Symposium on Improved Oil Recovery, SPE Paper 75246.

Liu, Y., 2003, Multiple-point geostatistics: Integration of geological concepts with numerical data: AAPG Mid-Continent Section Meeting, October 12–14, Tulsa, Oklahoma, Abstracts, http://www.searchanddiscovery.com/documents/abstracts/2003midcont/abs/liu.pdf (accessed June 22, 2006).

Liu, Y., A. Hardin, W. Abriel, and S. Strebelle, 2004, Multiple-point simulation integrating wells, 3-D seismic data and geology: AAPG Bulletin, v. 88, p. 905–921.

Lloyd, C. D., J. M. McKinley, and A. H. Ruffell, 2003, Conditional simulation of sandstone permeability, *in* J. Cubitt, ed., IAMG 2003, Proceedings of the 9th Annual Conference: International Association for Mathematical Geology, 5 p.: www.iamg.org/meetings/Proceedings_2003/papers/Lloyd.pdf (accessed April 19, 2005).

Long, J. C. S., and D. M. Billaux, 1986, The use of geostatistics to incorporate spatial variability in the modeling of flow through fracture networks: Berkeley, Lawrence Berkeley Laboratory, LBL-21439, U.S. Department of Energy Contract DE-AC03-76SF00098, 83 p.

Lozano, J. A., L. P. Costa, F. B. Alves, and A. da Costa e Silva, 1996, Upscaling of stochastic models for reservoir simulation—An integrated approach: Society of Petroleum Engineers Abu Dhabi International Petroleum Exhibition and Conference, SPE Paper 36205.

Lumb, P., 1975, Spatial variability of soil properties, *in* E. Schultze, Proceedings of the 2nd International Conference, Applications of Statistics and Probability in Soil and Structural Engineering (Aachen, Germany): Essen, Germany, Deutsche Gesellschaft für Erd- und Grundbau, p. 397–421.

Lunn, R. J., and R. Mackay, 1997, Geostatistical analysis of fractured rock characteristics, *in* A. Soares, J. Gomez-Hernandez, and R. Froidevaux, eds., GeoENV I, geostatistics for environmental applications: Dordrecht, The Netherlands, Kluwer Academic Publishers, p. 25–38.

Ma, Y. Z., and T. A. Jones, 2001, Modeling hole-effect variograms of lithology indicator variables: Mathematical Geology, v. 33, p. 631–648.

MacDonald, A. C., and J. O. Aasen, 1993, Parameter estimation for stochastic models of fluvial channel reservoirs (abs.): AAPG Bulletin, v. 77, p. 1643–1644.

MacDonald, A. C., and J. O. Aasen, 1994, A prototype procedure for stochastic modeling of facies tract distribution in shoreface reservoirs, *in* J. M. Yarus and R. L. Chambers, eds., Stochastic modeling and geostatistics: Principles, methods, and case studies: AAPG Computer Applications in Geology 3, p. 91–108.

Macdonald, A. C., L. M. Fält, and A.-L. Hektoen, 1998, Stochastic modeling of incised valley geometries: AAPG Bulletin, v. 82, p. 1156–1172.

MacMillan, J. R., and A. L. Gutjahr, 1986, Geologic controls on spatial variability for one-dimensional arrays of porosity and permeability normal to layering, *in* L. W. Lake and H. B. Carroll, eds., Reservoir characterization: San Diego, Academic Press, p. 265–292.

Mancini, E. A., 1990, Establishment of an oil and gas database for increased recovery and characterization of oil and gas carbonate reservoir heterogeneity: Washington, U.S. Department of Energy, DOE/BC/14425-T4, 6 p.

Mansoori, J., 1994, A review of basic upscaling procedures: Advantages and disadvantages, *in* J. M. Yarus and R. L. Chambers, eds., Stochastic modeling and geostatistics: Principles, methods, and case studies: AAPG Computer Applications in Geology 3, p. 65–74.

Mao, N. H., 1985, Interpreting geological structure using kriging: American Institute of Mining, Metallurgical and Petroleum Engineers Preprint No. 85-347, 7 p.

Maréchal, A., 1984, Kriging seismic data in presence of faults, *in* G. Verly, M. David, A. G. Journel, and A. Marechal, eds., Geostatistics for natural resources characterization: Dordrecht, The Netherlands, D. Reidel Publishing Company, p. 271–294.

Maslyn, R. M., 1986, Kriging to determine reservoir thickness and geometry (abs.), *in* D. C. Peters and S. A. Krajewski, eds., Denver geotech '86, Computer aided methods in geology and engineering: American Institute of Petroleum Geologists Colorado Section, p. 149–158.

Massonnat, G., 1999, Breaking of a paradigm: Geology can provide 3D complex probability fields for stochastic facies modeling: Society of Petroleum Engineers Annual Technical Conference and Exhibition, SPE Paper 56652.

Mateu, J., and P. J. Ribeiro, 1999, Geostatistical data versus point process data: Analysis of second-order characteristics, *in* J. Gomez-Hernandez, R. Froidevaux, and A. Soares, eds., Geostatistics for environmental applications: Dordrecht, The Netherlands, Kluwer Academic Publishers, p. 213–224.

Matheron, G., H. Beucher, C. De Fouquet, A. Galli, D. Guérillot, and C. Ravenne, 1987, Conditional simulation of the geometry of fluvio-deltaic reservoirs: Society of Petroleum Engineers Annual Technical Conference and Exhibition, SPE Paper 16753.

Mayer, L., M. Mergner-Keefer, and C. M. Wentworth, 1981, Probability models and computer simulation of landscape evolution: U.S. Geological Survey Open-file Report OFR-81-656, 37 p.

McBratney, A. B., and R. Webster, 1986, Choosing functions for semi-variograms of soil properties and fitting them to sampling estimates: Journal of Soil Science, v. 37, p. 617–639.

McKinley, J. M., C. D. Lloyd, and A. H. Ruffell, 2003, Variogram analysis of permeability in visibly homogeneous sandstone reservoirs, *in* J. Cubitt, ed., IAMG 2003, Proceedings of the 9th Annual Conference: International Association for Mathematical Geology, 6 p.: www.iamg.org/meetings/Proceedings_2003/papers/McKinley.pdf (accessed April 19, 2005).

Meddaugh, W. S., 2002, Pragmatic considerations for reservoir modeling using geostatistical techniques (abs.): AAPG Annual Meeting Program, v. 11, http://www.searchanddiscovery.com/documents/abstracts/annual2002/DATA/2002/13ANNUAL/SHORT/41231.pdf (accessed June 22, 2006).

Meddaugh, W. S., 2003, Assessing reservoir connectivity uncertainty in geostatistical models (abs.): AAPG Annual Meeting Program, v. 12, http://www.searchanddiscovery.com/documents/abstracts/annual2003/short/78199.PDF (accessed June 22, 2006).

Meo, N.-H., 1985, Interpreting geological structure using kriging: National Technical Information Service, No. B51054-3, 11 p.

Meunier, G., C. Coulomb, and J. P. Laille, 1987, Geostatistics applied to gas reservoirs: Society of Petroleum Engineers Annual Technical Conference and Exhibition, SPE Paper 16946.

Miesch, A. T., 1975, Variograms and variance components in geochemistry and ore evaluation: Geological Society of America Memoir 42, p. 333–340.

Miller, S. M., 1979, Geostatistical analysis for evaluating spatial dependence in fracture set characteristics, *in* T. J. O'Neil, ed., Proceedings of the 16th International Symposium on the Application of Computers and Operations Research in the Mineral Industry: New York, McGraw-Hill, p. 537–545.

Missman, R. A., 1990, Discussion of "Reservoir performance prediction methods based on fractal geostatistics:" SPE Reservoir Engineering, v. 5, p. 24.

Molz III, F. J., J. W. Castle, and S. Lu, 2003, The fractal/facies concept: An alternative to using variograms for generating subsurface heterogeneity (abs.): AAPG Annual Meeting Program, v. 12, http://www.searchanddiscovery.com/documents/abstracts/annual2003/short/78600.PDF (accessed June 22, 2006).

Morelon, I. F., B. Doligez, D. R. Guerillot, D. Rahon, and Y. Touffait, 1991, Application of a 3D geostatistical imaging to reservoir fluid flow simulations: Society of Petroleum Engineers Petroleum Computer Conference, Dallas, Texas, SPE Paper 22312.

Moss, B. P., 1990, Stochastic reservoir description—A methodology, *in* A. Hurst, M. A. Lovell, and A. C. Morton, eds., Geological applications of wireline logs: Geological Society (London) Special Publication 48, p. 57–75.

Mouliere, D., H. Beucher, L. Y. Hu, F. Fournier, P. Terdich, F. Mechiori, and G. Griffi, 1997, Integration of seismic derived information in reservoir stochastic modeling using truncated Gaussian approach, *in* E. Y. Baafi and N. A. Schofield, eds., Geostatistics Wollongong '96: Dordrecht, The Netherlands, Kluwer Academic Publishers, p. 374–385.

Mrugala, M., 1984, Application of statistical methods to the determination of mechanical model parameters for salt, *in* H. R. Hardy Jr. and M. Langer, eds., The mechanical behavior of salt, Proceedings of the 1st Conference: University Park, Pennsylvania State University, p. 339–354.

Mu, L.-H., and S.-M. Peng, 2004, Universal kriging method for quantitative sandstone prediction between wells (in Chinese): Petroleum Exploration and Development, v. 31, p. 73–75.

Muge, F., M. Vairinho, and P. Pina, 1994, A methodology for supervised classification of exploration areas based on the joint use of geostatistics and image analysis techniques, *in* C.-J. F. Chang, ed., Proceedings of IAMG '94: International Association for Mathematical Geology, p. 249–253.

Mukerji, T., G. Mavko, and P. Rio, 1997, Scales of reservoir heterogeneities and impact of seismic resolution on geostatistical integration: Mathematical Geology, v. 29, p. 933–950.

Mukerji, T., P. Avseth, G. Mavko, I. Takahashi, and E. F. Gonzalez, 2001, Statistical rock physics; combining rock physics, information theory, and geostatistics to reduce uncertainty in seismic reservoir characterization: The Leading Edge, v. 20, p. 313–319.

Mundim, E. C., P. Johann, and M. Remacre, 1999, Factorial kriging analysis; geostatistical filtering applied to reservoir characterization: The Leading Edge, v. 18, p. 787–788.

Murray, C. J., 1992, Stochastic simulation of hydrocarbon pore volume for risk assessment and economic planning: Society of Petroleum Engineers Annual Technical Conference and Exhibition, SPE Paper 25527.

Murray, C. J., 1993, Indicator simulation of petrophysical rock types, *in* A. Soares, ed., Geostatistics tróia '92: Dordrecht, The Netherlands, Kluwer Academic Publishers, p. 399–412.

Nasser, K. H., and A. K. Turner, 1995, Modeling the geometry of three-dimensional fracture networks using spherical statistics and fractals, *in* A. G. Fabbri and J. J. Royer, eds., Proceedings of the 3rd CODATA Conference on Geomathematics and Geostatistics: Nancy, Association Scientifique pour la Géologie et ses Applications, p. 81–114.

Neuman, S. P., 1984, Role of geostatistics in subsurface hydrology, *in* G. Verly, M. David, A. G. Journel, and A. Marechal, eds., Geostatistics for natural resources characterization: Dordrecht, The Netherlands, D. Reidel Publishing Company, p. 787–816.

Noble, A. C., and D. E. Ranta, 1984, Zoned kriging: A successful union of geology and geostatistics, *in* A. J. Erickson Jr., ed., Applied mining geology: American Institute of Mining, Metallurgical and Petroleum Engineers Special Publication, p. 115–128.

Norrena, K. P., and C. V. Deutsch, 2002, Automatic determination of well placement subject to geostatistical and economic constraints: Society of Petroleum Engineers International Thermal Operations and Heavy Oil Symposium and International Horizonal Well Technology Conference, SPE Paper 78996.

Norris, R. J., and G. Massonnat, 1993, The impact of geological interpretation on stochastic reservoir modeling and flow simulation (abs.): AAPG Annual Meeting Program, v. 2, p. 159.

Nosal, E. A., 1983, Statistical determination of geophysical well log response functions: Geophysics, v. 48, p. 1525–1535.

Nowak, M., and G. Verly, 2005, The practice of sequential Gaussian simulation, *in* O. Leuangthong and C. V. Deutsch, eds., Geostatistics Banff 2004, Proceedings of the Seventh International Geostatistics Congress: New York, Springer, p. 387–398.

Oliver, D. S., 2002, Conditioning channel meanders to well observations: Mathematical Geology, v. 34, p. 185–202.

Omre, H., 1988, Geostatistics applied to oil in place prediction, *in* J. McKenna and R. Teman, eds., ICIAM '87, Proceedings of the 1st International Conference on Industrial and Applied Mathematics: Philadelphia, Society of Industrial and Applied Mathematics, 5 p.

Omre, H., 2002, Stochastic reservoir models conditioned to non-linear production history observations, *in* W. J. Kleingeld and D. C. Krige, eds., Geostatistics 2000, Proceedings of the 6th International Geostatistics Congress: Capetown, Geostatistical Association of Southern Africa, p. 166–175.

Omre, H., and K. Sølna, 1992, Stochastic modelling and simulation of fault zones, *in* P. A. Dowd and J. J. Royer, eds., Proceedings of the 2nd CODATA Conference on Geomathematics and Geostatistics: Nancy, Association Scientifique pour la Géologie et ses Applications, p. 177–194.

Omre, H., and H. Tjelmeland, 1997, Petroleum geostatistics, *in* E. Y. Baafi and N. A. Schofield, eds., Geostatistics Wollongong '96: Dordrecht, The Netherlands, Kluwer Academic Publishers, p. 41–52.

Omre, H., K. Sølna, N. Dahl, and B. Tørudbakken, 1994, Impact of fault heterogeneity in fault zones on fluid flow, *in* J. O. Aasen, ed., North Sea oil and gas reservoirs III: Dordrecht, The Netherlands, Kluwer Academic Publishers, p. 185–200.

Osmani, S., T. Korini, and F. Kazazi, 1997, Finite elements, shape functions and elementary convariation matrices on reservoir characterization and geostatistics, *in* OMC '97 Proceedings: Ravenna, Italy, Offshore Mediterranean Conference and Exhibition, p. 1091–1096.

Overbey, W. K., T. K. Reeves, S. P. Salamy, C. D. Locke, and H. R. Johnson, 1991a, Novel geotechnical/geostatistical approach for exploration and production of natural gas from multiple geologic strata: Phase I—Volume I: Topical report: Washington, U.S. Department of Energy, DOE/MC/26026-3074, 35 p.

Overbey, W. K., T. K. Reeves, S. P. Salamy, C. D. Locke, and H. R. Johnson, 1991b, Novel geotechnical/geostatistical approach for exploration and production of natural gas from multiple geologic strata: Phase I—Volume II: Geology and engineering: Washington, U.S. Department of Energy, DOE/MC/26026-3074, 140 p.

Pan, G., 1997, Conditional simulation as a tool for measuring uncertainties in petroleum exploration: Nonrenewable Resources, v. 6, p. 285–293.

Panda, M. N., and L. W. Lake, 1993, Parallel simulated annealing for stochastic reservoir modeling: Society of Petroleum Engineers, 68th Annual Technical Conference and Exhibition, SPE Paper 26418.

Pathak, R., D. Ogbe, and J. Jensen, 2000, Application of geostatistical and fluid flow simulations to evaluate options for well placement (abs.): AAPG Bulletin, v. 84, p. 889.

Pawlowsky, V., R. A. Olea, and J. C. Davis, 1994, Additive log ratio estimation of regionalized compositional data: An application to calculation of oil reserves, *in* R. Dimitrakopoulos, ed., Geostatistics for the next century: Dordrecht, The Netherlands, Kluwer Academic Publishers, p. 371–382.

Pelletier, J. D., and D. L. Turcotte, 1999, Application of a stochastic deposition and erosion model to reservoir heterogeneity and stratigraphic data, *in* J. W. Harbaugh, W. L. Watney, E. C. Rankey, R. Slingerland, R. H. Goldstein, and E. K. Franseen, eds., Numerical experiments in stratigraphy: Recent advances in stratigraphic and sedimentologic computer simulations: SEPM Special Publication 62, p. 253–264.

Pendrel, J. V., and P. van Riel, 1997, Estimating porosity from 3D seismic inversion and 3D geostatistics (abs.): Society of Exploration Geophysicists Expanded Abstracts 16, p. 834.

Pendrel, J. V., and P. van Riel, 1998, Using 3D geostatistics and seismic inversion to estimate porosity (abs.): AAPG Annual Meeting Program, v. 2, p. A517.

Pereira, H. G., A. da Costa e Silva, A. Soares, L. Ribeiro, and J. de Carvalho, 1990, Improving reservoir description by using geostatistical and multivariate data analysis techniques: Mathematical Geology, v. 22, p. 879–914.

Pereira, H. G., A. da Costa e Silva, L. F. Guerreiro, and J. de Carvalho, 1992, The application of geostatistics to the description of oil reservoirs, *in* P. A. Dowd and J. J. Royer, eds., Proceedings of the 2nd CODATA Conference on Geomathematics and Geostatistics: Nancy, Association Scientifique pour la Géologie et ses Applications, p. 155–168.

Perez, G., and M. Kelkar, 1992, Effectiveness of conditional simulation to describe carbonate reservoir properties: Society of Petroleum Engineers, SPE Permian Basin Oil and Gas Recovery Conference, Midland, Texas, SPE Paper 23971.

Perez, G., and M. Kelkar, 1993a, Assessing distributions of reservoir properties using horizontal well data, *in* B. Linville, T. E. Burchfield, and T. C. Wesson, eds., Reservoir characterization III: Tulsa, PennWell Books, p. 399–436.

Perez, G., and M. Kelkar, 1993b, Three-dimensional reservoir description using conditional simulation, *in* K. S. Johnson and J. A. Campbell, Proceedings of the Petroleum-reservoir Geology in the Southern Mid-continent: Norman, Oklahoma Geological Survey Circular 95, p. 182–194.

Perez, G., A. K. Chopra, and C. D. Severson, 1995, Enhanced geostatistical mapping of reservoir fluids: Society of Petroleum Engineers International Meeting on Petroleum Engineering, Beijing, SPE Paper 29964.

Petit, F. M., P. Y. Biver, P. M. Calatayud, J.-L. Lesueur, and F. G. Alabert, 1994, Early quantification of hydrocarbon in place through geostatistical object modelling and connectivity: Society of Petroleum Engineers Annual Technical Conference and Exhibition, SPE Paper 28416.

Petit, F. M., M. Garcia, R. Froidevaux, and N. Noé, 1997, Application of object-simulation techniques to petroleum reservoir characterization: Practical considerations, *in* V. Pawlowsky-Glahn, ed., Proceedings of IAMG '97, 3rd Annual Conference of the International Association for Mathematical Geology: Barcelona, International Center for Numerical Methods in Engineering, p. 731–376.

Piazza, J. L., L. Sandjivy, and S. Legeron, 1997, Use of geostatistics to improve seismic velocities: Case studies (abs.): Society of Exploration Geophysicists Expanded Abstracts 16, p. 1293.

Pizarro, J. O. S., and L. W. Lake, 1999, A simple method to estimate interwell autocorrelation, *in* R. Schatzinger and J. Jordan, eds., Reservoir characterization—Recent advances: AAPG Memoir 71, p. 369–380.

Plotnick, R., and K. Prestegaard, 1995, Fractal methods for the analysis of stratigraphic sequences, *in* C. C. Barton and P. R. LaPointe, eds., Fractals in petroleum geology and earth sciences: New York, Plenum Press, p. 73–96.

Plotnick, R. E., 1986, A fractal model for the distribution of stratigraphic hiatuses: Journal of Geology, v. 94, p. 885–890.

Poletto, C. A., W. L. Lanzarini, and G. Tavares, 1999, Object based modeling and stochastic simulation of the geometry of fluvial reservoirs, *in* S. J. Lippard, A. Ness, and R. Sinding-Larsen, eds., IAMG '99, Proceedings of the 5th Annual Conference of the International Association for Mathematical Geology: Trondheim, Tapir, p. 647–652.

Poon, D. C., M. McCormack, and H. F. Thimm, 1993, The application of fractal geostatistics to oil and gas property evaluation and reserve estimates: Journal of Canadian Petroleum Technology, v. 32, p. 24–27.

Poquioma, W., and M. Kelkar, 1992, Application of geostatistics to forecast performance for waterflooding an oil field: Society of Petroleum Engineers Annual Technical Conference and Exhibition, SPE Paper 23677.

Pyrcz, M. J., and C. V. Deutsch, 2003, Stochastic surface models for integration of geologic information (abs.): AAPG Annual Meeting Program, v. 12, http://www.searchanddiscovery.com/documents/abstracts/annual2003/extend/77965.PDF (accessed June 22, 2006).

Raspa, G., R. Bruno, N. Kuka, and I. Manastiri-Hoxha, 1995, Borehole desurvey calculation using geostatistical methods, *in* A. G. Fabbri and J. J. Royer, eds., Proceedings of the 3rd CODATA Conference on Geomathematics and Geostatistics: Nancy, Association Scientifique pour la Géologie et ses Applications, p. 307–316.

Ravenne, C., and H. Beucher, 1988, Recent development of sedimentary bodies in a fluvio-deltaic reservoir and their 3D conditional simulations: Society of Petroleum Engineers 63rd Annual Technical Conference and Exhibition, SPE Paper 18310.

Ravenne, C., A. Galli, B. Doligez, H. Beucher, and R. Eschard, 2002, Quantification of facies relationships via proportion curves, *in* M. Armstrong, C. Bettini, N. Champigny, A. Galli, and A. Remacre, eds., Geostatstistics Rio 2000: Dordrect, The Netherlands, Kluwer Academic Publishers, p. 19–40.

Remacre, A. Z., A. A. Ferreira, and O. G. Souza Jr., 1989, Spatial distribution of erratic shales using conditional simulation, *in* M. Armstrong, ed., Geostatistics, Proceedings of the 3rd International Geostatistics Congress: Dordrecht, The Netherlands, Kluwer Academic Publishers, p. 507–516.

Rendu, J.-M., 1985, Geostatistical modelling and geological controls: Institute of Mining and Metallurgy Transactions, v. 94, sect. B, p. B166–B172.

Rendu, J.-M., and L. Ready, 1982, Geology and the semi-variogram—A critical relationship, *in* T. B. Johnson and R. J. Barnes, eds., Proceedings of the 17th International Symposium on the Application of Computers and Operations Research in the Mineral Industry: Society of Mining Engineers, p. 771–783.

Reynolds, A. C., N. He, and L. Chu, 1995, Reparameterization techniques for generating reservoir description conditioned variogram and well-test pressure data: Society of Petroleum Engineers Annual Technical Conference and Exhibition, SPE Paper 30588.

Reynolds, A. C., N. He, L. Chu, and D. S. Oliver, 1996, Reparametrization techniques for generating reservoir descriptions conditioned to variograms and well-test pressure data: Society of Petroleum Engineers Journal, v. 4, p. 413–426.

Reynolds, A. C., N. He, and D. S. Oliver, 1999, Reducing uncertainty in geostatistical description with well-testing pressure data, *in* R. Schatzinger and J. Jordan, eds., Reservoir characterization—Recent advances: AAPG Memoir 71, p. 149–162.

Rhea, L., M. Person, G. de Marsily, E. Ledoux, and A. Galli, 1994, Geostatistical models of secondary oil migration within heterogeneous carrier beds: A theoretical example: AAPG Bulletin, v. 78, p. 1679–1691.

Ripley, B. D., 1989, Stochastic geometry in reservoir modelling, in Proceedings, Statistics, Earth and Space Sciences: Bernoulli Society for Mathematical Statistics and Probability, p. 50–51.

Ritzi Jr., R. W., and D. F. Dominic, 1998, Quantitative models for permeability and facies distributions, and macrodispersive effects on transport (abs.): Geological Society of America Abstracts with Programs, v. 30, p. 103–104.

Rodionov, D. A., and B. V. Smirnov, 1995, Sequential estimation method of oil and gas deposits reserves, *in* A. G. Fabbri and J. J. Royer, eds., Proceedings of the 3rd CODATA Conference on Geomathematics and Geostatistics: Nancy, Association Scientifique pour la Géologie et ses Applications, p. 177–186.

Roth, C., 2000, Incorporating information about edge effects when simulating lithofacies: Mathematical Geology, v. 32, p. 277–300.

Rouleau, A., and J. E. Gale, 1985, Fracture size, interconnectivity and groundwater flow: Site specific stochastic analysis, *in* G. de Marsily, ed., Proceedings, The stochastic approach to subsurface flow: Paris, Centre Nationale de la Reserche Scientifique, p. 415–522.

Rovellini, M., F. Brega, and C. Monico, 1998, Facies related geological model; a reliable method to describe complex reservoirs: Petroleum Geoscience, v. 4, p. 227–235.

Rowbotham, P. S., P. Lamy, P. A. Swaby, O. Dubrule, and T. Cadoret, 1998, Geostatistical inversion for reservoir characterization (abs.): Society of Exploration Geophysicists Expanded Abstracts 17, p. 886.

Royle, A. G., F. L. Clausen, and P. Frederiksen, 1981, Practical universal kriging and automatic contouring: Geo-Processing, v. 1, p. 377–394.

Rudkiewicz, J. L., and R. Eschard, 1991, Geological knowledge in the geostatistical modeling of reservoir properties, *in* L. W. Lake, H. B. Carroll Jr., and T. C. Wesson, eds., Reservoir characterization II: San Diego, Academic Press, p. 686–688.

Ruseti, B., and G. Beqiraj, 1992, A geostatistical interpolation algorithm and its combination with spline-cubic interpolation functions for the calculation of the reserve of oil and solid minerals, *in* P. A. Dowd and J. J. Royer, eds., Proceedings of the 2nd CODATA Conference on Geomathematics and Geostatistics: Nancy, Association Scientifique pour la Géologie et ses Applications, p. 195–200.

Saad, N., V. Maroongroge, and C. T. Kalkomey, 1996, Ranking geostatistical models using tracer production data, *in* Proceedings of the European 3-D Reservoir Modeling Conference, Stavanger, Norway: Society of Petroleum Engineers, SPE Paper 35494, p. 131–142.

Sagar, R. K., B. G. Kelkar, and L. G. Thompson, 1993, Reservoir description by integrating well test data and spatial statistics: Society of Petroleum Engineers Annual Technical Conference and Exhibition, SPE Paper 26462.

Saleri, N. G., 1990, Discussion of stochastic modeling: Journal of Petroleum Technology, v. 42, p. 929.

Sampson, R. J., and R. A. Olea, 1993, Automatic contouring by triangulation and kriging (abs.), *in* R. B. McCammon, ed., Program and Abstracts, Silver Anniverary Meeting, 1968–1993: International Association for Mathematical Geology, p. 51.

Seifert, D., and J. L. Jensen, 1997, Using sequential indicator simulation as a tool in reservoir description: Issues and uncertainties, *in* V. Pawlowsky-Glahn, ed., Proceedings of IAMG '97, 3rd Annual Conference of the International Association for Mathematical Geology: Barcelona, International Center for Numerical Methods in Engineering, p. 641–646.

Seifert, D., J. J. M. Lewis, C. Y. Hern, and N. C. T. Steel, 1996, Well placement optimization and risking using 3-D stochastic reservoir modeling techniques: Society of Petroleum Engineers, SPE/NPF European 3-D Reservoir Modeling Conference (Stavanger, Norway), SPE Paper 35520, 12 p.

Seifert, D., J. D. H. Newbery, C. Ramsey, and J. J. M. Lewis, 1999, Evaluation of field development plans using 3-D reservoir modeling, *in* R. Schatzinger and J. Jordan, eds., Reservoir characterization—Recent advances: AAPG Memoir 71, p. 321–332.

Sen, M. K., G. A. Pope, A. Datta-Gupta, P. L. Stoffa, and L. W. Lake, 1995, Stochastic reservoir modeling using simulated annealing and genetic algorithms: SPE Formation Evaluation, v. 10, p. 49–55.

Sen, Z., 1990, Spatial simulation of geologic variables: Mathematical Geology, v. 22, p. 175–188.

Serra, J., 1987, Comments on "Geostatistics: Models and tools for the earth sciences" by A. G. Journel: Mathematical Geology, v. 19, p. 349–356.

Shibli, S. A. R., P. M. Wong, Y. Huang, and M. Maignan, 1997, Applications of AI-based techniques for spatial prediction, *in* V. Pawlowsky-Glahn, ed., Proceedings of IAMG '97, 3rd Annual Conference of the International Association for Mathematical Geology: Barcelona, International Center for Numerical Methods in Engineering, p. 675–680.

Sides, E. J., 1995, Hidden uncertainties in geostatistics, *in* A. G. Fabbri and J. J. Royer, eds., Proceedings of the 3rd CODATA Conference on Geomathematics and Geostatistics: Nancy, Association Scientifique pour la Géologie et ses Applications, p. 15–28.

Silliman, S. E., and B. Berkowitz, 2000, The impact of biased sampling on the estimation of the semivariogram within fractured media containing multiple fracture sets: Mathematical Geology, v. 32, p. 543–560.

Simlote, V. N., and S. A. Nikolayenko, 1998, Dealing with uncertainty: Geostatistics and the new era in reservoir studies, *in* Proceedings of the SPE India Oil and Gas Conference and Exhibition: Society of Petroleum Engineers, p. 383–398.

Skorstad, A., R. Hauge, and L. Holden, 1997, Well conditioning in a fluvial reservoir model, *in* V. Pawlowsky-Glahn, ed., Proceedings of IAMG '97, 3rd Annual Conference of the International Association for Mathematical Geology: Barcelona, International Center for Numerical Methods in Engineering, p. 725–730.

Springer, J., and J. Harff, 1992, A comparison of several regionalization methods, *in* P. A. Dowd and J. J. Royer, eds., Proceedings of the 2nd CODATA Conference on Geomathematics and Geostatistics: Nancy, Association Scientifique pour la Géologie et ses Applications, p. 43–50.

Srinivasan, S., and C. V. Deutsch, 1996, Improved reservoir management through ranking stochastic reservoir models: Society of Petroleum Engineers, 10th Society of Petroleum Engineers–Department of Energy Symposium on Improved Oil Recovery, SPE Paper 35411.

Srinivasan, S., and A. G. Journel, 1998, Simulation of permeability field conditioned to well test data: Society of Petroleum Engineers Annual Technical Conference and Exhibition, SPE Paper 49289.

Srivastava, R. M., 1992, Reservoir characterization with probability field simulation: Society of Petroleum Engineers Annual Technical Conference and Exhibition, SPE Paper 24753.

Srivastava, R. M., 1994a, An overview of stochastic methods for reservoir characterization, *in* J. M. Yarus and R. L. Chambers, eds., Stochastic modeling and geostatistics: Principles, methods, and case studies: AAPG Computer Applications in Geology 3, p. 1–16.

Srivastava, R. M., 1994b, The visualization of spatial uncertainty, *in* J. M. Yarus and R. L. Chambers, eds., Stochastic modeling and geostatistics: Principles, methods, and case studies: AAPG Computer Applications in Geology 3, p. 339–345.

Srivastava, R. M., 1996, An overview of stochastic spatial simulation, *in* H. T. Mowrer, R. L. Czaplewski, and R. H. Hamre, eds., Proceedings, 2nd International Symposium on Spatial Accuracy Assessment in Natural Resources and Environmental Sciences: Ft. Collins, U.S. Department of Agriculture, Forest Service, General Technical Report RM-GTR-277, p. 13–22.

Srivastava, R. M., P. Frykman, and M. Jensen, 2005, Geostatistical simulation of fracture networks, Journel, A. G., 2005, Beyond covariance: The advent of multiple-point geostatistics, *in* O. Leuangthong and C. V. Deutsch, eds., Geostatistics Banff 2004, Proceedings of the Seventh International Geostatistics Congress: New York, Springer, p. 295–304.

Starks, T. H., and J. H. Fang, 1982, On the estimation of the generalized covariance function: Mathematical Geology, v. 14, p. 57–64.

Strauss, D. J., and P. M. Sadler, 1989, Stochastic models for the completeness of stratigraphic sections: Mathematical Geology, v. 21, p. 37–59.

Strebelle, S., 2002, Conditional simulation of complex geological structures using multiple-point statistics: Mathematical Geology, v. 34, p. 1–22.

Strebelle, S., 2003, Integration of sequence stratigraphy concepts into multiple-point geostatistical models, *in* J. Cubitt, ed., IAMG 2003, Proceedings of the 9th Annual Conference: International Association for Mathematical Geology, 6 p.: www.iamg.org/meetings/Proceedings_2003/papers/Strebelle.pdf (accessed April 19, 2005).

Strebelle, S., and A. G. Journel, 2001, Reservoir modeling using multiple-point statistics: Society of Petroleum Engineers Annual Technical Conference and Exhibition, SPE Paper 71324.

Sturaro, J. R., and P. M. Barbosa Landim, 1997, Indicator kriging for GIS's maps integration, *in* V. Pawlowsky-Glahn, ed., Proceedings of IAMG '97, 3rd Annual Conference of the International Association for Mathematical Geology: Barcelona, International Center for Numerical Methods in Engineering, p. 699–704.

Sultan, A. J., J. P. Heller, and A. L. Gutjahr, 1990, Generation and testing of random fields and the conditional simulation of reservoirs, *in* Technology in the 90s, Prospering in a changing environment: Petroleum Society of Canadian Institute of Mining and the Society of Petroleum Engineers, CIM/SPE International Technical Meeting Preprints, Calgary, v. 3, p. 98-1–98-11.

Suro-Perez, V., and A. G. Journel, 1990, Stochastic simulation of lithofacies: An improved sequential indicator approach, *in* D. Guérillot and O. Guillon, eds., Proceedings of the 2nd European Conference on the Mathematics of Oil Recovery: Paris, Editions Technip, p. 3–10.

Svanes, T., P. I. Karstad, B. Myking, F. Hovland, and M. Madsen, 1994, Detailed stochastic modelling of a proximal fluvial reservoir with realizations used in flow simulation: Society of Petroleum Engineers, European Petroleum Conference (London), SPE Paper 28873.

Switzer, P., 1989, Non-stationary spatial covariances estimated from monitoring data, *in* M. Armstrong, ed., Geostatistics, Proceedings of the 3rd International Geostatistics Congress: Dordrecht, The Netherlands, Kluwer Academic Publishers, p. 127–138.

Syversveen, A. R., and H. Omre, 1997, Marked point models for facies units conditioned on well data, *in* E. Y. Baafi and N. A. Schofield, eds., Geostatistics Wollongong '96: Dordrecht, The Netherlands, Kluwer Academic Publishers, p. 415–423.

Szidarovsky, F., E. Y. Baafi, and Y. C. Kim, 1987, Kriging without negative weights: Mathematical Geology, v. 19, p. 549–559.

Taggart, I. J., M. Pitkin, and T. A. Gray, 1991, Application of fractal geostatistics to miscible flood performance in heterogeneous reservoirs: Society of Petroleum Engineers Annual Technical Conference and Exhibition, SPE Paper 22981.

Tamhane, D., L. Wang, and P. M. Wong, 1999, The role of geology in stochastic reservoir modelling: The future trends: Society of Petroleum Engineers, Asia Pacific Oil and Gas Conference and Exhibition, Jakarta, SPE Paper 54307.

Tang, R. W., R. A. Behrens, and A. S. Emanuel, 1991, Reservoir studies with geostatistics to forecast performance: SPE Reservoir Engineering, v. 6, p. 253–258.

Teles, V., G. de Marsily, and E. Perrier, 1998, A new approach for modelling sediment depostion in an alluvial plain in order to display its heterogeneity (in French): Sciences de la Terre et des Planetes, v. 327, p. 597–606.

Thadani, S. G., 1992, Stochastic imaging for reservoir characterization: Society of Petroleum Engineers Annual Technical Conference and Exhibition, SPE Paper 24741.

Thadani, S. G., 1993, Reservoir characterization with seismic data using pattern recognition and spatial statistics, *in* A. Soares, Geostatistics troia '92: Dordrecht, The Netherlands, Kluwer Academic Publishers, p. 519–542.

Thadani, S. G., F. Alabert, and A. G. Journel, 1987, An integrated geostatistical/pattern recognition technique for characterization of reservoir spatial variability (abs.): Society of Exploration Geophysicists Expanded Abstracts 6, p. 372–375.

Thiele, M. R., S. E. Rao, and M. J. Blunt, 1996, Quantifying uncertainty in reservoir performance using streamtubes: Mathematical Geology, v. 28, p. 843–856.

Tidwell, V. C., and J. L. Wilson, 1997, A comparison of simple upscaling models with multisupport permeability data measured on a block of Berea Sandstone, *in* V. Pawlowsky-Glahn, ed., Proceedings of IAMG '97, 3rd Annual Conference of the International Association for Mathematical Geology: Barcelona, International Center for Numerical Methods in Engineering, p. 737–742.

Tjelmeland, H., and H. Omre, 1997, A complex sand-shale facies model conditioned on observations from wells, seismics and production, *in* E. Y. Baafi and N. A. Schofield, eds., Geostatistics Wollongong '96: Dordrecht, The Netherlands, Kluwer Academic Publishers, p. 634–646.

Tomutsa, L., M.-M. Chang, and S. Jackson, 1993, Application of outcrop data for characterizing reservoirs and deriving grid-block scale values for numerical simulation, *in* B. Linville, T. E. Burchfield, and T. C. Wesson, eds., Reservoir characterization III: Tulsa, PennWell Books, p. 151–194.

Track, A., J. H. Meyer, S. Zurquiyah, and J. P. Delhomme, 1994, Combining geophisical data with geological prior knowledge in a geostatistical framework: The example of cross-well tomographic inversion, *in* R. Dimitrakopoulos, ed., Geostatistics for the next century: Dordrecht, The Netherlands, Kluwer Academic Publishers, p. 393–406.

Tran, T. T., C. V. Deutsch, and Y. Xie, 2001, Direct geostatistical simulation with multiscale well, seismic, and production data: Society of Petroleum Engineers Annual Technical Conference and Exhibition, SPE Paper 71323.

Tubman, K. M., and S. D. Crane, 1995, Vertical versus horizontal well log variability and applications to fractal reservoir modeling, *in* C. C. Barton and P. R. LaPointe, eds., Fractals in petroleum geology and earth sciences: New York, Plenum Press, p. 279–294.

Turcotte, D. L., 1989, Fractals in geology and geophysics: Pure and Applied Geophysics, v. 131, p. 171–196.

Tureyen, O. I., and J. Caers, 2003, A two-level optimization method for geostatistical integration of production data on non-uniform grids: Society of Petroleum Engineers Annual Technical Conference and Exhibition, SPE Paper 84358.

Tyler, K., T. Svanes, and S. Omdal, 1993, Faster history matching and uncertainty in predicted production profiles with stochastic modeling: Society of Petroleum Engineers Annual Technical Conference and Exhibition, Proceedings, SPE Paper 26420.

Tyler, K., A. Henriquez, and T. Svanes, 1994a, Modeling heterogeneities in fluvial domains: A review of the influence on production profiles, *in* J. M.

Yarus and R. L. Chambers, eds., Stochastic modeling and geostatistics: Principles, methods, and case studies: AAPG Computer Applications in Geology 3, p. 77–89.

Tyler, K., A. Henriquez, A. MacDonald, T. Svanes, L. Holden, and A. L. Hektoen, 1994b, MOHERES—A collection of stochastic models for describing heterogeneities in clastic reservoirs, *in* J. O. Aasen, ed., North Sea oil and gas reservoirs III: Dordrecht, The Netherlands, Kluwer Academic Publishers, p. 213–221.

Vasco, D. W., and A. Datta-Gupta, 1997, Integrating multiphase production history in stochastic reservoir characterization: SPE Formation Evaluation, v. 12, p. 149–156.

Verly, G., 1983, The multigaussian approach and its applications to the estimation of local reserves: Mathematical Geology, v. 15, p. 259–286.

Viseur, S., 1999, Stochastic Boolean simulation of fluvial deposits: A new approach combining accuracy with efficiency: Society of Petroleum Engineers Annual Technical Conference and Exhibition, SPE Paper 56688.

Viseur, S., 2004, Turbidite reservoir characterization: Object-based stochastic simulation meandering channels: Bulletin de la Société Géologique de France, v. 175, p. 11–20.

Viseur, S., A. Shtuka, and J.-L. Mallet, 1998, New fast, stochastic, Boolean simulation of fluvial deposits: Society of Petroleum Engineers Annual Technical Conference and Exhibition, SPE Paper 49281.

Vogely, W. A., E. Sani, and P. G. Monzon, 1978, The application of kriging methods to oil and gas resource estimation: Pennsylvania State University, Department of Mineral Economics, 105 p.

Waggoner, J. R., J. L. Castillo, and L. W. Lake, 1992, Simulation of EOR processes in stochastically generated permeable media: SPE Formation Evaluation, v. 7, p. 173–180.

Wang, J., and T. Zhang, 1995, Three-stage stochastic modelling method to characterize reservoir with fluvial facies: Society of Petroleum Engineers International Meeting on Petroleum Engineering, Beijing, SPE Paper 29965.

Wang, L., 1994, Animated visualization of multiple simulated realizations and assessment of uncertainty: Stanford, Stanford University Stanford Center for Reservoir Characterization, Report 7, 13 p.

Wang, L., 1996, Modeling complex reservoir geometries with multiple-point statistics: Mathematical Geology, v. 28, p. 895–907.

Wang, P. P., and D. Chen, 1994, Annealing using statistical inference and hierarchical structure for reservoir optimation models: Tuscaloosa, University of Alabama, Technical Report, 31 p.

Wang, R., and M. Yang, 1998, Some problems on the study of the fractal statistics for geological phenomena: Geoscience, v. 12, p. 91–96.

Watkins, A. J., 1993, Linking universal kriging and reservoir history matching: Society of Petroleum Engineers Annual Technical Conference and Exhibition, SPE Paper 26610.

Wattenbarger, R. C., K. Aziz, and F. M. Orr Jr., 1993, Optimal scales for representing reservoir heterogeneity, *in* B. Linville, T. E. Burchfield, and T. C. Wesson, eds., Reservoir characterization III: Tulsa, PennWell Books, p. 195–225.

Wen, J., Y.-N. Qiu, M.-X. Liu, and J.-X. Xiao, 1995, The construction and application of stochastic models in early evaluation phase of oil reservoirs: Society of Petroleum Engineers International Meeting on Petroleum Engineering, Beijing, SPE Paper 29966.

Wen, R., 1998, Multi-scale heterogeneity modeling in fluvial reservoirs; integrating geostatistics and conceptual models of geologists (abs.): Reservoir, v. 25, p. 8.

Wen, R., and R. Sinding-Larsen, 1994, Probability kriging of sub-seismic fault throws with multifractal distributions, *in* R. Dimitrakopoulous, Geostatistics for the next century: Dordrecht, The Netherlands, Kluwer Academic Publishers, p. 488–497.

Wen, R., and R. Sinding-Larsen, 1996, Stochastic modeling and simulation of small faults by marked point process and kriging, *in* E. Y. Baafi and N. A. Schofield, eds., Geostatistics Wollongong '96: Dordrecht, The Netherlands, Kluwer Academic Publishers, p. 398–414.

Wen, X.-H., A. S. Cullick, and C. V. Deutsch, 1997, Spatial permeability distribution parameters from a single-well transient pressure data, *in* V. Pawlowsky-Glahn, ed., Proceedings of IAMG '97, 3rd Annual Conference of the International Association for Mathematical Geology: Barcelona, International Center for Numerical Methods in Engineering, p. 761–766.

Wen, X.-H., T. T. Tran, and R. A. Behrens, 2002, Production data integration in sand/shale reservoirs using sequential self-calibration and geoMorphing: A comparison: SPE Reservoir Evaluation and Engineering, v. 5, p. 255–265.

Wen, X.-H., T. Yu, and S. Lee, 2005, Coupling sequential self-calibration and genetic algorithms to integrate production data in geostatistical reservoir modeling, *in* O. Leuangthong and C. V. Deutsch, eds., Geostatistics Banff 2004, Proceedings of the Seventh International Geostatistics Congress: New York, Springer, p. 691–702.

Whitten, E. H. T., 1977, Stochastic models in geology: Journal of Geology, v. 95, p. 321–330.

Whitten, E. H. T., 1981, Semivariograms and kriging; possible useful tools in fold description, *in* R. G. Craig, M. L. Labovitz, and J. C. Griffiths, eds., Future trends in geomathematics: London, Pion, p. 48–61.

Williams, N. H., A. R. Gardiner, and J. L. Jensen, 1999, Inter-relationship between measurement scale, sampling rate, and sedimentary heterogeneity in fluvio-aeolian succession permeability variograms, *in* S. J. Lippard, A. Naess, and R. Sinding-Larsen, eds., IAMG '99, Proceedings of the 5th Annual Conference of the International Association for Mathematical Geology, Trondheim, Tapir, p. 723–728.

Wilson, J. L., and A. Gutjahr, 1989, Synthetic generation of random permeability fields for heterogeneous reservoir simulation: Society of Petroleum Engineers, SPE Symposium on Reservoir Simulation, Houston, Texas, SPE Paper 18435.

Wist, H. T., H. Tjelmeland, and D. Lia, 1999, A stochastic facies model constrained by connectivity data, *in* S. J. Lippard, A. Naess, and R. Sinding-Larsen, eds., IAMG '99, Proceedings of the 5th Annual Conference of the International Association for Mathematical Geology, Trondheim, Tapir, p. 671–676.

Withers, K. D., D. J. Wolf, and M. D. Burnaman, 1994, Integration of well and seismic data using geostatistics (abs.): AAPG Annual Meeting Program, v. 3, p. 286.

Wolcott, D. S., and A. K. Chopra, 1993a, Incorporating reservoir heterogeneity with geostatistics to investigate waterflood recoveries: SPE Formation Evaluation, v. 8, p. 26–32.

Wolcott, D. S., and A. K. Chopra, 1993b, Investigating infill drilling performance and reservoir continuity using geostatistics, *in* B. Linville, T. E. Burchfield, and T. C. Wesson, eds., Reservoir characterization III: Tulsa, PennWell Books, p. 297–326.

Wolf, D. J., K. D. Withers, and M. D. Burnaman, 1994, Integration of well and seismic data using geostatistics, *in* J. M. Yarus and R. L. Chambers, eds., Stochastic modeling and geostatistics: Principles, methods, and case studies: AAPG Computer Applications in Geology 3, p. 177–199.

Xin, Q., Q. Du, and H. Xu, 1993, The application of geostatistics in the predetermination of reservoir parameters between wells (in Chinese): Geological Review, v. 39, p. 283–291.

Xu, J., and R. D. Jacobi, 1998, Characterizing natural fracture networks; density calculation and spatial structure analyses (abs.): Geological Society of America Abstracts with Programs, v. 30, p. 224–225.

Xu, W., T. Tran, R. M. Srivastava, and A. G. Journel, 1992, Integrating seismic data in reservoir modeling: the collocated cokriging alternative: Society of Petroleum Engineers Annual Technical Conference and Exhibition, SPE Paper 24742.

Xue, G., and A. Datta-Gupta, 1997, Structure preserving inversion: An efficient approach to conditioning stochastic reservoir models to dynamic data: Society of Petroleum Engineers Proceedings of the Annual Technical Conference and Exhibition, SPE Paper 38727.

Yang, A. P., 1992a, A geostatistical well correlation method that can identify faults: Society of Petroleum Engineers International Meeting on Petroleum Engineering, Beijing, SPE Paper 22448.

Yang, A. P., 1992b, A geostatistical interpolation method from only two or three wells: Society of Petroleum Engineers International Meeting on Petroleum Engineering, Beijing, SPE Paper 22346.

Yang, A. P., 1993, New geostatistical methods for mapping faults with well logs: Indonesian Petroleum Association, 22nd Annual Convention, October 12–13, Jakarta, p. 105–113.

Yang, A. P., 1994, Modelling faults with geostatistical gridding: ITC Journal, v. 1, p. 44–47.

Yang, A. P., 1995, Mapping temperature with geostatistics can help identify steam-chest and optimize oil production: Society of Petroleum Engineers International Heavy Oil Symposium, Calgary, SPE Paper 30254.

Yang, A. P., 1998, Integration of production history in geostatistical reservoir description: Society of Petroleum Engineers Asia Pacific Conference on Integrated Modeling for Asset Management, Kuala Lumpur, Malaysia, SPE Paper 39741.

Yao, T., and T. Mukerji, 1997, Application of factorial kriging to improve seismic data integration, *in* E. Y. Baafi and N. A. Schofield, eds., Geostatistics Wollongong '96: Dordrecht, The Netherlands, Kluwer Academic Publishers, p. 350–361.

Yao, T., T. Mukerji, A. Journel, and G. Mavko, 1999, Scale matching with factorial kriging for improved porosity estimation from seismic data: Mathematical Geology, v. 31, p. 23–27.

Yarus, J. M., C. R. Vestal, and C. L. Blauer, 1990, More information from computer-contoured maps: Regional component mapping and kriging-reading between lines (abs.): AAPG Bulletin, v. 74, p. 794–795.

Yarus, J. M., R. L. Chambers, and D. H. Zamora Guerrero, 2002, Principles of reservoir characterization and geostatistical modeling: Applications for early- and late-stage field development (abs.): AAPG Annual Meeting Program, v. 11, http://www.searchanddiscovery.com/documents/abstracts/annual2002/DATA/2002/13ANNUAL/SHORT/41269.pdf (accessed June 22, 2006).

Yarus, J. M., R. L. Chambers, and G. Jean, 2003, Identifying subtle stratigraphic and facies trends using advanced variography (abs.): AAPG Annual Meeting Program, v. 12, http://www.searchanddiscovery.com/documents/abstracts/annual2003/short/79089.PDF (accessed June 22, 2006).

Yates, S. R., and A. W. Warrick, 1999, Geostatistics and spatial variability of soil properties: Agronomy, v. 38, p. 1235–1260.

Yates, S. R., and M. V. Yates, 1988, Disjunctive kriging as an approach to management decision making: Ada, Robert S. Kerr Environmental Research Lab, EPA/600/J-88/431/ PB90-113473, 28 p.

Yeten, B., and F. Gumrah, 2000, The use of fractal geostatistics and artificial neural networks for carbonate reservoir characterization: Transport in Porous Media, v. 41, p. 173–197.

Young, D. A., 1987, Indicator kriging for unit vectors: Rock joint orientations: Mathematical Geology, v. 19, no. 6, p. 481–502.

Yuan, L.-P., and R. Strobl, 1993, Variograms and reservoir continuity, *in* B. Linville, T. E. Burchfield, and T. C. Wesson, eds., Reservoir characterization III: Tulsa, PennWell Books, p. 437–452.

Zeybek, A. D., and M. Onur, 2003, Conditioning fractal (fBm/FGn) porosity and permeability fields to multiwell pressure data: Mathematical Geology, v. 35, p. 577–612.

Zhang, F., A. C. Reynolds, and D. S. Oliver, 2002, Evaluation of the reduction in uncertainty obtained by conditioning a 3D stochastic channel to multiwell pressure data: Mathematical Geology, v. 34, p. 715–742.

Zhang, L., L. Cunha, and C. V. Deutsch, 2005, Local updating of reservoir properties for production data integration, *in* O. Leuangthong and C. V. Deutsch, eds., Geostatistics Banff 2004, Proceedings of the Seventh International Geostatistics Congress: New York, Springer, p. 769–776.

Zhanggui, L., J. Di, F. Jun, G. Jianchun, and Z. Kaihong, 1998, Integration of fuzzy methods into geostatistics for petrophysical property distribution simulation of a reservoir at exploration stage: Society of Petroleum Engineers Asia Pacific Oil and Gas Conference, SPE Paper 49964, p. 57–64.

Zhou, J., C. Wu, X. Mao, and X. Wang, 1998, Overview about reservoir model and simulation in petroliferous basin: Geological Science and Technology Information, v. 17, p. 67–72.

Zhuoxin, B., D. S. Oliver, and A. C. Reynolds, 1999, Conditioning 3D stochastic channels to pressure data: Society of Petroleum Engineers Annual Technical Conference and Exhibition, SPE Paper 56682.

Zingano, A. C., J. F. C. Costa, and J. C. Koppe, 1995, Oil content estimation in sedimentary rock assisted by geostatistics, *in* R. K. Singhal, ed., Proceedings of the 4th International Symposium on Mine Planning and Equipment Selection: Rotterdam, A. A. Balkema, p. 327–338.

## THEORETICAL DEVELOPMENTS

Ababou, R., A. C. Bagtzoglu, and E. F. Woods, 1994, On the condition number of covariance matrices in kriging, estimation, and simulation of random fields: Mathematical Geology, v. 26, p. 99–134.

Abbaspour, K. C., R. Schulin, M. T. van Genuchten, and E. Schläppi, 1998, An alternative to cokriging for situations with small sample size: Mathematical Geology, v. 30, p. 259–274.

Abrahamsen, P., 1997, A review of Gaussian random fields and correlation functions, 2nd ed.: Norsk Regnesentral, Norwegian Computer Center, Technical Report 917, 64 p.

Abrahamsen, P., and F. E. Benth, 1998, Gaussian field with unknown trend conditioned inequality data, *in* A. Buccianti, G. Nardi, and R. Potenza, eds., IAMG '98, Proceedings of the 4th Annual Conference of the International Association for Mathematical Geology: Naples, de Frede, p. 333–338.

Abrahamsen, P., and F. E. Benth, 2001, Kriging with inequality constraints: Mathematical Geology, v. 33, p. 719–744.

Adisoma, G. S., and Y. C. Kim, 1993, Jackknife kriging—A simple and robust new estimator, *in* X. Tang and J. Elbrond, eds., Proceedings of the 24th International Symposium on the Application of Computers and Operations Research in the

Mineral Industry: Canadian Institute of Mining, Metallurgy and Petroleum, p. 343–349.

Agterberg, F., 1976, New problems at the interface between geostatistics and geology, *in* M. Guarascio, M. David, and C. Huijbregts, eds., Advanced geostatistics in the mining industry: Dordrecht, The Netherlands, D. Reidel Publishing Company, p. 403–421.

Agterberg, F. P., 1994, Fractals, multifractals, and change of support, 1994, *in* R. Dimitrakopoulos, ed., Geostatistics for the next century: Dordrecht, The Netherlands, Kluwer Academic Publishers, p. 223–234.

Akin, H., 1983, Groupment of fundamental ideas of geostatistics (in German): Schriftenreihe der GDMB, Gesellschaft Deutscher Metallhütten- und Bergleute, v. 35, p. 159–161.

Alabert, F., 1987, The practice of fast conditional simulations through the LU decomposition of the covariance matrix: Mathematical Geology, v. 19, p. 369–386.

Alabert, F. G., and G. J. Massonat, 1990, Heterogeneity in a complex turbidic reservoir: Stochastic modeling of facies and petrophsyical variability: Society of Petroleum Engineers Annual Technical Conference and Exhibition, SPE Paper 20604, 16 p.

Aldworth, J. and N. Cressie, 1999, Sampling designs and prediction methods for Gaussian spatial processes, *in* S. Ghosh, ed., Multivariate analyses, design of experiments, and survey sampling: New York, Marcel Dekker, p. 1–54.

Alfaro, M., 1984, Statistical inference of the semivariogram and the quadratic model, *in* G. Verly, M. David, A. Journel, and A. Marechal, eds., Geostatistics for natural resources characterization: Dordrecht, The Netherlands, D. Reidel Publishing Company, p. 45–53.

Allard, D., and G. Guillot, 2000, Clustering geostatistical data, *in* W. J. Kleingeld and D. C. Krige, eds., Geostatistics 2000, Proceedings of the 6th International Geostatistical Congress: Geostatistical Association of Southern Africa, Cape Town, 15 p.

Almeida, A. S., and A. G. Journel, 1994, Joint simulation of multiple variables with a Markov-type coregionalization model: Mathematical Geology, v. 26, p. 565–588.

Ambroise, C., M. Dang, and G. Govaert, 1997, Clustering of spatial data by the EM algorithm, *in* A. Soares, J. Gomez-Hernandez, and R. Froidevaux, eds., GeoENV I, geostatistics for environmental applications: Dordrecht, The Netherlands, Kluwer Academic Publishers, p. 493–504.

Anis, A. A. and E. H. Lloyd, 1975, Stochastic reservoir theory: an outline of the state of the art as understood by applied probabilists: Laxenburg, International Institute for Applied Systems Analysis, Research Report RR-75-30, 22 p.

Arik, A., 2002, Area of influence kriging: Mathematical Geology, v. 34, p. 783–796.

Armstrong, M., 1984a, Improving the estimation and modelling of the variogram, *in* G. Verly, M. David, A. Journel, and A. Marechal, eds., Geostatistics for natural resources characterization: Dordrecht, The Netherlands, D. Reidel Publishing Company, p. 1–19.

Armstrong, M., 1984b, Testing variograms for positive definiteness: Mathematical Geology, v. 16, no. 4, p. 407–421.

Armstrong, M., 1984c, Problems with universal kriging: Mathematical Geology, v. 16, p. 101–108.

Armstrong, M., 1984d, Common problems seen in variograms: Mathematical Geology, v. 16, p. 305–313.

Armstrong, M., 1985, Comments on "The deterministic side of geostatistics" by A. G. Journel: Mathematical Geology, v. 17, p. 855–859.

Armstrong, M., and A. Boufassa, 1988, Comparing the robustness of ordinary kriging and lognormal kriging: Outlier resistance: Mathematical Geology, v. 20, p. 447–458.

Armstrong, M., and P. Diamond, 1984, Testing variograms for positive definiteness: Mathematical Geology, v. 16, p. 407–422.

Armstrong, M., and P. Diamond, 1986a, Disjunctive kriging revisited: Part I: Mathematical Geology, v. 18, p. 711–728.

Armstrong, M., and P. Diamond, 1986b, Disjunctive kriging revisited: Part II: Mathematical Geology, v. 18, p. 729–742.

Armstrong, M., and A. Galli, 1999, Derivative based plurigaussian simulations, *in* S. J. Lippard, A. Ness, and R. Sinding-Larsen, eds., IAMG '99, Proceedings of the 5th Annual Conference of the International Association for Mathematical Geology: Trondheim, Tapir, p. 591–596.

Armstrong, M., and R. Jabin, 1981, Variogram models must be positive definite: Mathematical Geology, v. 13, p. 455–460.

Armstrong, M., and H. Wackernagel, 1988, The influence of the covariance function on the kriged estimator: Nancy, Sciences de la Terre, Series Informatique Geologique, v. 27, p. 245–262.

Arpat, B. G., and J. Caers, 2005, A multiple-scale,

pattern-based approach to sequential simulation, Journel, A. G., 2005, Beyond covariance: The advent of multiple-point geostatistics, *in* O. Leuangthong and C. V. Deutsch, eds., Geostatistics Banff 2004, Proceedings of the Seventh International Geostatistics Congress: New York, Springer, p. 255–264.

Aug, C., J.-P. Chilès, G. Courrioux, and C. Lajaunie, 2005, 3D geological modelling and uncertainty: The potential field method, *in* O. Leuangthong and C. V. Deutsch, eds., Geostatistics Banff 2004, Proceedings of the Seventh International Geostatistics Congress: New York, Springer, p. 145–154.

Auñón, J., and J. J. Gómez-Hernandez, 2000, Dual kriging with local neighborhoods: Application to the representation of surfaces: Mathematical Geology, v. 32, p. 69–86.

Baafi, E. Y., and F. Szidarovszky, 1986, On nonnegative weights of linear kriging estimation: Mining Engineering, v. 38, p. 437–442.

Baczkowski, A. J., 1992, Testing for isotropy through the semi-variogram, *in* P. A. Dowd and J. J. Royer, eds., Proceedings of the 2nd CODATA Conference on Geomathematics and Geostatistics: Nancy, Association Scientifique pour la Géologie et ses Applications, p. 1–10.

Baczkowski, A. J., and K. V. Mardia, 1987, Approximate lognormality of the sample semi-variogram under a Gaussian process: Communications in Statistics—Simulation and Computation, v. 16, p. 571–585.

Bagtzoglou, A. C., and R. Ababou, 1997, Conditional simulation and estimation of Gauss-Markov random fields using the Bayesian nearest neighbor method, *in* A. Soares, J. Gomez-Hernandez, and R. Froidevaux, eds., GeoENV I: Geostatistics for environmental applications: Dordrecht, The Netherlands, Kluwer Academic Publishers, p. 477–492.

Bardossy, A., 1988, Notes on the robustness of the kriging system: Mathematical Geology, v. 20, p. 189–203.

Bardossy, A., I. Bogardi, and W. E. Kelly, 1988, Imprecise (fuzzy) information in geostatistics: Mathematical Geology, v. 20, p. 287–312.

Bardossy, A., I. Bogardi, and W. E. Kelly, 1990a, Kriging with imprecise (fuzzy) variograms: I. Theory: Mathematical Geology, v. 22, p. 63–80.

Bardossy, A., I. Bogardi, and W. E. Kelly, 1990b, Kriging with imprecise (fuzzy) variograms: II. Application: Mathematical Geology, v. 22, p. 81–94.

Barnes, R. J., 1991, The variogram sill and the sample variance: Mathematical Geology, v. 23, p. 673–678.

Barnes, R. J., and T. B. Johnson, 1984, Positive kriging, *in* G. Verly, M. David, A. Journel, and A. Marechal, eds., Geostatistics for natural resources characterization: Dordrecht, The Netherlands, D. Reidel Publishing Company, p. 231–244.

Barnes, R. J., and K. You, 1992, Adding bounds to kriging: Mathematical Geology, v. 24, p. 171–177.

Barrera, A., X. Liu, J. Ni, and S. Srinivasan, 2005, Multiple point geostatistics: Optimal template selection and implementation in multi-thread computational environments, *in* O. Leuangthong and C. V. Deutsch, eds., Geostatistics Banff 2004, Proceedings of the Seventh International Geostatistics Congress: New York, Springer, p. 621–630.

Barry, R. P., and R. K. Pace, 1997, Kriging with large data sets using sparse matrix techniques: Communications in Statistics—Simulation and Computation, v. 26, p. 619–629.

Barry, R. P,. and J. M. Ver Hoef, 1996, Blackbox krigining: Spatial prediction without specifying variogram models: Journal of Agricultural, Biological and Environmental Statistics, v. 1, p. 297–322.

Beckers, F., and P. Bogaert, 1997, Residual-based variogram in spatial linear model: Correction for bias, *in* V. Pawlowsky-Glahn, ed., Proceedings of IAMG '97, 3rd Annual Conference of the International Association for Mathematical Geology: Barcelona, International Center for Numerical Methods in Engineering, p. 629–634.

Beckers, F., and P. Bogaert, 1998, Nonstationarity of the mean and unbiased variogram estimation: Extension of the weighted least-squares method: Mathematical Geology, v. 30, p. 223–240.

Belfield, W. C., 1998, Incorporating spatial distribution into stochastic modeling of fractures: Multifractals and levy-stable statistics: Journal of Structural Geology, v. 20, p. 473–486.

Bell, G., M. J. Lechowitz, A. Appenzeller, M. Chandler, E. DeBlois, L. Jackson, B. Mackenzie, R. Preziosi, M. Scallenberg, and N. Tinker, 1993, The spatial structure of the physical environment: Oecologia, v. 96, p. 114–121.

Besag, J. E., 1974, Spatial interaction and the statistical analysis of lattice systems [with discussion]: Journal of the Royal Statistical Society, Series B, v. 36, p. 192–236.

Besag, J. E., 1977, Comment on "Modelling spatial patterns" by B. D. Ripley: Journal of the Royal Statistical Society, Series B, v. 39, p. 73–78.

Beucher, H., M. G. García-Morales, and F. Geffroy, 2005, Fitting the Boolean parameters in a non-stationarity case, *in* O. Leuangthong and C. V. Deutsch, eds., Geostatistics Banff 2004, Proceedings of the Seventh International Geostatistics Congress: New York, Springer, p. 569–578.

Biver, P., 2005, Efficient simulation techniques for uncertainty quantification on continuous variables: A process preserving the bounds, including uncertainty on data, uncertainty on average, and local uncertainty, *in* O. Leuangthong and C. V. Deutsch, eds., Geostatistics Banff 2004, Proceedings of the Seventh International Geostatistics Congress: New York, Springer, p. 205–214.

Bogaert, P., 1996, Comparison of kriging techniques in a space-time context: Mathematical Geology, v. 28, p. 17–24.

Bogaert, P., 1999, Assessing the variability of the variogram estimator, *in* J. Gomez-Hernandez, R. Froidevaux, and A. Soares, eds., Geostatistics for environmental applications: Dordrecht, The Netherlands, Kluwer Academic Publishers, p. 479–490.

Borgman, L. E., 1988, New advances in methodology for statistical tests useful in geostatistical studies: Mathematical Geology, v. 20, p. 383–404.

Borgman, L. E., K. Gerow, and G. T. Flatman, 1996, Cost-effective sampling for spatially distributed phenomena, *in* L. H. Keith, eds., Principles of environmental sampling, 2d ed.: Washington, American Chemical Society, p. 753–778.

Bosch, E. H., M. A. Oliver, and R. Webster, 2004, Wavelets and the generation of the variogram: Mathematical Geology, v. 36, p. 147–186.

Boufassa, A., and M. Armstrong, 1989, Comparison between different kriging estimators: Mathematical Geology, v. 21, p. 331–346.

Bourgault, G., 1994, Robust noise filtering by factorial kriging: Mathematical Geology, v. 26, p. 733–752.

Bourgault, G., 1996, Probability field for the post-processing of stochastic simulations: Mathematical Geology, v. 28, p. 723–734.

Bourgault, G., 1997, Using non-Gaussian distributions in geostatistical simulations: Mathematical Geology, v. 29, p. 315–334.

Bourgault, G., and A. G. Journel, 1994, Gaussian or indicator-based simulation? Which variogram is more relevant?, *in* C.-J. F. Chang, ed., Proceedings of IAMG '94: International Association for Mathematical Geology, p. 32–37.

Bourgault, G., and D. Marcotte, 1991, Multivariable variogram and its application to the linear model of coregionalization: Mathematical Geology, v. 23, p. 899–928.

Bourgault, G., and D. Marcotte, 1993, Spatial filtering under the linear coregionalization model, *in* A. Soares, ed., Geostatistics tróia '92: Dordrecht, The Netherlands, Kluwer Academic Publishers, p. 237–248.

Bradley, R., and J. Haslett, 1992, High-interaction diagnostics for geostatistical models of spatially referenced data: Statistician, v. 41, p. 371–380.

Bratvedt, K., E. Bolviken, T. Gimse, H. Holden, L. Holden, and R. Knarud, 1998, Comment on special issue on geostatistics with guest editor A. G. Journel: Mathematical Geology, v. 30, p. 245–246.

Brooker, P. I., 1975, Optimal block estimation by kriging: Australasian Institute of Mining and Metallurgy Proceedings, v. 253, p. 15–19.

Brooker, P. I., 1977, Robustness of geostatistical calculations: A case study: Australasian Institute of Mining and Metallurgy Proceedings, v. 264, p. 61–68.

Brooker, P. I., 1979, Kriging: Engineering and Mining Journal, v. 180, p. 148–153.

Brooker, P. I., 1980, Kriging, *in* A. G. Royle, ed., Geostatistics: New York, McGraw-Hill, p. 41–60.

Brooker, P. I., 1985, Two-dimensional simulation by turning bands: Mathematical Geology, v. 17, p. 81–90.

Brooker, P., 1986, A parametric study of robustness of kriging variance as a function of range and relative nugget effect for a spherical variogram: Mathematical Geology, v. 18, p. 477–488.

Bruining, J., D. van Batenburg, L. W. Lake, and A.-P. Yang, 1997, Flexible spectral methods for the generation of random fields with power-law semivariograms: Mathematical Geology, v. 29, p. 823–848.

Bruno, R., and G. Raspa, 1993, Towards a direct structural analysis of an IRF-k, *in* A. Soares, ed., Geostatistics tróia '92: Dordrecht, The Netherlands, Kluwer Academic Publishers, p. 49–60.

Bueso, M. C., and J. M. Angulo, 1999, Criteria for multivariate spatial sampling design based on covariance matrix perturbation, *in* J. Gomez-Hernandez, R. Froidevaux, and A. Soares, eds., Geostatistics for environmental applications: Dordrecht, The Netherlands, Kluwer Academic Publishers, p. 491–504.

Burger, H., and F. Birkenhake, 1994, Geostatistics and the polygonal method—A reexamination, *in* C.-J. F. Chang, ed., Proceedings of IAMG '94: International Association for Mathematical Geology, p. 50–55.

Burgess, T. M., and R. Webster, 1980, Optimal interpolation and isarithmic mapping of soil properties: I. The semi-variogram and punctual kriging: Journal of Soil Science, v. 31, p. 315–331.

Burridge, P., 1980, On the Cliff–Ord test for spatial correlation: Journal of the Royal Statistical Society, Series B, v. 42, p. 107–108.

Buxton, B. E., 1986, Global estimation variance: Mathematical Geology, v. 18, p. 693–696.

Caers, J., 2000a, Direct sequential indicator simulation, *in* W. J. Kleingeld and D. C. Krige, eds., Proceedings of the 6th International Geostatistics Congress: Cape Town, Geostatistical Association of South Africa, 12 p.

Caers, J., 2000b, Adding local accuracy to direct sequential simulation: Mathematical Geology, v. 32, p. 815–850.

Caers, J., 2001, Automatic histogram and variogram reproduction in simulated annealing simulation: Mathematical Geology, v. 33, p. 167–190.

Caers, J., 2005, Data integration using the probability perturbation method, *in* O. Leuangthong and C. V. Deutsch, eds., Geostatistics Banff 2004, Proceedings of the Seventh International Geostatistics Congress: New York, Springer, p. 13–22.

Caers, J., J. Gelders, A. Vervoot, and L. Rombouts, 1997, Non-conditional and conditional simulation of a spatial point process, *in* E. Y. Baafi and N. A. Schofield, eds., Geostatistics Wollongong '96: Dordrecht, The Netherlands, Kluwer Academic Publishers, p. 270–281.

Calle, E. O. F., and J. van Heteren, 1989, Statistical inference of trend and covariance of a random field with nonstationary mean and stationary covariance properties, *in* M. Armstrong, ed., Geostatistics, Proceedings of the 3rd International Geostatistics Congress: Dordrecht, The Netherlands, Kluwer Academic Publishers, p. 249–258.

Campbell, K., 1988, Bootstrap models for instrinsic random functions: Mathematical Geology, v. 20, p. 699–716.

Carle, S. F., and G. E. Fogg, 1994, Probability cokriging simulation, *in* C.-J. F. Chang, ed., Proceedings of IAMG '94: International Association for Mathematical Geology, p. 62–67.

Carle, S. F., and G. E. Fogg, 1996, Transition probability-based indicator geostatistics: Mathematical Geology, v. 28, p. 453–476.

Carr, J. R., 1990, Appliation of spatial filter theory to kriging: Mathematical Geology, v. 22, p. 1063–1080.

Carr, J. R., 1994, Order relation correction experiments for probability kriging: Mathematical Geology, v. 26, p. 605–622.

Carr, J. R., 2002, On visualization for assessing kriging outcomes: Mathematical Geology, v. 34, p. 421–434.

Carr, J. R., 2003, Comments about autokrieability in cokriging based on an empirical study: Mathematical Geology, v. 35, p. 217–225.

Carr, J. R., and E. D. Deng, 1987, Comparison of two techniques for applying disjunctive kriging: The Gaussian anamorphosis model versus the direct statistical inference of the bivariate distributions: Mathematical Geology, v. 19, p. 57–68.

Carr, J. R., and N.-H. Mao, 1993, Comparison of disjunctive kriging to generalized probability kriging in application to the estimation of simulated and real data, *in* A. Soares, ed., Geostatistics tróia '92: Dordrecht, The Netherlands, Kluwer Academic Publishers, p. 13–24.

Carr, J. R., and D. E. Myers, 1990, Efficiency of different equation solvers in cokriging: Computers and Geosciences, v. 16, p. 705–716.

Carr, J. R., R. E. Bailey, and E. D. Deng, 1985, Use of indicator variograms for an enhanced spatial analysis: Mathematical Geology, v. 17, p. 797–811.

Carrera, J., and F. J. Samper, 1992, On the similarities and differences between universal and residual kriging, *in* P. A. Dowd and J. J. Royer, eds., Proceedings of the 2nd CODATA Conference on Geomathematics and Geostatistics: Nancy, Association Scientifique pour la Géologie et ses Applications, p. 83–92.

Chautru, J. M., 1989, The use of Boolean random functions in geostatistics, *in* M. Armstrong, ed., Geostatistics, Proceedings of the 3rd International Geostatistics Congress: Dordrecht, The Netherlands, Kluwer Academic Publishers, p. 201–212.

Chauvet, P., 1982, The variogram cloud, *in* T. B. Johnson and R. J. Barnes, eds., Proceedings of the 17th International Symposium on the Application of Computers and Operations Research in the Mineral Industry: New York, Society of Mining Engineers, p. 757–764.

Chauvet, P., 1993, Processing data with a spatial support: Geostatistics and its methods: Fontainebleau, Centre de Géostatistique et de Morphologie Mathematique, Cahiers de Géostatistique 4, 57 p.

Cheng, Q., 1997, Fractal/multifractal modelling and spatial analysis, *in* V. Pawlowsky-Glahn, ed.,

Proceedings of IAMG '97, 3rd Annual Conference of the International Association for Mathematical Geology: Barcelona, International Center for Numerical Methods in Engineering, p. 57–72.

Cheng, Q., and F. P. Agterberg, 1996a, Multifractal modeling and spatial statistics: Mathematical Geology, v. 28, p. 1–16.

Cheng, Q., and F. P. Agterberg, 1996b, Multifractal modeling and spatial point statistics: Mathematical Geology, v. 27, p. 831–846.

Cherry, S., 1997, Non-parametric estimation of the sill in geostatistics: EnvironMetrics, v. 8, p. 13–27.

Cherry, S., J. Banfield, and W. Quimby, 1996, An evaluation of a nonparametric method of estimating semi-variograms of isotropic spatial processes: Journal of Applied Statistics, v. 23, p. 435–449.

Chilès, J. P., 1975, How to adapt kriging to nonclassical problems: Three case studies, *in* M. Guarascio, M. David, and C. Huijbregts, eds., Advanced geostatistics in the mining industry: Dordrecht, The Netherlands, D. Reidel Publishing Company, p. 69–90.

Chilès, J. P., 1979, Le variogramme généralisé: Fontainbleau, Ecole des Mines de Paris, Centre de Géostatistique et de Morphologie Mathematique, Report 612, 22 p.

Chilès, J. P., P. Delfiner, and B. Kedem, 2000, Geostatistics: Modeling spatial uncertainty: Society for Industrial and Applied Mathematics Review, v. 42, p. 530.

Christakos, G., 1984, On the problem of permissible covariance and variogram models: Water Resources Research, v. 20, p. 251–265.

Christakos, G., 1987, Stochastic simulation of spatially correlated geo-processes: Mathematical Geology, v. 19, p. 807–831.

Christensen, O. F., 2004, Monte Carlo maximum likelihood in model-based geostatistics: Journal of Computational and Graphical Statistics, v. 13, p. 702–718.

Christensen, R., 1990, The equivalence of predictions from universal kriging and intrinsic random-function kriging: Mathematical Geology, v. 22, p. 655–664.

Chu, J., 1996, Fast sequential indicator simulation: Beyond reproduction of indicator variograms: Mathematical Geology, v. 28, p. 909–922.

Chu, J., and A. G. Journel, 1994, Conditional fBm simulation with dual kriging, *in* R. Dimitrakopoulos, ed., Geostatistics for the next century: Dordrecht, The Netherlands, Kluwer Academic Publishers, p. 407–421.

Chu, J., W. Xu, and A. G. Journel, 1994, 3-D implementation of geostatistical analyses—The Amoco case study, *in* J. M. Yarus and R. L. Chambers, eds., Stochastic modeling and geostatistics: Principles, methods, and case studies: AAPG Computer Applications in Geology 3, p. 201–216.

Chung, C. F., 1984, Use of the jackknife method to estimate autocorrelation functions (or variograms), *in* G. Verly, M. David, A. G. Journel, and A. Marechal, eds., Geostatistics for natural resources characterization: Dordrecht, The Netherlands, D. Reidel Publishing Company, p. 55–71.

Clark, I., 1977a, Regularization of a semivariogram: Computers and Geosciences, v. 3, p. 341–346.

Clark, I., 1977b, Practical kriging in three dimensions: Computers and Geosciences, v. 3, p. 173–180.

Clark, I., 1979a, The semivariogram—Part I: Engineering and Mining Journal, v. 180, p. 90–94.

Clark, I., 1979b, The semivariogram—Part II: Engineering and Mining Journal, v. 180, p. 90–97.

Clark, I., 1980, Part 2: The semivariogram, *in* A. G. Royle, ed., Geostatistics: New York: McGraw-Hill, p. 29–40.

Clark, I., 1983, Regression revisited: Mathematical Geology, v. 15, p. 517–536.

Clark, I., K. L. Basinger, and W. V. Harper, 1989, MUCK: A novel approach to co-kriging, *in* B. E. Buxton, ed., Proceedings of the Conference on Geostatistical, Sensitivity, and Uncertainty Methods for Ground-water Flow and Radionuclide Transport Modeling: Columbus, Battelle Press, p. 473–493.

Cressie, N., 1979, Straight line fitting and variogram estimation: Bulletin of the International Statistical Institute, v. 38, bk. 3, p. 573–576.

Cressie, N., 1984, Towards resistant geostatistics, *in* G. Verly, M. David, A. G. Journel, and A. Marechal, eds., Geostatistics for natural resources characterization: Dordrecht, The Netherlands, D. Reidel Publishing Company, p. 21–44.

Cressie, N., 1985a, When are relative variograms useful in geostatistics?: Mathematical Geology, v. 17, p. 693–702.

Cressie, N., 1985b, Fitting variogram models by weighted least squares: Mathematical Geology, v. 17, p. 563–586.

Cressie, N., 1986, Kriging nonstationary data: Journal of the American Statistical Association, v. 81, p. 625–634.

Cressie, N., 1987a, A nonparametric view of generalized covariances for kriging: Mathematical Geology, v. 19, p. 425–450.

Cressie, N., 1987b, Exact interpolation: Geostatistics—An interdisciplinary geostatistics newsletter, v. 1, p. 11–13.

Cressie, N., 1987c, Mixing ergodicity and geostatistics: Geostatistics—An interdisciplinary geostatistics newsletter, v. 2, p. 9–12.

Cressie, N., 1988, Spatial prediction and ordinary kriging: Mathematical Geology, v. 20, p. 405–421.

Cressie, N., 1989a, The many faces of spatial prediction, *in* M. Armstrong, ed., Geostatistics, Proceedings of the 3rd International Geostatistics Congress: Dordrecht, The Netherlands, Kluwer Academic Publishers, p. 163–176.

Cressie, N., 1989b, Erratum: Spatial prediction and ordinary kriging: Mathematical Geology, v. 21, p. 493–494.

Cressie, N., 1990, The origins of kriging: Mathematical Geology, v. 22, p. 239–252.

Cressie, N., 1991, Geostatistical analysis of spatial data, *in* Spatial statistics and digital image analysis: Washington, National Academy Press, p. 87–108.

Cressie, N., 1993, Aggregation in geostatistical problems, *in* A. Soares, ed., Geostatistics troia '92: Dordrecht, The Netherlands, Kluwer Academic Publishers, p. 25–36.

Cresssie, N., 1994, Models for spatial processes, *in* J. Stanford and S. Vardeman, eds., Statistical methods for physical science: San Diego, Academic Press, p. 93–124.

Cressie, N., 1996, Change of support and the modifiable areal unit problem: Geographical Systems, v. 3, p. 159–180.

Cressie, N., 1997, Comment on "Model-based geostatistics" by P. J. Diggle, J. A. Tawn, and R. A. Moyeed: Applied Statistics, v. 47, p. 335.

Cressie, N., and J. Aldworth, 1997, Spatial statistical analysis and its consequences for spatial sampling, *in* E. Y. Baafi and N. A. Schofield, eds., Geostatistics Wollongong '96: Dordrecht, The Netherlands, Kluwer Academic Publishers, p. 126–137.

Cressie, N., and N. H. Chan, 1989, Spatial modeling of regional variables: Journal of the American Statistical Association, v. 84, p. 393–401.

Cressie, N., and G. Glonek, 1984, Median based covariogram estimators reduce bias: Statistics and Probability Letters, v. 2, p. 299–304.

Cressie, N., and D. M. Hawkins, 1980, Robust estimation of the variogram: I: Mathematical Geology, v. 12, p. 115–125.

Cressie, N., and J. D. Helterbrand, 1994, Multivariate spatial statistical models: Geographical Systems, v. 1, p. 179–188.

Cressie, N., and G. Johannesson, 2001, Kriging for cutoffs and other difficult problems, *in* P. Monestiez, D. Allard, and R. Froidevaux, eds., GeoENV III—Geostatistics for environmental applications: Dordrecht, The Netherlands, Kluwer Academic Publishers, p. 299–310.

Cressie, N., and M. Pavlikova, 2002, Calibrated spatial moving average simulations: Statistical Modeling, v. 2, p. 267–279.

Cressie, N., and M. Pavlikova, 2005, Lognormal kriging: Bias adjustment and kriging variances, *in* O. Leuangthong and C. V. Deutsch, eds., Geostatistics Banff 2004, Proceedings of the Seventh International Geostatistics Congress: New York, Springer, p. 1027–1036.

Cressie, N., and C. K. Wikle, 1998, The variance-based cross-variogram: You can add apples and oranges: Mathematical Geology, v. 30, p. 789–799.

Cressie, N., and D. L. Zimmerman, 1992, On the stability of the geostatistical methods: Mathematical Geology, v. 24, p. 45–59.

Cressie, N. A. C., and M. O. Grondona, 1992, A comparison of variogram estimation with covariogram estimation, *in* K. V. Mardia, ed., The art of statistical science: New York, John Wiley & Sons, p. 191–208.

Crozel, D., and M. David, 1985, Global estimation variance: Formulas and calculation: Mathematical Geology, v. 17, p. 785–796.

Curriero, F. C., and S. Lele, 1999, A composite likelihood approach to semivariogram estimation: Journal of Agricultural, Biological and Environmental Statistics, v. 4, p. 9–28.

Dahle, P., O. Kolbjørnsen, and P. Abrahamsen, 2005, When can shape and scale parameters of a 3D variogram be estimated?, *in* O. Leuangthong and C. V. Deutsch, eds., Geostatistics Banff 2004, Proceedings of the Seventh International Geostatistics Congress: New York, Springer, p. 949–958.

Daly, C., 2005, Higher order models using entropy, Markov random fields and sequential simulation, *in* O. Leuangthong and C. V. Deutsch, eds., Geostatistics Banff 2004, Proceedings of the Seventh International Geostatistics Congress: New York, Springer, p. 215–224.

David, M., 1973, Tools for planning variances and conditional simulations, *in* Proceedings of the 11th International Symposium on the Application of Computers and Operations Research on the Mineral Industry: Society of Mining Engineers, p. D10–D23.

David, M., 1976, The practice of kriging, *in* M. Guarascio, M. David, and C. J. Huijbregts, eds., Advanced geostatistics in the mining industry: Dordrecht, The Netherlands, D. Reidel Publishing Company, p. 31–48.

David, M., 1978, Sampling and estimation problems for three-dimensional spatial stationary and nonstationary stochastic processes as encountered in the mineral industry: Journal of Statistical Planning and Inference, v. 2, p. 211–244.

David, M., 1986, Global estimation variance: Formulas and calculation: Mathematical Geology, v. 18, p. 697–698.

David, M., D. Marcotte, and M. Soulie, 1984, Conditional bias in kriging and a suggested correction, *in* G. Verly, M. David, A. G. Journel, and A. Marechal, eds., Geostatistics for natural resources characterization: Dordrecht, The Netherlands, D. Reidel Publishing Company, p. 121–128.

Davis, B. M., 1982, A note on the symptotic distribution of the sample variogram: Mathematical Geology, v. 14, p. 189–193.

Davis, B. M., 1987, Uses and abuses of cross validation in geostatistics: Mathematical Geology, v. 19, no. 3, p. 241–248.

Davis, B. M., and L. E. Borgman, 1979, Some exact sampling distributions for variogram estimators: Mathematical Geology, v. 11, p. 643–653.

Davis, B. M., and L. E. Borgman, 1982, A note on the asymptotic distribution of the sample variogram: Mathematical Geology, v. 14, p. 189–193.

Davis, B. M., and G. J. Jalkanen, 1988, Nonparametric estimation of multivariate joint and conditional spatial distributions: Mathematical Geology, v. 20, p. 367–382.

Davis, G. J., and M. D. Morris, 1997, Six factors which affect the condition number of matrices associated with kriging: Mathematical Geology, v. 29, p. 669–683.

Davis, M. W., 1987a, Generating large stochastic simulations—The matrix polynomial approximation method: Mathematical Geology, v. 19, p. 99–107.

Davis, M. W., 1987b, Production of conditional simulations via the LU triangular decomposition of the covariance matrix: Mathematical Geology, v. 19, p. 91–98.

Davis, M. W., and C. Grivet, 1984, Kriging in a global neighborhood: Mathematical Geology, v. 16, p. 249–266.

Davis, M. W. D., and M. David, 1978a, Automatic kriging and contouring in the presence of trends (universal kriging made simple): Journal of Canadian Petroleum Technology, v. 17, p. 49–68.

Davis, M. W. D., and M. David, 1978b, The numerical calculation of integrals of an isotropic variogram function on hypercubes: Mathematical Geology, v. 10, p. 311–314.

Davis, M. W. D., M. David, and J.-M. Belisle, 1978, A fast method for the solution of a system of simultaneous linear equations—A method adapted to a particular problem: Mathematical Geology, v. 10, p. 369–374.

Deffeyes, K. S., B. D. Ripley, and G. S. Watson, 1982, Stochastic geometry in petroleum geology: Mathematical Geology, v. 14, p. 419–432.

de Fouquet, C., 1994, Reminders on the conditioning kriging, *in* M. Armstrong and P. A. Dowd, eds., Geostatical simulations: Dordrecht, The Netherlands, Kluwer Academic Publishers, p. 131–146.

de Fouquet, C., 2000, Joint simulation of a random function and its derivatives, *in* W. J. Kleingold and D. C. Krige, eds., Proceedings of the 6th International Geostatistics Congress: Cape Town, Geostatistical Association of South Africa, 10 p.

De Iaco, S., D. E. Myers, and D. Posa, 2003, The linear coregionalization model and the product-sum space-time variogram: Mathematical Geology, v. 35, p. 25–38.

Delay, F., and G. de Marsily, 1994, The integral of the semivariogram: A powerful method for adjusting the semivariogram in geostatistics: Mathematical Geology, v. 26, p. 301–322.

Delfiner, P., 1976, Linear estimation of non-stationary spatial phenomena, *in* M. Guarascio, M. David, and C. J. Huijbregts, eds., Advanced geostatistics in the mining industry: Dordrecht, The Netherlands, D. Reidel Publishing Company, p. 49–68.

Delfiner, P., 1994, Comments on "Image compression and kriging," *in* R. Dimitrakopoulos, ed., Geostatistics for the next century: Dordrecht, The Netherlands, Kluwer Academic Publishers, p. 162–167.

Delfiner, P., and J. P. Delhomme, 1973, Optimum interpolation by kriging, *in* J. C. Davis and M. J. McCullagh, eds., Proceedings of the NATO advanced study institute, display and analysis of spatial data: New York, John Wiley & Sons, p. 96–114.

Desmond, A. F., 1997, Prediction functions and geostatistics, *in* I. V. Basawa, V. P. Godambe, and R. L. Taylor, eds., Selected Proceedings of the Symposium on Estimating Functions: Institute of Mathematical Statistics Lecture Notes Monograph Series, v. 32, p. 353–367.

Deutsch, C., 1985, Estimating block effective permeability with geostatistics and power averaging: Society of Petroleum Engineers Annual Technical Conference and Exhibition, SPE Paper 15991.

Deutsch, C. V., 1991, The relationship between universal kriging, kriging with an external drift, and cokriging: Stanford, Stanford Center for Reservoir Forecasting Annual Report 4, 8 p.

Deutsch, C. V., 1993, Kriging in a finite domain: Mathematical Geology, v. 25, p. 41–52.

Deutsch, C. V., 1994a, Kriging with strings of data: Mathematical Geology, v. 26, p. 623–638.

Deutsch, C. V., 1994b, Algorithmically-defined random function models, *in* R. Dimirakopoulos, ed., Geostatistics for the next century: Dordrecht, The Netherlands, Kluwer Academic Publishers, p. 393–406.

Deutsch, C. V., 1996a, Constrained smoothing of histograms and scatterplots with simulated annealing: Technometrics, v. 38, p. 266–275.

Deutsch, C. V., 1996b, Correcting for negative weights in ordinary kriging: Computers and Geosciences, v. 22, p. 765–775.

Deutsch, C. V., and P. W. Cockerham, 1994, Practical considerations in the application of simulated annealing to stochastic simulation: Mathematical Geology, v. 26, p. 67–82.

Deutsch, C. V., and H. Kupfersberger, 1997, Geostatistical simulation with large-scale soft data, *in* V. Pawlowsky-Glahn, ed., Proceedings of IAMG '97, 3rd Annual Conference of the International Association for Mathematical Geology: Barcelona, International Center for Numerical Methods in Engineering, p. 73–90.

Deutsch, C. V., and X. H. Wen, 1998, An improved perturbation mechanism for simulated annealing simulation: Mathematical Geology, v. 30, p. 801–816.

Deutsch, C. V., and X. H. Wen, 2000, Integrating large-scale soft data by simulated annealing and probability constraints: Mathematical Geology, v. 32, p. 49–67.

Diamond, P., 1988, Interval-valued random functions and the kriging of intervals: Mathematical Geology, v. 20, p. 145–166.

Diamond, P., and M. Armstrong, 1984, Robustness of variograms and conditioning of kriging: Mathematical Geology, v. 16, p. 809–822.

Diblasi, A., 1998, Confidence intervals for the variogram of a spatial process via block-bootstrap, *in* A. Buccianti, G. Nardi, and R. Potenza, eds., IAMG '98, Proceedings of the 4th Annual Conference of the International Association for Mathematical Geology: Naples, de Frede, p. 364–369.

Dietrich, C. R., and G. N. Newsam, 1995, Efficient generation of conditional simulations by Chebyshev matrix polynomial approximations to the symmetric square root of the covariance matrix: Mathematical Geology, v. 27, p. 207–228.

Diggle, P. J., and P. J. Ribeiro Jr., 1999, Bayesian inference in Gausian model-based geostatistics: Geographical and Environmental Modelling, v. 6, p. 129–146.

Diggle, P. J., J. A. Tawn, and R. A. Moyeed, 1998, Model-based geostatistics (with discussion): Applied Statistics, v. 47, p. 299–350.

Diggle, P. J., P. J. Ribeiro Jr., and O. F. Christensen, 2003, An introduction to model-based geostatistics, *in* J. Møller, ed., Spatial statistics and computational methods: New York, Springer, p. 43–86.

Dimitrakopoulos, R., 1990, Conditional simulation of instrinsic random functions of order *k*: Mathematical Geology, v. 22, p. 361–380.

Dimitrakopoulos, R., and X. Luo, 1994, Spatiotemporal modeling: Covariances and ordinary kriging systems, *in* R. Dimitrakopoulos, ed., Geostatistics for the next century: Dordrecht, The Netherlands, Kluwer Academic Publishers, p. 88–93.

Dimitrakopoulos, R., and X. Luo, 2004, Generalized sequential Gaussian simulation on group size $\nu$ and screen-effect approximations for large field simulations: Mathematical Geology, v. 36, p. 567–592.

Dowd, P. A., 1982, Lognormal kriging—The general case: Mathematical Geology, v. 14, p. 475–500.

Dowd, P. A., 1984, The variogram and kriging: Robust and resistant estimators, *in* G. Verly, M. David, A. G. Journel, and A. Marechal, eds., Geostatistics for natural resources characterization: Dordrecht, The Netherlands, D. Reidel Publishing Company, p. 91–106.

Dowd, P. A., 1991, A review of recent developments in geostatistics: Computers and Geosciences, v. 17, p. 1481–1500.

Dowd, P. A., 1994, The use of neural networks for spatial simulation, *in* R. Dimitrakopoulos, ed., Geostatistics for the next century: Dordrecht, The Netherlands, Kluwer Academic Publishers, p. 173–184.

Dowd, P., and C. Saraç, 1994, An extension of the LU decomposition method of simulation, *in* M. Armstrong and P. A. Dowd, eds., Geostatistical simulations: Dordrecht, The Netherlands, Kluwer Academic Publishers, p. 23–36.

Dubrule, O., 1983a, Cross validation of kriging in a unique neighborhood: Mathematical Geology, v. 15, p. 687–699.

Dubrule, O., 1983b, Two methods with different objectives: Mathematical Geology, v. 15., p. 245–258.

Dubrule, O., 1989, A review of stochastic models for petroleum reservoirs, *in* M. Armstrong, ed., Geostatistics, Proceedings of the 3rd International Geostatistics Congress: Dordrecht, The Netherlands, Kluwer Academic Publishers, p. 493–506.

Dubrule, O., 1994a, Estimating or choosing a geostatistical model?, *in* R. Dimitrakopoulos, ed., Geostatistics for the next century: Dordrecht, The Netherlands, Kluwer Academic Publishers, p. 3–14.

Dubrule, O., 1994b, Reply to "Comment: Estimating or choosing a geostatistical model?," *in* R. Dimitrakopoulos, ed., Geostatistics for the next century: Dordrecht, The Netherlands, Kluwer Academic Publishers, p. 17.

Dunn, M. R., 1984, Conditions for a variogram model to yield positive variances under restrictions: Reply: Mathematical Geology, v. 16, p. 433–434.

Dunn, M. R., and J. R. Alldredge, 1982, General polygonal variogram functions: Evaluation of estimation variance integrals: Mathematical Geology, v. 14, p. 77–85.

Dunn, M. R., H. D. Tolley, and J. R. Alldredge, 1980, Exact formulas for approximating variogram function integrals: Mathematical Geology, v. 12, p. 577–588.

Dutilleul, P., 1993, Modifying the *t* test for assessing the correlation between two spatial processes: Biometrics, v. 49, p. 305–314.

Dutter, R., 1996, On robust estimation of variograms in geostatistics: Lecture Notes in Statistics, no. 109, p. 153–172.

Ecker, M. D., and A. E. Gelfand, 1997, Bayesian variogram modeling for an isotropic spatial process: Journal of Agricultural, Biological, and Environmental Statistics, v. 2, p. 347–369.

Emery, X., 2002, Conditional simulation of nongaussian random functions: Mathematical Geology, v. 23, p. 79–100.

Emery, X., 2003, Disjunctive kriging with hard and imprecise data: Mathematical Geology, v. 35, p. 699–718.

Emery, X., 2005, Variograms of order omega: A tool to validate a bivariate distribution model: Mathematical Geology, v. 37, p. 163–181.

Emery, X., and J. M. Ortiz, 2005a, Histogram and variogram inference in the multigaussian model: Stochastic Environmental Research and Risk Assessment, v. 19, p. 48–58.

Emery, X., and J. M. Ortiz, 2005b, Internal consistency and inference of change-of-support isofactorial models, *in* O. Leuangthong and C. V. Deutsch, eds., Geostatistics Banff 2004, Proceedings of the Seventh International Geostatistics Congress: New York, Springer, p. 1057–1066.

Eriksson, M., and P. P. Siska, 2000, Understanding anisotropy computations: Mathematical Geology, v. 32, p. 683–700.

Ersbøll, A. K., and B. K. Ersbøll, 1997, On spatiotemporal kriging, *in* V. Pawlowsky-Glahn, ed., Proceedings of IAMG '97, 3rd Annual Conference of the International Association for Mathematical Geology: Barcelona, International Center for Numerical Methods in Engineering, p. 617–622.

Fang, J. H., and P. P. Wang, 1997, Random field generation using simulated annealing vs. fractal-based stochastic interpolation: Mathematical Geology, v. 29, p. 849–858.

Fernandéz-Casal, R., W. González-Manteiga, and M. Febrero-Bande, 2003, Flexible spatio-temporal stationary variogram models: Statistics and Computing, v. 13, p. 127–136.

Filmoser, P., and C. Reimann, 2002, Robust multivariate methods in geostatistics, *in* W. Gaul and G. X. Ritter, eds., Classification, automation, and new media: New York, Springer, p. 429–436.

Francois-Bongarcon, D., 1986, Rank order kriging (R.O.K.): A powerful, robust, simple alternative to linear kriging, *in* D. Michel, ed., Ore reserve estimation, methods, models and reality: Montreal, Canadian Institute of Mining and Metallurgy, p. 313–322.

Francois-Bongarcon, D. M., 1994, Comments on F. Pitard's "Exploration of the nugget effect," *in* R. Dimitrakopoulos, ed., Geostatistics for the next century: Dordrecht, The Netherlands, Kluwer Academic Publishers, p. 137–141.

Fredericks, A. K., and K. B. Newman, 1998, A comparison of the sequential Gaussian and Markov-Bayes simulation methods for small samples: Mathematical Geology, v. 30, p. 1011–1032.

Freulon, X., 1994, Conditional simulation of a gaussian random vector with non linear and/or noisy observations, *in* M. Armstrong and P. A. Dowd, eds., Geostatistical simulations: Dordrecht, The Netherlands, Kluwer Academic Publishers, p. 57–71.

Freulon, X., and C. de Fouquet, 1993, Conditioning a Gaussian model with inequalities, *in* A. Soares, ed., Geostatistics troia '92: Dordrecht, The Netherlands, Kluwer Academic Publishers, p. 201–212.

Freund, M., 1986, Cokriging: multivariable analysis in petroleum exploration: Computers and Geosciences, v. 12, p. 485–489.

Froidevaux, R., 1984, Conditional estimation variances: An empirical approach: Mathematical Geology, v. 16, p. 327–350.

Froidevaux, R., 1993a, Probability field simulation, *in* A. Soares, ed., Geostatistics troia '92: Dordrecht, The Netherlands, Kluwer Academic Publishers, p. 73–84.

Froidevaux, R., 1993b, Constrained kriging as an estimator of local distribution functions, *in* V. Capasso, G. Girone, and D. Posa, eds., Statistics of spatial processes: Theory and applications: Bari, Italy, Istituto per Ricerche de Mathematica Applicata, p. 106–118.

Frykman, P., and C. V. Deutsch, 1998, Model-based declustering for deriving representative distributions prior to conditional simulations (abs.): 2nd Institute of Mathematics and its Applications (IMA) Conference on Quantification and Modeling of Spatial Patterns in Permeable Rocks, March 23–25, Cambridge, Churchill College.

Fuentes, M., 2001, Fixed-domain asymptotics for variograms using subsampling: Mathematical Geology, v. 33, p. 693–718.

Galli, A., E. Murillo, and J. Thomann, 1983, Dual kriging—Its properties and its uses in direct contouring, *in* G. Verly, M. David, A. G. Journel, and A. Marechal, eds., Geostatistics for natural resources characterization: Dordrecht, The Netherlands, D. Reidel Publishing Company, p. 621–634.

Galli, A., H. Beucher, G. Le Loc'h, and B. Doligez, 1994, The pros and cons of the truncated Gaussian method, *in* M. Armstrong and P. A. Dowd, eds., Geostatistical simulations: Dordrecht, The Netherlands, Kluwer Academic Publishers, p. 217–234.

Genton, M. G., 1998a, Highly robust variogram estimation: Mathematical Geology, v. 30, p. 213–222.

Genton, M. G., 1998b, Variogram fitting by generalized least squares using an explicit formula for the covariance structure: Mathematical Geology, v. 30, p. 323–346.

Genton, M. G., 1998c, Spatial breakdown point of variogram estimators: Mathematical Geology, v. 30, p. 853–872.

Genton, M. G., 2000, The correlation structure of Matheron's classical variogram estimator under elliptically contoured distributions: Mathematical Geology, v. 32, p. 127–138.

Genton, M. G., and D. J. Gorsich, 2003, Nonparametric variogram and covariogram estimation with Fourier-Bessel matrices: Computational Statistics and Data Analysis, v. 41, p. 47–58.

Glacken, C., 1997, Change of support and use of economic parameters for block simulation, *in* E. Y. Baafi and N. A. Schofield, eds., Geostatistics Wollongong '96: Dordrecht, The Netherlands, Kluwer Academic Publishers, p. 811–821.

Gneiting, T., Z. Sasvári, and M. Schlather, 2001, Analogies and correspondences between variograms and covariance functions: Advances in Applied Probability, v. 33, p. 617–630.

Gómez-Hernandez, J. J., and E. F. Cassiraga, 1994, Theory and practice of sequential simulation, *in* M. Armstrong and P. A. Dowd, eds., Geostatical simulations: Dordrecht, The Netherlands, Kluwer Academic Publishers, p. 111–124.

Gómez-Hernandez, J. J., and A. G. Journel, 1993, Joint sequential simulation of multigaussian fields, *in* A. Soares, ed., Geostatistics troia '92: Dordrecht, The Netherlands, Kluwer Academic Publishers, p. 85–94.

Gómez-Hernandez, J. J., and X. Wen, 1994, To be or not to be multiGaussian? That is the question: Stanford, Stanford Center for Reservoir Forecasting Report 7, 42 p.

Gómez-Hernandez, J. J., R. Froidevaux, and P. Biver, 2005, Exact conditioning to linear constraints in kriging and simulation, *in* O. Leuangthong and C. V. Deutsch, eds., Geostatistics Banff 2004, Proceedings of the Seventh International Geostatistics Congress: New York, Springer, p. 999–1006.

Goovaerts, P., 1993, Spatial orthogonality of the principal components completed from coregionalized variables: Mathematical Geology, v. 25, p. 281–302.

Goovaerts, P., 1994a, Comparison of CoIK, IK, mIK

performances for modeling conditional probabilities of categorical variables, *in* R. Dimitrakopoulos, ed., Geostatistics for the next century: Dordrecht, The Netherlands, Kluwer Academic Publishers, p. 18–29.

Goovaerts, P., 1994b, On a controversial method for modeling a coregionalization: Mathematical Geology, v. 26, p. 197–204.

Goovaerts, P., 1994c, Comparative performance of indicator algorithms for modeling conditional probability distribution functions: Mathematical Geology, v. 26, p. 389–412.

Goovaerts, P., 1994d, Study of spatial relationships between two sets of variables using multivariate geostatistics, *in* J. J. de Gruyter, R. Webster, and D. E. Myers, eds., Pediometrics '92: Developments in spatial statistics for soil science: Amsterdam, Elsevier, p. 93–108.

Goovaerts, P., 1996, Stochastic simulation of categorical variables using a classification algorithm and simulated annealing: Mathematical Geology, v. 28, p. 909–921.

Goovaerts, P., 1998a, Ordinary cokriging revisited: Mathematical Geology, v. 30, p. 21–42.

Goovaerts, P., 1998b, Accounting for estimation optimality criteria in simulated annealing: Mathematical Geology, v. 30, p. 511–534.

Goovaerts, P., 1999, Impact of the simulation algorithm, magnitude of ergodic fluctuations and number of realizations on the spaces of uncertainty of flow properties: Stochastic Environmental Research and Risk Assessment, v. 13, p. 161–182.

Gorsich, D. J., and M. G. Genton, 2000, Variogram model selection via nonparametric derivative estimation: Mathematical Geology, v. 32, p. 249–270.

Gorsich, D. J., and M. G. Genton, 2004, On the discretization of nonparametric isotropic covariogram estimators: Statistics and Computing, v. 14, p. 109–108.

Goovaerts, P., and A. G. Journel, 1996, Accounting for local probabilities in stochastic modeling of facies data: Society of Petroleum Engineers Journal, v. 1, p. 21–29.

Gorsich, D. J., M. G. Genton, and G. Strang, 2002, Eigenstructures of spatial design matrics: Journal of Multivariate Analysis, v. 80, p. 138–165.

Gotway, C. A., 1991, Fitting semivariogram models by weighted least squares: Computers and Geosciences, v. 17, p. 171–172.

Gotway, C. A., and A. H. Hartford, 1996, Geostatistical methods for incorporating auxiliary information in the prediction of spatial variables: Journal of Agricultural, Biological, and Environmental Statistics, v. 1, p. 17–39.

Gotway, C. A., and G. W. Hergert, 1997, Incorporating spatial trends and anisotropy in geostatistical mapping of soil properties: Soil Science Society of America Journal, v. 61, p. 298–312.

Gotway, C. A., and B. M. Rutherford, 1994, Stochastic simulation for imaging spatial uncertainty: Comparison and evaluation of available algorithms, *in* M. Armstrong and P. A. Dowd, eds., Geostatistical simulations: Dordrecht, The Netherlands, Kluwer Academic Publishers, p. 1–21.

Gotway, C. A., and L. J. Young, 2002, Combining incompatible spatial data: Journal of the American Statistical Association, v. 97, p. 632–648.

Goulard, M., 1989, Inference in a coregionalization model, *in* M. Armstrong, ed., Geostatistics, Proceedings of the 3rd International Geostatistics Congress: Dordrecht, The Netherlands, Kluwer Academic Publishers, p. 397–408.

Goulard, M., and M. Voltz, 1992, Linear coregionalization model: Tools for estimation and choice of cross-variogram matrix: Mathematical Geology, v. 24, p. 269–286.

Gray, T. A., F. X. Jian, and I. J. Taggart, 1993, A critical comparison of kriging, fractal, and indicator kriging techniques: Australian Petroleum Exploration Association Journal, v. 33, p. 330–342.

Gribov, A., and K. Krivoruchko, 2004, Geostatistical mapping with continuous moving neighborhood: Mathematical Geology, v. 36, p. 267–282.

Gringarten, E., 1998, Fracnet: Stochastic simulation of fractures in layered systems: Computers and Geosciences, v. 24, p. 729–736.

Grondona, M. O., and N. Cressie, 1995, Residuals based estimators of the covariogram: Statistics, v. 26, p. 209–218.

Guardiano, F., and M. Srivastava, 1993, Multivariate geostatistics: Beyond bivariate moments, *in* A. Soares, ed., Geostatistics tróia '92: Dordrecht, The Netherlands, Kluwer Academic Publishers, p. 133–144.

Guertin, K., 1984, Correcting conditional bias, *in* G. Verly, M. David, A. G. Journel, and A. Marechal, eds., Geostatistics for natural resources characterization: Dordrecht, The Netherlands, D. Reidel Publishing Company, p. 435–448.

Gunst, R. F., and M. L. Hartfield, 1997, Robust semivariogram estimation in the presence of influential spatial data values, *in* T. G. Gregoir, D. R.

Brillinger, P. J. Diggle, E. Russek-Cohen, W. G. Warren, and R. D. Wolfinger, eds., Modelling longitudinal and spatially correlated data: New York, Springer, p. 265–274.

Gutjahr, A., B. Bullard, and S. Hatch, 1994, Joint conditional simulation and flow modeling, *in* R. Dimitrakopoulos, ed., Geostatistics for the next century: Dordrecht, The Netherlands, Kluwer Academic Publishers, p. 185–196.

Gutjahr, A., B. Bullard, and S. Hatch, 1997, General joint conditional simulations using a fast Fourier transform method: Mathematical Geology, v. 29, p. 361–390.

Gutjahr, A. L., and L. W. Gelhar, 1981, Stochastic models of subsurface flows: Infinite versus finite domains and stationarity: Water Resources Research, v. 17, p. 337–350.

Gutjahr, A. L., and J. L. Wilson, 1985, Co-kriging and stochastic differential equations, *in* G. de Marsily, Proceedings of the International Symposium on the Stochastic Approach to Subsurface Flow: Montevillargenne, Centre Nationale de ala Reserche Scientifique, p. 180–191.

Gutjahr, A. L., L. W. Gelhar, A. A. Bakr, and J. R. Macmillan, 1978, Stochastic analysis of spatial variability in subsurface flows: 2. Evaluation and applications: Water Resources Research, v. 14, p. 953–960.

Handcock, M. S., 1994a, Comment on "Estimating or choosing a geostatistical model?," *in* R. Dimitrakopoulos, ed., Geostatistics for the next century: Dordrecht, The Netherlands, Kluwer Academic Publishers, p. 15–16.

Handcock, M. S., 1994b, Measuring the uncertainty in kriging, *in* R. Dimitrakopoulos, ed., Geostatistics for the next century: Dordrecht, The Netherlands, Kluwer Academic Publishers, p. 436–447.

Handcock, M. S., and M. L. Stein, 1993, A Bayesian analysis of kriging: Technometrics, v. 35, p. 403–410.

Haslett, J., 1989, Geostatistical neighbourhoods and subset selection, *in* M. Armstrong, ed., Geostatistics, Proceedings of the 3rd International Geostatistics Congress: Dordrecht, The Netherlands, Kluwer, p. 569–578.

Hawkins, D. M., and N. Cressie, 1984, Robust kriging—A proposal: Mathematical Geology, v. 16, p. 3–18.

Hazlett, R. D., 1997, Statistical characterization and stochastic modeling of pore networks in relation to fluid flow: Mathematical Geology, v. 29, p. 801–822.

Hegstad, B. K., H. Omre, H. Tjelmeland, and K. Tyler, 1994, Stochastic simulation and conditioning by annealing in reservoir description, *in* M. Armstrong and P. A. Dowd, eds., Geostatistical simulations: Dordrecht, The Netherlands, Kluwer Academic Publishers, p. 43–56.

Helterbrand, J. D., and N. Cressie, 1994, Universal cokriging under instinsic coregionalization: Mathematical Geology, v. 26, p. 205–226.

Henley, S., and D. F. Watson, 1998, Possible alternatives to geostatistics, *in* APCOM '98, International Symposium of the Application of Computers and Operations Research in the Mineral Industry: London, Institution of Mining and Metallurgy, p. 337–354.

Herzfeld, U. C., 1989, A note on programs performing kriging with nonnegative weights: Mathematical Geology, v. 21, p. 391–393.

Herzfeld, U. C., 1992, Least-squares collocation, geophysical inverse theory and geostatistics: A bird's eye view: Geophysical Journal International, v. 11, p. 237.

Heuvelink, G. B. M., 1992, An interative method for multi-dimensional simulation with nearest neighbour models, *in* P. A. Dowd and J. J. Royer, eds., Proceedings of the 2nd CODATA Conference on Geomathematics and Geostatistics: Nancy, Association Scientifique pour la Géologie et ses Applications, p. 43–50.

Heuvelink, G. B. M., and E. J. Pebesma, 2002, Is the ordinary kriging variance a proper measure of interpolation error, *in* G. Hunter, and K. Lowell, eds., Accuracy 2002: Proceedings of the 4th International Symposium on Spatial Accuracy Assessment in Natural Resources and Environmental Sciences: Melbourne, RMIT University, p. 179–186.

Hillmann, T., 2000, Estimation of the model parameters for geostatistical prediction (in German): Mathematische Geologie, v. 5, p. 49–57.

Hohn, M. E., and M. V. Fontana, 1990a, Dissecting variograms: Computers and Geology, v. 6, p. 147.

Hohn, M. E., and M. V. Fontana, 1990b, Dissecting variograms, *in* J. T. Hanley and D. F. Merriam, eds., Microcomputer applications in geology, II: Oxford, Pergamon Press, p. 147–160.

Hohn, M. E., and D. W. Neal, 1986, Geostatistical analysis of gas potential in Devonian shales of West Virginia: Computers and Geosciences, v. 12, p. 611–617.

Hjort, N. L., and H. Omre, 1994, Topics in spatial statistics: Scandinavian Journal of Statistics, v. 32, p. 289–358.

Hu, L. Y., 1989, Comparing gamma isofactorial disjunctive kriging and indicator kriging for estimating local spatial distributions, *in* M. Armstrong, ed., Geostatistics, Proceedings of the 3rd International Geostatistics Congress: Dordrecht, The Netherlands, Kluwer Academic Publishers, p. 335–346.

Hu, L. Y., 2000, Gradual deformation and iterative calibration of Gaussian-related stochastic models: Mathematical Geology, v. 32, p. 87–108.

Hu, L. Y., and M. le Ravalec-Dupin, 2005, On some controversial issues of geostatistical simulation, *in* O. Leuangthong and C. V. Deutsch, eds., Geostatistics Banff 2004, Proceedings of the Seventh International Geostatistics Congress: New York, Springer, p. 175–184.

Hu, L. Y., G. Blanc, and B. Noetinger, 1999, Gradual deformation and iterative calibration of sequential stochastic simulations, *in* S. J. Lippard, A. Naess, and R. Sinding-Larsen, eds., IAMG '99, Proceedings of the 5th Annual Conference of the International Association for Mathematical Geology: Trondheim, Tapir, p. 493–498

Hu, L. Y., M. le Ravalec, and G. Blanc, 2001, Gradual deformation and iterative calibration of truncated Gaussian simulations: Petroleum Geoscience, v. 7, p. 25–30.

Huijbregts, C., 1973, Regionalized variables and application to quantitatve analysis of spatial data, *in* Proceedings of the North Atlantic Treaty Organization Advanced Study Institute, display and analysis of spatial data: New York, John Wiley & Sons, p. 38–53.

Huijbregts, C. J., and G. Matheron, 1971, Universal kriging (an optimal method for estimating and contouring in trend surface analysis), *in* J. I. McGerrigle, ed., Proceedings of the 9th International Symposium on Techniques for Decision-making in the Mineral Industry: Montreal, Canadian Institute of Mining and Metallurgy, Special Volume 12, p. 159–169.

Issaks, E. H., 2005, The kriging oxymoron: Conditionally unbiased and accurate predictor, *in* O. Leuangthong and C. V. Deutsch, eds., Geostatistics Banff 2004, Proceedings of the Seventh International Geostatistics Congress: New York, Springer, p. 363–374.

Issaks, E. H., and R. M. Srivastava, 1988, Spatial continuity measures for probabilistic and deterministic geostatistics: Mathematical Geology, v. 20, p. 313–341.

Jeannée, N., and C. de Fouquet, 2000, Which experimental variogram for the structural inference?, *in* P. Monestiez, D. Allard, and R. Froidevaux, eds., GeoENV III, Geostatistics for environmental applications: Dordrecht, The Netherlands, Kluwer Academic, p. 357–368.

Jeulin, D., 1988, Sequential random function models, *in* M. Armstrong, ed., Geostatistics, Proceedings of the 3rd International Geostatistics Congress: Dordrecht, The Netherlands, Kluwer Academic Publishers, p. 189–200.

Jeulin, D., 1994, Change of scale in random media, *in* R. Dimitrakopoulos, ed., Geostatistics for the next century: Dordrecht, The Netherlands, Kluwer Academic Publishers, p. 235–244.

Jimenez-Gonzalez, S., V. Pawlowsky-Glahn, J. J. Egozcue, and R. Ramirez-Llorente, 1998, Ward's method adapted to spatially dependent data, *in* A. Buccianti, G. Nardi, and R. Potenza, eds., Proceedings of IAMG '98, 4th Annual Conference of the International Association for Mathematical Geology: Naples, de Frede, p. 445–450.

Johns, C. J., D. Nychka, T. G. F. Kittel, and C. Daly, 2003, Infilling sparse records of spatial fields: Journal of the American Statistical Association, v. 98, p. 796–806.

Jones, R. H., and A. V. Vecchia, 1993, Fitting continuous ARMA models to unequally spaced spatial data: Journal of the American Statistical Association, v. 88, p. 947–954.

Journal, A. G., and D. Posa, 1990, Characteristic behavior and order relations for indicator variograms: Mathematical Geology, v. 22, p. 1011–1026.

Journel, A. G., 1974, Geostatistics for conditional simulation of ore bodies: Economic Geology, v. 69, p. 673–687.

Journel, A. G., 1977, Kriging in terms of projections: Mathematical Geology, v. 9, p. 563–586.

Journel, A G., 1982, The indicator approach to estimation of spatial distributions, *in* T. B. Johnson and R. J. Barnes, eds., Proceedings of the 17th International Symposium of the Applications of Computers and Operations Research in the Mineral Industry: New York, Society of Mining Engineers, p. 793–806.

Journel, A. G., 1983, Nonparametric estimation of spatial distributions: Mathematical Geology, v. 15, p. 445–468.

Journel, A. G., 1984a, MAD and conditional quantile estimators, *in* G. Verly, M. David, A. G. Journel, and A. Marechal, eds., Geostatistics for natural resources characterization: Dordrecht, The Netherlands, D. Reidel Publishing Company, p. 261–270.

Journel, A. G., 1984b, The place of non-parametric geostatistics, *in* G. Verly, M. David, A. G. Journel, and A. Marechal, eds., Geostatistics for natural resources characterization: Dordrecht, The Netherlands, D. Reidel Publishing Company, p. 307–335.

Journel, A. G., 1985a, The deterministic side of geostatistics: Mathematical Geology, v. 17, p. 1–15.

Journel, A. G., 1985b, Answers to Margaret Armstrong and Robert F. Shurtz's comments on "The deterministic side of geostatistics:" Mathematical Geology, v. 17, p. 869.

Journel, A. G., 1986, Constrained interpolation and qualitative information—The soft kriging approach: Mathematical Geology, v. 18, p. 269–286.

Journel, A. G., 1988, New distance measures: The route towards truly non-Gaussian geostatistics: Mathematical Geology, v. 20, p. 459–475.

Journel, A. G., 1994a, Resampling from stochastic simulations: Environmental and Ecological Statistics, v. 1, p. 63–84.

Journel, A. G., 1994b, Modeling uncertainty: Some conceptual thoughts, *in* R. Dimitrakopoulos, ed., Geostatistics for the next century: Dordrecht, The Netherlands, Kluwer Academic Publishers, p. 30–43.

Journel, A. G., 1994c, Discussion: Technometrics, v. 36, p. 149.

Journel, A. G., 1995, Probability fields: Another look and a proof: Stanford, Stanford Center for Reservoir Forecasting Report 8, 7 p.

Journel, A. G., 1996, Modelling uncertainty and spatial dependence: Stochastic imaging: International Journal of Geographical Information Systems, v. 10, p. 517–522.

Journel, A. G., 1997, Deterministic geostatistics: A new visit, *in* E. Y. Baafi and N. A. Schofield, eds., Geostatistics Wollongong '96: Dordrecht, The Netherlands, Kluwer Academic Publishers, p. 292–301.

Journel, A. G., 2005, Beyond covariance: The advent of multiple-point geostatistics, *in* O. Leuangthong and C. V. Deutsch, eds., Geostatistics Banff 2004, Proceedings of the Seventh International Geostatistics Congress: New York, Springer, p. 225–234.

Journel, A. G., and F. G. Alabert, 1988, Focusing on spatial connectivity of extreme-valued attributes: Stochastic indicator models or reservoir heterogeneities: Society of Petroleum Engineers Annual Technical Conference and Exhibition, SPE Paper 18324, 12 p.

Journel, A. G., and F. G. Alabert, 1989, Non-Gaussian data expansion in the earth sciences: Terra Nova, v. 1, p. 123–134.

Journel, A. G., and R. J. Barnes, 1982, The indicator approach to estimation of spatial distributions, *in* T. B. Johnson and R. J. Barnes, eds., Proceedings of the 17th International Symposium on the Application of Computers and Operations Research in the Mineral Industry: Society of Mining Engineers, p. 793–806.

Journel, A. G., and C. V. Deutsch, 1993, Entropy and spatial disorder: Mathematical Geology, v. 25, p. 329–355.

Journel, A. G., and R. Froidevaux, 1982, Anisotropic hole effect modeling: Mathematical Geology, v. 14, p. 217–240.

Journel, A. G., and S. E. Rao, 1996, Deriving conditional distributions from ordinary kriging: Stanford, Stanford Center for Reservoir Forecasting Report 9, 25 p.

Journel, A. G., and M. R. Rossi, 1989, When do we need a trend model in kriging?: Mathematical Geology, v. 21, p. 715–739.

Journel, A. G., and W. Xu, 1994, Posterior identification of histograms conditional to local data: Mathematical Geology, v. 26, p. 323–360.

Journel, A. G., and Z. Ying, 2001, The theoretical links between sequential Gaussian simulation, Gaussian truncated simulation, and probability field simulation: Mathematical Geology, v. 33, p. 31–40.

Journel, A., C. Deutsch, T. Hewett, M. Blunt, and M. Thiele, 1998, Replies to comment by K. Bratveld, E. Bølviken, T. Gimse, H. Holden, L. Holden, and R. Knarud: Mathematical Geology, v. 30, p. 247–249.

Journel, A. G., P. C. Kyriakidis, and S. Mao, 2000, Correcting the smoothing effect of estimators: A spectral postprocessor: Mathematical Geology, v. 32, p. 787–815.

Kacewicz, M., 1992, On the problem of the kriging cost in a global neighborhood, *in* P. A. Dowd and J. J. Royer, eds., Proceedings of the 2nd CODATA Conference on Geomathematics and Geostatistics: Nancy, Association Scientifique pour la Géologie et ses Applications, p. 25–34.

Kacewicz, M., 1994, "Fuzzy" geostatistics—An integration of qualitative description into spatial analysis, *in* R. Dimitrakopoulos, ed., Geostatistics for the next century: Dordrecht, The Netherlands, Kluwer Academic Publishers, p. 448–463.

Kanevsky, M., M. Maignan, V. Demyanov, and M. F. Maignan, 1997, How neural network 2-D interpolations can improve spatial data analysis: Neural network residual kriging (NNRK), *in* V. Pawlowsky-Glahn, ed., Proceedings of IAMG '97, 3rd Annual Conference of the International Association for Mathematical Geology: Barcelona, International Center for Numerical Methods in Engineering, p. 549–554.

Kerner, C., 1992, Anisotropy in sedimentary rocks modeled as random media: Geophysics, v. 57, p. 564–576.

Kirkpatrick, S., C. D. Gellat Jr., and M. P. Vecchi, 1983, Optimization by simulated annealing: Science, v. 220, no. 4598, p. 671–680.

Kitanidis, P. K., 1985, Minimum variance unbiased quadratic estimation of covariances of regionalized variables: Mathematical Geology, v. 17, p. 195–208.

Kitanidis, P. K., 1987, Parametric estimation of covariances of regionalized variables: Water Resources Bulletin, v. 23, p. 557–567.

Kitanidis, P. K., 1996, Analytical expressions of conditional mean, covariance, and sample functions in geostatistics: Stochastic Hydrology and Hydraulics, v. 10, p. 279–295.

Kolbjørnsen, O., and P. Abrahamsen, 2005, Theory of the cloud transform for applications, *in* O. Leuangthong and C. V. Deutsch, eds., Geostatistics Banff 2004, Proceedings of the Seventh International Geostatistics Congress: New York, Springer, p. 45–54.

Krige, D. G., 1984, Geostatistics and the definition of uncertainty: Institute of Mining and Metallurgy Transactions, v. 93, sect. A, p. A41–A47.

Krige, K. G., 1999, Conditional bias and uncertainty of estimation in geostatistics, *in* K. Dagdelen, ed., APCOM '99, Proceedings of the 28th International Symposium on the Application of Computers and Operations Research in the Mineral Industry: Society of Mining Engineers, p. 3–14.

Krishnan, S., 2005, Experimental study of multiple-support, multiple-point dependence and its modeling, *in* O. Leuangthong and C. V. Deutsch, eds., Geostatistics Banff 2004, Proceedings of the Seventh International Geostatistics Congress: New York, Springer, p. 1109–1116.

Krishnan, S., and A. G. Journel, 2003, Spatial connectivity: from variograms to multiple-point measures: Mathematical Geology, v. 35, p. 915–926.

Krivoruchko, K., and A. Gribov, 2004, Geostatistical interpolation in the presence of barriers, *in* X. Sánchez-Vila, J. Carrera, and J. Gómez-Hernandez, eds., GeoENV IV, Geostatistics for environmental applications: Dordrecht, The Netherlands, Kluwer Academic Publishers, p. 331–342.

Krzanowski, R. M., 1986, Validation of geostatistical model: Canadian Institute or Mining, Metallurgy, and Petroleum, Annual Meeting Preprint 155, 18 p.

Kuka, N., and J. J. Royer, 1992, Improving algorithm for the solution of kriging equations in a global neighbourhood, *in* P. A. Dowd and J. J. Royer, eds., Proceedings of the 2nd CODATA Conference on Geomathematics and Geostatistics: Nancy, Association Scientifique pour la Géologie et ses Applications, p. 83–92.

Künsch, H., A. Papritz, and F. Bassi, 1997, Generalized cross-covariances and their estimation: Mathematical Geology, v. 29, p. 779–799.

Kupfersberger, H., C. V. Deutsch, and A. G. Journel, 1998, Deriving constraints on small-scale variograms due to variograms of large-scale data: Mathematical Geology, v. 30, p. 837–852.

Kyriakidis, P. C., 2005, Sequential spatial simulation using Latin hypercube sampling, *in* O. Leuangthong and C. V. Deutsch, eds., Geostatistics Banff 2004, Proceedings of the Seventh International Geostatistics Congress: New York, Springer, p. 65–74.

Kyriakidis, P. C., and A. G. Journel, 2000, Erratum: Geostatistical space-time models: A review: Mathematical Geology, v. 32, p. 893.

Lahiri, S. N., M. S. Kaiser, N. Cressie, and N.-J. Hsu, 1999a, Prediction of spatial cumulative distribution functions using subsampling: Journal of the American Statistical Association, v. 94, p. 86–97.

Lahiri, S. N., M. S. Kaiser, N. Cresie, and N.-J. Hsu, 1999b, Rejoinder: Prediction of spatial cumulative distribution functions using subsampling: Journal of the American Statistical Association, v. 94, p. 107–111.

Lajaunie, C., 1990, Comparing some approximate methods for building local confidence intervals for predicting regionalized variables: Mathematical Geology, v. 22, p. 123–144.

Lajaunie, C., 2000, Introduction to the modeling of the support effect, *in* H. S. Pandalai and P. K. Saraswati, eds., Geological data analysis—Statistical methods: New Delhi, Hindustan Publishing Corporation, p. 155–171.

Lamorey, G., and E. Jacobsen, 1995, Estimation of semivariogram parameters and evaluation of the effects of data sparsity: Mathematical Geology, v. 27, p. 327–358.

Langlais, V., 1989, On the neighborhood search procedure to be used when kriging with constraints, *in* M. Armstrong, ed., Geostatistics, Proceedings of the 3rd International Geostatistics Congress: Dordrecht, The Netherlands, Kluwer Academic Publishers, p. 603–614.

Lantuéjoul, C., 1993, Substitution random functions, *in* A. Soares, ed., Geostatistics trola '92: Dordrecht, The Netherlands, Kluwer Academic Publishers, p. 37–48.

Lantuéjoul, C., 1994, Non conditional simulation of stationary isotropic multigaussian random functions, *in* M. Armstrong and P. A. Dowd, eds., Geostatical simulations: Dordrecht, The Netherlands, Kluwer Academic Publishers, p. 147–177.

Lantuéjoul C., 1997, Iterative algorithms for conditional simulations, *in* E. Y. Baafi and N. A. Schofield, eds., Geostatistics Wollongong '96: Dordrecht, The Netherlands, Kluwer Academic Publishers, p. 27–40.

Lantuéjoul, C., 1999, Conditional simulation of random sets, *in* Proceedings of the Section on Statistics and the Environment: American Statistical Association, Joint Statistical Meetings, p. 37–45.

Laslett, G. M., 1994, Kriging and splines: An empirical comparison of their predictive performance in some applications: Journal of the American Statistical Association, v. 89, p. 391–409.

Le, N. D., and J. V. Zidek, 1992, Interpolation with uncertain spatial covariances: A Bayesian alternative to kriging: Journal of Multivariate Analysis, v. 43, p. 351–374.

Lee, Y. D., and S. N. Lahiri, 2002, Least squares variogram fitting by spatial subsampling: Journal of the Royal Statistical Society, Series B, v. 64, p. 837–854.

Lee, Y.-M., and J. H. Ellis, 1997, On the equivalence of kriging and maximum entropy estimators: Mathematical Geology, v. 29, p. 131–152.

Lele, S., 1997, Estimating functions for semivariogram estimation, *in* I. Basawa, V. P. Godambe, and R. L. Taylor, eds., Selected Proceedings of the Symposium on Estimating Functions: Institute of Mathematical Statistics Lecture Notes Monograph Series, v. 32, p. 381–396.

Lemmer, C, 1994a, Modelling in the presence of skewed distributions, *in* R. Dimitrakopoulos, ed., Geostatistics for the next century: Dordrecht, The Netherlands, Kluwer Academic Publishers, p. 108–119.

Lemmer, C., 1994b, Reply to "Comment on modelling in the presence of skewed distributions" by R. M. Srivastava, *in* R. Dimitrakopoulos, ed., Geostatistics for the next century: Dordrecht, The Netherlands, Kluwer Academic Publishers, p. 122–123.

Lemmer, I. C., 1986, Monomodal indicator variography: Part 1. Theory: Mathematical Geology, v. 8, p. 589–604.

le Ravalec, M., B. Noetinger, and L. Y. Hu, 2000, The FFT moving average (FFT-MA) generator: An efficient numerical method for generating and conditional Gaussian simulations: Mathematical Geology, v. 32, p. 701–725.

Leuangthong, O., 2005, The promises and pitfalls of direct simulation, Journel, A. G., 2005, Beyond covariance: The advent of multiple-point geostatistics, *in* O. Leuangthong and C. V. Deutsch, eds., Geostatistics Banff 2004, Proceedings of the Seventh International Geostatistics Congress: New York, Springer, p. 305–314.

Leuangthong, O., and C. V. Deutsch, 2003, Stepwise conditional transformation for simulation of multiple variables: Mathematical Geology, v. 35, p. 155–173.

Leuangthong, O., and C. V. Deutsch, 2004, Transformation of residuals to avoid artifacts in geostatistical modeling with a trend: Mathematical Geology, v. 36, p. 287–306.

Leuangthong, O., J. A. McLennan, and C. V. Deutsch, 2004, Minimum acceptance criteria for geostatistical realizations: Natural Resources Research, v. 13, p. 131–141.

Li, D., and L. W. Lake, 1994, A moving window semivariance estimator: Water Resources Research, v. 30, p. 1479–1489.

Liebers, R. T., 1993, Bayesian kriging compared with nonparametric estimation methods (abs.), *in* R. B. McCammon and D. F. Merriam, eds., Program and Abstracts, Silver Anniverary Meeting, 1968–1993: International Association for Mathematical Geology, p. 32.

Limíc, N., and A. Mikelíc, 1984, Constrained kriging using quadratic programming: Mathematical Geology, v. 16, p. 423–430.

Liu, N., and D. S. Oliver, 2003, Conditional simulation of truncated random fields using gradient methods, *in* J. Cubitt, ed., IAMG 2003, Proceedings of the 9th Annual Conference: International

Association for Mathematical Geology, 6 p.: www.iamg.org/meetings/Proceedings_2003/papers/Liu.pdf (accessed April 19, 2005).

Liu, Y., 2005, An information content measure using multiple-point statistics, *in* O. Leuangthong and C. V. Deutsch, eds., Geostatistics Banff 2004, Proceedings of the Seventh International Geostatistics Congress: New York, Springer, p. 1047–1056.

Liu, Y., and A. Journel, 2004, Improving sequential simulation with a structured path guided by information content: Mathematical Geology, v. 36, p. 945–964.

Liu, Y., A. Harding, and R. Gilbert, 2005, A workflow for multiple-point geostatistical simulation, Journel, A. G., 2005, Beyond covariance: The advent of multiple-point geostatistics, *in* O. Leuangthong and C. V. Deutsch, eds., Geostatistics Banff 2004, Proceedings of the Seventh International Geostatistics Congress: New York, Springer, p. 245–254.

Long, A. E., and D. E. Myers, 1997, A new form of the cokriging equations: Mathematical Geology, v. 29, p. 669–682.

Luo, J., and A. Thomsen, 1995, On direct estimation of the bivariate probability distribution of a regionalized variable form ITS spatial samples, *in* A. G. Fabbri and J. J. Royer, eds., Proceedings of the 3rd CODATA Conference on Geomathematics and Geostatistics: Nancy, Association Scientifique pour la Géologie et ses Applications, p. 115–124.

Ma, Y. Z., 1992, Spectral estimation by simple kriging in one dimension, *in* P. A. Dowd and J. J. Royer, eds., Proceedings of the 2nd CODATA Conference on Geomathematics and Geostatistics: Nancy, Association Scientifique pour la Géologie et ses Applications, p. 35–42.

Ma, Y. Z., and D. E. Myers, 1995, Simple and ordinary factorial cokriging: Sciences de la Terre Informatique Geologique, v. 1, no. 32, p. 49–62.

Mallet, J. L., 1989, Geometric modeling and geostatistics, *in* M. Armstrong, ed., Geostatistics, Proceedings of the 3rd International Geostatistics Congress: Dordrecht, The Netherlands, Kluwer Academic Publishers, p. 737–748.

Mallet, J. L., and A. Shtuka, 2000, Blending-based stochastic simulator: Mathematical Geology, v. 32, p. 367–380.

Manchuk, J., O. Leuangthong, and C. V. Deutsch, 2005, Direct geostatistical simulation on unstructured grids, *in* O. Leuangthong and C. V. Deutsch, eds., Geostatistics Banff 2004, Proceedings of the Seventh International Geostatistics Congress: New York, Springer, p. 85–94.

Mantoglou, A., and J. L. Wilson, 1981, Simulation of random fields with the turning band method: Cambridge, Massachusetts Institute of Technology, Department of Civil Engineering, Ralph M. Parsons Laboratory, Reprint 264, 199 p.

Mantoglou, A., and J. L. Wilson, 1982, The turning bands method for simulation of random fields using line generation by a spectral method: Water Resources Research, v. 18, p. 1379–1394.

Marbeau, J. H., and J.-P. E. Marbeau, 1989, Formal computation of integrals for 3-D kriging, *in* M. Armstrong, ed., Geostatistics, Proceedings of the 3rd International Geostatistics Congress: Dordrecht, The Netherlands, Kluwer Academic Publishers, p. 773–784.

Marchant, B. P., and R. M. Lark, 2004, Estimating variogram uncertainty: Mathematical Geology, v. 36, p. 867–898.

Marcotte, D., 1995, Conditional simulation with data subject to measurement error: Post-simulation filtering with modified factorial kriging: Mathematical Geology, v. 27, p. 749–762.

Marcotte, D., and M. David, 1988, Trend surface analysis as a special case of IRF-k kriging: Mathematical Geology, v. 20, p. 821–824.

Marcotte, D., and M. Powojowski, 2005, Covariance models with spectral additive components, *in* O. Leuangthong and C. V. Deutsch, eds., Geostatistics Banff 2004, Proceedings of the Seventh International Geostatistics Congress: New York, Springer, p. 115–124.

Mardia, K. V., and C. A. Gill, 1982, Some statistical inference problems in kriging: 1. Numerical applications: Geological Society (London) Miscellaneous Paper 14, p. 56–71.

Mardia, K. V., and J. T. Kent, 1996, Kriging and splines with derivative information: Biometrika, v. 83, p. 207–221.

Mardia, K. V., and R. J. Marshall, 1984, Maximum likelihood estimation of models for residual covariance in spatial regression: Biometrika, v. 71, p. 135–146.

Maréchal, A., 1974, Krigeage normal et lognormal: Fontainebleau, Centre de Geostatistique et de Morphologie Mathematique, no. N376, 10 p.

Maréchal, A., and J. Serra, 1970, Random kriging, *in* D. R. Merriam, ed., Geostatistics—A colloquium: New York, Penum Press, p. 91–112.

Marschallinger, R., 1993, Improving geostatistical

modeling by pre- and post-processing input and result data in a 3D CAD system (abs.), *in* R. B. McCammon and D. F. Merriam, eds., Program and Abstracts, Silver Anniverary Meeting, 1968–1993: International Association for Mathematical Geology, p. 34.

Marshall, R. J., and K. V. Mardia, 1985, Minimum norm quadratic estimation of components of spatial covariance: Mathematical Geology, v. 17, p. 517–525.

Mateu, J., and P. J. Ribeiro, 1999, Geostatistical data versus point process data: Analysis of second-order characteristics, *in* J. Gomez-Hernandez, R. Froidevaux, and A. Soares, eds., Geostatistics for environmental applications: Dordrecht, The Netherlands, Kluwer Academic Publishers, p. 213–224.

Matheron, G., 1962, Traité de géostatistique appliqué: Paris, Editions Technip, Mémoires du Bueau de Recherches Gélogiques et Miniéres, v. 14, 333 p.

Matheron, G., 1963, Principles of geostatistics: Economic Geology, v. 58, p. 1246–1266.

Matheron, G., 1965, Les variables régionalisées et leur estimation: Paris, Massson et Cie, Editeurs, 305 p.

Matheron, G., 1967, Kriging, or polynomial interpolation procedures?: Canadian Institute of Mining and Metallurgy Transactions, v. 70, p. 240–244.

Matheron, G., 1969, Le krigeage universel: Fontainebleau, École Nationale Supérieure de Paris, Les Cahiers du Centre de Morphologie Mathématique, Fasc. 1, 82 p.

Matheron, G., 1970, Random functions and their application in geology, *in* D. F. Merriam, ed., Geostatistics—A colloquium: New York, Plenum Press, p. 79–87.

Matheron, G., 1971, The theory of regionalized variables and its application: Fontainebleau, École Nationale Supérieure de Paris, Les Cahiers du Centre de Morphologie Mathematique, Fasc. 5, 211 p.

Matheron, G., 1973, The intrinsic random functions and their applications: Advances in Applied Probability, v. 5, p. 439–468.

Matheron, G., 1976a, A simple substitute for conditional expectation: Disjunctive kriging, *in* M. Guarascio, M. David, and C. Huijbregts, eds., Advanced geostatistics in the mining industry: Dordrecht, The Netherlands, D. Reidel Publishing Company, p. 221–236.

Matheron, G., 1976b, Le choix des modeles en géostatistique, *in* M. Guarascio, M. David, and C. Huibregts, eds., Advanced geostatistics in the mining industry: Dordrecht, The Netherlands, D. Reidel Publishing Company, p. 11–30.

Matheron, G., 1981, Splines and kriging: Their formal equivalence, *in* D. F. Merriam, ed., Down-to-earth statistics: Solutions looking for geological problems: Syracuse, Syracuse University Geological Contributions, p. 77–95.

Matheron, G., 1984, Isofactorial models and change of support, *in* G. Verly, M. David, A. G. Journel, and A. Marechal, eds., Geostatistics for Natural Resources Characterization: Dordrecht, The Netherlands, D. Reidel Publishing Company, p. 449–467.

Matheron, G., 1985, Change of support for diffusion-type random functions: Mathematical Geology, v. 17, p. 137–166.

Matheron, G., 1989, The internal consistency of models in geostatistics, *in* M. Armstrong, ed., Geostatistics, Proceedings of the 3rd International Geostatistics Congress: Dordrecht, The Netherlands, Kluwer Academic Publishers, p. 21–38.

Matheron, G., and W. J. Kleingeld, 1987, The evolution of geostatistics, *in* I. C. Lemmer, H. Schaum, and G. M. Canisani-Calzolari, eds., APCOM '87, Proceedings of the 20th International Symposium on the Application of Computers and Operations Research in the Mineral Industry, v. 3 (geostatistics): Johannesburg, South African Institute of Mining and Metallurgy, p. 9–12.

Matsuoka, T., 1998, Geostatistics: Geophysical Exploration, v. 51, p. 96–99.

McBratney, A. B., and R. Webster, 1981, The design of optimal sampling schemes for local estimation and mapping of regionalized variables: I. Theory and method: Computers and Geosciences, v. 7, p. 331–334.

McCarn, D. W., and J. R. Carr, 1992, Influence of numerical precision and equation solution algorithm on computation of kriging weights: Computers and Geosciences, v. 18, p. 1127–1167.

Mejía, E., 2001, The KTNVS method—Using a variogram surface instead of fitting theoretical model variograms to improve kriging, *in* IAMG 2001, Proceedings of the 7th Annual Conference of the International Association for Mathematical Geology (Cancun): Lawrence, Kansas Geological Survey, 27 p.

Menz, J., and S. Bian, 1998, Multiquadric method, collocation and kriging: A comparison (in German): Das Markscheidewesen, v. 105, p. 222–226.

Menz, J., and J. Pilz, 1997, Bayesian collocation: Equivalences between adjustment computations and geostatistics (in German): Das Markscheidewesen, v. 104, p. 85–90.

Menz, J., and K. Wälder, 2000, Calculating exceeding probabilities by conditional simulation or by indicator kriging? (in German): Mathematische Geologie, v. 5, p. 205–213.

Menz, J., D. Stoyan, and N. Kolesnikow, 2000, A method for estimating variogram parameters in the presence of trends: Mathematische Geologie, v. 5, p. 59–68.

Militino, A. F., and M. D. Ugarte, 1997, Analysing censored spatial data, *in* V. Pawlowsky-Glahn, ed., Proceedings of IAMG '97, 3rd Annual Conference of the International Association for Mathematical Geology: Barcelona, International Center for Numerical Methods in Engineering, p. 623–628.

Militino, A. F., and M. D. Ugarte, 2001, Assessing the covariance function in geostatistics: Statistics and Probability Letters, v. 52, p. 199–206.

Montoglou, A., and J. L. Wilson, 1982, The turning bands method for simulation of random fields using line generation by a spectral method: Water Resources Research, v. 18, p. 1379–1394.

Moritz, H., 2000, Least-squares collocation: Mathematische Geologie, v. 5, p. 35–42.

Morkoc, F., J. Biggar, D. Neilsen, and D. E. Myers, 1987, Kriging with generalized covariances: Journal of the American Soil Science Society, v. 51, p. 1126–1130.

Moyeed, R. A., and A. Papritz, 2002, An empirical comparison of kriging methods for nonlinear spatial point prediction: Mathematical Geology, v. 34, p. 365–386.

Mueller, W. G., 1999, Least squares fitting from the variogram cloud: Statistics and Probability Letters, v. 43, p. 93–98.

Mueller, W. G., and D. L. Zimmerman, 1995, An algorithm for sampling optimization for semivariogram estimation, *in* C. P. Kitsos and W. G. Mueller, eds., Proceedings, 4th International Workshop on Model-oriented Data Analysis (MODA): Heidelberg, Physica-Verlag, p. 173–180.

Mueller, W. G., and D. L. Zimmerman, 1999, Optimal designs for variogram estimation: EnvironMetrics, v. 10, p. 23–27.

Murcha, J., 1997, Some practical aspects of geostatistical modelling related to sampling errors and micro-variability, *in* V. Pawlowsky-Glahn, ed., Proceedings of IAMG '97, 3rd Annual Conference of the International Association for Mathematical Geology: Barcelona, International Center for Numerical Methods in Engineering, p. 653–658.

Murthy, P. S. N., and M. Makki, 1999, Possible alternatives to geostatistics—Some views, *in* K. Dagdelen, ed., APCOM '99, Proceedings of the 28th International Symposium on the Applications of Computers and Operations Research in the Mineral Industry: Society of Mining Engineers, p. 435–444.

Myers, D. E., 1982, Matrix formulation of co-kriging: Mathematical Geology, v. 14, p. 249–257.

Myers, D. E., 1983, Estimation of linear combinations and co-kriging: Mathematical Geology, v. 15, p. 633–638.

Myers, D. E., 1984a, Cokriging—New developments, *in* G. Verly, M. David, A. Journel, and A. Marechal, eds., Geostatistics for natural resources characterization: Dordrecht, The Netherlands, Reidel Publishing Company, p. 295–305.

Myers, D. E., 1984b, Conditions for a variogram model to yield positive variances under restrictions: Comment: Mathematical Geology, v. 16, p. 431–432.

Myers, D. E., 1985, Cokriging: Methods and alternatives, *in* P. S. Glaeser, ed., The role of data in scientific progress: Amsterdam, Elsevier, p. 425–428.

Myers, D. E., 1989a, To be or not to be... stationary? That is the question: Mathematical Geology, v. 21, p. 347–362.

Myers, D. E., 1989b, Vector conditional simulation, *in* M. Armstrong, ed., Geostatistics, Proceedings of the 3rd International Geostatistics Congress: Dordrecht, The Netherlands, Kluwer Academic Publishers, p. 283–293.

Myers, D. E., 1991a, Pseudo-cross-variograms, positive definiteness, and cokriging: Mathematical Geology, v. 23, p. 805–816.

Myers, D. E., 1991b, On variogram estimation, *in* P. R. Nelson, ed., Proceedings of the 1st International Conference on Statistical Computing: Columbus, American Sciences Press, p. 261–281.

Myers, D. E., 1991c, Interpolation and estimation with spatially located data: Chemometrics and Intelligent Laboratory Systems, v. 11, p. 209–228.

Myers, D. E., 1992, Cokriging, radial basis functions and the role of positive definiteness: Computers in Mathematical Applications, v. 24, p. 139–148.

Myers, D. E., 1993, Cross-validation and variogram estimation: Theory of Probability and Its Applications (translation of Teorija Verojatnostei i ee Primenenija), v. 37, p. 345–347.

Myers, D. E., 1994a, Change of support and transformations, *in* R. Dimitrakopoulos, ed., Geostatistics for the next century: Dordrecht, The Netherlands, Kluwer Academic Publishers, p. 253–258.

Myers, D. E., 1994b, Statistical methods for interpolation of spatial data: Journal of Applied Science and Computations, v. 1, p. 283–318.

Myers, D. E., 1994c, Spatial interpolation: An overview: Geoderma, v. 62, p. 17–28.

Myers, D. E., 1995a, The linear coregionalization model and simultaneous diagonalization of the variogram matrix function: Sciences de la Terre Iniformatique Geologique, v. 1, no. 32, p. 125–140.

Myers, D. E., 1995b, The linear coregionalization model and simultaneous diagonalization of the variogram matrix function, *in* A. G. Fabbri and J. J. Royer, eds., Proceedings of the 3rd CODATA Conference on Geomathematics and Geostatistics: Nancy, Association Scientifique pour la Géologie et ses Applications, p. 141–152.

Myers, D. E., and A. G. Journel, 1990, Variograms with zonal anisotropies and noninvertible kriging systems: Mathematical Geology, v. 22, p. 779–786.

Neprash, J. A., 1934, Some problems in the correlation of spatially distributed variables: Journal of the American Statistical Association Supplement, v. 29, p. 167–168.

Olea, R. A., 1974, Optimal contour mapping using universal kriging: Journal of Geophysical Research, v. 79, p. 695–702.

Olea, R. A., 1975, Optimum mapping techniques using regionalized variable theory: Lawrence, Kansas Geological Survey, Series on Spatial Analysis, no. 2, 137 p.

Olea, R. A., 1977, Measuring spatial dependence with semivariograms: Lawrence, Kansas Geological Survey, Series on Spatial Analysis, no. 3, 29 p.

Olea, R. A., 1984, Sampling design for spatial functions: Mathematical Geology, v. 16, p. 369–392.

Olea, R. A., 1994, Fundamentals of semivariogram estimation, modeling, and usage, *in* J. M. Yarus and R. L. Chambers, eds., Stochastic modeling and geostatistics: Principles, methods, and case studies: AAPG Computer Applications in Geology 3, p. 27–35.

Olea, R. A., and V. Pawlowsky, 1994, Conditionally de-smoothed kriging, *in* C.-J. F. Chang, ed., Proceedings of IAMG '94: International Association for Mathematical Geology, p. 273–278.

Olea, R. A., and V. Pawlowsky, 1996, Compensating for estimation smoothing in kriging: Mathematical Geology, v. 28, p. 407–417.

Oliver, D. S., 1995, Moving averages for Gaussian simulation in two and three dimensions: Mathematical Geology, v. 27, p. 939–960.

Oliver, D. S., 1996, On conditional simulation to inaccurate data: Mathematical Geology, v. 28, p. 811–818.

Oliver, D. S., 2003, Gaussian cosimulation: Modeling of the cross-covariance: Mathematical Geology, v. 35, p. 681–698.

Oliver, M. A., and R. Webster, 1986, Combining nested and linear sampling for determining the scale and form of spatial variation of regionalized variables: Geographical Analysis, v. 18, p. 227–242.

Oliver, M. A., and R. Webster, 1989, A geostatistical basis for spatial weighting in multivariate classification: Mathematical Geology, v. 21, p. 15–36.

Oliver, M. A., R. Webster, and K. Slocum, 1999, Image filtering by kriging analysis, *in* J. Gomez-Hernandez, R. Froidevaux, and A. Soares, eds., Geostatistics for environmental applications: Dordrecht, The Netherlands, Kluwer Academic Publishers, p. 549–560.

Omre, H., 1984, The variogram and its estimation, *in* G. Verly, M. David, A. G. Journel, and A. Marechal, eds., Geostatistics for natural resources characterization: Dordrecht, The Netherlands, D. Reidel Publishing Company, p. 107–125.

Omre, H., 1987, Bayesian kriging—Merging observations and qualified guesses in kriging: Mathematical Geology, v. 19, p. 25–39.

Omre, H., and K. B. Halvorsen, 1989, The Bayesian bridge between simple and universal kriging: Mathematical Geology, v. 21, p. 767–786.

Omre, H., K. B. Halvorsen, and V. Berteig, 1988, A Bayesian approach to kriging, *in* M. Armstrong, ed., Geostatistics, Proceedings of the 3rd International Geostatistics Congress: Dordrecht, The Netherlands, Kluwer Academic Publishers, p. 109–126.

Omre, H., K. Sølna, and H. Tjelmeland, 1993, Simulation of random functions on large lattices, *in* A. Soares, ed., Geostatistics troia '92: Dordrecht, The Netherlands, Kluwer Academic Publishers, p. 179–200.

Omre, K. H., 1984, Introduction to geostatistical theory and examples of practical applications: Norwegian Computing Centre, NCC-Note SAND/14/87, 20 p.

Ortiz, C. J., and C. V. Deutsch, 2000, Hierarchical indicator simulation, *in* S. Bandopadhyay, ed., Proceedings of the 29th International Symposium on the Applications of Computers and Operations Research in the Mineral Industry: Society of Mining, Metallurgy, and Exploration, p. 275–284.

Ortiz, C. J., and C. V. Deutsch, 2002, Calculation of uncertainty in the variogram: Mathematical Geology, v. 34, p. 169–183.

Ortiz, J. M., and C. V. Deutsch, 2004, Indicator simulation accounting for multiple-point statistics: Mathematical Geology, v. 36, p. 545–566.

Ortiz, J. M., and C. V. Deutsch, 2005, Integrating multiple-point statistics into sequential simulation algorithms, *in* O. Leuangthong and C. V. Deutsch, eds., Geostatistics Banff 2004, Proceedings of the Seventh International Geostatistics Congress: New York, Springer, p. 969–978.

Ortiz, J. M., B. Oz, and C. V. Deutsch, 2005, A step by step guide to bi-Gaussian disjunctive kriging, *in* O. Leuangthong and C. V. Deutsch, eds., Geostatistics Banff 2004, Proceedings of the Seventh International Geostatistics Congress: New York, Springer, p. 1097–1102.

Osamani, S., 1995, Finite elements on geostatistics calculation, *in* A. G. Fabbri and J. J. Royer, eds., Proceedings of the 3rd CODATA Conference on Geomathematics and Geostatistics: Nancy, Association Scientifique pour la Geologie et ses Applications, p. 1–14.

Palacios, M. B., 1997, Influence measures for a multivariate spatial linear model, *in* V. Pawlowsky-Glahn, ed., Proceedings of IAMG '97, 3rd Annual Conference of the International Association for Mathematical Geology: Barcelona, International Center for Numerical Methods in Engineering, p. 670–674.

Pan, G., 1994, Restricted kriging: A link between sample value and sample configuration: Mathematical Geology, v. 26, p. 135–155.

Papritz, A., H. R. Kunsch, and R. Webster, 1993, On the pseudo-cross-variogram: Mathematical Geology, v. 25, p. 1015–1026.

Pardo-Igúzquiza, E., and P. A. Dowd, 1998, The second-order stationary universal kriging model revisited: Mathematical Geology, v. 30, p. 347–378.

Pardo-Igúzquiza, E., and P. A. Dowd, 2001, Variance-covariance matrix of the experimental variogram: Assessing variogram uncertainty: Mathematical Geology, v. 33, p. 397–419.

Pardo-Igúzquiza, E., and P. A. Dowd, 2005, On the automatic inference and modelling of a set of indicator covariances and cross-covariances, *in* O. Leuangthong and C. V. Deutsch, eds., Geostatistics Banff 2004, Proceedings of the Seventh International Geostatistics Congress: New York, Springer, p. 185–194.

Park, J.-S., and J. Baek, 2001, Efficient computation of maximum likelihood estimators in a spatial linear model with power exponential covariogram: Computers and Geosciences, v. 27, p. 1–8.

Pawlowsky, V., 1989, Co-kriging of regionalized compositions: Mathematical Geology, v. 21, p. 513–522.

Pebesma, E. J., and G. B. M. Heuvelink, 1999, Latin hypercube sampling of Gaussian random fields: Technometrics, v. 41, p. 303–312.

Pelletier, B., P. Dutilleul, G. Larocque, and J. W. Fyles, 2004, Fitting the linear model of coregionalization by generalized least squares: Mathematical Geology, v. 36, p. 323–344.

Philip, G. M., and D. F. Watson, 1986a, Comment on "Comparing splines and kriging:" Computer and Geosciences, v. 12, p. 243–245.

Phillip, G. M., and D. F. Watson, 1986b, Matheronian geostatistics—Quo vadis?: Mathematical Geology, v. 18, p. 93–117.

Pilger, G. G., J. F. C. Leite Costa, and J. C. Koppe, 2005, Improving the efficiency of the sequential simulation algorithm using Latin hypercube sampling, *in* O. Leuangthong and C. V. Deutsch, eds., Geostatistics Banff 2004, Proceedings of the Seventh International Geostatistics Congress: New York, Springer, p. 989–998.

Pilkington, M., and J. P. Todoeschuck, 1990, Stochastic inversion for scaling geology: Geophysical Journal International, v. 102, p. 205–217.

Pilz, J., 1993, Bayesian kriging and optimal design (abs.), *in* R. B. McCammon and D. F. Merriam, eds., Program and Abstracts, Silver Anniverary Meeting, 1968–1993: International Association for Mathematical Geology, p. 46.

Pilz, J., 1994, Robust Bayes linear prediction of regionalized variables, *in* R. Dimitrakopoulos, ed., Geostatistics for the next century: Dordrecht, The Netherlands, Kluwer Academic Publishers, p. 464–475.

Pitard, F. F., 1994, Exploration of the "nugget effect," *in* R. Dimitrakopoulos, ed., Geostatistics for the next century: Dordrecht, The Netherlands, Kluwer Academic Publishers, p. 124–136.

Ploner, A., and R. Dutter, 2000, New directions in geostatistics: Journal of Statistical Planning and Inference, v. 91, p. 499–509.

Posa, D., 1989, Conditioning of the stationary kriging matrices for some well-known covariance models: Mathematical Geology, v. 21, p. 755–765.

Puente, C. E., 1989, Practical use of disjunctive kriging: Effects of two dimensional parameterizations, *in* M. Armstrong, ed., Geostatistics, Proceedings of the 3rd International Geostatistics Congress: Dordrecht, The Netherlands, Kluwer Academic Publishers, p. 359–370.

Puente, C. E., 1994, A fractal-multifractal approach to geostatistics, *in* R. Dimitrakopoulos, ed., Geostatistics for the next century: Dordrecht, The Netherlands, Kluwer Academic Publishers, p. 476–487.

Puente, C. E., and R. L. Bras, 1986, Disjunctive kriging, universal kriging, or no kriging: Small sample results with simulated fields: Mathematical Geology, v. 18, p. 287–305.

Pyrcz, M. J., and C. V. Deutsch, 2001, Two artifacts of probability field simulation: Mathematical Geology, v. 33, p. 775–800.

Pyrcz, M. J., and C. V. Deutsch, 2002, Debiasing for improved inference of the one-point statistic, *in* S. Bandopadhyay, Proceedings of the 30th International Symposium on the Application of Computers and Operations Research in the Mineral Industry: Littleton, Society for Mining, Metallurgy, and Exploration, p. 251–260.

Pyrcz, M. J., and C. V. Deutsch, 2005, Conditioning geostatistical simulation models in the presence of highly non-linear geologic structure, *in* O. Leuangthong and C. V. Deutsch, eds., Geostatistics Banff 2004, Proceedings of the Seventh International Geostatistics Congress: New York, Springer, p. 135–144.

Qureshi, S. E., and R. Dimitrakopoulos, 2005, Comparison of stochastic simulation algorithms in mapping spaces of uncertainty of non-linear transfer functions, *in* O. Leuangthong and C. V. Deutsch, eds., Geostatistics Banff 2004, Proceedings of the Seventh International Geostatistics Congress: New York, Springer, p. 959–968.

Rao, S., and A. G. Journel, 1997, Deriving conditional distributions from ordinary kriging, *in* E. Y. Baafi and N. A. Schofield, eds., Geostatistics Wollongong '96: Dordrecht, The Netherlands, Kluwer Academic Publishers, p. 92–102.

Rao, S. V. L. N., and J. Prasad, 1982, Definition of kriging in terms of fuzzy logic: Mathematical Geology, v. 14, no. 1, p. 37–42.

Raty, L., and M. Gilbert, 1998, Large-scale versus small-scale variation decomposition, followed by kriging based on a relative variogram, in presence of a non-stationary residual variance: Journal of Geographic Information and Decision Analysis, v. 2, p. 91–115.

Ren, W., L. B. Cunha, and C. V. Deutsch, 2005, Preservation of multiple point structure when conditioning by kriging, *in* O. Leuangthong and C. V. Deutsch, eds., Geostatistics Banff 2004, Proceedings of the Seventh International Geostatistics Congress: New York, Springer, p. 643–652.

Rendu, J.-M, 1979, Kriging, logarithmic kriging and conditional expectation, *in* T. J. O'Neil, ed., Proceedings of the 16th International Symposium on the Application of Computers and Operations Research in the Mineral Industry: New York, McGraw-Hill, p. 199–212.

Rendu, J.-M., 1980, Disjunctive kriging: comparison of theory with actual results: Mathematical Geology, v. 12, p. 305–320.

Ripley, B. D., 1976a, The second-order analysis of stationary point processes: Journal of Applied Probability, v. 13, p. 255–266.

Ripley, B. D., 1976b, On stationarity and superposition of point processes: Annals of Probability, v. 4, p. 999–1005.

Ripley, B. D., 1977, Modelling spatial patterns (with discussion): Journal of the Royal Statistical Society, Series B, v. 39, p. 172–212.

Ripley, B. D., 1984, Spatial statistics—Developments 1980–1983: International Statistical Review, v. 52, p. 141–150.

Ripley, B. D., 1987, Point processes for the earth sciences, *in* C. F. Chung, A. G. Fabbri, and R. Sinding-Larsen, eds., Quantitative analysis of mineral energy resources: Dordrecht, The Netherlands, D. Reidel Publishing Company, p. 301–322.

Ripley, B. D., 1992a, Classification and clustering in spatial and image data, *in* M. Schader, ed., Analyzing and modeling data and knowledge: New York, Springer, p. 93–105.

Ripley, B. D., 1992b, Stochastic models for the distribution of rock types in petroleum reservoirs, *in* A. Walden and P. Guttorp, eds., Statistics in the environmental and earth sciences: London, Edward Arnold Publishers Ltd., p. 247–282.

Rivoirard, J., 1987, Two key parameters when choosing the kriging neighborhood: Mathematical Geology, v. 19, p. 851–856.

Rivoirard, J., 1993, Relations between the indicators

related to a regionalized variable, *in* A. Soares, ed., Geostatistics tróia '92: Dordrecht, The Netherlands, Kluwer Academic Publishers, p. 273–284.

Rivoirard, J., 2000, Weighted variograms, *in* W. J. Kleingeld and D. C. Krige, eds., Proceedings of the 6th International Geostatistics Congress: Cape Town, Geostatistical Association of South Africa, 11 p.

Rivoirard, J., 2001, Which models for collocated kriging?: Mathematical Geology, v. 33, p. 117–132.

Rivoirard, J., 2002, On the structural link between variables in kriging with external drift: Mathematical Geology, v. 34, p. 797–808.

Rivoirard, J., 2004, On some simplifications of co-kriging neighborhood: Mathematical Geology, v. 36, p. 899–915.

Rivoirard, J., 2005, On simplifications of co-kriging, *in* O. Leuangthong and C. V. Deutsch, eds., Geostatistics Banff 2004, Proceedings of the Seventh International Geostatistics Congress: New York, Springer, p. 195–204.

Rivoirard, J., and N. Cressie, 1997, Introduction to disjunctive kriging and nonlinear geostatistics: Society for Industrial and Applied Mathematics Review, v. 39, p. 337–340.

Rossi, M. E., 1992, Assessment of uncertainty using geostatistics: EnvironMetrics, v. 3, p. 71–79.

Roth, C., 1998, Is lognormal kriging suitable for local estimation?: Mathematical Geology, v. 30, p. 999–1010.

Roth, C., M. Armstrong, A. Galli, and G. Le Loc'h, 1998, Using plurigaussian simulations to reproduce lithofacies with contrasting anisotropies, *in* APCOM '98, International Symposium on the Application of Computers and Operations Research in the Mineral Industry: London, Institution of Mining Metallurgy, p. 201–214.

Royer, J. J., 1983, Dual formalism of kriging, *in* G. Verly, M. David, A. G. Journel, and A. Marechal, eds., Geostatistics for natural resources characterization: Dordrecht, The Netherlands, D. Reidel Publishing Company, p. 691–702.

Royer, J. J., 1989, Multivariate geostatistics and the sampling problems, *in* M. Armstrong, ed., Geostatistics, Proceedings of the 3rd International Geostatistics Congress: Dordrecht, The Netherlands, Kluwer Academic Publishers, p. 823–836.

Royer, J. J., 1992, Recursive solution of kriging system application to the optimization of a moving neighbourhood, *in* P. A. Dowd and J. J. Royer, eds., Proceedings of the 2nd CODATA Conference on Geomathematics and Geostatistics: Nancy, Association Scientifique pour la Géologie et ses Applications, p. 65–76.

Royle, A. G., 1979, Why geostatistics?: Engineering and Mining Journal, v. 180, no. 5, p. 92–101.

Rubin, Y., J. J. Gómez-Hernández, and A. G. Journel, 1991, Analysis of upscaling and effective properties in disordered media, *in* L. W. Lake, H. B. Carroll Jr., and T. C. Wesson, eds., Reservoir characterization II: San Diego, Academic Press, p. 251–276.

Russell, B., and D. Hampson, 1997, Multivariate geostatistics: Canadian Society of Exploration Geophysicists Recorder, April, p. 5–9.

Russo, D., and W. A. Jury, 1987, A theoretical study of the estimation of the correlation scale in spatially variable fields: 1. Stationary fields: Water Resources Research, v. 7, p. 1257–1268.

Sabourin, R., 1976, Application of two methods for the interpretation of the underlying variogram, *in* M. Guarascio, M. David, and C. Huibregts, eds., Advanced geostatistics in the mining industry: Dordrecht, The Netherlands, D. Reidel Publishing Company, p. 101–109.

Salkauskas, K., 1982, Some relationships between surface splines and kriging, *in* W. Schempp and K. Zeller, eds., Multivariate approximation theory: II. Basel, Birkhauser, p. 313–325.

Salomão, M. C., and A. Z. Remacre, 2000, Experimental variogram in discrete Markovian and truncated Gaussian stochastic processes, *in* W. J. Kleingeld and D. C. Krige, eds., Proceedings of the 6th International Geostatistics Congress: Cape Town, Geostatistical Association of South Africa, p. 1-1.

Samper, F. J., J. C. Cuchi, and J. Carrera, 1992, Estimation of spatial cross-covariances by maximum likelihood cross-validation, *in* P. A. Dowd and J. J. Royer, eds., Proceedings of the 2nd CODATA Conference on Geomathematics and Geostatistics: Nancy, Association Scientifique pour la Géologie et ses Applications, p. 121–134.

Sampson, P. D., and P. Guttorp, 1992, Nonparametric estimation of nonstationary spatial covariance structure: Journal of the American Statistical Association, v. 87, p. 108–119.

Sandfur, R. L., 1982, Progressive kriging—A method of testing the geostatistical hypothesis, *in* T. B. Johnson and R. J. Barnes, eds., Proceedings of the 17th International Symposium on the Applications of Computers and Operations Research in the Mineral Industry: New York, American Institute of

Mining, Metallurgical, and Petroleum Engineers, p. 789–791.

Saraç, C., and P. A. Dowd, 1995, Conditional simulation by ring decomposition of the covariance matrix, *in* A. G. Fabbri and J. J. Royer, eds., Proceedings of the 3rd CODATA Conference on Geomathematics and Geostatistics: Nancy, Association Scientifique pour la Géologie et ses Applications, p. 1–14.

Schaaf, T., G. Chevert, and M. Mezghani, 2004, Refinement indicators for optimal selection of geostatistical realizations using the gradual deformation method: Mathematical Geology, v. 36, p. 425–446.

Schaap, W., and J. D. St. George, 1981, On weights of linear kriging estimator: Institute of Mining and Metallurgy Transactions, v. 90, sect. A., p. A25–A27.

Schaffrin, B., 1993, Biased kriging on the sphere, *in* A. Soares, ed., Geostatistics tróia '92: Dordrecht, The Netherlands, Kluwer Academic Publishers, p. 121–132.

Schaffrin, B., 2000, On the reliability of data obtained by kriging, *in* K. J. Beek and M. Molenaar, eds., Proceedings of the 19th ISPRS Congress: Amsterdam, International Society for Photogrammetry and Remote Sensing, p. 893–900.

Scharrin, B., 2001, Equivalent systems for various forms of kriging, including least-squares collocation: Zeitschrift für Vermessungswesen, v. 126, p. 87–94.

Schmitt, M., 2000, Random sets, sampling and connectivity, *in* W. J. Kleingeld and D. C. Krige, eds., Proceedings of the 6th International Geostatistics Congress: Cape Town, Geostatistical Association of South Africa, 9 p.

Schwartz, F. W., L. Smith, and A. S. Crowe, 1983, A stochastic analysis of macroscopic dispersion in fractured media: Water Resources Research, v. 19, p. 1253–1265.

Seifert, D., and J. L. Jensen, 2000, Object and pixel-based reservoir modeling of a braided fluvial reservoir: Mathematical Geology, v. 32, p. 581–603.

Sen, Z., 1989, Cumulative semivariogram models of regionalized variables: Mathematical Geology, v. 21, p. 891–904.

Shafer, J. M., and M. D. Varljen, 1990, Approximation of confidence limits on sample semivariograms from single realizations of spatially correlated random fields: Water Resources Research, v. 26, p. 1787–1802.

Shapiro, A., and J. Botha, 1991, Variogram fitting with a general class of conditionally nonnegative definite functions: Computational Statistics and Data Analysis, v. 11, p. 87–96.

Sharp, W. E., 1982a, Stochastic simulation of semivariograms: Mathematical Geology, v. 14, p. 445–456.

Sharp, W. E., 1982b, Estimation of semivariograms by the maximum entropy method: Mathematical Geology, v. 14, p. 457–474.

Sharp, W. E., 1982c, A statistical test for experimental semivariograms: American Statistical Association, Annual Meetings, Program with Abstracts, p. 55.

Shurtz, R. F., 1985, A critique of A. Journel's "The deterministic side of geostistics:" Mathematical Geology, v. 17, p. 861–868.

Shurtz, R. F., 1999, The role of quotient spaces and Lagrange multipliers in constrained kriging systems, *in* S. J. Lippard, A. Næss, and R. Sinding-Larsen, eds., IAMG '99, Proceedings of the 5th Annual Conference of the International Association for Mathematical Geology: Trondheim, Tapir, p. 261–266.

Sides, E. J., 1995, Hidden uncertainties in geostatistics, *in* A. G. Fabbri and J. J. Royer, eds., Proceedings of the 3rd CODATA Conference on Geomathematics and Geostatistics: Nancy, Association Scientifique pour la Géologie et ses Applications, p. 15–28.

Smith, L., and F. W. Schwartz, 1980, Mass transport: 1. Stochastic analysis of macrodispersion: Water Resources Research, v. 16, p. 303–313.

Smith, L., C. W. Mase, and F. W. Schwartz, 1985, A stochastic model for transport in networks of planar fractures, *in* G. de Marsily, ed., Proceedings, the stochastic approach to subsurface flow: Paris, Centre Nationale de la Recherche Scientifique, p. 523–536.

Soares, A., 1990, Geostatistical estimation of orebody geometry: Morphological kriging: Mathematical Geology, v. 22, p. 787–802.

Soares, A., 1998, Sequential indicator simulation with correction for local probabilities: Mathematical Geology, v. 30, p. 761–765.

Soares, A., 2001, Direct sequential simulation and cosimulation: Mathematical Geology, v. 33, p. 911–926.

Soares, A., and J. Almeida, 1996, Zonal control of geostatistical estimation: Lisbon, Instituto Superior Técnico, Caderno 16 do CMRP, 16 p.

Solow, A. R., 1983, The analysis of second-order stationary processes: Time series analysis, spectral analysis, harmonic analysis, and geostatistics, *in* G. Verly, M. David, A. G. Journel, and A. Marechal, eds., Geostatistics for natural resources characterization: Dordrecht, The Netherlands, D. Reidel Publishing Company, p. 573–585.

Solow, A. R., 1986, Mapping by simple indicator kriging: Mathematical Geology, v. 18, p. 335–352.

Solow, A. R., 1990, Geostatistical cross-validation: A cautionary note: Mathematical Geology, v. 22, p. 637–640.

Solow, A. R., 1993, On the efficiency of the indicator approach in geostatistics: Mathematical Geology, v. 25, p. 53.

Solow, A. R., and S. J. Ratick, 1994, Conditional simulation and the value of information, *in* R. Dimitrakopoulous, ed., Geostatistics for the next century: Dordrecht, The Netherlands, Kluwer Academic Publishers, p. 209–217.

Spease, C., and J. Carr, 1988, Biased kriging: A theoretical development, *in* D. F. Merriam, ed., Current trends in geomathematics: New York, Plenun Press, p. 105–116.

Srivastava, R. M., 1994a, Comment on "Conditional simulation and the value of information" by A. R. Solow, and S. J. Ratick, *in* R. Dimitrakopoulos, ed., Geostatistics for the next century: Dordrecht, The Netherlands, Kluwer Academic Publishers, p. 218–222.

Srivastava, R. M., 1994b, An annealing procedure for honouring change of support statistics in conditional simulation, *in* R. Dimitrakopoulos, ed., Geostatistics for the next century: Dordrecht, The Netherlands, Kluwer Academic Publishers, p. 277–292.

Srivastava, R. M., 1994c, Comment on "Modeling uncertainty: Some conceptual thoughts," *in* R. Dimitrakopoulos, ed., Geostatistics for the next century: Dordrecht, The Netherlands, Kluwer Academic Publishers, p. 44–45.

Srivastava, R. M., 1994d, Comment on "Modelling in the presence of skewed distributions," *in* R. Dimitrakopoulos, ed., Geostatistics for the next century: Dordrecht, The Netherlands, Kluwer Academic Publishers, p. 120–121.

Srivastava, R. M., and R. Froidevaux, 2005, P-field simulation: A retrospective, *in* O. Leuangthong and C. V. Deutsch, eds., Geostatistics Banff 2004, Proceedings of the Seventh International Geostatistics Congress: New York, Springer, p. 55–64.

Srivastava, R. M., and H. Parker, 1988, Robust measures of spatial continuity, *in* M. Armstrong, ed., Geostatistics, Proceedings of the 3rd International Geostatistics Congress: Dordrecht, The Netherlands, Kluwer Academic Publishers, p. 295–308.

Stanley, K. O., K. Jorde, N. Raestad, and C. P. Stockbridge, 1990, Stochastic modelling of reservoir sand bodies for input to reservoir simulation, Snorre field, northern North Sea, Norway, *in* A. T. Buller, E. Berg, O. Hjelmeland, J. Kleppe, O. Torsæter, and J. O. Aasen, eds., North Sea oil and gas reservoirs II: London, Graham and Trotman, p. 91–101.

Starks, T. H., and J. H. Fang, 1982a, The effect of drift on the experimental semivariogram: Mathematical Geology, v. 14, p. 309–320.

Starks, T. H., and J. H. Fang, 1982b, On the estimation of the generalized covariance function: Mathematical Geology, v. 14, p. 57–64.

Stein, A., and L. C. A. Corsten, 1991, Universal kriging and cokriging as a regression procedure: Biometrics, v. 47, p. 575–587.

Stein, M. L., 1987a, Minimum norm quadratic estimation of spatial variograms: Journal of the American Statistical Association, v. 82, p. 765–772.

Stein, M. L., 1987b, Gaussian approximations to conditional distributions for multi-Gaussian processes: Mathematical Geology, v. 19, p. 387–405.

Stein, M. L., 1988, Asymptotically efficient prediciton of a random field with a misspecified covariance function: Annals of Statistics, v. 16, p. 55–63.

Stein, M. L., 1989a, Asymptotic distributions of minimum norm quadratic estimators of the covariance function of a Gaussian random field: Annals of Statistics, v. 17, p. 980–1000.

Stein, M. L., 1989b, The loss of efficiency in kriging prediciton caused by misspecification of the covariance structure, *in* M. Armstrong, ed., Geostatistics, Proceedings of the 3rd International Geostatistics Congress: Dordrecht, The Netherlands, Kluwer Academic Publishers, p. 273–282.

Stein, M. L., 2002, The screening effect in kriging: The Annals of Statistics, v. 30, p. 298–323.

Stein, M. L., and M. S. Handcock, 1989, Some asymptotic properties of kriging when the covariance function is misspecified: Mathematical Geology, v. 21, p. 171–190.

Stephenson, J., C. Holmes, and K. Gallagher, 2005, A statistical technique for modelling non-stationary spatial processes, *in* O. Leuangthong and C. V. Deutsch, eds., Geostatistics Banff 2004, Proceedings of the Seventh International Geostatistics Congress: New York, Springer, p. 125–134.

Strebelle, S., and A. Journel, 2000, Sequential simulation drawing structures from training images, *in* W. J. Kleingeld and D. G. Krige, eds., Geostatistics 2000: Cape Town, Geostatistical Association of Southern Africa, p. 381–392.

Strebelle, S., and N. Remy, 2005, Post-processing of multiple-point geostatistical models to improve reproduction of training patterns, *in* O. Leuangthong and C. V. Deutsch, eds., Geostatistics Banff 2004, Proceedings of the Seventh International Geostatistics Congress: New York, Springer, p. 979–988.

Strebelle, S., and T. Zhang, 2005, Non-stationary multiple-point geostatistical models, A. G. Journel, 2005, Beyond covariance: the advent of multiple-point geostatistics, *in* O. Leuangthong and C. V. Deutsch, eds., Geostatistics Banff 2004, Proceedings of the Seventh International Geostatistics Congress: New York, Springer, p. 235–244.

Subramanyam, A., and H. S. Pandalai, 2004, On the equivalence of the cokriging and kriging systems: Mathematical Geology, v. 36, p. 507–524.

Sugiyama, T., and T. Matsuoka, 1994, New estimation of kriging in geostatistics, *in* M. Ichimura, S. Mao, and G. Fan, eds., Proceedings of the 5th Japan–China Symposium on Statistics: Okayama, University Education Press, p. 247–251.

Sullivan, J., 1984, Conditional recovery estimation through probability kriging—Theory and practice, *in* G. Verly, M. David, A. G. Journel, and A. Marechal, eds., Geostatistics for natural resources characterization: Dordrecht, The Netherlands, D. Reidel Publishing Company, p. 365–384.

Suro-Pérez, V., 1993, Indicator principal component kriging: the multivariate case, *in* A. Soares, ed., Geostatistics tróia '92: Dordrecht, The Netherlands, Kluwer Academic Publishers, p. 441–454.

Switzer, P., 1972, Application of random process models to the description of spatial distributions of qualitative geologic variables (abs.), *in* J. E. Gell, ed., Abstracts, 24th Session of the International Geological Congress, Montreal, p. 518.

Switzer, P., 1980, The practical relevance of the choice between a stationary and a nonstationary geostatistical model (abs.), *in* A. D. Miall, ed., Facts and principles of world petroleum occurrence: Calgary, Canadian Society of Petroleum Geologists, p. 1001.

Switzer, P., 1983, Inference for spatial autocorrelation functions, *in* G. Verly, M. David, A. G. Journel, and A. Marechal, eds., Geostatistics for natural resources characterization: Dordrecht, The Netherlands, D. Reidel Publishing Company, p. 127–140.

Switzer, P., 1989, Non-stationary spatial covariances estimated from monitoring data, *in* M. Armstrong, ed., Geostatistics, Proceedings of the 3rd International Geostatistics Congress: Dordrecht, The Netherlands, Kluwer Academic Publishers, p. 127–138.

Switzer, P., 1993, The spatial variability of prediction errors, *in* A. Soares, ed., Geostatistics tróia '92: Dordrecht, The Netherlands, Kluwer Academic Publishers, p. 261–272.

Szidarovsky, F., E. Y. Baafi, and Y. C. Kim, 1987, Kriging without negative weights: Mathematical Geology, v. 19, p. 549–559.

Taylor, C. C., and P. A. Burrough, 1986, Multiscale sources of spatial variation in soil: III. Improved methods for fitting the nested model to one-dimensional semivariograms: Mathematical Geology, v. 18, p. 811–821.

Tetzlaff, D. M., and E. Rodriquez, 1996, Probabilistic estimates from reservoir-scale sedimentation models, *in* W. L. Watney, E. C. Rankey, E. K. Franseen, R. H. Goldstein, conveners, Numerical experiments in stratigraphy: Lawrence, Kansas Geology Survey Open-file Report 96-27, p. 145–146.

Tjolsen, C. B., E. Damsleth, G. Johnsen, G. Halvorsen, and A. Ryseth, 1996, Seismic data can improve stochastic facies modeling: SPE Formation Evaluation, v. 11, p. 141–145.

Todini, E., and F. Pellegrini, 1999, A maximum likelihood estimator for semi-variogram parameters in kriging, *in* J. Gomez-Hernandez, R. Froidevaux, and A. Soares, eds., GeoENV II, Geostatistics for environmental applications: Dordrecht, The Netherlands, Kluwer Academic Publishers, p. 187–200.

Todoeschuck, J. P., O. G. Jensen, and S. Laboute, 1990, Gaussian scaling noise model of seismic reflection sequences: Evidence from well logs: Geophysics, v. 55, p. 480–484.

Tolosana-Delgado, R., and V. Pawlowsky-Glahn, 2003, A new approach to kriging of positive variables, *in* J. Cubitt, ed., IAMG 2003, Proceedings of the 9th Annual Conference: International Association for Mathematical Geology, 6 p.

Tompson, A. F. B., R. Ababou, and L. W. Gelhar, 1989, Implementation of the three-dimensional turning bands random field generator: Water Resources Research, v. 25, p. 2227–2242.

Tureyen, O. I., and J. Caers, 2005, A parallel scheme for multi-scale data integration, Journel, A. G., 2005, Beyond covariance: the advent of multiple-point geostatistics, *in* O. Leuangthong and C. V. Deutsch, eds., Geostatistics Banff 2004, Proceedings of the Seventh International Geostatistics Congress: New York, Springer, p. 275–284.

van Beers, W. C. M., and J. P. C. Kleijnen, 2001, Kriging for interpolation in random simulation: Tilburg, Tilburg University, Center for Economic Research, Report No. 2001-74, 22 p.

van den Boogaart, K. G., 2003, Odds and ends of variogram modelling, *in* J. Cubitt, ed., IAMG 2003, Proceedings of the 9th Annual Conference: International Association for Mathematical Geology, 6 p.: www.iamg.org/meetings/Proceedings_2003/papers/VanDenBoogaart2.pdf (accessed April 19, 2005).

van den Boogaart, K. G., and A. Brenning, 2001, Why is universal kriging better than IRFk-kriging: Estimation of variograms in the presence of trend, *in* S. J. Lippard, A. Naess, and R. Sinding-Larsen, eds., IAMG 2001, Proceedings of the 7th Annual conference: International Association for Mathematical Geology, 15 p.: www.geographie.uni-erlangen.de/abrenning/lit/trend.pdf (accessed April 19, 2005).

van den Boogaart, K. G., and H. Schaeben, 2002a, Kriging of regionalized directions, axes, and orientations: I. Directions and axes: Mathematical Geology, v. 34, p. 479–505.

van den Boogaart, K. G., and H. Schaeben, 2002b, Kriging of regionalized directions, axes, and orientations: II. Orientations: Mathematical Geology, v. 34, p. 671–678.

van Groenigen, J. W., 1997, Spatial simulated annealing for optimizing sampling: Different optimization criteria compared, *in* A. Soares, J. Gomez-Hernandez, and R. Froidevaux, eds., GeoENV I, geostatistics for environmental applications: Dordrecht, The Netherlands, Kluwer Academic Publishers, p. 351–362.

van Groenigen, J. W., 2000, The influence of variogram parameters on optimal sampling schemes for mapping by kriging: Geoderma, v. 97, p. 223–236.

van Groenigen, J. W., and A. Stein, 1998, Spatial simulated annealing for constrained optimization of spatial sampling schemes: Journal of Environmental Quality, v. 27, p. 1078–1086.

van Storch, H., 1999, Representation of conditional random distributions as a problem of ''spatial'' interpolation, *in* J. Gomez-Hernandez, R. Froidevaux, and A. Soares, eds., GeoENV II, geostatistics for environmental applications: Dordrecht, The Netherlands, Kluwer Academic Publishers, p. 13–24.

Vargas-Guzmán, J. A., 2004, Geostatistics for power models of Gaussian fields: Mathematical Geology, v. 36, p. 307–322.

Vargas-Guzmán, J. A., and R. Dimitrakopoulos, 2002, Conditional simulation of random fields by successive residuals: Mathematical Geology, v. 34, p. 597–612.

Vargas-Guzmán, J. A., and R. Dimitrakopoulos, 2003, Successive nonparametric estimation of conditional distributions: Mathematical Geology, v. 35, p. 39–52.

Vargas-Guzmán, J. A., D. E. Myers, and A. W. Warrick, 2000, Derivatives of spatial variances of growing windows and the variogram: Mathematical Geology, v. 32, p. 851–872.

Vargas-Guzmán, J. A., A. W. Warrick, and D. E. Myers, 2002, Coregionalization by linear combination of nonorthogonal components: Mathematical Geology, v. 34, p. 405–420.

Ver Hoef, J. M., and R. M. Barry, 1998, Constructing and fitting models for cokriging and multivariate spatial prediction: Journal of Statistical Planning and Inference, v. 69, p. 275–294.

Ver Hoef, J. M., and N. Cressie, 1993, Multivariable spatial prediction: Mathematical Geology, v. 25, p. 219–240.

Ver Hoef, J. M., N. Cressie, and R. P. Barry, 2004, Flexible spatial models for kriging and cokriging using moving averages and the fast Fourier transform: Journal of Computational and Graphical Statistics, v. 13, p. 265–282.

Verly, G., 1983, The multigaussian approach and its application to the estimation of local reserves: Mathematical Geology, v. 15, p. 259–286.

Verly, G., 1993, Sequential Gaussian co-simulation: A simulation method integrating several types of information, *in* A. Soares, ed., Geostatistics tróia '92: Dordrecht, The Netherlands, Kluwer Academic Publishers, p. 543–554.

Wackernagel, H., 1988, Geostatistical techniques for interpreting multivariate spatial information, *in* C. F. Chung, A. G. Fabbri, and R. Sinding-Larsen, eds., Quantitative analysis of mineral and energy resources: Dordrecht, The Netherlands, D. Reidel Publishing Company, p. 393–409.

Wackernagel, H., 1994, Cokriging versus kriging

in regionalized multivariate data analysis: Geoderma, v. 62, p. 83–92.

Wackernagel, H., and L. Bertino, 2005, Geostatistics and sequential data assimilation, *in* O. Leuangthong and C. V. Deutsch, eds., Geostatistics Banff 2004, Proceedings of the Seventh International Geostatistics Congress: New York, Springer, p. 893–898.

Wackernagel, H., P. Petitgas, and Y. Touffait, 1989, Overview of methods for coregionalization analysis, *in* M. Armstrong, ed., Geostatistics, Proceedings of the 3rd International Geostatistics Congress: Dordrecht, The Netherlands, Kluwer Academic Publishers, p. 409–420.

Wälder, O., 1997, On applications of point process statistics in geostatistics, *in* V. Pawlowsky-Glahn, ed., Proceedings of IAMG '97, 3rd Annual Conference of the International Association for Mathematical Geology: Barcelona, International Center for Numerical Methods in Engineering, p. 665–669.

Walvoort, D. J. J., and J. J. de Gruijter, 2001, Compositional kriging: A spatial interpolation method for compositional data: Mathematical Geology, v. 33, p. 951–966.

Wang, P. P., and D. Chen, 1996, Continuous optimization by a variant of simulated annealing: Computational Optimization and Applications, v. 6, p. 59–71.

Wang, R., 1993, Some combined applications of geostatistics and fractal theory (in Chinese): Geological Science and Technology Information, v. 12, p. 88–97.

Warnes, J. J., 1986, A sensitivity analysis for universal kriging: Mathematical Geology, v. 18, p. 653–676.

Warnes, J. J., and B. D. Ripley, 1987, Problems with likelihood estimation of covariance functions of spatial Gaussian processes: Biometrika, v. 74, p. 640–642.

Warrick, A. W., and D. E. Myers, 1987a, Optimization of sampling locations for variogram calculations: Water Resources Research, v. 23, p. 496–500.

Warrick, A. W., and D. E. Myers, 1987b, Calculations of error variances with standardized variograms: Journal of the American Soil Science Society, v. 51, p. 265–268.

Warrick, A. W., R. Zhang, M. K. El-Haris, and D. E. Myers, 1988, Direct comparisons between kriging and other interpolators, *in* P. J. Wierenga and D. Bachelet, eds., Proceedings of the International Conference and Workshop on the Validation of Flow and Transport Models in the Unsaturated Zone: Las Cruces, New Mexico State University, p. 505–510.

Watkins, A. J., and F. H. M. Al-Boutiahi, 1990, On maximum likelihood estimation of parameters in incorrectly specified models of covariance for spatial data: Mathematical Geology, v. 22, p. 151–174.

Watson, G. S., 1972, Trend surface analysis and spatial correlations: Geological Society of America Special Paper 146, p. 39–46.

Watson, G. S., 1984a, Smoothing and interpolation by kriging with splines: Mathmematical Geology, v. 16, p. 601–615.

Watson, G. S., 1984b, Smoothing and interpolation by kriging and with splines: Mathematical Geology, v. 16, p. 601–616.

Webster, R., and M. A. Oliver, 1992, Confidence intervals on variograms for samples of various sizes, *in* P. A. Dowd and J. J. Royer, eds., Proceedings of the 2nd CODATA Conference on Geomathematics and Geostatistics: Nancy, Association Scientifique pour la Géologie et ses Applications, p. 11–24.

Webster, R., and M. A. Oliver, 1993, How large a sample is needed to estimate the regional variogram adequately? *in* A. Soares, ed., Geostatistics tróia '92: Dordrecht, The Netherlands, Kluwer Academic Publishers, p. 155–166.

Whitten, E. H. T., and M. E. V. Koelling, 1973, Spline-surface interpolation, spatial filtering, and trend surfaces for geological mapper variables: Mathematical Geology, v. 5, p. 111–126.

Wikle, C. K., and L. M. Berliner, 2005, Combining information across spatial scales: Technometrics, v. 47, p. 80–91.

Wingle, W. L., and E. P. Poeter, 1998, Directional semivariograms; kriging anisotropy without anisotropy factors (abs.): AAPG Annual Meeting Program, v. 7.

Wong, D. W. S., 1996, Aggregation effects in georeferenced data, *in* D. Griffiths, ed., Advanced spatial statistics: Boca Raton, CRC Press, p. 83–106.

Xie, T., and D. E. Myers, 1995a, Fitting matrix-valued variogram models by simultaneous diagonalization: Part I. Theory: Mathematical Geology v. 27, p. 867–876.

Xie, T., and D. E. Myers, 1995b, Fitting matrix-valued variogram models by simultaneous diagonalization: Part II. Application: Mathematical Geology, v. 27, p. 877–888.

Xu, W., 1996, Conditional curvilinear stochastic simulation using pixel-based algorithms: Mathematical Geology, v. 28, p. 937–950.

Xu, W., and A. G. Journel, 1995, Histogram and scattergram smoothing using convex quadratic programming: Mathematical Geology, v. 27, p. 83–105.

Yakowitz, S. J., and F. Szidarovsky, 1985, A comparison of kriging with nonparametric regression methods: Journal of Multivariate Analysis, v. 16, p. 21–53.

Yamamoto, J. K., 2000, An alternative measure of the reliability of ordinary kriging estimates: Mathematical Geology, v. 32, p. 489–509.

Yamamoto, J. K., 2005, Correcting the smoothing effect of ordinary kriging estimates: Mathematical Geology, v. 36, p. 69–94.

Yao, T., 2000, Notes and corrections to "Automated modeling of (cross) covariance tables using fast Fourier transform:" Mathematical Geology, v. 32, p. 147–148.

Yao, T., 2001, Non-parametric covariance modeling using FFT, *in* P. Wong, M. Nikravesh, and F. Aminzadeh, eds., Soft computing for reservoir characterization and modeling: Heidelberg, Physica-Verlag, p. 559–584.

Yao, T., 2002, Integrating seismic data for lilthofacies modeling: A comparison of sequential indicator simulation algorithms: Mathematical Geology, v. 34, p. 387–404.

Yao, T., 2004, Reproduction of the mean, variance, and variogram model in spectral simulation: Mathematical Geology, v. 36, p. 487–506.

Yao, T., and A. G. Journel, 1998, Automatic modeling of (cross) covariance tables using fast Fourier transform: Mathematical Geology, v. 30, p. 589–615.

Yao, T., C. Calvert, G. Gishop, T. Jones, Y. Ma, and L. Foreman, 2005, Spectral component geologic modeling: A new technology for integrating seismic information at the correct scale, *in* O. Leuangthong and C. V. Deutsch, eds., Geostatistics Banff 2004, Proceedings of the Seventh International Geostatistics Congress: New York, Springer, p. 23–34.

Yates, S. R., and A. W. Warrick, 1999, Geostatistics and spatial variability of soil properties: Agronomy, v. 38, p. 1235–1260.

Yates, S., A.Warrick, and D. E. Myers, 1986, Disjunctive kriging I: An overview of estimation and conditional probability: Water Resources Research, v. 22, p. 615–621.

Yeh, T. C. J., L. W. Gelhar, and A. L. Gutjahr, 1985a, Stochastic analysis of unsaturated flow in heterogeneous soils: 1. Statistically isotropic media: Water Resources Research, v. 21, p. 447–456.

Yeh, T. C. J., L. W. Gelhar, and A. L. Gutjahr, 1985b, Stochastic analysis of unsaturated flow in heterogeneous soils: 2. Statistically anisotropic media with variable $\alpha$: Water Resources Research, v. 21, p. 457–464.

Yeh, T. C. J., L. W. Gelhar, and A. L. Gutjahr, 1985c, Stochastic analysis of unsaturated flow in heterogeneous soils: 3. Observations and applications: Water Resources Research, v. 21, p. 465–471.

Yfantis, E. A., 1994, Reply to "Comments on image compression and kriging," *in* R. Dimitrakoupolos, ed., Geostatistics for the next century: Dordrecht, The Netherlands, Kluwer Academic Publishers, p. 168–172.

Yfantis, E. A., and G. T. Flatman, 1988, On sampling nonstationary spatial autocorrelated data: Computers and Geosciences, v. 14, p. 667–686.

Yfantis, E. A., G. T. Flatman, and J. V. Behar, 1987, Efficiency of kriging estimation for square, triangular, and hexagonal grids: Mathematical Geology, v. 19, p. 183–205.

Yfantis, E. A., M. Au, and F. S. Makri, 1994, Image compression and kriging, *in* R. Dimitrakoupolos, ed., Geostatistics for the next century: Dordrecht, The Netherlands, Kluwer Academic Publishers, p. 156–161.

Ying, Z., 1993, Maximum likelihood estimation of parameters under a spatial sampling scheme: Annals of Statistics, v. 21, p. 1567–1590.

Young, D. S., 1982, Development and application of disjunctive kriging model: discrete Gaussian model, *in* T. B. Johnson and R. J. Barnes, Proceedings of the 17th International Symposium on the Application of Computers and Operations Research in the Mineral Industry: New York, Society of Mining Engineers, p. 544–556.

Yu, K., and J. Mateu, 2002, Nonparametric nearest-neighbour variogram estimation, *in* J. Mateu and F. Montes, eds., Spatial statistics through applications: Boston, WIT Press, p. 103–125.

Zanon, S., and O. Leuangthong, 2005, Implementation aspects of sequential simulation, *in* O. Leuangthong and C. V. Deutsch, eds., Geostatistics Banff 2004, Proceedings of the Seventh International Geostatistics Congress: New York, Springer, p. 543–550.

Zanon, S., T. Faechner, and C. V. Deutsch, 2002, Improved integration of secondary data using self-healing sequential Gaussian simulation, *in* S. Bandopadhyay, ed., Proceedings of the 30th International Symposium on the Application of

Computers and Operations Research in the Mineral Industry: Society for Mining, Metallurgy, and Exploration, p. 285–294.

Zhang, H., 2004, Inconsistent estimation and asymptotically equal interpolations in model-based geostatistics: Journal of the American Statistical Association, v. 99, p. 250–261.

Zhang, T., P. Switzer, and A. Journel, 2005, Journel, A. G., 2005, Beyond covariance: The advent of multiple-point geostatistics, *in* O. Leuangthong and C. V. Deutsch, eds., Geostatistics Banff 2004, Proceedings of the Seventh International Geostatistics Congress: New York, Springer, p. 265–274.

Zheng, L., and S. E. Silliman, 2000, Estimating the theoretical semivariogram from finite numbers of measurements: Water Resources Research, v. 36, p. 361–366.

Zhu, H., 1992, Dual kriging: Geostat Newsletter, v. 4, p. 4–5.

Zhu, H., and A. G. Journel, 1989, Indicator conditioned estimator: Society for Mining, Metallurgy and Exploration Transactions, v. 286, p. 1880–1886.

Zhu, H., and A. G. Journel, 1993, Formatting and integrating soft data: Stochastic imaging via the Markov-Bayes algorithm, *in* A. Soares, ed., Geostatistics troia '92: Dordrecht, The Netherlands, Kluwer Academic Publishers, p. 1–12.

Zimmerman, D. L., 1989a, Computationally exploitable structure of covariance matrices and generalized covariance matrices in spatial models: Journal of Statistical Computation and Simulation, v. 32, p. 1–15.

Zimmerman, D. L., 1989b, Computationally efficient restricted maximum likelihood estimation of generalized covariance functions: Mathematical Geology, v. 21, p. 655–672.

Zimmerman, D. L., 1993, Another look at anisotropy in geostatistics: Mathematical Geology, v. 25, p. 453.

Zimmerman, D. L., and N. Cressie, 1991, Mean squared prediction error in the spatial linear model with estimated covariance parameters: Annals of the Institute of Statistical Mathematics, v. 44, p. 27–44.

Zimmerman, D. L., and M. B. Zimmerman, 1991, A Monte Carlo comparison of spatial semivariogram estimators and corresponding ordinary kriging predictors: Technometrics, v. 33, p. 77–91.

Zimmerman, D. L., C. Pavlik, A. Ruggles, and M. P. Armstrong, 1999, An experimental comparison of ordinary and universal kriging and inverse distance weighting: Mathematical Geology, v. 31, p. 375–390.


## DISSERTATIONS AND THESES


Adisoma, G. S., 1993, The application of the jackknife in geostatistical resrouce estimation: Robust estimator and its measure of uncertainty: Ph.D. Dissertation, University of Arizona, Phoenix, 117 p.

Alabert, F. G., 1987, Stochastic imaging of spatial distributions using hard and soft information: M.Sc. Thesis, Stanford University, Stanford, 208 p.

Aldworth, W. J. K., 1998, Spatial prediction, spatial sampling, and measurement: Ph.D. Dissertation, Iowa State University, Ames, 186 p.

Alldredge, J. R., 1980, Nonparametric modeling of the covariance structure of regionalized variables: Ph.D. Dissertation, Texas A&M University, College Station, 82 p.

Almeida, A. S., 1993, Joint simulation of multiple variables with a Markov-type coregionalization model: Ph.D. Dissertation, Stanford University, Stanford, 199 p.

Aprilian, S. S., 1998, Integrating multiphase production data into reservoir characterization using stochastic inverse modeling: Ph.D. Dissertation, Texas A&M University, College Station, 130 p.

Baker, S, L., 1989, Numerical estimation of effective permeability: Use of non-conditional geostatistics to model non-stationary sand-shale sequences: M.S. Thesis, Southern Methodist University, Dallas, 116 p.

Bathija, A., 1998, Uncertainty evaluation and integration of dynamic data in reservoir characterization: M.S. Thesis, Stanford University, Stanford, 107 p.

Bhosale, J. V., 1992, Generation and analysis of stochastic fields by the source point method: M.S. Thesis, University of Texas, Austin, 92 p.

Campozana, F. P., 1997, Incorporating dynamic data into geostatistical reservoir modeling: Ph.D. Dissertation, University of Texas, Austin, 281 p.

Chessa, A. G., 1995, Conditional simulation of spatial stochastic models for reservoir heterogeneity: Ph.D. Dissertation, Technical University of Delft, Delft, 174 p.

Chilès, J. P., 1977, Géostatistique des phénomenès

non stationnaires (dans le plan): Ph.D. Dissertation, Universitié Nancy, Nancy, 152 p.

Chu, J., 1993, Conditional fractal simulation, fast sequential indicator simulation, and interactive variogram modeling: Ph.D. Dissertation, Stanford University, Stanford, 159 p.

Coakley, K. J., 1989, Spatial statistics for predicting flow through a rock fracture: Ph.D. Dissertation, Stanford University, Stanford, 185 p.

Collins, J. B., 1998, Geostatistical methods for analysis of multiple scales of variation in spatial data: Ph.D. Dissertation, Boston University, Boston, 129 p.

Curriero, F. C., 1996, The use of non-Euclidean distances in geostatistics: Ph.D. Dissertation, Kansas State University, Manhattan, 213 p.

Davis, B. M., 1978, Some tests of hypothesis concerning variograms: Ph.D. Dissertation, University of Wyoming, Laramie, 147 p.

de Lima, L. C., 1995, Large-scale conditional simulation: Domain and matrix decomposition and the moving template model: Ph.D. Dissertation, University of Texas, Austin, 320 p.

Desbarats, A. J., 1987, Stochastic modeling of flow in sand-shale sequences: Ph.D. Dissertation, Stanford University, Stanford, 185 p.

Deutsch, C. V., 1987, A probabilistic approach to estimate effective absolute permeability: M.S. Thesis, Stanford University, Stanford, 165 p.

Deutsch, C. V., 1992, Annealing techniques applied to reservoir modeling and the integration of geological and engineering (well test) data: Ph.D. Dissertation, Stanford University, Stanford, 325 p.

Dimitrakopoulos, R., 1989, Conditional simulation of IRF-k in the petroleum industry and the expert system perspective: Ph.D. Dissertation, Ecole Polytechnique, Montreal, 303 p.

dos Santos, M. B., 1998, Object based modeling and stochastic simulation of channelized turbidite reservoirs: M.S. Thesis, Universidade Estadual de Campinas, Campinas.

Dubrule, O., 1981, Krigeage et splines en cartogrphie automatique—Application a des exemples perolieres: Thesis, ENSMP, Centre de Geostatistique et de Morphologie Mathematique, Fontainbleau, 150 p.

Fredericks, A. K., 1995, A comparison of the sequential Gaussian and Markov-Bayes simulation methods: M.S. Thesis, University of Idaho, Moscow, 83 p.

Garaj, A., 1996, The inclusion of dynamic constraints in stochastic conditional simulation: Ph.D. Dissertation, University of Tulsa, Tulsa, 233 p.

Ghori, S. G. K., 1992, Reservoir rock geostatistics and well-to-well tracer tests by simulations and laboratory experiments: Ph.D. Dissertation, New Mexico Institute of Mining and Technology, Socorro, 291 p.

Glacken, I., 1996, Change of support by direct conditional block simulation: M.S. Thesis, Stanford University, Stanford, 127 p.

Gomez-Hernández, J. J., 1991, A stochastic approach to the simulation of block conductivity fields conditioned upon data measured at a smaller scale: Ph.D. Dissertation, Stanford University, Stanford, 351 p.

Gould, G. E., 1991, Geostatistical study and reservoir simulation of the Pen field, Graham Couty, Kansas: M.S. Thesis, University of Kansas, Lawrence, 163 p.

Gringarten, E. J., 1998, Geometric modeling of fracture networks: Ph.D. Dissertation, Stanford University, Stanford, 115 p.

Grzebyk, M., 1993, Ajustement d'une corégionalisation stationnaire: Ph.D. Dissertation, Ecole Nationale Supérieure des Mines de Paris, Paris, 154 p.

Guo, G., 1989, Applications of geostatistic in reservoir engineering: M.S. Thesis, University of Montana, Missoula, 110 p.

Harihara, S., 1998, Sampling fluctuation and its effect on variogram modeling and simulation: M.S. Thesis, Stanford University, Stanford, 63 p.

Honoré, M., 1998, Geostatistics of petroleum reserves: M.S. Thesis, University of Alberta, Edmonton, 102 p.

Idrobo Hurtado, E. A., 1999, Characterization and ranking of reservoir models using geostatistics and streamline simulation: Ph.D. Dissertation, Texas A&M University, College Station, 95 p.

Isaaks, E. H., 1990, The application of Monte Carlo methods to the analysis of spatially correlated data: Ph.D. Dissertation, Stanford University, Stanford, 213 p.

Ivanova, V. M., 1998, Geologic and stochastic modeling of fracture systems in rocks: Dissertation, Massachusetts Institute of Technology, Cambridge, 530 p.

Kark, M. J., 1985, A software approach to linear geostatistics: M.S. Thesis, Stanford University, Stanford, 144 p.

Laine, E., 1998, Geostatistical, geological and geophysical modeling of subsurface structures of

Precambrian bedrock in Finland: Ph.D. Dissertation, Helsinki University of Technology, Espoo, 104 p.

Lambert, M. E., 1981, A statistical study of reservoir heterogeneity: Ph.D. Dissertation, University of Texas, Austin, 181 p.

Liu, Y., 2003, Downscaling seismic data into a geological sound numerical model: Ph.D. Dissertation, Stanford University, Stanford, 202 p.

Malallah, A. H., 2001, Data integration into high resolution reservoir models using geostatistics and multiscale Markov random fields: Ph.D. Dissertation, Texas A&M University, College Station, 130 p.

Miller, S. M., 1979, Determination of spatial dependence in fracture set characteristics by geostatistical methods: M.S. Thesis, University of Arizona, Phoenix, 111 p.

Murray, C. J., 1992, Geostatistical applications in petroleum geology and sedimentary geology: Ph.D. Dissertation, Stanford University, Stanford, 301 p.

Olea, R. A., 1972, Application of regionalized variable theory to automatic contouring: M.S. Thesis, University of Kansas, Lawrence, 191 p.

Olea, R. A., 1982, Systematic approach to sampling of spatial functions: Ph.D. Dissertation, University of Kansas, Lawrence, 277 p.

Omre, K. H., 1985, Alternative variogram estimators in geostatistics: Ph.D. Dissertation, Stanford University, Stanford, 267 p.

Ouenes, A., 1992, Application of simulated annealing to reservoir characterization and petrophysics inverse problem: Ph.D. Dissertation, New Mexico Institute of Mining and Technology, Socorro, 205 p.

Paul, B. C., 1985, Geostatistical evaluation of the Mahogany oil shale zone in the eastern Uinta Basin of Utah using a rotating Cartesian coordinate system: M.S. Thesis, University of Utah, Salt Lake City, 124 p.

Perez, G., 1991, Stochastic conditional simulation for description of reservoir properties: Ph.D. Dissertation, University of Tulsa, Tulsa, 245 p.

Poletto, C. A., 1996, Object based modeling and stochastic simulation of the geometry of fluvial reservoirs: M.S. Thesis, Universidade Estadual de Campinas, Campinas, 141 p.

Poquioma-Shaffer, J. W., 1991, Application of geostatistics to forecast performance for waterflooding of an oil field: M.S. Thesis, University of Tulsa, Tulsa, 162 p.

Pyrcz, M. J., 2004, The integration of geologic information into geostatistical models: Ph.D. Dissertation, University of Alberta, Edmonton, 296 p.

Quimby, W. F., 1986, Selected topics in spatial statistical analysis: Nonstationary vector kriging, large scale conditional simulation of three dimensional Gaussian random fields, and hypothesis testing in a correlated random field: Ph.D. Dissertation, University of Wyoming, Laramie, 155 p.

Ratchkovski, A., 1997, Integration of geostatistics and conventional methods to derive flow units for improved reservoir description: M.S. Thesis, University of Alaska, Fairbanks, 147 p.

Remy, N., 2001, GsTL: The geostatistical template library in C++: M.S. Thesis, Stanford University, Stanford, 136 p.

Rivoirard, J., 1984, Le comportment des poids de krigeage: Dr.-Ing. Thesis, Ecole de Mines de Paris, Centre de Geostatistique et de Morphologie Mathematique, Fontainebleau, 72 p.

Robbana, E., 2002, A better understanding of a Uinta Basin channelized analog reservoir through geostatistics and reservoir simulation: M.S. Thesis, Texas A&M University, College Station, 112 p.

Rossi, M. E., 1988, Impact of spatial clustering on geostatistical analysis: M.S. Thesis, Stanford University, Stanford, 101 p.

Rouleau, A., 1984, Statistical characterization and numerical simulation of a fracture system: Application to groundwater flow in the Stripa Granite: Ph.D. Thesis, University of Waterloo, Waterloo, 416 p.

Shibli, S., 1991, An approach to generating reservoir properties distribution using stochastic conditional simulation method: M.S. Thesis, University of Tulsa, Tulsa, 219 p.

Solow, A. R., 1980, The multivariate Gaussian approach to geostatistics: M.S. Thesis, Stanford University, Stanford, 56 p.

Solow, A. R., 1985, Kriging under a mixture model: Ph.D. Dissertation, Stanford University, Stanford, 258 p.

Srivastava, R. M., 1987, A non-ergodic framework for variogram and covariance functions: M.S. Thesis, Stanford University, Stanford, 113 p.

Strebelle, S., 2000, Sequential simulation drawing structures from training images: Ph.D. Dissertation, Stanford University, Stanford, 187 p.

Sullivan, J., 1984, Nonparametric estimation of spatial distributions: Ph.D. Dissertation, Stanford University, Stanford, 367 p.

Sultan, A. J., 1992, Geostatistical characterization of reservoir heterogeneity and its applications: Ph.D. Dissertation, New Mexico Institute of Mining and Technology, Socorro, 157 p.

Suro-Peréz, V., 1988, Indicator kriging based on principal component analysis: M.S. Thesis, Stanford University, Stanford, 168 p.

Suro-Peréz, V., 1994, Estimation and stochastic simulation using indicator principal component kriging: Ph.D. Dissertation, Stanford University, Stanford, 191 p.

Syversveen, A. R., 1998, Spatial stochastic point models for reservoir characterization: Dr.-Ing. Thesis, Norwegian University of Science and Technology, Trondheim, 132 p.

Vivas Vejar, M. A., 1992, Technique for interwell description by applying geostatistics and fractal geometry methods to well logs and core data: Ph.D. Dissertation, University of Oklahoma, Norman, 363 p.

Wang, L., 2000, Integrating radial basis function neural networks and geostatistics in petroleum reservoir modelling: Ph.D. Dissertation, University of New South Wales, Sydney, 518 p.

Warnes, J., 1987, Applications of spatial statistics in petroleum geology: Ph.D. Dissertation, Strathclyde University, Glasgow, 202 p.

Weissmann, G. S., 1999, Toward new models of subsurface heterogeneity: An alluvial fan sequence stratigraphic framework with transition probability geostatistics: Ph.D. Dissertation, University of California, Davis, 279 p.

Xiao, H., 1985, A description of the behavior of indicator variograms for a bivariate normal distribution as a function of cut-off: M.S. Thesis, Stanford University, Stanford, 51 p.

Xu, W., 1995, Stochastic modeling of reservoir lithofacies and petrophsyical properties: Ph.D. Dissertation, Stanford University, Stanford, 214 p.

Yang, A. P., 1990, Stochastic heterogeneity and dispersion: Ph.D. Dissertation, University of Texas, Austin, 242 p.

Zanon, S. D. J., 2004, Advanced aspects of sequential Gaussian simulation: M.S. Thesis, University of Alberta, Edmonton, 65 p.

Zeybek, D. A., 2000, Application of fractals to reservoir characterization: Ph.D. Dissertation, Istanbul University, Istanbul, 191 p.

Zhu, H., 1991, Modeling mixture of spatial distributions with integration of soft data: Ph.D. Dissertation, Stanford University, Stanford, 310 p.

## SOFTWARE, SOFTWARE APPLICATIONS, AND USER MANUALS

Alabert, F., and J. L. Mallet, 1989, A local grid updating scheme for interpolation: Computers and Geosciences, v. 15, p. 1019–1024.

Araktingi, U. G., T. A. Hewett, and T. T. Tran, 1992, GEOLITH: An interactive geostatistical modeling program: Society of Petroluem Engineers, Computer Conference Proceedings, SPE Paper 24432.

Augedal, H. O., H. Omre, and K. O. Stanley, 1986, SISABOSA—A program for stochastic modeling and evaluation of reservoir geology: Oslo, Norwegian Computing Center, NR-Note SAND/18/86.

Barry, R., 1996, A diagnostic to assess the fit of a variogram model to spatial data: Journal of Statistical Software, v. 1, no. 1 (electronic), 6 p.

Berkowitz, B., and M. Ben-Zvi, 1991, An algorithm and Pascal program for geostatistical mapping: Computers and Geosciences, v. 16, p. 489–503.

Bivand, R. S., 1992, SYSTAT—Compatible software for modeling spatial dependence among observations: Computers and Geosciences, v. 18, p. 951–964.

Boger, L. W., 1988, GEOSTAT: A computer system for spherical semi-variogram modeling and kriging: Reston, U.S. Geological Survey, Open-file Report 88-864, 47 p.

Bradley, R., and J. Haslett, 1992, Interactive graphics for the exploratory analysis of spatial data—The interactive variogram cloud, *in* P. A. Dowd and J. J. Royer, eds., Proceedings of the 2nd CODATA Conference on Geomathematics and Geostatistics: Nancy, Association Scientifique pour la Géologie et ses Applications, p. 373–386.

Bratvold, R. B., L. Holden, T. Svanes, and K. Tyler, 1994, STORM: Integrated 3D stochastic reservoir modeling tool for geologists and reservoir engineers: Society of Petroleum Engineers Annual Technical Conference and Exhibition, SPE Paper 27563, p. 242–261.

Bridges, N. J., 1985, Kriging—An interactive program to determine the best linear unbiased estimation: Reston, U.S. Geological Survey Open-file Report 85-112, 46 p.

Buxton, B., A. Pate, and M. Morara, 2005, BATGAM geostatistical software based on GSLIB, *in* O. Leuangthong and C. V. Deutsch, eds., Geostatistics Banff 2004, Proceedings of the Seventh International Geostatistics Congress: New York, Springer, p. 1131–1136.

Carr, J. R., 1990, UVKRIG: A FORTRAN-77 program for universal kriging: Computers and Geosciences, v. 16, p. 211–236.

Carr, J. R., and K. Mela, 1998, Visual Basic programs for one, two, and three-dimensional geostatistial analysis: Computers and Geosciences, v. 24, p. 531–536.

Carr, J. R., and D. E. Myers, 1985, COSIM: A FORTRAN IV program for co-conditional simulation: Computers and Geosciences, v. 11, p. 675–705.

Carr, J. R., D. E. Myers, and C. E. Glass, 1985, Cokriging—A computer program: Computers and Geosciences, v. 11, p. 111–127.

Chen, Y., and X. Jiao, 2001, Semivariogram fitting with linear programming: Computers and Geosciences, v. 27, p. 71–76.

Chen, Z., K. G. Osadetz, H. Gao, and P. K. Hannigan, 2004, SuperSD: An object-based stochastic simulation program for modeling the locations of undiscovered petroleum accumulations: Computers and Geosciences, v. 30, p. 281–290.

Chu, J., 1993, Xgam: A 3-D interactive graphic software for modeling variograms and crossvariograms under conditions of positive definiteness: Stanford, Stanford University, Center for Reservoir Forecasting Report 6.

Clark, I., 1977, Regularization of a semivariogram: Computers and Geosciences, v. 3, p. 341–346.

Clark, I., K. L. Basinger, and W. V. Harper, 1989, MUCK: A novel approach to co-kriging, *in* B. E. Buxton, ed., Proceedings of the conference on geostatistical, sensitivity, and uncertainty methods for ground-water flow and radionuclide transport modeling: Columbus, Battelle Press, p. 473–493.

Clayton, C. M., 1994, Public domain geostatistics programs: STATPAC, Geo-EAS, GEOPACK, Geostatistical Toolbox, and GSLIB, *in* J. M. Yarus and R. L. Chambers, eds., Stochastic modeling and geostatistics: Principles, methods, and case studies: AAPG Computer Applications in Geology 3, p. 349–367.

Clemetsen, R., L. Holden, H. Omre, R. Solberg, and T. Taxt, 1987, SISABOSA, user's guide and technical documentation: Oslo, Norwegian Computing Center, NR-Note SAND/12/87.

David, M., R. Dimitrakopoulos, and D. Marcotte, 1987, GEOSTAT1—A prototype expert system for the explicit knowledge approach to geostatistics, *in* I. C. Lemmer, H. Schaum, and G. M. Canisani-Calzolari, eds., APCOM '87, Proceedings of the 20th International Symposium on the Application of Computers and Operations Research in the Mineral Industry, v. 3 (geostatistics): Johannesburg, South African Institute of Mining and Metallurgy, p. 121–128.

Davis, B. M., R. Hagan, and L. E. Borgman, 1981, A program for the finite Fourier transform simulation of realizations from a one-dimensional random function with known covariance: Computers and Geosciences, v. 7, p. 199–206.

Delfiner, P., J. P. Delhomme, J. P. Chilès, D. Renard, and F. Irigoir, 1980, BLUEPACK-3D Manual: Fontainebleau, France, Centre de Geostatistique et de Morphologie Mathematique, Ecole des Mines, 128 p.

Deutsch, C. V., 1989, DECLUS: A FORTRAN 77 program for determining optimum spatial declustering weights: Computers and Geosciences, v. 15, p. 325–332.

Deutsch, C. V., 1998, FORTRAN programs for calculating connectivity of three-dimensional numerical models and for ranking multiple realizations: Computers and Geosciences, v. 24, p. 69–76.

Deutsch, C. V., and A. G. Journel, 1992, GSLIB—Geostatistical software library and user's guide: New York, Oxford University Press, 340 p.

Deutsch, C. V., and A. G. Journel, 1998, GSLIB, geostatistical software library and user's guide, 2d ed.: New York, Oxford University Press, 369 p.

Deutsch, C. V., and T. T. Tran, 2002, FLUVSIM: A program for object-based stochastic modeling of fluvial depositional systems: Computers and Geosciences, v. 28, p. 525–535.

Dowd, P. A., 1991, A review of recent developments in geostatistics: Computers and Geosciences, v. 17, p. 1481–1500.

Englund, E., and A. Sparks, 1988, Geo-EAS user's guide: Washington, U.S. Environmental Protection Agency, Environmental Monitoring Systems Laboratory, EPA/600/4-88/033, 172 p.

Englund, E., and A. Sparks, 1991, GEO-EAS 1.2.1 user's guide: Las Vegas, U.S. Environmental Protection Agency, Environmental Monitoring Systems Laboratory, EPA/600/8-91/008, 147 p.

Franklin, D. J., H. J. Crofts, and A. J. Watkins, 1994, Geostatistical toolkit: Using advanced visualisation to interpret geostatistical data: Society of Petroleum Engineers European Petroleum Computer Conference, SPE Paper 27562, p. 233–242.

Freund, M., 1986, Cokriging: Multivariable analysis in petroleum exploration: Computers and Geosciences, v. 12, p. 485–489.

Froidevaux, R., 1990, Geostatistical toolbox primer, version 1.30: Troinex, Switzerland, FSS International.

Glatzer, E., and W. G. Mueller, 2004, Residual diagnostics for variogram fitting: Computers and Geosciences, v. 30, p. 859–866.

Gomez-Hernandez, J. J., and R. M. Srivastava, 1990, ISIM3D: An ANSI-C three-dimensional multiple indicator conditional simulation program: Computers and Geosciences, v. 16, p. 395–440.

Gringarten, E., 1998, Fracnet: Stochastic simulation of fractures in layered systems: Computers and Geosciences, v. 24, p. 729–736.

Gundesø, R., and O. Egeland, 1990, SESIMIRA—A new geological tool for 3-D modeling of heterogeneous reservoirs, *in* A. T. Buller, E. Berg, O. Hjelmeland, J. Kleppe, O. Torsæter, and J. O. Aasen, eds., North Sea oil and gas reservoirs II: London, Graham & Trotman, p. 361–371.

Haslett, J., R. Bradley, M. Dillon, and G. Wills, 1992, Interactive graphics for the exploratory analysis of spatial data—Applications to data of different support, *in* P. A. Dowd and J. J. Royer, eds., Proceedings of the 2nd CODATA conference on geomathematics and geostatistics: Nancy, Association Scientifique pour la Géologie et ses Applications, p. 361–372.

Hervada-Sala, C., and E. Jarauta-Bragulat, 2001, Modifications to KB2D program in GSLIB to allow use of tabulated covariances calculated with fast Fourier transform method: Computers and Geosciences, v. 27, p. 887–889.

Herzfeld, U. C., 1989, A note on programs performing kriging with nonnegative weights: Mathematical Geology, v. 21, p. 391–394.

Hessami, M., F. Anctil, and A. A. Viau, 2001, Delaunay implementation to improve kriging computing efficiency: Computers and Geosciences, v. 27, p. 237–240.

Holden, L., H. Omre, R. Solberg, and T. Taxt, 1986, Technical documentation of the SISABOSA system: Oslo, Norwegian Computing Center, NR-Note SAND/12/86.

Jian, X., R. A. Olea, and Y.-S. Yu, 1996, Semivariogram modeling by weighted least squares: Computers and Geosciences, v. 22, p. 387–398.

Kaluzny, S. P., S. C. Vega, T. P. Cardoso, and A. A. Shelly, 1998, S+ spatial stats: users manual for Windows and UNIX: New York, Springer, 327 p.

Kirk, K. G., 1991, Residual analysis for evaluating the robustness of inverse distance, kriging, and minimum tension gridding algorithms: Lakewood, GeoTech/GeoChautaugua '91 Conference Proceedings, p. 15.

Kyriakidis, P. C., C. V. Deutsch, and M. L. Grant, 1999. Calculation of the normal scores variogram used for truncated Gaussian lithofacies simulation: Theory and FORTRAN code: Computers and Geosciences, v. 25, p. 161–171.

Littell, R. C., G. A. Milliken, W. W. Stroup, and R. D. Wolfinger, 1996, SAS system for mixed models: Cary, SAS Institute, 633 p.

Lloyd, C. D., and P. M. Atkinson, 2001, Assessing uncertainty in estimates with ordinary and indicator kriging: Computers and Geosciences, v. 27, p. 929–937.

Ma, X., and A. G. Journel, 1999, An expanded GSLIB cokriging program allowing for two Markov models: Computers and Geosciences, v. 25, p. 627–641.

Ma, X., and T. Yao, 2001, A program for 2D modeling (cross) correlogram tables using fast Fourier transform: Computers and Geosciences, v. 27, p. 763–774.

Mao, S., and A. G. Journel, 1999, Conditional 3D simulation of lithofacies with 2D seismic data: Computers and Geosciences, v. 25, p. 845–863.

Marcotte, D., 1991, Cokriging with Matlab: Computers and Geosciences, v. 17, p. 1265–1280.

Marcotte, D., 1996, Fast variogram computation with FFT: Computers and Geosciences, v. 22, p. 1175–1186.

McBratney, A. B., and R. Webster, 1981, The design of optimal sampling schemes for local estimation and mapping of regionalized variables: II. Program and examples: Computers and Geosciences, v. 7, p. 335–365.

McCarn, D. W., and J. R. Carr, 1992, Influence of numerical precision and equation solution algorithm on computation of kriging weights: Computers and Geosciences, v. 18, p. 1127–1167.

Morisette, J., 1997, Examples using SAS to fit the model of linear coregionalization: Computers and Geosciences, v. 23, p. 317–324.

Nienhaus, P. R., 1986, CROSSV, a simple FORTRAN 77 program for calculating 2-dimensional experimental cross-variograms: Computers and Geosciences, v. 12, p. 375–387.

Olea, R. A., 1996, XVAN: A computer program for the analysis of spatial estimation errors: Computers and Geosciences, v. 22, p. 445–449.

O'Leary, J., and A. Monge, 1980, VAR: A program to calculate the semivariance function in three orthogonal planes: Computers and Geosciences, v. 6, p. 413–499.

Ouenes, A., and N. Saad, 1993, A new, fast parallel simulated annealing algorithm for reservoir characterization: Society of Petroleum Engineers Annual Technical Conference and Exhibition, SPE Paper 26419.

Oz, B., C. V. Deutsch, and P. Frykman, 2002, A visual-basic program for histogram and variogram scaling: Computers and Geosciences, v. 28, p. 21–31.

Pan, G., K. Moss, T. Heiner, and J. R. Carr, 1992, A FORTRAN program for three-dimensional cokriging with case demonstration: Computers and Geosciences, v. 18, p. 557–578.

Pannatier, Y., 1994, MS-WINDOWS programs for exploratory variography and variogram modeling in 2D, *in* V. Capasso, G. Girone, and D. Posa, eds., Statistics of spatial processes: Theory and applications: Bari, Italy, Istituto per Ricerche de Mathematica Applicata (IRMA), p. 165–170.

Pannatier, Y., 1996, VARIOWIN: Software for spatial data analysis in 2D: New York, Springer, 91 p.

Pardo-Igúzquiza, E., 1998a, Inference of spatial indicator covariance parameters by maximum likelihood using MLREML: Computers and Geosciences, v. 24, p. 453–464.

Pardo-Igúzquiza, E., 1998b, MLREML4: A program for the inference of the power variogram model by maximum likelihood and restricted maximum likelihood: Computers and Geosciences, v. 24, p. 537–544.

Pardo-Iqúizguiza, E., 1999, VARFIT: A FORTRAN-77 program for fitting variogram models by weighted least squares: Computers and Geosciences, v. 25, p. 251–262.

Pardo-Iqúizguiza, E., and P. A. Dowd, 2001, VARIOG2D: A computer program for estimating the semi-variogram and its uncertainty: Computers and Geosciences, v. 27, p. 549–561.

Pardo-Iqúizguiza, E., and P. A. Dowd, 2005, EMLK2D: A computer program for spatial estimation using empirical maximum likelihood kriging: Computers and Geosciences, v. 31, p. 361–370.

Pardo-Igúzquza, E., M. Chica-Olmo, J. Delgado-García, 1992, SICON1D: A FORTRAN-77 program for conditional simulation in one dimension: Computers and Geosciences, v. 18, p. 665–688.

Pebesma, E. J., 2004, Multivariable geostatistics in S: The GSTAT package: Computers and Geosciences, v. 30, p. 683–691.

Pebesma, E. J., and C. G. Wesseling, 1998, GSTAT, a program for geostatistical modelling, prediction and simulation: Computers and Geosciences, v. 24, p. 17–32.

Remy, N., 2005, The geostatistical earth modelling software (GEMS): A tool for new algorithms, *in* O. Leuangthong and C. V. Deutsch, eds., Geostatistics Banff 2004, Proceedings of the Seventh International Geostatistics Congress: New York, Springer, p. 865–872.

Remy, N., A. Shtuka, B. Levy, and J. Caers, 2002, Gs TL: The geostatistical template library in C++: Computers and Geosciences, v. 28, p. 971–979.

Renard, D., 1990, BLUEPACK 3-D and its use in the petroleum industry: Society of Petroleum Engineers, 5th Petroleum Computer Conference, SPE Paper 20352.

Ribeiro Jr., P. J., and P. J. Diggle, 2001, GeoR: A package for geostatistical analysis: R News, v. 1, p. 14–18.

Ricci, K., 1999, Geostatistics using SAS software, *in* Proceedings of the 24th Annual Conference of the SAS Users Group International: Cary, SAS Institute, p. 1614–1619.

Richmond, A., 2002a, An alternative implementation of indicator kriging: Computers and Geosciences, v. 28, p. 555–565.

Richmond, A., 2002b, Two-point declustering for weighting data pairs in experimental variogram calculations: Computers and Geosciences, v. 28, p. 231–241.

Romesburg, H. C., 1989, ZORRO: A randomization test for spatial pattern: Computers and Geosciences, v. 15, p. 1011–1018.

Röttig, D. T., F. Tonn, and H.-J. Wendel, 2000, SAFARI—An extended geostatistical software package (in German): Mathematische Geologie, v. 5, p. 23–33.

SAS/STAT technical report: Spatial prediction using the SAS ststem, 1996: Cary, SAS Institute, 72 p.

SAS/STAT user's guide, version 8, v. 1, 2 & 3: Cary, SAS Institute, 3809 p.

Skrivan, J. A., and M. R. Karllinger, 1980, Semi-variogram estimation and universal kriging program: NTIS Report No. Pb-81 120 560, 100 p.

Tavares, G., H. Lopes, S. Pesco and C. Poletto, 2001, PETBOOL: A software for stochastic modeling of geometric objects, *in* U. Bayer, H. Burger, and W. Skala, eds., IAMG 2002, Proceedings of the Annual Conference of the International Association for Mathematical Geology: Berlin, Terra Nostra, p. 203–208.

Tran, T. T., 1994, Improving variogram reproduction

on dense simulation grids: Computers and Geosciences, v. 20, p. 1161–1168.

Xiaodong, J., R. A. Olea, and Y. Y.-Sheng, 1996, Semivariogram modeling by weighted least squares: Computers and Geosciences, v. 22, p. 387–399.

Yates, S. R., 1986, A disjunctive kriging program for two dimensions: Ada, Robet S. Kerr Environmental Research Lab, EPA/600/J-86/247, 28 p.

Yates, S. R., A. W. Warrick, and D. E. Myers, 1986, A disjunctive kriging program for two dimensions: Computers and Geosciences, v. 12, p. 281–313.

Yfantis, E. A., and G. T. Flatman, 1988, On sampling nonstationary spatial autocorrelated data: Computers and Geosciences, v. 14, p. 667–686.

Ying, Z., 2000, Iksim: A fast algorithm for indicator kriging and simulation in the presence of inequality constraints and soft data: Computers and Geosciences, v. 26, p. 493–508.

Zhang, X. F., J. C. Van Eijkeren, and A. W. Helmink, 1995, On the weighted least-squares method for fitting a semivariogram model: Computers and Geosciences, v. 20, p. 605–608.

## GENERAL INTEREST PAPERS, DISCUSSIONS, AND COMMENTARIES

Alldredge, J., and N. G. Alldredge, 1978, Geostatistics: A bibliography: International Statistics Review, no. 46, p. 77–88.

Beliveau, D., 1995, Heterogeneity, geostatistics, horizontal wells, and blackjack poker: Journal of Petroleum Technology, v. 47, p. 1068–1075.

Bell, G. D., and M. Reeves, 1979, Kriging and geostatistics: A review of the literature available in English: Australian Institute of Mining and Metallurgy Proceedings, v. 269, p. 17–27.

Brooker, P. I., 1975, Avoiding unnecessary drilling: Australian Institute of Mining and Metallurgy Proceedings, v. 253, p. 21–23.

Bryant, I. D., and S. S. Flint, 1993, Quantitative clastic reservoir geological modelling: Problems and perspectives, *in* S. S. Flint and I. D. Bryant, eds., The geological modelling of hydrocarbon reservoirs and outcrop analogues: Oxford, Blackwell Scientific Publications, p. 3–20.

Chambers, R. L., J. M. Yarus, and K. B. Hird, 2000a, Petroleum geostatistics for nongeostatisticians: Part 1: The Leading Edge, v. 19, p. 474–480.

Chambers, R. L., J. M. Yarus, and K. B. Hird, 2000b, Petroleum geostatistics for nongeostatisticians: Part 2: The Leading Edge, v. 19, p. 592–600.

Clark, I., 1979, Does geostatistics work?, *in* T. J. O'Neil, ed., Proceedings of the 16th International Symposium on the Application of Computers and Operations Research in the Mineral Industry: New York, McGraw-Hill, p. 213–225.

Coburn, T. C., 1994, Reflections on the proliferation of geostatistics in petroleum exploration and production, *in* J. M. Yarus and R. L. Chambers, eds., Stochastic modeling and geostatistics: Principles, methods, and case studies: AAPG Computer Applications in Geology 3, p. 21–23.

Cressie, N., 1989, Geostatistics: The American Statistician, v. 43, p. 197–202.

Cressie, N., 1990, Reply to comments on "Geostatistics:" The American Statistician, v. 44, p. 256–258.

Damsleth, E., H. Haldorsen, A. Henriquez, and K. Tyler, 1993, Letter to the editor: How well are we doing with reservoir description technology?: Journal of Petroleum Technology, v. 45, p. 957.

David, M., 1975, The practice of kriging, in Geostat 75, *in* M. Guarescio, M. David, and C. Huijbregts, eds., Proceedings of the North Atlantic Treaty Organization Advanced Study Institute Congress: Dordredcht, The Netherlands, D. Reidel Publishing Corporation, p. 31–48.

Deutsch, C. V., and T. A. Hewett, 1996, Challenges in reservoir forecasting: Mathematical Geology, v. 28, p. 829–843.

Deutsch, C. V., and D. N. Meehan, 1996, Geostatistical techniques improve reservoir management: Petroleum Engineer International, v. 69, p. 21–28.

Dowd, P., 1991, A review of recent developments in geostatistics: Computers and Geosciences, v. 17, p. 1481–1500.

Dowds, J. P., 1972, Geostatistics aid gas, oil discovery and exploration: World Oil, v. 75, p. 417–455.

Dubrule, O., 1989, A review of stochastic models for petroleum reservoirs, *in* M. Armstrong, Geostatistics: Proceedings of the 3rd International Geostatistics Congress: Dordrecht, The Netherlands, Kluwer Academic Publishers, p. 493–506.

Dubrule, O., and E. Damsleth, 2001, Achievements and challenges in petroleum geostatistics: Petroleum Geoscience, v. 7, p. 1–7.

Dubrule, O., A. Haas, and J. C. Herweijer, 1994, Geostatistics: A way towards data integration and uncertainty quantification, *in* Proceedings of the 14th World Petroleum Congress: New York, John Wiley & Sons, p. 265–276.

Englund, E. J., 1990, A variance of geostatisticians: Mathematical Geology, v. 22, p. 417–456.

Etris, N., 1998, Conventional geological modeling methods versus geostatistical (abs.): Reservoir, v. 25, p. 8.

Galli, A., and H. Beucher, 1997, Stochastic models for reservoir characterization: A user-friendly review: Society of Petroleum Engineers 5th Latin American and Caribbean Petroleum Engineering Conference and Exhibition, SPE Paper 38999.

Gringarten, E., and C. V. Deutsch, 2001, Variogram interpretation and modeling: Mathematical Geology, v. 33, p. 507–534.

Haldorsen, H. H., and E. Damsleth, 1993, Challenges in reservoir characterization: AAPG Bulletin, v. 77, p. 541–551.

Haldorsen, H. H., P. J. Brand, and C. J. Macdonald, 1988, Review of the stochastic nature of reservoirs, *in* S. Edwards and P. R. King, eds., Mathematics in oil production: Oxford, Clarendon Press, p. 109–209.

Harper, W. V., 1986, Inside the black box of geostatistics, *in* Proceedings of the section on statistical education: Alexandria, American Statistical Association, p. 173–175.

Henley, S., 1987, Kriging—Blue or pink?: Mathematical Geology, v. 19, p. 155–158.

Henley, S., and D. F. Watson, 1998a, Possible alternatives to geostatistics, *in* APCOM '98 International Symposium on the Application of Computers and Operations Research in the Mineral Industry: London, Institution of Mining and Metallurgy, p. 337–354.

Henley, S., and D. F. Watson, 1998b, Geostatistics: A critical review, *in* G. N. Panagiotou and T. N. Michalakopoulos, eds., Proceedings of the 1st International Conference on Information Technologies in the Minerals Industry: Rotterdam, A. A. Balkema, p. 51.

Hirsche, K., S. Boerner, and C. Astaldi, 1998, Avoiding pitfalls in geostatistical reservoir characterization: A survival guide: Canadian Society of Exploration Geophysicists Recorder, June, p. 34–56.

Hirsche, K. L., L. Mewhort, J. Porter-Hirsche, and R. Davis, 1996, Geostatistical reservoir characterization of a Canadian reef or the use and abuse of geostatistics: Canadian Society of Exploration Geophysicists Recorder, December, p. 26–38.

Hirsche, K. L., J. Porter-Hirsche, L. Mewhort, and R. Davis, 1997, The use and abuse of geostatistics: The Leading Edge, v. 16, p. 253–260.

Journel, A., 1989, A democratic research impetus: Geostatistics (an Interdisciplinary Newsletter), v. 3, no. 3, p. 5–6.

Journel, A. G., 1983, Geostatistics, *in* D. L. Banks, C. B. Read, and S. Kotz, eds., Encyclopedia of statistical sciences: New York, John Wiley & Sons, v. 3, p. 424–431.

Journel, A. G., 1984, Geostatistics, simple tools applied to difficult problems, *in* H. A. David and H. T. David, eds., Statistics: An appraisal: Ames, Iowa State University Press, p. 237–256.

Journel, A. G., 1993, Geostatistics: Roadblocks and challenges, *in* A. Soares, ed., Geostatistics tróia '92: Dordrecht, The Netherlands, Kluwer Academic Publishers, p. 213–224.

Kelkar, M., 2000, Application of geostatistics for reservoir characterization—Accomplishments and challenges: Journal of Canadian Petroleum Technology, v. 39, p. 25–30.

Kerbs, L., 1986, Geo-statistics: The variogram: Computer Oriented Geological Society Newsletter, v. 2, p. 54–59.

Lake, L. W., H. B. Carroll, and T. Wesson, 1991, Second International Reservoir Characterization Conference, *in* L. W. Lake, H. B. Carroll Jr., and T. C. Wesson, eds., Reservoir characterization II: San Diego, Academic Press, p. 478–491.

Lieber, B., 1996, Geostatistics: The next step in reservoir modeling: PetroSystems World, March–April, p. 28–29.

Marcotte, D., 1991, Cokriging with Matlab: Computers and Geosciences, v. 17, p. 1265–1280.

Matheron, G., 1986, Philipian/Watsonian high (flying) philosophy: Mathematical Geology, v. 18, p. 501–502.

Matheron, G., 1987, A simple answer to an elementary question: Mathematical Geology, v. 19, p. 455–458.

Murthy, P. S. N., and M. Makki, 1999, Possible alternatives to geostatistics—Some views, *in* K. Dagdelen, ed., APCOM '99, Proceedings of the 28th International Symposium on the Application of Computers and Operations Research in the Mineral Industry: Society of Mining Engineers, p. 435–444.

Myers, D. E., 1986, Matheronian geostatistics—Quo vadis?: Comment: Mathematical Geology, v. 18, p. 699–700.

Neidell, N., and R. Mullin, 2000, Geologic information and seismic data; how much do we need? (abs.): AAPG Bulletin, v. 84, p. 1686–1687.

Olea, R. A., 1992, Kriging—Understanding allays intimidation: Geobyte, v. 7, p. 12–17.

Parker, H. M., and R. L. Sandfur, 1976, A review of recent developments in geostatistics: Redwood City, California, Fluor Mining & Metals Inc., 30 p.

Parker, W. C., 1989, Kriging applications and pitfalls: A layman's discussion (abs.): Geological Society of America Abstracts with Programs, v. 21, p. 54.

Pawar, R. J., 2003, Introduction to geostatistics: Developments in Petroleum Science, v. 51, p. 85–96.

Philip, G. M., and D. F. Watson, 1986a, Geostatistics and spatial data analysis: Mathematical Geology, v. 18, p. 505–510.

Phillip, G. M., and D. F. Watson, 1986b, Matheronian geostatistics—Quo vadis?: Mathematical Geology, v. 18, p. 93–117.

Philip, G. M., and D. F. Watson, 1987, An open letter to G. Matheron: Mathematical Geology, v. 19, p. 453–454.

Ricard, O., 1990, How geostatistics helps decisions on complementary development, *in* A. T. Buller, E. Berg, O. Hjelmeland, J. Kleppe, O. Torsæter, and J. O. Aasen, eds., North Sea oil and gas reservoirs II: London, Graham & Trotman, p. 407–414.

Ripley, B. D., 1988, Spatial data analysis *in* S. Kotz, N. L. Johnson, and C. B. Read, eds., Encyclopedia of Statistical Science, v. 8: New York, John Wiley & Sons, p. 570–573.

Royle, A. G., 1980, Why geostatistics? *in* A. G. Royle, ed., Geostatistics: New York, McGraw-Hill, p. 1–16.

Russell, B., and D. Hampson, 1997, Multivariate geostatistics: Canadian Society of Exploration Geophysicists Recorder, April, p. 5–9.

Srivastava, R. M., 1986, Philip and Watson—Quo vadunt?: Mathematical Geology, v. 18, p. 141–146.

Srivastava, R. M., 1987, Minimum variance or maximum profitability?: Canadian Institute of Mining and Metallurgy Bulletin, v. 80, no. 901, May, p. 63–68.

Statistical Inference for Spatial Processes, 1987: Institute of Mathematical Statistics Bulletin, v. 16, p. 113.

Vogt, J., 1993, Reservoir property grids improve with geostatistics: World Oil, v. 214, p. 67–75.

Wahba, G., 1990, Comments on "Geostatistics" (N. Cressie, The American Statistician, 1989): The American Statistician, v. 44, p. 255–256.

Yarus, J. M., 1992, 3-D reservoir modeling—A lot of noise can go a long way: Geobyte, v. 7, p. 76.

## BOOKS, REVIEWS OF BOOKS, AND COURSE NOTES

Agterberg, F. P., 1974, Geomathematics: Mathematical background and geoscience applications: Amsterdam, Elsevier, 596 p.

Arlinghaus, S. L., and D. A. Griffith, 1996, Practical handbook of spatial statistics: Boca Raton, CRC Press, Inc., 307 p.

Armstrong, M., 1989, Geostatistics: Proceedings of the 3rd International Geostatistics Congress: Dordrecht, The Netherlands, Kluwer Academic Publishers, 1038 p.

Armstrong, M., 1998, Basic linear geostatistics: New York, Springer, 153 p.

Armstrong, M., and P. A. Dowd, 1994, Geostatistical simulations: Proceedings of the Geostatistical Simulation Workshop: Dordrecht, The Netherlands, Kluwer Academic Publishers, 255 p.

Armstrong, M., C. Bettini, N. Champigny, A. Galli, and A. Remacre, 2002, Geostatistics Rio 2000: Dordrecht, The Netherlands, Kluwer Academic Publishers, 235 p.

Armstrong, M., A. Galli, G. Le Loc'h, F. Geoffroy, and R. Eschard, 2003, Plurigaussian simulations in the geosciences: New York, Springer, 149 p.

Baafi, E. Y., and N. A. Schofield, 1997, Geostatistics Wollongong '96: Dordrecht, The Netherlands, Kluwer Academic Publishers, 1333 p.

Bailey, T. C., and A. C. Gatrell, 1995, Interactive spatial data analysis: Harlow, Longman Scientific & Technical, 436 p.

Barton, C. C., and P. R. La Pointe, 1995, Fractals in petroleum geology and earth processes: New York, Plenum Press, 317 p.

Benes, M., 1978, A spatial-directional analysis of geological objects: Praha, Czechoslovakia, Ústrednf ústav geologický, 106 p.

Bilonick, R. A., and S. L. Collins, 1980, Geostatistics for the practicing geologist: The first confrontation: Pittsburg Geological Society, Short Course Notes, 189 p.

Bloom, L. M., 1997, Review: Multivariate geostatistics: An introduction with applications (H. Wackernagel, Springer, 1995): Metrika, v. 45, p. 185–186.

Bohling, G., 2001, Review: Variowin—Software for spatial data analysis in 2D (Y. Pannatier, Springer, 1996): Computers and Geosciences, v. 27, p. 125.

Caers, J., 1999, Review: Basic linear geostatistics (M. Armstrong, Springer, 1998): Earth Sciences Review, v. 47, p. 146–147.

Carr, J. R., 1991, A short course for geostatistics and

multivariate data analysis: Utah Geological Survey Course Notes, 192 p.

Carr, J. R., 1995, Numerical analysis for the geological sciences: Englewood Cliffs, New Jersey, Prentice Hall, 592 p.

Carr, J. R., 2002a, Review: Practical geostatistics: Modeling and spatial analysis (S. W. Houlding, Springer, 2000): Computers and Geosciences, v. 28, p. 431.

Carr, J. R., 2002b, Review: Geostatistics for engineers and earth scientistists (R. A. Olea, Kluwer Academic Publishers, 1999): Mathematical Geology, v. 34, p. 613–616.

Chang, T., 2001, Review: Interpolation of spatial data: Some theory for kriging (M. L. Stein, Springer, 1999): Journal of Statistical Planning and Inference, v. 96, p. 431–436.

Chauvet, P., 1999, Aide-mémoirè de géostatistique linéaire: Paris, Les Presses de l'Ecole des Mines, 367 p.

Chilès, J.-P., and P. Delfiner, 1999, Geostatistics: Modeling spatial uncertainty: New York, John Wiley & Sons, 695 p.

Christakos, G., 1992, Random field models in earth sciences: San Diego, Academic Press, 474 p.

Christakos, G., 2000, Modern spatiotemporal geostatistics: Oxford, Oxford University Press, 304 p.

Clark, I., 1979, Practical geostatistics: London, Applied Science Publishers, 129 p.

Clark, I., 1992, Geostatistics applied to real data: Reno, University of Nevada, Mackay School of Mines, Course Notes, 28 p.

Clark, I., and W. V. Harper, 2000, Practical geostatistics 2000: Columbus, Ecosse North America, 416 p.

Cliff, A. D., and J. K. Ord, 1973, Spatial autocorrelation: London, Pion, 178 p.

Cliff, A. D., and J. K. Ord, 1981, Spatial processes: Models and applications: London, Pion, 266 p.

Coburn, T. C., 2000, Review: Geostatistics for natural resources evaluation (P. Goovaerts, Oxford University Press, 1997): Technometrics, v. 42, p. 437–438.

Coburn, T. C., 2004, Review: Geostatistics with applications in earth sciences (D. D. Sarma, Capital Publishing Company, 2002): Mathematical Geology, v. 36, p. 987–989.

Craig, R. G., 1988, Review: Stochastic simulation (B. D. Ripley, John Wiley & Sons, 1987): Mathematical Geology, v. 20, p. 893–898.

Cressie, N., 1991, Statistics for spatial data: New York, John Wiley & Sons, 900 p.

Cressie, N., 1993, Statistics for spatial data, revised edition: New York, John Wiley & Sons, 900 p.

Cressie, N., 1997a, Review: Introduction to disjunctive kriging and non-linear geostatistics (J. Rivoirard, Oxford University Press, 1994): Society for Industrial and Applied Mathematics Review, v. 39, p. 337–340.

Cressie, N., 1997b, Review: Spatial data analysis by example, v. 1 (G. Upton and B. Fingleton, John Wiley & Sons, 1985): Technometrics, v. 29, p. 114–116.

Cressie, N., and J. Symanzik, 1998, Review: Variowin—Software for spatial data analysis in 2D (Y. Pannatier, Springer, 1996): Computational Statistics, v. 13, p. 419–422.

David, M., 1974a, A short course in geostatistics: Golden, Colorado School of Mines, 100 p.

David, M., 1974b, A course in geostatistics: Reno, University of Nevada, Mackay School of Mines, 303 p.

David, M., 1977, Geostatistical ore reserve estimation: Amsterdam, Elsevier, 364 p.

David, M., 1988, Handbook of applied advanced geostatistical ore reserve estimation: Amsterdam, Elsevier, 216 p.

Davis, J. C., 1986, Statistics and data analysis in geology, 2d ed: New York, John Wiley & Sons, 646 p.

de Oliveira, V., and B. Kedem, 1997, Review: Multivariate geostatistics: An introduction with applications (H. Wackernagel, Springer, 1995): Society for Industrial and Applied Mathematics Review, v. 39, p. 340.

Desbarats, A., 1996, Review: Stochastic modeling and geostatistics: Principles, methods, and case studies (J. M. Yarus, and R. L. Chambers, AAPG, 1994): Computers and Geosciences, v. 22, p. 951.

de Smedt, F., 1991, Multivariate geostatistics: Brussels, Vrije Universiteit Brussel, Laboratory of Hydrology, ERASMUS Course Textbook, 31 p.

Deutsch, C. V., 2002, Geostatistical reservoir modeling: New York, Oxford University Press, 386 p.

Diggle, P. J., 1983, Statistical analysis of spatial point patterns: New York, Academic Press, 148 p.

Diggle, P. J., 2003, Statistical analysis of spatial point patterns, 2d ed.: London, Arnold, 159 p.

Dimitrakopoulos, R., 1994, Geostatistics for the next century: An international forum in honour of Michel David's contribution to geostatistics: Dordrecht, The Netherlands, Kluwer Academic Publishers, 497 p.

Dowd, P. A., and J. J. Royer, 1992, The 2nd CODATA

Conference on Geomathematics and Geostatistics: Nancy, Association Scientifique pour la Géologie et ses Applications, 498 p.

Dubrule, O., 1998, Geostatistics in petroleum geology: AAPG Continuing Education Course Notes 38, 52 p.

European Association of Geoscientists and Engineers–Society of Petroleum Engineers International Symposium on Petroleum Geostatistics, 1999: London, Society of Petroleum Engineers, Extended Abstracts.

Fabbri, A. G., and J. J. Royer, 1995, Proceedings of the 3rd CODATA Conference on Geomathematics and Geostatistics: Nancy, Association Scientifique pour la Géologie et ses Applications, 666 p.

Gomez-Hernandez, J. J., A. Soares, and R. Froidevaux, 1999, Proceedings: GeoENV II—Geostatistics for environmental applications: Dordrecht, The Netherlands, Kluwer Academic Publishers, 584 p.

Goovaerts, P., 1997, Geostatistics for natural resources evaluation: New York, Oxford University Press, 483 p.

Griffith, D. A., 1988, Advanced spatial statistics: Special topics in the exploration of quantitative spatial data series: Dordrecht, The Netherlands, Kluwer Academic Publishers, 273 p.

Griffith, D. A., L. J. Layne, J. K. Ord, and A. Sone, 1999, Casebook for spatial statistical data analysis: A compilation of analyses of different thematic data sets (spatial information systems): New York, Oxford University Press, 524 p.

Guarascio, M., M. David, and C. Huijbregts, 1976, Advanced geostatistics in the mining industry: Dordrecht, The Netherlands, D. Reidel Publishing Company, 461 p.

Henley, S., 1981, Nonparametric geostatistics: London, Applied Science Publishers, 145 p.

Hohn, M. E., 1988, Geostatistics and petroleum geology: New York, Van Nostrand Rienhold, 264 p.

Hohn, M. E., 1999, Geostatistics and petroleum geology, 2d ed: Dordrecht, The Netherlands, Kluwer Academic Publishers, 235 p.

Hohn, M. E., 2003, Review: Geostatistics: Modeling spatial uncertainty (J.-P. Chilès and P. Delfiner, John Wiley & Sons, 1999): Mathematical Geology, v. 35, p. 353–356.

Houlding, S., 2000, Practical geostatistics: Modeling and spatial analysis: New York, Springer, 159 p.

Issaks, E. H., and R. M. Srivastava, 1989, An introduction to applied geostatistics: New York, Oxford University Press, 561 p.

Jensen, J. L., L. W. Lake, P. W. M. Corbett, and D. J. Goggin, 2000, Statistics for petroleum engineers and geoscientists, 2d ed.: Amsterdam, Elsevier, 338 p.

Johnston, K., J. M. Ver Hoef, and K. Krivoruchko, 2001, Using ArcGIS geostatistical analyst: Redlands, Environmental Systems Research Institute Press, 316 p.

Jones, T. A., 1990, Review: Fundamentals of geostatistics in five lessons (A. G. Journel, American Geophysical Union, 1989): Mathematical Geology, v. 22, p. 876–878.

Jones, T. A., 2003, Review: Statistics for petroleum engineers and geoscientists, 2d ed. (J. L. Jensen, L. W. Lake, P. W. Corbett, and D. J. Goggin, Elsevier, 2000): Mathematical Geology, v. 35, p. 357–359.

Journel, A. G., 1989a, Short course in geology: Fundamentals of geostatistics in five lessons: Washington, 28th International Geological Congress, 36 p.

Journel, A. G., 1989b, Fundamentals of geostatistics in five lessons: Washington, American Geophysical Union Course Notes, 40 p.

Journel, A. G., 1989c, Review: Geostatistics and petroleum geology (M. E. Hohn, New York, Van Nostrand Rienhold, 1988): Mathematical Geology, v. 21, p. 803–804.

Journel, A. G., 2000, Review: Geostatistics and petroleum geology (M. E. Hohn, Dordrecht, Kluwer Academic Publishers, 1999): Mathematical Geology, v. 32, p. 139–142.

Journel, A. G., and C. J. Huijbregts, 1978, Mining geostatistics: New York, Academic Press, 600 p.

Kedem, B., 2000, Review: Geostatistics: Modeling spatial uncertainty (J.-P. Chilès and P. Delfiner, John Wiley & Sons, 1999): Society for Industrial and Applied Mathematics Review, v. 42, p. 530.

Kelkar, M., G. Perez, and A. Chopra, 2002, Applied geostatistics for reservoir characterization: Richardson, Society of Petroleum Engineers, 264 p.

Kitanidis, P. K., 1997, Introduction to geostatistics: Applications to hydrology: Cambridge, Cambridge University Press, 249 p.

Kork, J. O., 1993, Review: Statistics for spatial data (N. Cressie, John Wiley & Sons, 1991): Nonrenewable Resources, v. 2, p. 62.

Laarhoven, P. V., and Aarts, E., 1987, Simulated annealing: Theory and applications: Dordrecht, The Netherlands, D. Reidel Publishing Company, 186 p.

Lantuéjoul, C., 2002, Geostatistical simulation: Models and algorithms: New York, Springer, 255 p.

Leuangthong, O., and C. V. Deutsch, 2005, Geostatistics Banff 2004, Proceedings of the Seventh International Geostatistics Congress: New York, Springer, 1167 p.

Majewski, E., 1997, Review: Fractals in petroleum geology and earth processes (C. C. Barton and P. R. LaPointe, Plenum Press, 1995): Pure and Applied Geophysics, v. 150, p. 357–359.

Matérn, R., 1966, Spatial variation: New York, Springer, 151 p.

Matheron, G., and M. Armstrong, 1987, Geostatistical case studies: Dordrecht, The Netherlands, Kluwer Academic Publishers, 248 p.

Monestiez, P., D. Allard, and R. Froidevaux, 2001, Proceedings: GeoENV III—Geostatistics for environmental applications: Dordrecht, The Netherlands, Kluwer Academic Publishers, 555 p.

Moore, M., 2001, Spatial statistics: Methodological aspects and applications: New York, Springer, 282 p.

Myers, D. E., 1993, Review: Random field models in the earth sciences (G. Christakos, Academic Press, 1992): Mathematical Geology, v. 25, p. 407–410.

Myers, D. E., 1995, Review: Geostatistical simulations (M. Armstrong and P. Dowd, Kluwer Academic Publishers, 1994): Journal of Geological Education, v. 43, p. 285–286.

Myers, D. E., 1999, Review: GSLIB: Geostatistical software library and user's guide, 2d ed. (C. V. Deutsch and A. G. Journel, Oxford University Press, 1998): Computers and Geoscieces, v. 25, p. 309–312.

Myers, D. E., 2000, Review: Basic linear geostatistics (M. Armstrong, Springer, 1998): Technometrics, v. 42, p. 437.

Myers, D. E., 2002a, Review: Practical geostatistics 2000 (I. Clark and W. V. Harper, Ecosse North America, 2000): Mathematical Geology, v. 34, p. 777–779.

Myers, D. E., 2002b, Review: Interpolation of spatial data: Some theory for kriging (M. L. Stein, Springer, 1999): International Journal of Geographic Information Science, v. 16, p. 205–207.

Myers, D. E., 2002c, Review: Modern spatiotemporal geostatistics (G. Christakos, Oxford University Press, 2000): Natural Resources Research, v. 11, p. 315–318.

Myers, D. E., 2003, Review: Spatial statistics: Methodological aspects and applications (M. Moore, Springer, 2001): Mathematical Geology, v. 35, p. 647–648.

Olea, R., 1995, Review: Geostatistics for the next century (R. Dimitrakopoulos, Kluwer Academic Publishers, 1994): Nonrenewable Resources, v. 4, p. 353–354.

Olea, R. A., 1975, Optimum mapping techniques using regionalized variable theory: Lawrence, Kansas Geological Survey, Series on Spatial Analysis 2, 137 p.

Olea, R. A., 1977, Measuring spatial dependence with semivariograms: Lawrence, Kansas Geological Survey, Series on Spatial Analysis 3, 29 p.

Olea, R. A., 1990, Review: An introduction to applied geostatistics (E. H. Isaaks and R. M. Srivastava, Oxford University Press, 1989): Mathematical Geology, v. 22, p. 625–630.

Olea, R. A., 1991, Geostatistical glossary and multilingual dictionary: New York, Oxford University Press, 177 p.

Olea, R. A., 1999, Geostatistics for engineers and earth scientists: Dordrecht, The Netherlands, Kluwer Academic Publishers, 303 p.

Oliver, M., 2002, Review: Modern spatiaotemporal geostatistics (G. Christakos, Oxford University Press, 2000): Geoderma, v. 108, p. 149–150.

Otten, R., and L. Van Ginneken, 1989, The annealing algorithm: Dordrecht, The Netherlands, Kluwer Academic Publishers, 224 p.

Parker, H. M., 1988, Review: Geostatistical case studies (G. Matheron and M. Armstrong, Kluwer Academic Publishers, 1987): Mathematical Geology, v. 20, p. 899–904.

Pawlowsky, V., 1997, Review: Multivariate geostatistics: An introduction with applications (H. Wackernagle, Springer, 1995): Computers and Geosciences, v. 23, p. 135.

Pawlowsky-Glahn, V., 2000, Review: Multivariate geostatistics: An introduction with applications, 2d ed. (H. Wackernagle, Springer, 1998): Computers and Geosciences, v. 26, p. 1069.

Reyment, R., 1989, Review: Handbook of applied advanced geostatistical ore reserve estimation (M. David, Elsevier, 1988): Mathematical Geology, v. 21, p. 579–580.

Ripley, B. D., 1981, Spatial statistics: New York, John Wiley & Sons, 252 p.

Ripley, B. D., 1984, Review: Spatial processes: Models and applications (A. D. Cliff and J. K. Ord, Pion, 1981): Journal of the American Statistical Association, v. 79, p. 283.

Ripley, B. D., 1986a, Review: Statistical analysis of spatial point patterns (P. J. Diggle, Academic

Press, 1983): Mathematical Geology, v. 18, 353–354.

Ripley, B. D., 1986b, Review: Spatial data analysis by example, v. 1 (G. J. G. Upton & B. Fingelton, John Wiley & Sons, 1985): Mathematical Geology, v. 18, 863–864.

Ripley, B. D., 1987, Stochastic simulation: New York, John Wiley & Sons, 237 p.

Ripley, B. D., 1988, Statistical inference for spatial processes: Cambridge, Cambridge University Press, 148 p.

Ripley, B. D., 1989, Review: Spatial variation (B. Matérn, Springer, 1986): Journal of Official Statistics, v. 5, 103–104.

Ripley, B. D., 1993, Review: Statistics for spatial data (N. Cressie, John Wiley & Sons, 1991): Mathematical Geology, v. 25, 415–417.

Rivoirard, J., 1994, Introduction to disjunctive kriging and nonlinear geostatistics: New York, Oxford University Press, 180 p.

Rogers, A., 1974, Statistical analysis of spatial dispersion: London, Pion, 164 p.

Royle, A. G., 1979a, A practical introduction to geostatistics: University of Leeds, Department of Mining and Mineral Sciences, 103 p.

Royle, A. G., 1979b, Exercises in geostatistics: University of Leeds, Department of Mining and Mineral Sciences, 115 p.

Royle, A. G., 1980, Geostatistics: New York, McGraw-Hill, 168 p.

Sarma, D. D., 2002, Geostatistics with applications in earth sciences: New Delhi, Capital Publishing Company, 170 p.

Schabenberger, O., and C. A. Gotway, 2004, Statistical methods for spatial data analysis: Boca Rotan, Chapman & Hall/CRC, 488 p.

Scheunemeyer, J. H., 1987, Review: Statistics and data analysis in geology, 2d ed. (J. C. Davis, John Wiley & Sons, 1986): Mathematical Geology, v. 19, p. 165–168.

Sharp, W. E., 1987, Review: Statistics and data analysis in geology, 2d ed. (J. C. Davis, John Wiley & Sons, 1986): Mathematical Geology, v. 19, p. 161–164.

Skjaeveland, S. M., and J. Kleppe, 1992, Recent advances in improved oil recovery methods for North Sea sandstone reservoirs: Stavanger, Norwegian Petroleum Directorate, State Programme for Improved Oil Recovery Monograph, 325 p.

Soares, A., 1993, Geostatistics troia '92: Dordrecht, The Netherlands, Kluwer Academic Publishers, 1038 p.

Soares, A., J. J. Gomez-Hernandez, and R. Froidevaux, 1997, Proceedings: GeoENV I—Geostatistics for environmental applications: Dordrecht, The Netherlands, Kluwer Academic Publishers, 516 p.

Stein, M. L., 1999, Interpolation of spatial data: some theory for kriging: New York, Springer, 247 p.

Stoyan, D., W. S. Kendall, and J. Mecke, 1987, Stochastic geometry and its applications: New York, John Wiley & Sons, 345 p.

Upton, G. J. G., and B. Fingleton, 1985, Spatial data analysis by example, v. 1 (point pattern and quantitative data): New York, John Wiley & Sons, 410 p.

van Laarhoven, P. J. M., and E. H. L. Aarts, 1987, Simulated annealing: Theory and applications: Dordrecht, The Netherlands, D. Reidel Publishing Company, 186 p.

Vanmarcke, E., 1983, Random fields: Analysis and synthesis: Cambridge, MIT Press, 382 p.

Vargas–Guzmán, J. A., 2004, Review: Geostatistics Rio 2000 (by M. Armstrong, C. Bettini, N. Champiguy, A. Galli, and A. Remacre, Kluwer Academic Publishers, 2002): Mathematical Geology, v. 36, p. 759–760.

Verly, G., M. David, A. G. Journel, and A. Marechal, 1984, Geostatistics for natural resources characterization: Dordrecht, The Netherlands, D. Reidel Publishing Company, 1092 p.

Voortman, L., 1990, Review: An introduction to applied geostatistics (E. H. Isaaks and R. M. Srivastava, Oxford University Press, 1989): Economic Geology, v. 85, p. 653–657.

Wackernagel, H., 1995, Multivariate geostatistics: An introduction with applications: New York, Springer, 256 p.

Wackernagel, H., 1998, Multivariate geostatistics: An introduction with applications, 2d ed.: New York, Springer, 291 p.

Wackernagel, H., 2003, Multivariate geostatistics: An introduction with applications, 3d ed.: New York, Springer, 387 p.

Wrigley, N., 1979, Statistical applications in the spatial sciences: London, Pion, 310 p.

Woodard, R., 2000, Review: Interpolation of spatial data: Some theory for kriging (M. L. Stein, Springer, 1999): Technometrics, v. 42, p. 436–437.

Yarus, J. M., and R. L. Chambers, 1994, Stochastic modeling and geostatistics: Principles, methods, and case studies: AAPG Computer Applications in Geology 3, 379 p.

Ziegel, E., 1990, Review: Geostatistics and petroleum

geology (M. E. Hohn, Van Nostrand Reinhold, 1988): Technometrics, v. 32, p. 231.

Ziegel, E., 1998, Editor's report: Geostatistical software library and user's guide, 2d ed. (C. Deutsch and A. Journel, Oxford University Press, 1998): Technometrics, v. 40, p. 357.

Ziegel, E., 2000a, Review: Geostatistics: Modeling spatial uncertainty (J.-P. Chilès and P. Delfiner, John Wiley & Sons, 1999): Technometrics, v. 42, p. 444.

Ziegel, E., 2000b, Review: Geostatistics and petroleum geology, 2d ed. (M. E. Hohn, Kluwer Academic Publishers, 1999): Technometrics, v. 42, p. 444.

Ziegel, E., 2000c, Review: Geostatistics for engineers and earth scientists (R. Olea, Kluwer Academic Publishers, 1999): Technometrics, v. 42, p. 444.